Wake Conference 2015

Journal of Physics: Conference Series Volume 625

Visby, Sweden
9-11 June 2015

ISBN: 978-1-5108-0798-3
ISSN: 1742-6588

Printed from e-media with permission by:

Curran Associates, Inc.
57 Morehouse Lane
Red Hook, NY 12571

Some format issues inherent in the e-media version may also appear in this print version.

Printed by Curran Associates, Inc. (2015)

For permission requests, please contact the Institute of Physics
at the address below.

Institute of Physics
Dirac House, Temple Back
Bristol BS1 6BE UK

Phone: 44 1 17 929 7481
Fax: 44 1 17 920 0979

techtracking@iop.org

Additional copies of this publication are available from:

Curran Associates, Inc.
57 Morehouse Lane
Red Hook, NY 12571 USA
Phone: 845-758-0400
Fax: 845-758-2634
Email: curran@proceedings.com
Web: www.proceedings.com

Table of contents

Volume 625

Wake Conference 2015
9–11 June 2015, Visby, Sweden

Accepted papers received: 1 June 2015
Published online: 18 June 2015

Aerodynamics, Wakes

Experiments, Wind Tunnels

MexNext contributions

Simulations, LES

Simulations, RANS

Simulations, Wind farms

Vortex dynamics, Stability

Wake modeling

WakeBench contributions

The 44 papers in this volume constitute the proceedings of the 2015 Wake Conference, held in Visby on the island of Gotland in Sweden. It is the fourth time this conference has been held. The Wake Conference series started in Visby, where it was held in 2009 and 2011. In 2013 it took place in Copenhagen where it was combined with the International Conference on Offshore Wind Energy and Ocean Energy. In 2015 it is back where it started in Visby, where it takes place at Uppsala University Campus Gotland, June 9th–11th.

The global yearly production of electrical energy by wind turbines has grown tremendously in the past decade and it now comprises more than 3% of the global electrical power consumption. Today the wind power industry has a global annual turnover of more than 50 billion USD and an annual average growth rate of more than 20%. State-of-the-art wind turbines have rotor diameters of up to 150 m and 8 MW installed capacity. These turbines are often placed in large wind farms that have a total production capacity corresponding to that of a nuclear power plant. In order to make a substantial impact on one of the most significant challenges of our time, global warming, the industry's growth has to continue for a decade or two yet. This in turn requires research into the physics of wind turbine wakes and wind farms.

Modern wind turbines are today clustered in wind farms in which the turbines are fully or partially influenced by the wake of upstream turbines. As a consequence, the wake behind the wind turbines has a lower mean wind speed and an increased turbulence level, as compared to the undisturbed flow outside the farm. Hence, wake interaction results in decreased total production of power, caused by lower kinetic energy in the wind, and an increase in the turbulence intensity. Therefore, understanding the physical nature of the vortices and their dynamics in the wake of a turbine is important for the optimal design of a wind farm.

This conference is aimed at scientists and PhD students working in the field of wake dynamics. The conference covers the following subject areas: Wake and vortex dynamics, instabilities in trailing vortices and wakes, simulation and measurements of wakes, analytical approaches for modeling wakes, wake interaction and other wind farm investigations.

Many people have been involved in producing the 2015 Wake Conference proceedings. The work by the more than 60 reviewers ensuring the quality of the papers is greatly appreciated. The timely evaluation and coordination of the reviews would not have been possible without the work of the section editors: Christian Masson, *ÉTS*, Fernando Porté-Agel, *EPFL*, Gerard Schepers, *ECN Wind Energy*, Gijs Van Kuik, *Delft University*, Gunner Larsen, *DTU Wind Energy*, Jakob Mann, *DTU Wind Energy*, Javier Sanz Rodrigo, *CENER*, Johan Meyers, *KU Leuven*, Rebecca Barthelmie, *Cornell University*, Sandrine Aubrun-Sanches, *Université d'Orléans* and Thomas Leweke, *IRPHE-CNRS.*

We are also immensely indebted to the very responsive support from the editorial team at IOP Publishing, especially Sarah Toms, during the review process of these proceedings.

Visby, Sweden, June 2015

Andrew Barney, Jens Nørkær Sørensen and Stefan Ivanell
Uppsala University – Campus Gotland

Journal of Physics: Conference Series **625** (2015) 011002 doi:10.1088/1742-6596/625/1/011002

Peer review statement

All papers published in this volume of *Journal of Physics: Conference Series* have been peer reviewed through processes administered by the proceedings Editors. Reviews were conducted by expert referees to the professional and scientific standards expected of a proceedings journal published by IOP Publishing.

Wake Conference 2015 IOP Publishing
Journal of Physics: Conference Series **625** (2015) 012001 doi:10.1088/1742-6596/625/1/012001

First comparison of LES of an offshore wind turbine wake with dual-Doppler lidar measurements in a German offshore wind farm

L Vollmer[1], M van Dooren[1], D Trabucchi[1], J Schneemann[1], G Steinfeld[1], B Witha[1], J Trujillo[1] and M Kühn[1]

[1] ForWind, Carl von Ossietzky Universität Oldenburg, Ammerländer Heerstr.136, Oldenburg, DE

E-mail: `lukas.vollmer@forwind.de`

Abstract. Large-Eddy Simulations (LES) are more and more used for simulating wind turbine wakes as they resolve the atmospheric as well as the wake turbulence. Considering the expenses and sparsity of offshore measurements, LES can provide valuable insights into the flow field in offshore wind farms. However, for an application of LES wind fields to assess offshore wind farm flow, a proper validation with measured data is necessary. Such a proper validation requires that the LES can closely reproduce the atmospheric conditions during the measurement. For this purpose, a representation of the large-scale features that drive the wind flow is required. Large-scale-forcing and nudging of the LES model PALM is tested with reanalysis data of the COSMO-DE model for a case study during one particular day in the beginning of 2014 at a German offshore wind farm. As wind and temperature profiles of the LES prove to follow the large-scale features closely, the wake of a single wind turbine is simulated with an advanced version of an actuator disc model. Measurement data is provided by processed dual-Doppler lidar measurements during the same day in the same wind farm. Several methods have been investigated at the University of Oldenburg to compare LES wind fields and lidar measurements. In this study a dual-Doppler algorithm was applied in order to estimate the horizontal stationary wind field. The raw data originate from Plan Position Indicator (PPI) measurements, which have been performed with two long-range wind lidars installed at different opposing platforms at the border of the wind farm.

1. Introduction

The planning of offshore wind farms is still tainted with high risks due to unknown power losses and a higher level of fatigue loads related to wake effects. Thus, a good knowledge of the atmospheric boundary layer and its interaction with the wake flow is essential to predict wind farm power production as well as structural loads on the wind turbines. In recent years, Large-Eddy Simulations (LES) have become a common tool to simulate the influence of atmospheric turbulence on the turbine structure [1, 2] and to study the wakes inside wind farms [3, 4]. In general, measurements do not provide the information necessary for the setup of LES to create comparable atmospheric conditions. So in practice, LES are conducted for idealized situations and are normalized for comparison with results from measurement campaigns. This procedure naturally has several shortcomings. One of them is, that the scale and magnitude of turbulence influenced by e.g. atmospheric stability and the local topographic features at the turbine site are

expected to have significant influence on shape and magnitude of the wake and on the turbine performance [5, 6]. Recently, efforts have been made to establish more realistic conditions [7] in LES or to focus on short-term phenomena like gusts [8].
In the presented study we introduce an approach to establish LES with closely comparable conditions to the atmospheric state recorded during a lidar measurement campaign of wind turbine wakes at the German offshore wind farm *alpha ventus* that is situated approximately 45 km north of the German island Borkum in the North Sea. The lidar measurements provide the rare possibility to have access to wake wind fields from an offshore location. To establish the atmospheric conditions in the model, the variables of the LES are forced towards the state of a COSMO-DE [9] reanalysis. In so doing, meso-scale mean wind and temperature fluctuations are object of the LES. The large distance of the wind farm to the coast should further improve the comparability of the LES with measured conditions as influences of topography and land use can be neglected. The LES wind field is generated for one particular day in the beginning of 2014 and is used for a simulation of a single turbine of the wind farm. Wake velocities from the LES are compared to horizontal wind fields behind the same turbine that were computed from lidar measurements, taken during the same time interval.
We start this paper with an introduction to the lidar data that is used for comparison. In the following, the measured wind data from a met mast close to the wind farm site is compared to the reanalysis data that is used for the large-scale-forcing of the LES. In the last part, the LES using a turbine model is presented and compared to results from the lidar campaign.

2. Measurement scenario

The basis for this study is a lidar measurement campaign that took place during the period from August 2013 until March 2014 at the German offshore wind farm *alpha ventus*. During the campaign three long-range lidars were positioned around the wind farm. Two of the systems were placed on the met mast (FINO1) support platform to the west of the wind farm, the third one was placed on the transformer platform in the south-east corner (see Fig.1). Line-of-sight velocities from PPI (Plan Position Indicator) scans from two of the pulsed lidars were combined to evaluate the horizontal velocity components u and v in a well defined control volume inside the wind farm. The algorithm used is the Multiple-lidar Wind Field Evaluation Algorithm (MuLiWEA), developed at ForWind. The approach is based on a wind vector evaluation on a Cartesian grid and has an adjustment imposed by the continuity equation. For more information the reader is referred to [10].
The setup of the lidar system in *alpha ventus* has some consequences for the measurements. Hard targets in the wind farm, e.g. masts and the wind turbines themselves obstruct the view

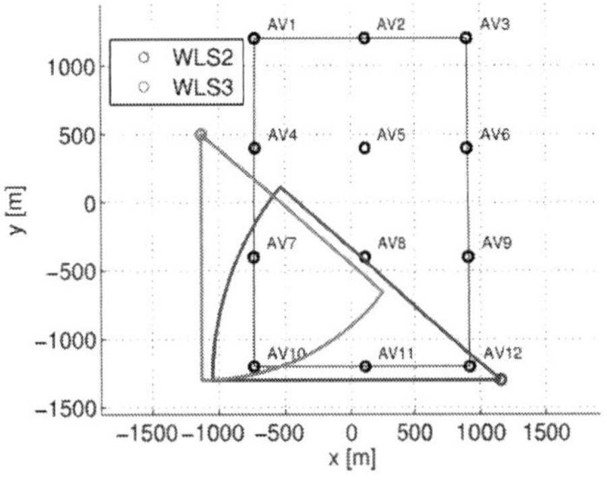

Figure 1. Layout of alpha ventus and position of the two lidars (WLS2, WLS3) that were used for the analysis and the scan area for measurements of the wake of AV10.

of the lidars. The optimal accuracy of the algorithm is reached when scanning under a relative angle of 90° between the individual scans, e.g. scanning the wake of AV3 or AV10. On the other hand the large error at a relative angle of 180° causes that the wake of e.g. AV8 cannot be characterized with sufficient accuracy. Results shown in this study are from selected cases measured at 20 February 2014 behind turbine AV10. The wakes are calculated at hub height of the wind turbine. Only fields that contain enough measurements within a reasonable altitude range centered around hub height are used. In addition, measurements are filtered at both sides of the Carrier-to-Noise level. That means highly scattered signals with a high noise level and reflexions from hard targets, that produce a very high signal, are removed. Velocities are averaged on $(20\text{ m})^3$ cubes with the center at hub height and over a time interval of 10 min to have more measurements for the calculation of the hub height velocities. Both lidars scan at a relatively small azimuth angle, so 5-7 sweeps over the scan area contribute to the calculation of 10 min mean velocities.

3. Modeling the selected case

3.1. Large-scale-forcing routines

For the modeling of the scenario on 20 February 2014, the LES model PALM was used with the default numeric routines of version 4.0 [11, 12]. The goal of the large-scale-forcing routines is to force the horizontal mean variables of the model towards a time-dependent predefined state. The forcing is done in PALM in two different ways: The first is by a variation of the boundary conditions and the second is by nudging. The time-dependent, large-scale pressure gradients that drive the flow in the model are given by prescribing the vertical profile of the geostrophic wind vector $\vec{v}_g(t)$ for each time step. In addition, the change of potential temperature Θ by horizontal advection is prescribed:

$$\frac{\partial\Theta(z)}{\partial t}|_{LS} = -\left(u(z)_{LS}\frac{\partial\Theta(z)_{LS}}{\partial x} + v(z)_{LS}\frac{\partial\Theta(z)_{LS}}{\partial y}\right)$$

Terms with the subscript $_{LS}$ are the large-scale values of the variables. The change of temperature by horizontal advection acts as a steady sink or source of heat at every grid point of the model. At the lower boundary the bottom temperature, pressure and heat flux is provided by the large-scale-forcing during the simulation .

The term nudging describes a relaxation of the spatial mean LES state $< \phi >$ towards a prescribed state ϕ_{LS} by a time constant τ at every time step.

$$\frac{\partial\phi(z)}{\partial t}|_{Nudge} = -\frac{<\phi(z)> - \phi(z)_{LS}}{\tau}$$

Profiles for the large-scale variables can be set at discrete times and on its own vertical grid. The input is afterwards linearly interpolated on the PALM model vertical grid and on the discrete time steps of the simulation.

3.2. Input data

The forcing input data was taken from a reanalysis with the COSMO-DE model. It is available on a $2.8\,\text{km} \times 2.8\,\text{km}$ horizontal grid and a non-equidistant vertical grid. Instantaneous values of the first order variables are available once per hour. Higher order terms like the surface heat flux are output as an hourly mean. The necessary variables were averaged horizontally over an area of approximately $(120\,\text{km})^2$ around *alpha ventus* to get the large-scale-forcing input for the LES (see Fig.2).

To evaluate the accuracy of the input data, it was compared to measurements at the met mast FINO1. For the predominant south-west wind direction during the scenario, these measurements

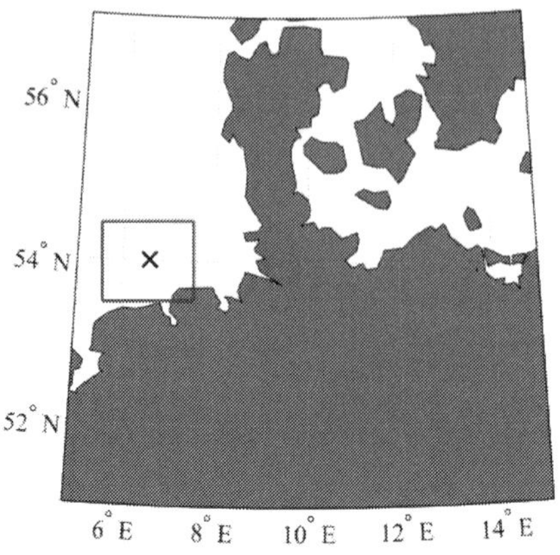

Figure 2. Position of FINO1 (cross) and averaging domain of the COSMO-DE data (box).

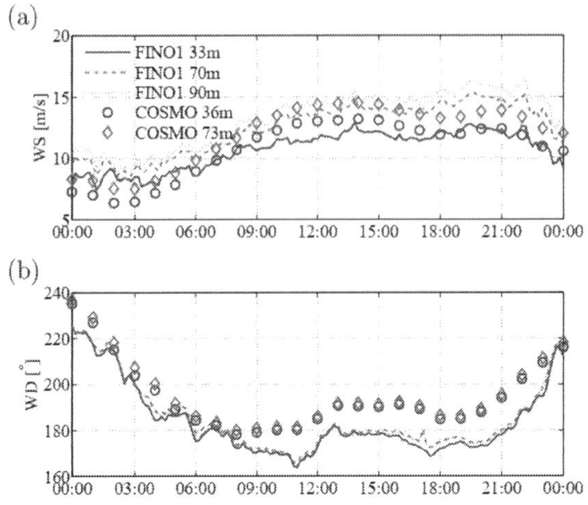

Figure 3. Time series of the wind speed (a) and the wind direction (b) at different heights from FINO1 measurements (lines) and from spatially averaged COSMO-DE data (circles) for 20 February 2014.

are not affected by the adjacent wind farm. 10 min wind speeds from anemometers and wind directions from wind vanes are available for this day at heights of 33 m, 50 m, 60 m, 70 m and 90 m, temperature data is unfortunately not available.

The comparison of the FINO1 measurements with the spatially averaged COSMO-DE data for 20 February 2014 shows some good agreement, especially between the wind measurements at 70 m height and the meso-scale data at 73 m height (see Fig.3). The general trend of wind speed during the day is a speed up from around 8 m/s to 14 m/s from the early morning to noon and a constant strong wind afterwards. The wind direction turns only slightly from south-west to south during the morning and back towards south-west in the evening. There seems to be a quite constant offset of about 10 degrees between the wind direction of the two data sets. Differences are largest for southerly wind during day time. However for this wind direction an influence of *alpha ventus* on the FINO1 measurements can not be ruled out.

3.3. Results from the LES with large-scale-forcing and nudging

The forcing with meso-scale data was tested in PALM for a horizontal and vertical resolution of 5 m. The domain size was 3200 m x 3200 m horizontally and 2200 m vertically with cyclic lateral boundary conditions. At the lower boundary, Monin-Obhukhov similarity theory was applied between the surface and the first grid level. A roughness length of $z_0 = 2 \cdot 10^{-3}$ m was used, corresponding to offshore conditions. Before the forcing was applied, a spin-up period of 6 hours was conducted for atmospheric turbulence to evolve. The process was triggered with random perturbations of the velocity field during the initial phase of the simulation. Initial profiles were the meso-scale profiles at 20 February 2014 0:00 am. After the 6 hour spin up, the time dependent nudging and large-scale-forcing were applied. The relaxation time τ for the nudging of the horizontal velocities and the potential temperature was set to $\tau = 1.5$ h.

Domain-averaged vertical profiles of the PALM output and the forcing profiles are shown in Fig.4. The difference for the initial profile at 0:00 am is caused by the divergence of the wind field during the spin-up before the actual forcing is applied. Afterwards the profiles of the

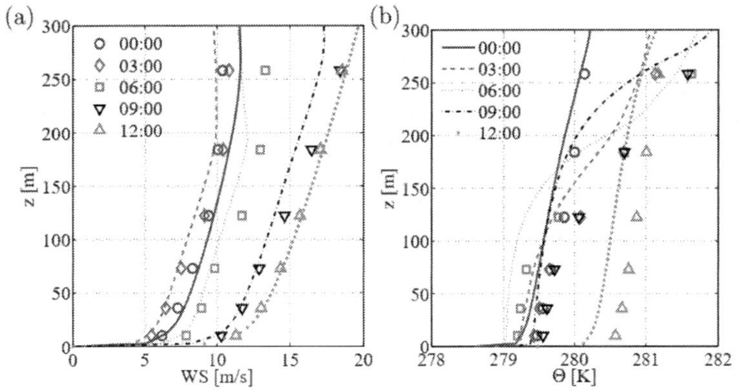

Figure 4. Vertical profiles of wind speed (a) and potential temperature (b) of the COSMO-DE input data (symbols) and the PALM output (lines) during the first 12 h of the large-scale-forcing of the model

LES are close to the forcing data. For the potential temperature a vertically constant offset of about 0.25 K persists. Overall the differences between the LES velocities and the COSMO-DE velocities are small compared to the differences between FINO1 and COSMO-DE.

4. Wind turbine simulation

A LES utilizing a wind turbine model was conducted for the time between 5.30 am and 7.00 am. For this time-span multiple 2D wakes from the lidar-campaign exist for comparison, and furthermore the wind profile of the meso-scale data and of the LES simulation are quite close to the measured profile from the met mast evaluation. The setup was the same as in the aforementioned simulation. Using cyclic boundaries in wake simulations often modifies the results as the wake reenters the boundary and modifies the inflow. In the presented case this effect is attenuated by the slight diagonal inflow, so that the wake is advected through the domain multiple times and vanishes before serving as inflow for the turbine again. As wind turbine model a modified actuator disc model (ADMR) was used that includes non-uniform thrust, a rotational component of the wake and yawing of the turbine. The influence of tower and nacelle on the flow were represented by constant drag coefficients. Further details con be found in [13].

The ADMR is based on blade element momentum theory and requires information about the structure and the lift- and drag coefficients for the different airfoils of the turbine blade. As this information is not available for the turbine AV10, properties of the freely available NREL 5MW turbine of approximately the same size and rated power were used as replacement. To control the rotational speed of the rotor and to maximize power capture below the rated operation point, the generator torque was adjusted for every time step of the simulation (approximately every 0.3 s). The low-pass filtered generator speed was used as the only feedback input, as explained in the turbine definition [14]. Average data was recorded for each 10 min interval starting at 5:40 am. The yaw alignment of the turbine was corrected at the beginning of each interval to the mean wind direction during the interval.

Fig.5 shows the horizontal velocity at hub height for three different time intervals of the LES and the lidar measurements. Velocities from the LES were averaged on $(20\ \text{m})^3$ cubes for better comparison. Results show particularly good agreement in the direction of the wake propagation, indicating that the large-scale-forcing of the LES model delivers a quite accurate wind direction. The magnitude of the wind speed is normalized by the domain averaged wind speed in case of the LES and by the average wind speed measured at 90 m height at FINO1 in case of the lidar measurements. Deviations from unity outside of the wake are probably due to the nonhomogeneity of the wind field during the short averaging interval and due to the

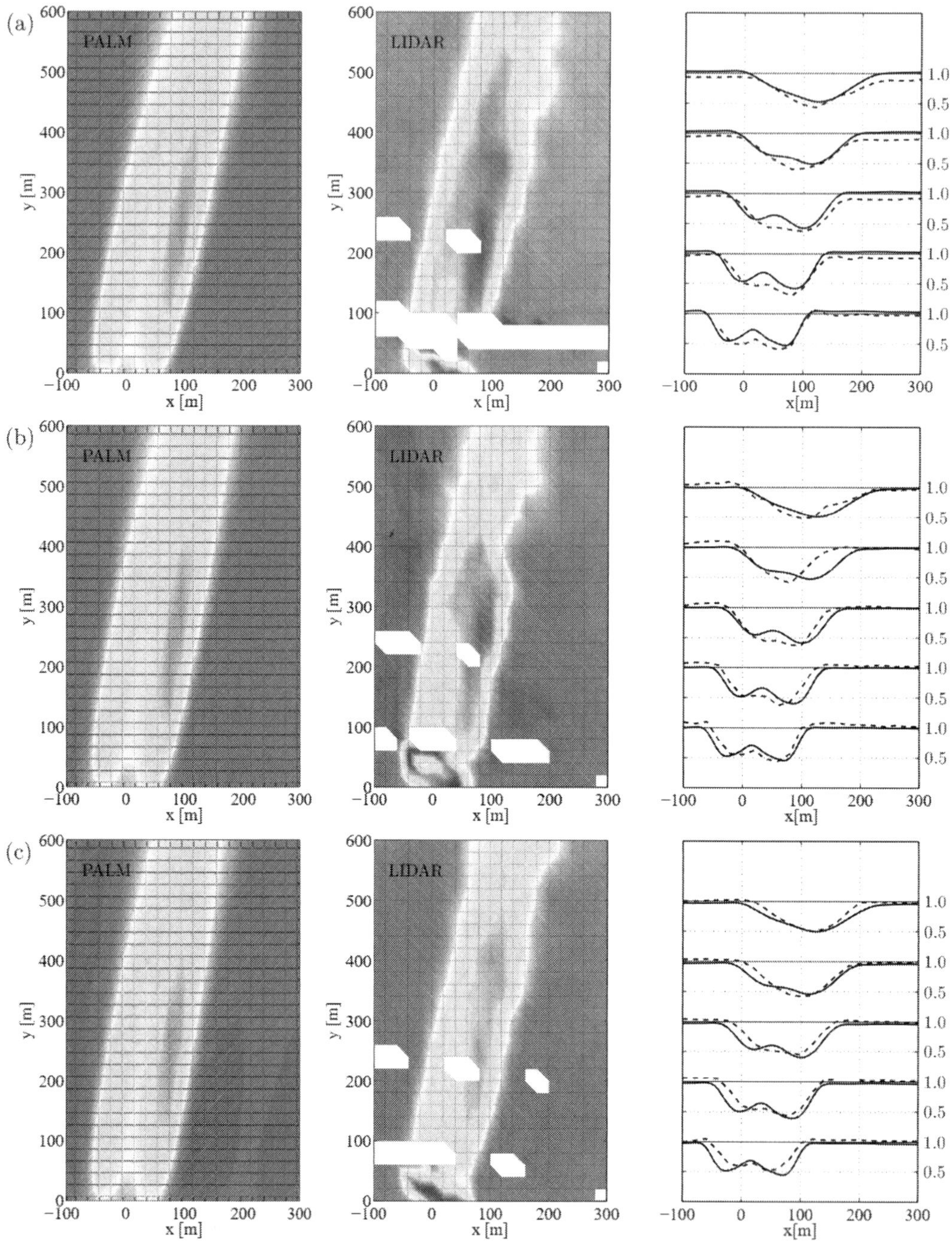

Figure 5. Hub height wind speed during three different 10 min measurement and simulation intervals starting at (a) 5:40 am, (b) 6:10 am and (c) 6:30 am. In the rightmost panels cross-sections of PALM wind speed (solid lines) and measured wind speed (dashed lines) at different distances are compared. Wind speeds of simulations and measurements are normalized by the domain-averaged wind speed at hub height and by the 10 min average wind speed at 90 m measured at FINO1, respectively.

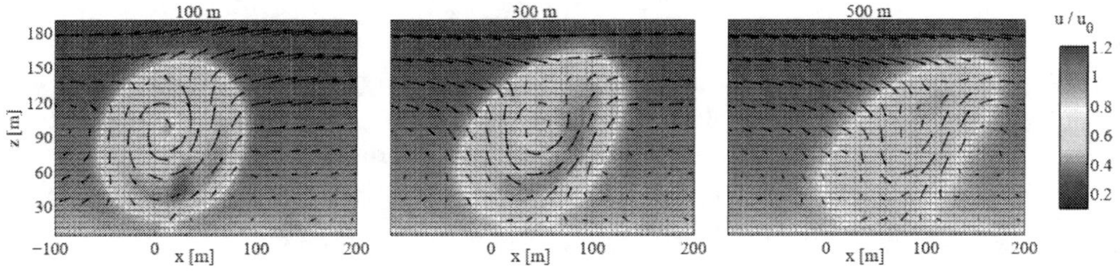

Figure 6. Streamwise (contours) and cross stream component (vectors) of the normalized simulated wind speed at different distances behind the wind turbine for the interval starting at 6:30 am.

distance of the met mast to AV10 (see Fig.1). The deficit is generally slightly larger in the lidar measurements. Both results show a development from a double Gaussian-shaped deficit to a single Gaussian-shaped deficit in the first 500 m downstream of the wind turbine. The double Gaussian shape is more pronounced and the transition appears to occur further downstream in the LES. A clearly visible feature of the wakes is the higher deficit on the right side of the wake, where the clockwise rotation of the rotor causes an upward flow. This upward flow advects lower momentum due to the vertical shear of the wind field. The effect is more clear in the vertical cross sections shown in Fig.6, where also the impact of the vertical wind veer on the shape of the wake is evident.

5. Discussion

In the scope of this paper we show that the LES model PALM delivers realistic results, when driven by large-scale profiles from a COSMO-DE reanalysis using the implemented large-scale-forcing and nudging routines. As base for comparison, wind data from the FINO1 met mast was used. The diurnal cycle of wind speed and direction measured during one particular day of operation was quite well reproduced with the model. The modeled wind field was subsequently used for a LES study with a wind turbine model. Results from the simulation were compared to horizontal wind fields computed from lidar wake measurements, that were taken at the exact same time. We find that in particular the wind direction and thus the propagation of the wake is well reproduced in the model. Furthermore wind turbine wakes from the lidar measurements and the LES show the same characteristic features, even though the turbines were not exactly the same. This encourages to continue working with the wind turbine model as well as with the usage of the large-scale forcing routines.

For further use, the sensitivity to the relaxation time constant of the nudging procedure and the procedure to develop initial turbulence before starting the forcing needs to be evaluated in more detail. Also the capability of the LES to develop realistic atmospheric turbulence, when forced towards an external state, needs further evaluation.

As manner to compare LES results with lidar measurements the recently developed dual-Doppler algorithm MuLiWEA, that combines two long-range lidar measurements to estimate a stationary 2D wind field, was used. This approach is quite intuitive as a realization for averaged measurements because the differences are well visible. Other methods are also studied at ForWind, e.g. a direct comparison of line-of-sight velocities with projected LES velocities, that allows for a more detailed evaluation of atmospheric and wake turbulence.

Acknowledgments
The work presented in this study is funded by the national research projects CompactWind and Parallelrechner-Cluster für CFD und WEA-Modellierung (FKZ 0325492B, 0325220; Federal Ministry for Economic Affairs and Energy) and the Ministry for Education, Science and Culture of Lower Saxony. Computer resources have been partly provided by the North German Supercomputing Alliance (HLRN). We thank Deutscher Wetterdienst (DWD) for providing analysis data. The authors further thank R. Heinze from IMUK Hannover for assistance on the execution and preparation of the large-scale-forcing routines of the LES model PALM.

References

[1] Lee S, Churchfield M J, Moriarty P J, Jonkman J and Michalakes J 2013 *J. Sol. Energy Eng.* **135** doi: 10.1115/1.4023319
[2] Park J, Basu S and Manuel L 2013 *Wind Energy* **17** 359–385 doi: 10.1002/we.1580
[3] Lu H and Porté-Agel F 2011 *Phys. Fluids* **23** 065101 doi: 10.1063/1.3589857
[4] Witha B, Steinfeld G, Dörenkämper M and Heinemann D 2014 *J. Phys.: Conference Series* **555** 012108 doi: 10.1088/1742-6596/555/1/012108
[5] Wharton S and Lundquist J K 2010 Tech. Rep. LLNL-TR-424425 Lawrence Livermore National Laboratory
[6] Dörenkämper M, Tambke J, Steinfeld G, Heinemann D and Kühn M 2014 *J. Phys.: Conference Series* **555** 012029 doi: 10.1088/1742-6596/555/1/012029
[7] Lundquist J K, Mirocha J D and Kosovic B 2010 *The 5th Int. Symp. on Computational Wind Engineering, Chapel Hill, North Carolina, USA*
[8] Storey R, Norris S and Cater J 2013 *Proc. of the World Congress on Engineering 2013 Vol III, WCE 2013, July 3 - 5, 2013, London, U.K.*
[9] Baldauf M, Förstner J, Klink S, Reinhardt T, Schraff C, Seifert A, Stephan K and Wetterdienst D 2009 *Deutscher Wetterdienst, Geschäftsbereich Forschung und Entwicklung, Offenbach, Germany*
[10] v Dooren M 2014 Master's thesis ForWind - University of Oldenburg
[11] Raasch S and Schröter M 2001 *Meteorol. Z.* **10** 363–372 doi: 10.1127/0941-2948/2001/0010-0363
[12] Maronga B, Gryschka M, Heinze R, Hoffmann F, Kanani-Sühring F, Keck M, Ketelsen K, Letzel M O, Sühring M and Raasch S 2015 *Geosci. Model Dev. Discuss.* **8** 1539–1637 doi: 10.5194/gmdd-8-1539-2015
[13] Dörenkämper M, Witha B, Steinfeld G, Heinemann D and Kühn M 2015 *J. Wind Eng. Ind. Aerod.* accepted for publication doi : 10.1016/j.jweia.2014.12.011
[14] Jonkman J M, Butterfield S, Musial W and Scott G 2009 Technical Report NREL/TP-500-38060 National Renewable Energy Laboratory

The role of atmospheric stability/turbulence on wakes at the Egmond aan Zee offshore wind farm

R.J. Barthelmie[1], M.J. Churchfield[2], P.J. Moriarty[2], J.K. Lundquist[2,3], G.S. Oxley[4], S. Hahn[4], S.C. Pryor[1]

[1]Cornell University, Ithaca, NY 14853, USA
[2]National Renewable Energy Laboratory, Golden, CO 80401, USA
[3]University of Colorado, Boulder, CO 80309, USA
[4]Vestas Wind Systems A/S, Aarhus 8200, Denmark, USA

rb737@cornell.edu

Abstract. The aim of the paper is to present results from the NREL SOWFA project that compares simulations from models of different fidelity to meteorological and turbine data from the Egmond aan Zee wind farm. Initial results illustrate that wake behavior and impacts are strongly impacted by turbulence intensity [1]. This includes both power losses from wakes and loading illustrated by the out of plane bending moment. Here we focus on understanding the relationship between turbulence and atmospheric stability and whether power losses due to wakes can effectively be characterized by measures of turbulence alone or whether atmospheric stability as a whole plays a fundamental role in wake behavior. The study defines atmospheric stability using the Monin-Obukhov length estimated based on the temperature difference between 116 and 70 m. The data subset selected using this method for the calculation of the Monin-Obukhov length indicate little diurnal or directional dependence of the stability classes but a dominance of stable classes in the spring/unstable classes in fall and of near-neutral classes at high wind speeds (Figure 2). The analysis is complicated by the need to define turbulence intensity. We can select the ratio of the standard deviation of wind speed to mean wind speed in each observation period using data from the meteorological mast, in which case a substantial amount of data must be excluded due to the presence of the wind farm. An alternative is to use data from the wind turbines which could provide a larger data set for analysis. These approaches are examined and compared to illustrate their robustness. Finally, power losses from wakes are categorized according to stability and/or turbulence in order to understand their relative importance in determining the behavior of wind turbine wakes.

1. Introduction

Previous research has indicated that wind turbine wakes are responsible for significant power losses and increased wind turbine fatigue loading in both onshore and offshore wind farms (where the 'deep array' effect is particularly marked) [2]. The goals of this research are thus to: (1) Optimize techniques associated with characterizing the atmospheric parameters responsible for varying wake behavior. (2) Improve physical understanding of wake development and wake merging under different wind speed, turbulence and atmospheric stability conditions, and thus reduce uncertainty in wind farm power prediction by better quantification, modeling and description of wakes.

In prior research [2], we analyzed data from two offshore wind farms in Denmark – Horns Rev and Nysted and showed that total wind farm efficiency appeared to exhibit a clearer variation (and greater absolute magnitude of variability) with stability (as measured using the Monin-Obukhov length) than with turbulence intensity (as described using the standard deviation of wind speed to the mean). Herein we investigate whether the degree to which one can differentiate the effects of ambient turbulence versus stability metrics on wake recovery and behavior is a function of the precise metrics used and which of these properties provides a clearer forensic diagnostic of wake behavior using data from the Egmond aan Zee (OWEZ) wind farm in the Netherlands.

There are different indices that can be used to describe the atmospheric stability (e.g. the gradient (or bulk) Richardson number, and the Monin-Obukhov length) and depending on which metric is used and the approach applied to compute the index, they can generate quite different stability climates [3], [4]. Here we employ the Monin-Obukhov length (L) in part because it has a strong theoretical foundation that can be used quantitatively to correct wind speed profiles in the surface layer [5]:

$$L = -\frac{\theta_v u_*^3}{kg\left(\overline{w'\theta'_v}\right)} \quad (1)$$

where θ_v is the virtual potential temperature, u_* is the friction velocity, $(\overline{w'\theta'_v})$ is the mean virtual heat flux, k is the von Karman constant (0.4) and g is acceleration due to gravity.

Where measurements of the virtual heat flux are not available, a range of techniques (see section 3) can be applied to derive estimates of L, though the method used to determine each value of L does to some extent determine the result and thus the resulting stability climate. For example, if L is calculated from the Richardson number (see Section 3) then it is strongly impacted by wind shear - and is sensitive to measurement error therein and the depth of the layer over which it is computed. Using L determined from an estimate of heat flux as in equation (1) [6] tends to produce a large number of observations in the stable or unstable classes, likely because it is more sensitive to any measurement errors in the high-frequency temperature observations. L is function of the cube of the friction velocity and therefore tends towards high values (neutral conditions) at high wind speeds (and $|L| \to \infty$). In unstable conditions then buoyant production of turbulent mixing increases, although shear production may be declining (and L is negative and $\to 0$). In a stable atmosphere turbulent motions are acting against gravity hence turbulence is supressed by buoyancy while turbulence is generated mechanically by wind shear (L is positive and $\to 0$) [5]. Because L is unbounded, it is often (as here) presented normalized by measurement height (z) (i.e. as z/L). z/L is a surface layer scaling parameter.

While stable conditions tend to be predominately associated with low turbulence, and higher turbulence tends to occur in unstable conditions [2], the occurrence of low turbulence conditions does not necessarily determine that conditions are stable. Equally higher turbulence conditions can occur in stable conditions [7]. Thus far it is unclear whether stability or turbulence intensity is most important to wake propagation and recovery or even whether the two quantities act in different ways. For example, it is possible that turbulence intensity at large length scales determines the degree of wake meander (i.e. lateral movement of the wake centreline) that acts to apparently widen wakes in unstable conditions while stability controls the extent of downward mixing of momentum (and thus wake dispersion) which is an important control of wake recovery in large wind farms. Here we use observations from the OWEZ wind farm to determine the interplay between stability and turbulence and wake properties, and thus address one of the key objectives of the Simulator for Wind Farm Applications (SOWFA) project.

2. The OWEZ wind farm

The OWEZ wind farm is located between 10 and 18 km off the Dutch North Sea coast, at 52°36'N, 4°23'E (Figure 1). It comprises 36 Vestas V90 3.0 MW turbines in 4 rows oriented from the NW to the SE. The rotor diameter is 90 m and the hub-height 70 m. The turbine spacing is 7.1 D (rotor diameters) within the row and 11.5 D between rows. Meteorological data are available from a 116 m meteorological mast located southwest of the wind farm (Figure 1). Meteorological data are available for 2005-2010 and turbine data from 2007-2010. Note that water temperature is missing for 2009 and 2010. Data availability for each turbine, after applying quality control and related data screening criteria (see Section 3) are shown in Figure 1. Consistent with analyses of data from other offshore wind farms, prior analyses of OWEZ data indicated; (i) a strong dependence of wake losses on wind direction and wind speeds, and a dominance of nearest wakes [8] and (ii) evidence of amplification of wake losses under stable stratification [1].

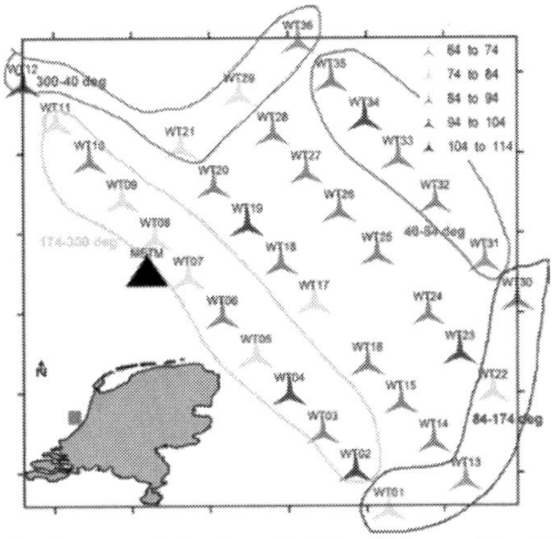

Figure 1. Wind farm location (red square in bottom left inset map of the Netherlands) and layout of the turbines at the Offshore Wind Farm Egmond aan Zee (OWEZ). The spacing shown on each axis is 1 km. The groups of turbines shown are used in the freestream calculation. The scale indicates the percentage of observations available after the quality control described in Section 3 is applied and is normalized to the wind turbine with the most observations leading to a value greater than 100% for the meteorological mast data

3. Turbulence intensity and stability

3.1 Determining stability from meteorological data

Prior analysis of stability conditions at OWEZ in [4] suggested:

- avoiding the wind farm sector (135-315°) and using the freestream sector 225-315°
- using wind speeds > 4 ms^{-1}
- using the bulk Richardson number method to characterize stability
- subtracting 0.82 K from the measured water temperature when used to compute stability.

As is the case for most offshore wind farms, direct measurements of the sensible and latent heat fluxes are not available from OWEZ. Thus, a number of approaches for determining the Monin-Obukhov length were tested here including:

- Using combinations of the sea surface temperature, the air temperature difference between measurements at different heights 116-70 m, 116-18 m and 70-18 m. Note that temperature differences were not measured directly and so can be impacted by sensor calibration errors.

- Using different methods for the calculation including:
  1. the AMOK tool [9] which uses the Richardson number based approach
  2. the Beljaars [6] routines (here denoted as BJx-y, where xy are the heights of the temperature measurements used) that estimate surface fluxes based on the temperature and wind speed profiles and find an iterative solution for L
  3. calculation of a bulk Richardson number which can then be used to estimate an Obukhov Length based on [10] (Here denoted as GRx-y where x and y are the heights of the wind speed and temperature).

Using AMOK most of the observations fall into the neutral class with very few in either stable or unstable classes. This is an indication that there is a temperature offset (likely a calibration issue) between the air and water temperature. Applying the temperature correction suggested in [4] did not improve results from AMOK significantly but improved the results from BJ in terms of the distribution of stability classes by wind speed, direction, time of day and month (Figure 2). A prior analysis based on the gradient Richardson number indicated a dominance of the near-neutral and unstable classes with very few stable observations except at low wind speeds. Since no unbiased estimate of L was available, the 'best' method was determined by using each of the L estimates to predict wind speeds at 116 m and the functions given in [6]. Using BJ116-70 minimized the root mean square error (RMSE) in the prediction of the wind speed relative to observations and thus in the following we use the BJ116-70 method for classifying stability. This approach was used previously in e.g.[11]. It is worth noting that calculating L using data from these heights is also subject to errors since Monin-Obukhov similarity theory applies only in the surface layer (see discussion in [4]).

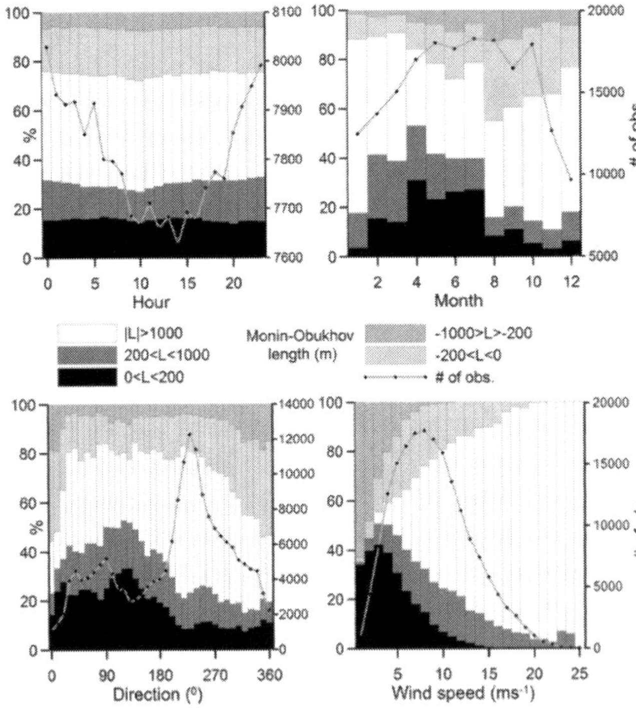

Figure 2. Distribution of observations by stability class where L is calculated using the Beljaars [6] routines with data from 116-70m (i.e. BJ116-70) conditionally sampled by hour of the day, month, wind direction and wind speed.

Journal of Physics: Conference Series **625** (2015) 012002 doi:10.1088/1742-6596/625/1/012002

Table 1. Percentage of observations in different stability classes based on OWEZ data for 2007-2010 and the different approximation approaches listed above.

% of observation in each stability class	Very stable	Stable	Slightly stable	Near-neutral	Slightly unstable	Unstable	Very Unstable
L in m	$10 \le L \le 50$	$50 \le L \le 200$	$200 \le L \le 500$	$\lvert L \rvert > 500$	$-500 \le L \le -200$	$-200 \le L \le -100$	$-100 \le L \le -50$
Estimates from [4]	1	9	11	23	15	23	18
GR116-70. For 2007	2	8	17	67	1	1	1
GR116-21. For 2007	14	19	25	10	1	3	3
GR70-21. For 2007	3	5	12	60	4	6	4
L in m		$0 \le L \le 200$	$200 \le L \le 1000$	$\lvert L \rvert > 1000$	$-1000 \le L \le -200$	$-200 \le L \le 0$	
BJ116-18. 2007-2010		35	27	26	8	5	
BJ116-70. 2007-2010		15	15	44	19	6	

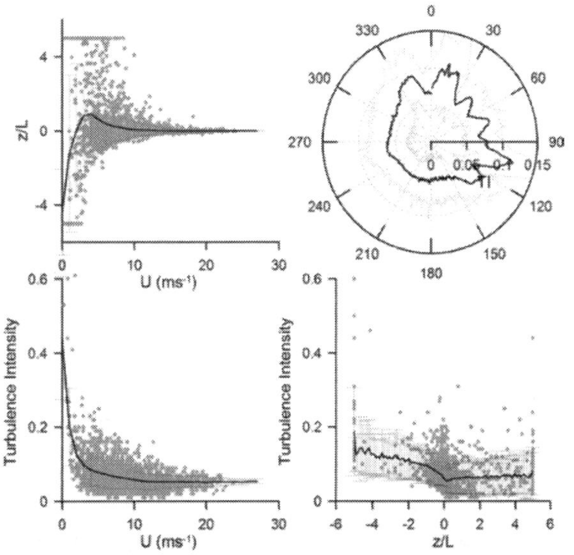

Figure 3. Relationship between *TI*, z/L and U with *TI* and U using OWEZ data from 116 m height. The black solid lines depict the mean value, the vertical bars show ±one standard deviation (±1sd), and the grey dots show the individual data points. The *TI* rose (top right) is also from 116 m height and is for all wind speeds. The black line shows the mean by 1° direction bin and the gray lines ±1sd.

3.2 Estimating turbulence intensity from meteorological data

An estimate of ambient turbulence intensity *TI* can be derived using the ratio of the standard deviation of wind speed in each 10 minute observation period, σ, divided by the mean of the wind speed U in the period, presuming that the meteorological data are available that are representative of the freestream conditions. The relationships between *TI* and z/L and U derived using data from the 116 m height follows patterns that have been observed in other offshore environments [2] (Figure 3). *TI* is high at low wind speeds, decreases with wind speed and then increases slightly as wind speeds continue to increase and there is an increase in mechanical generation of turbulence due to the increasingly rough sea-surface. Consistent with the definitions of stability indices higher mean turbulence intensity is observed in unstable conditions, and lower *TI* occurs in stable conditions. However, as shown by the error bars, lower turbulence intensity can occur in unstable conditions and vice versa. In the freestream direction *TI* is less than 6%. Higher *TI* to the north and east of the mast is driven by flow through the wind farm, and hence is a motivation behind sections 3.3 and 3.4 wherein we determine whether robust estimates can be derived from the turbine power output.

3.3 Estimating the freestream wind and power

As indicated above, the meteorological mast at OWEZ is very close to WT06 and WT07 and frequently in the 'wind farm wake' and thus data from the mast cannot be used to determine the freestream wind speed or ambient turbulence intensity for use in wake analyses or to estimate whole wind farm efficiency (see Figure 4, right). In the absence of these data we thus determined the degree to which data from the wind turbine nacelle-mounted anemometers and the variability of wind turbine power output can be used as proxies of freestream wind speed and turbulence intensity respectively using the approach described in [12]. Data from groups of turbines are used to determine the freestream wind speed by direction (Figure 1) and the arithmetic mean of the data from nacelle-mounted anemometers is used to approximate the freestream wind speed. The results (Figure 4) show this approach gives a reasonable estimate of the freestream power curve as a function of freestream wind speed derived as the average of the nacelle-mounted anemometers.

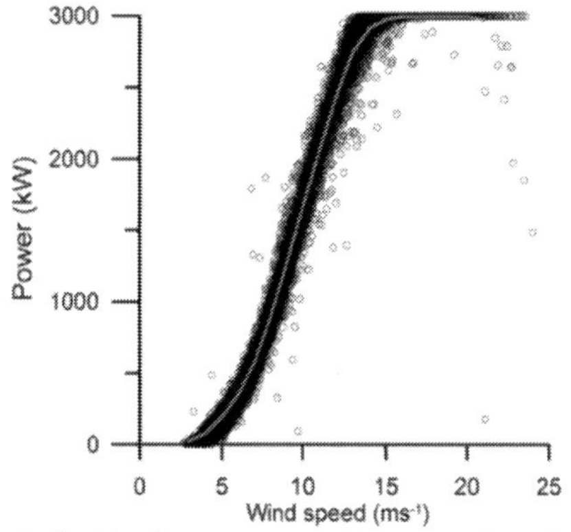

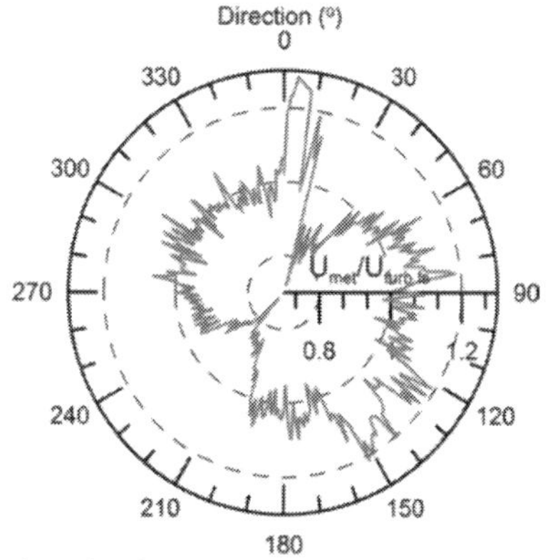

Figure 4. Left: The freestream power curve plotted against the freestream wind speed as derived from the nacelle-mounted anemometers. Right: Ratio of wind speeds at 70 m from the meteorological mast (U_{met}) to those selected as the freestream from the nacelle anemometer (U_{turb}).

3.4 Estimating turbulence intensity from power data

To estimate turbulence intensity at each turbine two measures were tested. One in which *TI* is derived as above but using data from the nacelle anemometer and the second in which the power output and its standard deviation are used as in [13]:

$$\sigma_p = B\sigma_U\left(\frac{\partial P}{\partial U}\right) \qquad (2)$$

where σ_p is the standard deviation of power output, σ_u is the standard deviation of wind speed, B is a constant which is in the range of 0.8–0.9, depending on mean wind speed [13], P is power output and U is the wind speed.

The former approach may be deficient because of turbulence induced by the nacelle and the blades. The latter (i.e. Equation 2) also has some limitations, namely there is uncertainty regarding the value of the coefficient B. It cannot be used when the turbine is not producing power or if there is no change in the power output in a given period. Use of the power and the power change mean that the rotor itself may be acting to filter turbulence at smaller scales. To evaluate the different measures of *TI*: TI_{na} from the nacelle anemometer measurements and TI_{pw} from the wind turbine power standard deviation were compared with TI_{met} from the anemometer mounted on the meteorological mast (Figure 5). There is good agreement between TI_{met} and TI_{na} but using B=0.9 (as in [13]) gave values that were too high for TI_{pw} so this technique has been disregarded here. If it is to be utilized a technique for determining the value of B in Equation 2 needs further investigation. Although in the following section TI_{na} is used to describe ambient turbulence intensity it is worth noting that TI_{met} is a better indicator of atmospheric conditions, and when plotted against z/L, TI_{na} does not vary. TI_{na} may be influenced by flow around the nacelle and smoothing by the rotor resulting in a value that does not vary by direction or wind speed. However, it appears from results in the northeast sector that it could be an acceptable alternative in the directions where the meteorological mast is influenced by direct wakes (339-29° and 98-117°).

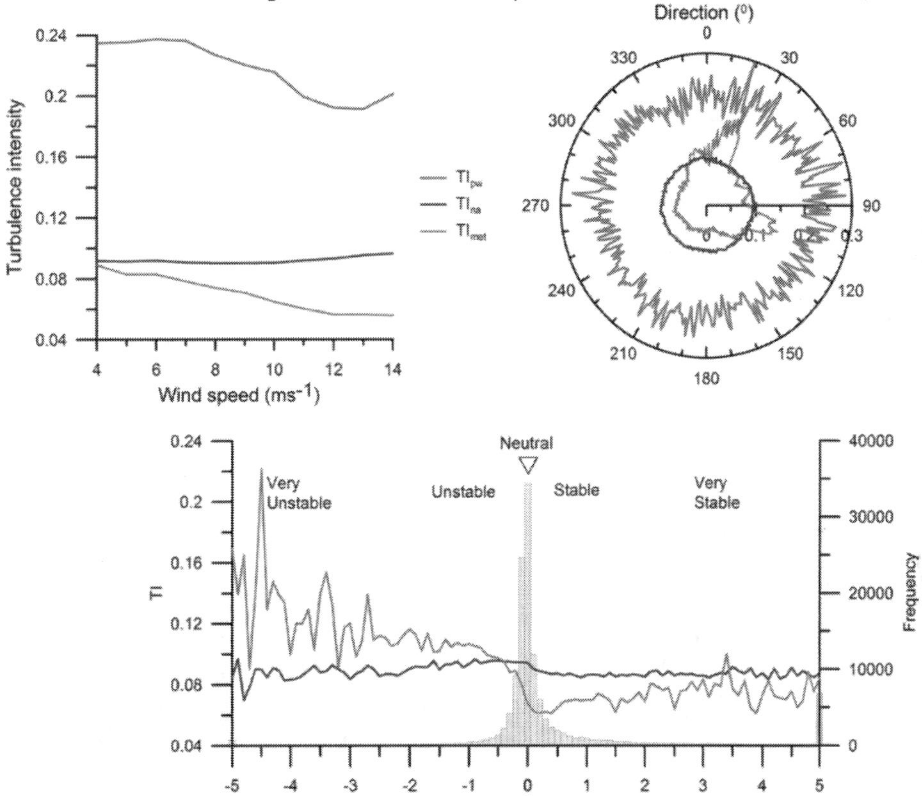

Figure 5. Comparison of *TI* determined by different techniques. Top left: *TI* binned by wind speed. Top right: *TI* binned by wind direction. Bottom: *TI* binned by z/L. Note that TI_{met} has not been screened for wakes but TI_{pw} and TI_{na} are both freestream values.

4. Wake analysis

4.1 Wake behavior conditionally sampled by turbulence intensity

As noted previously [2], the main driver of wake magnitudes in approximate order of impact are 1) wind speeds (via their impact on the thrust coefficient), 2) the wind direction (as a proxy for the number of upstream turbines, the distance to the nearest turbine, any roughness change or stability issues) 3) turbulence intensity which is itself a function of wind speed and stability and 4) stability which is directly related to the previous 3 variables.

In the following the freestream wind, power and turbulence intensity are taken from the turbine data while the stability measure z/L is from the meteorological mast (BJ116-70). To assess the impact of turbulence intensity and stability conditions on wake losses we use the average power output for the wind farm (normalized by the freestream power output by direction) calculated for each 1° directional bin. Recall that, as shown in Figure 1, the turbines are aligned in a regular grid with turbine spacing of 11.5 D between the rows and 7.1 D within the rows. It is also worth noting that in these analyses the direction is taken from the meteorological mast which may have an offset of 2° [14], and that wind directions can (and should) be verified [8].

Figure 6 shows the normalized power output (total average power from all wind turbines in the array / freestream power calculated using the power curve and freestream wind speed) by direction. It clearly indicates the main wake directions at 139 and 319° (7-12 turbines upstream with the closest at 7.1 D). The secondary wind turbines wakes expected at 229° and 49° based on the wind farm layout (there are 1 to 4 turbines upstream at 11.5 D and beyond in these directions) are – however – not visible, or at least are not discernible above the general 'wind farm wake'.

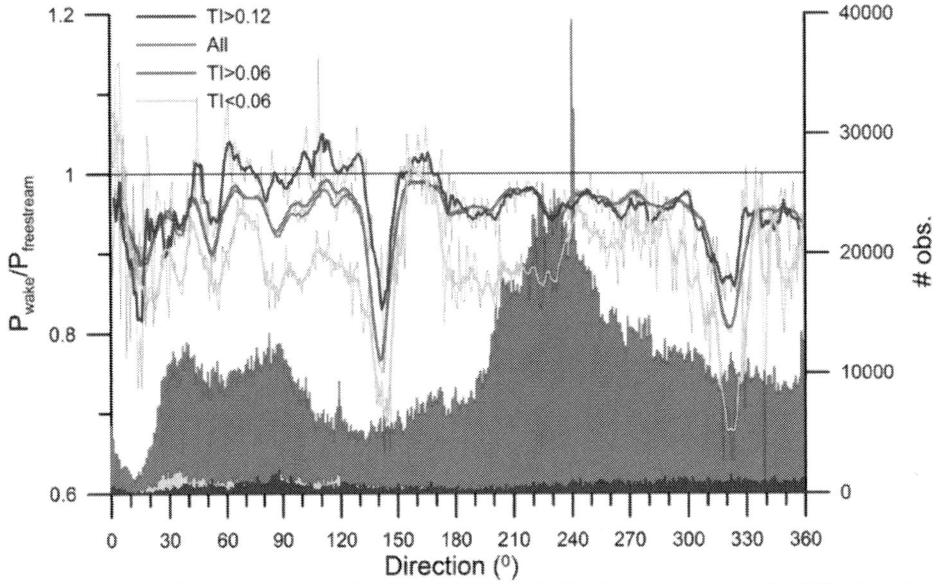

Figure 6. Normalized power output by direction classified by turbulence intensity (*TI*) (from the nacelle anemometer). Note the data are binned by 1° (thin gray lines) and a running mean of 3 is applied (thicker colored lines).

If the normalized power is conditionally sampled by both wind direction and *TI* then it is apparent that wake losses are larger in low turbulence conditions and vice versa. The impact on the wind farm power output is that normalized power is slightly higher (~0.5-1.0%) if $TI > 0.12$ and normalized power is 7% lower if $TI < 0.06$. This is a similar value to that given for efficiency changes (relative normalized power) at Horns Rev and Nysted wind farms of 0.98-1.4% per 1% change in turbulence intensity [2], suggesting this may be a generalizable response function for offshore wind farms in relatively high wind speed regimes (e.g. in the seas of northern Europe).

Figure 7 shows the normalized power output (total average power from all wind turbines in the array / freestream power calculated using the power curve and freestream wind speed) conditionally sampled by both direction and stability class. This analysis indicates that observations in the stable class typically have the highest wake losses, and the near-neutral class the lowest wake losses. Unexpectedly, the unstable class appears to experience higher wake losses than the near-neutral class. However, this is not completely consistent by direction. It is possible that from 290° over north, the measurements at the meteorological mast are being impacted by wakes from the closest wind turbines.

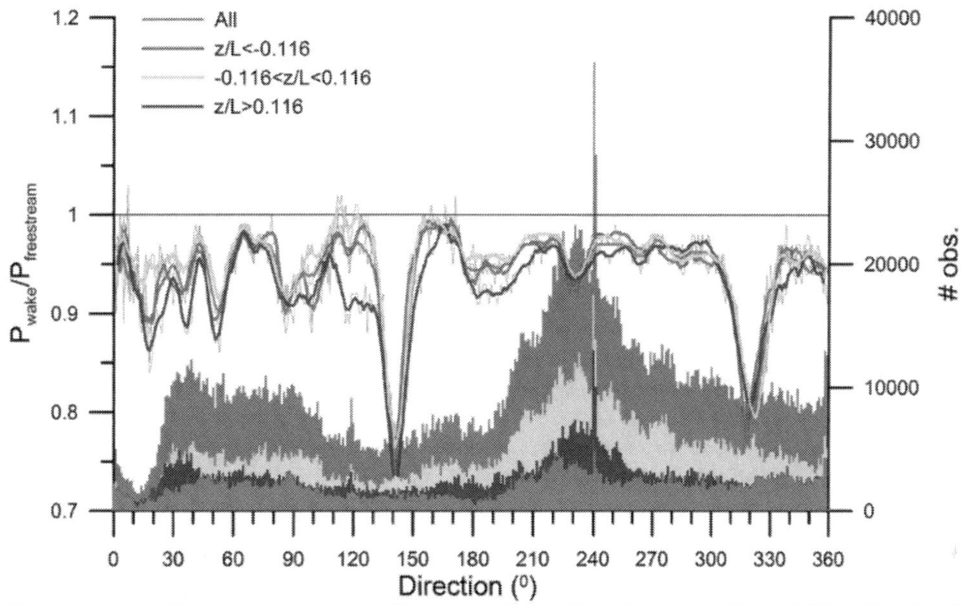

Figure 7. Normalized power output by direction and conditionally sampled by stability *z/L* derived from the meteorological mast using the Beljaars approach (BJ116-70). Stable conditions are shown in blue, unstable in red and near-neutral in green.

Deconvoluting the impacts of wind speed, *TI* and stability on wake depth, width and dissipation is confounded by their co-dependence. For example, different stability classes tend to have different distributions of wind speeds and turbulence intensity. To examine this, the data were selected for a single wind speed bin (8-10 ms^{-1}). Wake losses were larger due to the higher thrust coefficients in this wind speed range but the overall difference in normalized power varied by only 1% overall between the stable and near-neutral class and the unstable and near-neutral classes. Mean wind speed is close to 9 ms^{-1} in all three cases, while mean *TI* is 8.0, 8.7 and 8.5% in the stable, near-neutral and unstable classes respectively. This result is thus in contrast to our previous analyses of data from Horns Rev and Nysted that had indicated a stronger relationship between efficiency (normalized power output) changes of -8.1% to +9.7% over a range of stabilities [2].

5. Conclusions

Observations at the OWEZ wind farm have been analyzed by stability and *TI* classes. Following previous analysis it is clear that wakes are influenced primarily by wind speeds and then by wind direction as a proxy for both spacing and for stability. *TI* is also shown to have a major impact on normalized power output that is of similar magnitude (about a 1% change in normalized power output for a 1% change in *TI*) as found at other offshore wind farms [2]. Analysis indicates that while *TI* is generally lower in the stable class this assumption cannot be uniformly applied due to the wide range of *TI* in both stable and unstable classes. Stability is also shown to have an impact on the normalized power output but the impact of stability on power losses due to wakes appears to be of lesser magnitude than *TI*, and less distinct than found for other offshore wind farms [2] which might be partly due to the height of measurements used to define stability or use of the nacelle anemometers to define turbulence intensity.

6. Acknowledgements

This research was supported by the SOWFA project subcontract XFC-5-42084-01 and also in part by NSF#1464383, DE-EE0005379. We acknowledge data provision from the OWEZ data set by Vestas.

7. References

1. Churchfield, M.J., et al. *A comparison of the dynamic wake meandering model, large-eddy simulation, and field data at the Egmond aan Zee Offshore wind plant.* in *AIAA Science and Technology Forum and Exposition (SciTech 2015)* 2015. Florida 5-9 January 2015.
2. Barthelmie, R.J., K.S. Hansen, and S.C. Pryor, *Meteorological controls on wind turbine wakes.* Proceedings of the IEEE, 2013. **101(4)**: p. 1010-1019.
3. Barthelmie, R.J., *The effects of atmospheric stability on coastal wind climates.* Meteorological Applications, 1998. **6**: p. 39-47.
4. Sathe, A., S.-E. Gryning, and A. Peña, *Comparison of the atmospheric stability and wind profiles at two wind farm sites over a long marine fetch in the North Sea.* Wind Energy, 2011. **14(6)**: p. DOI: 10.1002/we.456.
5. Stull, R.B., *An introduction to boundary layer meteorology*. ISBN 90-277-2768-6 ed1988, Dordrecht: Kluwer Publications Ltd. 666.
6. Beljaars, A.C.M., A.A.M. Holtslag, and R.M. van Westrhenen, *Description of a software library for the calculation of surface fluxes.*, 1989, KNMI: De Bilt, Netherlands.
7. Sun, J., et al., *Turbulence Regimes and Turbulence Intermittency in the Stable Boundary Layer during CASES-99.* Journal of the Atmospheric Sciences, 2012. **69**(1): p. 338-351.
8. Larsen, T.J., et al., *Validation of the Dynamic Wake Meander Model for Loads and Power Production in the Egmond aan Zee Wind Farm* Wind Energy, 2012. **DOI: 10.1002/we.**
9. Hansen, K.S., G.C. Larsen, and S. Ott, *Dependence of offshore wind turbine fatigue loads on atmospheric stratification.* Journal of Physics: Conference Series 2014. **524**: p. doi:10.1088/1742-6596/524/1/012165.
10. Grachev, A.A. and C.W. Fairall, *Dependence of Monin-Obukhov stability parameter on the bulk Richardson number for unstable conditions over the ocean.* Journal of Applied Meteorology 1997. **36**: p. 406-415.
11. Motta, M., R.J. Barthelmie, and P. Vølund, *The influence of non-logarithmic wind speed profiles on potential power output at Danish offshore sites.* Wind Energy, 2005. **8**: p. 219-236.
12. Barthelmie, R.J. and L.E. Jensen, *Evaluation of power losses due to wind turbine wakes at the Nysted offshore wind farm.* Wind Energy, 2010. **13**: p. 573-586.
13. Barthelmie, R.J., et al., *Modelling and measurements of power losses and turbulence intensity in wind turbine wakes at Middelgrunden offshore wind farm.* Wind Energy, 2007. **10**(DOI: 10.1002/we.238): p. 217-228.
14. Alblas, L.M., *Power output of offshore wind farms in relation to atmospheric stability*, in *Technical University of Delft in cooperation with Vestas Wind Systems A/S*2012: Delft. p. 190.

A spectral model of stably stratified surface-layer turbulence

A. Segalini[1], J. Arnqvist[2], I. Carlén[3], H. Bergström[2] & P. H. Alfredsson[1]

[1] Linné FLOW Centre, KTH Mechanics, Osquars Backe 18, 10044 Stockholm
[2] Dept. Earth Sc., Uppsala University, Villavagen 16, 75236 Uppsala
[3] TG Teknikgruppen AB, Kemistvägen 6, 18379 Täby

E-mail: segalini@mech.kth.se

Abstract. A new model to determine the spectral velocity tensor in a stably stratified flow is proposed. This model is complementary to the Mann model as it solves the stratified inviscid Rapid Distortion Theory equations analytically, allowing for the determination of the single and two-point velocity spectra as well as the temperature-velocity cross-spectra. The model has been here calibrated and validated against field measurements conducted over a forested area with measurements up to 140 m, therefore covering a region of interest for wind-energy applications.

1. Introduction

The knowledge of the instantaneous velocity field characteristics of the atmospheric boundary layer has crucial importance for an efficient wind-energy utilisation. However, the flow over the rotor is highly turbulent and depends on a large number of parameters, like the boundary-layer height, the geostrophic-wind variation and the surface inhomogeneities, to mention some. Wind turbines must be designed to operate in and withstand such a turbulent environment. The current physical understanding of turbulent motions comes from experiments (in real fields or wind tunnels) and numerical simulations. In the latter case the high Reynolds number of the flow becomes challenging because of the increasing scale separation between large and small eddies both inside the domain and at the upstream domain boundaries. In order to obtain a realistic turbulent velocity field, we must force a realistic turbulent field at the inlet of the computational domain with prescribed characteristics or, alternatively, use a precursor simulation that serves as input for the main simulation. The latter approach provides a realistic field but it has the cost of additional computational power that is needed to provide the input for the main simulation. The former method, on the other hand, is quite efficient but necessitates of some information about the boundary-layer velocity field and its temporal characteristics. The model developed by Mann [1], for instance, provides a description of the spectral velocity tensor (Φ_{ij}, where i and j correspond to the fluctuating velocity components u', v' and w', indicating the streamwise, lateral and vertical velocity components, respectively) that can be used to determine the single- and two-point spectra of velocity time series inside and at the boundaries of the numerical domain. This approach provides some information about the frequency distribution of the energy so that a synthetic velocity field can be created that has the same one- and two-point statistics of the observed one. Since 1994, the model has become the standard tool to synthesise turbulence time series at the inlet of many numerical simulations aimed at the characterisation of unsteady loads and aeroelastic effects.

However, the Mann model provides results in absence of density stratification and empirical approaches must be used to extend its application to non-neutral conditions (see for instance [2]). The absence of a density equation in the Mann model limits its use as an in-flow tool since, in addition to the initial velocity spectrum, models with non-neutral stratification need a description of the initial temperature spectrum.

The Mann model is based on Rapid Distortion Theory (RDT) where the turbulence is driven by the mean-flow distortion rather than by the non-linear distortions by turbulence itself. As the theory is time-dependent, a characteristic time scale is introduced to stop the structures evolution at a time related to their expected lifetime. In order to extend the model to stratified conditions, a solution of the inviscid stratified RDT equations with homogeneous shear is proposed, extending the solution of Hanazaki and Hunt [3] to the two horizontal velocity components. The analytical solution allows for the determination of the spectra at any given time, starting from a prescribed initial condition. Following the same approach adopted by Mann [1], a model for the spectral velocity tensor in the atmospheric surface layer is obtained where the spectral tensor, assumed to be isotropic at the initial time, evolves until the break-up time, where the spectral tensor is supposed to achieve its equilibrium state observed in surface-layer turbulence. A similar approach was developed by Chougule [4] by numerically integrating the stratified RDT equations.

The present paper aims at assessing whether or not the new extended model (discussed in section 2) can be used to simulate realistic atmospheric turbulence (as the Mann model does) with the additional information that regards the heat fluxes and, in general, the temperature-velocity correlations for different stability classes. Here only stably stratified conditions will be discussed, although the model can be extended to unstable stratification as well. Data from a measurement campaign over a forested site (section 3) will be used to calibrate and validate the model (section 4). Section 5 will conclude the paper with some final remarks.

2. Mathematical model

Let us consider a homogeneously stratified flow (with constant $\mathrm{d}\rho_0/\mathrm{d}z$) with a uniform mean velocity shear in the z-direction. By normalising physical variables with the characteristic length and velocity scale L and U_s (for the moment left unspecified), the dimensionless inviscid RDT governing equations can be written in Fourier space (where the $\hat{\cdot}$ indicates the Fourier transform operator) as

$$\frac{\mathrm{d}\hat{u}}{\mathrm{d}t} = \alpha\left(\frac{2k_x^2}{k^2} - 1\right)\hat{w} + \frac{k_x k_z}{k^2}\hat{\rho}\,, \tag{1}$$

$$\frac{\mathrm{d}\hat{v}}{\mathrm{d}t} = \alpha\frac{2k_x k_y}{k^2}\hat{w} + \frac{k_y k_z}{k^2}\hat{\rho}\,, \tag{2}$$

$$\frac{\mathrm{d}\hat{w}}{\mathrm{d}t} = \alpha\frac{2k_x k_z}{k^2}\hat{w} + \left(\frac{k_z^2}{k^2} - 1\right)\hat{\rho}\,, \tag{3}$$

$$\frac{\mathrm{d}\hat{\rho}}{\mathrm{d}t} = N^2\hat{w}\,, \tag{4}$$

where $\alpha = \mathrm{d}U/\mathrm{d}zL/U_s$ is the dimensionless mean wind shear, $N = \left[-\left(g/\rho_0\right)\left(\mathrm{d}\rho_0/\mathrm{d}z\right)\right]^{1/2} L/U_s$ is the scaled Brunt-Väisälä frequency, g is the gravitional acceleration, $k = \left(k_x^2 + k_y^2 + k_z^2\right)^{1/2}$ is the wavenumber magnitude and the density is normalised as $\rho/(\rho_0\mathrm{Fr}^2)$ where $\mathrm{Fr} = U_s/\left(gL\right)^{1/2}$ is the characteristic Froude number.

The equation for the evolution of the wavenumber vector, $\mathbf{k}(t) = (k_x, k_y, k_z)$, is [5]

$$\frac{\mathrm{d}k_x}{\mathrm{d}t} = \frac{\mathrm{d}k_y}{\mathrm{d}t} = 0 \quad , \quad \frac{\mathrm{d}k_z}{\mathrm{d}t} + \alpha k_x = 0 \quad \longrightarrow \quad \mathbf{k} = (k_{x0}, k_{y0}, k_{z0} - \alpha k_{x0} t) \; , \tag{5}$$

where $(k_{x,0}, k_{y,0}, k_{z,0})$ is the initial wavenumber vector with magnitude k_0. As a notation, the horizontal wavenumber $k_h = \left(k_x^2 + k_y^2\right)^{1/2}$ is introduced, a quantity that, according to equation (5), is independent of time.

As discussed in the introduction, Hanazaki & Hunt [3] provided an analytical solution of the stratified RDT equations for equations (3) and (4). The solution is based on the introduction of the variable $\zeta = ik_z/k_h$ in the equation which results from the combination of the density and vertical-velocity equations, leading to the differential equation

$$\left(1 - \zeta^2\right) \frac{\mathrm{d}^2 \hat{\rho}}{\mathrm{d}\zeta^2} - 2\zeta \frac{\mathrm{d}\hat{\rho}}{\mathrm{d}\zeta} + \lambda \left(1 + \lambda\right) \hat{\rho} = 0 \, , \tag{6}$$

where

$$\lambda \left(1 + \lambda\right) = -\frac{N^2 k_h^2}{\alpha^2 k_x^2} \, . \tag{7}$$

Equation (6) can be solved in terms of Legendre functions of first and second order as

$$\hat{\rho} = A P_\lambda \left(\zeta\right) + B Q_\lambda \left(\zeta\right) \, , \tag{8}$$

while the vertical-velocity Fourier transform is

$$\hat{w} = -\frac{i\alpha k_x}{N^2 k_h} \left[A P'_\lambda \left(\zeta\right) + B Q'_\lambda \left(\zeta\right)\right] \, , \tag{9}$$

in which the primes indicate the derivative of the Legendre functions with respect to their argument ζ.

After some algebra [6], it is possible to get the analytical solution for the other two velocity components as

$$\hat{u} = \hat{u}_0 + \frac{\alpha}{N^2 k_h^2} \left\{ A \left[k_x^2 \zeta P'_\lambda \left(\zeta\right) - k_y^2 P_\lambda \left(\zeta\right)\right]_{\zeta_0}^{\zeta} + B \left[k_x^2 \zeta Q'_\lambda \left(\zeta\right) - k_y^2 Q_\lambda \left(\zeta\right)\right]_{\zeta_0}^{\zeta} \right\} \, , \tag{10}$$

$$\hat{v} = \hat{v}_0 + \frac{\alpha k_x k_y}{N^2 k_h^2} \left\{ A \left[\zeta P'_\lambda \left(\zeta\right) + P_\lambda \left(\zeta\right)\right]_{\zeta_0}^{\zeta} + B \left[\zeta Q'_\lambda \left(\zeta\right) + Q_\lambda \left(\zeta\right)\right]_{\zeta_0}^{\zeta} \right\} \, , \tag{11}$$

where $\zeta_0 = ik_{z,0}/k_h$ indicates the value of ζ at the initial time.

The constants A and B can be determined through the initial-density and vertical-velocity fluctuations Fourier transforms by solving the linear system

$$\begin{bmatrix} P_\lambda \left(\zeta_0\right) & Q_\lambda \left(\zeta_0\right) \\ P'_\lambda \left(\zeta_0\right) & Q'_\lambda \left(\zeta_0\right) \end{bmatrix} \begin{bmatrix} A \\ B \end{bmatrix} = \begin{bmatrix} \hat{\rho}_0 \\ \left(iN^2 k_h \hat{w}_0\right) / \left(\alpha k_x\right) \end{bmatrix} \, . \tag{12}$$

The time evolution of the velocity and density perturbation vector $\mathbf{q} = [\hat{u}, \hat{v}, \hat{w}, \hat{\rho}]^T$ from its initial state $\mathbf{q}_0 = [\hat{u}_0, \hat{v}_0, \hat{w}_0, \hat{\rho}_0]^T$ can be written as $\mathbf{q} = \mathbf{M}\mathbf{q}_0$ where the matrix $\mathbf{M}$ is given by

$$\mathbf{M}\left(t; \mathbf{k}_0\right) = \begin{bmatrix} 1 & 0 & F_{uw_0} & F_{u\rho_0} \\ 0 & 1 & F_{vw_0} & F_{v\rho_0} \\ 0 & 0 & F_{ww_0} & F_{w\rho_0} \\ 0 & 0 & F_{\rho w_0} & F_{\rho\rho_0} \end{bmatrix} \, , \tag{13}$$

with the coefficients obtained from equations (10), (11), (9) and (8) and reported in [6].

The generic time-dependent spectral matrix $\mathbf{\Phi}(t;\mathbf{k})$ is now introduced as

$$\mathbf{\Phi}(t;\mathbf{k}) = \Re\left[\overline{\mathbf{q}(t)\mathbf{q}^{T*}(t)}\right] = \begin{bmatrix} \Phi_{xx} & \Phi_{xy} & \Phi_{xz} & \Phi_{x\rho} \\ \Phi_{xy} & \Phi_{yy} & \Phi_{yz} & \Phi_{y\rho} \\ \Phi_{xz} & \Phi_{yz} & \Phi_{zz} & \Phi_{z\rho} \\ \Phi_{x\rho} & \Phi_{y\rho} & \Phi_{z\rho} & \Phi_{\rho\rho} \end{bmatrix}, \tag{14}$$

where the $*$ superscript indicates the complex conjugate operator. The evolution of the spectral tensor (14) is given by

$$\mathbf{\Phi}(t;\mathbf{k}) = \mathbf{M}(t;\mathbf{k}_0)\,\mathbf{\Phi}(0;\mathbf{k}_0)\,\mathbf{M}^T(t;\mathbf{k}_0)\,. \tag{15}$$

Equation (15) is the key result of this work since it determines the evolution of the spectral tensor of velocity and density fluctuations starting from a generic initial condition $\mathbf{\Phi}(0;\mathbf{k}_0)$ after a generic time t.

The dimensionless initial velocity field can be assumed to be isotropic and characterised by means of the von-Kármán spectrum [1] that, according to the present scaling, becomes

$$\Phi_{ij}(0;\mathbf{k}_0) = \frac{Q_u}{4\pi}\frac{\delta_{ij}k_0^2 - k_{i,0}k_{j,0}}{\left(1+k_0^2\right)^{17/6}} \qquad \text{with} \qquad Q_u = \frac{\alpha_1 \epsilon^{2/3} L^{2/3}}{{U_s}^2}, \tag{16}$$

where ϵ is the turbulent kinetic energy dissipation rate and α_1 is the Kolmogorov constant [7]. The corresponding isotropic dimensionless density spectrum can be also written as a power law [8] as

$$\Phi_{\rho\rho}(0;\mathbf{k}_0) = \frac{Q_\rho}{\left(1+k_0^2\right)^{11/6}} \qquad \text{with} \qquad Q_\rho = \frac{5\beta_1}{6\pi\alpha_1}\frac{1}{\mathrm{Fr}^4}\frac{\epsilon_\rho {U_s}^2}{{\rho_0}^2\epsilon}Q_u\,, \tag{17}$$

where ϵ_ρ is the dissipation rate of half the density variance and β_1 is a universal constant [7]. Equations (16) and (17) characterise the initial spectral tensor and allow some flexibility by means of the parameters Q_u and Q_ρ that represent two important properties of the initial isotropic field: Q_u is the normalised turbulent kinetic energy dissipation of the velocity field, while Q_ρ/Q_u is the ratio between the density and turbulent kinetic energy dissipation. Starting with isotropic conditions brings the important property that the initial density-velocity covariances are all zero, which reduces the addition in non-neutral stratification to the density spectrum alone.

Following [1], the dimensionless eddy break-up time scale under neutral stratification can be written in terms of the hypergeometric function

$$\tau_{M94} = \frac{\Gamma}{\alpha}k^{-2/3}\left[{}_2F_1\left(\frac{1}{3},\frac{17}{6};\frac{4}{3};-k^{-2}\right)\right]^{-1/2}, \tag{18}$$

where Γ is the anisotropy parameter. The presence of the oscillation time scale introduced by the Brunt-Väisälä frequency suggests a new time scale, τ, here assumed to be of the form

$$\frac{1}{\tau} = \frac{1}{\tau_{M94}} + cN \longrightarrow \tau = \frac{\tau_{M94}}{1+cN\tau_{M94}}\,. \tag{19}$$

The advantage of this assumption is that the neutral time scale is kept exactly as in [1] and the stability dependence comes from an independent parallel coupling of τ with N by the constant c. Besides the scaling quantities U_s and L and the initial spectral tensor parameters Q_u and Q_ρ, two other parameters are therefore needed to complete the model by determining the break-up time scale, namely Γ and c. By assuming that U_s is equal to the streamwise velocity standard deviation σ_u, and according to the observations done in Ryningsnäs, some of these parameters can be assumed to be constant (Γ and c) or function of the Richardson number (Q_ρ/Q_u and z/L), as discussed in [6].

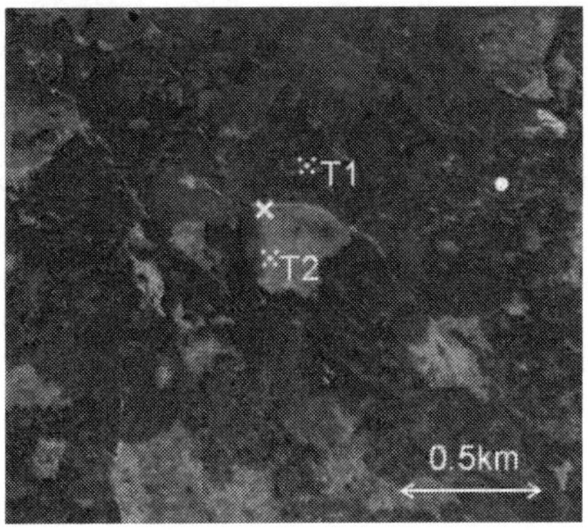

Figure 1. Landscape of the Ryningsnäs site.

3. Experimental setup

The experiment took close to the inland mansion of Ryningsnäs in Sweden, located 30 km inland from the Swedish south-eastern coast. A complete description of the site and the equipment has been reported by Arnqvist *et al.* [9] and here only some brief details are recalled. A 138 m tall meteorological tower was located in the northwestern corner of a 200 m by 250 m large clearing surrounded by a forested area consisting of predominantly Scots Pine trees (with a mean canopy height of $h_c = 20$ m). Two wind turbines, labeled T1 and T2 in figure 1, are located approximately 200 m (corresponding to 2.2 rotor diameters) from the tower in the southern and north-eastern directions, respectively. The hub height is $z_{\mathrm{hub,T1}} = 100$ m and $z_{\mathrm{hub,T2}} = 80$ m, for the turbine T1 and T2, respectively.

The experiment ran between November 2010 and February 2012, yielding a total of 10560 hours of available measurements. 6 tri-axial Metek sonic anemometers were placed at $z =$ 40, 59, 80, 98, 120, 137.7 m complemented by additional Risø PT-100 temperature sensors at $z =$ 40, 98, 120, 137.7 m. The data were first sorted into 30-min blocks and, for each 30-min time series, statistics were computed. At each measurement height the velocity components were also rotated to a Cartesian reference frame xyz where the average lateral velocity, V, was zero. The data were sorted according to the velocity at $z = 100$ m (in bins of 1 m/s) and to the ratio h_c/L_o, where L_o is the Obukhov length, evaluated by means of the conventional definition

$$L_o = -\frac{u_*^3\Theta_0}{\kappa g \langle w'\theta' \rangle}, \tag{20}$$

where $\kappa = 0.4$ is the von Kármán constant, $\langle w'\theta'\rangle$ is the vertical temperature flux determined from the sonic anemometer at the lowest height, Θ_0 is the reference temperature taken from the temperature sensor at the lowest height, and u_* is the friction velocity that is consistently evaluated at the lowest height ($z = 40$ m), by $u_* = \left[\langle w'u'\rangle^2 + \langle v'w'\rangle^2\right]^{1/4}$.

The western sector (with direction within [240°, 280°]) was defined as the sector where the influence from the two turbines and the clearing was minimal (see [9] for a discussion regarding the nearby clearing effects). Luckily, the western sector was also coincident with the most probable mean wind direction as well as being the direction with the longest upwind forest cover.

4. Results

The model parameters have been calibrated against single-point spectra measured at Ryningsnäs in the velocity bin where $U(z = 100\ \mathrm{m}) \approx 8\ \mathrm{ms}^{-1}$. Figure 2 shows a comparison between the measured and modelled pre-multiplied spectra. Obviously the agreement here is a result of the fitting method (which determines the parameters L, Q_u, Q_ρ, Γ and c by maximising the agreement between the measured and modelled pre-multiplied single-point velocity and temperature spectra at some frequencies), although it is worth to notice that the $u\theta$ spectrum was not used in the parameters determination, which can be considered here as an indicator of the quality of the model. The single-point temperature spectra have not been shown as the measured spectra for near-neutral to moderately stable conditions shows limited quality with large high frequency scatter. This is a known problem of sonic anemometers and does not affect the co-variances [10]. Due to the high frequency scatter causing upward tails of the spectra and the subsequent problem with model comparison the temperature spectra have not been considered in the present analysis.

At this point, it is worth computing the integral of the spectra providing the measured velocity and temperature covariances, shown in figure 3. It is visible that the model is able to compute the velocity variances and the shear stress $\langle u'w' \rangle$ as well as the heat fluxes $\langle u'\theta' \rangle$ and $\langle w'\theta' \rangle$. All these integral quantities appear to be well estimated, underlining the fact that the present model complements [1, 2, 4]. The only information going in to the model is the background stratification and shear as well as horizontal velocity variance. It seems thus that the mean distortion by temperature and velocity gradients is enough to reproduce the same distribution of energy between different Reynold stress components and heat fluxes that is observed in nature.

Both single-point spectra and associated integrals are however a direct result of the fitting of the spectra and their agreement with the measurements is expected. The fitting process was performed by using only single-point data with no information about two-point spectra. Therefore a comparison of the model prediction with two-point spectra represents a way to validate the proposed model. Figures 4 and 5 show two-point data that assess how well the model performs. Figure 4 shows the covariance between two different heights for the uu, vv, ww cross spectra in three different stability conditions (from nearly neutral to stable). The covariances are here normalised with the product of the standard deviations of the relative velocity components at the heights z_1 and z_2, and plotted against the height difference $\Delta z = |z_2 - z_1|$ normalised by the average wind speed $\overline{U} = [U(z_1) + U(z_2)]/2$. The advantage of this normalisation is discussed in [11] where it is demonstrated that the data collapse well. As visible in the figure, the modelled covariances have the same qualitative behaviour of the measured data, although with a slight overestimation of the covariances (especially in uu). The stability trend is correctly predicted for all the velocity components.

Figure 5 shows the comparison between the phase of the cross-spectra between the two different heights (for a frequency of 0.01 Hz) against the variable $\Delta z/\overline{U}$ where a monotonic trend is visible (except for ww that is scattered around a zero-phase shift). Here the model is able to qualitatively follow the data trend but the phase shift is slightly underestimated. No significant stability trend can be discerned in both measurements and modelled phases. Noteworthy, the phases follow the inequality $\phi_w < \phi_u < \phi_v$ proposed by [12] for neutral stratification.

5. Conclusions

In the present paper a new model to estimate the velocity and temperature spectral tensor of the atmospheric boundary layer is proposed. The model is based on an analytical solution of the stratified RDT equations for a homogeneous turbulent flow subjected to homogeneous shear and stable stratification. This approach allows to determine the time-wise evolution of the Fourier transform of the velocity and density fluctuations starting from an arbitrary initial

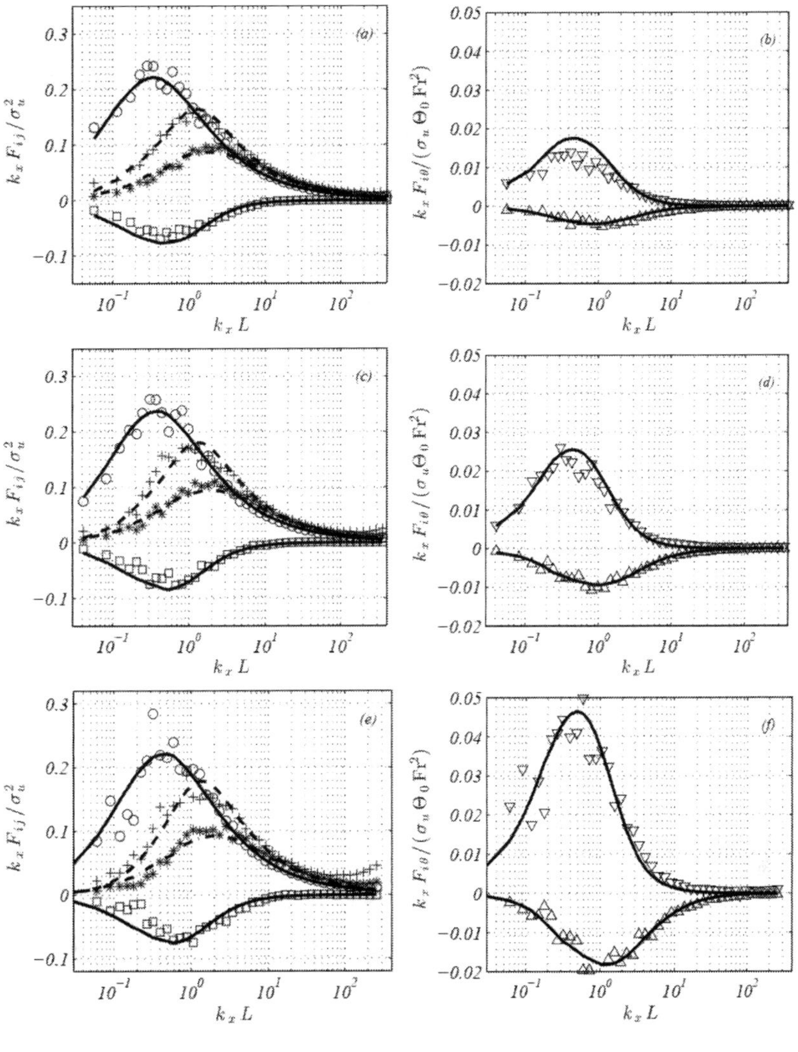

Figure 2. Comparison of the pre-multiplied spectra at $z \approx 138$ m and $U(z = 100 \text{ m}) \approx 8 \text{ ms}^{-1}$ for $(a-b)$ Neutral conditions (Ri ≈ 0.07), $(c-d)$ Near-Neutral Stable conditions (Ri ≈ 0.10) and $(e-f)$ Stable conditions (Ri ≈ 0.14). Experimental data: ($\circ$) F_{uu}, (+) F_{vv}, ($*$) F_{ww}, ($\square$) F_{uw}, ($\bigtriangledown$) $F_{u\theta}$, ($\triangle$) $F_{w\theta}$. Model: *(solid line)* F_{uu}, F_{uw}, $F_{u\theta}$, $F_{w\theta}$. *(dashed line)* F_{vv}. *(dash-dotted line)* F_{ww}.

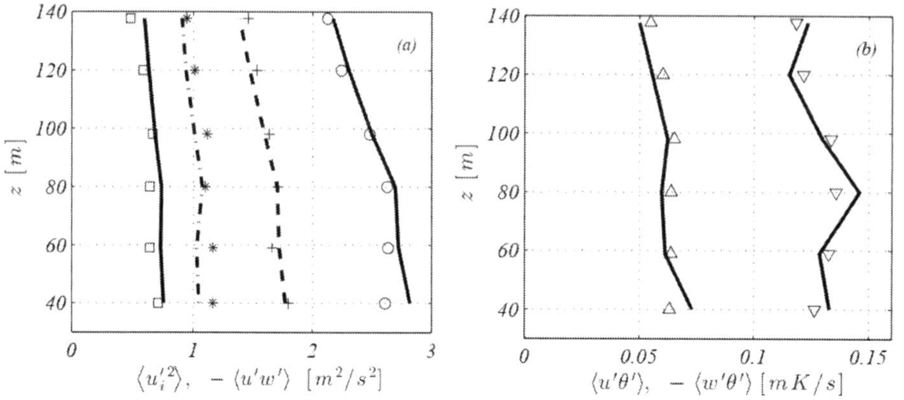

Figure 3. Comparison of the integral of the single-point spectra when $U(z = 100\ \text{m}) \approx 8\ \text{ms}^{-1}$ for Near-Neutral Stable conditions (Ri $\approx$ 0.10) for zero separation distance. Experimental data: ($\circ$) $\langle u'^2 \rangle$, (+) $\langle v'^2 \rangle$, ($*$) $\langle w'^2 \rangle$, ($\square$) $-\langle u'w' \rangle$, ($\bigtriangledown$) $\langle u'\theta' \rangle$, ($\triangle$) $-\langle w'\theta' \rangle$. Model: (*solid line*) $\langle u'^2 \rangle$, $-\langle u'w' \rangle$, $\langle u'\theta' \rangle$, $-\langle w'\theta' \rangle$. (*dashed line*) $\langle v'^2 \rangle$. (*dash-dotted line*) $\langle w'^2 \rangle$.

state, as long as the rapid-distortion closure is valid. By following the same approach of Mann [1], the equilibrium state of the time-dependent spectral tensor is determined by introducing a wavenumber-dependent time scale.

The model is based upon a number of scaling quantities and parameters. These have all been determined by optimising the comparison between model outputs and measurements over a forested area [2, 9]. According to our observations, the proposed model is able to determine reasonably well the single-point velocity spectra and the single-point velocity and temperature cross-spectra (including $F_{u\theta}$, which was purposefully kept out of the optimisation approach) for different Richardson numbers. Although the model is a function of only one initial velocity scale and one initial temperature scale it predicts the same distribution of energy between all the different covariances as is observed in the measurements. This suggests that in turbulent flows at the height span and velocity range that is interesting to wind energy the rapid distortion equations provide a valid approximation to the flow in a statistical sense. When evaluated against vertically separated two-point spectra, the model showed qualitative agreement with the available data both in terms of decay with separation distance and the tendency of decreasing correlation with increasing stratification.

References

[1] Mann J 1994 *J. Fluid Mech.* **273** 141–168
[2] Chougule A, Mann J, Segalini A and Dellwik E 2014 *Wind Energy* **18** 469–481
[3] Hanazaki H and Hunt J C R 2004 *J. Fluid Mech.* **507** 1–42
[4] Chougule A 2013 *Influence of atmospheric stability on the spatial structure of turbulence* Ph.D. thesis DTU Wind Energy
[5] Townsend A A 1976 *The structure of turbulent shear flow* (Cambridge University press)
[6] Segalini A and Arnqvist J 2015 *J. Fluid Mech.* **submitted**
[7] Kaimal J C, Wyngaard J C, Izumi Y and Coté O R 1972 *Q. J. Roy. Meteor. Soc.* **98** 536–589
[8] Goedecke G H, Ostashev V E, Wilson K D and Auvermann H J 2004 *Bound.-Lay. Meteor.* **112** 33–56
[9] Arnqvist J, Segalini A, Dellwik E and Bergström H 2015 *Bound.-Lay. Meteor.* DOI 10.1007/s10546-015-0016-x
[10] Nielsen M and Larsen S E 2002 *The influence of pulse-firing delays on sonic anemometer response characteristics* (American Meteorological Society) pp 139–142

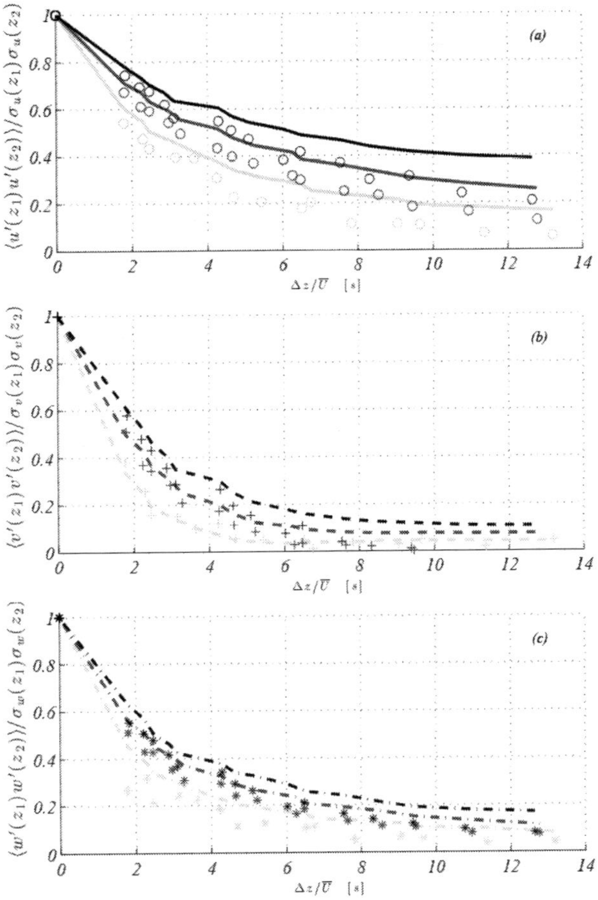

Figure 4. Comparison of the integral of the two-point cross-spectra when $U(z = 100 \text{ m}) \approx 8 \text{ ms}^{-1}$ for different stability conditions (*black* Ri ≈ 0.07, *blue* Ri ≈ 0.10 and *cyan* Ri ≈ 0.14) between different points at heights z_1 and z_2 between the u-component (a), v-component (b) and w-component (c). Experimental data: ($\circ$) $\langle u'(z_1)u'(z_2)\rangle$, (+) $\langle v'(z_1)v'(z_2)\rangle$, ($*$) $\langle w'(z_1)w'(z_2)\rangle$. Model prediction: (*solid line*) $\langle u'^2\rangle$, (*dashed line*) $\langle v'^2\rangle$, (*dash-dotted line*) $\langle w'^2\rangle$.

[11] Bergström H, Alfredsson P, Arnqvist J, Carlén I, Dellwik E, Fransson J, Ganander H, Mohr M, Segalini A and Söderberg S 2013 Wind power in forests: wind and effects on loads Tech. rep. Elforsk Rapport, 13:09
[12] Chougule A, Mann J, Kelly M, Sun J, Lenschow D H and Patton E G 2012 *J. of Turbulence* **13** 1–13

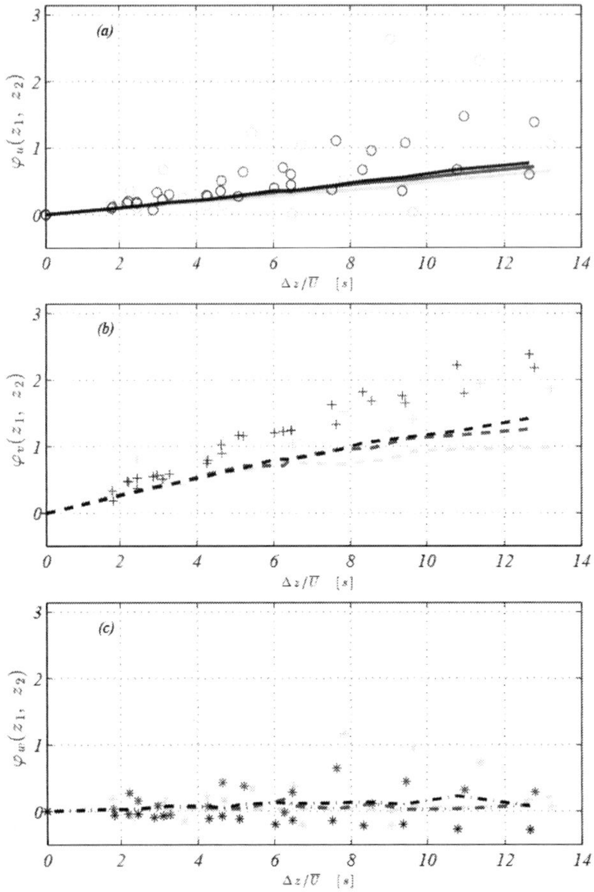

Figure 5. Comparison of the spectral phase when $U(z = 100\ \text{m}) \approx 8\ \text{ms}^{-1}$ and $f = 0.01$ Hz for different stability conditions (*black* Ri ≈ 0.07, *blue* Ri ≈ 0.10 and *cyan* Ri ≈ 0.14) between different points at heights z_1 and z_2 between the u-component (a), v-component (b) and w-component (c). Experimental data: (∘) $\varphi_u(z_1, z_2)$, (+) $\varphi_v(z_1, z_2)$, (∗) $\varphi_w(z_1, z_2)$. Model prediction: (*solid line*) $\langle u'^2 \rangle$, (*dashed line*) $\langle v'^2 \rangle$, (*dash-dotted line*) $\langle w'^2 \rangle$.

Using the coupled wake boundary layer model to evaluate the effect of turbulence intensity on wind farm performance

Richard J.A.M. Stevens[1,2], Dennice Gayme[1], Charles Meneveau[1]

[1]Department of Mechanical Engineering, The Johns Hopkins University, 3400 North Charles Street, Baltimore MD 21218, USA

[2] Department of Science and Technology and J.M. Burgers Center for Fluid Dynamics, University of Twente, P.O Box 217, 7500 AE Enschede, The Netherlands

E-mail: `r.j.a.m.stevens@jhu.edu`

Abstract. We use the recently introduced coupled wake boundary layer (CWBL) model to predict the effect of turbulence intensity on the performance of a wind farm. The CWBL model combines a standard wake model with a "top-down" approach to get improved predictions for the power output compared to a stand-alone wake model. Here we compare the CWBL model results for different turbulence intensities with the Horns Rev field measurements by Hansen *et al.*, Wind Energy 15, 183196 (2012). We show that the main trends as function of the turbulence intensity are captured very well by the model and discuss differences between the field measurements and model results based on comparisons with LES results from Wu and Porté-Agel, Renewable Energy 75, 945-955 (2015).

1. Introduction

For the design of wind farms it is important to understand the effect of the relative turbine positioning on the overall power output [1]. Two analytical approaches have been used for such evaluations. The first approach, is the use of classical wake models [2–8] to estimate the wind farm performance. Such models have been used in several wind farm optimization studies [9–12]. Wake models predict the wake deficits in the entrance region of the wind farm well, but have difficulty predicting the performance further downstream where many wakes interact [13–15]. "Top-down" models [16–18] can capture the vertical structure of the atmospheric boundary layer and the associated wake-atmosphere interactions in this fully developed regime of the wind farm better than wake models. Therefore "top-down" models have been used for the evaluation of the optimal spacing in very large wind farms [19, 20]. However, "top-down" models do not consider the effect of the relative turbine positioning [21, 22] on the wind farm performance. Previous works have used a one-way coupling between wake and "top-down" models [6, 17, 23] to improve the performance.

In this paper we use the recently introduced coupled wake boundary layer (CWBL) model [15, 22] to predict the influence of the turbulence intensity on the performance of Horns Rev. In particular, to investigate how well computationally inexpensive analytical models can predict the influence of this flow feature. The CWBL model combines the Calaf *et al.* [18, 24] "top-down" model and a classical wake model [3, 4, 6] through two-way coupling. The effect of the relative turbine positioning is captured by the wake model part and the interaction between the wind farm and the atmospheric boundary layer by the "top-down" part. It has been shown that the CWBL model [15, 22] gives improved predictions over either of its constitutive parts for the turbine power production in the fully developed regime of the wind farm. In this proceeding we start with a very short review of the basic concepts of the CWBL model in section 2. In section 3 the CWBL model results for different turbulence intensities are compared with field measurement data for Horns Rev [25] and corresponding LES results by Wu and Porté-Agel [26]. Section 4 concludes the paper.

2. The generalized CWBL model

In the CWBL model both the wake (section 2.1) and the "top-down" (section 2.2) model have one parameter that needs to be determined. An iterative procedure (section 2.3) is used to calculate these parameters from the complementary part of the CWBL model to make sure that a consistent CWBL system is obtained. For a detailed description of the (generalized) CWBL model we refer the reader to Refs. [15, 22].

2.1. Wake model part

The classic wake model [2–4, 6, 7] assumes that wind turbine wakes grow linearly with downstream distance and uses the following expression for the velocity (which is simplified as a top-hat distribution) in the turbine wakes [2, 3, 5]:

$$u = u_0 \left(1 - \frac{1-\sqrt{1-C_{\mathrm{T}}}}{(1+k_{\mathrm{w}}x/R)^2}\right) = u_0\left(1 - \frac{2a}{(1+k_{\mathrm{w}}x/R)^2}\right). \tag{1}$$

Here u_0 is the incoming free stream velocity, k_{w} is the wake expansion coefficient, R is the rotor radius, $C_{\mathrm{T}} = 4a(1-a)$ is the thrust coefficient with flow induction factor a, and x is the downstream distance with respect to the turbine, which is modeled as an actuator disk. Wake interactions are accounted for by adding the squared velocity deficits of interacting wakes [4]. The predicted normalized turbine power P_{T}/P_1 is given by

$$\frac{P_{\mathrm{T}}}{P_1} = \left(\frac{1}{N_{\mathrm{d}}}\sum_{k=1}^{N_{\mathrm{d}}}\frac{u(\mathbf{x}_{T,k})}{u_0}\right)^3. \tag{2}$$

Here the summation k is over all points (N_{d}) on the turbine disk and the velocity ratio is obtained by calculating the effect of all wind turbine wakes, including the wake interaction effects, at these locations. We use a uniform inflow, which is common in wake models. Computing the sum over the entire disk, instead of just considering the value at hubheight, ensures that the ground effect modeled by the image turbines sets in incrementally, i.e. there are no discontinuous changes in the wake behavior with increasing the streamwise distance.

The wake decay parameter at the entrance of the wind farm $k_{\mathrm{w},0}$ can be modeled as the ratio of friction velocity to the mean velocity at hubheight, which gives

$$k_{\mathrm{w},0} = \frac{\kappa}{\ln(z_{\mathrm{h}}/z_{0,\mathrm{lo}})}, \tag{3}$$

where $z_{0,\mathrm{lo}}$ is the roughness length of the ground surface, z_{h} the turbine hubheight, and $\kappa = 0.4$ is the von Kármán constant [2, 17, 27]. To match the wake expansion coefficient to the turbulence intensity, expressed based on the streamwise velocity fluctuations, we use the following logarithmic laws for the mean

$$\langle\overline{u}\rangle/u_* = \kappa^{-1}\ln(z/z_{0,\mathrm{lo}}) \tag{4}$$

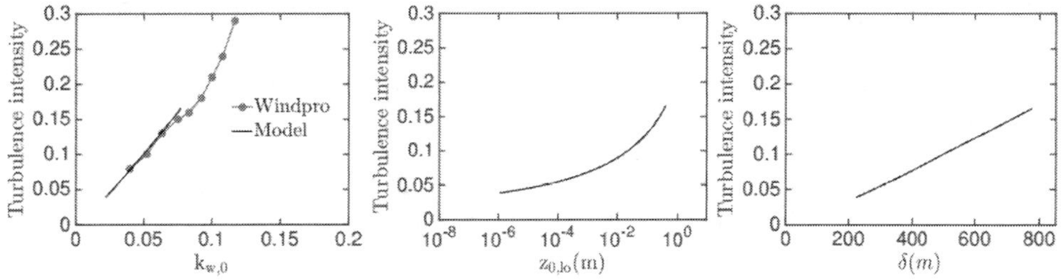

Figure 1. (a) The turbulence intensity at hubheight as a function of the wake expansion coefficient at the entrance of the wind-farm, $k_{w,0}$. The data points give the recommended relationship by Windpro [5], which is based on comparisons with field measurement data. The black line is based on the model equations (3), (6), and (7). The turbulence intensity as function of (b) the roughness length $z_{0,lo}$, and as function of (c) the boundary layer height for the data shown in panel (a).

and variance

$$\langle \overline{(u'^{+})^2} \rangle = B_1 - A_1 \log(z/\delta), \tag{5}$$

in the boundary layer [28–30].

Equation (4) gives the mean velocity profile in an atmospheric boundary layer, while equation (5) gives the streamwise velocity fluctuations in the boundary layer. Combining both equations gives the turbulence intensity at hubheight as

$$\mathrm{TI}_{\mathrm{hub}} = \frac{[\langle \overline{(u'^{+})^2} \rangle]^{1/2}}{\langle \overline{u} \rangle} = \frac{[B_1 - A_1 \log(z_{\mathrm{h}}/\delta)]^{1/2}}{\kappa^{-1} \ln(z_{\mathrm{h}}/z_{0,\mathrm{lo}})}. \tag{6}$$

The constants A_1 and B_1, are based on those measured in different high Reynolds number turbulent boundary layer experiments, see Marusic *et al.* [28] for an overview. They concluded that $A_1 \approx 1.25$ is an universal constant, while $B_1 \approx 1.5 - 2.1$ depends slightly on the flow geometry. Based on comparisons of our atmospheric boundary layer simulations with experiments we concluded that $B_1 \approx 1.6$ is an appropriate value for this case [30]. According to equation (6) the turbulence intensity at hubheight depends on the roughness length $z_{0,\mathrm{lo}}$ and the boundary layer height δ, while equation (3) states that the wake expansion coefficient at hubheight only depends on $z_{0,\mathrm{lo}}$.

In addition, we use that the atmospheric boundary layer height is given by [31]

$$\delta = Cu_*/f \tag{7}$$

where $f = 2\Omega \sin(\psi)$ with $\Omega = 2\pi/(24 \times 3600s) = 7.27 \times 10^{-5}\ 1/s$, $\psi = 55^\circ$ (the latitude of Horns Rev), and $C = 0.15$, which is typical for neutral atmospheric boundary layers [32–34]. We note that stratification affects the very largest scales the most, i.e. things happening at a height $z = \delta$, but that these effects are much weaker at hubheight, and below (i.e. where $z_{0,\mathrm{lo}}$ enters). We verified in the model calculations that small changes in C do not significantly influence the results. Using equations (3), (6), and (7) we obtain for each turbulence intensity the corresponding boundary layer height δ, the roughness height of the ground $z_{0,\mathrm{lo}}$, and wake expansion coefficient $k_{w,0}$. Figure 1 shows that the results agree very well with the recommended values by Windpro [5], which are based on comparisons with field measurement data.

2.2. The "top-down" model

In the CWBL model we use the "top-down" approach introduced by Calaf *et al.* [18]. This model is used to obtain the ratio of the mean velocity in the fully develop regime to the reference incoming velocity at hubheight

$$\frac{\langle \overline{u} \rangle (z_{\rm h})}{\langle \overline{u_0} \rangle (z_{\rm h})} = \frac{\ln \left(\delta_{\rm IBL}/z_{0,\rm lo} \right)}{\ln \left(\delta_{\rm IBL}/z_{0,\rm hi} \right)} \ln \left[\left(\frac{z_{\rm h}}{z_{0,\rm hi}} \right) \left(1 + \frac{D}{2z_{\rm h}} \right)^{\beta} \right] \left[\ln \left(\frac{z_{\rm h}}{z_{0,\rm lo}} \right) \right]^{-1}. \tag{8}$$

Here $\delta_{\rm IBL}$ indicates the height of the internal boundary layer in the fully developed regime of the wind farm, D the turbine diameter, $\beta = \nu_w^*/(1+\nu_w^*)$ and $\nu_w^* \approx 28\sqrt{\pi C_{\rm T}/(8 s_{\rm x} s_{\rm ye})}$, and $z_{0,\rm hi}$ denotes the roughness length of the wind farm, which is defined as

$$z_{0,\rm hi} = z_{\rm h} \left(1 + \frac{D}{2z_{\rm h}} \right)^{\beta} \exp \left(- \left[\frac{\pi C_{\rm T}}{8 s_{\rm x} s_{\rm ye} \kappa^2} + \left(\ln \left[\frac{z_{\rm h}}{z_{0,\rm lo}} \left(1 - \frac{D}{2z_{\rm h}} \right)^{\beta} \right] \right)^{-2} \right]^{-1/2} \right). \tag{9}$$

Here $s_{\rm ye}$ is the effective spanwise turbine spacing, which is obtained from the two-way coupling with the wake model part of the CWBL model. Due to the two-way coupling between the wake and the "top-down" model $s_{\rm ye}$ depends on parameters such as the streamwise distance between the turbines $s_{\rm x}$, the relative positioning of the turbines, and the wake coefficient in the fully developed regime of the wind farm $k_{\rm w,\infty}$.

The reason for changing the effective spanwise spacing $s_{\rm ye}$ in the "top-down" model is that this model considers a momentum balance averaged over the entire horizontal plane. In that model the horizontally averaged velocity thus depends on the friction velocity, which depends on the stresses that are imposed by the turbines in the wind farm. However, when the spanwise spacing between the turbines is very large this assumption is not longer valid. This fact can be clearly seen from cases with very small streamwise spacing, $s_{\rm x}$, and very large spanwise spacings, $s_{\rm y}$, which in the "top-down" model (that only depends on s) will result in limited wake effects. This assumption is obviously unrealistic, as even for a single line of turbines aligned in the wind direction for which $s_{\rm y} = \infty$ a significant power reduction is observed for downstream turbines. In the CWBL model we assume that the momentum analysis in the "top-down" part of the model should be performed over the control area that is representative of the region that is directly influenced by the wind turbine wakes. Since the wakes progress very far in the downstream direction, the streamwise spacing, $s_{\rm x}$, is unadjusted in the model, while the wake expansion is limited in the spanwise direction. In the CWBL modeling approach the effective spanwise spacing, $s_{\rm ye}$, that should be used is obtained from a two-way coupling procedure with a wake model. For further details about this procedure we refer the reader to section IV of Ref. [15].

2.3. Coupling

The wake and "top-down" parts of the CWBL model are coupled by demanding that both models give the same prediction for the turbine velocity in the fully developed regime of the wind farm. An iterative procedure is used to obtain the effective spanwise spacing, $s_{\rm ye}$, in equation (9) and the wake expansion rate, $k_{\rm w,\infty}$, in equation (1) in the fully developed regime. The effective spanwise spacing, $s_{\rm ye}$, is the spanwise dimension of the control volume size that is used to account for the large scale interactions with the atmosphere as explained above. Because in the wake model the turbine velocity in the fully developed regime depends on $k_{\rm w,\infty}$, while in the "top-down" model this velocity depends on $s_{\rm ye}$, these values need to be iterated until convergence is reached and this iteration is accomplished through the two-way coupling in the CWBL model. This procedure is described in detail for aligned and staggered wind farms in Stevens *et al.* [15]. To make sure that the effect of the wind farm geometry is taken into account this two-way coupling should be enforced for each wind direction separately, which is described in detail in Ref. [22]. Here we in addition, iterate the internal boundary later thickness in the fully developed regime as

$$\delta_{\mathrm{IBL}} = \delta \frac{u_{\mathrm{hi},*}}{u_*}, \tag{10}$$

where

$$\frac{u_{*\mathrm{hi}}}{u_*} = \frac{\ln\left(\delta_{\mathrm{IBL}}/z_{0,\mathrm{lo}}\right)}{\ln\left(\delta_{\mathrm{IBL}}/z_{0,\mathrm{hi}}\right)}. \tag{11}$$

is obtained from the CWBL model. We limit the internal boundary layer thickness to 500 meters when we compare with the LES data (as this was the boundary layer thickness used in the corresponding simulations by Porté-Agel *et al.* [35]) and 1000 meters (height of the thermal inversion) for the comparison with the field measurements.

In order to capture the entrance effects the CWBL model assigns a wake coefficient

$$k_{\mathrm{w,T}} = k_{\mathrm{w},\infty} + (k_{\mathrm{w},0} - k_{\mathrm{w},\infty}) \exp(-\zeta m), \tag{12}$$

to each turbine in the wind farm by interpolating between $k_{\mathrm{w},0}$ (the wake expansion coefficient at the entrance of the wind farm) and $k_{\mathrm{w},\infty}$ (the wake expansion coefficient in the fully developed regime, which is found using the iterative procedure). Here $\zeta = 1$ and m is the number of turbine wakes that interact with the turbine of interest. Thus the wake model part dominates in the entrance region of the wind farm, while the wake development further downstream is determined by the two-way coupling between the wake and "top-down" models that comprise the CWBL model. Therefore the CWBL predictions in the fully developed regime of the wind farm depend much less on the quadratic superposition of the wakes [4] than those of the stand alone wake model.

In this proceeding we assumed that all turbines operate with the same thrust coefficient C_T. In the field measurements [25] and the LES results [26] to which we compare the model predictions the wind speeds considered correspond to turbines operating in region II for which this assumption is reasonable. Because the measurement consider a very narrow range of wind speeds we neglected the variation of the power coefficient C_P with wind speed as this effect cancels out when relative powers are considered.

3. Results

Figure 2a shows the layout of the Horns Rev wind farm, which consists of 80 turbines with a hubheight of 70 meters and a turbine diameter of 80 meters, in a rectangular pattern. The streamwise spacing (East-West direction) is 7 turbine diameters and the spanwise spacing (North-South) is 6.95 turbine diameters. Figure 2 shows the power deficits as function of the downstream distance averaged for the wind directions $270° \pm 15°$, $222° \pm 15°$, and $312° \pm 15°$, and a wind speed of 8 ± 0.5 m/s with an average turbulence intensity of 7.0% (6.3% for $222° \pm 15°$) (see figure 4 of Hansen *et al.* [25]). The corresponding model results capture these measurements quite accurately. The figure also shows that the CWBL model agrees well with LES results from Wu and Porté-Agel [26].

The study of Hansen *et al.* [25] also analyzed the effect of turbulence intensity and atmospheric stability on the downstream power development. Hansen *et al.* [25] defined different stability classes ($cL = 3$, $cL = 2$, and $cL \leq 1$) based on the Obukhov length, see table 1 of Ref. [25]. In the remainder of the proceeding we will refer to these different cases using the naming convention introduced by Hansen *et al.* [25], i.e. "very stable" (case label $cL = 3$), "stable" (case label $cL = 2$), and "remaining" (case label $cL \leq 1$). From figure 7 of their paper we obtain that the average turbulence intensity of these three main stability classes they consider are 7.1%, 5.1%, and 3.9%, respectively, when the wind speed is 8 ± 0.5 m/s. We use these reported

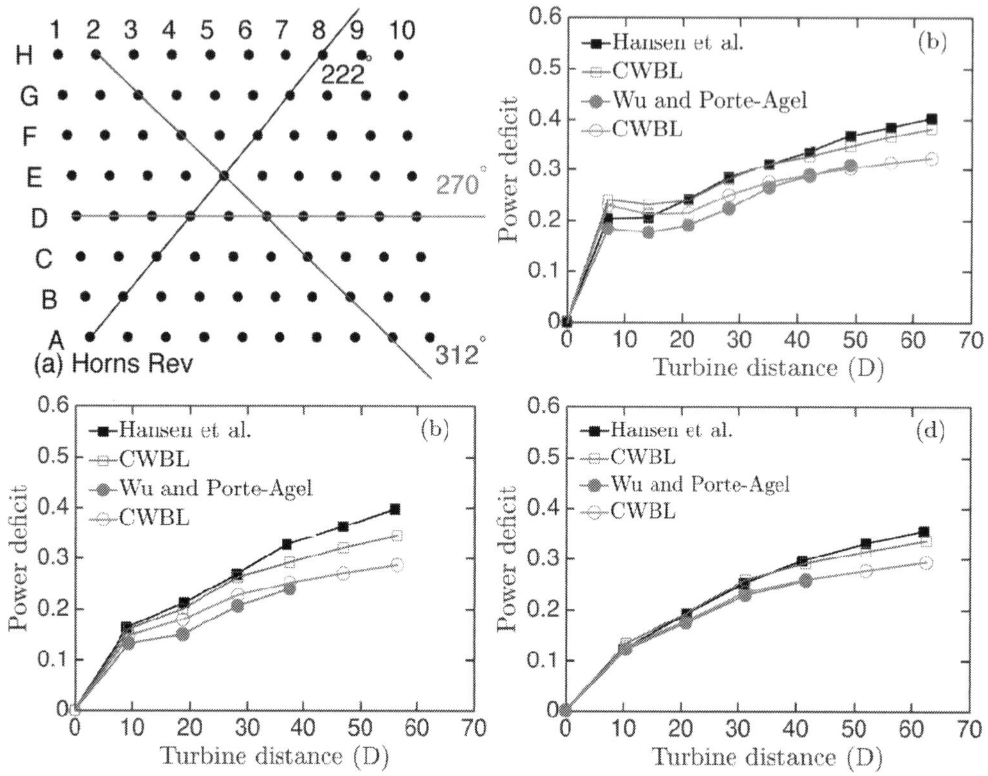

Figure 2. (a) Sketch of the Horns Rev layout indicating the wind directions 270°, 222°, and 312°. (b) The power deficit along the rows of wind turbines for the wind direction sectors $270 \pm 15°$, (c) 222 ± 15, and (d) $312 \pm 15°$. For the Hansen *et al.* [25] data the wind speed range is 8 ± 0.5 m/s with an average turbulence intensity of 7.0% [25] for $270 \pm 15°$ and $312 \pm 15°$, and 6.3% for $222 \pm 15°$. The corresponding CWBL predictions have been computed using a turbine thrust coefficient $C_{\mathrm{T}} = 0.80$. For the LES data of Wu and Porté-Agel [26] the turbulence intensity is 7.7% and we used $C_T = 0.78$ for the corresponding CWBL calculations [22, 26]. The squares provide a comparison between the data from Hansen *et al.* [25] and the CWBL model and the circles between the data from Wu and Porté-Agel [26] and the CWBL model. The data from Hansen *et al.* [25] and Wu and Porté-Agel [26] are digitally extracted from their figures.

turbulence intensities to distinguish the different atmospheric stability classes. We note that the effect of the atmospheric stability is not included in the generalized CWBL model, but that the effect of increasing turbulence intensity is included. Figure 3a shows the measured power deficits for these three stability classes as a function of the downstream direction for the $270 \pm 15°$ direction. Panel (b) of that figure shows the model results for the three corresponding turbulence intensities. The lower panels of figure 3 compare the field experiments and the model results for the three cases separately. In agreement with the experimental observations, the model predicts an increased velocity deficit for decreasing turbulence intensity. The reason is that a higher turbulence intensity results in a wake that better mixes with the surrounding flow. The figure shows that the effect of the turbulence intensity on the wind turbine performance is more pronounced further inside the wind farm.

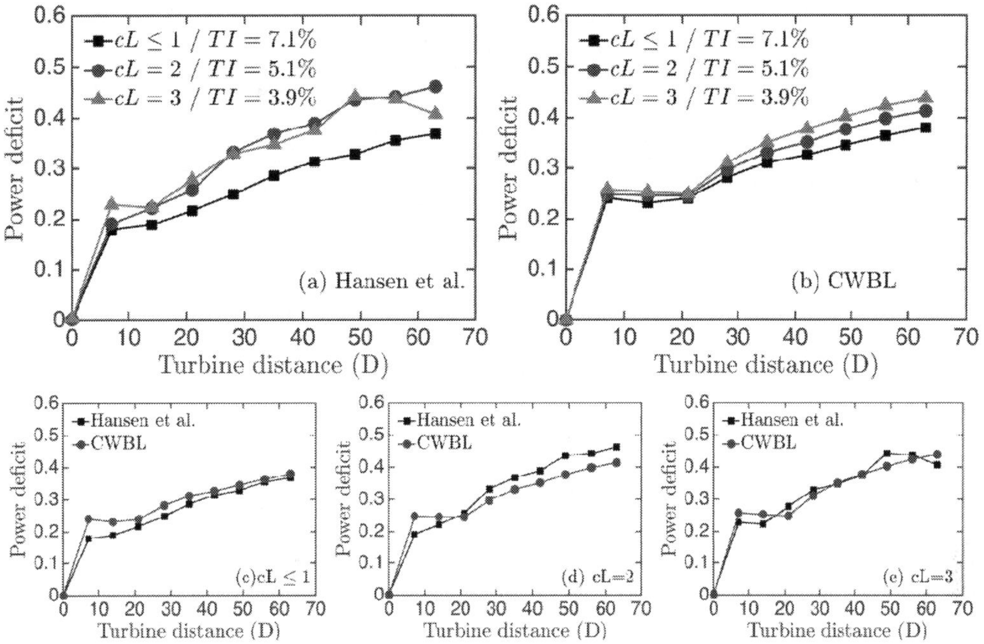

Figure 3. The power deficit along rows of wind turbines with a streamwise spacing of $7D$ averaged for the wind directions $270\pm15°$ and a wind speed of 8 ± 0.5 m/s. The field measurement results are grouped in the very stable ($cL = 3$), stable ($cL = 2$) and remaining ($cL \leq 1$) stability classes. The CWBL predictions have been computed for three different turbulence intensities, i.e. 7.1%, 5.1%, and 3.9%, which correspond to the average turbulence intensity for these three different stability classes. Panels (a) and (b) show a comparison of the three different cases for the field measurements and the CWBL model separately, and panels (c), (d), and (e) compare the field measurements and the corresponding CWBL predictions. The data from Hansen *et al.* [25] are digitally extracted.

Figure 4 shows the same results as in figure 3 for the wind-directions $222\pm15°$ and $312\pm15°$. A comparison with figure 3 reveals that the effect of changes in the turbulence intensity result in similar changes in the wind farm performance for the different wind directions. Considering that the effective streamwise distance is "similar" for these different wind directions (222°, 270°, 312°), i.e. $9.4D$, $7.0D$, and $10.4D$, respectively, and the overall wind turbine density is the same for these cases, this is reasonable. However, the corresponding experimental data seem to indicate a different trend, specifically a bigger influence of the turbulence intensity on the performance for the $222 \pm 15°$ and $312 \pm 15°$ cases than for the $270 \pm 15°$. We do not know the reason for this discrepancy. However, we do note that it is well known in the literature that the comparison with the Horns Rev data to model results tends to be more favorable for some wind-directions than for others [36].

Here we also emphasize that, apart from the turbulence intensity that is included in the model, the effect of the atmospheric stability is not accounted for in the model. So, for example, the effect of the atmospheric thermal stability on the wind shear is not taken into account. Generalizations to this approach can be made by including stability correction functions. Such a direction has been explored in a recent paper dealing with another topic (Sescu & Meneveau

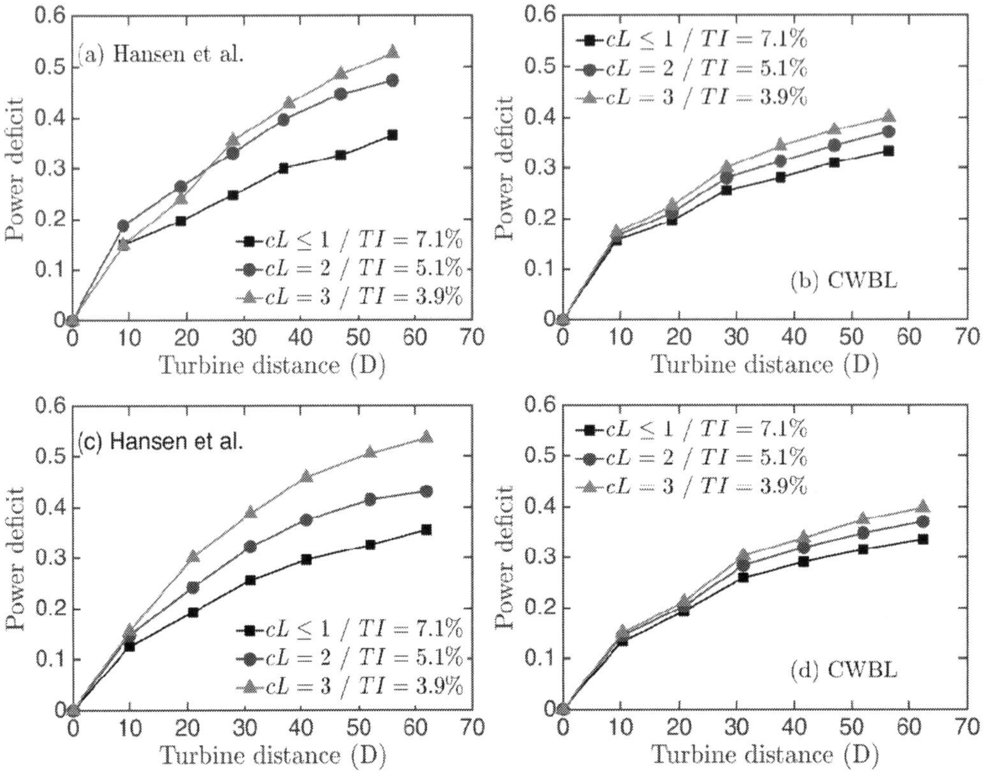

Figure 4. The power deficit for the wind directions $222° \pm 15°$ (top panels) and $312° \pm 15°$ (lower panels) for a wind speed of 8 ± 0.5 m/s. The field measurement results are grouped in the very stable ($cL = 3$), stable ($cL = 2$) and remaining ($cL \leq 1$) stability classes. The CWBL predictions have been computed for three different turbulence intensities, i.e. 7.1%, 5.1%, and 3.9%, which correspond to the average turbulence intensity for these three different stability classes. The left panels indicate the field measurement by Hansen *et al.* [25] and the right panels the CWBL predictions. The data from Hansen *et al.* [25] are digitally extracted.

[37]). Nevertheless the CWBL model is able to predict some of the main trends observed in the field data, which seems to indicates that the change of the turbulence intensity due to the thermal effects is one of the effects that influences the wind farm performance. In the field measurements the effect of the different atmospheric conditions seems stronger for some wind directions.

4. Summary

This paper demonstrates that the predicted performance trends by the coupled wake boundary layer (CWBL) model as a function of the turbulence intensity compare well with trends observed in field measurement data of Horns Rev [25]. The model has been compared with the field experiments using the reported turbulence intensities for the different atmospheric stability classes. We find that the model captures the trend that an increase in the turbulence intensity leads to decreased wake defect velocities due to enhanced mixing. The effect of the turbulence intensity is bigger further downstream. In the model predictions the change in the wind farm performance due to changes in the turbulence intensity are similar for different wind directions. Considering the similarity between these cases this is reasonable. However, in the field measurements the effect of the different atmospheric conditions seems to be stronger for some wind directions. Further research will be necessary to clarify this behavior. Further research will be necessary to clarify this behavior.

Acknowledgments

RJAMS's work is supported by the program 'Fellowships for Young Energy Scientists' (YES!) of the Foundation for Fundamental Research on Matter (FOM), which is financially supported by the Netherlands Organization for Scientific Research (NWO). Further partial support has been provided by the US National Science Foundation from grant IIA 1243482 (the WINDINSPIRE project).

References

[1] J. F. Herbert-Acero, O. Probst, P.-E. Réthoré, G. C. Larsen, and K. K. Castillo-Villar, *A Review of Methodological Approaches for the Design and Optimization of Wind Farms*, Energies **7**, 6930 (2014).
[2] P. B. S. Lissaman, *Energy Effectiveness of Arbitrary Arrays of Wind Turbines*, J. of Energy **3**, 323 (1979).
[3] N. O. Jensen, *A note on wind generator interaction*, Risø-M-2411, Risø National Laboratory, Roskilde (1983).
[4] I. Katić, J. Højstrup, and N. O. Jensen, *A simple model for cluster efficiency*, European Wind Energy Association Conference and Exhibition, 7-9 October 1986, Rome, Italy 407 (1986).
[5] J. Choi and M. Shan, *Advancement of Jensen (PARK) wake model*, EWEA Conference, Vienna, February 2013 (2013).
[6] A. Peña and O. Rathmann, *Atmospheric stability-dependent infinite wind-farm models and the wake-decay coefficient*, Wind Energy **17**, 1269 (2014).
[7] M. Bastankhah and F. Porté-Agel, *A new analytical model for wind-turbine wakes*, Renewable Energy **70**, 116 (2014).
[8] N. G. Nygaard, *Wakes in very large wind farms and the effect of neighbouring wind farms*, J. Phys. Conf. Ser. **524**, 012162 (2014).
[9] G. Marmidis, S. Lazarou, and E. Pyrgioti, *Optimal placement of wind turbines in a wind park using Monte Carlo simulation*, Renewable Energy **33**, 1455 (2008).
[10] A. Emami and P. Noghreh, *New approach on optimization in placement of wind turbines within wind farm by genetic algorithms*, Renewable Energy **35**, 1559 (2010).
[11] A. Kusiak and Z. Song, *Design of wind farm layout for maximum wind energy capture*, Renewable Energy **35**, 685 (2010).
[12] B. Saavedra-Moreno, S. Salcedo-Sanz, A. Paniagua-Tineo, L. Prieto, and A. Portilla-Figueras, *Seeding evolutionary algorithms with heuristics for optimal wind turbines positioning in wind farms*, Renewable Energy **36**, 2838 (2011).
[13] R. J. Barthelmie, K. Hansen, S. T. Frandsen, O. Rathmann, J. G. Schepers, W. Schlez, J. Phillips, K. Rados, A. Zervos, E. S. Politis, and P. K. Chaviaropoulos, *Modelling and Measuring Flow and Wind Turbine Wakes in Large Wind Farms Offshore*, Wind Energy **12**, 431 (2009).
[14] E. Son, S. Lee, B. Hwang, and S. Lee, *Characteristics of turbine spacing in a wind farm using an optimal design process*, Renewable Energy **65**, 245 (2014).
[15] R. J. A. M. Stevens, D. F. Gayme, and C. Meneveau, *Coupled wake boundary layer model of wind-farms*, J. Renewable and Sustainable Energy **7**, 023115 (2015).
[16] S. Frandsen, *On the wind speed reduction in the center of large clusters of wind turbines*, J. Wind Eng. and Ind. Aerodyn. **39**, 251 (1992).
[17] S. Frandsen, R. Barthelmie, S. Pryor, O. Rathmann, S. Larsen, J. Højstrup, and M. Thøgersen, *Analytical modelling of wind speed deficit in large offshore wind farms*, Wind Energy **9**, 39 (2006).
[18] M. Calaf, C. Meneveau, and J. Meyers, *Large eddy simulations of fully developed wind-turbine array boundary layers*, Phys. Fluids **22**, 015110 (2010).
[19] J. Meyers and C. Meneveau, *Optimal turbine spacing in fully developed wind farm boundary layers*, Wind Energy **15**, 305 (2012).
[20] R. J. A. M. Stevens, *Dependence of optimal wind-turbine spacing on wind-farm length*, Wind Energy DOI: 10.1002/we.1857 (2015).
[21] R. J. A. M. Stevens, D. F. Gayme, and C. Meneveau, *Effects of turbine spacing on the power output of extended wind-farms*, Wind Energy DOI: 10.1002/we.1835 (2015).

[22] R. J. A. M. Stevens, D. F. Gayme, and C. Meneveau, *Generalized coupled wake boundary layer model: applications and comparisons with field and LES data for two real wind farms*, Submitted to Wind Energy (2015).

[23] X. Yang and F. Sotiropoulos, *Analytical model for predicting the performance of arbitrary size and layout wind-farms*, preprint (2014).

[24] C. Meneveau, *The top-down model of wind farm boundary layers and its applications*, J. of Turbulence **13**, 1 (2012).

[25] K. S. Hansen, R. J. Barthelmie, L. E. Jensen, and A. Sommer, *The impact of turbulence intensity and atmospheric stability on power deficits due to wind turbine wakes at Horns Rev wind farm*, Wind Energy **15**, 183 (2012).

[26] Y.-T. Wu and F. Porté-Agel, *Modeling turbine wakes and power losses within a wind farm using LES: An application to the Horns Rev offshore wind farm*, Renewable Energy **75**, 945 (2015).

[27] X. Yang, S. Kang, and F. Sotiropoulos, *Computational study and modeling of turbine spacing effects in infinite aligned wind farms*, Phys. Fluids **24**, 115107 (2012).

[28] I. Marusic, J. Monty, M. Hultmark, and A. Smits, *On the logarithmic region in wall turbulence*, J. Fluid Mech. **716**, R3 (2013).

[29] C. Meneveau and I. Marusic, *Generalized logarithmic law for high-order moments in turbulent boundary layers*, J. Fluid Mech. **719**, R1 (2013).

[30] R. J. A. M. Stevens, M. Wilczek, and C. Meneveau, *Large eddy simulation study of the logarithmic law for high-order moments in turbulent boundary layers*, J. Fluid Mech. **757**, 888 (2014).

[31] C. Rossby and R. Montgomery, *The layers of frictional influence in wind and ocean currents*, Pap. Phys. Oceanogr. Meteor. **3**, 1 (1935).

[32] P. Seibert, F. Beyrich, S.-E. Gryning, S. Joffre, A. Rasmussen, and P. Tercier, *Mixing height determination for dispersion modelling*, Cost Action 710: Preprocessing of Meteorological data for dispersion modeling. Report of Working Group 2 (1997).

[33] A. Peña, S.-E. Gryning, and C. Hasager, *Measurements and Modelling of the Wind Speed Profile in the Marine Atmospheric Boundary Layer*, Boundary-Layer Meteorol **129**, 479 (2008).

[34] A. Peña, S.-E. Gryning, and C. Hasager, *Comparing mixing-length models of the diabatic wind profile over homogeneous terrain*, Theor Appl Climatol **100**, 325 (2010).

[35] F. Porté-Agel, Y.-T. Wu, and C.-H. Chen, *A Numerical Study of the Effects of Wind Direction on Turbine Wakes and Power Losses in a Large Wind Farm*, Energies **6**, 5297 (2013).

[36] R. J. Barthelmie, S. T. Frandsen, O. Rathmann, K. Hansen, E. Politis, J. Prospathopoulos, J. G. Schepers, K. Rados, D. Cabezón, W. Schlez, A. Neubert, and M. Heath, *Flow and wakes in large wind farms: Final report for UpWind WP8*, Report number Risø-R-1765(EN) (2011).

[37] A. Sescu and C. Meneveau, *Large Eddy Simulation and single column modeling of thermally stratified wind-turbine arrays for fully developed, stationary atmospheric conditions*, J. Atmospheric and Oceanic Technology, http://dx.doi.org/10.1175/JTECH-D-14-00068.1 (2015).

Journal of Physics: Conference Series **625** (2015) 012005 doi:10.1088/1742-6596/625/1/012005

Predicting wake meandering in real-time through instantaneous measurements of wind turbine load fluctuations

S Aubrun[1], YA Muller[1,2], C Masson[2]

[1] Univ. Orléans, INSA-CVL, PRISME EA-4229, 8, rue Léonard deVinci, 45072 Orléans Cedex 2, France

[2]Ecole de Technologie Supérieure, 1100, rue Notre-Dame Ouest, Montréal H3C 1K3, Canada

E-mail: sandrine.aubrun@univ-orleans.fr

Abstract. The present study deals with the real-time estimation of the wind turbine wake meandering and the determination of the best indicators to measure in order to build a real-time wake meandering model. Results are obtained through experiments performed in an atmospheric boundary layer wind tunnel, where a specific set-up enables to measure simultaneously the incoming lateral velocity fluctuations, the lateral force fluctuations applied to the wind turbine model and to track in real-time the lateral position of its wake. It is an extension of a previous work about the determination of good candidates to build some real-time predictors of the wake meandering. The strong correlations between the incoming transverse velocity fluctuations, the lateral force fluctuations and the lateral position of the wake farther downstream are quantified, confirming that the large-scale turbulent eddies impact directly the wake meandering. Three different versions of a wake meandering predictor model are compared. It leads to the conclusion that the monitoring of the global force fluctuations applied to a wind turbine could be used to predict in real-time the meandering of the generated wake.

1. Introduction

Wind turbine wake interactions are one source of power losses in wind farms. Their predictions are inaccurate since the wake models implemented in commercial wind resource assessment models are very basic. For instance, they do not take into account the atmospheric flow conditions and its unsteady features [1-3]. However, the large scale turbulent eddies contained in the atmospheric boundary layer (ABL) represent one source of the meandering process of the wind turbine wake, which is characterized mainly by low frequency horizontal oscillations of the whole wake [4-6]. The present study is exclusively focused on this source of meandering. Some meandering wake models were developed based on this hypothesis [7-9]. PRISME Laboratory from the University of Orléans already performed several experiments in its atmospheric boundary layer test section of the Malavard wind tunnel in order to prove the existence of this phenomenon and characterize it [10-12]. It had already obtained a transfer function between the incoming flow turbulent properties and the wake meandering, giving the possibility to predict in real-time the main meandering behavior of the wake just by monitoring the incoming flow [13]. This result could be used in numerical models of meandering but one could also think of applying this procedure in real time and in real-configurations in order to improve the wind farm monitoring and control. On the other hand, it is not yet common to have access to the measurements of the incoming flow of a wind turbine (even if the LIDAR technology is mature enough to enable the development of LIDAR-assisted control systems [14]) and our goal in this present work is to determine a better candidate than the incoming flow velocity as meandering predictor.

It is expected that the global unsteady aerodynamic loads applied to the wind turbine could be adapted to this application and an experiment had been designed on in order to prove this expectation.

The results showed that the global drift force fluctuations measured on the actuator disc, which are a signature of the integrated transverse velocity fluctuations all over the disc and can be interpreted as an unsteady misalignment of the rotor according to the incoming flow, are also correlated with the meandering process. A correlation between the vertical torque (yaw) fluctuations and the meandering is identified when the modeled wind turbine interacts with an upstream wind turbine wake. It was concluded that the yaw fluctuations could be a good indicator of the presence of an upstream wake [13]. The present work will be dedicated to combine these results with a model dealing with the horizontal wake deviation versus the misalignment angle in order to determine whether this unsteady misalignment can be a source of amplification of the meandering.

2. Experimental set-up

The present experiments are based on the physical modelling in a wind tunnel of the ABL and of the wind turbine at a very reduced geometric scale (1:400). This ABL modelling approach is extensively used in atmospheric dispersion applications for 50 years and more recently in wind energy applications. Several referenced guidelines and articles described precisely the rules to properly reproduce an ABL in a wind tunnel and the limitations of this modelling [15-19]. For instance, the fact that the characteristic Reynolds number of the model is much smaller than in full scale must be particularly studied. Even if a critical Reynolds number based on the obstacle dimensions of 10,000 had been established to ensure the Reynolds independency of obtained results, this independency must be checked by reproducing experiments at different reference velocities. Additionally, any element composing the model must be smooth in order to avoid the re-laminarization process. These constraints were taken into account in the present set-up. The strategy to model the wind turbine was also driven by these constraints: the rotor is modelled by a porous disc made of metallic mesh. The flow through a porous grid (i.e. a turbulence grid) is known to be Reynolds independent [20]. Additionally, this modelling concept eliminates the unsteady flow signatures due to the blade motion (blade wake, rotational momentum, tip vortex, root vortex) and due to the massive separation (Van Karman-type vortices as in a bluff-body wake). It therefore ensures that, if a meandering process is detected, it is only due to the interactions between the incoming turbulent length scales and the wind turbine [11, 12]. The remaining discrepancies related to the relatively low Reynolds number is the modification of the dissipation scales but they do not play a role in the present scenario of meandering process, based on the large turbulent scales.

The experiments were carried out in the "Lucien Malavard" wind tunnel located at the PRISME Laboratory from the University of Orléans (Fig. 1). The wind tunnel is equipped to model a neutrally stratified ABL flow at a geometric scale of 1:400. In order to obtain a model of a moderately rough ABL, the 14m development plate is covered with perforated steel sheet. Furthermore a turbulence grid, combined with turbulence generators, is located at the entrance of the wind tunnel section in order to initiate appropriate mean velocity and turbulence intensity profiles. Although the return test section is 4m×4m, the effective test area is 3m wide and 1.5m high.

Properties of the obtained boundary layer (mean velocity, turbulence intensity and integral length scale profiles) are checked and are in agreement with the standards for ABL flows [15-19]. The ABL is representative of a moderately rough terrain, with a roughness length of $\boldsymbol{z_0} = 5.10^{-5}$m (equivalent full scale roughness length of 0.02m), a power law exponent of $\alpha = 0.14$ and a friction velocity of $\boldsymbol{u^*} = 0.29 \text{m.s}^{-1}$. The mean velocity measured at hub height is $U_\infty = 6.9\ \text{m.s}^{-1}$. The turbulence intensity profiles are within the expected range for this type of terrain and the expected ratio between the turbulence intensities $Iw/Iu = 0.5$ is obtained. The streamwise integral length scale $\boldsymbol{Lu_x}$ is relatively constant at the height of interest and is 2 to 4 times larger than the rotor diameter. This ensures that the meandering process due to the presence of large turbulent eddies in the incoming flow can exist.

Journal of Physics: Conference Series **625** (2015) 012005 doi:10.1088/1742-6596/625/1/012005

The modelling concept of the wind turbine is based on the actuator disc theory as described by Aubrun et al [10]. The rotor is reproduced with a disc of metallic mesh of 200mm of diameter (***D***) (80m at full scale). The solidity of the porous disc is 45% (wire diameter of 1mm and thread spacing of 3.2mm), leading to an equivalent axial induction factor of a = 0.19 (defined in $3.2) and a thrust coefficient of C_T = 0.62. The center of the porous disc is located at a variable hub height H_{hub} with a threaded rod with a 0.05***D*** diameter.

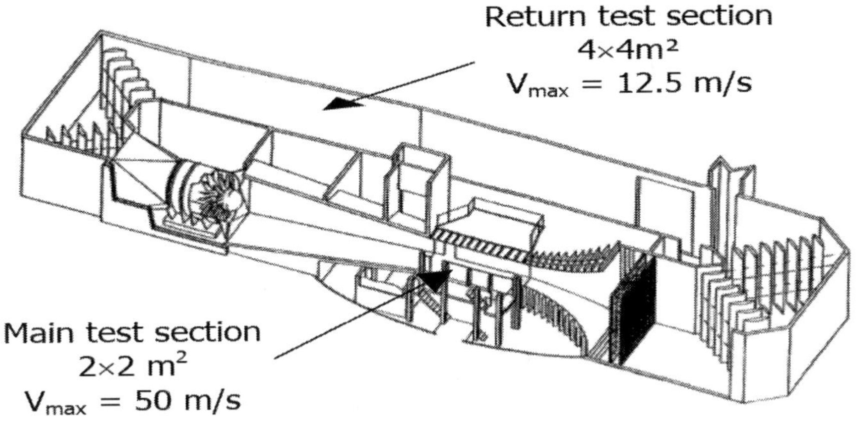

Figure 1. The "Lucien Malavard" wind tunnel of PRISME Laboratory, University of Orléans. Measurements were performed in the return test circuit, adapted to model a neutrally stratified atmospheric boundary layer

A horizontal hot wire rake composed of 5 hot wire probes (four 55P11 single probes, one 55P61 X-probe and anemometers from Dantec Dynamics) located at a distance ***L***=5***D*** downstream of the wind turbine is used to determine the instantaneous horizontal wake position (Fig. 2).

The choice is made to focus the whole study only on the horizontal meandering since it has been observed that it is the main meandering direction due to the fact that the horizontal turbulent velocity fluctuations in a turbulent boundary layer are larger than the vertical ones, which is itself a consequence of the blockage effect by the ground [11, 18].

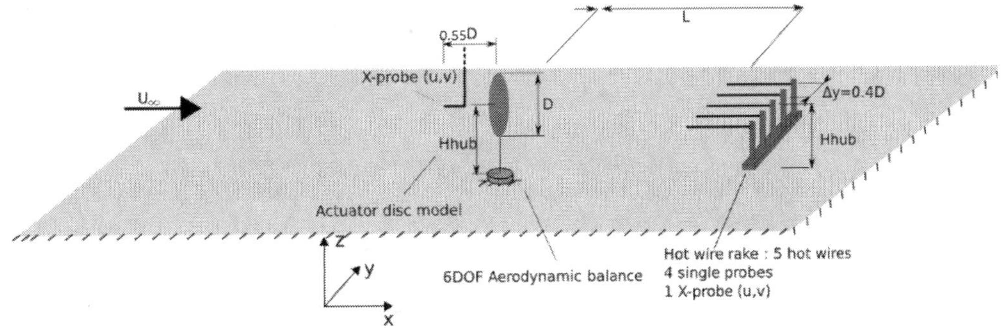

Figure 2. Experimental set-up and dimensions.

On one side of the rake, the X hot-wire probe is used in order to measure the local transverse velocity fluctuations. The uniform spacing between probes is Δy = 0.4***D***. The methodology to calculate the horizontal wake position is based on the computation of a weighted average of the locations of each probe (which are fixed), where the weighting is chosen as the exponential of the instantaneous local velocity deficit (see details and validation in [13]).

A 55P61 DANTEC hot wire X-probe is located 0.55D upstream of the wind turbine model of interest, measuring the horizontal components of the velocity.

The wind turbine model is fixed on a 6-axis aerodynamic balance (ATI Mini40), which is located underneath the floor. The three components of the aerodynamic force (F_x, F_y, F_z) and the three components of the associated torque (T_x, T_y, T_z) computed according to the center of the rotor model are measured. The 6 channels of the aerodynamic balance are simultaneously acquired with the hot wire time series so as to allow time correlations.

3. Results and discussion

3.1. Coherence and phase shift

In order to assess the relationship between the upstream large scale turbulence, the global load fluctuations and wake meandering, the coherences and the associated phase shifts are computed between the upstream transverse velocity $v_{upstream}$, the lateral force F_y and the wake position y_w (Fig. 3). The coherence is defined as a dimensionless value deduced from the cross-spectrum and in the range between 0 (no correlation) and 1 (full correlation), as a function of the frequency.

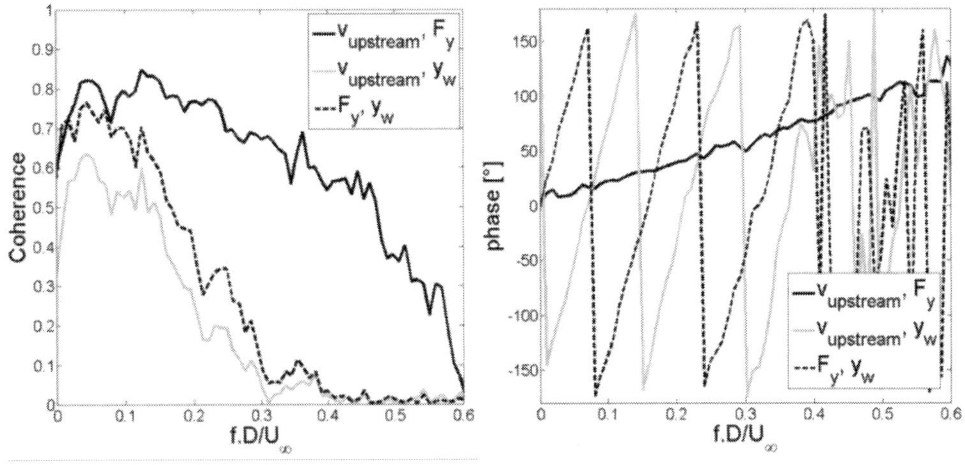

Figure 3. Coherence and phase shift between physical values.

The coherence between the upstream transverse velocity $v_{upstream}$ and the lateral force F_y shows a very high level of coherence over a broad frequency range and reaches the value of 0.8 for low frequencies. The associated phase shift presents a linear law versus the frequency that is representative of a convective phenomenon. The slope of the linear law is directly linked to the time delay between both signals. It is assessed to $\tau_{F/v}$= 16.4 ms, which leads to a convection velocity $Uc_{F/v}$=0.55 $\boldsymbol{D}/\tau_{F/v}$ = $0.98U_\infty$. It illustrates that the lateral load fluctuations measured on the wind turbine model are directly driven by the incoming transverse velocity fluctuations, with a time delay due to the streamwise advection.

The coherence between the upstream transverse velocity $v_{upstream}$ and the lateral wake position y_w shows a lower level of coherence since it reaches only a value of 0.6 and the frequency range of correlation is also narrower. Indeed, the coherence tends to zero for $f.D/U_\infty = 0.3$. One can associate this cut-off frequency to length scales of 0.33$\boldsymbol{D}$, leading to the conclusion that only the turbulent eddies larger than 0.33$\boldsymbol{D}$ contribute to the wake meandering signature that is captured in the present experiment.

On the other hand, previous results showed that the spectral content of the meandering process was assessed to be significant up to $f.D/U_\infty = 0.5$ [13]. The long distance between both signals $\boldsymbol{L}+0.55\boldsymbol{D} = 5.55\boldsymbol{D}$, which is higher than 2 times $\boldsymbol{Lu_x}$ is the reason of the correlation loss for the intermediary turbulent eddies. They are more rapidly distorted and renewed than the larger ones during the advection process.

The associated phase shift between upstream transverse velocity $v_{upstream}$ and the lateral wake position y_w presents also a linear law versus the frequency. The time delay between signal is assessed to $\tau_{v/y}$= 202.2 ms, which leads to a convection velocity $Uc_{v/y}$=5.55 $\boldsymbol{D}/\tau_{v/y}$ = 0.8$\boldsymbol{U_\infty} \approx$ (1-$\boldsymbol{a}$) $\boldsymbol{U_\infty}$. This information confirms that the meandering process, illustrated here by the lateral wake position fluctuations y_w, is mainly driven by the transverse velocity fluctuations that impact the wind turbine model.

The coherence between the lateral force F_y and the lateral wake position y_w, the associated phase shift and convection velocity $\tau_{F/y}$ are similar as the previous one. The maximum level of coherence is slightly higher with 0.7 instead of 0.6. This information is of interest since lateral force fluctuations are an image of the interactions between the yawed flow and the actuator disc. Having a higher coherence level between the position of the wake and the force than between the wake and the velocity tends to indicate that theses interactions should not be neglected into the wake meandering models.

3.2. *Unsteady yaw effects.*

The presence of transverse velocity fluctuations can be considered as an unsteady yaw effect, where $\gamma(t)$ is the angle between the instantaneous horizontal velocity vector and the streamwise direction. This instantaneous yaw angle is assessed with the following equation:

$$\gamma(t) = tan^{-1}\left(\frac{v_{upstream(t)}}{U_\infty}\right) \quad (1)$$

According to the vortex cylinder theory of a yawed actuator disc [21], the induced velocity that actually impacts the disc is skewed from an angle χ, with χ the skew angle of the wake, which is related to the axial induction factor a and the yaw angle γ:

$$\chi = (0.6a + 1)\gamma \quad (2)$$

With $a = 1 - \frac{U_{disc}}{U_\infty} = \frac{1}{2}\left(1 - \frac{U_{wake}}{U_\infty}\right)$, U_{disc} is the streamwise velocity at the disc location and U_{wake} is the streamwise velocity in the far-wake.

This formula is demonstrated for a steady yaw but it is attempted to extrapolate it to the present unsteady analysis, considering that a hypothesis of quasi-steadiness (succession of steady states) is acceptable for yaw modifications at low frequency.

If one expects that the lateral force fluctuations $F_y(t)$ are directly related to the induced velocity fluctuations that impact the disc, and that the instantaneous angle between the horizontal force vector and the streamwise direction $\theta_F(t)$ is equal to the angle between the instantaneous induced velocity vector and the streamwise direction:

$$\theta_F(t) = \chi(t) = (0.6a + 1)\gamma(t - \tau_{F/v}) = G_{\theta_F/\gamma_{theo}}.\gamma(t - \tau_{F/v}) \quad (3)$$

With $G_{\theta_F/\gamma_{theo}} = 1.11$ in the present configuration.

Fig. 4 shows the gain of the transfer function between the angles $\gamma(t)$ and $\theta_F(t)$ deduced from the experiments, $G_{\theta_F/\gamma_{exp}}(f)$. It is relatively constant all over the presented frequency range at a value of 0.4. Even if this is the right order of magnitude, it shows that the relationship between the unsteady yaw angle and the unsteady drift (lateral force $\boldsymbol{F_y}$) needs to be improved. One explanation might be that the velocity probe measuring the incoming flow is to close to the disc (0.55***D***) and that the local flow is already distorted due to the whole stream tube deflection or that the force vector is not aligned with the averaged induced velocity at the disc location. This point must be checked in a future measurement campaign.

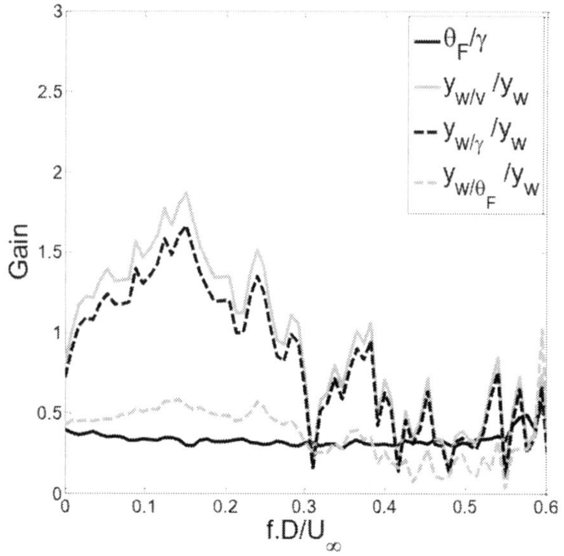

Figure 4. Gain obtained from the transfer function between physical values.
Example: $G_{Y/X}$ corresponds to the gain obtained from the transfer function estimate between the time series $X(t)$ (input) and $Y(t)$ (output)

3.3. Wake meandering models

This experimental set-up gives the opportunity to test and validate some assumptions that are used in wake meandering models as the passive transport of the wake "emitted" from the wind turbine by the incoming turbulent flow [7]. The lateral wake position at an instant t_i and at a fixed downstream distance L is expected to be totally driven by the transverse velocity fluctuations that impact the disc at the instant $(t_i - U/L)$, with U the mean streamwise velocity. In the present study, the convection velocity has been assessed to $Uc_{v/y}=0.8U_\infty$, thanks to the phase shift between the upstream transverse velocity fluctuations and the wake position fluctuations. It will be therefore used in the tested wake meandering model:

$$y_{w/v}(t) = \frac{L}{Uc_{v/y}} v_{upstream}\left(t - \tau_{v/y}\right) \tag{4}$$

In their model, the notion of passive transport of the wake is a very strong hypothesis that ignores the effects of interactions between the flow and the wind turbine model. Indeed, there is a strong coupling between the wind turbine misalignment characterized by the yaw angle and the skewed wake characterized by the skew angle. This can be taken into account into an alternative wake meandering model, considering that the meandering is directly due to the skewness of the wake:

$$\begin{aligned} y_{w/\gamma}(t) &= L.tan\left(\chi\left(t - \tau_{v/y}\right)\right) = L.tan\left((0.6a + 1)\,\gamma\left(t - \tau_{v/y}\right)\right) \\ &= L.tan\left((0.6a + 1)\, tan^{-1}\left(\frac{v_{upstream}\left(t-\tau_{v/y}\right)}{U_\infty}\right)\right) \end{aligned} \tag{5}$$

The same hypothesis of strong coupling between the flow and the actuator disc is now used to find a relationship between the lateral force fluctuations and the wake meandering, by considering that the force vector is aligned with the induced velocity at the disc location :

$$y_{w/\theta_F}(t) = L.tan\left(\chi(t-\tau_{F/y})\right) = L.tan\left(tan^{-1}\left(\frac{F_{y(t-\tau_{F/y})}}{F_x}\right)\right) \quad (6)$$

Fig. 4 shows the gain of the transfer functions between the measured instantaneous lateral wake position $y_w(t)$ and the assessed ones from the three models:

- the model 1, based on the passive transport hypothesis $y_{w/v}(t)$, $G_{y_{w/v}/y_w}(f)$,
- the model 2, based on the yaw effect to the skewness of the wake $y_{w/Y}(t)$, $G_{y_{w/Y}/y_w/}(f)$.
- The model 3, based on the yaw effect to the skewness of the wake $y_{w/\theta_F}(t)$, $G_{y_{w/\theta_F}/y_w/}(f)$

If one model is correct, it is expected that its gain is close to 1 in the frequency range of interest ($f.D/U_\infty < 0.3$).

The model 1 presents a gain between 1 and 2. It means that the model proposed by Larsen et al. [7] overestimates the lateral transport of the wake. For the model 2, the gain is slightly lower than for the passive transport hypothesis but still overestimates the meandering. For the model 3, the gain is close to 0.5 for the whole frequency range of interest In that case, the meandering is overestimated. It seems that this latter model based on the measurement of the instantaneous lateral force fluctuations could also be a good candidate to predict the wake meandering since these fluctuations are very much correlated with the incoming flow fluctuations. But the gain of the transfer function is not more appropriate than the ones based on the velocity fluctuations.

4. Conclusion

The present study was performed through atmospheric boundary layer wind tunnel testing, where a specific set-up enabled to measure simultaneously the incoming lateral velocity fluctuations, the lateral force fluctuations applied to the wind turbine model and to track in real-time the lateral position of its wake. It is an extension of the work presented in [13] about the determination of good candidates to build some real-time predictors of the wake meandering. The strong correlations between the incoming transverse velocity fluctuations, the lateral force fluctuations and the lateral position of the wake farther downstream were quantified, showing that, at 5D downstream of the wind turbine model, the large-scale turbulent eddies (larger than 3***D***) impact directly the wake meandering. Three different versions of a wake meandering predictor model were compared. The first one was developed by Larsen et al. [7] and was based on the classical hypothesis of passive transport of the wake by the large turbulent eddies. The second and third ones were proposed in the present study. They were based on the hypothesis that the interactions between the incoming flow and the wind turbine model must not be neglected and took into account the skewness of the wake due to the unsteady yaw effect. The second one used as indicator the incoming velocity fluctuations, whereas the third one used the lateral force fluctuations. It was observed that none of them was able to reproduce the right magnitude of meandering. It leads to the conclusion that the monitoring of the global force fluctuations applied to a wind turbine could also be used to predict in real-time the meandering of the generated wake but the function transfer between this information and the meandering magnitude must be improved.

This study must be extended to multiple configurations through a parametrical study, in order to confirm these conclusions. The influence of the disc porosity (and so the velocity deficit and the wake skewness), the incoming wind properties (turbulence intensity and turbulence length scales) and the streamwise distance between the wind turbine model and the location where the meandering is predicted will be tested. The model sensitivity to a steady misalignment (static yaw errors encountered on real wind turbines) will be also quantified. This parametrical study might be also accompanied by LES computations.

References

[1] Ainslie JP (1988) Calculating the flow field in the wake of wind turbines, J. Wind Eng. Ind. Aerodyn. 27:213-224.

[2] Jensen NO (1983) A note on wind generator interaction. Risø Report M-2411

[3] Frandsen S, Barthelmie R, Pryor S, Rathmann O, Larsen S, Højstrup J, Thøgersen M (2006) Analytical modelling of wind Speed deficit in large offshore wind farms. Wind Energy 9:39–53

[4] Taylor GJ, Milborrow DJ, McIntosh DN, Swift-Hook DT (1985) Wake measurements on the Nibe windmills. Proceedings of the 7th British Wind Energy Association Conference March 27-29 1985, Oxford

[5] Bingöl F, Mann J and Larsen GC, (2010) Light detection and ranging measurements of wake dynamics. Part 1: one-dimensional scanning, Wind Energy 13:51–61

[6] Trujillo JJ, Bingöl F, Larsen GC, Mann J (2011) Light detection and ranging measurements on wake dynamics, Part II: Two-dimensional Scanning. Wind Energy 14:61-75

[7] Larsen G.C., Madsen H. A, Thomsen K. and Larsen T. J., (2008), Wake meandering – a pragmatic approach, Wind Energy, 11:377-395

[8] Larsen TJ, Madsen HA, Larsen GC, Hansen KS (2013) Validation of the dynamic wake meander model for loads and power production in the Egmond aan Zee wind farm. Wind Energy 16/4: 605–624

[9] Trujillo JJ, Kühn M (2009) Adaptation of a Lagrangian Dispersion Model for Wind Turbine Wake Meandering, in Proceedings of the EWEA Conference, March 16-19, Marseille, France

[10] Aubrun S, Devinant P, España G, (2007) Physical modelling of the far wake from wind turbines. Application to wind turbine interactions, Proceedings of the European Wind Energy Conference, Milan, Italy.

[11] España G, Aubrun S, Loyer S, Devinant P (2011) Spatial study of the wake meandering using modelled wind turbines in a wind tunnel. Wind Energy 14:923-937

[12] España G, Aubrun S, Loyer S, Devinant P (2012) Wind tunnel study of the wake meandering downstream of a modelled wind turbine as an effect of large scale turbulent eddies. J. Wind Eng. Ind. Aerodyn. 101:24-33

[13] Muller YA, Aubrun S, Masson C (2015) Determination of real-time predictors of the wind turbine wake meandering, Exp Fluids DOI 10.1007/s00348-015-1923-9.

[14] Davoust S, Jehu A, Bouillet M, Bardon M, Vercherin B, Scholbrock A, Fleming P, Wright A (2014) Assessment and Optimization of LiDAR Measurement Availability for Wind Turbine Control. Scientific. Proceedings of EWEA Conference March 10-13, 2014, Barcelona, Spain

[15] Counihan J (1975) Adiabatic atmospheric boundary layers: a review and analysis of data from the period 1880-1972. Atmos. Environ. 9:871-905

[16] Snyder WH (1981) Guideline for fluid modelling of atmospheric diffusion. US Environmental Protection Agency. Report EPA-600/8-81-009

[17] Engineering Sciences Data Unit (1985) Characteristics of atmospheric turbulence near the ground. Item N° 85020

[18] Kaimal JC, Finnigan JJ (1994) Atmospheric boundary layer flows, their structure and measurements. Oxford University Press

[19] VDI guideline 3783/12 (2000) Physical modelling of flow and dispersion processes in the atmospheric boundary layer - Application for wind tunnels. Beuth Verlag, Berlin

[20] Mohamed, M.S., LaRue, J.C., 1990. The decay power law in grid generated turbulence. Journal of Fluid Mechanics 219, 195–214

[21] Burton T, Sharpe D, Jenkins N, Bossanyi E (2001) Wind Energy Handbook. John Wiley & Sons, Chichester

Wake Conference 2015 IOP Publishing
Journal of Physics: Conference Series **625** (2015) 012006 doi:10.1088/1742-6596/625/1/012006

Characterizing Wake Turbulence with Staring Lidar Measurements

D Bastine[1], M Wächter[1], J Peinke[1]

[1] AG TWiSt, ForWind, University of Oldenburg, Ammerländer Herrstr. 136, 26111 Oldenburg

D Trabucchi[2], M Kühn[2]

[2] AG WE-Sys, ForWind, University of Oldenburg, Ammerländer Herrstr. 136, 26111 Oldenburg

E-mail: david.bastine@uni-oldenburg.de

Abstract.
Lidar measurements in the German offshore wind farm Alpha Ventus were performed to investigate the turbulence characteristics of wind turbine wakes. In particular, we compare measurements of the free flow in the surroundings of the wind turbines with measurements in the inner region of a wake flow behind one turbine. Our results indicate that wind turbines modulate the turbulent structures of the flow on a wide range of scales. For the data of the wake flow, the power spectrum as well as the multifractal intermittency coefficient reveal features of homogeneous isotropic turbulence. Thus, we conjecture that on scales of the rotor a new turbulent cascade is initiated, which determines the features of the turbulent wake flow quite independently from the more complex wind flow in the surroundings of the turbine.

1. Introduction

It is a well-known fact that the power output and loads of wind turbines are strongly influenced by wakes of neighboring turbines [1, 2]. These wakes are highly turbulent flows which even show an increased turbulence intensity compared to the atmospheric inflow [3, 1]. Since the response times of wind turbines are typically in the order of seconds the small scale properties of turbulent flows have a strong impact on the turbines [4, 5, 6, 7, 8]. Hence, to understand the impact of wakes on neighboring turbines, the turbulence present in a wake flow needs to be investigated in detail.

Recently, laboratory experiments with model wind turbines have been performed [9, 10, 11] showing that the turbulent statistics in the wake flow differ strongly from the inflow characteristics.

Since laboratory experiments cannot fully emulate the atmospheric boundary layer turbulence and since they are performed for relatively small Reynolds numbers, field measurements play an essential role for the investigation of wake flows. Measurements performed on met masts offer only a limited number of measurement locations and are therefore not sufficient to characterize flows in wind farms. Thus, the use of remote sensing techniques, such as sodars or Doppler lidars, has increased rapidly in the last years [12, 13, 14, 15, 16, 17]. However, field measurements focusing on turbulent features of wakes[16, 17, 18] are still rare and have to the authors knowledge not yet been performed in an offshore environment. Furthermore, they mostly do not investigate higher order statistics. Therefore, this study aims for a characterization of wake turbulence using lidar measurements in an offshore wind farm. By comparing the statistics of the wind field in a wake region with the statistics of a region with approximately free atmospheric flow, we investigate the modulation of the turbulent structures through the wind turbine.

We start with a concise description of the measurements in Sec. 2. The statistical analysis is performed in Sec. 3 including a description of power spectral densities and higher order statistics of temporal increments. In Sec. 4 the results of our analysis are discussed and conclusions are drawn in Sec. 5.

2. Measurements

The measurements analyzed in this work stem from a measurement campaign in the offshore wind farm "Alpha Ventus" (Fig. 1) located in the German North Sea. In this campaign, three scanning long range Doppler lidar systems of type Leosphere Windcube200S were used for a detailed analysis of the flow in the wind farm. One lidar was positioned on the wind farm's sub station while two others were positioned on the nearby research platform and meteorological mast "Fino 1".

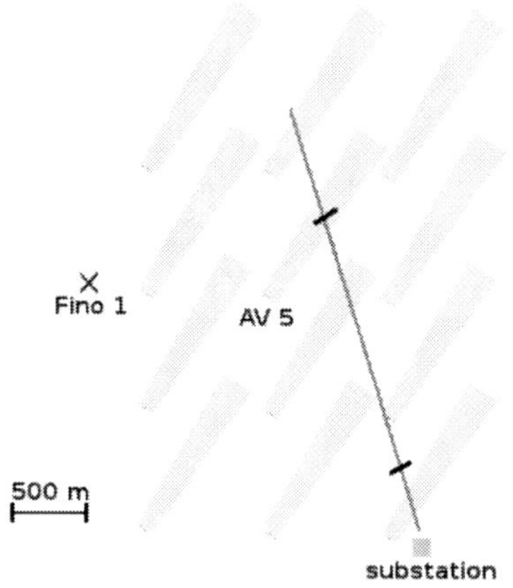

Figure 1. Illustration of the measurement scenario in the wind park "Alpha Ventus". The blue color represents the wakes of the different turbines in the farm. Note that the wakes do also exist in larger distances to the turbine than indicated by the blue color. The black cross represents the position of the research platform "Fino 1". The red line represents the lidar beam. The black lines through the lidar beam show approximate positions of the center of the measurement regions, defined in Sec. 3.1.

In this work, we analyze data from a measurement scenario using only the lidar on the substation platform and an ultrasonic anemometer located at a height of z = 60 m on the nearby met mast Fino 1. Measurements were taken for approx. 5 days but only 10 hours of this data are selected for the analysis presented in this work. In the selected time interval the wind speed and the wind direction indicate a relatively stationary atmospheric velocity field, as illustrated by the sonic measurements shown in Fig. 2. The average wind speed, noted as U

in the following, is given by approx. 9.9 $\frac{\text{m}}{\text{s}}$ and the average wind direction by approx. 213°. The approximate stationarity for this relatively long period is essential for a reliable statistical analysis.

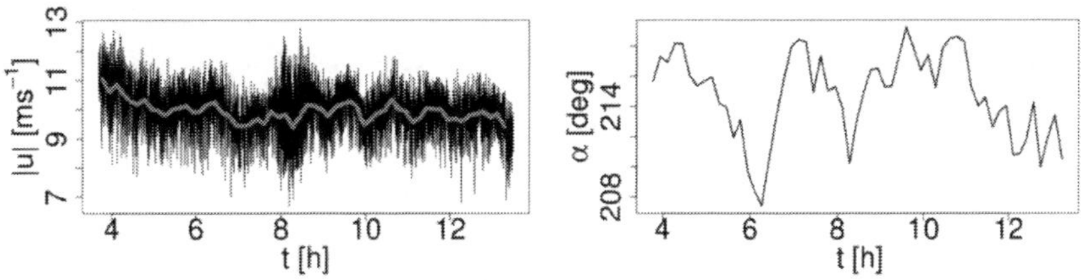

Figure 2. Ultrasonic Anemometer measurements on "Fino 1" (z = 60 m). Left: Absolute value of the wind speed measured with a sampling rate of 10 Hz (black line) and its ten minute average (red line). Right: Ten minute average of the wind direction α.

To facilitate the investigation of short time scales, the lidar was set to a staring mode with a fixed direction of the laser beam, as illustrated in Fig. 1. In the following, the distance along the laser beam will be noted as r and the time as t. The angle between the average wind vector and the lidar beam is approx. 51° and the vertical position of the laser beam is shown in Fig. 3. The lidar measures the velocity component along the laser beam, which will be called line of sight (los) velocity in the following and be noted as u_{los}. Further technical settings for the lidar can be found in table 1. With these settings u_{los} could be measured instantaneously at different positions with a relatively high temporal resolution, as shown in Fig. 4. Measurements with a carrier to noise ratio (CNR) lower than −22.5 dB were omitted from the analysis which were less than 4% of the data in the analyzed measurement regions, defined in Sec. 3.1. The measurements were not taken at equidistant times and the average time step was given by approximately 1.5 s corresponding to an average sampling rate of approx. 0.6 Hz. Since in the following 2-point statistics, such as moments of temporal increments are investigated, a resampling procedure was applied. Linear interpolation was used to obtain values at equidistant time steps of 1.5 s. Values stemming from an interpolation over a time of more than 2 s were omitted.

Table 1. Lidar settings for the measurement scenario described in Sec. 2 .

Setting	Value
Azimuth angle	−17.62 °
Elevation angle	1.80 °
Pulse length (Full Width Half Maximum)	200 ns
Range gate size	64 points
Sampling rate of the photodiode	250 Mhz
Accumulation time	0.60 s
Physical resolution	30 m − 40 m
Range gate distance	4 m

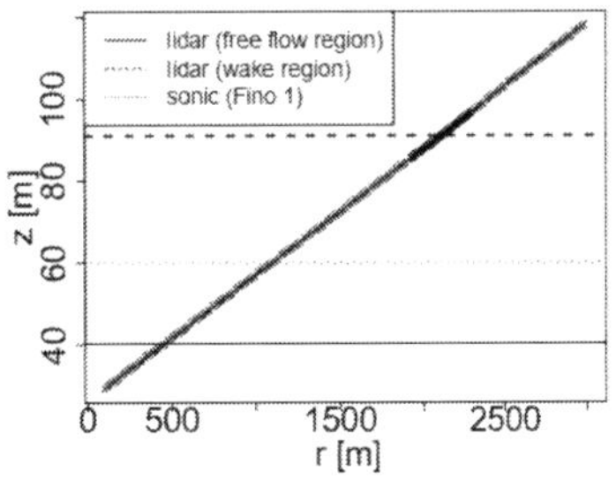

Figure 3. Height z of the lidar beam dependent on the distance r along the laser beam. The horizontal lines represent the approximate heights at the analyzed regions defined in Sec. 3.1.

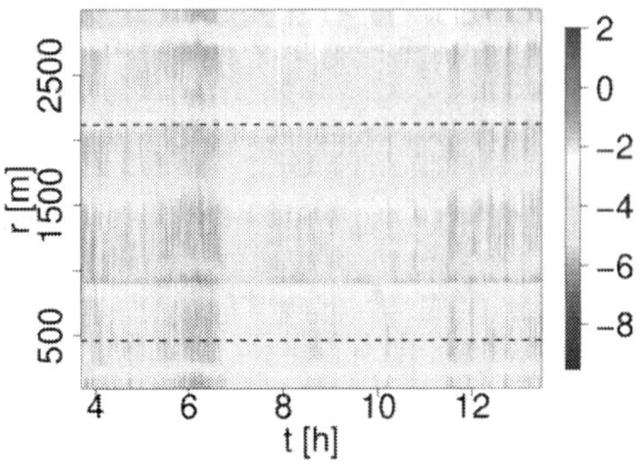

Figure 4. Los velocity, denoted by color, dependent on the distance along the laser beam r and time t. The dashed lines show the approximate positions of the measurement regions analyzed in this paper. The lower line corresponds to a region almost uninfluenced by the park while the upper line is a wake region corresponding to the wake of AV5 as shown in Fig. 1.

3. Results

3.1. Defining Regions of Interest

To compare the statistics of the lidar measurements in a wake and a free flow situation, we choose two different regions for our analysis. The first region is centered at $r = 460$ m (lower black line crossing the laser beam in Fig. 1 and lower dashed line in Fig. 4) and contains all range gates with $r \in [430, 490]$ m. For the wind directions occurring in the selected measurement period this region is assumed to be only weakly influenced by the presence of the wind farm (see Fig. 1). Thus, it will be referred to as the free flow region the following.

The second region corresponds to the wake of AV 5 (Fig. 1 and upper dashed line in Fig. 4) and contains all range gates with $r \in [r_c(t) - 30 \text{ m}, r_c(t) + 30 \text{ m}]$, as illustrated in Fig. 6. $r_c(t)$ is an approximation of the position of the wake center given by the maxima of the 10-min averages of the measured los velocities (Fig. 6). Note that these maxima are the minima of the absolute wind speed due to $u_{los} < 0$. The defined region is approximately $5D$ away from the turbine, where $D = 126$ m is the rotor diameter. Due to the described tracking procedure we expect that almost all measurements in this region correspond to the inner region of the wake and we will refer to it as the wake region or inner wake region in the following. Note that we do not aim to analyze the outer region of the wake in this work.

For both defined regions, statistical averages will be taken over all enclosed range gates and the complete time domain.

Additionally, we compare the lidar measurements to the ultrasonic anemometer measurements done on the platform "Fino 1" with a sampling frequency of 10 Hz. To do this, the three dimensional sonic measurements are projected on the line of sight direction of the lidar. After this procedure, we see that, both, the lidar measurements for the free flow case and the projected sonic data, show the same large scale dynamics (Fig. 5). The obvious offset is likely to be caused by the height difference between sonic and the center of the free flow region. In Sec. 3.2 the sonic data is used for a simplified correction of the power spectra of the lidar measurements.

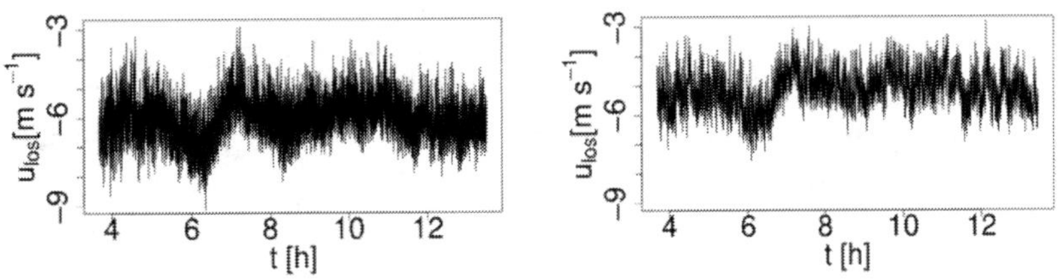

Figure 5. Left: Ultrasonic anemometer measurements (10 Hz) projected on the line of sight of the lidar. Right: Lidar data (approx. 0.6 Hz) in the center of the free flow region at $r = 460$ m.

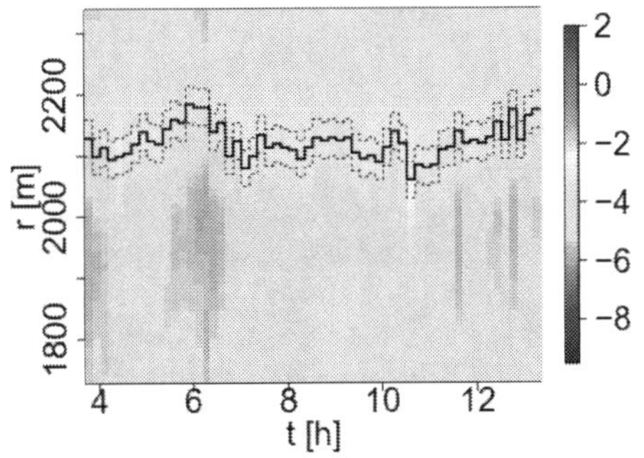

Figure 6. Tracking the wake center: The colors show the ten minute averages of the measured los velocities. The solid line represents the maxima of these ten minute averages and is used as a rough approximation of the wake center, noted as $r_c(t)$. The region of $[r_c(t) - 30\text{ m}, r_c(t) + 30\text{m}]$ marked by the dashed lines is used for the statistical analysis and is referred to as the (inner) wake region in this work.

3.2. Spectral Analysis and Correlations

As expected, we find a reduction of the absolute mean los velocity in the wake region to 3.3ms^{-1} compared to $5.1\,\text{m}\,\text{s}^{-1}$ in the free flow case. The turbulent kinetic energy (variance) in the los direction is increased in the wake situation to $1.15\,\text{m}^2\text{s}^{-2}$ from $0.62\,\text{m}^2\text{s}^{-2}$ in the free flow, corresponding to a turbulence intensity of u_{los} of 34% and 12% respectively. This indicates that the mean kinetic energy of the free flow gets partially converted into turbulent kinetic energy.

To investigate the scale dependence of this conversion process we now compare the spectral characteristics in the inner wake region and the free flow region. To estimate the power spectral density (PSD) the averaging of the power spectrum is done over 5−min windows using every range gate in the respective region, as defined in Sec. 3.1. In the wake case, the 10−min windows defined by the wake tracking (Fig. 6) are split in two while the same wake center position is used for both windows.

The power spectra S multiplied by the frequency f in Fig. 7 reveal an enhanced energy in the wake for all investigated frequencies ($0.05\frac{U}{D} < f < 5\frac{U}{D}$). The strongest energy enhancement is found at around $0.35\frac{U}{D}$ (Fig. 7). This maximum of $S \cdot f$ in the wake region corresponds to a sharp bend in the double logarithmic plot of the PSD (Fig. 8). This bend is followed by an approximate power law behavior with an exponent of approx. −2.3. In free flow case no clear peak or bend can be found but a power law behavior for higher frequencies ($f > 2.5\frac{U}{D}$) is indicated. The exponent is approx. −2.1. These rather unusual exponents could be related to the volume average effect of the lidar [19, 20, 21]. Therefore, we additionally investigate the PSD of the sonic anemometer measurements which shows a common $f^{-\frac{5}{3}}$ scaling for higher frequencies ($f > 2.5\frac{U}{D}$). Assuming that the sonic anemometer shows the correct spectrum we obtain a simplified correction of the PSD of the lidar in the wake case by multiplying it with the ratio

$$\frac{S_{\text{sonic}}(f)}{S_{\text{lidar(free flow)}}(f)} .$$

This procedure reveals an approx. $f^{-\frac{5}{3}}$ scaling almost directly after the bend in the wake spectrum and thus for much lower frequencies than for the sonic measurements representing the free flow case.

In addition to the spectral analysis, we investigate the corresponding autocorrelation functions defined as

$$c(\tau) := \frac{\langle (u_{los}(t+\tau) - \langle u_{los}\rangle) \cdot (u_{los}(t) - \langle u_{los}\rangle)\rangle}{\langle (u_{los}(t) - \langle u_{los}\rangle)^2\rangle} . \tag{1}$$

Due to our wake tracking procedure, first the autocorrelation function $c(\tau)$ is calculated for every 10-min window instead of using a complete time series. The mean of the autocorrelation functions of all windows is then used as a final estimate for $c(\tau)$. As shown in Fig.9, the correlations decrease much faster in the wake region than in the free flow case. Even though large-scale correlations are partially suppressed since we calculated the autocorrelation functions of 10-min windows only, these results indicate a smaller integral time scale in the wake region in the order of $\tau \approx 12\text{ s} \approx \frac{D}{U}$.

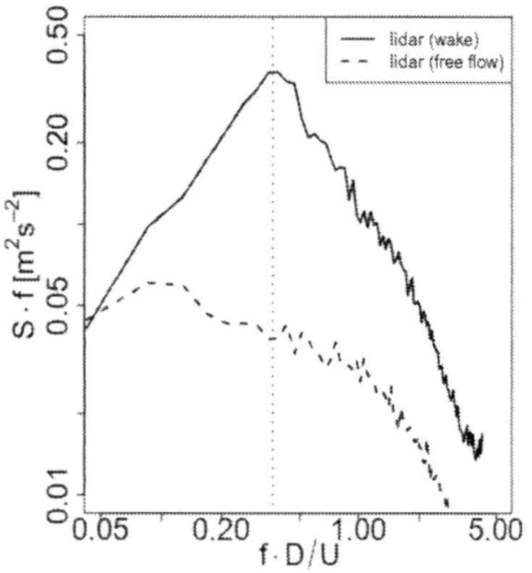

Figure 7. Frequency multiplied with power spectral densities of the lidar measurements in the wake region and the free flow. The dashed line marks the maximum of wake spectrum which also corresponds to the highest enhancement through the turbine. The frequency is normalized by $\frac{U}{D} \approx 0.08$ Hz.

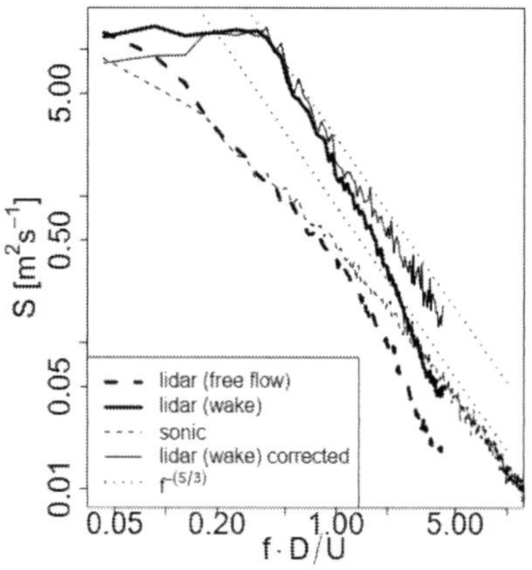

Figure 8. Power Spectra of lidar measurements in the wake, of lidar measurements in the free flow and of ultrasonic measurements. Additionally, a corrected wake spectrum, as described in Sec. 3.2, is plotted. The frequency is normalized by $\frac{U}{D} \approx 0.08$ Hz.

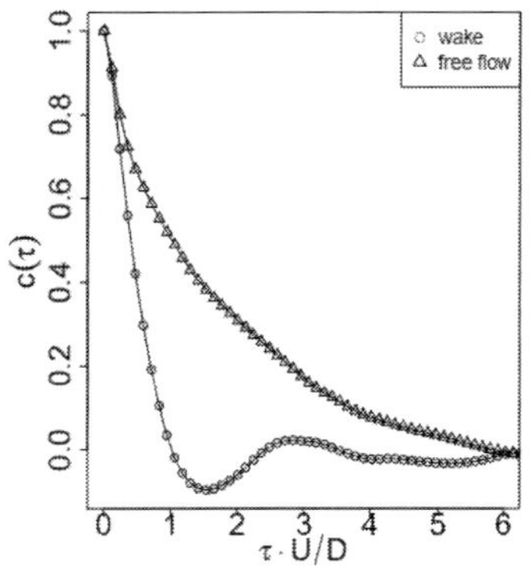

Figure 9. Autocorrelation Function of the los velocities of the lidar measurements in the wake and free flow regions respectively. The time scale is normalized by $\frac{D}{U} \approx 12$ s.

3.3. Higher Order Statistics

To investigate higher order two point statistics, we analyze the statistics of velocity increments, defined as

$$\Delta_\tau u_{los} := u_{los}(t+\tau) - u_{los}(t)$$

for a certain time lag τ. The power spectral density and the autocorrelation function, presented in the last subsection, are related to the variance of such increments. Additional information about the turbulent structures can be found in higher order cumulants of increments [22, 23]. Therefore, the normalized increment pdfs of the lidar data corresponding to different time scales τ are shown in Fig. 10. For smaller scales (3 s), the pdfs in the wake and free flow situation show heavy-tails. While the pdf in the wake case becomes Gaussian between 6 s and 12 s the free flow pdf looses it heavy tails only for large time scales (higher than 30 s). These results are consistent with the idea of a reduced integral time scale in the wake in the order of $\frac{D}{U} \approx 12$ s.

One possibility to quantify the structural change of the pdfs and their heavy tailed structure is to determine the flatness, defined as

$$F(\tau) := \frac{\langle \Delta_\tau {u_{los}}^4 \rangle}{\langle \Delta_\tau {u_{los}}^2 \rangle^2} \ . \tag{2}$$

This flatness depending on the scale τ is presented in Fig. 11. As expected from Fig. 10, we find a higher flatness in the free flow case compared to the wake situation. The flatness in both regions obviously varies with the scale τ indicating a multifractal scaling behavior as typical for turbulent flows [22, 23]. As for homogeneous isotropic turbulence (HIT), the flatness in the wake region approximately behaves like a power law for small τ ($\tau < \frac{D}{U}$). Following K62-theory [24, 23] for HIT, we would expect:

$$F(\tau) \propto \tau^{-\frac{4}{9}\mu} \ , \tag{3}$$

where μ is is often called intermittency factor. The dashed line in Fig. 11 corresponds to a value of $\mu \approx 0.27$, which is a realistic value for HIT (see e.g.[25, 23]). For the free flow case, no clear power law behavior for the small scales is found.

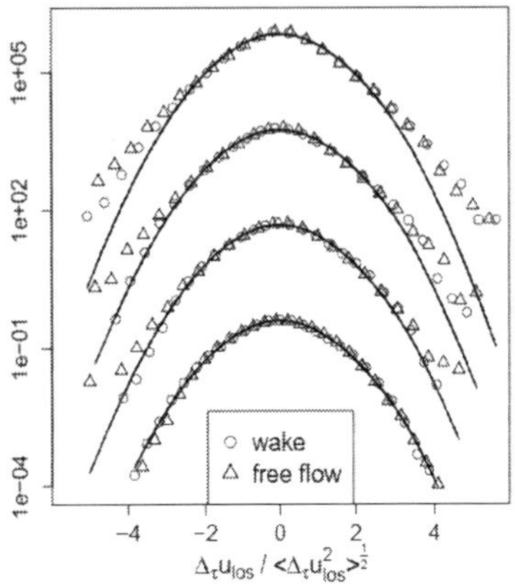

Figure 10. Normalized increment pdfs of the lidar measurements in the wake and free flow region. For clarity, the pdfs are multiplied by different factors for different temporal increments. Time lags used from top to bottom: 3 s, 6 s, 12 s, 45 s.

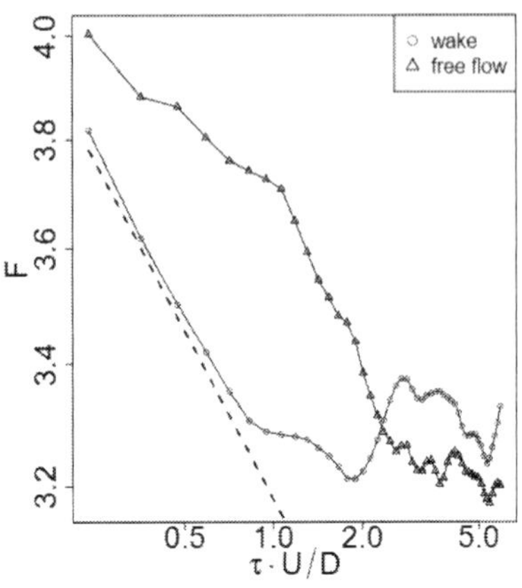

Figure 11. Double logarithmic plot of the flatness of the increments dependent on the temporal lag τ for the wake and the free flow region. The dashed line corresponds to a power law as given in Eq. (3) with $\mu = 0.27$. The time scale is normalized by $\frac{D}{U} \approx 12$ s.

4. Discussion

In contrast to the results for the free flow region, we find clear indications of homogeneous isotropic turbulence (HIT) in the inner far wake region of a wind turbine. The power spectra indicate that energy injection on the scales of the rotor diameter seems to be the starting point of an energy cascade. Features of HIT such as the approximate $f^{-\frac{5}{3}}$ scaling of the spectrum and multifractal scaling behavior with an intermittency factor in the same order of magnitude as known from HIT are found. For the undisturbed flow, the HIT-like properties are much less clearly present, may be seen only in the higher frequency part. One possible explanation of this phenomenon is that the anisotropic features of the ambient atmospheric flow become less relevant in the wake due to the mixing by the rotor. An alternative interpretation is offered by the works of [26, 27] which indicate that HIT cannot be seen in the atmospheric flow due to its non-stationary character. From this point of view, our results indicate that the wake flow might be in some sense more stationary than its surroundings.

As in our case of the wind turbine wake, properties of HIT are also found in the center of free jet flows or flows behind bluff bodies such as a cylinder (see e.g. [28]). When moving further away from the center of these flows the velocity field obviously becomes less isotropic (see e.g. [29]). Analogously, we only expect statistical behavior similar to HIT in the inner region of a wind turbine wake and not in the outer region where the wake strongly interacts with the ambient field. In contrast to most laboratory experiments concerning turbulence behind bluff bodies, the inflow for a wind turbine in the atmosphere is strongly anisotropic and furthermore the rotational direction of the turbine also breaks the rotational symmetry. This is particularly important in the near wake [30]. Therefore, our findings of statistical properties similar to HIT in the far wake are quite remarkable.

It is important to note that the wind turbine seems to modulate the turbulent structures of the atmosphere on a wide range of scales possibly even for frequencies smaller than $0.05\frac{U}{D}$

also supported by the results from Singh et al.[9] for model turbines in laboratory experiments. This might be relevant when modeling the wake as a passive tracer driven by the large scale atmospheric structures [31, 32]. In agreement with our results, Singh et al.[9] also found a maximum energy enhancement for frequencies on the order of $0.4\frac{U}{D}$. Since our resolution in the frequency domain was comparably coarse we could not confirm that the wake flow has a reduced energy for $0.01\frac{U}{D} < f < 0.1\frac{U}{D}$. As Singh et al. [9] in the lab and Wessel [18] in the atmosphere, we also observed a less heavy-tailed distribution of increments in the wake compared to the ambient flow. A homogenization in the sense of a reduced intermittency factor in the wake [9] could not be found here.

It should be noted that the free flow and wake region compared here, are located at two different heights. Thus, the statistical differences can partially stem from this height difference and should be interpreted with care. However, we do not expect that our major findings in the wake region, such as the integral length scale related to the rotor diameter together with a scaling behavior similar to HIT, could also be found in the free flow when measured in the same height as the wake.

5. Conclusions

Our results and their partial agreement with lab experiments clearly show that staring line of sight lidar measurements can be a useful tool to qualitatively investigate the turbulent flow inside a wind farm. A further and more detailed investigation of corrections due to the volume averaging effect, particularly for higher order statistics would be very useful.

Since a wind turbine operates in a complex anisotropic flow field, statistical properties similar to HIT in the inner region of a far wake are a remarkable result. It would be very interesting to investigate whether further statistical quantities also show HIT-like behavior. Multiple lidars could for example be used to measure other velocity components providing information about the turbulent momentum fluxes in the wake.

Acknowledgements

We would like to thank the whole "Lidar and Wake" group at ForWind (University of Oldenburg) for the realization of this measurement campaign. Furthermore, we would like to acknowledge fruitful discussions with Philip Rinn from the working group TWiSt (University of Oldenburg).

This work has been funded by the Bundesministerium für Wirtschaft und Energie (BMWi) due to a decision of the German Bundestag (FKZ0325397A).

References

[1] Frandsen S 2005 Turbulence and turbulence generated fatigue in wind turbine clusters Risø-R 1188 Risø National Laboratory: Roskilde
[2] Barthelmie R J, Pryor S C, Frandsen S T, Hansen K S, Schepers J G, Rados K, Schlez W, Neubert A, Jensen L E and Neckelmann S 2010 *Journal of Atmospheric and Oceanic Technology* **27** 1302–1317
[3] Burton T, Sharpe D, Jenkins N and Bossanyi E 2001 *Wind energy handbook* (New York: Wiley)
[4] Tavner P, Qiu Y, Korogiannos A and Feng Y 2011 *Proceedings of EWEA 2011 (Brussels)* URL `http://proceedings.ewea.org/annual2011/proceedings/allfiles2/594_EWEA2011presentation.pdf`
[5] Wächter M, Heißelmann H, Hölling M, Morales A, Milan P, Mücke T, Peinke J, Reinke N and Rinn P 2012 *Journal of Turbulence* **13** N26 (*Preprint* `http://www.tandfonline.com/doi/pdf/10.1080/14685248.2012.696118`)
[6] Milan P, Wächter M and Peinke J 2013 *Phys. Rev. Lett.* **110**(13) 138701
[7] Mücke T A, Wächter M, Milan P and Peinke J 2014 *Wind Energy* ISSN 1099-1824
[8] Lind P G, Herráez I, Wächter M and Peinke J 2014 *Energies* **7** 8279–8293
[9] Singh A, Howard K B and Guala M 2014 *Physics of Fluids* **26** ISSN 1070-6631
[10] Chamorro L P, Guala M, Arndt R E A and Sotiropoulos F 2012 *Journal of Turbulence* **13** ISSN 1468-5248
[11] Melius M S, Tutkun M and Cal R B 2014 *Physica D: Nonlinear Phenomena* **280** 14–21

[12] Barthelmie R, Folkerts L, Ormel F, Sanderhoff P, Eecen P, Stobbe O and Nielsen N 2003 *Journal of Atmospheric and Oceanic Technology* **20** 466–477
[13] Barthelmie R, Larsen G, Frandsen S, Folkerts L, Rados K, Pryor S, Lange B and Schepers G 2006 *Journal of atmospheric and oceanic technology* **23** 888–901
[14] Bingöl F, Mann J and Larsen G C 2010 *Wind Energy* **13** 51–61
[15] Trujillo J J, Bingöl F, Larsen G C, Mann J and Kühn M 2011 *Wind Energy* **14** 61–75
[16] Iungo G V, Wu Y T and Porté-Agel F 2012 *Journal of Atmospheric and Oceanic Technology* **30** 274–287
[17] Smalikho I N, Banakh V A, Pichugina Y L, Brewer W A, Banta R M, Lundquist J K and Kelley N D 2013 *J. Atmos. Oceanic Technol.* **30** 2554–2570 ISSN 0739-0572 URL `http://dx.doi.org/10.1175/JTECH-D-12-00108.1`
[18] Wessel A 2008 *Entwicklung Eines Physikalischen Modells Der Im Windpark Generierten Turbulenz: Development of a Physical Model for Calculation of the Turbulence Inside Wind Farms* Ph.D. thesis
[19] Wächter M, Rettenmeier A, Kühn M and Peinke J 2009 *Meteor. Z.* **18** 277–280 URL `http://www.ingentaconnect.com/content/schweiz/mz/2009/00000018/00000003/art00005`
[20] Sjöholm M, Mikkelsen T, Mann J, Enevoldsen K and Courtney M 2009 *Meteorologische Zeitschrift* **18** 281–287
[21] Mann J, Cariou J P, Courtney M S, Parmentier R, Mikkelsen T, Wagner R, Lindelöw P, Sjöholm M and Enevoldsen K 2008 *IOP Conference Series: Earth and Environmental Science* **1** 012012 URL `dx.doi.org/10.1088/1755-1315/1/1/012012`
[22] Frisch U 1995 *Turbulence* (Cambridge University Press)
[23] Argyris J H, Faust G, Friedrich R and Haase M 2015 *An Exploration of Dynamical Systems and Chaos* (Springer)
[24] Kolmogorov A N 1962 *J. Fluid Mech.* **13** 82–85
[25] Pope S B 2000 *Turbulent flows* (Cambridge university press)
[26] Böttcher F, Barth S and Peinke J 2007 *Stoch Environ Res Ris Assess* **21** 299–308
[27] Morales A, Wächter M and Peinke J 2012 *Wind Energy* **15** 391–406
[28] Mathieu J and Scott J 2000 *An introduction to Turbulent Flow* (Cambridge University Press)
[29] Tritton D J 1998 *Physical Fluid Dynamics* (Oxford University Press)
[30] Hamilton N and Cal R B 2015 *Physics of Fluids (1994-present)* **27** 015102
[31] Larsen G C, Madsen Aagaard H, Bingöl F, Mann J, Ott S, Sørensen J N, Okulov V, Troldborg N, Nielsen N M and Thomsen K 2007 Dynamic wake meandering modeling Risø-R 1607 Risø National Laboratory
[32] Larsen G C, Madsen H A, Thomsen K and Larsen T J 2008 *Wind Energy* **11** 377–395

Journal of Physics: Conference Series **625** (2015) 012007 doi:10.1088/1742-6596/625/1/012007

A wake detector for wind farm control

C L Bottasso[1,2], S Cacciola[1] and J Schreiber[1]
[1] Technische Universität München, München, Germany
[2] Politecnico di Milano, Milano, Italy

E-mail: carlo.bottasso@tum.de

Abstract. The paper describes an observer capable of detecting the impingement on a wind turbine rotor of the wake of an upstream machine. The observer estimates the local wind speed and turbulence intensity on the left and right parts of the rotor disk. The estimation is performed based on blade loads measured by strain gages or optical fibers, sensors which are becoming standard equipment on many modern machines. A lower wind speed and higher turbulence intensity on one part of the rotor, possibly in conjunction with other information, can then be used to infer the presence of a wake impinging on the disk. The wake state information is useful for wind plant control strategies, as for example wake deflection by active yawing. In addition, the local wind speed estimates may be used for a rough evaluation of the vertical wind shear.

1. Introduction and motivation

Wind turbines operating in a wind plant may be affected by the wake of neighboring wind turbines. When this happens, the affected machine experiences reduced power output and increased fatigue loading. The problem of how to control the machines within a wind power plant to minimize the undesirable effects of wake interactions is nowadays the subject of intensive research [1-3], enabled by the development of suitable wake models [3,4].

The implementation of control strategies to address this problem may require the ability to detect such an interference condition. For example, figure 1 depicts the case of wind plant control by wake deflection; in such a case, the detection of a wake interference condition may be used to yaw the affecting wind turbine until a clean condition on the affected machine is achieved.

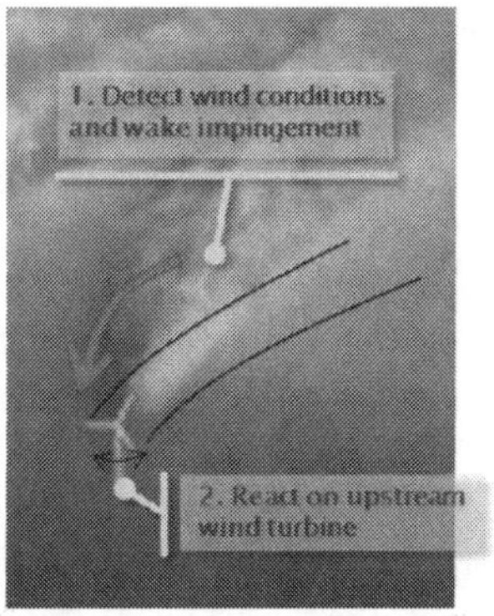

Figure 1. Wake interference detection followed by deflection.

Journal of Physics: Conference Series **625** (2015) 012007 doi:10.1088/1742-6596/625/1/012007

This paper describes a methodology to detect wake impingement on a wind turbine caused by an upstream machine. LiDARs (Light Detection and Ranging) may be able to detect the location of a wind turbine wake [5], and therefore in principle they could be used to achieve such a goal. However, their use is still confined to research applications, and they are not yet routinely deployed in the field on production machines. In this work, the wake interference detection is based on the use of rotor loads. In fact, as rotor load sensors are becoming commonly available on modern wind turbines, for example to enable individual blade pitch control, no additional sensors or equipment are necessary for the implementation of the present method. The use of rotor loads for the detection of wind conditions is a technology that has been proposed and demonstrated in [6-10]. The present work extends the technology to the estimation of the wake state.

The proposed method is based on the estimation of the wind speed by the out-of-plane blade root bending moment. Similarly to the thrust coefficient, a cone root bending coefficient may be defined, which depends on the tip speed ratio and blade pitch. Using the three blades together, knowledge of the loads in addition to rotor speed and blade pitch setting allows one to estimate from the cone coefficient a rotor-equivalent wind speed. On the other hand, when using the information for each single blade, one may sense the wind at the azimuthal location occupied by that blade. Averaging over an azimuthal interval, an estimate of the local wind speed in a rotor sector can be readily obtained, from which in turn one may also obtain an estimate of the turbulence intensity.

2. Methods

At first, a new rotor-effective wind speed estimator is developed, based on the out-of-plane cone (i.e. averaged over the number of blades) bending rotor loads. The cone coefficient is defined as

$$C_{m_0}(\lambda_{RE},\beta,V_{RE}) = \frac{\frac{1}{2\pi}\int_{\psi=0}^{2\pi}\sum_{i=1}^{B} m_i(\psi_i)\,\mathrm{d}\psi}{\frac{1}{2}\rho ARV_{RE}{}^2}, \tag{1}$$

where λ_{RE} is the rotor-effective tip speed ratio, β the pitch angle, m_i the out-of-plane bending moment of blade i, which occupies the azimuthal position ψ_i, ρ is the density of air, A the rotor disc area, R the rotor radius and V_{RE} the unknown rotor-effective wind speed. This latter quantity can be now interpreted as that constant wind speed that induces the same out-of-plane bending moment as the one being currently measured. With this definition, the blade collective bending moment (per blade) $m_0 = \sum_{i=1}^{B} m_i/B$ is computed as

$$m_0(t) = \frac{1}{2B}\rho AR\, V_{RE}{}^2(t)\, C_{m_0}\big(\lambda_{RE}(t),\beta(t),V_{RE}(t)\big), \tag{2}$$

where B is the number of blades. At each instant of time, m_0 is measured by blade load sensors. As also the rotor speed Ω can be easily measured together with the blade pitch angle β, the sole unknown in the equation is the effective wind speed V_{RE}, which can therefore be readily computed.

To increase robustness of the estimates in the face of measurement and process noise, an Extended Kalman filter is used for the computation of V_{RE}. The wind speed update is defined as

$$V_{RE_k} = V_{RE_{k-1}} + w_{k-1}, \tag{3}$$

w_k being the process noise with covariance $\boldsymbol{Q}$, while the non-linear output equation is defined as

$$z_k = \frac{1}{2}\rho AR\, V_{RE_k}^2\, C_{m0}(\lambda_{RE},\beta) - \widehat{m}_0 + v_k, \tag{4}$$

where v_k is the measurement noise with covariance $\boldsymbol{R}$, $\widehat{m}_0$ are the measured loads, while the output z_k is set to 0 to enforce the desired equation at each step.

Such a method delivers estimates of the wind speed and turbulence intensity that are similar, if not slightly superior, to the classical rotor-effective wind speed estimators based on the torque balance equation using the power coefficient C_P [11]. In fact, the present approach may be somewhat less limited in frequency than the classical one, which is slowed by the significant inertia of the rotor. An example of the quality of the resulting estimates of the wind speed in the time domain is shown in figure 2. The effective wind speed reference, shown in the figure using a black solid line, was computed from the simulation input as the spatial mean of the longitudinal wind speed within the rotor disk area, as already done in Østergaard et al. [12].

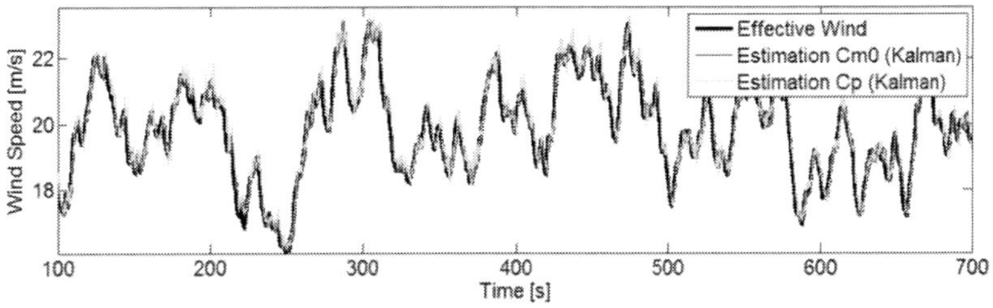

Figure 2. Effective wind speed time history, and its estimates obtained by the cone estimator using the C_{m_0} coefficient, or the classical formulation based on the power coefficient C_P [11].

The estimates provided by the present method are of good accuracy and robustness with respect to the tuning parameters of the filter, as shown in figure 3.

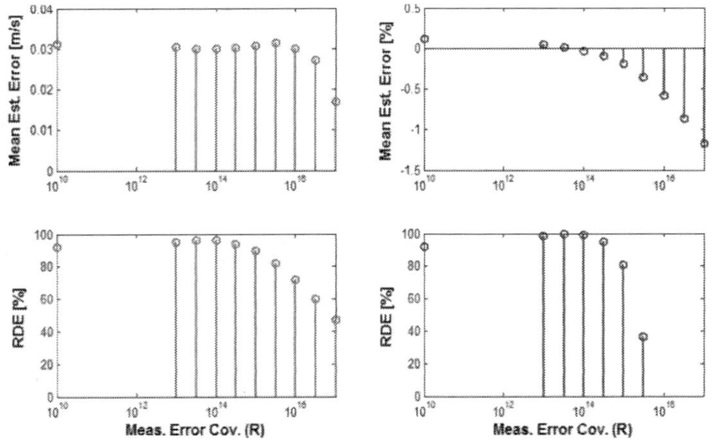

Figure 3. Estimation of rotor-effective wind speed (left) and turbulence intensity (right): mean estimation error (top), and relative degree of explanation (RDE) [11] (bottom).

Next, the method is specialized to the estimation of wind speed on sectors of the rotor disk. To this end, considering the ith blade, the previous moment equation is modified as

$$m_i(t) = \frac{1}{2B} \rho A R \, V_{LE}{}^2(t,\psi) \, C_{m_0}\big(\lambda_{LE}(t,\psi), \beta(t), V_{LE}(t)\big), \tag{5}$$

where the blade local-effective wind speed is defined as

$$V_{LE}(\psi) = \frac{1}{A_B} \int_{A_B} V \mathrm{d}A_B(\psi), \tag{6}$$

A_B being the planform area of the rotor blade. As clear from (6), the ith blade local-effective wind speed can be seen as that constant wind speed that would lead to the same out-of-plane-bending moment of the ith blade. From the blade local-effective wind speed, a sector-effective wind speed is obtained by averaging over an azimuthal interval of interest. The concept is illustrated in figure 4, which shows how the passage of a blade over a disk sector can be used for estimating a sector-effective wind speed:

$$V_{SE}(\bar{t}) = \frac{1}{A_S} \int_{A_S} V_{LE}(t,\psi) \, \mathrm{d}A_S = \frac{1}{\psi_2 - \psi_1} \int_{\psi_1}^{\psi_2} V_{LE}(t,\psi) \mathrm{d}\psi. \tag{7}$$

Finally, from the knowledge of the wind speed, the sector turbulence intensity can be readily computed.

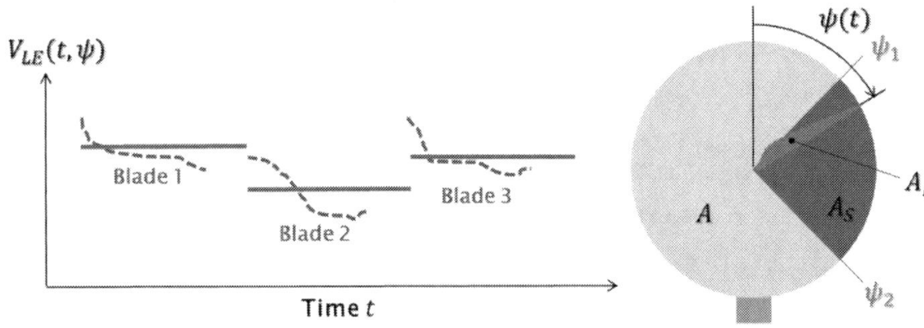

Figure 4. Estimation of local-effective and sector-effective wind speed and turbulence intensity, from the loads of a blade passing through a rotor disk sector.

3. Results

At first, the new method was tested using field data of the NREL CART 3 wind turbine [13]. As no known wake interference condition of that machine with other wind turbines was available for this study, the local wind estimation method was used to estimate the different velocities in the top and bottom quadrants of the rotor, which gives an idea of the vertical wind shear.

Using a nearby met-mast, a comparison with anemometer measurements can be performed, the position of the anemometers being shown in figure 5. To compare with these two point measurements, it is useful to derive a point measurement from the sector-effective wind speed estimated by the present method. To this end, the blade bending moment averaged over the sector, m_{iS}, can be alternatively computed as

$$m_{iS} = \frac{1}{2B} \rho \int_{\psi_1}^{\psi_2} \int_0^R V^2(\psi,r) \, C_T(\psi,r) r^2 \mathrm{d}r \, \mathrm{d}\psi, \tag{7}$$

where C_T is the local thrust coefficient. Assuming a constant speed and thrust coefficient over the rotor disk, the bending moment is readily expressed as

$$m_{iS} = \frac{1}{2B}\rho A_S V_{SE}^2 C_T \frac{2}{3}R. \tag{8}$$

This expression shows that the bending loads used for estimating the sector-effective wind speed can be seen as generated by the sector thrust applied at $2/3\ R$; in this sense, the estimator can be interpreted as sampling the wind field at about 66% of the blade span. Consequently, the anemometer measurements were interpolated at hub height $\pm 2/3\ R$ and used as reference measurements in the following plot to judge the quality of the load-based estimates.

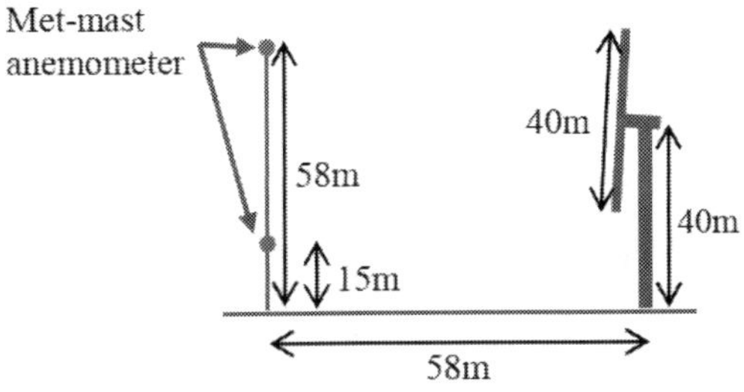

Figure 5. Sketch of met-mast anemometer position wrt the NREL CART 3 wind turbine.

Figure 6 shows a good agreement between the estimated sector-effective wind speeds with the interpolated point-wise anemometer measurements. The trends are reasonably well followed and a higher wind speed is detected in the top than in the bottom quadrant.

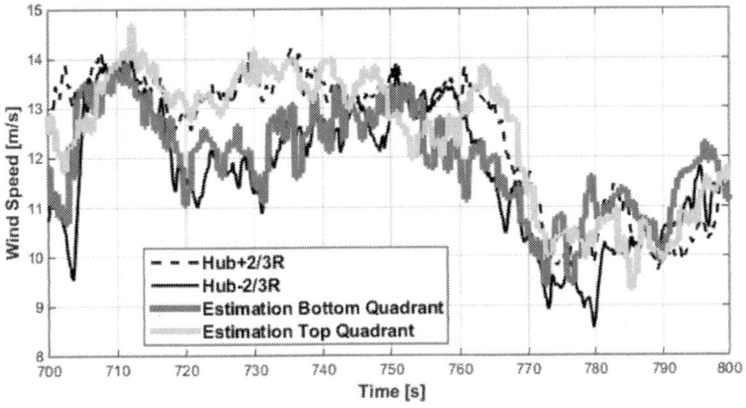

Figure 6. Estimation of top and bottom quadrant effective wind speeds, and comparison with interpolated met-mast data for the NREL CART 3 wind turbine.

Finally, the new method was used for estimating the local wind speed separately on the left and right parts of the rotor, thereby detecting the possible presence of an area of reduced speed and increased turbulence intensity, which may indicate the presence of a wake. As no experimental data was available for this case, a simulation study was conducted, by using a high-fidelity aeroservoelastic model [14] of a multi-MW wind turbine operating in different partial and full wake conditions. A representative image of the wind field at the rotor disk in waked and unwaked conditions is shown if figure 7.

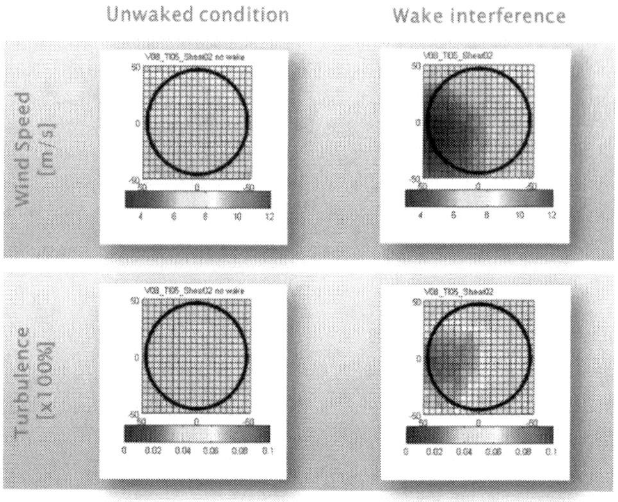

Figure 7. Wind field at the rotor disk in unwaked (left) and waked (right) conditions.

The results of the estimation are summarized by figures 8 and 9, which show the actual and estimated local wind speeds and turbulence intensities in two lateral quadrants of the rotor. Each subplot refers to a different overlap indicated by the lateral distance between the rotor and the wake center.

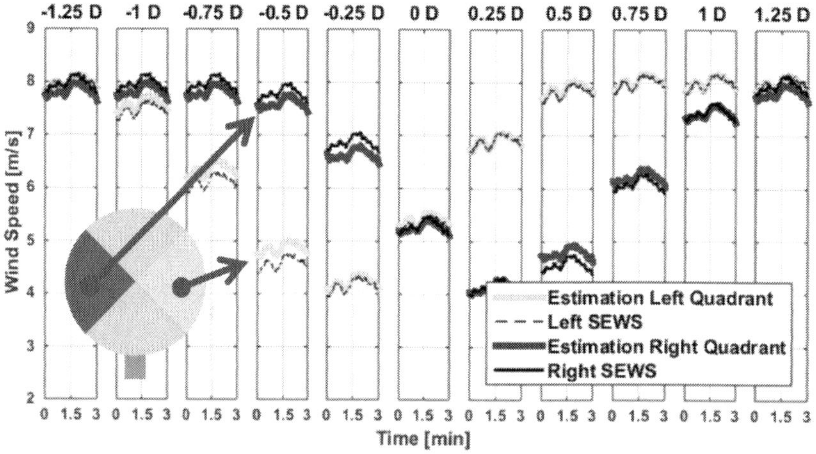

Figure 8. Estimation of local wind speed on two lateral rotor quadrants.

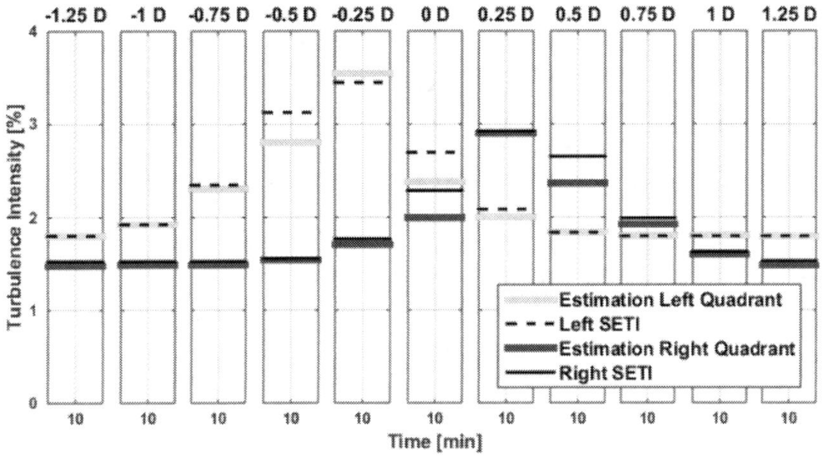

Figure 9. Estimation of local wind turbulence intensity on two lateral rotor quadrants.

4. Conclusions

As well illustrated by the results shown here, the proposed method is capable of estimating with good accuracy the local wind speed and turbulence intensity, and in particular it is able to detect variations of these quantities on the two sides of the rotor that may be indicative of a wake interference condition. Similar results can be derived for various wind conditions, demonstrating the robustness of the local wind speed estimator. The new wake detector is currently being used for driving wake deflection strategies by active wind turbine yaw. In addition, the same approach may be used for a rough estimation of the vertical wind shear.

References

[1] Lund B F, Oma P N and Jakobsen T 2014 Wind farm production maximization through joint optimization of turbine efficiency and yaw. Simulated case studies. *Poster EWEA* (Barcelona, Spain)

[2] Park J, Kwon S and Law K H 2013 Wind farm power maximization based on a cooperative static game approach *Proceedings of the SPIE Active and Passive Smart Structures and Integrated Systems Conference* (San Diego, USA)

[3] Fleming P, Gebraad P, Lee S, van Wingerden J, Johnson K, Churchfield M, Michalakes J, Spalart P, Moriarty P 2013 High-fidelity simulation comparison of wake mitigation control strategies for a two-turbine case *Proceedings of ICOWES Conference* (Lyngby, Denmark)

[4] Hao Y, Lackner M, Keck R E, Lee S, Churchfield M, Moriarty P 2013 Implementing the dynamic wake meandering model in the NWTC design codes *AIAA*

[5] Bingöl F, Mann J, Larsen G. 2010 Light detection and ranging measurements of wake dynamics. Part I: one-dimensional scanning. *Wind Energy* **13** 51–61

[6] Bottasso C L and Riboldi C E D 2015 Validation of a wind misalignment observer using field test data *Renewable Energy* **74** 298–306

[7] Bottasso C L and Riboldi C E D 2014 Estimation of wind misalignment and vertical shear from blade loads *Renewable Energy* **62** 293–302

[8] Bottasso C L, Campagnolo F and Petrović V 2014 Wind tunnel testing of scaled wind turbine models: beyond aerodynamics *J Wind Eng Ind Aerod* **127** 11–28

[9] Bottasso C L, Croce A and Riboldi C E D 2010 Spatial estimation of wind states from the aeroelastic response of a wind turbine *Proc. TORQUE* (Heraklion, Greece)

[10] Bottasso C L, Croce A, Riboldi C E D and Bir G 2009 Real-time estimation of structural and wind states for wind turbine advanced control *Proc. EWEC* (Marseille, France)

[11] Soltani M and Wisniewsky R 2013 Estimation of rotor effective wind speed: a comparison *IEEE Trans. Contr. Syst. Technol.* **21**(4) 1155–1167

[12] Østergaard KZ, Brath P, Stoustrup J 2007 Estimation of effective wind speed *Journal of Physics: Conference Series 75, EWEA*

[13] Fleming P A, Wright A D, Fingersh L J and van Wingerden J W 2011 Resonant vibrations resulting from the re-engineering of a constant-speed 2-bladed turbine to a variable-speed 3-bladed turbine *Proc. 49th AIAA Aerospace Sciences Meeting* (Orlando, Florida)

[14] Bottasso C L and Croce A 2006–2015 *Cp-Lambda – User's Manual* Politecnico di Milano (Milano, Italy)

Simulation of wake effects between two wind farms.

KS Hansen[1], P-E Réthoré[1], J Palma[2], BG Hevia[3], J Prospathopoulos[4], A Peña[1], S Ott[1], G Schepers[5], A Palomares[6], MP van der Laan[1] and P Volker[1]

[1]DTU Wind Energy-DK, [2]UPorto-PT, [3]CENER-ES, [4]CRES-GR, [5]ECN-NL and [6]Ciemat-ES

E-mail: kuhan@dtu.dk

Abstract. SCADA data, recorded on the downstream wind farm, has been used to identify flow cases with visible clustering effects. The inflow condition is derived from a partly undisturbed wind turbine, due to lack of mast measurements. The SCADA data analysis concludes that centre of the deficit for the downstream wind farm with disturbed inflow has a distinct visible maximum deficit zone located only 5-10D downstream from the entrance. This zone, representing 20-30% speed reduction, increases and moves downstream for increasing cluster effect and is not visible outside a flow sector of 20-30°. The eight flow models represented in this benchmark include both RANS models, mesoscale models and engineering models. The flow cases, identified according to the wind speed level and inflow sector, have been simulated and validated with the SCADA results. The model validation concludes that all models more or less are able to predict the location and size of the deficit zone inside the downwind wind farm.

1. Introduction

The EERA-DTOC project (European Energy Research Alliance – Design Tools for Offshore wind farm clusters) is aiming at creating a robust, easy to use and flexible tool to facilitate the optimized design of individual or cluster of offshore wind farms [1, 2].

The wake modeling part of EERA-DTOC has been used to improve the fundamental understanding of wind turbine wakes and modeling. A number of wake cases for large offshore wind farms have been identified. These cases cover a range of conditions, likely to be experienced in offshore wind farms and provide sufficient high-quality input data to form a basis for a systematic evaluation. This includes models from wind farm scale models with straightforward parameterizations for wake behavior through Computational Fluid Dynamics (CFD) models.

This investigation is focusing on benchmarking existing wake modes and in particular their ability to handle the upscaling from wind farm to cluster scale. All benchmark flow cases are defined with reference to existing wind farm measurements, which is used in the validation.

Datasets with potential wake effects within wind farms and between wind farms have been identified and used for the verification. Some of the preliminary flow cases have been reported [3] based on the template from the WakeBench project [4]. The wake models are compared to measurements and possible improvements are identified through three benchmarks:

1) The first model validation has been focused on a wind farm with constant spacing (7-10 D), combined with sensitivity analysis of power deficit as function of inflow direction, turbulence or stratification on the Horns Rev offshore wind farm [3, 5].

2) The second model validation has been performed on an offshore wind farm, characterized with a small spacing (3.3 – 4.3 D) and two "missing" turbines. Results of this benchmark have been presented [6].

3) The third validation has been performed on a large wind farm with variable spaced wind turbines, arranged on 5 distinct arches. The last part of the validation encompass a complete wind farm cluster, consisting of two large offshore wind farms with 90 & 72 wind turbines and separated with more than 30D.

This article will present the different models, the wind farm cluster layout, comparison of the measured and simulated wake deficits and the sector wise park efficiency from the third validation.

2. Models

A range of different models used in the benchmarking and their affiliations are listed in Table 1. The models are divided into three groups: engineering models, WRF mesoscale model and CFD models.

Table 1. Models used in the EERA-DTOC benchmark.

DTOC	Models	Affiliation	Mail
1	SCADA/BA	DTU Wind Energy/K.S.Hansen	kuhan@dtu.dk
2	FUGA/SO	DTU Wind Energy/S. Ott	sqot@dtu.dk
3	NOJ/GU	DTU Wind Energy/A. Peña	aldi@dtu.dk
4	NOJ/Peña	DTU Wind Energy/A. Peña	aldi@dtu.dk
5	WRF/UPM	CIEMAT/A.Palomares	ana.palomares@ciemat.es
6	WRF/PV	DTU Wind Energy/P.Volker	pvol@dtu.dk
7	AD/RANS	UPORTO/J.L. Palma	jpalma@fe.up.pt
8	CFDWake	CENER/B.G. Hevia	bgarcia@cener.com
9	CRESflowNS	CRES/ J. Prospathopoulos	jprosp@cres.gr
10	FarmFlow	ECN Wind Energy/J.G Scheepers	schepers@ecn.nl
11	RANS/f$_P$C	DTU Wind Energy/P.van der Laan	plaa@dtu.dk

A short summary of the models are given, but detailed information have been published previously and will only be referred below.

1. SCADA represent the processed wind farm measurements, which are compared to the simulated wake model results. Wind farm SCADA data are usually not referred as a model result in the literature. However, considering the amount of assumptions and processing methods that have to be applied in order to produce comparable results with a wind farm wake model, a processed SCADA data should in all fairness be treated as a model result. This point is further detailed in the succeeding chapters.
2. FUGA is a linearized flow solver based on the steady-state RANS equations, currently only applicable to flat and homogeneous terrain. In FUGA the flow is assumed to be incompressible and lid-driven at the chosen inversion height. It uses a simple eddy viscosity turbulence closure and the Actuator Disk (AD) approach. The model description and validation results are available in [7].
3. NOJ/Peña is a mass-conserving engineering wake model, which is used to estimate the hub-height wind speed downstream of a wind turbine at a given distance. The estimation is based on the hub-height inflow wind speed, known thrust coefficient, rotor radius, wake decay coefficient [8] in combination with the PARK wind farm flow model [9]. The NOJ results are calculated for a distinct, center line direction of the inflow sectors.
4. NOJ/GU is also based on the NOJ model and the results labeled (GU= Gausian Uncertainty) are calculated for distinct directions and post-processed according to the methodologies described in [10], as function of the directional variability. The GU method is applicable for each of the listed methods.
5. WRF/PV is the Weather Research and Forecasting (WRF) model [11], combined with an Explicit Wake Parametrization (EWP) developed for mesoscale models. A description is given in [12]. The wind turbine is treated as a drag device, which executes a force in the opposite flow direction. In the parametrization, it is assumed that the grid cell pressure field is in average, uninfluenced by the wind turbine and that a wind turbine will cause only additional turbulence due to the enhanced shear production [13].
6. WRF/UPM, also based on WRF, is used to downscale the outputs from the reanalysis data to the domain. The WRF model is coupled to the UPMPARK model [14], which describes the diffusion of multiple wakes in the atmospheric surface layer parameterized by Monin-Obukhov scaling, in order to model the wind flow inside the wind farm.
7. AD/RANS represents the VENTOS®/2 code, which is built around a segregated finite volume implicit method for the solution of the Reynolds averaged Navier-Stokes (RANS) equations for non-

stratified flows, with a two-equation k−ε turbulence model. This Navier-Stokes solver is specifically geared towards the solution of wind flow problems over complex terrain, with a stable core [15, 16] upon which improvements have been added over the years. Modeling the momentum drag associated with the presence of a wind turbine is done implicitly in VENTOS®/2 [17], using a uniformly loaded AD model.

8. CFDWake 1.0 is an elliptic CFD wind farm model, implemented in the open source OpenFOAM, which allows simulating wake effects inside big wind farms through the actuator disk concept. The model is designed for offshore sites where met mast measurements are representative for the freestream flow of all wind turbine positions in the wind farm. CFDWake is implemented as a module of the CFDWind1.0 wind model, designed for the simulation of wind in the Superficial Boundary Layer [18,19,20]. Only simulation results for westerly inflow has been forwarded.

9. CRESflowNS, [21] is a full 3D Navier–Stokes solver using the Wilcox k–ω turbulence model [22] for turbulence closure. Simulation of the embedded wind turbines is made through the actuator disk theory where the rotor of each wind turbine is simulated as a disk discretized by a number of control volumes. In order to determine the reference velocity for thrust calculation, a parabolic procedure is applied, [23]. The simulation starts without turbines and activates the first row when a certain convergence criterion is fulfilled for the velocities at the wind turbine positions of that row. This procedure is successively repeated until all wind turbine rows are included. Only simulation results for westerly inflow has been forwarded.

10. FarmFlow wake model was originally based on the UPMwake code as developed by the Universidad Polytecnica de Madrid (UPM) [24]. FarmFlow is a 3-dimensional parabolized RANS code with a k−ε turbulence model [25]. The free stream wind field in Farmflow is modelled by taking into account the effect of atmospheric stability through the Monin-Obukhov length and the rotor is modelled as an actuator disk characterized by the thrust curve. The k−ε turbulence parameter adjustments are used to correct for the actuator disk assumptions, based on local field measurements [26].

11. RANS/f_PC is based on the flow solver EllipSys3D [27-28], where the RANS equations are employed, together with the k-ε-f_P [29] turbulence model, capable of handling the complex near wake flow through a variable eddy-viscosity coefficient. The wind turbines are represented as ADs using a variable force method suited for interacting ADs [30]. Velocity and power deficits predicted by the k-ε-f_P model combined with the variable AD force method compare well with large eddy simulations and measurements, both for single wind turbine wakes [29] and complete wind farms [30]. Furthermore the effect of the Coriolis force on wind turbine wake and wind farm wake interaction has been included by modeling the entire atmospheric boundary layer (ABL) [31].

3. Cluster layout and measurements

The offshore wind farm cluster modelled in EERA-DTOC is located in the southern part of Denmark, 5-10 km outside the coastline, as shown in Figure 1.
The wind farm cluster consists of two large wind farms separated with a distance of 3 km equal to 30 diameters, as shown in Figure 2.

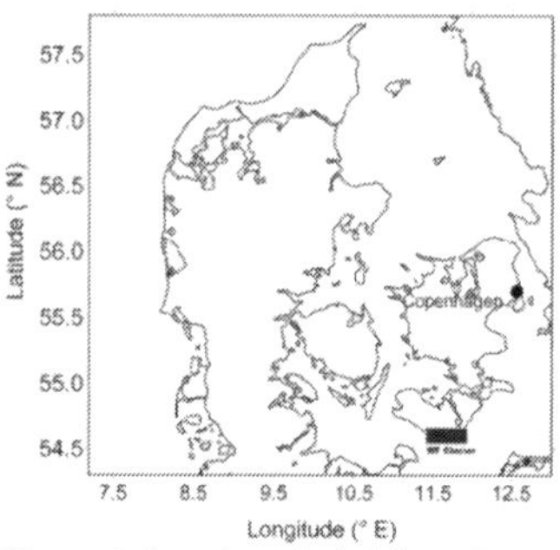

Figure 1. Location of wind farm cluster.

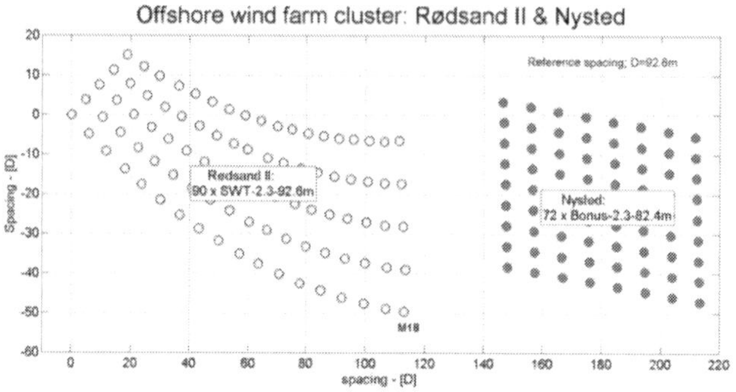

Figure 2. Layout of the wind farm cluster, where the reference spacing is based on SWP-2.3-92.6

3.1 Wind farm layout

Rødsand II offshore wind farm consists of 90 SWP 2.3MW wind turbines with a 92.6 m rotor diameter and a hub height of 68.5 m, partly owned by E·ON. The turbines, which are controlled with a combination of variable speed and pitch, are located on 5 distinct arcs with increased spacing, as shown in Figure 2. The spacing increases from 5D to 10D for changing wind direction. The turbines in Rødsand II wind farm are identical to the turbines presented in the previous EERA-DTOC analysis performed on the Lillgrund offshore wind farm [6]. The wind farm is only equipped with one met mast, located in the south western corner and the wind measurements are very disturbed by the two wind farms for easterly winds.

Nysted offshore wind farm consists of 72 Bonus (later SWP) 2.3MW/82.4m wind turbines, which are active stall controlled, with a hub height of 69 m. This wind farm was commissioned in 2003 and is owned (partly) by DONG Energy A/S. The wind farm includes 4 met masts, located west and east for wind farm. The SCADA dataset from this wind farm has previously been used to intensively wake studies [32]. The internal spacing along the two main directions is 180°/5.7D and 97°/10.3D. Nysted wind farm has been in operation for more than 10 years and obtained an average of 3300 full load hours annually.

3.2 Data sharing and qualification

The SCADA data from Rødsand II includes two periods without time stamp, representing westerly and easterly inflow to the wind farm, for a wind speed range 5-11 m/s. Easterly inflow represent an inflow sector of 70 – 110°, as shown in Figure 3. The wind farm data is recorded with the (supervisory control and data acquisition) SCADA system as 10-minute statistics for each individual wind turbine in the wind farm. The SCADA signals are: electric power, pitch angle, rotor speed, yaw position and nacelle wind speed.

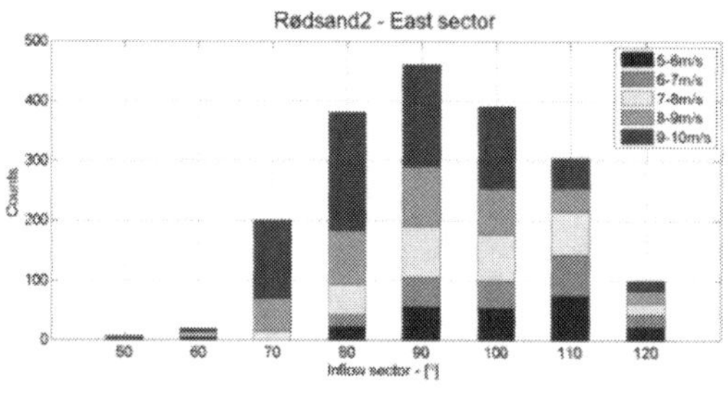

Figure 3. Rødsand II, qualified dataset for easterly inflow.

The SCADA data has been quality-controlled according to the guidelines given in [33, 34]. While the qualification only disqualified 3% of the data and it was decided not to apply further constraints. Due to a lack of measured inflow conditions, it has been necessary to derive the inflow conditions directly from the wind turbines:

- The inflow wind speed at hub height is derived from the measured electrical power combined with the official power curve.
- The inflow direction at hub height is derived from the wind turbine yaw position.

The model comparison is based on wind speed values derived from power values with reference to the official power curve. Unfortunately it has not been possible to verify any of the wind turbine power curves against the official power curve.

The SCADA data from Nysted wind farm includes 2½ years of operation, which has been used in previous flow deficit analysis [34]. The dataset consists of wind farm SCADA data synchronized with meteorological measurements.

Unfortunately there is no synchronized SCADA data representing the whole wind farm cluster. The model validation will focus on the periods where Rødsand II wind farm is online, assuming the Nysted offshore wind farm is online and operating with 100% availability. The Danish wind turbine production register hosted by the Danish Energy Agency reports that the energy production for the Nysted wind farm throughout the period 2011-2014 has been reliable and large.

3.3 Ambient flow conditions

The wind conditions have been measured before the installation of the Rødsand II wind farm and displays a sector wise turbulence intensity (TI) of 7% at hub height for easterly wind at 8 ± 0.5 m/s for the wind direction sector [75-105°].

The average speed recovery behind the Nysted wind farm for easterly inflow has been extracted as function of the inflow angle, and the estimated wind speed reduction is 15% for the inflow sector [80-105°] at wind speed of 8 m/s.

3.4 SCADA data analysis

The normalized speed deficit[1] inside the wind farm is determined with reference to the wind speed at the reference wind turbine M18, which is located in the south eastern corner of Rødsand II wind farm, Figure 2. Unfortunately this wind turbine is operating partly in the wake of Nysted wind farm, as all other Rødsand turbines for easterly wind. The inflow wind speed is derived from the power level, corresponding to 8±0.5 m/s and the inflow wind direction is equal to the wind turbine yaw position.

The normalized speed deficit distribution for the "variable" spaced turbines in the Rødsand II wind farm is difficult to visualize – compared to previous analysis of wind farms with regular layout and fixed spacing. The normalized speed deficit values are therefore illustrated as colored contour plot, as show in Figure 4. The normalized wind speed deficit distribution for Nysted wind farm is calculated for the wind speed interval 8±0.5 m/s with reference to the measured wind speed at hub height on a mast located 2 km east of the wind farm, for 10 degree sectors. The Rødsand II wind farm reference turbine (=M18) is partly disturbed for WD=77-97° by the Nysted wake. The figures 4b, 4c and 4e demonstrates how the distinct Rødsand II speed deficit zone develops at 87°, increases in size and moves downstream for increasing inflow angle until 107° when the Nysted wake only covers a tiny part of the Rødsand II wind farm.

The integrated effect of the speed deficit distribution is expressed as the sector wise park efficiency, calculated with reference to the top-3 producing wind turbines, using a moving window averaging technique, for a window size of 4°. Figure 5 shows a decreasing park efficiency when the Nysted wake partly covers Rødsand II wind farm. The decrease is a combined wake effect and cluster effect. This effect seems to reach a minimum at 107°, when the Nysted wake has passed the Rødsand II wind farm.

[1] The normalized speed deficit is equal to (1-speed ratio).

Due to the asymmetric layout of the Rødsand II wind farm it is not possible directly to conclude on the park efficiency obtained for western and eastern inflow sector.

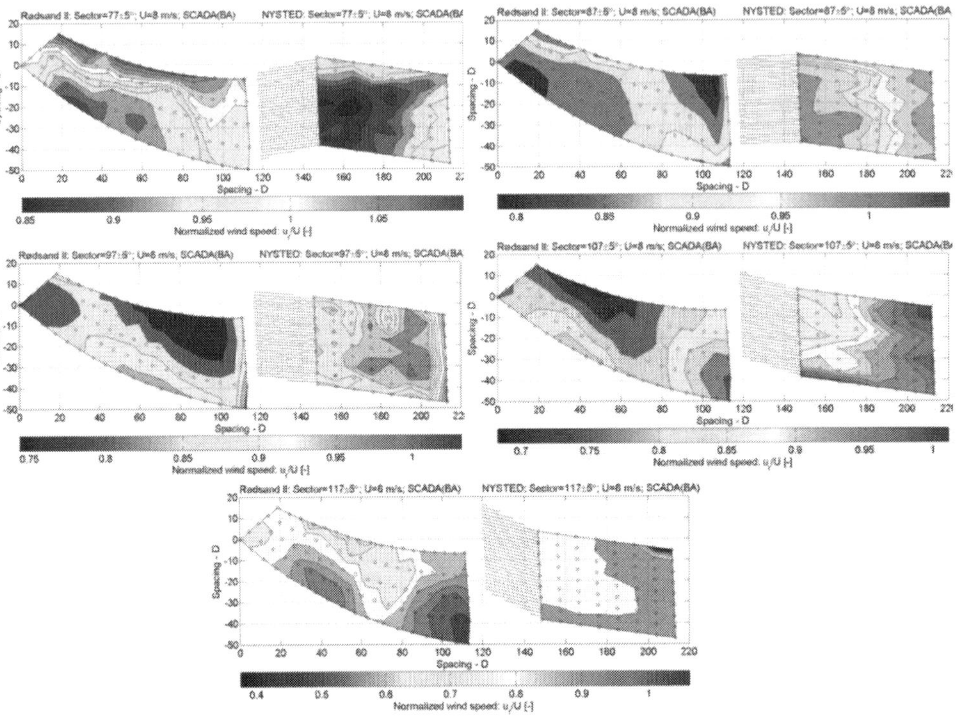

Figure 4. Measured normalized speed deficit distribution on Rødsand II & Nysted wind farm for WD=77, 87, 97,107, 117 ±5° and U=8 ±0.5 m/s sectors respectively.

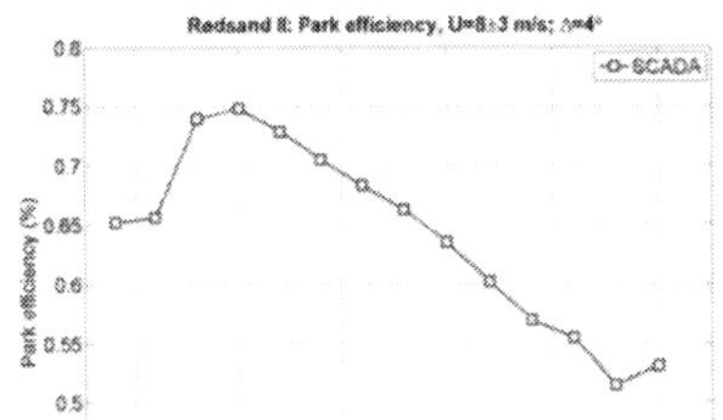

Figure 5. Measured park efficiency for the Rødsand II wind farm.

The SCADA data analysis includes a high uncertainty related to operational wind turbine parameters, which can be summarized as:

- Disturbed inflow to the reference wind turbine, M18;
- Lack of power curve validation and signal calibration;
- Transformation from electric power to wind speed without taking pitch angle and rotor speed into account;
- Wind turbine yaw position depends on the yaw controller, time constants and yaw threshold values;

- The spatial variability of both wind speed and wind direction over very large distance are difficult to take into account.

4. Flow cases

In the easterly wind direction, both wind farms need to be modelled to predict the cluster effect. While there is a limited amount of information available on Rødsand II wind farm and nothing about the operational status of Nysted wind farm, the modellers should assume that all 162 turbines were operating under neutral atmospheric conditions.

The wind farm and cluster wake models are simulated for the inflow conditions applied to the Nysted wind farm, summarized as:

- Inflow wind speeds at hub height is 8±0.5° m/s;
- Turbulence intensity at hub height is 7%;
- 5 different inflow sectors, represented with: [87±5°, 97±5°, 107±5°, 77±5°, 117±5°] degrees;

The modellers were asked to model the five distinct inflow directions or the sectors. All modellers should provide the simulated power values for each of the 162 wind turbines for the 5 distinct flow sectors.

5. Results

The model results were divided into two parts; 4 partners submitted results covering the whole wind farm cluster while the remaining 4 modellers submitted results covering only the Rødsand II wind farm - but still taking the influence into account the wake from Nysted wind farm. The model contributions are listed in Table 2.

Table 2. Wind farm cluster simulation matrix

	Rødsand II wind farm					Nysted wind farm				
WD=	77°	87°	97°	107°	117°	77°	87°	97°	107°	117°
SCADA(BA)	x	x	x	x	x	x	x	x	x	x
AD/RANS	x	x	x	x	x	x	x	x	x	x
FarmFlow	x	x	x	x	x	x	x	x	x	x
NOJ/Peña	x	x	x	x	x					
NOJ(GU)	x	x	x	x	x					
FUGA/SO	x	x	x	x	x					
Mesoscale	x	x	x	x	x					
RANS/f_PC	x	x	x	x	x	x	x	x	x	x
WRF/UPM	x	x	x	x					x	

5.1. Discussion

The whole wind farm was modelled as a single grid-cell in the Mesoscale model and the resulting power/wind speed is represented with one single value.

Sector WD=77° demonstrates that the model agrees well, but predicts larger speed reduction compared to the SCADA results with NMAE[2] values equal to 0.1±0.02 for all models.

The discussion below refers to the presented results compared to SCADA results, shown in Figure 4.

Sector WD=87° on Figure 6, demonstrates that most of the models predict the triangular speed reduction zone in the NE part initiated by Nysted wind farm wake quite well. The NMAE values are equal to 0.04±0.02.

Sector WD=97° on Figure 7, represents the aligned flow in the Nysted wind farm and demonstrates how the triangular speed reduction zone moves [downstream] towards W for increasing inflow direction

[2] NMAE is the Normalized Mean Averaged Error from WakeBench [4] and has been calculated with reference to the SCADA values.

for most of the simulated distributions. All the models predicts the deficit well with comparable NMAE values equal to 0.06±0.02.

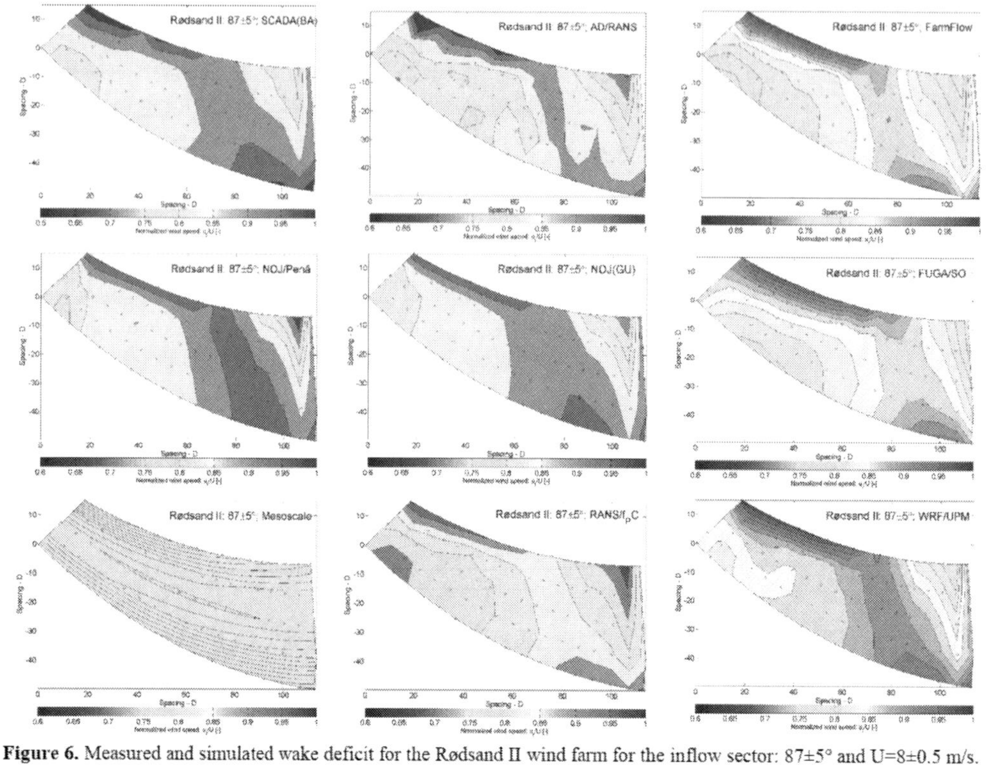

Figure 6. Measured and simulated wake deficit for the Rødsand II wind farm for the inflow sector: 87±5° and U=8±0.5 m/s.

Sector WD=107° (plot is not included); despite that most of the Nysted wake influence has vanished, some models still predict a distinct N-S deficit area around 90D with a visible speed reduction. All models predict the increased speed around the SE wind farm "corner", with undisturbed inflow.

Sector WD=117° (plot is not included), some of the models predicts a large, distinct deficit area, 30D from the inlet. All models predict the increased speed around the SE wind farm "corner" with undisturbed inflow. The NMAE values equal to 0.1±0.02 demonstrates a limited scatter compared to the previous sector results.

Figure 8 shows the complete, simulated cluster effect for 2 inflow sectors equal to 87° & 97°. The three models seems to predict very well both the undisturbed area and the location and size of the peak deficit zone.

The park efficiency has been calculated for each of the five flow cases (U=8±0.5 mm/s) and compared to the measured (U=5-11 m/s) park efficiency in Figure 9. The modelled park efficiency levels varies quite much but all models predicts the minimum efficiency to be located in the sector [97-107°].

5.2. Summary

Previous SCADA analysis concludes that centre of the deficit for a wind farm with variable spacing and undisturbed inflow is located 80-90 diameters downstream from the inflow turbines. Furthermore, the location of the deficit zone is not very sensitive to the inflow direction and the maximum deficit inside the zone is 20 – 25 %. The present analysis of disturbed inflow situations concludes that the maximum deficit zone is distinct and located only 5-10D downstream from the wind farm entrance. The size of the deficit zone increases and moves downstream for increasing inflow direction e.g. where the wind farm operates in partly wake conditions.

The eight models in the benchmark includes both RANS models, mesoscale models and engineering models. The flow cases, identified with wind speed and inflow sector, have been simulated and compared with the SCADA results. The validation confirms that a distinct triangular deficit zone appears 5-10D into the wind farm, when the upstream wind farm wake encompass the downwind wind farm. The deficit zone, representing 20-30% speed reduction, increases and moves downstream for increasing inflow direction (partial cluster effect), and the external wake effect disappears outside a flow sector of ±15°.

The benchmark demonstrates that the models are able to predict cluster performance, but it is difficult to draw a conclusion about which models are in error because of the very large uncertainty of the measurements for this test case. Most models use different physical assumptions to model the wake effect. These differences in assumption have an influence on the prediction accuracy that becomes greater as the wind farms become larger. When there are two wind farms in the wake of one another this effect is even more pronounced.

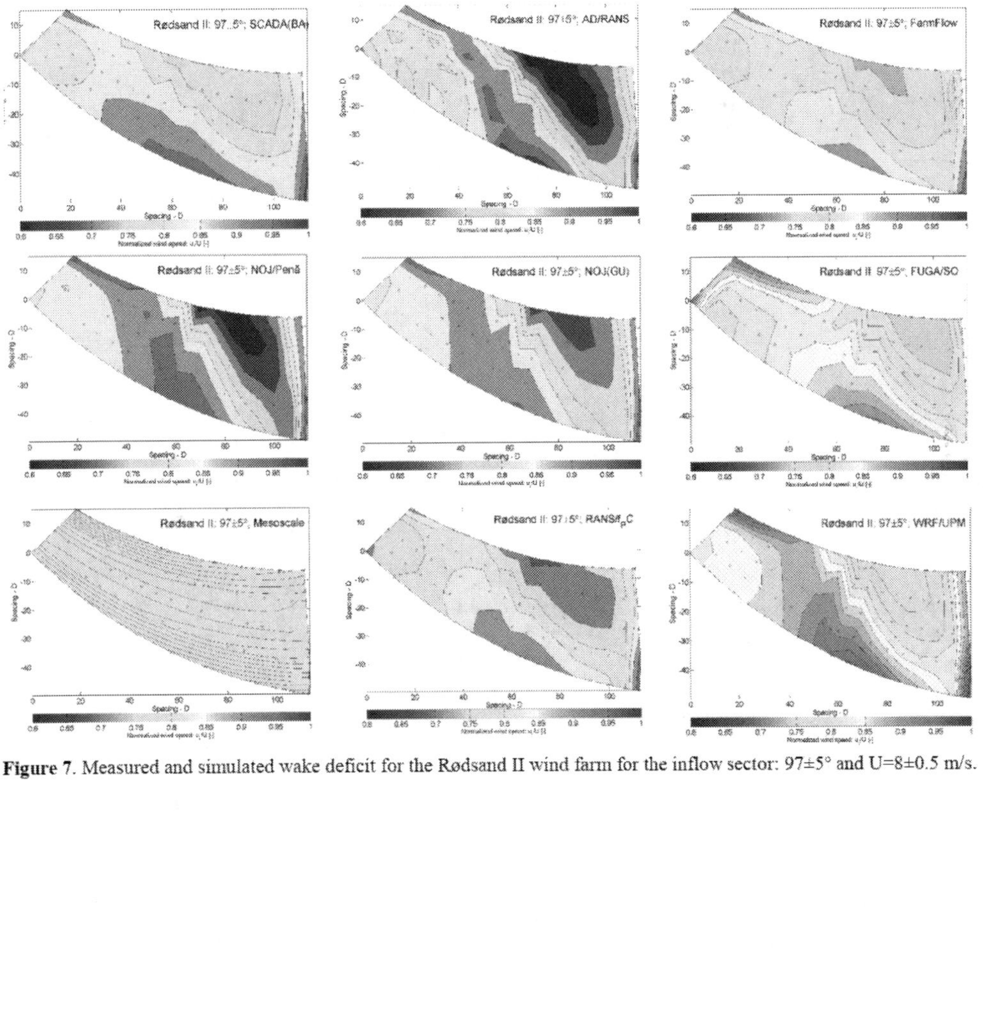

Figure 7. Measured and simulated wake deficit for the Rødsand II wind farm for the inflow sector: 97±5° and U=8±0.5 m/s.

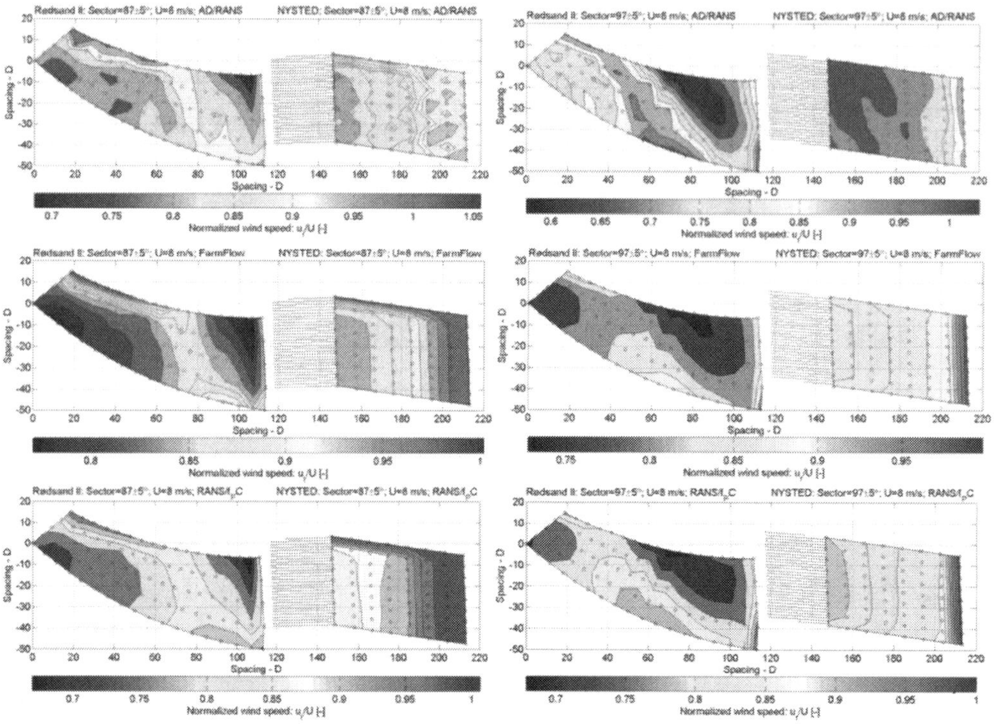

Figure 8. Cluster effect WD=87 & 97±5° inflow sectors.

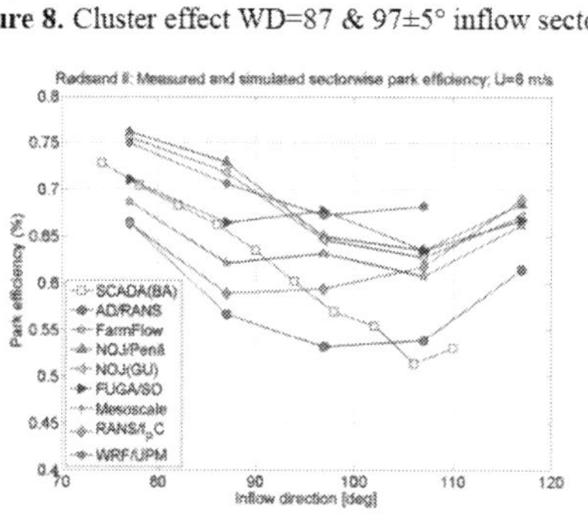

Figure 9. Measured and simulated park efficiency.

6. Conclusion

This is the first investigation that included two wind farms, and the main apparent conclusion is that the models give a significantly large spread, which indicates that there is a need for investigating further these types of flow cases to obtain more robust predictions. The current data we obtained from the wind farm owners are not of sufficient quality in order to draw more meaningful conclusion.

Another way of looking at these comparisons is that the spread of the models illustrate our current predictive capability as the wind energy community. It does encompass the results, which is a good thing, but we should strive at reducing this spread in order to reduce the risks in designing offshore wind farms close to each-other.

7. Acknowledgments

This work was supported by the EU EERA-DTOC project nr. FP7-ENERGY-2011/n 282797. We acknowledge E•ON for having access to the SCADA data from the Rødsand II offshore wind farm and we acknowledge DONG Energy A/S for having access to the SCADA data from Nysted offshore wind farm.

8. References

[1] http://EERA-DTOC.EU project homepage
[2] Schepers G, et.al. *EERA-DTOC: Design tools for off-shore wind farm clusters including new results on wake bench*; Presented at: European Wind Energy Conference & Exhibition 2013, 2013, Vienna
[3] Réthoré P-E, et.al. *Benchmarking of wind farm scale wake models in the EERA - DTOC project*, Presented at: International Conference on aerodynamics of Offshore Wind Energy Systems and wakes (ICOWES 2013), 2013, Lyngby
[4] Moriaty P, et.al. *IEA-Task 31 WAKEBENCH: Towards a protocol for wind farm flow model evaluation. Part 2: Wind farm wake models*, IOP online (DOI: http://dx.doi.org/10.1088/1742-6596/524/1/012185), vol: 524, issue: 1, 2014
[5] Hansen KS, et.al. *The impact of turbulence intensity and atmospheric stability on power deficits due to wind turbine wakes at Horns Rev wind farm*; Wind Energy (ISSN: 1095-4244) DOI: http://dx.doi.org/10.1002/we.512, vol: 15, issue: 1, pages: 183-196, 2012
[6] Hansen KS,et.al. *Benchmarking of Lillgrund offshore wind farm scale wake models*. Presented at: EERA DeepWind 2014 - 11th Deep Sea Offshore Wind R&D Conference, 2014, Trondheim
[7] Ott S, Berg J, Nielsen M *Linearised CFD Models for Wakes*. Technical Univ. of Denmark, Risoe National Lab. for Sustainable Energy. Wind Energy Div. Risoe-R-1772(EN) ISBN 978-87-550-3892-9. Available online.
[8] Jensen NO *A note on wind generator interaction*. Technical Report Risoe-M-2411(EN), Risø National Laboratory, Roskilde 1983.
[9] Peña A, Réthoré PE, Rathmann O *Modeling large offshore wind farms under different atmospheric stability regimes with the park wake model*. Renew. Energ. 2014; 70:164–171.
[10] Gaumond M et.al. *Evaluation of the wind direction uncertainty and its impact on wake modelling at the Horns Rev offshore wind farm*. Wind Energy 2014; 17:1169–1178.
[11] Skamarock WC et.al. *A description of the advanced research WRF version 3*. Technical Report TN-475+STR, NCAR, 2008.
[12] Volker PJ et.al. *The Explicit Wake Parameterization: Development, implementation in the WRF model and evaluation against mast measurements*. Submitted to Geoscientic Model Development.
[13] Ainsley JF *Calculating the flow field in the wake of wind turbines*. J Wind Eng Ind Aerodyn, 27:213-224, 1988.
[14] Crespo A, et.al. *UPMPARK: a parabolic 3D code to model wind farms*. Proc. EWEC 1994, European Wind Energy Conference, Thessaloniki, p 454-459.
[15] Castro F et.al. Simulation of the Askervein Flow. Part 1: Reynolds Averaged Navier–Stokes Equations (k−ε Turbulence Model), Boundary-Layer Meteorology, 2003, 107, 501-530.
[16] Palma J et.al. *Linear and nonlinear models in wind resource assessment and wind turbine micrositing in complex terrain*, Journal of Wind Engineering and Industrial Aerodynamics, 2008, 96, 2308 – 2326.
[17] Gomes VMMGC et.al. *Improving actuator disk wake model, Journal of Physics*, Conference Series, 2014, 524, 012170.

[18] Politis E et.al. *CFD modeling issues of wind turbine wakes under stable atmospheric conditions*, Proceedings of the European Wind Energy Conference, March 2009, Marseille (France).
[19] Cabezón D et.al. *CFD modelling of the interaction between the Surface Boundary Layer and rotor wake. Comparison of results obtained with different turbulence models and mesh strategies*, Proceedings of the European Wind Energy Conference, March 2009, Marseille (France)
[20] Cabezón D et.al. *Analysis and validation of CFD wind farm models in complex terrain. Effects induced by topography and wind turbines*, EWEC 2010 scientific proceedings, Warsaw, Poland, April 2010
[21] Chaviaropoulos PK, Douvikas, DI *Mean-flow-field Simulations over Complex Terrain using a 3D Reynolds-averaged Navier–Stokes Solver*, Proceedings of ECCOMAS '98, Vol. I, part II. Papailiou KD, Tsahalis D, Périaux J, Hirsch C, Pandolf M (eds.). Wiley: West Sussex, England, pp. 842-848.
[22] Wilcox DC *Turbulence Modelling for CFD*, DCW Industries Inc., La Canada, California, 1993, IBSN 0-9636051-0-0.
[23] Prospathopoulos JM, Chaviaropoulos PK *Numerical simulation of offshore wind farm clusters*, European Wind Energy Association, Conference proceedings 2013.
[24] Crespo A et. Al *Numerical Analysis of wind turbine wakes*, Proceedings of Delphi Workshop on "Wind turbine applications, 1985
[25] Rodi W *Turbulent buoyant jets and plumes* Pergamon Press, 1982
[26] Schepers JG *Engineering models in Wind Turbine Aerodynamics*, TUDelft, November 2012 ISBN 9789461915078"
[27] Sørensen NN *General purpose flow solver applied to flow over hills*, 1994 Ph.D. thesis Risø National Laboratory Roskilde, Denmark
[28] Michelsen JA *Basis3d - a platform for development of multiblock PDE solvers*. 1992 Tech. Rep. AFM 92-05 Technical University of Denmark, Lyngby, Denmark
[29] Van der Laan MP et.al. *An improved k- ϵ model applied to a wind turbine wake in atmospheric turbulence* 2014 Wind Energy Published online: DOI: 10.1002/we.1736
[30] Van der Laan MP et.al. *The k-ϵ-fP model applied to wind farms* 2014 Wind Energy Published online: DOI: 10.1002/we.1816
[31] Van der Laan MP et.al. *Predicting wind farm wake interaction with RANS: an investigation of the Coriolis force*, 2015 Submitted to Wake Conference 2015 in Visby, Sweden
[32] Barthelmie RJ, Jensen LE *Evaluation of wind farm efficiency and wind turbine wakes at the Nysted offshore wind farm 2010 Wind Energy Published online: DOI: 10.1002/we.408*
[33] Barthelmie et.al. *Flow and wakes in large wind farms: Final report for UpWind WP8*; Appendix 7: Guideline to wind farm wake analysis, Report Risø-R-1765, January 2011
[34] Réthoré et al. *Systematic wind farm measurement data reinforcement tool for wake model calibration*, EOW Conference 2009, 14 – 16 September 2009 Stockholm

Wake Conference 2015 IOP Publishing
Journal of Physics: Conference Series **625** (2015) 012009 doi:10.1088/1742-6596/625/1/012009

Data-driven Reduced Order Model for prediction of wind turbine wakes

G.V. Iungo[1], C. Santoni-Ortiz[1], M. Abkar[2], F. Porté-Agel[2], M.A. Rotea[1] and S. Leonardi[1]

[1] The University of Texas at Dallas, Mechanical Engineering Department, 75080 Richardson, TX

[2] Ecole Polytechnique Fédérale de Lausanne (EPFL), Wind Engineering and Renewable Energy (WIRE) Lab, Lausanne, Switzerland

E-mail: valerio.iungo@utdallas.edu

Abstract. In this paper a new paradigm for prediction of wind turbine wakes is proposed, which is based on a reduced order model (ROM) embedded in a Kalman filter. The ROM is evaluated by means of dynamic mode decomposition performed on high fidelity LES numerical simulations of wind turbines operating under different operational regimes. The ROM enables to capture the main physical processes underpinning the downstream evolution and dynamics of wind turbine wakes. The ROM is then embedded within a Kalman filter in order to produce a time-marching algorithm for prediction of wind turbine wake flows. This data-driven algorithm enables data assimilation of new measurements simultaneously to the wake prediction, which leads to an improved accuracy and a dynamic update of the ROM in presence of emerging coherent wake dynamics observed from new available data. Thanks to its low computational cost, this numerical tool is particularly suitable for real-time applications, control and optimization of large wind farms.

1. Introduction

Computational fluid dynamics (CFD) techniques with different level of complexity and accuracy have been developed over the last decade in order to simulate wind turbine wake flows [1, 2]. Although a large improvement has been achieved to predict wake fluid dynamic features and power production, the computational cost connected to high fidelity LES simulations of wind turbine arrays still remains far too high, which makes them inapplicable for operational purposes and real-time applications. For this reason, simple analytical wake models are still considered as a standard procedure for predictions of power harvesting from a wind power plant, and for siting of new wind farms. Recently improved wake models, such as the dynamic wake meandering model, have been proposed in order to capture the large-scale downstream motion of wind turbine wakes [3, 4].

An in-depth comprehension and characterization of the physical processes occurring in intra-wind farm wake flows is the substrate to develop new concepts for wind turbine wake prediction. The fluid motion is an infinite-dimensional dynamic system governed by the well known Navier-Stokes equations, which are nonlinear partial differential equations. Therefore, the analysis and prediction of fluid dynamics and instabilities result to be quite challenging tasks. In the domain of wind turbine wakes, in [5, 6] wake instabilities were predicted through linear stability analysis performed on the time-averaged velocity field obtained from wind tunnel experiments. For this technique, the Navier-Stokes equations are linearized on the base flow, and small perturbations are applied in order to analyze their spatial and temporal evolution.

The dynamic modes of a fluid motion can also be detected by analyzing time-resolved observations, obtained through experiments or numerical simulations [7, 8], and by applying different modal decomposition techniques, such as proper orthogonal decomposition (POD). The POD modes are characterized by spatial orthogonality, but they have multi-frequential temporal content. This means that different wake dynamics acting at completely separated time scales, thus connected to different phenomena, cannot be distinguished if they have comparable energy. In wind energy, POD has been used to investigate wind turbine wake dynamics in [9, 10].

In the present work, LES simulation data of wind turbine wakes are analyzed through the dynamic mode decomposition (DMD) [11]. For DMD time-resolved data are processed in order to evaluate the dominant physical processes, which are denoted as DMD modes. The latter may be non-orthogonal, but each DMD mode captures a single spectral contribution, thus an isolated physical phenomenon. DMD modes are characterized as for their spectral content (i.e. frequency and spatial wavelength), growth rate (i.e. a stable mode is going to be damped by proceeding in space and time, conversely an unstable mode will be amplified and add kinetic energy to the system), and energy. All those information are essential to highlight the main contributions to wake morphology and dynamics, which in turn affect power production and fatigue loads of wind turbines. DMD modes can be sorted accordingly to different objective functions, rather then their merely energy contribution to wake dynamics, as for POD. DMD modes can be sorted as a function of their spectral contribution and characteristic length scale. This feature can turn out to be very effective for investigations on wake interactions and added fatigue loads of waked wind turbines. Indeed, the overall life cycle of a wind turbine is mainly affected by dynamic loads acting within a well defined spectral range, which neglects very small and very large scales of the turbulent atmospheric wind with respect to the rotor diameter. In the same spirit, DMD modes can be sorted as a function of their growth rate. This criterion is useful for design of wind farm layout, for which only the DMD modes with a relevant energy at a certain downstream distance are considered. Therefore, accordingly to the specific criterion adopted, only a subset of DMD modes is selected in order to formulate a reduced order model (ROM).

The proposed tool for wind turbine wake prediction is meant to be suitable for real-time applications, thus it should be characterized by a relatively low computational cost, but at the same time by an adequate accuracy. For this reason a data-driven approach is considered, where new available data are injected into the simulation tool in order to provide a more accurate wake prediction. This operation is performed by embedding the ROM obtained via DMD in a Kalman filter [12]. The Kalman filer provides an efficient computational recursive algorithm to estimate the state of a process by minimizing the mean square error.

The paper is organized as follows. The database obtained from high fidelity LES simulations of wind turbines is described in section 2. The DMD formulation is recalled in section 3, together with a description of the Kalman filter. Two different applications are then presented in this paper: for section 4 the proposed numerical tool is applied to highly detailed LES data, which are obtained through the actuator line model and simulating also the presence of the wind turbine tower and nacelle. In section 5, wind turbine wake is simulated through the actuator disc model, but a more realistic atmospheric wind field under neutral thermal stability regime is generated through precursor simulations. Final discussion and conclusions are reported in section 6.

2. Description of the LES database

Two different databases have been used to developed the reduced order model, one produced at UT Dallas and another one at EPFL. For the former, the flow around the wind turbine blades are modeled using the actuator line model. The wind turbine tower and nacelle are simulated via an immersed boundary technique described in [14]. This method allows the flexibility of imposing complex geometries inside the domain without the need of body fitted grids, thus reducing the computational time. The computational box is $12.5d \times 2.1d \times 3d$ (d is the rotor diameter of 0.89 m), i.e. $1024 \times 512 \times 512$ grid points, in the streamwise, wall normal and spanwise direction, respectively. A uniform velocity profile is used as inflow condition. For these simulations, the rotational frequency of the rotor is equal to 10.7 Hz, while the time-averaged incoming wind at hub height, U_∞, is 10 m/s, which produces a tip speed ratio equal to 3. The axial velocity field is evaluated over the vertical symmetry plane of the wake for 214 snapshots acquired with a sampling frequency of 1124 Hz, which represent about 2 full rotations of the rotor disc.

A second set of LES simulations performed by the WIRE Lab at EPFL, which will be analyzed in section 5, employed the scale-dependent Lagrangian dynamic model and to parameterize the turbine-induced forces, the actuator-disk model with rotation is used. It is worth mentioning that the grid resolution is chosen such that 10 points in the spanwise direction and 16 points in the vertical direction cover the turbine rotor diameter. In these simulations, the Vestas V80-2MW wind turbine, with a rotor diameter of 80 m and a hub-height of 70 m, is immersed in a neutral boundary layer flow. The wind turbine operates with a mean angular velocity of about 0.27 Hz. The code is run for a long-enough time to guarantee quasi-steady conditions [15].

3. Dynamic mode decomposition and reduced order model

The DMD algorithm is briefly outlined in the following. At each timestamp, the measurements ,ψ denoted as snapshot, collected with a fixed sampling frequency, F_{samp}, are arranged as columns of a data matrix. The number of measurement points for each snapshot is equal to M, and the total number of snapshots is equal to N. From the snapshot sequence, two data matrices are generated:

$$\Psi_0 := [\psi_0, \psi_1, ..., \psi_{N-1}]\,; \qquad \Psi_1 := [\psi_1, \psi_2, ..., \psi_N]\,. \tag{1}$$

The DMD is based on the assumption that temporal consecutive snapshots are generated through a linear, time-invariant operator applied to the the previous snapshot. In other therms:

$$\psi_{t+1} = A\ \psi_t; \quad t = [0, ..., N-1]. \tag{2}$$

It is worth to point out that the assumption of a linear system for the DMD time-marching algorithm is not equivalent to perform a linearization of the dynamic system, as performed for instance within linear stability analysis. For DMD all the non-linearities in the evolution of the system dynamics are still kept, only the advancement in time is linearized. By combining Eq. 1 with Eq. 2, it is obtained:

$$\Psi_1 = [\psi_1, \psi_2, ..., \psi_N] = [A\psi_0, A\psi_1, ..., A\psi_{N-1}] = A\Psi_0. \tag{3}$$

If r is the dimension of the aimed reduced order model (ROM), with $r < N$, a matrix F can be defined through DMD in order to be the optimal representation of the matrix A in the basis spanned by the POD modes of Ψ_0,

$$A \approx U\ F\ U^*. \tag{4}$$

U^* denotes the complex-conjugate-transpose of the POD modes of the snapshot matrix Ψ_0, obtained with an economic-size singular value decomposition:

$$\Psi_0 = U\ \Sigma\ V^*, \tag{5}$$

where Σ is the $r \times r$ diagonal matrix of the singular values. The matrix F can be determined by minimizing the Frobenius norm of the difference between Ψ_1 and $A\ \Psi_0$, thus by injecting Eqs. 4 and 5 in Eq. 3,

$$\underset{F}{\text{minimize}} \|\Psi_1 - U\ F\ \Sigma\ V^*\|_F^2. \tag{6}$$

The matrix F can be expressed in a diagonal form, by means of its eigenvalues, μ_i, and eigenvectors, y_i,

$$F = \begin{bmatrix} y_1 & \cdots & y_r \end{bmatrix} \begin{bmatrix} \mu_1 & & \\ & \ddots & \\ & & \mu_r \end{bmatrix} \begin{bmatrix} z_1^* \\ \vdots \\ z_r^* \end{bmatrix} = Y D_\mu Z^*, \tag{7}$$

where z_i are the eigenvectors of F^*, corresponding to the eigenvalues $\overline{\mu_i}$ (the eigenvectors are bi-orthogonal $z_i^* y_j = \delta_{ij}$). It is possible to evaluate each snapshot in the form:

$$\psi_t = A\ \psi_{t-1} = U\ F\ U^*\ \psi_{t-1} = U\ Y\ D_\mu\ Z^*\ U^*\ \psi_{t-1} = U\ Y\ D_\mu^t\ Z^*\ U^*\ \psi_0. \tag{8}$$

The DMD modes, ϕ_i, are then obtained:

$$\Phi = UY. \tag{9}$$

Thus, Eq. 8, can be reformulated as a linear combination of the DMD modes:

$$\psi_t = \sum_{i=1}^{r} \phi_i\ \mu_i^t\ \alpha_i, \tag{10}$$

and α_i are the amplitude of each DMD mode. In matrix form:

$$\begin{bmatrix} \psi_0 & \psi_1 & \cdots & \psi_{N-1} \end{bmatrix} \approx \begin{bmatrix} \phi_1 & \phi_2 & \cdots & \phi_r \end{bmatrix} \begin{bmatrix} \alpha_1 & & & \\ & \alpha_2 & & \\ & & \ddots & \\ & & & \alpha_r \end{bmatrix} \begin{bmatrix} 1 & \mu_1 & \cdots & \mu_1^{N-1} \\ 1 & \mu_2 & \cdots & \mu_2^{N-1} \\ \vdots & \vdots & \ddots & \vdots \\ 1 & \mu_r & \cdots & \mu_r^{N-1} \end{bmatrix} = \tag{11}$$
$$= \Phi\ D_\alpha\ V_{and},$$

where V_{and} is the Vandermonde matrix of the eigenvalues of the matrix F. The determination of the DMD amplitudes, α_i, is obtained through the following optimization problem:

$$\underset{\alpha}{\text{minimize}} \|\Psi_0 - \Phi D_\alpha V_{and}\|_F^2. \tag{12}$$

The solution of this problem is detailed in [13]. The different dynamics of a wind turbine wake flow are then properly characterized through the DMD modes, ϕ_i, defined in Eq. 9, whose amplitudes are α_i, frequency equal to $F_{samp}/(2\pi)\Im[log(\mu_i)]$, and growth rate equal to $F_{samp} \times \Re[log(\mu_i)]$ (μ_i are the eigenvalues of the matrix F). A reduce order model (ROM) for prediction of the wind turbine wake flow can be formulated by using the matrix F as linear operator of a time-marching algorithm. However, a ROM can be formulated by only selecting a subset of DMD modes that fulfill a particular criterion, such as for their energy content, frequency, growth rate, or a combination of multiple criteria. Thus, the columns of the matrix Y corresponding to the selected DMD modes produce the matrix Y^{ROM}. The same procedure is applied to the matrix Z of the eigenvectors of F^*, producing the matrix Z^{ROM}. Finally, the respective eigenvalues constitute the diagonal matrix D_μ^{ROM}. Therefore, the linear operator used for the ROM is:

$$F^{ROM} = Y^{ROM}\ D_\mu^{ROM}\ Z^{*\ ROM}. \tag{13}$$

The main feature of the ROM is to enable operating on a lower dimensional space R^r, rather than for the DMD operating with the matrix A, which has size $M \times M$, and most probably impossible to calculate on standard computers because of memory requirements. Therefore, the ROM allows achieving very fast computational algorithms, and the DMD is now applied as

$$\widetilde{\psi}_{t+1} = F^{ROM}\widetilde{\psi}_t \quad (14)$$

with $\widetilde{\psi}_i \in R^r$, which is obtained

$$\widetilde{\psi}_i = U^*\psi_i. \quad (15)$$

The ROM is then embedded in a Kalman filter in order to produce a time-marching algorithm for wake prediction, and it also enables data-assimilation of new available observations. The Kalman filter addresses the general problem to estimate the state $\widetilde{\psi} \in R^r$ of a process governed by the linear stochastic difference equation

$$\widetilde{\psi}_{t+1} = F^{ROM}\widetilde{\psi}_t + B\widetilde{w}_t. \quad (16)$$

The variables $\widetilde{w}$ represents the process noise, and the respective linear operator B is evaluated by applying the DMD to the residual of the analyzed data not extracted through the ROM. With a Kalman filter, a measurement of the system $\widetilde{z} \in R^r$ can be considered

$$\widetilde{z}_t = H\ \widetilde{x}_t + \widetilde{v}_t; \quad (17)$$

thus, when a new observation of the system is available, $\widetilde{z}_t$, it is embedded in the system in order to improve the accuracy of the ROM. The variable $\widetilde{v}$ is the measurement noise, which is neglected by assuming an ideal measuring system. The observation matrix, H, is the $r \times r$ identity matrix, because all the LES data obtained from each snapshot are used for data-assimilation. The Kalman filter allows to evaluate the so-called Kalman gain, M, which is an $r \times r$ matrix. Each time a new observation, $\widetilde{z}_t$ is available, the forecast of the process, ψ_{t+1} is obtained as follows:

$$\widetilde{\psi}_{t+1} = F^{ROM}(\widetilde{\psi}_t + M(\widetilde{z}_t - \widetilde{\psi}_t)). \quad (18)$$

4. ROM for wind turbine wakes simulated with the actuator line model

In this section the ROM is evaluated for the wind turbine wake obtained from LES simulations carried out by the CFD Lab at UT Dallas by means of the actuator line model. The time-averaged axial velocity field evaluated over the wake vertical symmetry plane is shown in Fig. 1. The velocity deficit produced by the presence of the turbine tower and nacelle is evident in the very near-wake. Then, the whole wake, which is generated by the blade rotation, the turbine tower and nacelle, gradually recovers by proceeding downstream.

DMD has been applied to the full data set, and 210 DMD modes were extracted in order to achieve a detailed analysis of the wake dynamics. Their respective frequency, growth rate and energy are summarized in Fig. 2. Specifically, frequency is reported on the horizontal axis, the opposite value of the growth rate on the vertical axis, and energy is represented by both size of the dots and the colorbar reported therein. All the DMD modes have a negative growth rate, which characterizes stable modes with a reducing energy by proceeding downstream. The

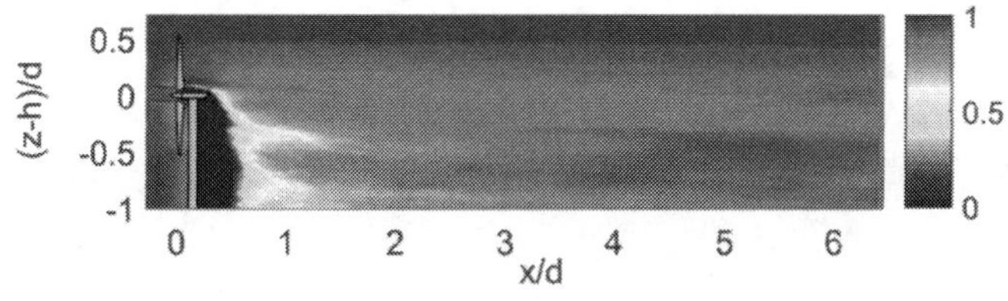

Figure 1. Time-averaged axial velocity field, U/U_∞.

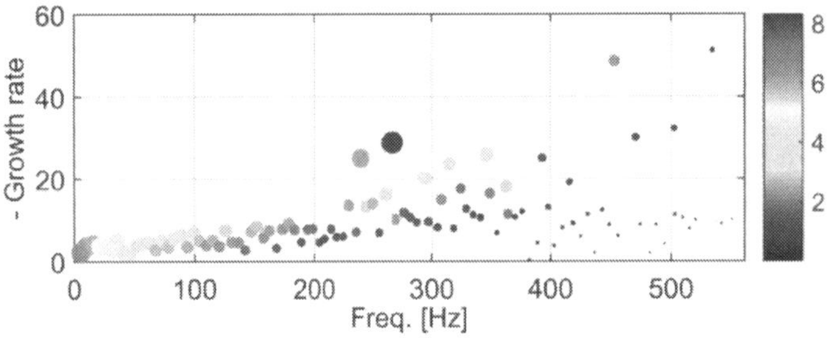

Figure 2. DMD spectrum obtained from the axial velocity field.

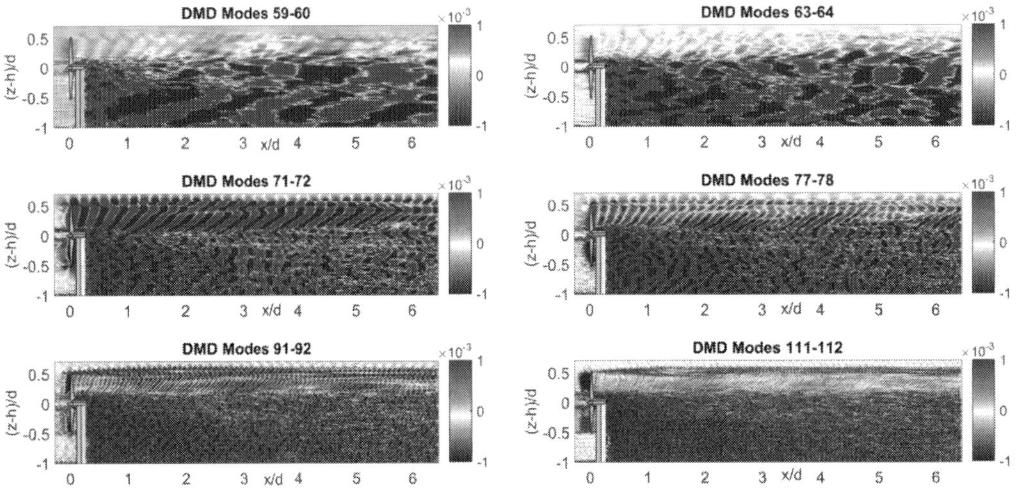

Figure 3. DMD modes selected for the reduced order model.

general damping of the DMD modes is a consequence of the uniform incoming wind, and thus wake dynamics are dissipated as they evolve downstream. The scenario is different for the case of a realistic atmospheric boundary layer flow presented in the next section, for which unstable modes are triggered by the incoming wind.

DMD highlights the presence of dominant modes, which are sorted as function of their physical origin, i.e. connected to the instability of the helicoidal tip vortices, the shedding of the trailing vortices, or to the vortex shedding from the turbine tower. In Fig. 3, the four most energetic DMD modes for each of those clusters are shown, and their characteristics are reported in Table 1.

The DMD modes 59 and 60 are connected to the vortex shedding from the wind turbine tower, and they are characterized by a frequency of 5.2 Hz. The DMD modes 63 and 64 are also connected to the same phenomenon, but they are characterized by a frequency of 10.4 Hz, which is double than the previous one. This harmonic contribution can be connected to the alternate vortex shedding, which produces a double frequency at the wake vertical symmetry plan. It is interesting to observe that this harmonic frequency is roughly equal to the rotational frequency of the rotor, f_{hub}, which leads to suppose that the blade rotation triggers, or at least, affects the vortex shedding from the turbine tower. This remark is also endorsed by the slanted structure of the vortices, which indicates a perturbation produced by the rotor.

Table 1. Characteristics of the DMD modes selected for the reduced order model.

DMD modes	Energy	Freq. [Hz]	Growth rate
59-60	6.6	5.2	-2.2
63-64	6.4	10.4	-3.8
71-72	4.6	31.5	-2.8
77-78	4.6	36.1	-5.1
91-92	3.3	62.6	-4.6
111-112	3.3	99	-6.8

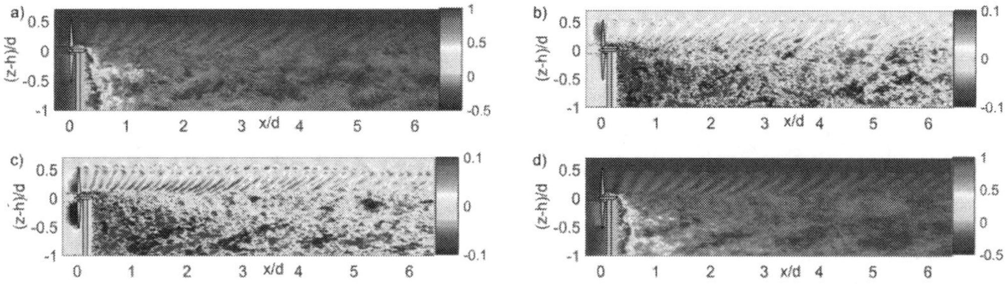

Figure 4. Flow reconstruction through the selected DMD modes: a) instantaneous axial velocity field; b) instantaneous velocity minus the time-averaged axial velocity; c) extracted velocity field; d) extracted velocity plus the time-averaged velocity field.

The DMD modes 71, 72, 77, and 78, are characterized by frequencies roughly equal to 3 f_{hub}, which are connected with the blade rotation, thus to the shedding of trailing and tip vortices. Finally, the DMD modes 111 and 112, are characterized by a frequency about 9 f_{hub}, and the DMD modes 91 and 92 by a frequency of about 6 f_{hub}. Their main energy contribution is localized at the rotor top tip height, which corresponds to the area where the helicoidal tip vortices are shed. All those features suggest that these DMD modes are associated with instabilities of the helicoidal tip vortices.

In order to evaluate the effectiveness of the selected DMD modes to constitute a ROM, the projection of the velocity field on those DMD modes is considered. Indeed, in Fig. 4(a) a snapshot of the axial velocity at a specific instant is reported, while in Fig. 4(b) the time-averaged velocity field is subtracted from the instantaneous one in order to highlight the wake dynamics. The instantaneous velocity field is projected on each of the selected DMD modes, then all the projections are added in order to produce the reconstructed velocity field. The latter is shown in Fig. 4(c), which clearly enables to capture the main flow dynamics shown in Fig. 4(b). Moreover, the accuracy in the flow reconstruction is even more evident when the time-averaged velocity field is added, as for Fig. 4(d). Therefore, it has been shown that the selected DMD modes are suitable to produce a ROM in order to represent the main dynamics present within the turbine wake.

The ROM is then embedded in a Kalman filter, as described above in section 3. The performance of the ROM is evaluated by calculating the global percentage error between the predicted velocity field, $\hat{\psi}$, and the original data, ψ:

$$E(t) = \frac{1}{M} \sum_i \sum_j \frac{\left|\psi(x_i, z_j, t) - \hat{\psi}(x_i, z_j, t)\right|}{< \psi(x_i, z_j) >}, \tag{19}$$

where $< \psi(x_i, z_j) >$ is the time-averaged velocity at a specific location. The global error is first calculated for the projection of the original data on the selected DMD modes, which is reported in Fig. 5 with black squares. This parameter is a reference for the approximation performed with the ROM. Then, the ROM embedded in the Kalman filter is used to predict the wind turbine wake flow by providing as only input the projection of the first snapshot on the selected DMD modes. Therefore, the velocity field is predicted for all the timestamps from 2 till N=214. The percentage global error for the wake prediction is reported in Fig. 5 with red dots, and it overlaps to close approximation with the error performed with the projection of the original data on the selected DMD modes. This result highlights the high accuracy obtained with the time-marching algorithm obtained by embedding the ROM in the Kalman filter.

In case the original velocity field is provided as initial condition of the ROM, then the ROM performance are further enhanced. Indeed, the global error, reported in Fig. 5 with a blue circles, shows that an even better prediction of the wake velocity field is obtained by using as initial condition the original snapshot of the axial velocity field.

5. ROM for wind turbine wake simulated with the actuator disc model and a neutral atmospheric boundary layer

The analysis of the LES data provided by the WIRE Lab at EPFL has been performed for the axial velocity field evaluated over the horizontal plane at hub height. The time-averaged velocity field is shown in Fig. 6. A total number of 2000 snapshots acquired with a sampling frequency of 9.7 Hz has been processed through DMD, producing the DMD spectrum presented in Fig. 7. For this investigation only unstable mode with a positive growth rate have been considered for the ROM in order to investigate the dominant wake dynamics that might affect fatigue loads on possible downstream wind turbines. The selected unstable DMD modes are shown in Fig. 8, and their characteristics are listed in Table 2. It interesting to observe that all the unstable DMD modes have a wavelength equal or larger than the rotor diameter. This result suggests that the large turbulent scales present in the incoming atmospheric wind excite the wind turbine wake flow producing far-wake dynamics, such as wake meandering [16].

The unstable DMD modes are now used to evaluate the ROM as presented in the previous sections. The first snapshot of the LES data is used as initial condition of the time-marching algorithm obtained through the Kalman filter, and a closer-look to a specific timestamp is presented in the following. The instantaneous axial velocity field obtained from the LES data

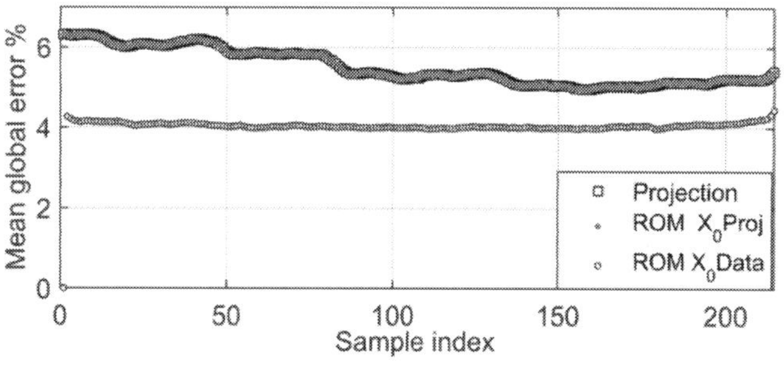

Figure 5. Global error in the wake flow prediction: *black square*) projection on the selected DMD modes; *red dot*) prediction by using as initial condition the projection of the first snapshot; *blue circle*) prediction by using as initial condition the first snapshot.

Journal of Physics: Conference Series **625** (2015) 012009 doi:10.1088/1742-6596/625/1/012009

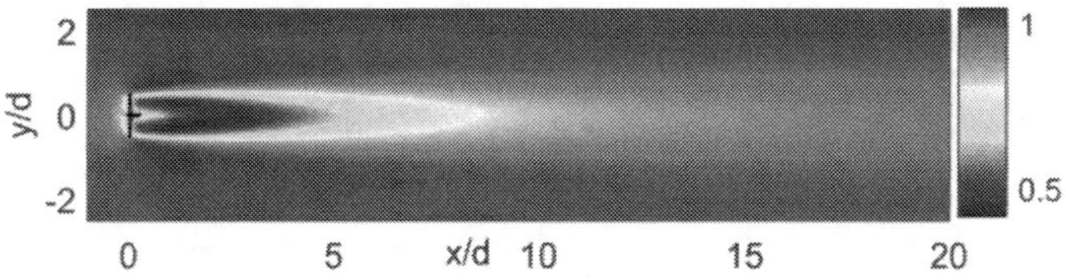

Figure 6. Time-averaged axial velocity field, U/U_∞, over the horizontal plane at hub height.

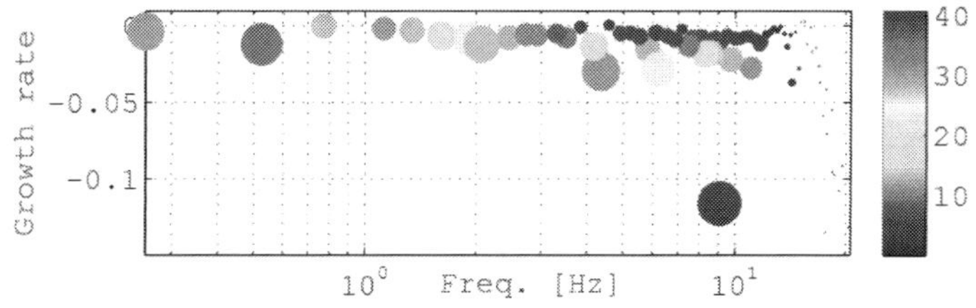

Figure 7. DMD spectrum obtained from the axial velocity field.

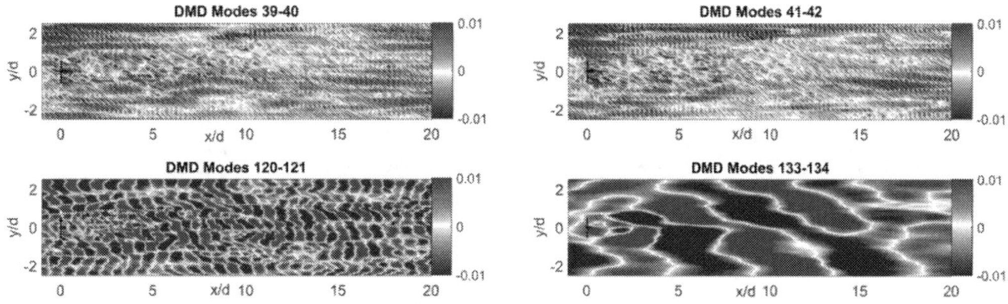

Figure 8. Unstable DMD modes selected for the reduced order model.

Table 2. Characteristics of the DMD modes selected for the reduced order model.

DMD modes	Energy	Freq. [Hz]	Growth rate
39-40	0.02	0.27	0.003
41-42	0.03	0.25	0.003
120-121	2.81	0.08	0.001
133-134	13.73	0.013	0.001

is reported in Fig. 9(a), while the same one minus the time-averaged velocity field is in Fig. 9(b). For the same timestamp, the velocity field predicted through the ROM shows in Fig. 9(c) the presence of coherent structures that evolve moving downstream with an increasing cross-width. Finally, in Fig. 9(d) the predicted velocity field clearly shows that the ROM formulated through the unstable DMD modes enables to predict the far-wake meandering excited by the large coherent structures present in the incoming atmospheric boundary layer.

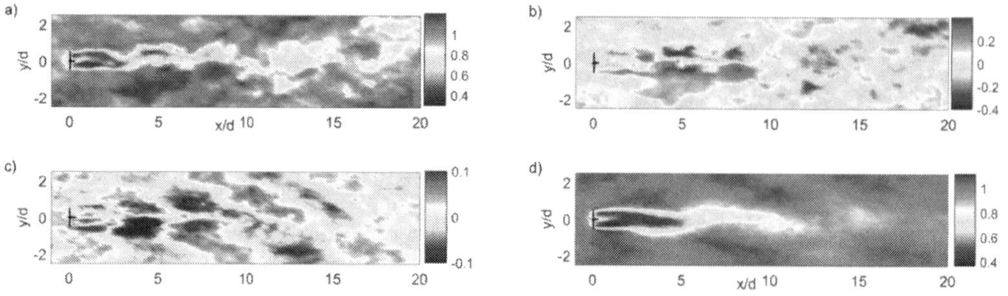

Figure 9. Flow prediction with a ROM formulated with only unstable DMD modes: a) instantaneous axial velocity field; b) instantaneous velocity minus the time-averaged axial velocity; c) predicted velocity minus the time-averaged velocity; d) predicted velocity.

6. Conclusions

A novel approach for fast and accurate prediction of wind turbine wakes has been presented, which consists in a reduced order model (ROM) embedded in a Kalman filter. The ROM is evaluated by means of dynamic mode decomposition performed on high fidelity LES experiments of wind turbines. This data-driven algorithm enables data assimilation of new measurements simultaneously to the wake prediction, which leads to an improved accuracy and a dynamic update of the ROM in presence of emerging coherent wake dynamics observed from new available data. It has been shown that the entire wake flow can be predicted with a global accuracy of about 4% by only providing time-resolved data of the wake flow, and then using as initial condition only for the prediction only one snapshot. It has been shown that only specific phenomena can be predicted by formulating opportunely the ROM, as for instance the prediction of far-wake meandering. The low computational cost of the proposed algorithm makes it suitable for real-time applications, control and optimization of large wind farms.

Acknowledgments

C. Santoni-Ortiz and S. Leonardi were supported by NSF PIRE Award IIA 1243482. TACC is acknowledged for providing computational time.

References

[1] Martinez-Tossas L A, Churchfield M J and Leonardi S 2014 *Wind Energy*, published online
[2] Porté-Agel F, Wu Y-T and Chen C-H 2013 *Energies*, **6** (10) 5297-5313
[3] Larsen G C, Madsen H A, Thomsen K and Larsen T J 2008 *Wind Energy*, **11** 377-395
[4] Madsen H A, Larsen G C, Larsen T J, Troldborg N and Mikkelsen R 2010 *J. Solar energy Eng.*, **132** 041014
[5] Iungo G V, Viola F, Camarri S, Porté-Agel F and Gallaire F 2013 *J. Fluid Mech.* **737** 499-526
[6] Viola F, Iungo G V, Camarri S, Porté-Agel F and Gallaire F 2014 *J. Fluid Mech.* **750** R1
[7] Iungo G V, Wu Y-T and Porté-Agel F 2013 *J. Atmos. Ocean. Technol.* **30** 274-287
[8] Iungo G V and Porté-Agel F 2014 *J. Atmos. Ocean. Technol.* **31** (10) 2035-2048
[9] Hamilton N, Tutkun M and Cal R B 2015 *Wind Energy.* **18** 297-315
[10] Bastine D, Witha B, Wachter M and Peinke J 2015 *Energies.* **8** 895-920
[11] Schmid P 2010 *J. Fluid Mech.* **656** 5-28
[12] Kalman R E 1960 *Trans. ASME J. Basic Eng.* **82** Series D 35-45
[13] Jovanovic M R, Schimd P J and Nichols J W 2014 *Physics of Fluids* **26** 024103
[14] Orlandi P and Leonardi S 2006 *J. of Turbulence* **7** 53
[15] Abkar M and Porté-Agel F 2015 *Physics of Fluids* in press
[16] Espana G, Aubrun S, Loyer S and Devinant P 2011 *Wind Energy* **14** 923-937

Local blockage effect for wind turbines

Takafumi Nishino[1] **and Scott Draper**[2]

[1] Centre for Offshore Renewable Energy Engineering, Cranfield University, Cranfield, UK
[2] Centre for Offshore Foundation Systems, University of Western Australia, Crawley, Australia

E-mail: t.nishino@cranfield.ac.uk

Abstract. This paper presents a combined theoretical and CFD study on the fluid-mechanical limit of power extraction by a closely-spaced lateral array of wind turbines. The idea of this study originates in recent studies on the array optimisation of tidal/marine turbines, for which the power coefficient of each turbine is known to increase significantly if the lateral spacing between turbines, or the local blockage, is optimised. The present study, using 3D Reynolds-averaged Navier-Stokes (RANS) simulations of a boundary-layer flow over a closely-spaced lateral array of up to 9 actuator discs, suggests that a similar—albeit less significant—power increase due to the effect of local blockage can be achieved even for wind turbines. A possible theoretical approach to estimating this power increase is also discussed.

1. Introduction

"How to maximise the total power of a large number of turbines?" is a common question asked by both wind and tidal farm designers. From the viewpoint of a fluid dynamicist, these two problems are not identical but closely related to each other. Recent studies on the array optimisation of tidal/marine turbines [1, 2] have suggested that the best strategy to maximise the total power of a given number of turbines is to arrange them in a closely-spaced lateral array and thereby utilise the blocking effect of neighbouring turbines. For example, it has been shown theoretically by Nishino and Willden [3] that the power coefficient of a large number of turbines arrayed in the middle of an infinitely wide water channel may increase significantly from the Lanchester-Betz-Joukowsky limit of 0.593 up to another limit of 0.798 if the spacing between turbines is optimised.

A similar—albeit less significant—power increase due to such "local" or "in-field" blockage has been reported for wind turbines as well [4]; however neither the exact mechanism nor the upper limit of this power increase has been clarified yet. Hence the present study aims to explore the mechanism of such a power increase and thus clarify how significant (or not so significant) this power increase can possibly be for wind turbines. To study this systematically, a number of 3D Reynolds-averaged Navier-Stokes (RANS) simulations of an array of up to 9 turbines (represented as actuator discs) have been performed first, followed by some theoretical considerations. Although we consider a lateral array of horizontal-axis turbines for the sake of convenience, the majority of our discussion and conclusions given in this paper are equally applicable to vertical-axis turbines.

2. Blockage effect and "local" blockage effect

In this section we briefly review some of the recent theoretical models developed in the tidal energy community to understand the effect of blockage on the efficiency of ideal turbines. These models are essentially extended versions of the classical (Lanchester-Betz type) actuator disc momentum theory. We review the basic "single-scale" blockage effect model of Garrett and Cummins [5] first and then the "two-scale" (i.e. combined local and global) blockage effect model of Nishino and Willden [3, 6]; the latter of which is closely related to the case of wind turbines discussed later in this paper.

2.1. *Blockage effect*

The blockage effect model of Garrett and Cummins [5] (see also Houlsby et al. [7] and Whelan et al. [8]) considers an actuator disc placed in the middle of a uniformly confined flow passage, e.g. duct and tube. Similarly to the classical actuator disc theory, this is essentially a one-dimensional model and hence the effect of flow confinement is described solely by the blockage ratio B (the ratio of the disc area to the cross-sectional area of the flow passage); i.e. neither the cross-sectional shape of the passage nor the exact position of the turbine within the passage can be taken into account. The key difference from the classical actuator disc theory is that this model considers the conservation of energy not only for the flow going through the disc (core flow) but also for the flow "bypassing" the disc (bypass flow). The model also considers the mixing of core and bypass flows but only in the far wake region, i.e. downstream of the location where pressure equalises between the core and bypass flows; hence the entire flow is considered to become uniform again far downstream. By using this model, it is analytically shown that the upper limit of power extraction increases exponentially with the blockage ratio B. If we define the power coefficient C_P in the traditional way, i.e. based on the naturally available kinetic energy:

$$C_P = \frac{\text{Power}}{\frac{1}{2}\rho U_0^3 A} \tag{1}$$

where ρ is the fluid density, U_0 is the upstream (or undisturbed) flow speed and A is the disc area, the maximum power coefficient $C_{P\,\max}$ is shown to be $16/27 \times (1-B)^{-2}$. Note that when $B = 0$ this $C_{P\,\max}$ goes back to the Betz limit.

This theoretical model is useful, for example, when estimating the effect of wind-tunnel blockage on the output power of a turbine. However, it should be noted that this additional power due to the blockage effect essentially comes from the pressure difference induced between far upstream and far downstream of the turbine. When $B = 0$, pressure far upstream and far downstream should be the same (as assumed in the classical actuator disc theory), which means that the turbine extracts energy only from the naturally available kinetic energy and hence C_P defined in (1) is an appropriate measure of the efficiency of power extraction in this case. When $B > 0$, however, the turbine extracts energy not only from the naturally available kinetic energy but also from the pressure head induced between far upstream and far downstream of the turbine, explaining why C_P can surpass the Betz limit and even become higher than unity. In fact, this C_P can increase to infinity as B approaches to unity, in which case the power depends solely on the pressure head (as the entire flow must go through the turbine) and hence C_P defined in (1) is no longer meaningful. Clearly, this blockage effect studied by Garrett and Cummins [5] is relevant only when the entire flow passage is confined (by wind tunnel walls, for example) and cannot explain the power increase for wind turbine arrays operating outside, where there is no confinement above the turbines (and hence the overall/global blockage is zero).

2.2. *Local blockage effect*

The "two-scale" blockage effect model recently proposed by Nishino and Willden [3] is an extended version of the blockage effect model described above. This new model considers a large number of actuator discs arrayed laterally like a fence in the middle of a wide flow passage, thereby addressing the combined effect of "local" and "global" blockage. The original model proposed by Nishino and Willden in 2012 (hereafter referred to as NW12) assumes a complete "scale separation" between the local (device-scale) and global (array-scale) flow events including the "far-wake" mixing for each scale [3]; this assumption holds strictly and thus the model does not require any empirical or *ad-hoc* inputs when the number of discs arrayed is sufficiently large. When the number of discs is not large

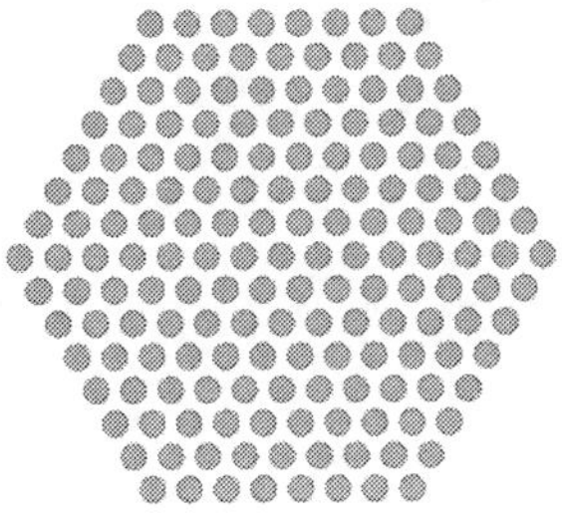

Figure 1. Schematic of a hypothetical array of a number of wind turbines (arrayed in both lateral and vertical directions) with the local blockage ratio $B_L \approx 0.4$. The NW12 model [3] suggests that this B_L is optimal and yields the theoretical upper limit, $C_{P\,\max} = 0.798$.

Figure 2. Cross-section of the computational domain with a lateral array of 9 actuator discs. The height and width of the domain are 25*d* and 50*d*, respectively, the lateral gap between each disc is 0.5*d*, and the vertical gap between the discs and the ground is also 0.5*d*.

enough, however, this scale separation will not be complete and the model will need to incorporate, for example, the rate of mixing downstream of each disc empirically [6]. For the case of tidal/marine applications, i.e. discs in the middle of a wide (but shallow) water channel, recent CFD studies have suggested that the scale separation assumption employed in the NW12 model is approximately valid when the number of discs is more than 10 or so [6].

Of particular interest here, for wind energy applications, is that this two-scale model is valid even when the entire flow passage is not confined, i.e. the global blockage ratio B_G (the ratio of the total area of discs to the cross-sectional area of the entire flow passage) is zero. In this particular case, the model describes only the local blockage effect. Importantly, in this case the array of discs extracts energy only from the naturally available kinetic energy (and hence C_P defined in (1) is still a proper measure of the efficiency of power extraction) because the pressure far upstream and far downstream of the entire disc array must be the same when $B_G = 0$. It should be noted that, although the original paper by Nishino and Willden [3] considered an infinitely wide water channel as an example of the zero-global-blockage situation for tidal applications, the model itself is essentially one-dimensional and may therefore be applied to wind applications as well.[3] If we consider a hypothetical situation where a number of wind turbines are arrayed in both lateral and vertical directions, as illustrated in figure 1, these turbines (except for those near the edges of the array) would also experience the local blockage effect due to neighbouring turbines and hence, according to the NW12 model, would have the same theoretical limit $C_{P\,\max} = 0.798$ (i.e. the Nishino-Willden limit) when the local blockage B_L (the ratio of the single disc area to the cross-sectional area of each local flow passage) is about 0.4. In practice, however, wind turbines cannot usually be arrayed in the vertical direction and therefore the power increase due to the local blockage effect must be less than this hypothetical case (also, B_L can no longer be defined unambiguously). This will be discussed further in Section 4.

[3] Nishino and Willden [3] introduced the height and width of a wide water channel as well as the lateral spacing between turbines for the sake of convenience, but then showed that this two-scale blockage effect could be described solely by two independent blockage ratios; hence the model is still essentially one-dimensional [6]. Theoretically, the only key difference between the tidal application discussed in [3] and the wind application illustrated in figure 1 would be the number of turbines required to satisfy the "scale separation" assumption, since the rate of array-scale flow expansion would be different between the tidal/marine case (two-dimensional expansion) and the wind case (three-dimensional expansion). If we consider a large enough number of turbines to satisfy the scale separation assumption, these two cases are theoretically identical.

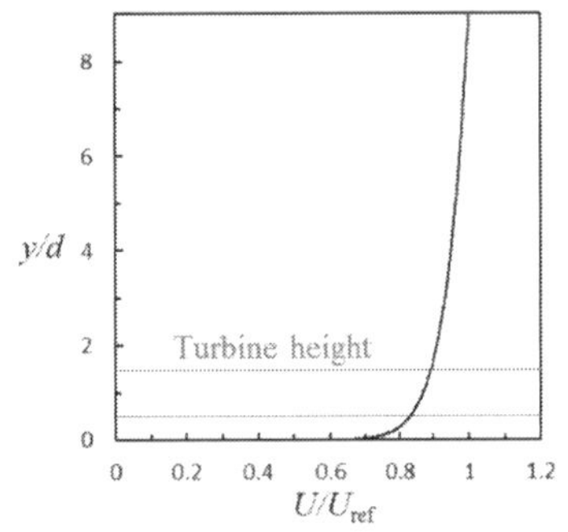

Figure 3. Streamwise velocity profile for the vertically sheared inflow case.

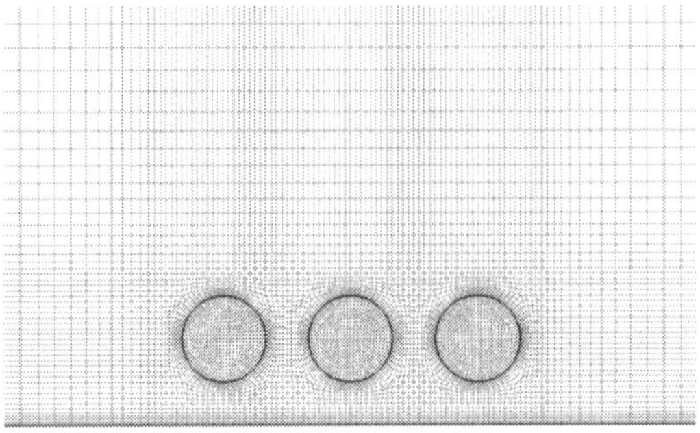

Figure 4. Cross-sectional view of the computational mesh around three actuator discs (normal resolution).

3. RANS simulations of a lateral array of wind turbines

In this section we present a series of 3D RANS simulations of a lateral array of up to 9 wind turbines (modelled as simple actuator discs without any rotations) operating near the ground, as illustrated in figure 2. Also studied is a lateral array of an infinite number of turbines, simulated using a narrower computational domain with periodic boundary conditions. Further details of the array configuration and flow conditions are described first, followed by computational methods and results.

3.1. Computational domain and array configuration

Figure 2 depicts the cross-section of our computational domain with 9 actuator discs. Except for the "infinite array" case, we employ this large computational domain of 25*d* height, 50*d* width and 100*d* length (where *d* is the disc diameter) throughout the study to ensure that the effect of global blockage is insignificant. The disc array consisting of either 1, 3, 5, 7 or 9 discs is located at the centre of the domain near the ground. The lateral gap between each disc and the vertical gap between the ground and the discs are both fixed at 0.5*d*. Note that the global blockage ratio B_G (which is the ratio of the total cross-sectional area of all discs to the cross-sectional area of the computational domain) slightly increases from 0.0006 to 0.006 as the number of discs increases from 1 to 9.

For the "infinite array" case, only one disc is simulated in a much narrower domain of 1.5*d* width (with periodic boundary conditions applied to the lateral direction). The height of the domain is the same as the other cases (25*d*) and therefore the global blockage ratio is still sufficiently small (B_G = 0.02) in this case. The vertical gap between the ground and the disc is again fixed at 0.5*d*.

3.2. Flow conditions

We consider three different inflow conditions: (i) uniform inflow with a very low level of ambient turbulence; (ii) vertically sheared inflow, i.e. boundary-layer flow; and (iii) uniform inflow with the same level of ambient turbulence as the vertically sheared inflow case. For the sake of convenience, we consider a constant air density and viscosity of 1.2kg/m^3 and 1.8×10^{-5}kg/m-s, respectively, and a disc diameter d = 100m. These, together with a reference velocity U_{ref} = 10m/s, result in the Reynolds number (based on the disc diameter) of about 67 million.

For the uniform inflow cases, uniform streamwise velocity $U = U_{ref}$ (10m/s) is applied to the inlet boundary of the domain. Inlet conditions for turbulence quantities (k and ε) will be described later in Section 3.3. Symmetry conditions are applied not only to the top and side boundaries but also to the bottom (ground) boundary of the domain; thus the uniform velocity profile is maintained throughout the domain unless the actuator discs disturb the flow.

For the vertically sheared inflow case, the mass flow through the entire computational domain is fixed such that the vertically or cross-sectionally averaged streamwise velocity $U_{avg} = U_{ref}$ (10m/s).

Symmetry conditions are again applied to the top and side boundaries, but no-slip (i.e. smooth wall) conditions are applied to the bottom (ground) boundary in this case. The streamwise velocity profile given at the inlet is shown in figure 3. This velocity profile and also the inlet profiles of turbulence quantities (not shown here) are "fully developed" profiles for the present computational domain and conditions (i.e. obtained from a preliminary simulation with periodic boundary conditions applied to the streamwise direction); hence the inlet profiles are maintained throughout the domain unless the actuator discs disturb the flow. Here the "undisturbed" streamwise velocity averaged across the disc area is U_{d0} = 8.69m/s. A summary of the flow conditions will be shown later in table 1.

3.3. Computational methods

The RANS actuator disc method used here is similar to that previously used by Nishino and Willden [6, 9] for tidal/marine applications. We solve numerically the 3D incompressible RANS equations, where the Reynolds stress terms are modelled using the standard k-ε model of Launder and Spalding [10] (with the standard wall functions applied to the bottom boundary for the sheared inflow case). Computations are performed using a commercial CFD solver ANSYS FLUENT 14 together with its user-defined function module for modifications. The solver is based on a finite volume method and is nominally second-order accurate in space. All computations are performed as steady state.

Similarly to Nishino and Willden [6, 9], each actuator disc is modelled as a stationary permeable (or porous) disc. The effect of each disc on the (Reynolds-averaged or mean) flow is considered as a loss of momentum in the streamwise (x) direction at the disc plane. The change in momentum flux (per unit area) is locally calculated as

$$M_x = K\left(\frac{1}{2}\rho U_d^2\right) \qquad (2)$$

where U_d is the local (rather than disc-averaged) streamwise velocity at the disc plane and K is the momentum loss factor, which is assumed to be uniform across the entire surface of all discs for each simulation. It should be noted that $K = 2$ is a theoretically optimal value (corresponding to the Betz limit) for an isolated disc, but for closely-arrayed discs the optimal value can be larger than 2 due to the local blockage effect; hence in this study we change the value of K between 1 and 5.

It has been shown in [9] that this RANS actuator disc method, if applied to a single disc placed in the middle of a rectangular channel with a very low level of ambient turbulence, agrees well with the analytical blockage effect model of Garrett and Cummins [5] (in terms of the power predicted for an ideal turbine). To predict the performance of real turbines, we would need a different method taking into account the aerodynamic characteristics of turbine blades (which could be modelled based on the blade element theory) and also the effect of blade-induced turbulence [11]. In this study we do not consider these issues for simplicity; however we still need to consider carefully the effect of ambient turbulence. When used with a high level of ambient turbulence, this RANS actuator disc method tends to predict strong mixing immediately downstream of the disc edges (where the shear induced by the disc is strongest) and importantly, such strong mixing in the near wake is known to increase the limit of power extraction by the disc [9, 12]. The reason for testing two different "uniform inflow" cases (with low and high levels of inlet turbulence) in addition to the "sheared inflow" case is basically to differentiate this effect (namely the near-wake-mixing effect) from the blockage effect to be studied. See table 1 for a summary of the flow conditions including the inlet conditions for k and ε.

3.4. Computational grids

Figure 4 shows a cross-sectional view of the computational mesh around three actuator discs (as an example). In this study we prepared seven different grids in total (for the cases with 1, 3, 5, 7 and 9 discs, the "infinite array" case, and the "infinite array" case with a finer resolution; see table 2 for a summary). These grids are used for both uniform inflow cases and sheared inflow cases, i.e. we use the same set of grids for different inflow conditions. As can be seen from figure 4, these grids are multi-block structured grids; we created a 2D grid first for one cross-section of the domain and then

Table 1. Summary of flow conditions.

Case	U_{avg} [m/s]	U_{d0} [m/s]	k_{in} [m^2/s^2]	ε_{in} [m^2/s^3]	$(\mu_T/\mu)_{d0}$
Uniform (low FST)	10	10	0.00015	3.02×10^{-7}	420
Vertically sheared	10	8.69	n/a	n/a	5.8×10^5
Uniform (high FST)	10	10	0.67335	4.18×10^{-3}	5.8×10^5

(FST: freestream turbulence. k_{in}: inlet value of k. ε_{in}: inlet value of ε. $(\mu_T/\mu)_{d0}$: turbulent viscosity ratio averaged across the disc area for the case without discs.)

Table 2. Summary of computational grids.

Grid	Domain width	Total number of cells
1 disc	50d	1,749,760
3 discs	50d	2,625,280
5 discs	50d	3,500,800
7 discs	50d	4,376,320
9 discs	50d	5,251,840
"infinite"	1.5d	437,760
"infinite" (F)	1.5d	3,502,080

(F: finer grid, for which the resolution was doubled in all directions.)

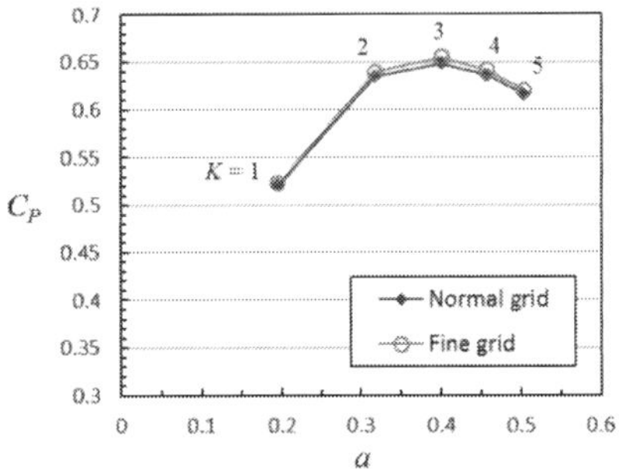

Figure 5. Effect of grid resolution on C_P (for the infinite array with low FST uniform inflow).

stacked the same 2D grid in the streamwise direction (non-equidistantly) to create a 3D grid, which consists of only hexahedral cells.

For the normal resolution grids, the minimum cell size is about 0.02d near the disc edges (in both disc-radial and disc-normal directions) and also near the bottom wall (in the wall-normal direction). For the finer resolution grid, the resolution was doubled in all directions; hence the total number of cells is 8 times larger than the normal resolution grid. As will be described in Section 3.5 the effect of grid resolution seems insignificant and therefore, unless specified, all results presented in this paper are those computed on the normal resolution grids.

3.5. *Results*

3.5.1. Grid sensitivity. A grid sensitivity study was performed first for the "infinite array" case with low FST uniform inflow. The results are summarised in figure 5, showing the (disc-averaged) power coefficient C_P plotted against the (disc-averaged) axial induction factor a. Note that in this study C_P and a are defined based on U_{d0} (in order to make a fair comparison between the uniform inflow and sheared inflow cases later):

$$C_P = \frac{\int M_x U_d \, dA}{\frac{1}{2}\rho U_{d0}^3 A} = K\frac{\langle U_d^3 \rangle}{U_{d0}^3} \tag{3}$$

$$a = 1 - \frac{\langle U_d \rangle}{U_{d0}} \tag{4}$$

where $\langle \phi \rangle$ denotes the average of a variable ϕ over the disc area A (which can be the area of one disc or of all discs, depending on whether C_P and a are for one particular disc or for the average of all discs). It can be seen that the differences between the normal and fine resolution grids are very small for all K values tested (between 1 and 5). The difference in the maximum power coefficient $C_{P\,max}$ (obtained at K = 3 for this particular case) is ~1% between the two grids, suggesting that the normal resolution is sufficient.

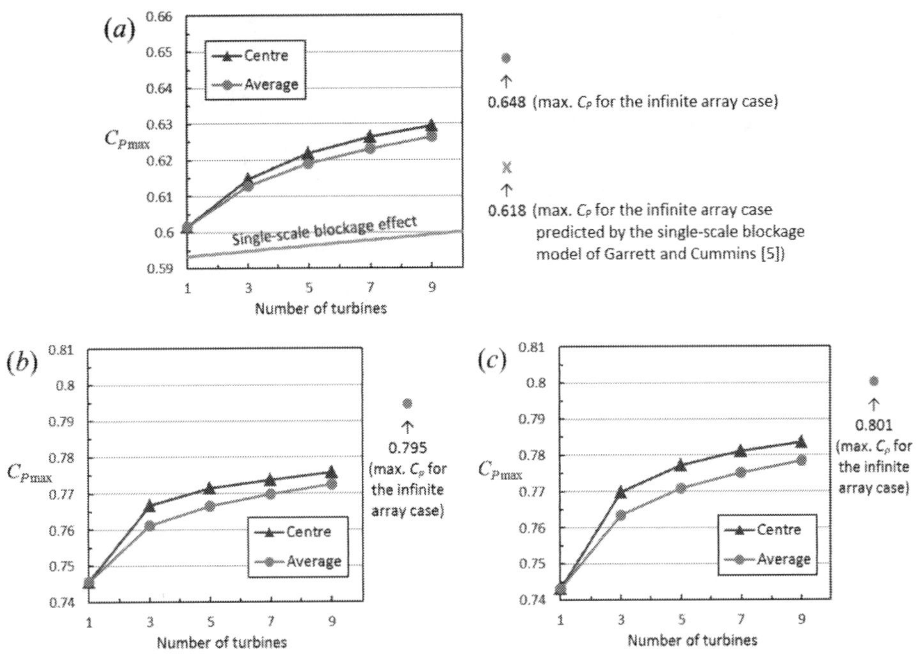

Figure 6. Effect of the number of turbines on $C_{P\,\max}$ for three different inflow conditions: (*a*) low FST uniform inflow; (*b*) vertically sheared inflow; and (*c*) high FST uniform inflow.

3.5.2. Effect of the number of turbines on $C_{P\,\max}$. Figure 6 shows the effect of the number of discs on the maximum power coefficient $C_{P\,\max}$ for the three different inflow conditions. The values of $C_{P\,\max}$ plotted here are those for the disc at the centre of the array (blue symbols) and those averaged for all discs in the array (red symbols). Also shown in figure 6(*a*) for comparison is the effect of "single-scale" blockage predicted by the theoretical model of Garrett and Cummins [5] (green line). Here the global blockage ratio B_G (varying from 0.0006 up to 0.02 as the number of discs increases from 1 to infinity) has been taken as the single-scale blockage ratio for the theoretical model.

For the case with low FST uniform inflow, the value of $C_{P\,\max}$ obtained for a single disc is 0.602, which is close to the Betz limit of 0.593 for a completely isolated disc. However, the value of $C_{P\,\max}$ increases as the number of discs increases, and importantly, the rate of increase (especially when the number of discs increases from 1 to 5 or so) is clearly higher than that predicted by the single-scale blockage effect model, indicating that this increase in $C_{P\,\max}$ is primarily due to the local blockage effect. As the number of discs increases further, the rate of increase reduces and becomes similar to that predicted by the single-scale blockage model, suggesting that about 7 to 9 discs are sufficient to take full advantage of the local blockage effect (which is about 5% increase in $C_{P\,\max}$). Note that the value of $C_{P\,\max}$ increases up to 0.648 for the "infinite array" case (corresponding to about 33 discs arrayed in the original computational domain of 50*d* width) but this additional increase in $C_{P\,\max}$ is most likely due to the slight increase in the global blockage ratio.

For the vertically sheared inflow case, the effect of the number of discs on $C_{P\,\max}$ is similar to that for the low FST uniform inflow case described above, although the values of $C_{P\,\max}$ are much higher than the previous case. The main reason of these high C_P values can be explained by comparing the results with the high FST uniform inflow case, for which the inlet values of turbulence quantities (k and ε) have been adjusted such that their values at the disc location (50*d* downstream of the inlet) are the same as those at the disc centre in the sheared inflow case. Comparing figures 6(*b*) and (*c*), it can

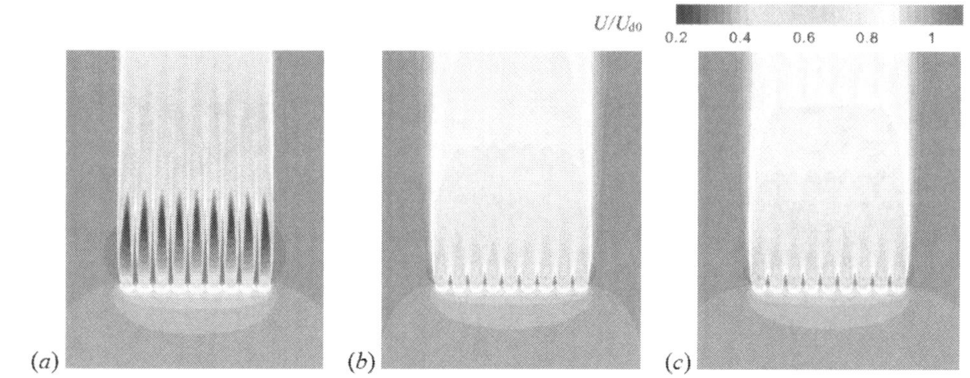

Figure 7. Streamwise velocity contours around 9 actuator discs, plotted on the horizontal plane through the disc centres: (*a*) low FST uniform inflow; (*b*) vertically sheared inflow; and (*c*) high FST uniform inflow. The axial induction factor $a = 0.37$ for all three cases.

Figure 8. Streamwise velocity contours plotted on the longitudinal plane through the centre of 9 actuator discs: (*a*) low FST uniform inflow; (*b*) vertically sheared inflow; and (*c*) high FST uniform inflow. The axial induction factor $a = 0.37$ for all three cases.

be seen that the results of these two cases are very similar to each other. This suggests that the high values of $C_{P\,\max}$ for the sheared inflow case are primarily due to the high level of ambient turbulence around the discs, causing the near wake mixing effect [9] mentioned earlier in Section 3.3. It should also be noted that the increase in $C_{P\,\max}$ due to the local blockage effect in these two cases is 4~5%, which is similar to the low FST uniform inflow case. This increase is not as significant as that for the hypothetical array shown earlier in figure 1 (more than 30% increase) but is still not insignificant.

3.5.3. Two-scale flow expansion and mixing. Figures 7 and 8 show the streamwise velocity contours plotted on the horizontal and longitudinal planes, respectively, through the centre of the array of 9 actuator discs. Note that the velocity on the horizontal plane (figure 7) has been normalised by U_{d0}, whereas that on the longitudinal plane (figure 8) has been normalised by U_{avg}. Comparing the three different inflow cases, it can be seen that the level of ambient turbulence affects the prediction of the rate of mixing behind the discs significantly.

Of particular interest here is that, on the horizontal plane, we can see a clear separation of scales between the flow around each disc and the flow around the entire array (similarly to what has been discussed in [6] for tidal applications). The flow starts decelerating upstream of the entire array first due to the "array-scale" flow expansion, and then decelerates further just in front of each disc due to the "device-scale" flow expansion. Also, the wake mixing behind each disc takes place much faster than the mixing behind the entire array. This scale separation of the flow patterns (visualised on the horizontal plane) can be seen in all three different inflow cases, suggesting that the basic mechanism of the local blockage effect for wind turbines is indeed similar to that for tidal turbines.

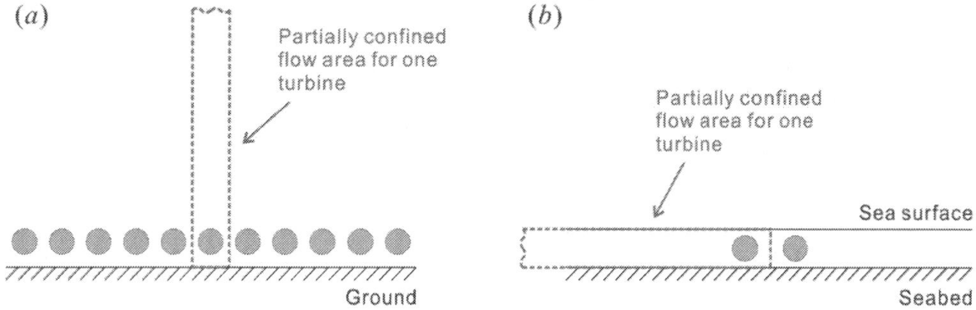

Figure 9. Configurational similarity between (*a*) an infinitely long lateral row of wind turbines and (*b*) two side-by-side tidal/marine turbines in an infinitely wide water channel.

On the longitudinal plane, it is difficult to recognise such a scale separation of the flow patterns. However, we can still see that the flow expansion and mixing takes place faster in the vicinity of the disc (where the flow is expanded/mixed three-dimensionally at the scale of the disc) compared to the regions far upstream/downstream of the disc (where the flow is expanded/mixed only vertically at the scale of the disc and only horizontally at the scale of the array).

4. Discussion and conclusions

The results of RANS actuator disc simulations have indicated that the limit of power extraction by a closely-spaced lateral array of wind turbines can be up to about 5% higher than that by the turbines placed far apart, due to the local blockage effect. This increase in power is less significant than that theoretically predicted for the hypothetical (lateral *and* vertical) array of turbines depicted earlier in figure 1, but is still potentially beneficial to wind farm developers.

It should be noted that the RANS simulations presented in this paper are only for one particular array configuration (i.e. $0.5d$ spacing laterally between each turbine and also vertically between the turbines and the ground), although we have tested three different inflow conditions to examine their influence on the local blockage effect. Since these computer simulations require a decent amount of computational resources, it would be desirable if this type of local blockage effect (for a lateral, but not vertical, array of turbines) can also be predicted theoretically for different configurations. If we assume that the vertical shear of the inflow is not directly related to the basic mechanism of the local blockage effect (which is a fair assumption based on the RANS results presented earlier in figure 6) and consider the "infinite array" of wind turbines with uniform inflow, the local blockage effect on these turbines should be essentially identical to that on two side-by-side tidal/marine turbines in the middle of a wide but shallow water channel with uniform inflow (note that in both cases each turbine is, effectively, locally blocked by three nearby boundaries with one open end; see figure 9). This suggests that the local blockage effect for a large number of wind turbines could be modelled in the same way as that for a (very) short array of tidal/marine turbines discussed in [6].

The difficulty here, however, is that the theoretical model of a short array of tidal/marine turbines tends to over-predict the limit of power extraction especially when the number of turbines arrayed is very small; this is essentially because the scale separation assumption does not hold for such a short array [6]. For example, if we apply this short tidal array model (with $n = 2$, where n is the number of turbines) to the low FST uniform inflow case of "infinite array" of wind turbines, the model predicts $C_{P\,\mathrm{max}} = 0.71$, which is substantially higher than the corresponding RANS actuator disc prediction of $C_{P\,\mathrm{max}} = 0.65$ shown earlier in figure 6(*a*). As discussed in [6], what we need to know here (in order to correct the theoretical model for such a short array of tidal/marine turbines) are essentially the rates of "device-scale" flow expansion and mixing relative to the rates of "array-scale" flow expansion and mixing. From the viewpoint of an infinitely long lateral row of wind turbines, this may be rephrased as follows: we need to know the rates of three-dimensional flow expansion and mixing (taking place in the vicinity of each turbine) relative to the rates of vertical flow expansion and mixing (taking place more upstream and downstream of the row of turbines).

In conclusion, the basic mechanism of the local blockage effect for wind turbines seems closely related to that for tidal/marine turbines. The concept of two-scale flow modelling recently proposed for tidal/marine turbines seems useful to understand the local blockage effect for wind turbines as well. Further communications between wind energy researchers and tidal/marine energy researchers would be beneficial to both research areas.

References

[1] Draper S and Nishino T 2014 Centred and staggered arrangements of tidal turbines *J. Fluid Mech.* **739** 72-93

[2] Hunter W, Nishino T and Willden R H J 2015 Investigation of tidal turbine array tuning using 3D Reynolds-averaged Navier-Stokes simulations *Int. J. Marine Energy* **10** 39-51

[3] Nishino T and Willden R H J 2012 The efficiency of an array of tidal turbines partially blocking a wide channel *J. Fluid Mech.* **708** 596-606

[4] McTavish S, Rodrigue S, Feszty D and Nitzsche F 2014 An investigation of in-field blockage effects in closely spaced lateral wind farm configurations *Wind Energy* (published online)

[5] Garrett C and Cummins P 2007 The efficiency of a turbine in a tidal channel *J. Fluid Mech.* **588** 243-51

[6] Nishino T and Willden R H J 2013 Two-scale dynamics of flow past a partial cross-stream array of tidal turbines *J. Fluid Mech.* **730** 220-44

[7] Houlsby G T, Draper S and Oldfield M L G 2008 Application of linear momentum actuator disc theory to open channel flow *Tech. Rep.* OUEL 2296/08 University of Oxford

[8] Whelan J I, Graham J M R and Peiró J 2009 A free-surface and blockage correction for tidal turbines *J. Fluid Mech.* **624** 281-91

[9] Nishino T and Willden R H J 2012 Effects of 3-D channel blockage and turbulent wake mixing on the limit of power extraction by tidal turbines *Int J. Heat Fluid Flow* **37** 123-35

[10] Launder B E and Spalding D B 1974 The numerical computation of turbulent flows *Comput. Methods Appl. Mech. Eng.* **3** 269-89

[11] Nishino T and Willden R H J 2014 Low-order modelling of blade-induced turbulence for RANS actuator disk computations of wind and tidal turbines *Wind Energy - Impact of Turbulence (Research Topics in Wind Energy* vol 2*)* eds M Hölling, J Peinke and S Ivanell (Berlin: Springer) pp 153-58

[12] Nishino T and Willden R H J 2013 The efficiency of tidal fences: a brief review and further discussion on the effect of wake mixing *Proc. ASME 32nd Int. Conf. on Ocean, Offshore and Arctic Engineering (Nantes)* OMAE2013-10207

Wake effect on a uniform flow behind wind-turbine model

V L Okulov [1,2] I V Naumov[2], R F Mikkelsen[1], and J N Sørensen[1]

[1]Department of Wind Energy, Technical University of Denmark, 2800 Lyngby, Denmark
[2] Institute of Thermophysics, SB RAS, Novosibirsk 630090, Russia

E-mail: okulov@dtu.dk

Abstract. LDA experiments were carried out to study the development of mean velocity profiles of the very far wake behind a wind turbine model in a water flume. The model of the rotor is placed in a middle of the flume. The initial flume flow is subjected to a very low turbulence level, limiting the influence of external disturbances on the development of the inherent wake instability. The rotor is three-bladed and designed using Glauert's optimum theory at a tip speed ratio λ =5 with a constant of the lift coefficient along the span, C_L= 0.8. The wake development has been studied in the range of tip speed ratios from 3 to 9, and at different cross-sections from 10 to 100 rotor radii downstream from the rotor.

By using regression techniques to fit the velocity profiles it was possible to obtain accurate velocity deficits and estimate length scales of the wake attenuation. The data are compared with different analytical models for wind turbine wakes.

1. Introduction

Due to the fast growth in the number of wind farms around the world extensive investigations have been carried out during the past two decades to better understand and predict wake flows behind wind turbines. The velocity deficit and turbulence levels in the wakes are parameters that define the available power and structural stability of wind turbines located downwind of other turbines in the farms. In spite of the extensive investigations, the prediction of wake evolution and decay are still open questions [1].

Various analytical approaches have been applied to describe the wind-turbine wakes, but a general model for the wake based on a common solution to predict the wake behavior under different wind conditions has still not been found. Among them, so-called kinematic models were obtained from experimental and theoretical works [2]. The kinematic models are based on self-similar velocity deficit profiles obtained from global momentum conservation, using as input the thrust coefficient of the wind turbine. One of the pioneering kinematic models of the far wake behind wind turbines assumed a top-hat shape for the velocity deficit [3]. Hereinafter we will refer all models to the first letter of the author's name, as such the J-model for the model of Jensen [3]. Lately, yet an analytical wake model was proposed and validated to predict the wind velocity distribution in the wake [4]. This BP-model is derived by applying conservation of mass and momentum and assuming a Gaussian distribution for the velocity deficit in the wake.

Out of these considerations of the wind-turbine wakes, George [5] proposed a new methodology for the far wake behind axisymmetric bluff bodies. This G-model of wake was called "equilibrium similarity" analysis when the axisymmetric profiles at a self-similarity state wake asymptotically attenuates at the same rate independent of the way it is generated. He also argued that the Reynolds number is the main factor in determining the rate of the axisymmetric wake. This theory was successfully confirmed by comparison to experimental data for the axisymmetric wake behind a streamline disk [6-8] in a wind tunnel. Based on these studies, the two different rates for the G-model were confirmed for very high and low local Reynolds numbers in the disk wake.

The current paper investigates the far wake of a three-bladed HAWT rotor, with a blade design based on the Glauert optimum rotor [9]. The purpose of the present investigation is to determine the wake development and decay properties observed in the far wake behind the rotor to compare it with the wake behind disk to estimate a possibility of the G-model for the wind-turbine wake. Compared to the flow around a bluff body, it is, however, rather difficult to discriminate between the various decay properties of the far field behind a rotor. A difference in the strength of the wake deficit behind a wind turbine and a disk of the same radius is not fully understood. Indeed the various types of the decay properties for both disk and rotor wakes may be of quite different nature. Moreover some difference may also exist in the wake development for different operating regimes of the wind turbine (with different thrust coefficient or tip speed ratio). The last question did not else clear even for the basic J- and BP models mentioned above [4].

Therefore, in the present study, a special attention has been paid to detect and categorize the various types of wake development behind the wind turbine under the different operation in a range of the tip speed ratio (TSR) from 2 to 10. The LDA-measurements at different cross-sections of the far wake up to 50 rotor diameters downstream from the rotor are accomplished by a comparison with the three basic analytical models mentioned above.

2. Experimental Method

The experiment is carried out in a water flume of length 35m, 3m width and an operative height of 0.9m. The 3m wide test section is fitted with transparent walls at a distance of 20 m from the channel inlet. The free flow speed in the flume was $U = 0.64$m/s. In order to filter out disturbances, the water was led through an inlet equipped with honeycomb. Prior to the experiment, the velocity profiles were measured at various positions in the flume in order to determine a cross section with as constant velocity as possible (figure 1). The velocity profile variation of the incoming water flow is less than 1% during all experiments, measured with LDA and an independent OTT Z400 velocimeter. This diagram indicates that the velocity profile is constant 0.25m from the edges and across the flume in the region 0.25-0.65m above the bottom with a minimal oscillation of the free flow (figure 2). Based on the measured velocity the rotor axis was positioned at a height of 0.5m from the channel bottom and 0.7m away from the flume walls [9, 11].

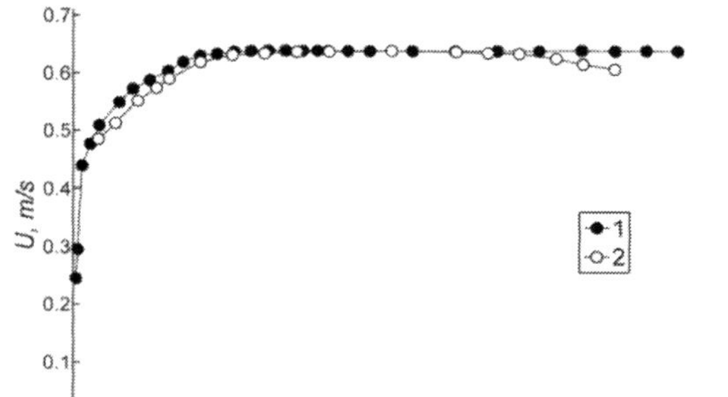

Figure 1. Free stream conditions (mean velocity) in the rotor cross section: average velocity profiles: 1 – from wall to rotor, 2 – from flume bottom to rotor.

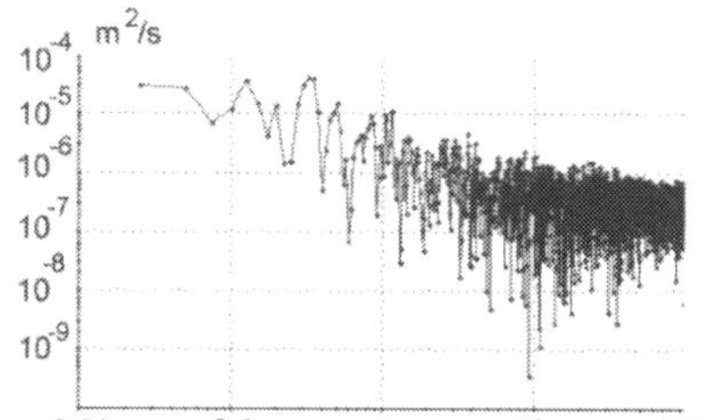

Figure 2. Spectrum of the free flow oscillations in the rotor area.

Figure 3. Photo of the model rotor and the LDA measurements.

The three-bladed model rotor has a diameter $D = 0.376$m with blades of length 0.159m, consisting of SD7003 airfoil sections (figure 3). The shape and pitch setting of the blade were calculated using the theory of Glauert [10] for optimum operating conditions at a tip speed ratio $\lambda = 5$, where $\lambda = \Omega R/U$, and Ω is angular speed of the rotor. The Reynolds numbers based on rotor diameter and free stream was varied in the range $140.000 < Re < 240.000$. As working fluid, tap water at a temperature of 20°C was used.

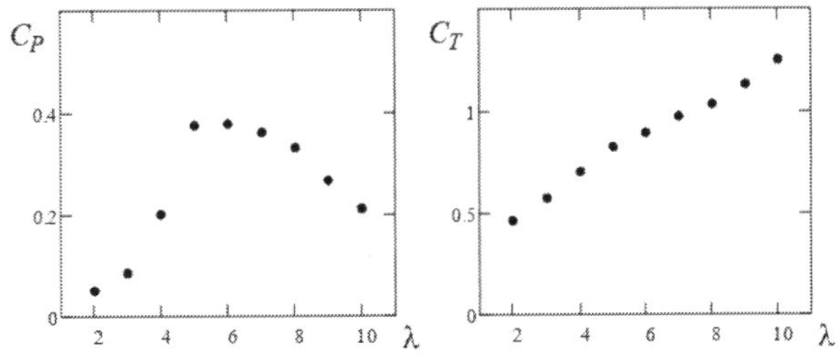

Figure 4. Measured power and trust coefficients as functions of tip speed ratio [9].

The rotor was driven by a JVL Industri Elektronik MAC400 servo motor which was operated at a constant rotational speed within 2% accuracy. The angle of attack of the blade was fixed for each test case. The torque of the motor was transferred to the rotor axis via a rigid gear transmission. The power (C_P) and thrust (C_T) coefficients were measured for different tip speed ratios (figure 4 and table 1).

Table 1. Power and thrust coefficients as function of tip speed ratio

λ	C_P	C_T
2	0.05	0.46
5	0.37	0.83
8	0.33	1.01
10	0.21	1.25

The flow downstream of the rotor develops into two different structures of the near and far wakes. The near wake defines as well-defined 3D structure consisting of slender vortices in which the helical tip vortices are undoubtedly visualized by fluorescence dye (figure 5) [12-13].

Figure 5. Dye visualization of the 3-D vortex structure in the near wake.

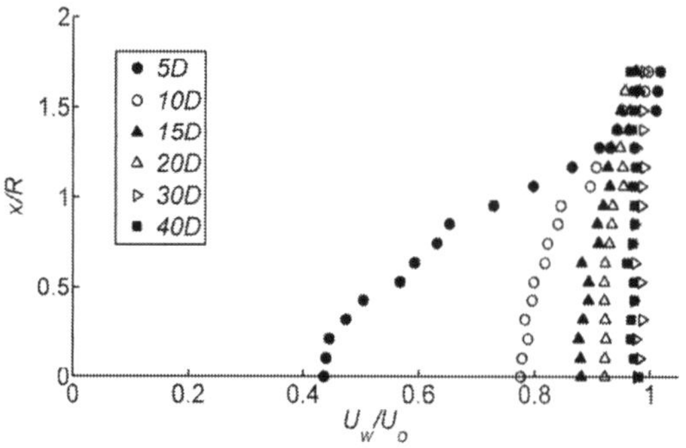

Figure 6. LDA-test of the far wake deficit of wind turbine with an optimal value of the tip speed ratio (λ = 5).

The visualizations are however not useable to detect the flow structures in the far wake of the rotor because the initial helical structure disappears due to mixing when the vortices become unstable and starts pairing up [9, 13]. A more accurate recognition of the flow structures was obtained using LDA (figure 6). To obtain more information about the far wake development, the flow velocities were measured with high temporal accuracy using LDA. These measurements were carried out through all wake area with stream-wise steps of $10R$ and at different radial positions from the rotor axis with the step of $0.1R$. In each point the local axial velocity history was obtained using a Dantec 2-D Fiber flow LDA, based on a 1W Argon laser with a differential optical configuration with a frequency shift of 40 MHz. The diameter of the optical gauge is 112mm and the focal length is 600mm with a beam diameter of 1.35mm. The wavelength of the laser beam is 514.5 nm (green light). The size of the probing optical field was $0.12\times0.12\times1.52\text{mm}^3$. The signal from the light-scattering particles is collected and processed by a Burst Specter Analyzer (BSA57N2 commercial signal processor) to detect the characteristic frequencies in the flow. Each time history was recorded in a period of 60s, corresponding to about 20 - 50 rotor rotation.

3. The analytical models of the far wake development

Based on the three basic models mentioned in the Introduction we will write the following expression for the maximal values of the velocity deficit

$$\frac{\Delta U}{U_0} = \frac{\max\left(U_0 - U_w\right)}{U_0}, \tag{1}$$

where U_0 is the free wind speed; U_w is velocity in the wake cross-sections and ΔU is the maximal velocity deficit. In the axisymmetric wake the maximum of the deficit lies on the rotor axis.

For the deficit maximum of (1) the J-model [3] can be expressed by a simple formula:

$$C_J\left(x\right) \equiv \frac{\Delta U}{U_0} = \frac{1-\sqrt{1-C_T}}{\left(1-\dfrac{2k_w x}{D}\right)^2}, \tag{2}$$

where C_T is the thrust coefficient from the table 1; k_w is the rate of the wake expansion and D is the rotor diameter. Jensen [3] assumed a top-hat shape for the velocity deficit and considered a constant value of the wake expansion ($k_w = 0.1$). However, $k_w = 0.075$ is the value usual used in literature [14].

The BP-model [4] is proposed to predict the maximum deficit of the streamwise velocity in the wake by the formula:

$$C_{BP}(x) \equiv \frac{\Delta U}{U_0} = 1 - \sqrt{1 - \frac{C_T}{8\left(\frac{k^* x}{D} - 0.2\sqrt{\beta}\right)^2}}, \quad (3)$$

where k^* represents the wake growth rate and β is obtained by the equation

$$\beta = \frac{1 + \sqrt{1 - C_T}}{2\sqrt{1 - C_T}}. \quad (4)$$

The author of the BP-model [4] assumed a Gaussian shape for the velocity deficit

$$\frac{U_0 - U_w}{U_0} = C_{BP}(x)\exp\left(-\frac{r^2}{2\sigma^2}\right), \quad (5)$$

where r is radial distance from the rotor axis and $\sigma = k^* x + \varepsilon$ is the standard deviation of the Gaussian-like velocity deficit profiles (5) at each x. The authors of [4] have considered a constant value of the wake growth rate ($k^*/D = 0.023$ and 0.031) in which the small deviations depend from operating regimes with different values of the thrust coefficient. However, $k^*/D = 0.015$ is best to fit the wake on the wind turbine optimum for our experience.

The G-model [5] was originally proposed to predict the wake decay behind a solid disk. Using similarity theory the maximum deficit of the streamwise velocity in the wake for high-Reynolds-number is described by the formula [8]:

$$C_G(x) \equiv \frac{\Delta U}{U_0} = h(x - x_0)^{-2/3}, \quad (6)$$

where x_0 is a virtual origin and the parameters a and x_0 depend on the type of the bluff body generating the wake, which, however, must keep the power $-2/3$ for any wake generators. Downstream evolution of the centreline velocity deficit behind his disk conditions was found excellent for the constants $h = 0.77$ and $x_0 = -2.4$.

Instead of the Gaussian shape for the velocity deficit (5), the G-model [6] assumed to fit it by a curve of the form

$$\frac{U_0 - U_w}{U_0} = C_G(x)\left(1 + a\eta^2 + b\eta^4\right)\exp\left(-c\eta^2 - d\eta^4\right), \quad (7)$$

where $a = 0.049$; $b = 0.128$; $c = 0.345$; $d = 0.134$ and $\eta = r/L$ is a non-dimension radial distance scaling by position where the maximum of the velocity deficit reduced to a half value $\frac{1}{2}\Delta U$

4. Development of the far wake behind a 3-blade rotor

Initially we try to separate the zones of the near and the far wake. The dye tracers in figure 5 give a good impression of the structure of the tip vortices in the near wake [9, 13], which shows the tip vortices as continuous curves giving a nearly perfect helical shape with a slight expansion. These photos also indicate an onset of initial disturbances for all operating regimes in the early development of the wake (approximately 2*D* downstream) as well as the initial point of onset of the transitional regime to far wake. This happens approximately 4*D* downstream from the rotor, where a full mixing of the dye tracers is seen to take place. The mixing connects with the wake instability and may be explained as a transition from the onset of initial disturbances to a fully developed far wake at 6*D* distance behind rotor. Due to the mixing in the far wake, the visualizations gave no information about the far wake behavior. As a consequence, to achieve a more accurate picture of the wake development, the visualizations were supplemented by LDA measurements in a number of points downstream of the rotor (see figure 6). Analyzing the LDA-spectra of the axial velocity in the rotor wake also confirmed the existence of two different zones of the wake with different main frequencies [9]. In the tested region the main flow oscillations have been determined to be associated with the blade and rotor frequencies in the near wake up to 6*D* distance behind rotor. Only a single dominating oscillation at a very low frequency has been indicated in far wake area after a distance of 6*D*.

Afterward we have analyzed the various LDA-profiles of the axial velocity in different cross-sections of the far wake at distances from 5*D* to 40*D* behind the rotor (see figure 6) and for the rotor operating under different tip speed ratios (see table 1). The self-similar velocity-deficit profiles at different downwind wake distances are shown in figure 7 for the optimum operating regime with $\lambda = 5$ and in figure 8 the different rotor operations. At each x the wake's half-width defined as

$$\left.\frac{U_0 - U_w}{U_0}\right|_{r=\frac{1}{2}L} = \frac{1}{2}\frac{\Delta U}{U_0}, \tag{8}$$

Figures 7 and 8 show that all wake profiles scaled by (8) approximately coinside with the Gaussian curve (5) from [4] except at the edge of the wake but approximation by the curve (7) introduced in [7] is excellent for fitting this data. It means that the velocity deficit profile can be assumed to be self-similar in far wakes. For $\lambda = 2$ (figure 8) the velocity profile is Gaussian instead of the one proposed by G-model. This is most likely explained by the lack of axial symmetry of the wake at very slow rotation of the three blades.

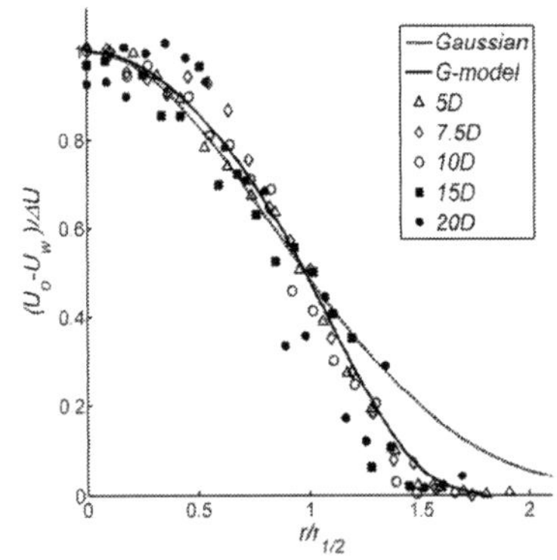

Figure 7. The self-similar behavior of the velocity deficit profiles in the wake on the optimum $\lambda = 5$.

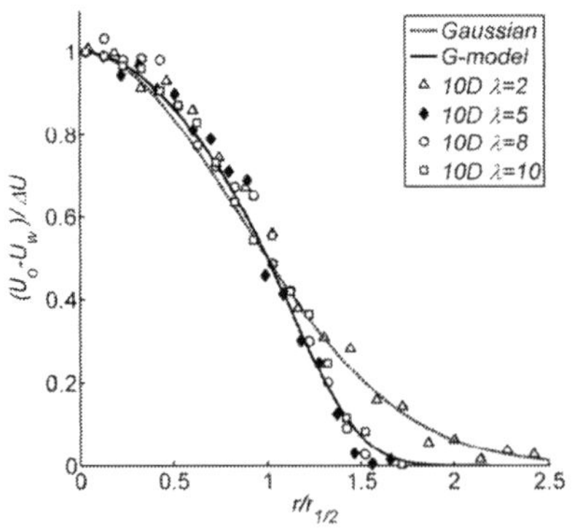

Figure 8. The self-similar behavior of the velocity deficit profiles in the wakes with $\lambda = 2, 5, 8, 10$.

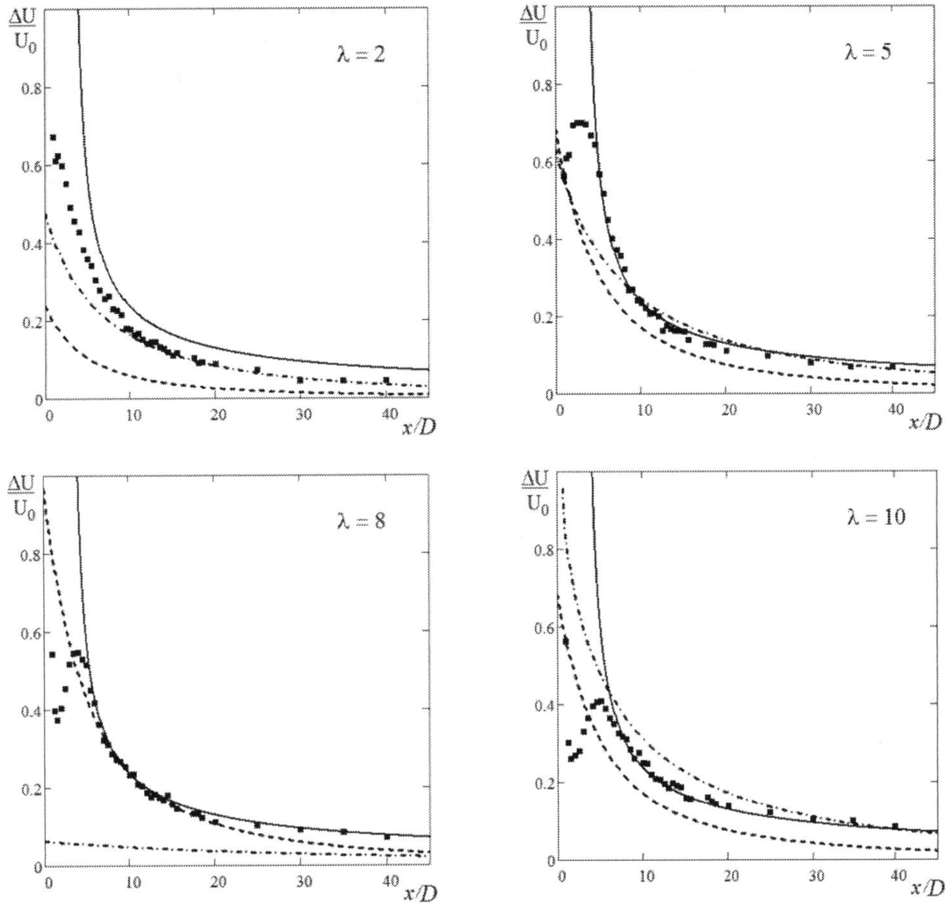

Figure 9. Downstream evolution of the maximal velocity deficit for different λ = 2, 5, 8, 10: Points – experiment; solid lines - C_G ; dashed lines - C_J ; dot-dashed lines - C_{BP}

The streamwise variation of the maximum velocity deficit in the rotor wake under the different tip speed ratios is shown in figure 9. It is clear that this value decreases monotonically and in a smooth manner for the different rotor operations. The solid line shows the high-Reynolds-number equilibrium similarity solution discussed below in equation (6) which was proposed in [5]. For cases of the wake evolution, the same constants, $h = 0.85$ and $x_0 = 3.2$, in (6) was used for the solid curve fitting. The first plot for the very low tip speed ratio (λ = 2) gives different values of the experimental data with (6), but next ones indicate the same behaviour with (6) which does not depend from the rotor thrust for all values of λ > 2. The first case has a different value only because the very slow rotation of the three blades could not form a symmetrical far wake.

The dashed and dot-dashed lines show the curves represented by (2) and (3) of the J- and BP-models. Unfortunately these models disagree with the experimental data.

5. Conclusions

The flow behind the model of a horizontal axis wind turbine rotor was investigated experimentally in a water flume. The initial flume flow was subject to a very low turbulence level and a uniform velocity profile, limiting the influence of external disturbances on the development of the far wake. The wake has been studied at different cross-sections from the near wake up to 40 rotor diameters downstream from the rotor. In the result of the present investigation the far wake development and decay properties have been determined and observed behind the rotor under different operating regimes for $\lambda = 2\text{-}10$. The data was used to compare the different wake models by Jensen [3] (J-model); Bastankhan & Porte-Agel [4] (BP-model) and the disk wake model by George [5] (G-model) which was extended to the rotor wake here.

The comparison with our high-resolution LDA measurements shows that the velocity profiles obtained with the disk-wake G-model after an expansion to the rotor wake are in acceptable agreement with the experimental data. By contrast, the J- and BP models generally underestimate the velocity deficit at the centerline of the wake. Our results also reveal that the G-model is consistent and acceptably accurate also in terms of the power estimation for the long distance of the wake development whereas J- and BP-models are less accurate.

6. Acknowledgments

The present work has been carried out with the support of the Danish Council for Strategic Research for the project COMWIND (grant 2104-09-067216/DSF) and the Russian Science Foundation (grant 14-19-00487). The authors also express their thanks to K.E. Meyer for useful discussions and help in calibration of LDA equipment during experiments.

References

[1] Vermeer L, Sørensen J, Crespo A. Wind turbine wake aerodynamics. *Prog Aerosp Sci* 2003; 39: 467-510.

[2] Lissaman PBS. Energy effectiveness of arbitrary arrays of wind turbines. In: *AIAA Paper* 79-0114, 1979. p. 1–7.

[3] Jensen N. A note on wind turbine interaction. Technical report Ris-M-2411. Roskilde, Denmark: Risø National Laboratory; 1983.

[4] Bastankhan M, Porte-Agel F. A new analytical model for wind turbine wakes. *Renewable Energy* 2014; 70: 116-123.

[5] George WK. The self-preservation of turbulent flows and its relation to initial conditions and coherent structures. *Advances in Turbulence*, 1989: p. 39-73.

[6] Johansson PBV, George WK, Gourlay MJ. Equilibrium similarity, effects of initial conditions and local Reynolds number on the axisymmetric wake. *Physics of Fluids*, 2003; 15(3): 603-617.

[7] Johansson PBV. The axisymmetric turbulent wake. PhD thesis of Department of Thermo and Fluid Dynamics, Chalmers University of Technology, 2002.

[8] Johansson PBV, George WK. The far downstream evolution of the high-Reynolds-number axisymmetric wake behind a disk. Part 1. Single-point statistics. *J. Fluid Mech.* 2006; 555: 363-386.

[9] Okulov VL, Naumov IV, Mikkelsen RF, Kabardin IK, Sørensen JN. A regular Strouhal number for large-scale instability in the far wake of a rotor. *J. Fluid Mech.* 2014; 747: 369-380.

[10] Glauert H 1935 Airplane propellers *Aerodynamic Theory* ed. W.F.Durand (Berlin: Springer) **IV**

[11] Naumov IV, Rahmanov VV, Okulov VL, Velte CM, Meyer KE, Mikkelsen RF 2012 Flow diagnostics downstream of a tribladed rotor model. *Thermophysics & Aeromechanics* **19(2)** 253-263

[12] Okulov VL, Naumov IV, Mikkelsen RF, Kabardin IK, Sørensen JN. Experimental investigation of the wake behind a model of wind turbine in a water flume. *Journal of Physics: Conference Series* 2014; 555: 012080. DOI: 10.1088/1742-6596/555/1/012080

[13] Naumov IV, Mikkelsen RF, Okulov VL, Sørensen JN. PIV and LDA measurements of the wake behind a wind turbine model. *Journal of Physics: Conference Series* 2014, vol. 524, 012168; doi:10.1088/1742-6596/524/1/012168

[14] Barthelmie RJ, Folkerts L, Larsen GC, Rados K, Pryor SC, Frandsen ST, et al. Comparison of wake model simulations with offshore wind turbine wake profiles measured by sodar. *J Atmospheric Ocean Technol* 2005; 23: 881-901.

Journal of Physics: Conference Series **625** (2015) 012012 doi:10.1088/1742-6596/625/1/012012

Wind-tunnel study of the wake behind a vertical axis wind turbine in a boundary layer flow using stereoscopic particle image velocimetry

V Rolin, F Porté-Agel

Wind Engineering and Renewable Energy Laboratory, Ecole Polytechnique Fédérale de Lausanne, Switzerland

E-mail: vincent.rolin@epfl.ch,fernando.porte-agel@epfl.ch

Abstract. Stereo particle image velocimetry is used in a wind-tunnel to study boundary layer effects in the wake behind a vertical axis wind turbine. The turbine is a three-bladed giromill with a solidity of 1.18. The wake is studied for a tip speed ratio of 2 and an average chord Reynolds number of 1.6×10^4. The velocity deficit and turbulence levels in the horizontal plane are observed to be strongly asymmetrical with two strong peaks corresponding to the two halves of the rotor where blades move either towards the oncoming flow or away from it. The stronger peak is measured behind the blades moving upstream, however this region also benefits from a greater rate of re-energization. Due to the incoming boundary layer profile, momentum is also entrained downwards into the wake from above and aids with the recovery of the core of the wake.

1. Introduction

In response to the growing need for renewable energy, vertical axis wind turbines are enjoying a period of renewed interest. Horizontal axis wind turbines (HAWTs) have benefited from a larger volume of research than vertical axis wind turbines (VAWTs). Nonetheless, VAWTs offer some advantages over HAWTs which merit further investigation of the technology. For instance, the generator of a VAWT is placed at ground level, as opposed to in the nacelle of a HAWT which is placed at the rotor hub-height. This reduces the loads on the structure, thereby diminishing capital and maintenance costs, and simplifying the construction of the turbine making it an attractive option for off-shore wind arrays [1]. These turbines are also insensitive to the direction of the incoming wind and therefore require no yaw equipment [2]. The simpler design of the VAWT has also been shown to reduce the noise produced which is an important commercial advantage [3].

Due to the fact that a VAWT rotates about an axis perpendicular to the incoming wind, the blades are subjected to a flow with continuously changing magnitude and direction. This behavior introduces some phenomena unique to VAWTs: the thrust and torque forces produced by the rotor oscillate as a function of the azimuthal position of the blades, the downstream half of the rotor will be subjected to the wake of the upstream half, and the high angles of attack encountered when the blade is perpendicular to the incoming flow can lead to dynamic stall conditions and induce vortex shedding. These effects have been shown not only to alter the flow downstream of the turbine but also to influence its performance. Many studies have been conducted using a variety of numerical and experimental techniques to examine this unsteady behavior. [4–7]

Possibly the first wake measurement behind a VAWT was performed by Muraca and Guillote (1976) [8], who profiled the wake velocity upstream and downstream of the turbine at different elevations. Bergeles et al.(1988) [9] carried out a more comprehensive study by using two-component hot-wire measurements to observe the velocity vectors and turbulence intensity in the wake at various downstream locations and tip speed ratios. They reported an easily identifiable wake, characterized by an abrupt change in velocity between the wake and unperturbed flow, and high turbulence intensities near the edges of the wake. They also reported a deflection of the wake in the direction of rotation and generally stronger wake characteristics behind the blades moving upwind at low tip speed ratios. More recently, Battisti et al.(2011) [10] performed a similar investigation using hot-wire probes over a vertical plane behind the VAWT. Battisti reported behavior similar to what has been previously observed, but by phase-locking the measurements, they related some of the asymmetrical characteristics of the wake to the unsteady behavior of the VAWT. The present study will aim to get a better picture of the wake and its re-energization by increasing the spatial resolution of measurements and reporting three-component statistics using PIV techniques. A key difference between the current study and those discussed in the literature, however, is that the present examination is conducted in a boundary layer flow, whereas previously they had been performed exclusively in uniform flows.

If VAWTs are to be deployed in an array, whether onshore or offshore, it is vital to understand and characterize the behavior of their wakes. Turbine-wake interaction in arrays dramatically reduces the effectiveness of the array, and increases the need for maintenance due to vibrational loading [11]. The atmospheric boundary layer plays an important role in wake flow, altering the rate of re-energization and entrainment of momentum. Many studies have been carried out in order to understand the interaction between the boundary layer and the wake of a HAWT. Chamorro and Porté-Agel (2009) [12] carried out a comprehensive examination of the wake behind a HAWT in boundary layer flow using hot wire measurements. Their findings indicate that turbulent mixing effects are stronger around the upper edges of HAWT wakes, where velocity gradients are steeper, and more momentum is entrained downwards from the faster moving winds of the unperturbed flow. At the time of writing there is little to no literature regarding a similar characterization of VAWT wakes in boundary layer flow.

The present study intends to analyze the interaction between the turbine and the boundary layer flow and to understand how it influences the structure of the wake. Stereoscopic particle image velocimetry (S-PIV) is deployed in the WIRE wind tunnel at the EPFL to observe the time-averaged wake behind a scaled-down model VAWT immersed in a boundary layer flow. The velocity deficit, turbulence kinetic energy (TKE), and the crosswind and vertical turbulent fluxes of momentum are quantified to characterize the wake and the entrainment of momentum.

2. Experimental setup

2.1. Model vertical axis wind turbine

The turbine used in this experiment is a three-bladed giromill, shown in figure 1. The blade profile is a 5% thick plate with a 5% circular arc camber, shown in figure 2. This shape has been shown to have better lift-to-drag ratios and delayed stall at low Reynolds numbers compared to the more conventional NACA profiles used at larger scales [13–15]. The turbine is designed with a blade span of 15.24 cm (6 in) in order to maintain a realistic scale relative to the height of boundary layer, which will be quantified in the following section. A 12 W Maxon EC 45 motor is used in order to maintain a tip speed ratio of 2. The geometry of the turbine is summarized in table 1.

Figure 1. A graphical representation of the model turbine used in the experiment.

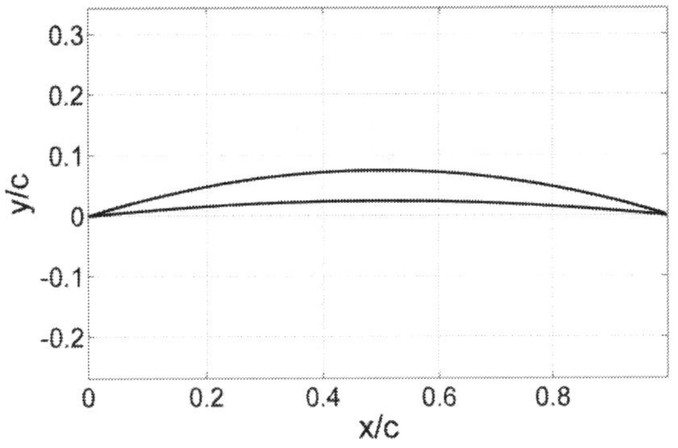

Figure 2. Thin plate with 5% circular arc camber.

Table 1. List of important rotor and airfoil design parameters

Parameter	Symbol	Value
Blade chord	C	5 cm
Rotor radius	R	7.62 cm
Rotor height	H	15.24 cm
Midspan height	H_m	12.7 cm
Number of blades	N	3
Rotor solidity	$\sigma = {}^{NC}/_{R}$	1.18
Velocity at midspan	U_∞	4ms^{-1}
Rotational velocity	ω	1,000RPM
Tip speed ratio	$\lambda = {}^{\omega R}/_{U_\infty}$	2
Average chord Reynolds number	$Re = \rho\lambda U_\infty C/\mu$	1.60×10^4

2.2. Wind tunnel

The EPFL boundary layer wind tunnel is a closed-looped facility at the WIRE Laboratory. The test section is 28 m long with a 2.57 m × 2 m cross section. The facility is equipped with a 130 kW fan. Ahead of the test section the flow passes through a contraction with a 5:1 area ratio.

The turbulence intensity in the center of the wind tunnel, or freestream, is approximately 0.5%. A turbulent boundary layer develops naturally over the long test section. This experiment is performed within the boundary layer approximately 22m from the inlet of the test section. The boundary layer velocity and turbulence intensity profiles are shown in figure 3. The boundary layer has a total height of roughly 40 cm surface roughness, z_0, of 2.1612×10^{-2}mm, and friction velocity, u^*, of 0.1319ms^{-1}. The power density spectra of the incoming flow is plotted for varying heights in figure 4. The steeper slope of spectra at large wave-numbers is due to the greater dissipation of energy near the wind tunnel floor. The incoming wind at the midspan height of

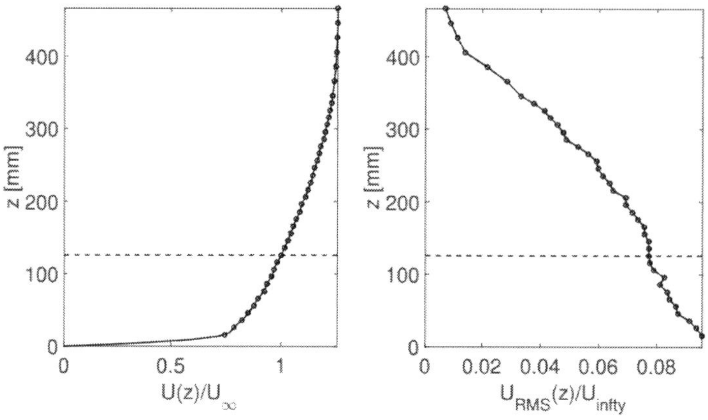

Figure 3. Normalized mean velocity (left) and turbulence intensity of the incoming wind as a function of distance from the wind tunnel floor, z. U_∞ denotes the wind velocity of the incoming wind at the midspan height. The dashed line represents the midspan height

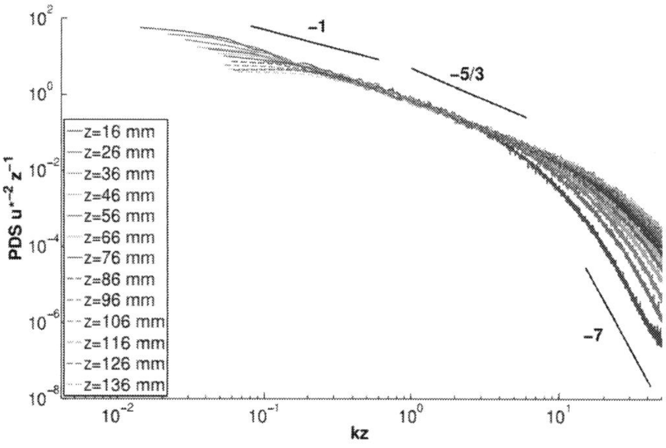

Figure 4. Normalized power density spectra of the incoming wind at various heights with respect to the wind tunnel floor.

the turbine, has a speed, U_∞, of $4.0\,\mathrm{ms}^{-1}$, and a turbulence intensity, I, of 7.71%.

2.3. Stereo-PIV configuration

The flow was continuously seeded with olive oil particles on the order of one micrometer in diameter. The particles were illuminated by a thin horizontal laser sheet at the equatorial height of the rotor. The laser sheet was produced by a dual head 435 mJ ND:YAG laser. A high-resolution stereoscopic PIV system developed by LaVision was used to measure the three velocity components in the illuminated plane. The imaging system comprised two 29MP 12-bit CCD cameras (6600 × 440 pixels) each with a 105 mm lens installed on scheimflug adapters to improve the focus quality over the field-of-view (FOV). The imaging system was installed above the wind tunnel, in a forward-backward configuration, such that the cameras are oriented

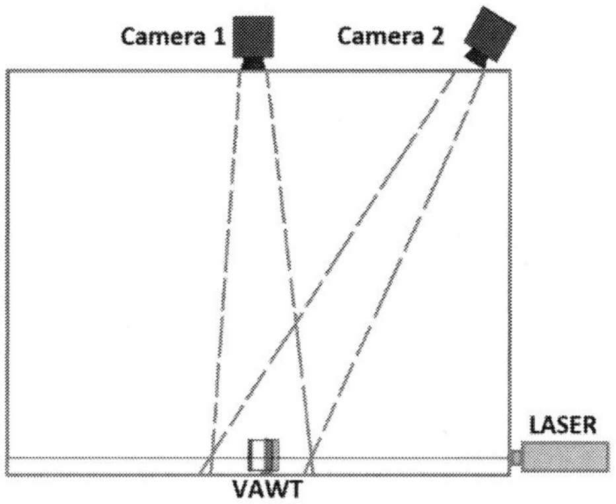

Figure 5. Schematic representation of the PIV set-up as seen from downwind of the turbine.

perpendicular to the direction of flow. A schematic illustration of the set-up is shown in figure 5.

This allows the FOV to approach the back edge of the rotor. Due to the large distance between the measurement plane and the camera, as well as the low photo-sensitivity of the CCD array, the image captured is dimly illuminated. This is especially egregious in the case of the backwards camera, which receives mostly backscatter laser radiation. To compensate, this camera is moved so that it is directly perpendicular to the measurement plane, as in a 2D PIV configuration.

The captured image has an area of 676 mm × 451 mm, with a resolution of approximately 0.1 mm. This corresponds to 4.4 × 3.0 rotor diameters. In order to capture a larger region of the wake, two sets of measurements are taken by keeping the PIV system stationary and changing the streamwise position of the turbine. The time delay between exposures was set such that particles in the core of the wake region were displaced by roughly five pixels. Exactly 1,200 samples were acquired for each streamwise location. Images were processed using 32-pixel interrogation windows with 50% overlap.

3. Experimental results

Due to the turbulent nature of the wake, a statistical approach is adopted in this analysis. From this perspective, the wind is considered as a stochastic quantity composed of a mean and a fluctuating quantity. By convention the following notation is implemented:

$$U = \overline{U} + u'. \tag{1}$$

In equation 1, U represents the total velocity, $\overline{U}$ denotes the mean velocity, and u' represents the fluctuating velocity. This representation is used for streamwise, crosswind, and vertical velocity, which are denoted as U, V, and W, respectively. The figures presented in this section illustrate the horizontal plane at the midspan of the rotor as seen from above. The rotor is centered at the origin and rotates counter-clockwise in this reference frame. Note that some figures have regions which have been cut; this is done to remove numerical inaccuracies on the edges of the image in the out-of-plane measurements.

3.1. Velocity deficit

Two clear minima can be observed in the wake velocity shown in figure 6. The profiles of

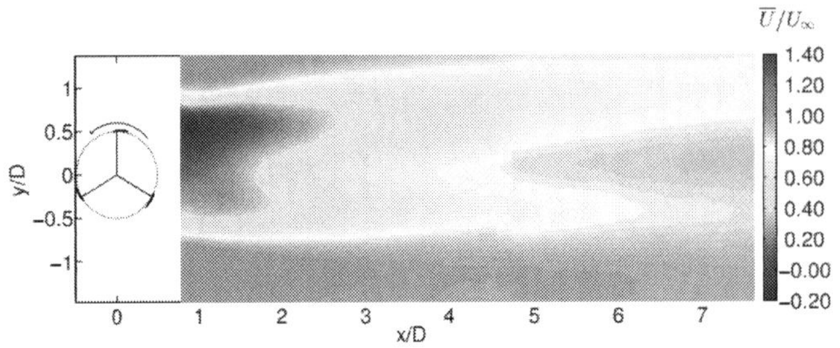

Figure 6. Contours of normalized time-averaged streamwise velocity ($\overline{U}/U_\infty$) of the wake in the symmetry plane behind the VAWT.

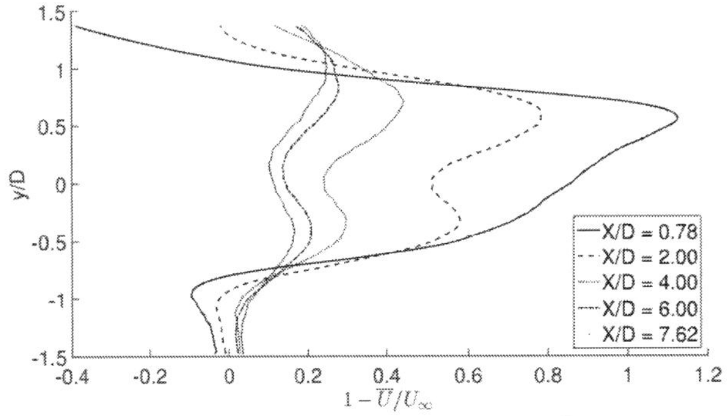

Figure 7. Crosswind profiles of normalized streamwise velocity deficit in the wake ($1-\overline{U}/U_\infty$) at various downstream locations

the velocity deficit at various downstream locations, shown in figure 7, show that the wake of the blades moving upwind is significantly stronger than that of those moving downwind. This is due to the greater difference between the incoming wind speed and the tangential velocity of the blades as they move upwind. These peak regions will hereinafter be referred to as the stronger and weaker sides of the wake, respectively. The profiles also show that the wake on the stronger side tends to spread and diffuse into the unperturbed flow to a greater extent than on the weaker side. This indicates a faster recovery of the wake due to turbulent mixing related to the mechanical production of TKE, given by $-\overline{u_i'u_j'}\frac{\partial \overline{U}_i}{\partial x_j}$ [16]. The mechanical production of TKE represents the contribution to the TKE produced by the wind large shears near solid objects. From this term, it is understood that the steeper velocity gradient on the stronger side of the wake is associated with higher levels of TKE in that region.

The crosswind velocity presented in figure 8 reveals motion tangential to the rear edge of the rotor. This suggests that the rotation of the turbine imparts some momentum to the wake. This may also play a role in the expansion of the wake on the stronger side into the unperturbed flow, as the crosswind momentum convects the wake turbulence outwards.

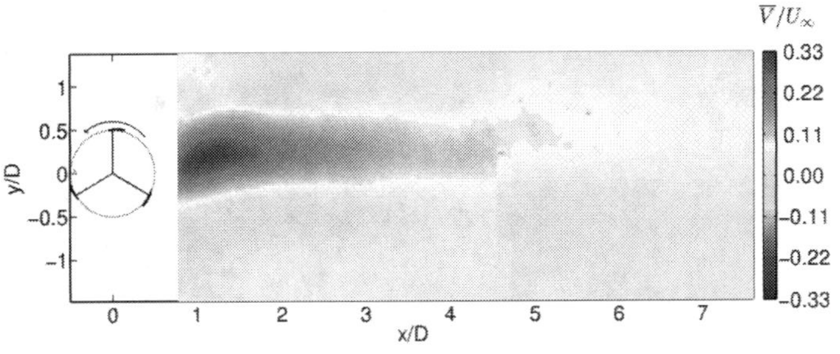

Figure 8. Contours of normalized time-averaged crosswind velocity ($\overline{V}/U_\infty$) of the wake in the symmetry plane behind the VAWT.

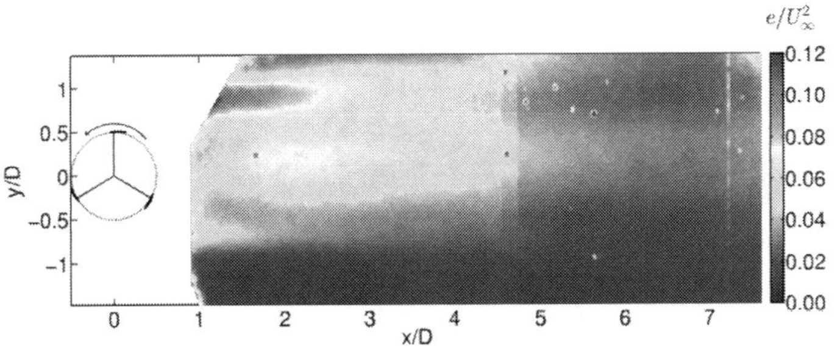

Figure 9. Normalized contours of turbulence kinetic energy ($e/U_\infty^2 = \frac{1}{2}\left[\overline{u'^2}/U_\infty^2 + \overline{v'^2}/U_\infty^2 + \overline{w'^2}/U_\infty^2\right]$) in the symmetry plane in the wake of the VAWT

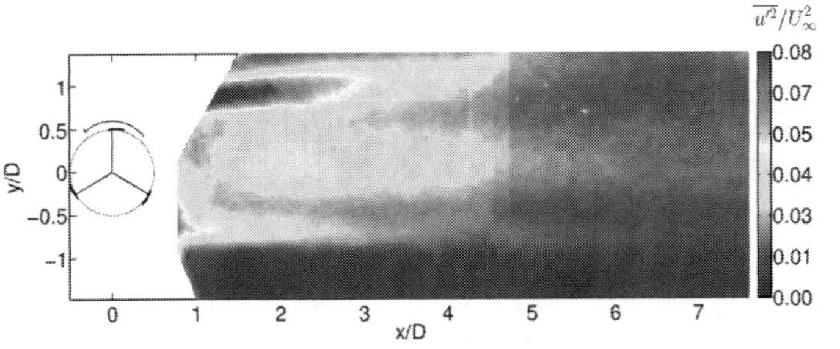

Figure 10. Normalized contours of the streamwise component of turbulence intensity ($\sqrt{\overline{u'^2}}/U_\infty^2$) in the symmetry plane in the wake of the VAWT.

3.2. Wake Turbulence

Examining the turbulence kinetic energy in figure 9 reveals a highly turbulent region in the core of the wake. This core turbulence may be due to the large periodic blockage of the center of the rotor as a blade turns perpendicular to the incoming wind. By decomposing the turbulence kinetic energy into streamwise, crosswind, and vertical components of turbulence intensity shown in figures 10, 11, and 12, respectively, it is clear that the fluctuations in the core of the wake are largely due to fluctuations in vertical velocity. This result is somewhat puzzling. Typically

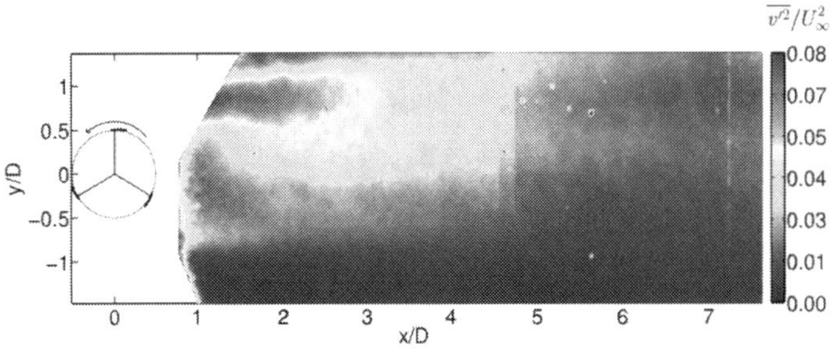

Figure 11. Normalized contours of the crosswind component of turbulence intensity ($\sqrt{\overline{v'^2}}/U_\infty^2$) in the symmetry plane in the wake of the VAWT.

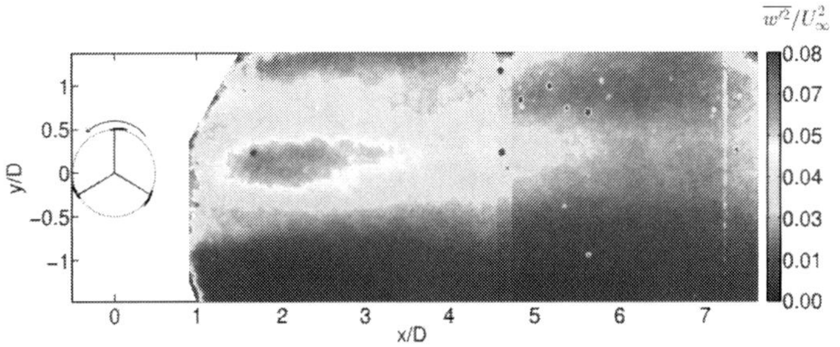

Figure 12. Normalized contours of the vertical component of turbulence intensity ($\sqrt{\overline{w'^2}}/U_\infty^2$) in the symmetry plane in the wake of the VAWT.

entrainment is strongest where the velocity gradients are greatest. However, in the core of the wake, where the vertical entrainment seems significant, the momentum is greater than in the surrounding wake. The presence of vertical entrainment is a sign of boundary layer effects. In a uniform flow, it would be expected that the momentum entrained downwards would be equivalent to that entrained upwards, and that vertical motion would not be observable at the rotor midspan.

In order to understand why vertical momentum is entrained into the core of the wake, an investigation of vertical planes in is required. We can, however, provide some hypotheses; one possible explanation could be attributed to vortex shedding. In HAWTs, tip-vortices tend to shield the wake from entrainment and turbulent mixing. Similarly, it is possible that tip-vortices shed from the top and bottom of the VAWT rotor act as a barrier against vertical entrainment in the stronger and weaker sides of the wake, and instead allow entrainment only in the core of the wake. Another possibility is that the rotor exerts some crosswind force on the flow, and causes the flow to diverge in the wake. Due to continuity, this would induce out-of-plane motion to conserve mass within the wake. This idea is reinforced by the contours of crosswind velocity, which indicate motion in both the positive and negative y direction in the core of the wake. These theories will require more investigation.

As mentioned previously, the mechanical production of turbulence kinetic energy is the product of the velocity gradient and turbulent flux of momentum. The turbulence flux of momentum represents the momentum transported by turbulent motion. Of particular interest are the crosswind and vertical components of turbulent flux of streamwise momentum, denoted as $\overline{u'v'}$ and $\overline{u'w'}$, respectively. The crosswind turbulent flux of momentum in figure 13 shows a

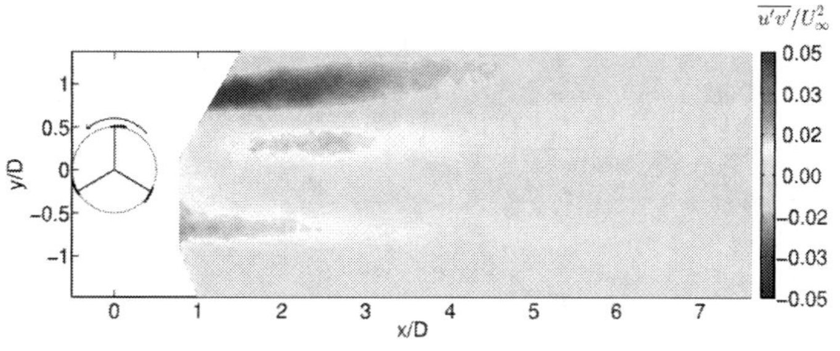

Figure 13. Normalized contours of the crosswind component of the streamwise turbulent flux of momentum $(\overline{u'v'}/U_\infty^2)$ in the symmetry plane in the wake of the VAWT.

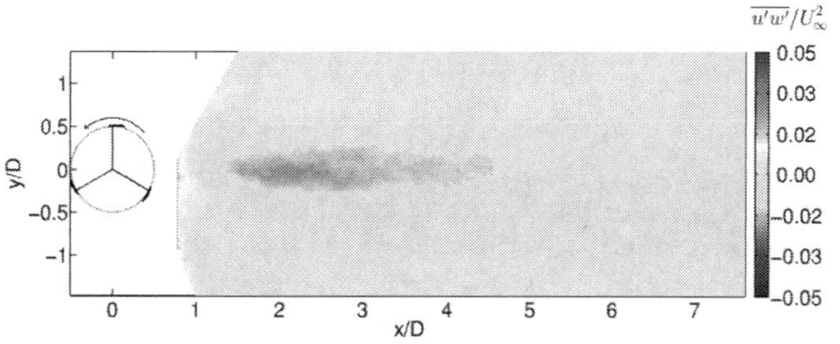

Figure 14. Normalized contours of the vertical component of the streamwise turbulent flux of momentum $(\overline{u'w'}/U_\infty^2)$ in the symmetry plane in the wake of the VAWT.

higher level of entrainment on the stronger side of the wake than the weaker one. It also shows some entrainment between the two sides with momentum being transfered from the weaker to the stronger side of the wake. This result is expected due to the previously observed shear on the stronger side of the wake which is associated with turbulent fluxes. The vertical turbulent flux of momentum in figure 14 shows entrainment downwards into the wake in agreement with the vertical fluctuations previously observed in figure 12. The stronger downwards entrainment of momentum is likely due to the higher momentum present above the rotor in the boundary layer. It remains unclear, however, why this momentum should be entrained directly into the core of the wake, rather than into the stronger and weaker sides of the wake where the velocity deficit is greater.

4. Summary

The wake behind a VAWT is complex and unique. In this study, the wake behind a VAWT in a turbulent boundary layer flow at low Reynolds numbers and a low tip speed ratio is examined. Two important regions of the wake can be distinguished: one behind the blades of the rotor as they move upwind, and the other as they move downwind. These two regions are termed the stronger and weaker side, respectively. The stronger side of the wake is associated with a significantly larger velocity deficit and a higher level of turbulence than the weaker side. The stronger side of the wake also tends to expand and diffuse into the unperturbed flow to a greater extent than the weaker one. This is primarily the result of the higher levels of turbulence produced by large wind shear on the stronger side, but may be due in part to the crosswind momentum imparted on the flow by the rotation of the turbine. In the core of the wake, momentum is transferred from the weaker side of the wake to the stronger side, such that far downstream the wake becomes more symmetric.

The effect of the boundary layer in the core is also evident as it re-energizes this region with momentum entrained downwards. If instead the flow were uniform, the wake would be symmetric about the midspan plane of the rotor and no vertical momentum would be evident. The boundary layer effects may be responsible for the lower velocity deficit in the core of the wake. It remains unclear why vertical entrainment is almost exclusive to the core of the wake. Typically, entrainment should be greatest where there are larger velocity deficits, such as in the stronger and weaker sides of the wake. It is possible that blade-tip vortices at the top and bottom edges of the rotor are shielding the outer perimeter of the wake from vertical entrainment, or that aerodynamic forces produced by the rotor cause the flow to diverge in the wake and induce out-of-plane motion. This phenomenon will require investigation of the wake at different heights to be better understood.

A comprehensive understanding of the behavior of the wake and how it is influenced by the rotor geometry and operating conditions is also crucial for the successful planning and deployment of VAWT arrays and will be the target of future studies.

References

[1] Sutherland H J, Berg D E and Ashwill T D 2012 A Retrospective of VAWT Technology Tech. Rep. January Sandia National Laboratories Albequerque, New Mexico
[2] Eriksson S, Bernhoff H and Leijon M 2008 *Renewable and Sustainable Energy Reviews* **12** 1419–1434
[3] Möllerström E, Larsson S, Ottermo F, Hylander J and Bååth L 2014 Noise Propagation from a Vertical Axis Wind Turbine *43rd International Congress on Noise Control Engineering* (Melbourne: Australian Acoustical Society) pp 1–8
[4] Dixon K R 2008 *The Near Wake Structure of a Vertical Axis Wind Turbine* Ph.D. thesis Delft University of Technology
[5] Ferreira C S 2009 *The near wake of the VAWT 2D and 3D views of the VAWT aerodynamics* Ph.D. thesis Universidade Tecnica de Lisboa
[6] Shamsoddin S and Porte-Agel F 2013 *Large Eddy Simulation of Flow through Vertical Axis Wind Turbines: Evaluation of Turbine Parameterizations* Master's thesis EPFL
[7] Tescione G, Ragni D, He C, Simão Ferreira C and van Bussel G 2014 *Renewable Energy* **70** 47–61
[8] Muraca R J and Guillotte R J 1976 Wind tunnel investigation of a 14ft veritcal axis windmill Tech. rep. NASA Langley Research Center Hampton, Virginia
[9] Bergeles G, Michos a and Athanassiadis N 1991 *Journal of Wind Engineering and Industrial Aerodynamics* **37** 87–101
[10] Battisti L, Zanne L, DellAnna S, Dossena V, Persico G and Paradiso B 2011 *Journal of Energy Resources Technology* **133** 031201
[11] Alfreddson P H and Dahlberg J A 1981 Measurements of wake interaction effects on the power output from small wind turbine models Tech. rep. The Aeronautical Research Institute of Sweden Structures Department Stockholm, Sweden
[12] Chamorro L P and Porté-Agel F 2009 *Boundary-Layer Meteorology* **132** 129–149
[13] Laitone E V 1997 *Experiments in Fluids* **23** 405–409
[14] Sunada S, Sakaguchi A and Kawachi K 1997 Airfoil Section Characteristics at a Low Reynolds Number
[15] Pelletier A and Mueller T J 2000 Low Reynolds Number Aerodynamics of Low-Aspect-Ratio, Thin/Flat/Cambered-Plate Wings
[16] Stull R B 1988 *An Introduction to Boundary Layer Meteorology* vol 13 (Dordrecht, The Netherlands: Kluwer Academic Publishers)

Comparison of actuator disc and Joukowsky rotor flows, to explore the need for a tip correction

Gijs van Kuik[1], **Wei Yu**[1], **Sasan Sarmast**[2] **and Stefan Ivanell**[2]
[1]DUWIND, Faculty of Aerospace Engineering, TU-Delft, Kluyverweg 1, 2629HS Delft, the Netherlands
[2]Wind Energy Campus Gotland, Dept. of Earth Sciences, Uppsala University, 621 67 Visby, Sweden

E-mail: g.a.m.vankuik@tudelft.nl

Abstract. In van Kuik & Lignarolo (2015, *Wind Energy* accepted with modifications) potential flow solutions have been obtained for classical actuator discs (axisymmetric, uniform load) presenting a model for the non-uniformity of the axial velocity at the disc. Current rotor design models proceed from a disc with a uniform axial velocity, modified by a tip correction like the one of Prandtl-Glauert-Shen. A comparison shows that this correction leads to a similar distribution as obtained from the potential flow solutions. A next comparison is made with the velocity distribution at the blade position of a Joukowsky rotor with constant bound circulation, calculated by an Actuator Line and a Lifting Line model. The resulting distributions correspond reasonably well to the potential flow disc distribution, in magnitude as well as shape. This implies that this non-uniform distribution is relevant for a rotor with a finite number of blades, and could be the basis for rotor design instead of the uniform but tip-corrected distribution.

1. Introduction

Most Blade Element Momentum codes (BEM) proceed from the assumption that the optimal induction at the rotor plane is uniform, with a correction for the difference between a disc and a rotor. Most used is the Prandtl-Glauert tip correction F which is analysed and expanded by Shen *et al.* [1, 2]. Also aerodynamic optimizations using CFD (computational fluid dynamics) like Johansen *et al.* [3] assume that a uniform induction is best. However, as BEM codes apply actuator disc momentum theory, the assumption of uniform induction should find its basis in actuator disc flow solutions. In van Kuik & Lignarolo [4] the potential flow solutions for actuator discs with a uniform normal load have been presented. It is shown that not the axial component v_z is uniform but the absolute velocity $|\boldsymbol{v}|$. The non-uniform distribution for v_z denoted as G is a function of the local radius r and the thrust T. It shows slightly higher values than the disc-averaged value $\overline{v_{z,disc}}$ for relative radius $r/R < 0.8$ and lower values near the disc edge. This 'edge effect' is qualitatively the same as the 'tip effect' of F. This raises two questions:

(i) is it better to optimize a rotor based on the non-uniform distribution G instead of a uniform induction and correcting it by F?
(ii) how representative is the actuator disc velocity distribution for a rotor?

The rotor to be studied has B blades carrying a constant blade-bound circulation. Such a rotor was first analysed by Joukowsky [5] by which it is called a 'Joukowsky rotor'. This

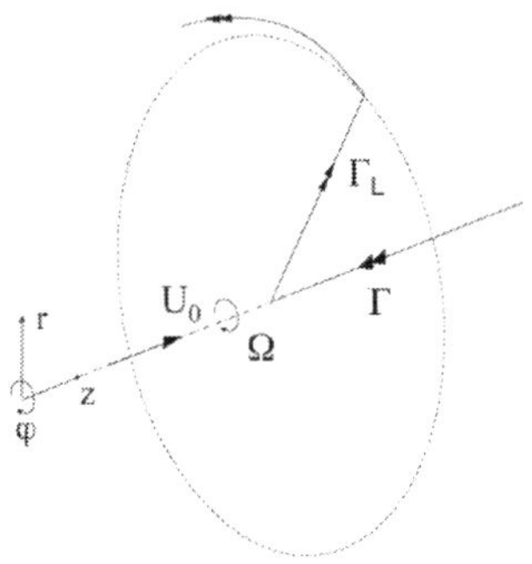

Figure 1. The coordinate system and orientation of the vortices, shown for one lifting line L

rotor becomes the disc with a uniform load after the limits $B \to \infty$ and tip speed ratio $\lambda \to \infty$ meanwhile keeping the converted power constant, see van Kuik [6]. The tip speed ratio is the tip speed ΩR divided by the wind speed U_0. The blades are modelled as lifting lines with prescribed constant bound circulation Γ_L, see figure 1, also showing the coordinate system (z, r, φ)[1]. The bound vortex continues as a root vortex at the axis and as a tip vortex at the wake boundary. All vortices have a kernel radius δ. The root vortex becomes $\Gamma = -B\Gamma_L$ or, dimensionless:

$$q = \frac{B\Gamma_L}{2\pi R U_0} \tag{1}$$

Note that the sign of Γ and q is different in order to obtain positive thrust and power coefficients.

First some general expressions for the thrust and power coefficients of a Joukowsky rotor are derived. Then the first question is discussed in section 3, by evaluating F for the three load cases in table 1, all for $B = 3$, and G for the disc with the same thrust coefficient. For the second question, two numerical models are used: a potential flow lifting line (LL) model and a computational fluid dynamic model using actuator lines (AL), both of which are briefly described in the appendix. Section 4 compares the results of both models with emphasis on the velocities at the rotor plane, where after section 5 compares all methods.

Table 1. Flow cases

Case #	λ	2λq
1	7	0.888
2	7	0.970
3	20	0.888

2. Expressions for the thrust and power of a Joukowsky rotor

Since the bound vorticity Γ_L of the lifting lines (L) is constant, simple expressions for the thrust and power are possible. The thrust and torque originate from the axial and azimuthal component of the Kutta-Joukowsky load L at Γ_L. The product of torque and Ω gives the power:

$$C_p = \frac{1}{\frac{1}{2}\rho U^3 \pi R^2} \rho \Omega B \Gamma_L \int_0^R v_{z,L} r dr = 2\lambda q \frac{\overline{v_{z,L}}}{U_0} \tag{2}$$

[1] In [4] the axis has the x-coordinate; the system is identical otherwise.

with $\frac{\overline{v_{z,L}}}{U_0} = \int_0^1 \frac{v_{z,L}}{U_0} d\left(\frac{r}{R}\right)^2$. The thrust is determined by the total axial load at the B lifting lines, $L_z = \rho B \Gamma_L (\Omega r - v_{\varphi,L})$. The expression for the thrust coefficient is:

$$C_T = \frac{\rho B \Gamma_L \int\limits_0^R (\Omega r - v_{\varphi,L}) dr}{\frac{1}{2}\rho U^2 \pi R^2} = 2\lambda q - 4q \int\limits_0^1 \frac{v_{\varphi,L}}{U} d\frac{r}{R} \tag{3}$$

The local thrust coefficient C_t defined as $dL_z / \left(\rho U_0^2 \pi r dr\right)$ becomes:

$$C_t = \frac{\rho B \Gamma_L (\Omega r - v_{\varphi,L})}{\rho U^2 \pi r} = 2\lambda q - 2q \frac{v_{\varphi,L}}{U} \frac{R}{r} \tag{4}$$

In the wake the azimuthally averaged value $\overline{v_\varphi} = \Gamma / (2\pi r)$ but in the rotor plane it is half this value: $\overline{v_{\varphi,x=0}} = \Gamma / (4\pi r)$. The azimuthal distribution of v_φ will be uniform for low values of r/R as the induction by the root vortex dominates. However, for larger r/R values the tip vortices will add a harmonic distribution. When the non-uniformity of v_φ is neglected, $v_{\varphi,L}/U \approx \Gamma / (4\pi r U)$ for $r > \delta$ and $\Gamma r / \left(4\pi \delta^2\right)$ for $r < \delta$. As the contribution of the area $\pi \delta^2$ to thrust and power is negligible, this contribution is omitted. The result is:

$$\left.\begin{aligned} C_T &= C_{T,\Delta P} + C_{T,\Delta\varphi} \\ C_{T,\Delta P} &= 2\lambda q \\ C_{T,\Delta\varphi} &= -4q \int\limits_{\delta/R}^{1} \frac{v_{\varphi,L}}{U} d\frac{r}{R} \approx -2q^2 \ln \frac{\delta}{R} \end{aligned}\right\} \tag{5}$$

and

$$\left.\begin{aligned} C_t &= C_{t,\Delta P} + C_{t,\Delta\varphi} \\ C_{t,\Delta P} &= 2\lambda q \\ C_{t,\Delta\varphi} &= -2q \frac{v_{\varphi,L}}{U} \frac{R}{r} \approx q^2 \left(\frac{R}{r}\right)^2 \end{aligned}\right\} \tag{6}$$

The first term at the right hand side of C_T and C_t is denoted $C_{T,\Delta P} = 2\lambda q$ since only this part of the thrust contributes to the conversion of power, see (2). The second term represents the contribution to the thrust by the radial pressure gradient balancing the radial distribution of v_φ. The singular logarithmic appears in C_p but not in C_T. However, it may have an indirect impact at C_p by $v_{z,L}$. With $\delta \ll R$ the singular term is positive, so it adds up to the thrust. For the hypothetical case of a potential flow model with a singular vortex when $\delta \to 0$, $C_{T,\Delta\varphi}$ becomes infinite. With the data of flow case 1: $2\lambda q = 0.8889$, $\lambda = 7$ and $\delta/R = 0.03$ which are values used in current rotor designs, $C_{T,\Delta\varphi} \approx 0.028$ so the thrust coefficient is somewhat higher than estimated in absence of $v_{\varphi,L}$. Apart from $C_{t,\Delta\varphi}$ the local thrust coefficient is constant.

The numerical results presented in the next section enable evaluation of the approximation $v_{\varphi,L} \approx q\frac{R}{2r}$ used in the expressions for $C_{T,\Delta\varphi}$ and $C_{t,\Delta\varphi}$. This appears to give a deviation $\Delta C_T = 0.0002$, and $\Delta C_t = 0.007$ close to the tip where $v_{\varphi,L}$ reaches a local minimum, see figure 5. The conclusion is that the approximation is well in place.

With (2), (5) and (6) the thrust and power coefficients are expressed in the independent parameters λ and q, and in the only unknown parameter $\overline{v_{z,L}}$. It is interesting to observe that (2) and (5) are identical to eq. 57 of Sørensen & van Kuik [7] except for the missing term $q^2/2$ accounting for the contribution of the vortex kernel. For the example values used here, this term contributes to C_T only 0.002 so is negligible indeed.

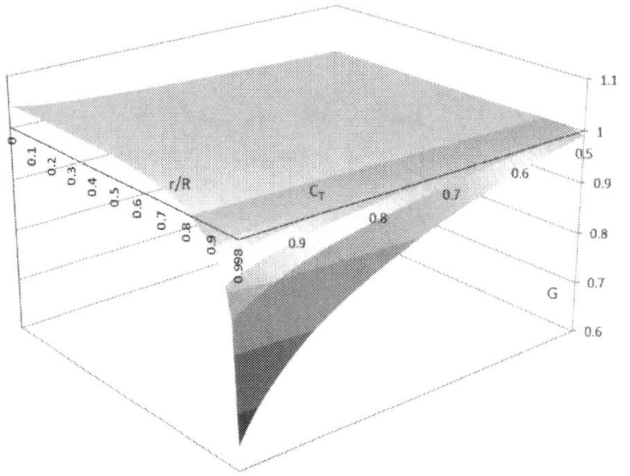

Figure 2. The velocity distribution function G (10) at the uniformly loaded actuator disc. As $\int_0^1 G d(r/R)^2 = 1$ the plane $G = 1$ corresponds to $\overline{v_{z,disc}}$ as obtained by momentum theory.

3. Momentum theory with two non-uniform v_z distribution functions

In BEM models without any tip correction, the distribution of the axial velocity is determined by applying the local momentum equation $C_t = 4v_{z,L}(1-v_{z,L})$, where the local thrust coefficient $C_t = L_z/\left(\rho U_0^2 \pi r\right)$. Branlard *et al.* [8] and Shen *et al.* [1] have evaluated existing tip corrections with emphasis on the Prandtl-Glauert correction F given by eq. (2) of [1]. Shen *et al.* [2] have adapted this correction resulting in the Prandtl-Glauert-Shen (PGS) method with an additional correction F_1 to account for local tip shape properties:

$$\left.\begin{aligned} F_1 &= \frac{2}{\pi}\cos^{-1}\left[\exp\left(-g\frac{B}{2}\left(\frac{R}{r}-1\right)^n \frac{1}{\sin(\theta)}\right)\right] \\ g &= \exp\left[\frac{0.125\,(B\lambda - 21)}{1-2k}\right] + 0.1 \\ n &= 1 + 0.5k \end{aligned}\right\} \tag{7}$$

where θ is the local inflow angle with $\sin(\theta) = v_{z,L}/\sqrt{(\Omega r - v_{\varphi,L})^2 + v_{z,L}^2}$. For $n = 1$ and $g = 1$ F_1 is identical to F. k is the minimum value of chord-to-radius derivative at the tip, with $k = -0.45$ used in Shen's tuning. The same value is used here. With B, λ and q given in table 1 F and F_1 are known except for the induction $a_L = (1 - v_{z,L}/U_0)$. As in BEM the second relation between the induction and F, F_1 is the local momentum equation, to be solved for a_L :

$$C_t F_1 = 4a_L F(1 - a_L F), \tag{8}$$

with C_t now given by (6) instead of blade element theory. The q^2 term in (6) is neglected.

Van Kuik & Lignarolo [4] have derived a function G representing the radial distribution of v_z of actuator discs with an axisymmetric uniform load, based on a potential flow model for the wake. Classical momentum theory gives the average value of $\overline{v_{z,disc}}$ but not the distribution. Figure 2 shows that the local axial velocity decreases towards the disc edge, instead of being constant as often assumed in momentum balances. The distribution is given by:

$$v_{z,disc}(r, C_T) = G(r, C_T)\overline{v_{z,disc}(C_T)} \tag{9}$$

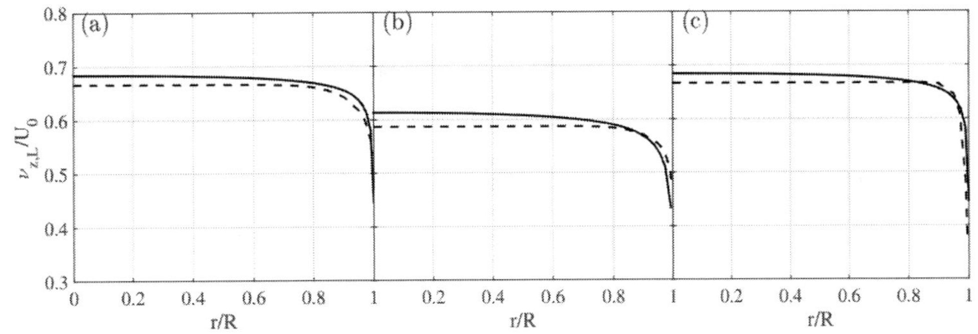

Figure 3. The axial velocity distribution obtained from the momentum theory + Prandtl-Glauert-Shen correction (dash line) and actuator disc function G (solid line). (a) Case 1, (b) Case 2 and (c) Case 3

with $\int_0^1 Gd(r/R)^2 = 1$ so the disc-averaged momentum theory result remains the same. G is approximated by the surface-fit to the numerical results of [4]:

$$\left.\begin{aligned} G(r, C_T) &= 1 + \alpha_1 \left(1 - 1.00076\left(1 - \left(\frac{r}{R}\right)^{\alpha_2}\right)^{0.0015}\right) \\ \text{with} \quad \alpha_1 &= 62.05(1 - C_T)^{0.42} - 47.56 \\ \alpha_2 &= 7 - 2.5C_T \end{aligned}\right\} C_T \geq 0.5 \tag{10}$$

$$G(r, C_T) = 1 \quad \} \, C_T < 0.5 \tag{11}$$

The local momentum equation to be solved is:

$$\left.\begin{aligned} C_t &= 4a(1 - a) \\ 1 - a_L &= G(1 - a) \end{aligned}\right\} \tag{12}$$

C_t is given by (6) with the q^2 term omitted as before in solving (8). The first equation gives the induction a as if the local and disc-averaged momentum equations are the same, the second equation gives the local value. Unlike the PGS correction model the axial velocity is higher than predicted by the unmodified momentum equation for $r/R < 0.8$. This is shown in figure 3 presenting $v_{z,L}$ as resulting from the PGS correction and G function for the 3 load cases considered. For $r/R > 0.8$ the actuator disc line G corresponds reasonably well with the PGS tip correction model.

4. Comparison of the Actuator Line and Lifting Line results

Before the relevance of $G(r, C_T)$ for a rotor with a finite number of blades is investigated, the Actuator Line and Lifting Line codes used for this purpose will be compared and validated. Figure 4 shows the wake development as calculated by both methods for cases 1 and 2. The wake contour and the downstream positions of the tip vortices agree reasonably well.

Figure 5 compares the velocity components at the actuator or lifting line. For all three cases the azimuthal and radial components are identical except near the tip vortex where the differences in de-singularization become visible. Also the axial velocity distributions in case 3 match very well, while these differ slightly in case 1 and 2. This is caused by the discretization of

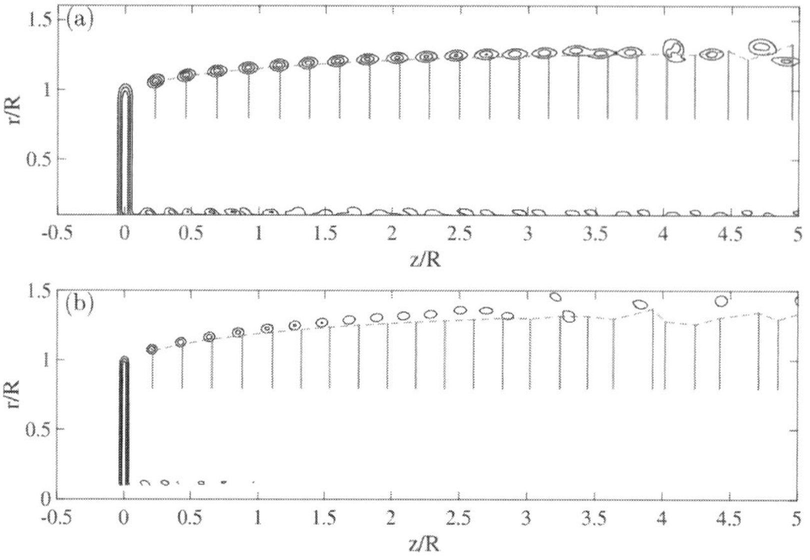

Figure 4. Azimuthal plane cut at the blade position $\phi = 0°$ for (a) case 1, (b) case 2. Case 3 is not shown as the AL-calculated vortices are not graphically discernible any more. Contours indicate the vorticity magnitude calculated from the AL approach. The dashed and solid red lines indicate the wake envelope and position of the vortex cores estimated from the LL approach.

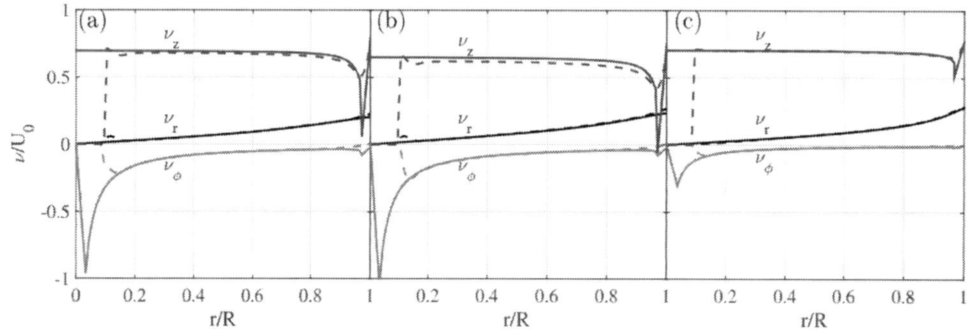

Figure 5. The velocity components at the actuator line (dash line) and lifting line (solid line) for (a) case 1, (b) case 2 and (c) case 3.

the tip vortices in the LL method, which is finer for increasing λ, as discussed in the appendix. As the correspondence of the results from both codes is quite good, no LL calculations with a finer discretization have been done.

Figure 6 shows the distribution of the axial velocity at the rotor plane. Again these distributions agree reasonably well except for $r < 0.1R$ and near the tip, due to differences in the desingularization of the vortices. The LL method assumes a Rankine vortex kernel with diameter δ, whereas in the AL approach the vortex structures are formed from a vortex sheet as a result of lift force distribution along the actuator line and the velocity fields are used to evaluate the vortex properties such as the vortex core size. With the LL method the sensitivity of the

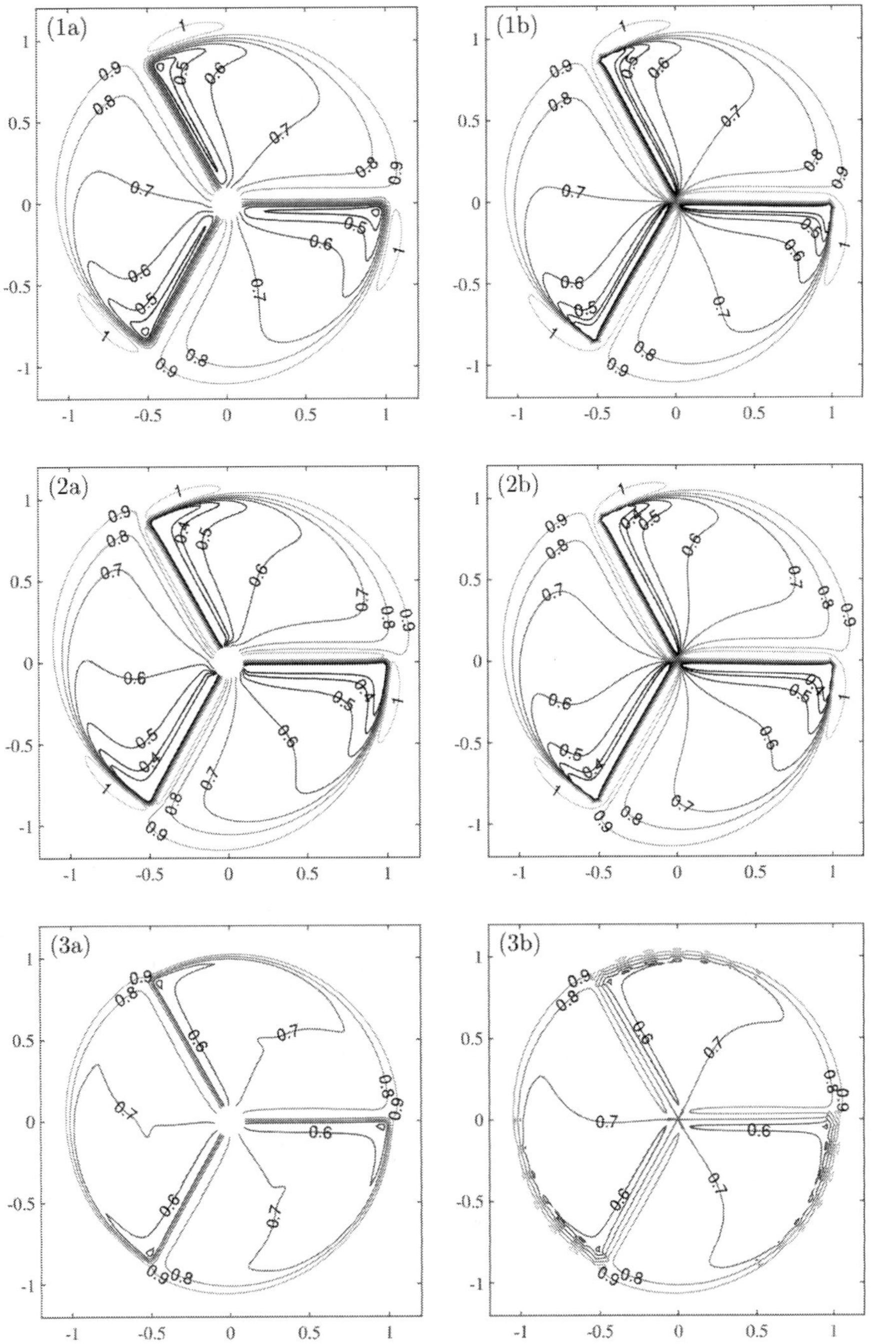

Figure 6. Distribution of v_z/U_0 at the rotor plane z=0 according to the actuator line (a), and lifting line (b) model, for flow case 1 (top), 2 (middle) and 3 (bottom).

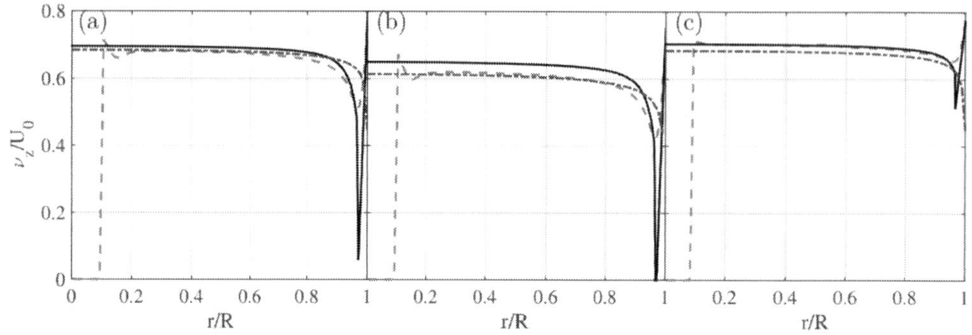

Figure 7. The axial velocity distribution at the actuator line (red dashed line), lifting line (black solid line) and actuator disc (blue dashed-dot line) for cases, (a) case 1, (b) case 2 and (c) case 3.

results for different δ has been tested. The distribution itself is invariant except for the area covered by the vortex kernel, as might be expected. The conclusion is that the de-singularization near the centre has no impact on the level and shape of the distributions. Figure 6, top row, corresponds well with figure 4(b) of Segalini *et al.* [9], showing the axial velocity for the same load case but with a different lifting line model.

5. Comparison of disc and rotor results

In figure 7 the distribution of the axial velocity of the actuator disc is compared with the distribution calculated by the AL and LL models at the blade position. For the inboard part, $r < 0.6R$, the disc distribution equals the AL distribution in cases 1 and 2, but is a bit lower than the AL and LL distributions in case 3. For this high λ case, the disc and AL-LL distributions match very well near the tip. For the lower λ cases 1 and 2, the AL and LL distributions correspond to the disc distribution at a slightly lower radius. This is to be expected since for a high λ the tip vortices are less strong compared to the low λ flow with equal thrust. The power and thrust coefficients according to the several methods are shown in table 2. The LL C_p values are consistently higher than the AL values, in accordance with the differences in axial induction.

Table 2. C_p and C_T as determined by eq.(2) respectively (5) with $\overline{v_{z,L}}$ from the AL and LL method, momentum theory plus PGS and G distributions, and the AL and LL simulation results, in % deviation from the actuator disc momentum theory values.

C_p	(2)+AL	(2)+LL	MT+PGS	MT+G	AL	LL
Case 1	-2.1	-2.0	-2.9	-0.3	-4.1	-2.0
Case 2	-0.4	2.2	-2.5	-1.4	-5.2	2.2
Case 3	4.0	3.2	-1.8	-0.3	1.3	3.4
C_T	(5)	(5)	MT	MT	AL	LL
Case 1	2.3	3.2	0.0	0.0	1.9	3.6
Case 2	3.0	3.5	0.0	0.0	2.1	3.9
Case 3	2.7	2.7	0.0	0.0	-0.1	0.4

The same holds for the momentum theory results obtained with the *PGS* or *G* function. The latter values are consistently higher since the average induction using the *PGS* function is higher compared to the *G* function values. With respect to C_T the analytical and AL-LL values are higher than the momentum theory values. The reason for this is the $C_{T,\Delta\varphi}$ term in (5). In particular case 2 shows $C_T > 1$ for the LL model. The AL and LL C_p values for case 3 are higher than the Betz-Joukowsky limit. Some of these observations are in line with the results obtained by Segalini *et al.* [9]. Figure 8 of this paper shows Lifting Line results for case 1 with $\delta = 0.03$, with C_T 3% higher than momentum theory gives, and C_p 4% lower. The results for C_p and C_T merit further investigation. However, this has no impact on the analysis of the present paper as this focuses on the non-uniformity of the velocities in the rotor plane as illustrated by the AL and LL results.

6. Evaluation and conclusions

The 2 research questions mentioned in the introduction lead to the following answers:

(i) The non-uniform induction at the actuator disc is accurately captured by the combination of momentum theory based on a uniform induction plus the correction by the Prandtl-Glauert-Shen model. In other words: when the non-uniform disc distribution is used as basis for BEM rotor design, the tip correction models do not need to correct for a higher induction - lower axial velocity near the tip. For the inboard region always $v_{z,disc} > v_{z,PGS}$.

(ii) The non-uniform distribution of $v_{z,disc}$ is reasonably well reproduced by the v_z distribution at the position of the bound vortex of a Joukowsky rotor, with the following remarks to be taken into account: the Actuator Line model and Lifting Line model show a very minor difference for the design thrust coefficient $C_T = 8/9$, $\lambda = 7$ and no difference for the same C_T, $\lambda = 20$ except at the tip itself. For $C_T = 0.97$, $\lambda = 7$ the difference in v_z is 5%. The shape of the disc distribution corresponds to the AL-LL distribution for $\lambda = 20$ while it corresponds to the AL-LL distribution for $\lambda = 7$ with a corrected, lower, radius.

A third conclusion is that the axial velocity field at the rotor disc calculated by the CFD and Potential Flow models are very similar, which confirms that the non-uniform distribution of the velocity does not depend on the type of model, but is an inherent feature of the flow.

The results for the Joukowsky rotor have shown that the non-uniform $v_{z,disc}$ distribution and the v_z distribution at the position of the blade of a rotor are similar, by which it is worth to consider the non-uniform $v_{z,disc}$ as basis for BEM, instead of a uniform $v_{z,disc}$ which needs a tip correction.

The physical or fluid dynamic origins of the Prandt-Glauert-Shen tip correction and the non-uniform velocity distribution G are different, so the good correspondence is remarkable. The tip correction as formulated by Prandtl is a correction of actuator disc results for the finite number of blades, by considering the non-uniform velocity field of the wake represented by helical vortex sheets. The effect of it is most noticeable near the edge of the streamtube, by which it has been named tip correction. Thereafter Prandtl's model has been adapted and calibrated by Glauert and Shen so it has become a calibrated engineering model for the flow non-uniformity. The good correspondence with the flow non-uniformity as calculated by actuator disc potential flow field solutions and with the Actuator-Lifting Line calculations may be considered as a confirmation of this calibration.

Acknowledgments

The Lifting Line simulations of TU-Delft were performed with financial support from the China Scholarship Council. The Actuator Line simulations of Uppsala University, campus Gotland, were performed on resources provided by the Swedish National Infrastructure for Computing (SNIC) within the project SNIC 2014/8-5.

Appendix

The Lifting Line model

The model is structured as shown in figure 1: the blades are modelled as bound vortices with constant strength Γ_L rotating with angular velocity Ω, the root vortex $\Gamma = B\Gamma_L$ is located at $r = 0$, the tip vortices with strength Γ_L are released at $r = R$ and are discretized by piecewise straight vortex elements. Using the Biot-Savart law for the induction by each element, the total induction at each point in space is calculated. The convection of the tip vortices is driven by the velocity acting at the elements, being the sum of the total induction and the free stream velocity. The kernel of all vortices is a Rankine vortex with radius δ.

At time $t = 0$ the first element of the B tip vortices is released, with new elements following after each time step $\Delta t = \Delta\varphi/\Omega$. The calculation proceeds until the wake has developed so far that the induced velocities at the rotor-plane do not change any more. This is checked by the convergence of C_T and C_p. The values for δ and $\Delta\varphi$ are given in table 3.

The Actuator Line model

The computational fluid dynamic computations are carried out using a three-dimensional flow solver developed by Sørensen and Michelsen [10, 11] called EllipSys3D. It solves the discritized incompressible Navier–Stokes equations using a finite volume approach. In computations, the flow around the rotor is simulated using the actuator line approach [12]. In this approach, the presence of the actual rotor is approximated by the body forces which are distributed radially among the computational grid points in a 3D Gaussian manner. In order to impose the constant circulation $\Gamma = L/\rho v_{rel}$ (L indicates the lift while v_{rel} is the relative velocity), the distribution of L along the actuator line is evaluated manually. The drag is considered to be zero.

The computations are performed using an axisymmetric polar grid with approximately 49 million grid points. In the vicinity of the actuator lines around 35 million grid points are used in order to resolve the near-wake dynamics. Each blade is represented by 91 grid points along the actuator lines. The simulations are conducted in a rotating reference frame.

Table 3. Numerical parameters in the computations. Lifting line: column 2 & 3 , Actuator Line 4 & 5

Case #	$\Delta\varphi$ (°)	δ/R	$\epsilon/\Delta r$	δ/R
1	5	0.03	4	0.08
2	5	0.03	2.5	0.05
3	10	0.05	2.5	0.05

References

[1] WZ Shen, R Mikkelsen, JN Sørensen, and C Bak. Tip loss corrections for wind turbine computations. *Wind Energy*, 8(4):457–475, 2005.
[2] WZ Shen, WJ Zhu, and JN Sørensen. Study of tip loss corrections using CFD rotor computations. *J. Phys. Conf. Ser.*, 555:012094, 2014.
[3] J Johansen, M Gaunaa, and C Bak. Design of a Wind Turbine Rotor for Maximum Aerodynamic Efficiency. *Wind Energy*, 12:261–273, 2009.
[4] GAM. van Kuik and LEM Lignarolo. Potential flow solutions for energy extracting actuator disc flows. *Wind Energy*, accepted pending a minor revision, 2015.
[5] NE Joukowsky. *Theory Tourbillionnaire de l'Helice Propulsive, Article I.* Gauthier-Villars, Paris, 1912.
[6] GAM van Kuik. The relationship between loads and power of a rotor and an actuator disc. *J. Phys. Conf. Ser.*, 555:012101, 2014.
[7] JN Sørensen and GAM van Kuik. General momentum theory for wind turbines at low tip speed ratios. *Wind Energy*, 14:821–839, 2011.
[8] E Branlard, K Dixon, and M Gaunaa. Vortex methods to answer the need for improved understanding and modelling of tip-loss factors. *IET Renewable Power Generation*, 7(4):311–320, 2013.
[9] A Segalini and PH Alfredsson. A simplified vortex model of propeller and wind-turbine wakes. *J. Fluid Mech.*, 725:91–116, 2013.
[10] JA Michelsen. Block structured multigrid solution of 2D and 3D elliptic PDE's, AFM 94-06. Technical report, Dept. Fluid Mechanics, Technical University Denmark, Lyngby, 1994.
[11] NN Sørensen. General purpose flow solver applied to flow over hills. *PhD thesis, Risø*, 1995.
[12] JN Sørensen and WZ Shen. Num. Modeling of Wind Turbine Wakes. *J. Fluids Eng.*, 124(2):393–399, 2002.

A wind-tunnel investigation of wind-turbine wakes in yawed conditions

Majid Bastankhah, Fernando Porté-Agel

Wind Engineering and Renewable Energy Laboratory (WIRE), École Polytechnique Fédérale de Lausanne (EPFL), EPFL-ENAC-IIE-WIRE, 1015 Lausanne, Switzerland

E-mail: majid.bastankhah@epfl.ch, fernando.porte-agel@epfl.ch

Abstract. Wind-tunnel experiments were performed to study the performance of a model wind turbine and its wake characteristics in a boundary layer under different operating conditions, including different yaw angles and tip speed ratios. High-resolution particle image-velocimetry (PIV) was used to measure the three velocity components in a horizontal plane at hub height covering a broad streamwise range from upstream of the turbine to the far-wake region. Additionally, thrust and power coefficients of the turbine were measured under different conditions. These power and thrust measurements, together with the highly-resolved flow measurements, enabled us to systematically study different wake properties.

The near-wake region is found to have a highly complex structure influenced by different factors such as tip speed ratio and wake rotation. In particular, for higher tip speed ratios, a noticeable speed-up region is observed in the central part of near wake, which greatly affects the flow distribution in this region. In this regard, the behavior of the near wake for turbines with similar thrust coefficients but different tip speed ratios can vary widely. In contrast, it is shown that the mean streamwise velocity in the far wake of the turbine with zero yaw angle has a self-similar Gaussian distribution, and the strength of wake in this region is consistent with the magnitude of the thrust coefficient.

With increasing yaw angle, as expected, the power and thrust coefficients decrease, and the wake deflection increases. The measurements also reveal that, in addition to turbulent momentum flux, lateral mean momentum flux boosts the flow entrainment in only one side of the wake, which results in a faster wake recovery in that side. It is also found that the induced velocity upstream of a yawed turbine has a non-symmetric distribution, and its distribution is in agreement with the available model in the literature. Moreover, the results suggest that in order to accurately predict the load distribution in yawed conditions, both normal and tangential (with respect to the rotor plane) components of the induced velocity upstream of the turbine should be taken into account.

1. Introduction

Wind-turbine performance and its wake characteristics depend on many factors such as the incoming atmospheric boundary-layer (ABL) and wind-turbine operating conditions [1]. Two important features of turbines' operation are yaw angle and tip speed ratio (i.e., the ratio between the speed of the tip blade and the velocity of the incoming flow), which both significantly affect the turbine performance and its wake structure.

Turbines working in yawed conditions have recently received considerable attention both as: (i) an unfavorable practical issue in the operation of wind turbines, (ii) and more importantly, as a favorable method to increase the power production of whole wind farm. The former reason is mainly associated with the inaccuracy of wind-direction measurements by the sensor installed on the turbine nacelle [2]. Consequently, it is quite common for turbines to work in yaw misalignments. Apart from this undesirable yaw situation, as mentioned earlier, controlling the yaw angle of turbines can be potentially used as a control method to improve the performance of the whole wind farm by deflecting the wake away from downwind turbines [3].

Several studies have been performed to investigate the turbine performance under these different operating conditions. Grant et al. [4] and later Grant and Parkin [5] did one of the first wind-tunnel experiments on the near wake (up to one rotor diameter) of a yawed turbine. Using flow visualisation methods, they observed helical vortex tubes downstream of the scaled-down turbine, and they also quantified the skew angle of the near wake for different yaw angles. Later, Medici and Alfredsson [6] characterized the wake of miniature yawed turbine in a few downstream positions. They found that a yaw turbine deflects the wake flow to one side, so it can be used as a potential way to control the turbine wake. Large-eddy simulation (LES) has been used by Jimenez et al. [3] to investigate the deflection of wakes of yawed turbines with different thrust coefficients. They pointed out that, for a given yaw angle, the wake deflects more for turbines with higher thrust coefficients. Recently, Krogstad and Adaramola [7] have controlled both the yaw angle and tip speed ratio of a turbine in their wind-tunnel studies, and examined the performance of the yawed turbine and its near-wake axial velocity. They found that the near wake is strongly affected by both the tip speed ratio and yaw angle.

It is important to note that the experimental studies mentioned above are limited to uniform inflow conditions. In the present work, the interaction between the wake of turbines under different conditions (e.g., yaw angle) and the turbulent boundary layer is studied, to our knowledge, for the first time. The results of this study will help us better understand and more accurately predict this complex interaction. To achieve this goal, detailed wind-tunnel measurements were carried out under different tip speed ratios and yaw angles for a model wind turbine placed in a neutrally stratified boundary layer. Highly spatially-resolved velocity measurements were performed in a broad streamwise range from upstream of the turbine to the far-wake region. Moreover, the main turbine characteristics (power and thrust coefficients) were measured.

In Section 2, the experimental setup is presented. The results are then shown and analyzed for different cases in Section 3. Finally, the summary and future research are presented in Section 4.

2. Experimental Set-up

Experiments were performed in the new closed-loop boundary-layer wind tunnel at the WIRE Laboratory of EPFL. The test section, designed for atmospheric boundary layer studies, is 2 m high, 2.57 m wide, and 28 m long. There is a contraction with a 5:1 area ratio upwind of the test section, and the tunnel is driven by a 130 kW fan. The turbulence intensity in the center of the wind tunnel (free stream) is lower than 0.1%. The turbulent boundary layer is naturally developed over the wind-tunnel floor thanks to the long test section. In the current study, the measurements were performed approximately 22 meters downstream of the test section entrance. The time-averaged incoming velocity at the hub-height of the turbine ($\overline{u}_h$) is kept approximately constant at 4.88 m/s. Figure 1 shows the vertical profiles of the mean velocity and turbulence intensity in the boundary layer in absence of the turbine. The boundary layer thickness is $\delta \approx 0.4m$ at the turbine location. The aerodynamic surface roughness length and the friction velocity were found to be $z_0 = 0.022mm$ and $u_* = 0.19m/s$, respectively, based on fitting a logarithmic velocity profile to the measured velocity profile in the surface layer (approximately lowest 15% of the boundary layer).

Two PIV setups were used to acquire detailed flow information. A high-resolution stereoscopic PIV (S-PIV) system from the LaVision company, called "*PIV-setup I*" in the following, was used to measure three velocity components downstream of the turbine in the horizontal plane at hub height. Two 29MP 12-bit CCD cameras (6600 × 4400 pixels) together with 105 mm lenses were installed on Scheimpflug mountings to maximize the focused area in the FOV. The FOV and spatial resolution are $4d \times 2.5d$ and $0.015d$, respectively, where d is the diameter of the wind turbine.

In addition to the mentioned S-PIV measurements, a high-resolution 2-Dimensional, 2-Components ($2D2C$) PIV system, called "*PIV-setup II*" later on in this paper, was used to

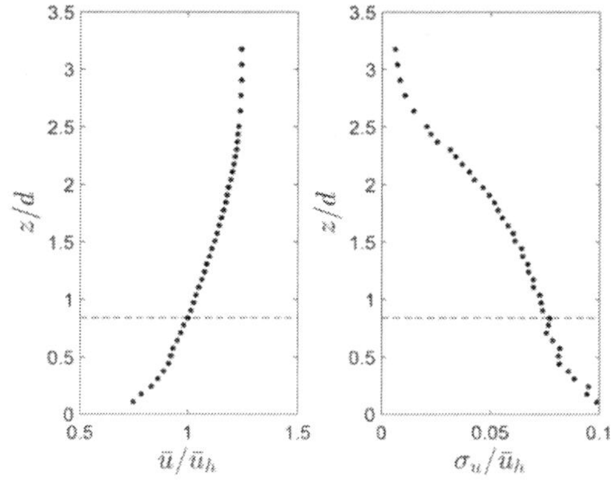

Figure 1. Characteristics of the turbulent boundary layer: normalized mean velocity (left), turbulence intensity (right). The *horizontal lines* represent the turbine axis.

measure two velocity components in a horizontal plane at hub height upstream and downstream of the model turbine. A 16-bit sCMOS camera (2560 × 2160 pixels) with a 105 mm lens were used to capture particle images in the horizontal plane. The field of view (FOV) and spatial resolution obtained by this PIV setup are $2d \times 1d$ and $0.013d$, respectively.

The model wind turbine used in this experiment was designed and built at the WIRE laboratory of EPFL. This horizontal-axis turbine is 3-bladed, with a diameter of $15cm$. The blade profile is a 5% thick plate with a 5% circular arc camber. The height of the turbine hub above the floor is $12.5cm$. The turbine rotor drives a small DC-generator to extract the energy from the wind.

The tower of the turbine was mounted on a strain gauge sensor to measure the thrust force exerted by the wind on the model turbine. The tip speed ratio was also varied by applying different electrical loads on the generator attached to the turbine rotor.

3. Results

3.1. Turbine characteristics

One of the key factors in turbine performance is the power coefficient (C_p) defined as: $C_p = Q\Omega/\left(0.5\rho\left(\frac{\pi}{4}d^2\right)\bar{u}_h^3\right)$, where Q is the torque generated by the rotor, Ω is the rotational velocity of turbine, ρ is the density of air and $\bar{u}_h$ is the mean incoming velocity at hub height. The graph of power coefficient versus tip speed ratio (λ) for different yaw angles (γ) is shown in Fig 2.a. For $\gamma = 0$, the maximum power coefficient is approximately 0.35 and occurs at $\lambda \approx 4$. The figure shows that a small yaw angle (e.g., $\gamma = 10$) slightly decreases the power production of the turbine. However, higher yaw angles (e.g., $\gamma = 30$) cause a dramatic drop in the power production. This is due to the reduction in the effective incoming velocity as well as the rotor frontal area for a yawed turbine [7]. To better compare the highest achievable power production in different yaw angles, Fig. 3 shows the maximum C_p for each yaw angle normalized with the maximum C_p at $\gamma = 0$. The figure includes, for comparison, the graphs of $cos^2(\gamma)$ and $cos^3(\gamma)$. One cannot find a unanimous statement on $C_p - \gamma$ curve in the literature. Based on the momentum theory [8] and some experimental studies (e.g., [7]), the maximum C_p versus γ has a shape similar to $cos^3(\gamma)$. On the other hand, other studies (e.g., [9]) proposed $cos^2(\gamma)$ shape for $C_p - \gamma$ curve. Fig. 3 shows that the maximum power production approximately varies as $cos^3(\gamma)$ in our wind-tunnel measurements.

Another key factor which is important to characterize the turbine performance is the thrust coefficient (C_T) defined as: $C_T = T/\left(0.5\rho\left(\frac{\pi}{4}d^2\right)\bar{u}_h^2\right)$, where T is the perpendicular force exerted on a turbine by the incoming wind. The variation of the thrust coefficient versus tip speed ratio for different yaw angles is shown in Fig. 2.b. The force measurements reveal that the thrust force decreases as the yaw angle increases, like the trend earlier seen for power production, which is in agreement with previous studies [7]. In general, it also shows that the value of thrust coefficient

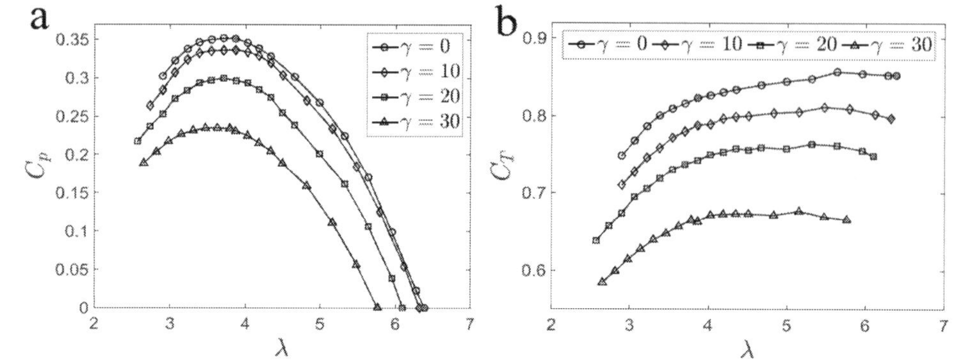

Figure 2. Turbine performance versus tip speed ratio for different yaw angles: (a) power coefficient (C_p), and (b) thrust coefficient (C_T). For red-colored points, PIV measurements were performed.

gradually increases by increasing the tip speed ratio.

3.2. Mean flow characteristics

The effect of tip speed ratio on turbine wake properties will be first presented, and then the effect of yawing the turbine on the flow both upstream and downstream of the model turbine will be shown. For each yaw angle, the flow measurements were performed for two different tip speed ratios: (i) The tip speed ratio of the turbine when the electrical circuit connected to the generator is open, so the turbine rotates freely. This tip speed ratio is called λ_f later on in this paper. (ii) The tip speed ratio at which the turbine has the maximum power production, and it is called the optimal tip speed ratio (λ_o) in the following. λ_f and λ_o for each yaw angle are indicated in Fig. 2 with red-colored points. As mentioned in section 2, two PIV systems were employed in this experiment. The *PIV-setup I* was used in three FOVs with some overlapping to capture the wake flow in a broad streamwise range from the near-wake ($0.4d$) to the far-wake region ($12d$). Note that in all the contour plots showing the results of *PIV-setup I*, overlapped locations are indicated by dashed lines. Additionally, the *PIV-setup II* was employed to quantify the wind flow in a smaller area around the turbine including upstream and very near wake of the turbine (from $-2d$ to $2d$). All the PIV measurements presented in this paper were carried out in the horizontal plane at hub height.

3.2.1. Effect of tip speed ratio on turbine wakes: Fig. 4 shows contours of normalized mean streamwise velocity ($\bar{u}/\bar{u}_h$) downwind of the turbine with zero yaw for the two aforementioned tip speed ratios. As seen in the figure, the center of the near wake region for the optimal tip speed ratio (λ_o) has higher velocity deficit compared to the one for the higher tip speed ratio

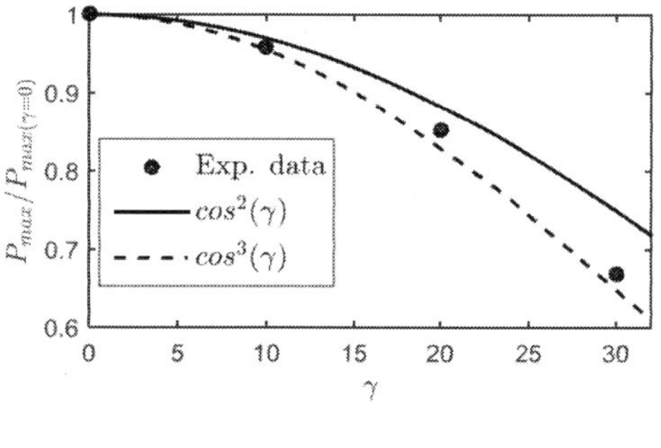

Figure 3. The maximum power production at different yaw angles normalized with the value corresponding to zero yaw angle. The *solid line* and *dashed line* represent $cos^2(\gamma)$ and $cos^3(\gamma)$ curves, respectively.

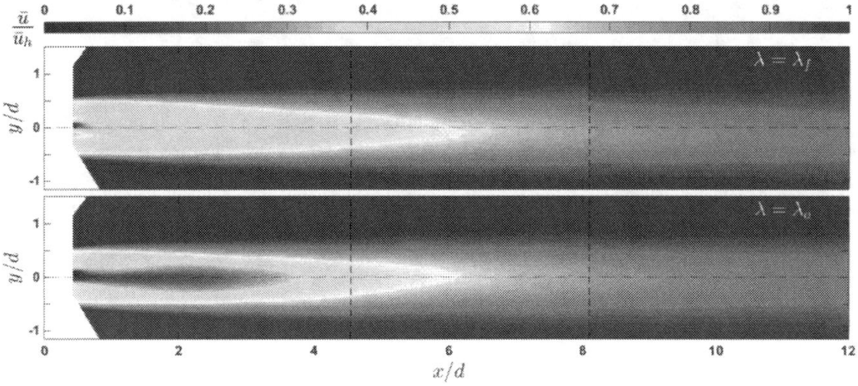

Figure 4. Contours of the normalized mean streamwise velocity ($\bar{u}/\bar{u}_h$) in the horizontal plane at hub height downwind of the zero-yawed turbine at: tip speed ratio of free-rotating turbine $\lambda_f = 6.34$ (top), the optimal tip speed ratio $\lambda_o = 3.9$ (bottom).

(λ_f). In other words, the near wake is surprisingly less pronounced for the case of higher tip speed ratio λ_f (i.e. higher blockage ratio and slightly higher C_T (see Fig. 2.b)). Consideration of the flow just behind the turbine might help us to explain the reason of this apparent paradox between the wake and force measurements. To do this, Fig. 5 shows contours of the normalized mean streamwise velocity ($\bar{u}/\bar{u}_h$) in the vicinity of the turbine for both tip speed ratios. From the figure, a speed-up region is found just downstream of the inner part of the rotor for both cases, although it is considerably stronger for the higher tip speed ratio (λ_f). Firstly, this is due to the fact that the wind velocity around the turbine hub increases based on conservation of mass. Secondly and more importantly, the angle of attack for the inner part of the blade in the case of higher tip speed ratio (λ_f) is very low or even negative [7]; this substantially intensifies the speed-up region in this case as shown in the figure. Further downstream, this high velocity region lessens the wake of the nacelle. As a result, the central part of the near wake region seems weaker for the higher tip speed ratio (λ_f) (Figs. 4 and 5) although C_T of the turbine is slightly higher in this case. It is also important to note that the central part of the rotor does

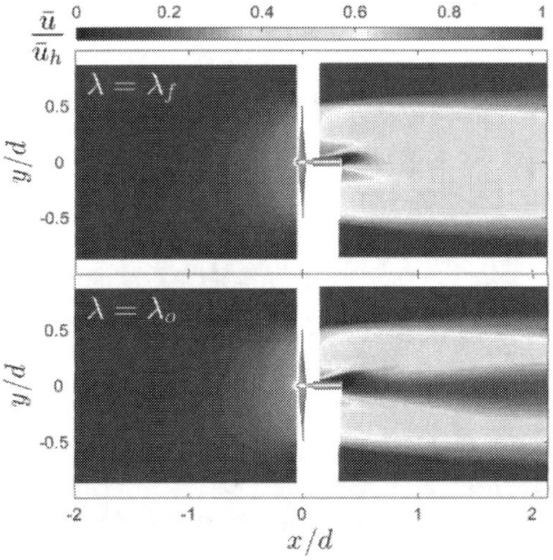

Figure 5. Contours of the normalized mean streamwise velocity ($\bar{u}/\bar{u}_h$) in the horizontal plane at hub height in the vicinity of the zero-yawed turbine at: tip speed ratio of free-rotating turbine $\lambda_f = 6.34$ (top), the optimal tip speed ratio $\lambda_o = 3.9$ (bottom).

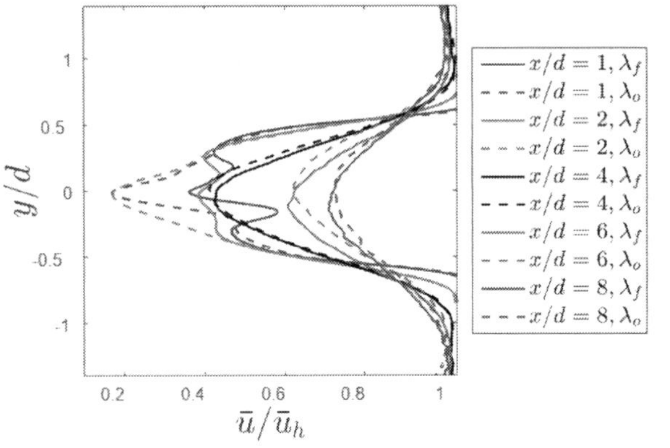

Figure 6. Lateral profiles of the normalized mean streamwise velocity ($\bar{u}/\bar{u}_h$) in the wake of zero-yawed turbine: solid lines correspond to the free-rotating turbine $\lambda_f = 6.34$, and dashed lines correspond to the optimal tip speed ratio $\lambda_o = 3.9$.

not make a significant contribution to the overall thrust force compared to outer part due to the large difference in their swept areas.

Despite the near-wake region, as seen in Fig. 4, the wake of the turbine in the far-wake region ($x/d > 4$), which is less sensitive to the rotor dynamics, is similar for the two considered tip speed ratios. To facilitate the quantitative comparison of the results, lateral profiles of the normalized mean streamwise velocity ($\bar{u}/\bar{u}_h$) in different positions downstream of the turbine (x/d = 1, 2, 4, 6 and 8) are shown in Fig. 6 for both λ_f and λ_o. As expected, the figure shows that the velocity profiles in the far-wake region ($x/d \geq 4$), unlike the near wake, have a Gaussian distribution. Furthermore, it can be seen in this figure that the minimum velocity in the far-wake region is slightly lower for the higher tip speed ratio (λ_f) despite having higher velocities in the center of near wake. This is consistent with our force measurements as C_T is found to be slightly higher in this case. Note that for the higher tip speed ratio (λ_f), the wake is slightly deflected to one side. This might be due to the interaction between the rotating wake of turbine and the incoming turbulent boundary layer.

In general, these wind-tunnel measurements confirm the well-known fact that although the overall thrust coefficient, along with the incoming ABL conditions, can represent the general characteristics of far-wake regions [10], it is not sufficient for near-wake predictions. In fact, our results show that relying only on the thrust coefficient can even yield completely untrue near-wake predictions. Detailed information regarding the rotor characteristics such as the operating tip speed ratio, blade geometry and twist distribution is indeed needed to systematically analyze and predict this very complex region [11].

One important feature of the near-wake region of wind turbines is the wake rotation. In a horizontal plane, the vertical component of velocity (w) shows rotation of the wake. Fig. 7 shows contours of normalized mean vertical velocity ($\bar{w}/\bar{u}_h$) in the same horizontal plane for two different tip speed ratios. As expected, the wake rotates in the opposite direction to that of the

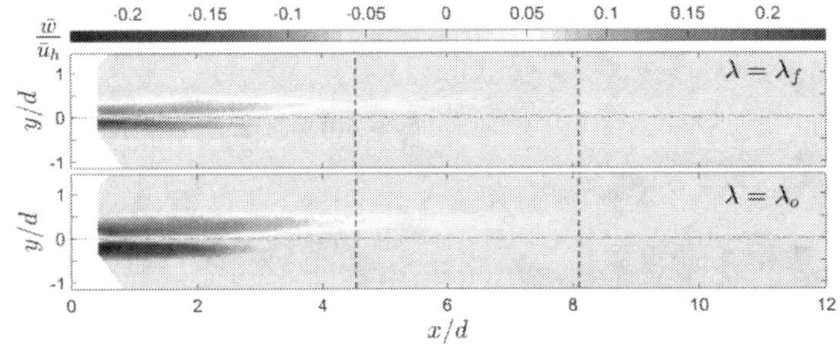

Figure 7. Contours of the normalized mean vertical velocity ($\bar{w}/\bar{u}_h$) in the horizontal plane at hub height downwind of a zero-yawed turbine at: tip speed ratio of free-rotating turbine λ_f = 6.34 (top), the optimal tip speed ratio λ_o = 3.9 (bottom).

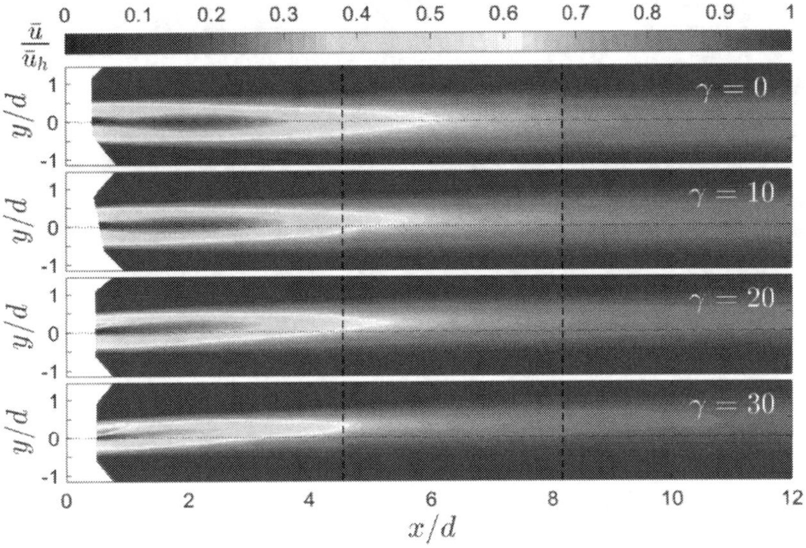

Figure 8. Contours of the normalized mean streamwise velocity ($\bar{u}/\bar{u}_h$) in the horizontal plane at hub height downwind of a turbine for different yaw angles ($\gamma = 0$, 10, 20 and 30).

blades due to to conservation of angular momentum [12]. Note that the turbine blades rotate in the counterclockwise direction, viewed from upwind of the turbine. This strong wake rotation, along with the presence of the turbulent boundary layer, makes the near wake region deviate considerably from axisymmetric behavior (see Fig. 6). With increasing downwind distance, however, the vertical velocity decreases due to wake recovery as shown in Fig. 7. In addition, it can be seen that the wake rotation is noticeably stronger for the optimal tip speed ratio (λ_o). This is due to the fact that the torque generated from the turbine is definitely higher in this case, so the turbine induces stronger rotation in the wake compared to the other case.

3.2.2. Effect of yaw angle on the flow upstream and downstream of the turbine: For each yaw angle, only the flow measurements for the optimal tip speed ratio (λ_o) are presented in this section. Figure 8 shows contours of normalized mean streamwise velocity ($\bar{u}/\bar{u}_h$) for different yaw angles ($\gamma = 0$, 10, 20 and 30) in the same horizontal plane. Based on momentum conservation, the wake deflection is caused by the lateral component of the force that turbine exerts on the air flow. As expected, it is clear that the wake deflection increases with increasing the yaw angle. In

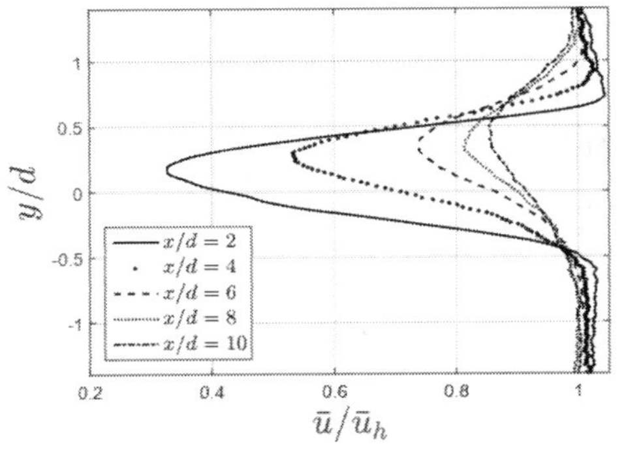

Figure 9. Lateral profiles of the normalized mean streamwise velocity ($\bar{u}/\bar{u}_h$) in the wake of a turbine with the yaw angle of 30° ($\gamma = 30°$).

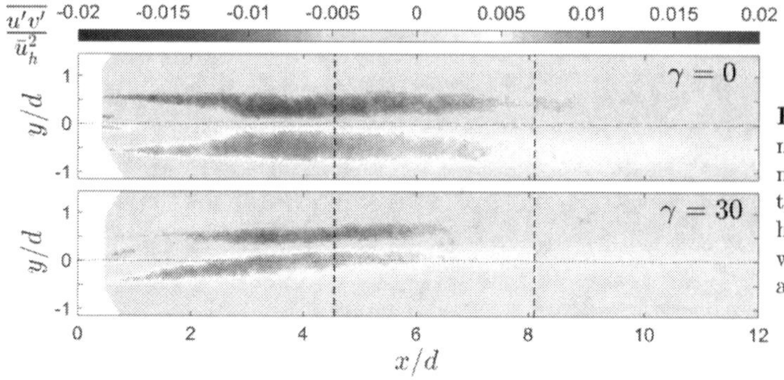

Figure 10. Contours of the normalized lateral turbulent momentum flux ($\overline{u'v'}/\bar{u}_h^2$) in the horizontal plane at hub height downwind of a turbine with: zero yaw angle (top), and 30° yaw angle (bottom).

addition, the figure shows that the wake is less pronounced as the yaw angle increases because of the reduction in thrust force of the turbine (Fig. 2.b).

To quantitatively investigate the wake of the yawed turbine, lateral profiles of normalized mean streamwise velocity ($\bar{u}/\bar{u}_h$) at different positions ($x/d = 2$, 4, 6, 8 and 10) downstream of a turbine with the yaw angle of 30° are shown in Fig. 9. As seen in the figure, one side of the wake (lower side in Fig. 9) is wider than the other side, and this difference is more obvious with increasing the downwind distance. This means that the wake recovery in one side is faster than the other side. Generally, the wake recovery for a turbine with zero yaw angle is associated with the turbulent momentum flux ($\overline{u'v'}$ in the horizontal plane) that transfers the energy from the outer flow into the wake region [13, 14]. For a yawed turbine, however, the mean components of velocity can also play a significant role as the wake is not parallel to the incoming flow. Thus, in addition to the turbulent momentum flux, the lateral mean momentum flux ($\bar{u}\bar{v}$) should be considered as an effective mechanism for the flow entrainment.

Figs. 10 and 11 show contours of normalized lateral turbulent ($\overline{u'v'}/\bar{u}_h^2$) and mean ($\bar{u}\bar{v}/\bar{u}_h^2$) momentum fluxes, respectively, for a turbine with yaw angles of 0° and 30° . For both yaw angles, as expected, Fig. 10 shows a strong enhancement in the magnitude of $\overline{u'v'}$ on both sides of the wake owing to the strong shear between the wake and the outer flow. Apart from the turbulent momentum flux, Fig. 11 shows that substantial entrainment flux caused by mean values of velocity ($\bar{u}\bar{v}$) exists only in one side of the wake of the turbine with the yaw angle of 30°. This can explain the non-axisymmetry of turbine wakes in yawed conditions shown in Fig. 9.

To illustrate the flow field in the vicinity of a yawed turbine, Fig. 12 shows the contours of normalized mean streamwise velocity ($\bar{u}/\bar{u}_h$) around the turbine with the yaw angle of 30°. Downstream of the turbine, the figure shows that the presence of the nacelle profoundly affects the central part of the wake. This can be explained by the increase in the frontal area of the

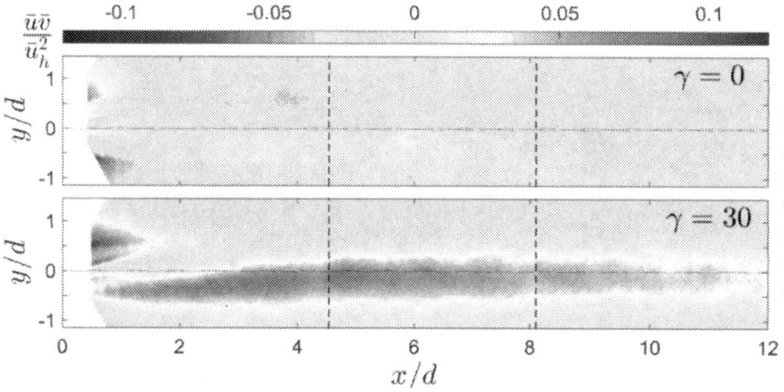

Figure 11. Contours of the normalized lateral mean momentum flux ($\bar{u}\bar{v}/\bar{u}_h^2$) in the horizontal plane at hub height downwind of a turbine with: zero yaw angle (top), and 30° yaw angle (bottom).

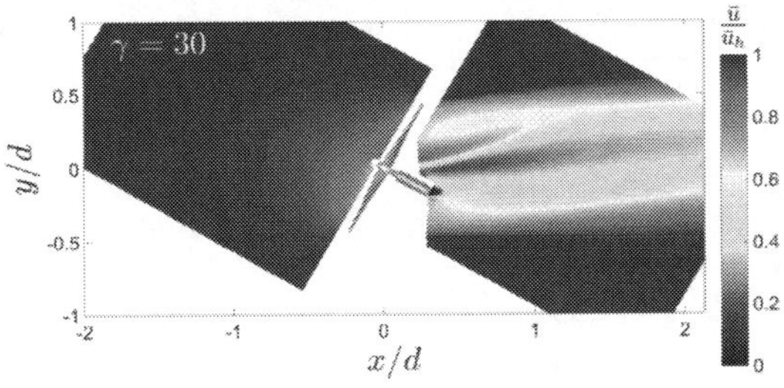

Figure 12. Contours of the normalized mean streamwise velocity ($\bar{u}/\bar{u}_h$) in the horizontal plane at hub height in the vicinity of the turbine with 30° yaw angle.

nacelle for a yawed turbine compared with the case of zero yaw.

Additionally, it can be seen that yawing the turbine induces a non-symmetric flow distribution upstream of the turbine. Glauert [15] addressed this non-symmetric distribution for the first time by studying the performance of an autogyro. Based on his model, the magnitude of induced velocity normal to the rotor plane is higher in front of the downwind side of a yawed rotor (in this paper, the side in the positive spanwise region) compared to the upwind one. Fig. 13 shows the vectors of the velocities induced by the turbine both in the normal and tangential directions (with respect to the rotor plane) for the turbine with the yaw angle of 0° (Fig. 13.a) and 30° (Fig. 13.b). The normalized mean streamwise velocity ($\bar{u}/\bar{u}_h$) is also shown as a contour plot. From the figure, the induced velocity in both normal and tangential directions for a turbine with zero yaw angle is approximately symmetric. For a yawed turbine, on the other hand, the induced velocities clearly have a non-symmetric distribution.

Next, these velocity measurements upstream of the turbine are used to assess the validity of Glauert's model [15]. Note that, however, the flow upstream of the central part of the rotor cannot be considered for this purpose as the flow is influenced by the presence of nacelle in this region. By considering the outer half of the rotor, it can be seen that the magnitude of the normal induced velocity is higher in front of the downwind side of the turbine, which is consistent with Glauert's theory. On the other hand, the contour plot in the figure shows that the streamwise velocity is lower in front of the upwind side of rotor (i.e., bigger projected induced velocity in the streamwise direction). This can be explained by the fact that the tangential induced velocity in front of turbine's upwind side decelerates the incoming flow, whereas it accelerates the incoming flow in front of the downwind side. This causes the higher induced velocity projected in the streamwise direction in front of the upstream side of rotor, although the normal induced velocity is lower there. This finding suggests that, in addition to the normal induced velocity, the induced velocity in the tangential direction must be taken into account in models that try to simulate turbines in yawed conditions (e.g., blade element-momentum (BEM) models).

4. Summary and future research

Wind-tunnel measurements were carried out to study the performance of a wind turbine and its wake characteristics under different yaw angles and tip speed ratios. In these experiments, a three-bladed horizontal-axis turbine was placed in the turbulent boundary-layer generated in the recirculating wind tunnel at the WIRE Laboratory of EPFL. High-resolution PIV was used to quantify the wind flow in the horizontal plane at hub height from 2 rotor diameters upstream of the turbine to 12 rotor diameters downstream. In addition to flow measurements, the thrust and power coefficients of the model turbine were measured under different conditions.

It is found that the near wake region is considerably influenced by different factors such as the tip speed ratio of turbine and the wake rotation. For higher tip speed ratios, as previously reported for bigger turbines in free-stream flows [7], a speed-up region exists downwind of the inner part of rotor. Due to the presence of this high-velocity region, the near wake of the turbine working at higher tip speed ratios seems weaker, although the thrust coefficient is higher in this case. This finding indicates that the overall characteristics of turbines such as thrust coefficient are not sufficient for near-wake predictions. The mean streamwise velocity in the far wake of the turbine with zero yaw angle, in contrast to the near wake, has a self-similar Gaussian distribution

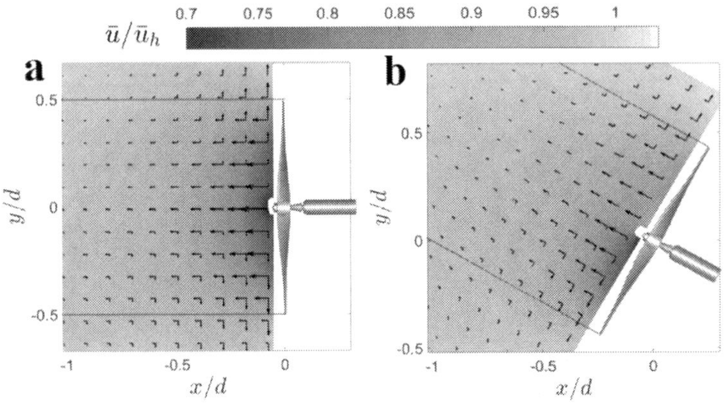

Figure 13. Vectors of mean normal and tangential velocities induced by the turbine in its upstream for: (a) zero yaw angle, and (b) 30° yaw angle. The background shows the contour plot of normalized mean streamwise velocity ($\bar{u}/\bar{u}_h$).

in the horizontal plane. The wake rotation is also found to be stronger when the turbine rotates at the optimal tip speed ratio with respect to the case without any electrical loads (i.e., freely spinning).

For a yawed turbine, as expected, both power production and thrust force of the model turbine decrease with increasing yaw angle. In particular, the maximum power production versus tip speed ratio varies approximately as $cos^3(\gamma)$, where γ is the yaw angle. As expected, the wake deflection increases with increasing yaw angle. Moreover, it is found that, in addition to turbulent momentum flux, lateral mean momentum flux greatly increases the wake entrainment only on one side of the wake of the yawed turbine. As a result, the wake recovers faster on that side.

Upstream of the yawed turbine, the induced velocities are non-symmetrically distributed in both normal and tangential directions (with respect to the rotor plane). Qualitative comparison between the measurements and the model proposed by Glauert [15], which addresses this non-symmetric distribution, shows an agreement.

In our future research, hot-wire anemometry will be used to obtain highly time-resolved measurements. That will enable us to study dynamic characteristics of turbine wakes such as tip vortices and wake meandering in different operating conditions. In addition, we will extend the analytical model proposed by Bastankhah and Porté-Agel [10] to include the effect of yaw angle on turbine-wake predictions.

References

[1] Vermeer L, Sørensen J and Crespo A 2003 *Progress in Aerospace Sciences* **39** 467–510
[2] Kragh K A, Hansen M H and Mikkelsen T 2013 *Wind Energy* **16** 353–366
[3] Jiménez Á, Crespo A and Migoya E 2010 *Wind energy* **13** 559–572
[4] Grant I, Parkin P and Wang X 1997 *Experiments in fluids* **23** 513–519
[5] Grant I and Parkin P 2000 *Experiments in Fluids* **28** 368–376
[6] Medici D and Alfredsson P 2006 *Wind Energy* **9** 219–236
[7] Krogstad P Å and Adaramola M S 2012 *Wind Energy* **15** 743–756
[8] Burton T, Sharpe D, Jenkins N and Bossanyi E 1995 *Wind energy handbook* 1st ed (Wiley)
[9] Pedersen T 2004 *Wind Energy* **7** 163–176
[10] Bastankhah M and Porté-Agel F 2014 *Renewable Energy* **70** 116–123
[11] Segalini A and Alfredsson P H 2013 *Journal of Fluid Mechanics* **725** 91–116
[12] Zhang W, Markfort C D and Porté-Agel F 2012 *Experiments in fluids* **52** 1219–1235
[13] Wu Y and Porté-Agel F 2012 *Energies* **5** 5340–5362
[14] Chamorro L and Porté-Agel F 2009 *Boundary-Layer Meteorol.* **132** 129–149
[15] Glauert H 1926 *A general theory of the autogyro* (HM Stationery Office)

Offshore wind farm flow measured by complementary remote sensing techniques: radar satellite TerraSAR-X and lidar windscanners

J. Schneemann[1], J. Hieronimus[1], S. Jacobsen[2], S. Lehner[2] and M. Kühn[1]

[1] ForWind-University of Oldenburg, Institute of Physics, Ammerländer Heerstr. 136, 26129 Oldenburg, Germany

[2] DLR-Maritime Safety and Security Lab, Henrich-Focke-Str. 4, 28199 Bremen, Germany

E-mail: [1]`j.schneemann@uni-oldenburg.de`, [2]`sven.jacobsen@dlr.de`

Abstract. Scanning Doppler lidar systems offer continuous wind measurements with some kilometres of range and a spatial distribution of concurrent measurements down to some metres. The synthetic aperture radar (SAR) satellite TerraSAR-X is capable to cover offshore areas of hundreds of square kilometres and to obtain wind data spatially distributed with some tens of metres. Images can be taken up to twice a day when the satellite passes the measurement site.

Simultaneous wind speed measurements with ground based scanning Doppler lidar and TerraSAR-X in the region of the offshore wind farm "alpha ventus" in the German North Sea were collected. A comparison of both systems in free stream conditions is performed by extrapolating the lidar data to the measurement height of the radar satellite assuming a logarithmic wind profile. In wake conditions the wake tracks obtained by lidar and TerraSAR-X are compared. In free stream conditions the comparison reveals a mean absolute wind velocity difference ≤ 0.4 m/s in two of the four considered cases and 1.1 m/s in one case. The fourth case shows a bad agreement due to a unusually low radar backscatter in the satellite's measurement. In wake conditions the wind turbine wakes could be tracked in the lidar and the satellite data. The comparison for the considered case reveals similar wake tracks in principle, but no matching due to the time difference of the measurements and the lower spatial resolution of the radar measurements.

1. Introduction

Remote sensing technologies offer the opportunity to study atmospheric quantities like the wind velocity over large areas. Especially Doppler lidar has become an important tool for wind energy related research in the last years with application onshore [1, 2] and offshore [3, 4]. Ground-based lidar windscanners enable the measurement of vertical profiles or horizontal planar scans of the radial wind speed within a range of several kilometres. Typical lengths of sampling volumes in radial direction (range gates) are about 30 m. Dynamic flows can be resolved with a spatial distribution of range gates of some metres. Scan durations take seconds to minutes depending on the used settings. Investigations on wind turbine wakes using scanning lidars onshore [5, 6] and offshore [7] have been reported.

Another remote sensing technology to measure wind speeds is space-born synthetic aperture radar. The German synthetic aperture radar satellite TerraSAR-X (TS-X) scans the surface of the earth with image resolutions of down to one meter. Using the intensity of the backscattered signal from the sea surface the wind field at 10 m height is calculated using an empirical geophysical model function. The obtained wind fields can have resolutions as high as a few tens of meters.
Results have already been validated against buoys equipped with anemometers at low measurement heights and the results of the German Weather Service (DWD) atmospheric model and have proven an accuracy of 1.5 m/s [8]. Wakes of offshore wind farms have already been investigated using SAR satellites [9, 10].
To our knowledge ground based lidar and radar satellite have been validated only against single point measurements or weather models, a direct comparison of the methods against each other has not been performed so far.
The contribution presents the comparison of concurrent measurements of the wind field in and around the offshore wind farm "alpha ventus" in the German North Sea taken with TerraSAR-X and a lidar system. The analysis is done separately for areas with undisturbed flow and those influenced by wakes. To obtain comparable data, the lidar measurements are extrapolated to the 10 m height level of the TS-X wind speed, using a logarithmic wind profile. Both data sets are interpolated to a reference grid for comparison.

2. Methods

2.1. Lidar measurements

Three scanning long range Doppler lidar systems of type Leosphere Windcube200S were operated in the offshore wind farm "alpha ventus" in the German North Sea. One lidar was located on the wind farm's sub station, two were positioned on the nearby research platform and meteorological mast "Fino1". The layout of the wind farm is shown in Figure 1. The lidar systems are described in [6].

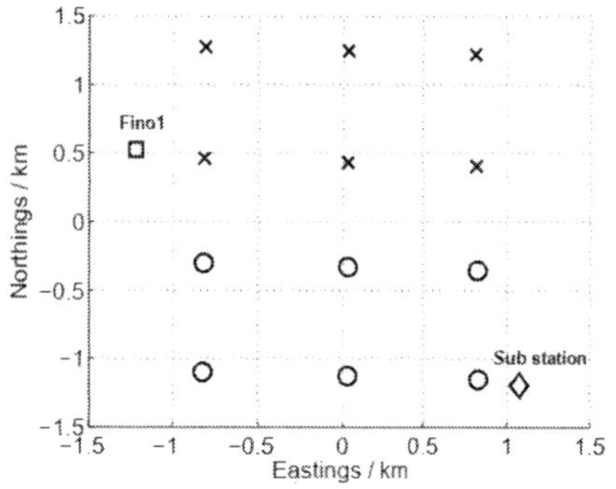

Figure 1. Layout of the offshore wind farm "alpha ventus". The 60 MW wind farm consists of six 5 MW wind turbines Senvion 5M (×) with a rotor diameter D_S = 126 m (two northerly rows) and six 5 MW wind turbines AREVA Wind M5000 (○) with D_A = 116 m (two southerly rows). One lidar was operated on the sub station (◇) and two lidars on the research platform "Fino1" (□).

All lidar systems were operated in the so-called PPI (plan position indicator) mode, scanning a wide azimuth range using a small fixed elevation angle $< 1°$ leading to increasing measurement heights with distance. The scanning speed was set to 1°/s with an averaging time of 1 s. All systems scanned the whole azimuthal range excluding some sectors with blocked sight. Each lidar recorded 198 range gates equally spaced from 100 m to 6000 m range. The pulse length of the lidar systems was set to 400 ns corresponding to a probed length of approx. 75 m.
The filtered line of sight wind (LoS) speeds of one lidar PPI scan are plotted in Figure 2, the distribution of the LiDAR measurement points over height are shown in Figure 4. For the

comparison with TS-X in free stream conditions data from the lidar located on the sub station of the wind farm is considered.

2.2. *Radar satellite TerraSAR-X measurements*

The X-band synthetic aperture radar (SAR) satellite TerraSAR-X sends radar pulses to the earth's surface. From the ocean, the radar pulse is returned to the sensor mainly by Bragg-Scattering from capillary waves in the centimetre scale. These small scale waves are directly related to the wind flow and therefore it is possible to relate the radar back scatter and the 10 m surface wind empirically via geophysical model functions (GMF). However, the GMF coefficients need to be tuned for every satellite, which is an extensive and cumbersome process and which demands a large and representative dataset. The empirical wind algorithm for TS-X is called XMOD2 and it has been successfully validated against buoy measurements [8]. An example of a wind field calculated from a TS-X image at "alpha ventus" is shown in Figure 3. For the measurement campaign radar images were taken simultaneously to the lidar measurements on four different dates in January 2014. As TS-X operates on a dawn/dusk orbit, measurements were done around 6:00 (UTC) for three dates and 17:30 (UTC) for one date, respectively. From the radar images, wind fields with a spatial resolution of 60 m have been calculated using the XMOD2 algorithm.

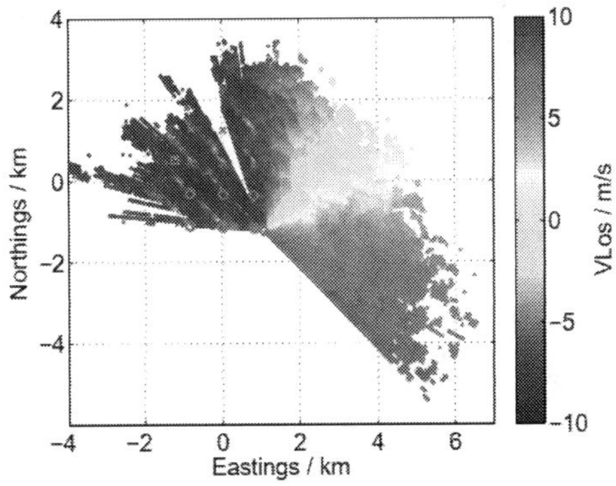

Figure 2. Line of sight wind speeds scanned by one long range lidar located on the sub station in "alpha ventus".

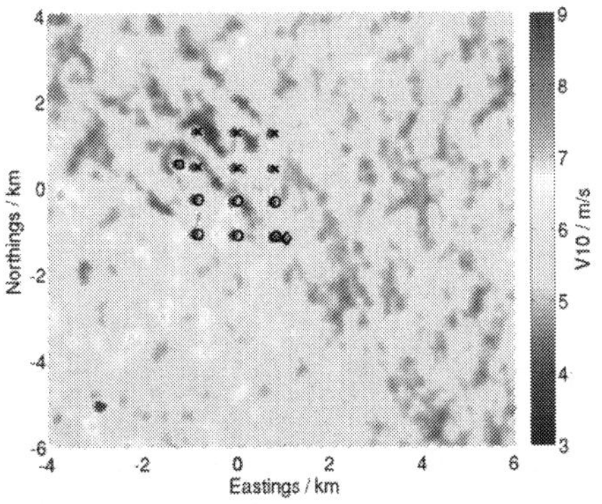

Figure 3. Image of the 10 m wind field in the "alpha ventus" region calculated with XMOD2 from an image of the radar satellite TerraSAR-X.

2.3. *Data analysis*

To compare both measurement systems the data has to be preprocessed. Particularly a calculation of absolute horizontal wind speeds from the lidar data and an extrapolation to the 10 m height of the TS-X data is performed. Before the comparison a linear interpolation of both data sets to a standard grid is done.

All lidar data is filtered for bad CNR (carrier to noise ratio) values and a maximum time lag to the measurement of TS-X of 120 s first. Some corrupt data points e.g. at the location of hard targets are removed by hand. Furthermore the field is divided to be in free stream or to be affected by the wind farm wake dependant on the wind direction.

For the data analysis, the atmospheric conditions around the wind farm "alpha ventus" are

needed. Since the meteorological mast "Fino1" was in the wake of the wind farm in all situations considered here, the parameter for the assumed logarithmic atmospheric boundary layer z_0 (c.f. Equation 2), the ambient wind direction and wind speed were taken from the operational analysis data from the COSMO-DE numerical weather prediction model. COSMO-DE uses a time step of one hour and a spatial resolution of 2.8 km [11], so just one grid point located in the wind farm region is considered.
To calculate the absolute horizontal wind from the lidar's line of sight measurement, all measured LoS wind speeds are transformed to the mean horizontal wind direction obtained from COSMO-DE using

$$v = v_{LoS} / \cos\alpha \tag{1}$$

with α being the angle between the LoS direction and the wind direction. The error of this method increases with the angle α and therefore sectors with $|\alpha| > 40°$ are not used in the further analysis. A correction for the small elevation angle is not performed. Next all measured lidar data is transferred to a planar polar grid with centre at the lidar's position and a resolution of 60 m in radial and 1° in azimuthal direction. In each grid point the average height of the measured range gates and the corresponding average horizontal wind speed v are calculated. Then all wind speeds are extrapolated to the height $z = 10$ m using the logarithmic wind profile for neutral conditions

$$v(z) = v_m \frac{\ln z/z_0}{\ln z_m/z_0} \tag{2}$$

with height z, the measurement height z_m, the wind speed in measurement height v_m and the surface roughness z_0. Offshore, the surface roughness is dependant on the swell. Typical values range from $z_0 = 1 \cdot 10^{-5}$ m to $z_0 = 1 \cdot 10^{-3}$ m [12].
The model used in Equation 2 is valid for a neutral stratification of the atmosphere. Due to a lack of dependable data neither of the actual stratification nor the predominant wind profile, it was chosen for all considered cases. To compare this assumption against measured data Figure 4 illustrates suitable lidar data available for heights from approx. 30 m to 90 m obtained from lidar and TS-X for free stream conditions on the 19.01 and 22.01. Lidar values are grouped in bins with 5 m height and for each bin mean value and standard deviation are calculated. The neutral logarithmic profile (c.f. Equation 2) is plotted using the bin average at 57.5 m height as reference.
The high variation in wind speed of the lidar data is supposed to result mainly from the spatially distributed measurements in the variable flow field. The spreading is increased by the error occurring from the transformation of the LoS measurement to the absolute wind speed v according to Equation 1 which increases with α.
Of course the used model could just be a first approximation. Nevertheless, in this case a check of the used logarithmic model is not possible since no data in the most critical heights regarding wind shear was available. In future applications the predominant wind profile should be measured directly especially at lower heights.

For comparing wind speeds of lidar and TerraSAR-X in free stream conditions, the spatial average and standard deviation of the extrapolated wind speeds of each system on the standard grid are calculated. Additionally the mean difference of lidar and TS-X wind speeds is determined. Furthermore lidar and radar satellite are compared to the meso scale weather model COSMO-DE.
In wake conditions the comparison has to be performed differently. The logarithmic wind profile used to transfer the wind speeds of the lidar to the height of the TS-X values does not hold valid in wake conditions. So a comparison of wind speeds is not meaningful. Therefore the analysis is limited to the comparison of wake positions. For that purpose lidar and TS-X data are interpolated to a Cartesian grid covering the wake area downstream of the turbine. The

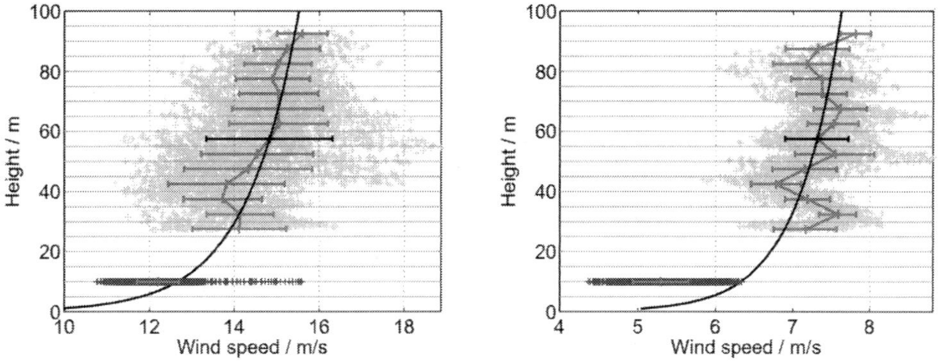

Figure 4. Vertical distribution of all available wind speed values measured during one lidar scan (green) and the corresponding TS-X measurement (blue) on the 19.01 (left) and 22.01 (right). Only data from free stream sectors are considered. Mean values and standard deviations of the lidar wind speed values are calculated in 5 m height intervals (red markers and lines). A neutral logarithmic profile according to equation 2 with z_0 taken from COSMO-DE and mean value at 57.5 m height taken as reference is plotted (black line).

maximum deficit of the wake is searched in several distances down stream fitting Gaussians to lines perpendicular to the wake.

3. Results and discussion

3.1. Free stream

Figure 5 illustrates the absolute horizontal wind speed in 10 m height on the standard grid obtained from lidar and radar satellite. It can be directly seen, that the mean wind speed calculated from the lidar measurement (6.4 m/s) is about 1.1 m/s higher than the one from TS-X (5.3 m/s). This is assumed to result mainly from the extrapolation of the lidar measurements to 10 m height, since no information on the predominant wind profile was available. The standard deviations of lidar and TS-X compare well with 0.39 m/s and 0.36 m/s respectively. Aside the speed difference most of the structures in the flow can be found to be very similar in both plots. The TS-X wind field looks a little bit smoother than the one from the lidar due to the lower spatial resolution of the original data being interpolated on the standard grid. These promising results benefit from the low time difference in the regarded case of ≤ 30 s.

Figure 6 summarizes the results found in four different concurrent measurements from the lidar system located on the sub station in "alpha ventus" and TS-X. On the 16.01. and 19.01. a very good agreement of less than 0.5 m/s mean difference between lidar and TS-X is found. The difference obtained on the 22.01. is still on a very good level of approximately 1.1 m/s. On the 17.01. the mean difference reaches a value of almost 4 m/s. The mean difference between lidar and TS-X is positive in all cases, meaning that the lidar measures higher averaged wind speeds in all cases.

The comparison of the mean lidar wind speed to the meso scale weather model COSMO-DE reveals negative values in all four cases, meaning higher wind speeds obtained by COSMO-DE than with the lidar. Differences are about 1.5 to 1.8 m/s, on the 19.01. the difference is lower than 0.5 m/s. Reasons for the deviations are seen in the time lag between lidar and COSMO-DE and in the assumption of a logarithmic wind profile in all cases.

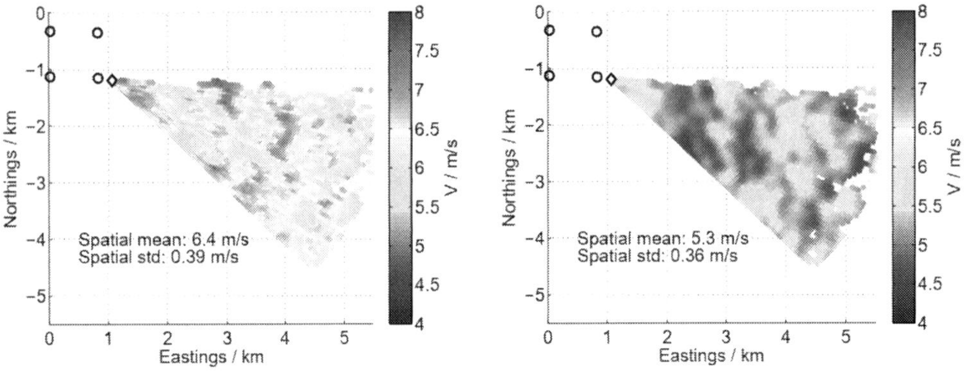

Figure 5. Absolute wind fields in 10 m height upstream of "alpha ventus" with south eastern wind conditions on the standard grid measured on the 22.01.2014. The time difference between lidar and radar measurement was ≤ 30 s. *Left*: Lidar-LoS measurements are transformed back to be the absolute wind speed in the ambient wind direction taken from the meso scale weather model COSMO-DE and extrapolated to 10 m height using a logarithmic wind profile. *Right*: TerraSAR-X wind speeds. The wind direction passed to the algorithm XMOD2 is taken from the meso scale weather model COSMO-DE.

A big deviation between lidar and TS-X and between TS-X and COSMO-DE on the 17.01. is observed, while lidar and COSMO-DE compare well. This seems to originate from an unusually low radar backscatter, which results in a lower derived wind speed. The reasons for the low backscatter level might be related to the ambient atmospheric conditions, but need to be further investigated in detail.

The remaining offset between lidar and TS-X can result from the transformation of the measured lidar data to the height of the TS-X measurement. The used logarithmic wind profile works in neutral conditions but introduces an error in stable or unstable stratification of the atmosphere. As a future application long range scanning lidar data measured on 10 m height or alternatively with the knowledge of the prevailing wind profile could serve to further optimize the geophysical model used to calculate the 10 m wind from the TS-X radar backscatter and to reduce the systematic error of the SAR measurement.

3.2. Wake

The comparison of wake tracks observed with lidar and TS-X is presented in Figure 7. The magnitude of the horizontal wind speed of the lidar is shown together with the wake tracks calculated from the lidar data and the corresponding TS-X data. Measurement times of both data sets on 22.01.2014 differ $[90 \pm 4]$ s from each other. The height of the lidar's measurement points reduces from 40 m at 4 D to 28 m at 16 D. The wake of turbine AV12 measured by the lidar can be nicely seen with a speed deficit of up to 60% at 4 D to 7 D and a recovery to a deficit of approximately 25% in 9.5 D. At approximately 10 D the wake partly hits turbine AV08 and merges with the new wake. The wake tracks of lidar and TS-X deviate from each other up to 2 D but lay in the same region in principle. With the resolution of 60 m in the horizontal plane TS-X resolves just about two data points per rotor diameter D at the regarded turbines ($D_A = 116$ m and $D_S = 126$ m). However, TS-X can yield wind fields of higher spatial resolution, but with drawbacks in the wind speed accuracy. So the localization of the wake tracks is limited

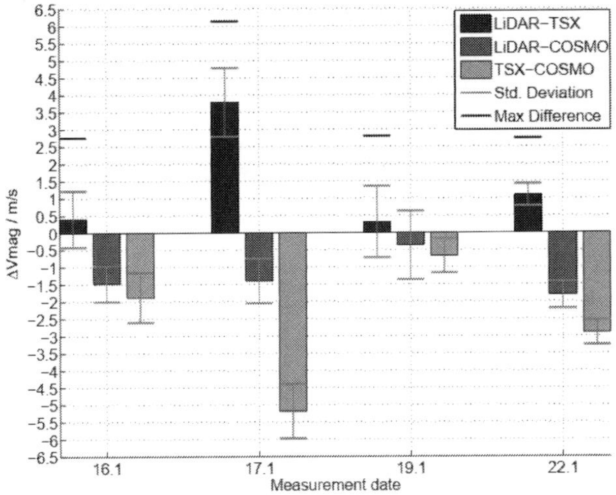

Figure 6. Comparison of absolute wind speeds obtained by lidar, TerraSAR-X and COSMO-DE. Mean difference, standard deviation and maximum deviation are plotted.

by the spatial resolution of the chosen dataset. Another reason for the differences is the time lag between lidar measurement and TS-X image of about 90 s. Furthermore the fundamental differences in the measurement methods have to be considered. While the lidar measures one component of the actual wind speed in a specific location, TS-X uses a tuned model function to calculate a wind speed from the capillary wave induced backscattered intensity on the sea surface. This makes the radar image prone for other influences on the sea surface like strong rain, oil spills, etc. While TS-X takes an image in about 1 s, a scan of the lidar takes up to some minutes depending on the area coverage. So there are considerable differences in the location, the duration and time difference and the principle itself of the measurements. Regarding this, the results obtained are quite promising.

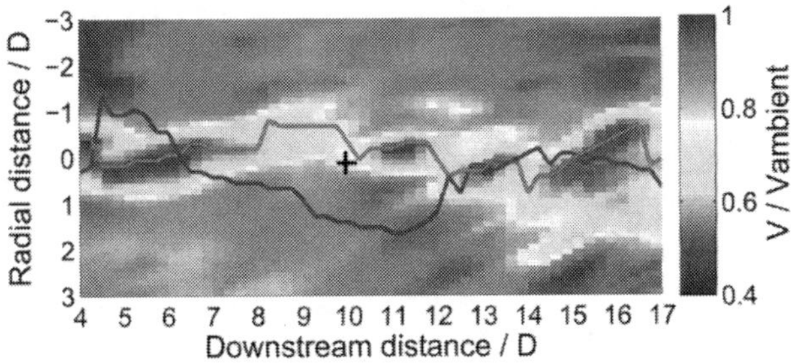

Figure 7. Normalized horizontal wind speed obtained from lidar measurements (color coded). The centre of the wake track is marked for the lidar measurement (red line) and the corresponding TS-X measurement (blue line). The time lag between both measurements is [90 ± 4] s. The position of turbine AV08 is marked (+). Both axis are normalized with the rotor diameter D_A of the wake generating turbine AV12 located at 0 D.

4. Conclusions

In free stream conditions a comparison of wind speeds between lidar and radar satellite is performed. Good agreement in the difference of the mean wind speed of both systems in the magnitude of less than 0.5 m/s is observed for two of the regarded four cases. This remaining offset of higher lidar wind speeds could be related to the assumption of the neutral stratification of the atmosphere for extrapolation. With information about the real wind profile obtained from meteorological measurements or from a lidar profiler another model can be used to transform the lidar wind speed to the level of the TS-X wind field in future campaigns. The comparison of the variability of the wind field characterized by the standard deviation reveals a good agreement. A parallel view of the wind fields of lidar and TS-X shows similarity of most flow structures. The comparison of wake tracks revealed low spatial agreement, but looked similar in principle. Due to the lower resolution of the TS-X wind field and the possibility to cover large areas, this method appears to be better suited to study the wake of whole offshore wind farms on a bigger scale. Future measurement campaigns will be used to compare the wake of a whole offshore wind farm measured with lidars to the ones calculated from TS-X data.

5. Acknowledgements

The lidar measurements in the offshore wind farm "alpha ventus" were carried out in the frame of the RAVE (Research at Alpha VEntus) research project GW Wakes, funded by the German Federal Ministry for Economic Affairs and Energy (BMWi) based on a decision of the Parliament of the Federal Republic of Germany (grant number 0325397A).
We thank Deutscher Wetterdienst (DWD) for providing analysis data.

References

[1] Trujillo J J, Bingöl F, Larsen G C, Mann J and Kühn M 2011 *Wind Energy* **14** 61–75 ISSN 1099-1824
[2] Käsler Y, Rahm S, Simmet R and Kühn M 2010 *J. Atmos. Oceanic Technol.* **27** 1529–1532
[3] Peña A, Hasager C B, Gryning S E, Courtney M, Antoniou I and Mikkelsen T 2009 *Wind Energy* **12** 105–124
[4] Cañadillas B, Westerhellweg A and Neumann T 2011 *DEWI Magazin* **38** 58–64
[5] Aitken M L, Banta R M, Pichugina Y L and Lundquist J K 2014 *J. Atmos. Oceanic Technol.* **31** 765 – 787
[6] Trabucchi D, Trujillo J J, Schneemann J, Bitter M and Kühn M 2014 *Meteorologische Zeitschrift*
[7] van Dooren M, Trabucchi D, Beck H, Schneemann J, Friedrichs W and Kühn M 2015 *Proceedings of EWEA Offshore, Copenhagen*
[8] Li X and Lehner S 2014 *Geoscience and Remote Sensing, IEEE Transactions on* **52** 2928–2939 ISSN 0196-2892
[9] Li X and Lehner S 2013 *Selected Topics in Applied Earth Observations and Remote Sensing, IEEE Journal of* **6** 1757–1768 ISSN 1939-1404
[10] Christiansen M B and Hasager C B 2005 *Remote Sensing of Environment* **98** 251 – 268 ISSN 0034-4257
[11] Baldauf M, Förstner J, Klink S, Reinhardt T, Schraff C, Seifert A and Stephan K 2014 Kurze Beschreibung des Lokal-Modells Kürzestfrist COSMO-DE (LMK) und seiner Datenbanken auf dem Datenserver des DWD. Techreport Deutscher Wetterdienst Version 2.3
[12] Emeis S 2013 *Wind Energy Meteorology* Green Energy and Technology (Springer Berlin Heidelberg) pp 95–133 ISBN 978-3-642-30522-1

Journal of Physics: Conference Series **625** (2015) 012016 doi:10.1088/1742-6596/625/1/012016

Study of wake meandering by means of fixed point lidar measurements: Spectral analysis of line-of-sight wind component

Davide Trabucchi, Gerald Steinfeld, David Bastine, Juan-José Trujillo, Jörge Schneemann and Martin Kühn

ForWind-University of Oldenburg, Institute of Physics, Ammerländer Heerstr. 136, 26129 Oldenburg

E-mail: davide.trabucchi@uni-oldenburg.de

Abstract. The validation of dynamic wake meandering models by full field measurements is an important but challenging task. Recently a new approach has been proposed, where a long range lidar was employed to analyse the power spectral density of the line of sight wind component measured across the wake trajectory. The method is promising, but the number of useful time series within a measurement campaign depends to a large extent on a favourable geometrical setup of the lidar position and the wind direction. In the first part of the paper the approach is further investigated. To this avail, lidar simulations based on large eddy simulation results are analysed. In the second part, the approach is applied to real measurement data from a campaign with a long range lidar windscanner in the offshore wind farm »alpha ventus«. Eventually results about the wake dynamics are discussed.

1. Introduction

When a wind turbine is operating, it extracts energy from the wind passing through its rotor. In this process a wind deficit is generated which slowly recovers while it moves downstream. The area of the wind field affected by the wind deficit with increased local turbulence is commonly known as wake.

In a wind farm wind turbines often operate in the wake of other turbines. In these conditions, not only is the available energy of the wind less than in the free stream, but more severe turbulent fluctuations affect the turbine. A precise knowledge of the turbulence properties of the wake is therefore necessary to correctly predict the dynamic loading of the turbine structure in the design process. Hence, for instance, longer life-cycle or optimised structure design can derive from the reduction of the uncertainty on turbulence fluctuations in wakes.

The wake models applied in aero-elastic simulations can be divided in steady and dynamic models. The former consider a straight propagation of the wake where turbulence fluctuations are given by the combination of ambient and aero-dynamic turbulence [1]. The latter describe the wake as a series of deficit disks released at the turbine rotor and moving downstream transported by the atmospheric turbulent structures with scales larger than the rotor diameter [2; 3]. In these models the contribution of the so-called wake meandering, i.e. the lateral displacement of the wake deficit from the rotor axis, is taken into account along with the ambient and aero-dynamic turbulence.

A wind tunnel can be a suitable facility to study the dynamics of the wake [4; 5]. Synthetic wind fields generated by means of Large Eddy Simulation (LES) have been also applied to investigate the wake meandering [6; 7]. Nacelle based lidar have been implemented in full field experiments, for instance in [8; 9], to track dynamically the position of the wake center and to compare the results with the expectations from meandering models.

According to wake meandering models, the points near the lateral edges of the deficit are alternately inside and outside the wake. This results in a fluctuation of the wind component aligned with the rotor axis which is related to the wake dynamics. An approach to characterize the described fluctuations is proposed in [10], where long range lidar measurements are performed staring the laser beam at about hub height, diagonally across the wake. With this configuration, the necessity of an inclination about the axial wind direction has the drawback that only the wind turbulent fluctuations aligned with the lidar line of sight (LOS) can be measured. This aspect is investigated in the first part of this paper by means of lidar simulations, where a varying offset is applied between the LOS and the wind direction. In the central part, spectral characteristics of the LOS measurements previously simulated are evaluated for the study of the wake meandering. Eventually, fixed point lidar measurements collected in the offshore wind farm »alpha ventus« are considered. In particular, the power spectral density (PSD) of the data sampled across the wake of a wind turbine is analyzed.

2. Influence of the lidar beam inclination on the power spectral density of line of sight measurements

2.1. Lidar simulations

An unsteady, three dimensional wind field generated by the LES implementation PALM [11] is used as virtual environment for lidar simulations. The wake of a 2 MW Tjaereborg wind turbine with a rotor diameter D_T of 62 m and a hub height of 61 m is simulated by means of an actuator line (ACL) model [12]. The resulting wind field has a spatial and temporal resolution of 4 m and 0.4 s respectively.

The target of the lidar simulations was a point located at hub height in the wake of the described turbine, about 3 D_T downstream with a lateral offset from the rotor axis of about 1 rotor radius R_T. The measurements were simulated with a 0.4 s sampling time. Four different LOS inclinations from the mean wind direction were considered, namely from 0° to 60° with a 20° increment. The layout of the simulations is illustrated in figure 1. In order to only estimate the influence of projection of the wind vector on the LOS direction, the volume average of the lidar measurements was not applied to the lidar simulations.

2.2. The PSD of the LOS wind component across the wake

The PSD of the simulated LOS measurements was evaluated by means of the Welch algorithm, applying 90 % overlapping rectangular windows of about 3 min duration and is presented in

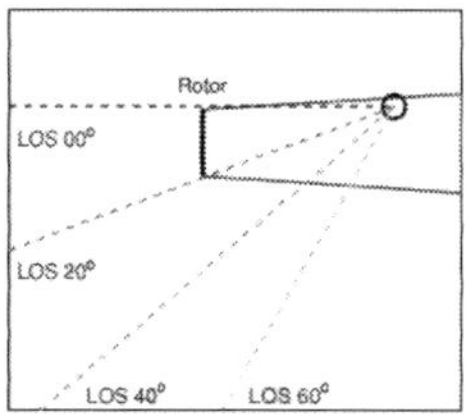

Figure 1. Layout of the lidar simulations (top view)

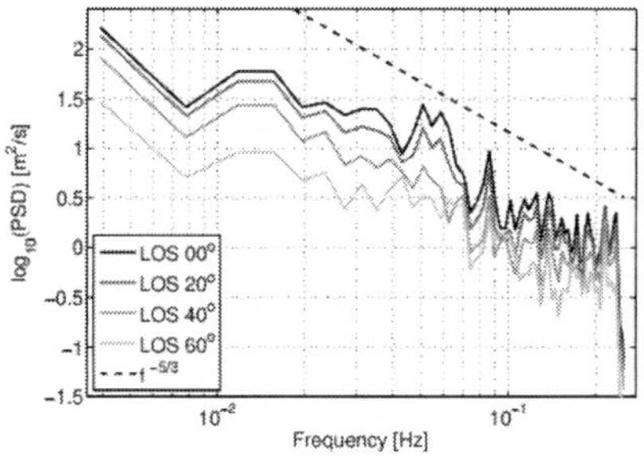

Figure 2. Power spectral density of the simulated LOS wind speed.

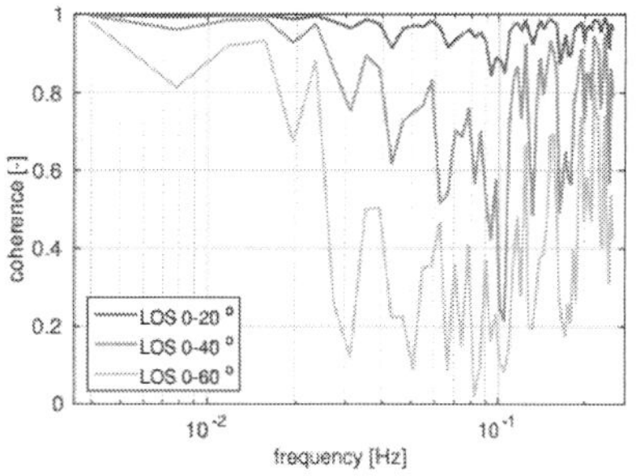

Figure 3. Variation of the spectral coherence between the simulated LOS measurements for an increasing anlgle of misalignment

figure 2. The 95 % confidence level conservatively estimated for this settings according to [13] corresponds to the range from −36 % to +79 % of the PSD value. From the graph it is evident that the energy content captured by the LOS projection decreases when the inclination angle increases. The spectral coherence function (SCF) can help to better interpret the results from the PSD analysis. It represents the normalized cross-spectral density between two signals [14] and indicates how well their spectral content is correlated. In particular, the SCF is constant and equal to 1 or 0 if the two considered signals are linearly correlated or completely uncorrelated. The SCF of the LOS simulated measurements aligned with the wind direction in relation to the the ones with an increasing offset is depicted on the graph of figure 3. A coherence close to one is found for a deviation angle of 20°. For larger deviations a coherence value above 0.8 characterizes only the low frequency region, up to 0.02 − 0.03 Hz. Then it progressively decays and eventually starts fluctuating randomly. This result indicates that for long time fluctuations the vertical and transverse wind components are only weekly affecting the LOS projection even for larger direction offset.

3. Application of fixed point lidar measurements to the study of the wake meandering

A wake tracking procedure [8] has been applied to the wind field at the cross-section 3 D_T downstream. The corresponding PSD of the lateral displacement of the wake deficit was evaluated to better understand whether a hidden relation exists between the lateral wake meandering and the fluctuations of the longitudinal wind speed at the lateral edge of the wake deficit. The normalized PSD of the wake lateral displacement and the one of the LOS measurements simulated with no offset with respect to the wind direction are compared in figure 4.

In both PSD curves, the energy is concentrated in the low frequency band and, at about 0.02 Hz, it starts decaying with a rate of about $-5/3$. While the PSD of the LOS simulations does not present any slope change in the high frequency range, the one of the wake lateral displacement sharply drops down. This may result from a low-pass filtering effect implicit in the wake-tracking procedure. Furthermore, the position of the wake center is supposed to be driven by the large-scale turbulent structures of the atmosphere.

It could be speculated on a resemblance to spectral peaks in the PSD of the LOS simulations near to 0.05 Hz and 0.09 Hz. This signature is not found in the spectral analysis of the wake tracking data, therefore no link with the wake meandering can be asserted. To better understand the nature of these possible peaks, further research will deal with longer datasets which could

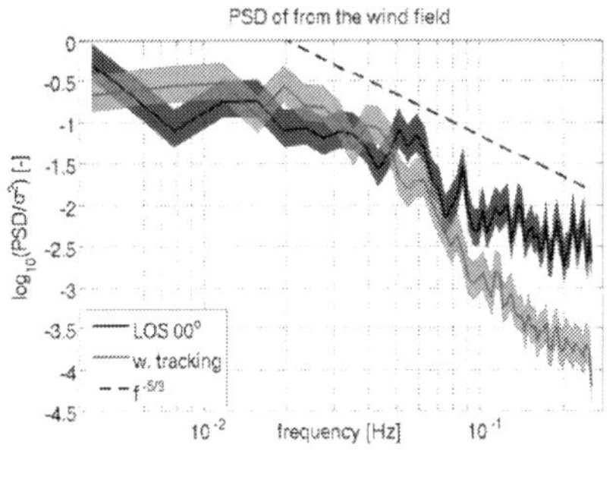

Figure 4. Normalized power spectral density and corresponding 95 % confidence level of the LOS simulations aligned with the wind direction and of the displacement of the wake center.

improve the statistical significance of the PSD.

4. Offshore lidar measurements of the wake meandering

Three long range scanning lidars Windcube WLS200S were deployed at the offshore wind farm »alpha ventus« in the German North Sea for one year from spring 2013. Twelve 5 MW wind turbines from two different manufacturers are installed as illustrated in figure 5 from West to East in rows of four units. The six turbines in the northern half of the wind farm are characterized by a rotor diameter D of 126 m mounted on a 92 m high tower, while the six in the southern half have a 116 m rotor at an hub height of 90 m. Two lidars were positioned on the platform of the meteorological mast FINO1, while the third unit was installed on the transformer station AV0 near the south-east corner of the wind farm.

Table 1. Technical sheet of the Windcube WLS200S.

Properties		Acquisition	
Wave length	1.54 μm	Photodiode sampling rate	250 MHz
Pulse length (FWHM)	0.1 – 0.4 μs	FFT length	64 – 128 – 256 points
Max laser power	5 mW	LOS component accuracy	0.2 m/s
Pulse repetition rate	10 – 20 kHz	max # range gates	250
Max range	6500 m		

Table 2. Measurement setup.

Setting	Value
Azimuth angle	−17.62°
Elevation angle	1.80°
Pulse length (Full Width Half Maximum)	0.20 μs
FFT points	64
Accumulation/Sampling time	0.60 s
Physical resolution in line of sight	~40 m

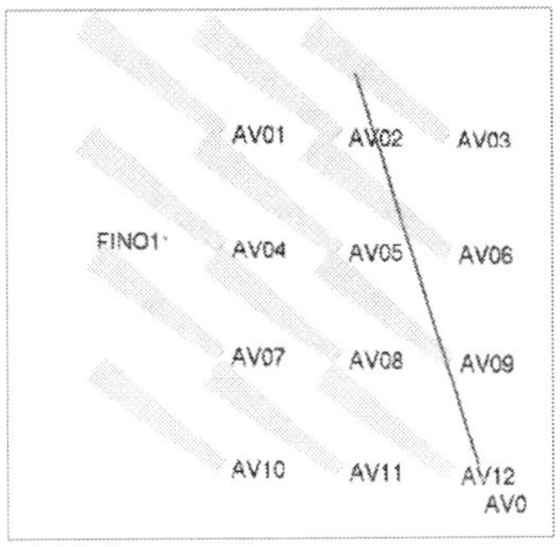

Figure 5. Position of the measurement beam (blue line) of the lidar installed on the transformer platform AV0 in the wind farm »alpha ventus«. Wakes are drawn downstream the turbines AV01-AV12 for a wind direction of 127°.

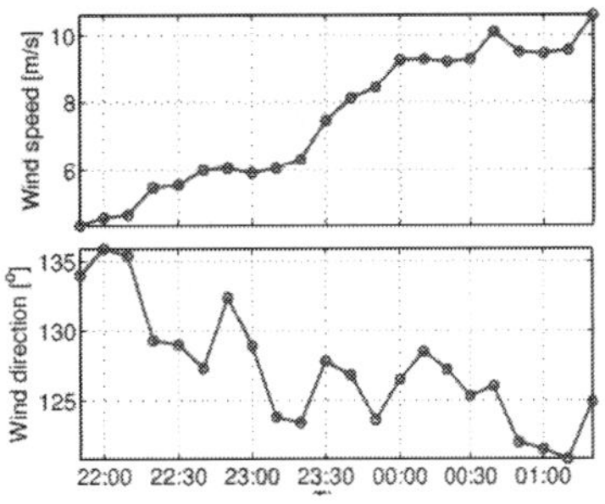

Figure 6. Ten-minute wind speed (top panel) and direction (bottom panel) collected by the top cup anemometer and the vane installed on the FINO1 meteorological mast at 100 m and 90 m respectively.

4.1. Measurement setup
Staring lidar measurements were performed with the lidar installed on the transformer station. The target point was located at hub height in the middle position among the four turbines in the north-east corner of the wind farm. The 244 range gates implemented in this experimental campaign were distributed equally every 15 m between 100 m and 2992 m. A finer separation of 5 m was applied along a 3 D distance centered to the aimed point. The objective of the measurements was to study the wake dynamics of the wind turbine AV06 by means of spectral analysis. The layout of the measurement is sketched in figure 5 for a wind direction of 127°, while the technical data and settings applied on the lidar are summarized in tables 1 and 2.

4.2. Dataset description
The measurements took place for about 3 hours on 17/12/2013 at night. The ten minute average of the wind speed and direction measured during the campaign by the top cup anemometer and the wind vane installed on FINO1 at 100 m and 90 m respectively [15] is reported in figure 6. Figure 7 shows the full dataset of measurements. The wake of the wind turbines AV06 and AV09 are clearly visible within the shorter and further ranges respectively. Since only the measurements which took place in the wake of AV06 are at about hub height, only the further ranges were considered. Among all the data, only six ten minute intervals were analysed for this study. For sake of concision only the interval measured on 18/12/2013 at 01:04 is presented here. Data with a carrier to noise ratio (CNR) below −25 dB or higher than −5 dB was filtered out. Before the analysis, also outliers identified with the Chauvenet criterion [16] were excluded from the filtered time series. The remaining data was re-sampled applying a linear interpolation with a constant time step of 0.6 s.

In order to represent the LOS measurements in the fixed frame of reference of the wake of AV06, the orientation of the wake axis was determined using the center of the average deficit measured by the lidar. The derived wind direction, i.e. 127°, has a 6° offset with respect to the wind direction provided by the FINO1 instrumentation. The fact that for this direction FINO1 is in the wake of the wind farm could explain this deviation. In figures 8 and 9 the position of the measurement points along with the data considered in this study are represented in the

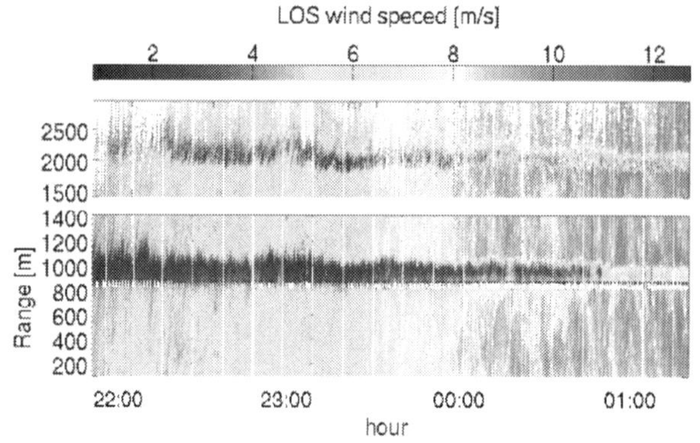

Figure 7. LOS measurements conducted on 17/12/2013 at night. In the top panel the measurements cross the wake of AV06, while in the bottom panel the one of AV09.

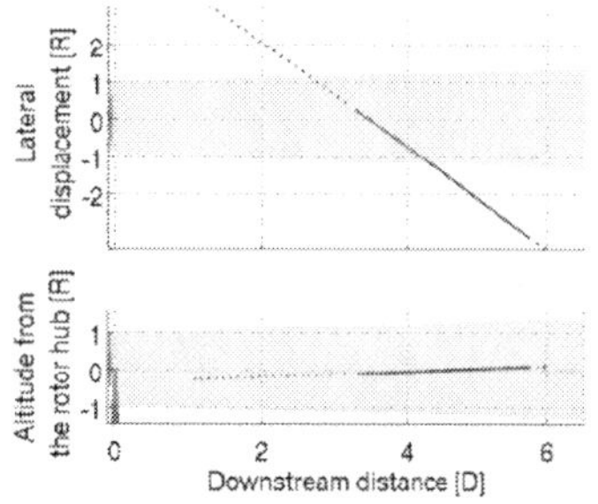

Figure 8. Top and side view of the measurement points represented in AV06 rotor frame of reference.

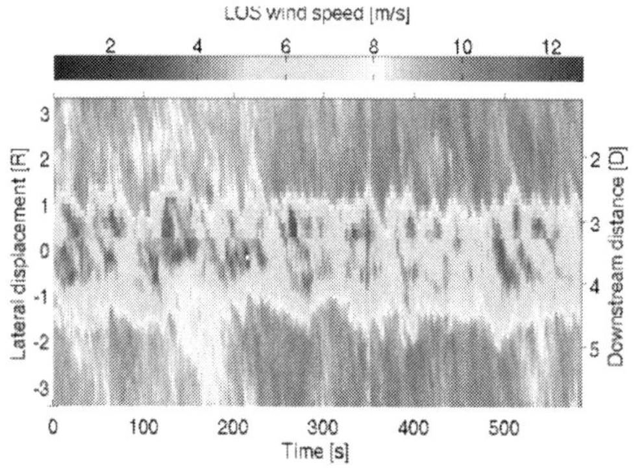

Figure 9. Ten minute time interval at 1:04 selected from the full dataset and represented in the AV06 rotor frame of reference.

reference frame of AV06 wake. Here, the x dimension is normalized with the rotor diameter D, while y and z with the corresponding radius R. Considering this setup, the inclination of the LOS from the wind direction is about 35°, which is an acceptable value according to the results of section 2.2.

4.3. Spectral analysis

The PSD of the lidar measurements was performed applying the same procedure described in section 2.2. The related results are illustrated in figure 10 where the lateral boundaries of the wake can be easily identified. In fact, in this region the energy content is particularly high in relation to other positions over the rotor, especially in the low frequency range. Around 2.7 D downstream, the area characterized by a higher level of the spectral energy covers a range of ± 0.5 R across the edge of the wake. At the opposite edge, about 4 D downstream, the range characterized by the increased PSD is almost twice as wide. This results are in accordance with [10], where a similar behaviour has been observed between 1.5 D and 4 D downstream. Figure 11 offers a closer view of the PSD of the LOS measurements evaluated at the center of

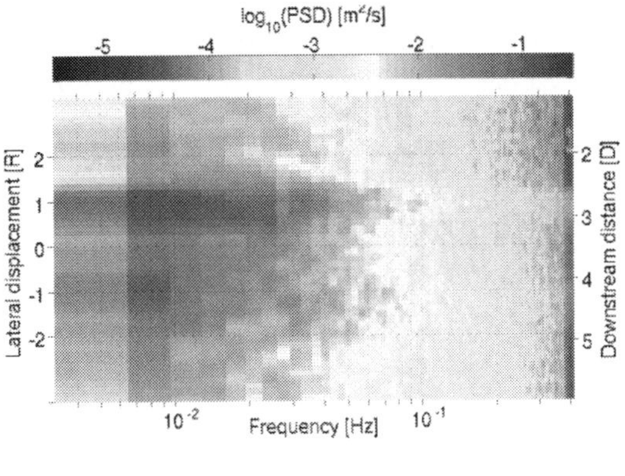

Figure 10. Power spectral density of the LOS wind component as a function of the frequency (abscissa), of the lateral displacement (left ordinate) and of the downstram position (right ordinate).

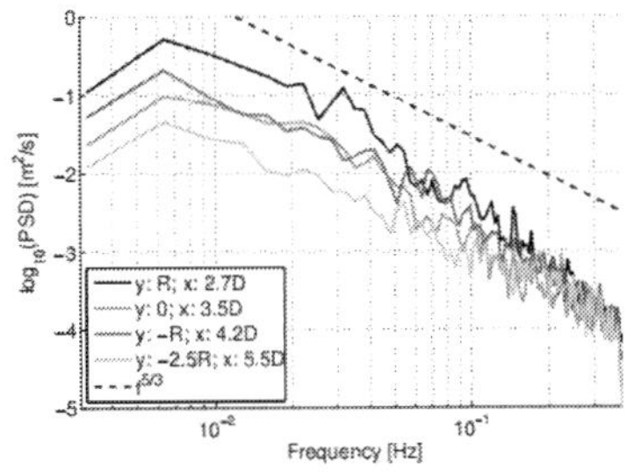

Figure 11. Power spectral density of the LOS wind component measured at different downstream distances, respectively at the lateral edges of the wake, in its center and in the free stream.

the wake 3.5 D downstream and at the two opposite sides of the wake located at 2.7 D and 4.2 D respectively. The same graphic includes the PSD calculated in the free stream.

A comparison of the curves confirms that the spectral energy content outside the wake is remarkably lower than in the area affected by it. Moreover, the diagram reveals that the PSD values in the center and at the edge of the deficit are very similar in the far wake, while they are considerably higher in the transition between the near and far wake. The high spectral energy content found in the near wake at the boundary of the deficit seems to spread radially towards the center of the wake and towards the free stream while moving downstream, until eventually a quite uniform and lower level of turbulent energy is reached.

Considering the wake meandering models, these effect could be in part related to the amplitude of the displacement of the wake deficit, which generates the alternation of the measurement point position between in and outside the wake. In the transition between near and far wake smaller horizontal displacements characterize the wake due to the short time characterized by lateral transport, while, further downstream, the displacements could be larger than 1 R as observed also in [17].

However, the increase of low frequency turbulence at the wake lateral boundaries could also be explained by turbulent diffusion. Only if an anti-correlation was found between the fluctuations measured at opposite sides of the wake, the relation between the fluctuations of the measurements at the edge of the wake and the wake meandering could be confirmed [5]. The applied measurement setup does not allow to perform the anti-correlation analysis for two points at the same downstream distance. The cross-correlation coefficient [14] has therefore been evaluated for the LOS time series sampled at the opposite edges of the wake deficit, but 2.8 D and 4.2 D downstream, respectively. The resulting curve visible in figure 12 as a function of the imposed time lag between the considered signals shows a minimum of -0.34 for a time lag of 26 s. The corresponding peak is not necessarily statistically significant, for this reason a moderate anti-correlation can only be assumed. This hypothesis is sustained by the delay which is in the order of the expected travelling time of the deficit between the two considered points.

A dominant peak was probably not found because of the large longitudinal distance between the two considered points. To support this explanation, the cross-correlation analysis was applied

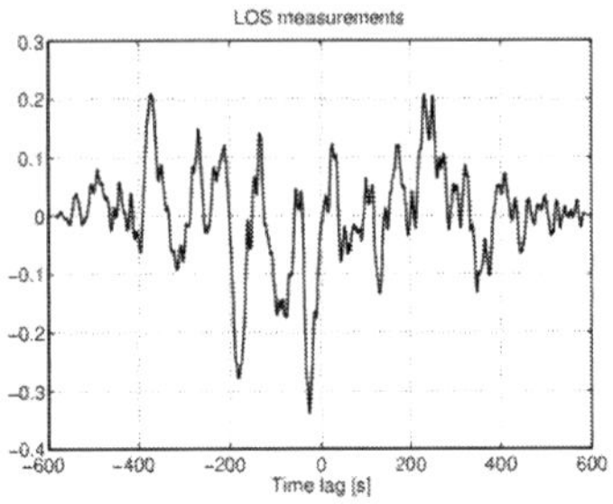

Figure 12. Cross-correlation of two LOS time series measured around hub height, at the opposite edges of AV06 wake deficit 2.8 D and 4.2 D downstream, respectively.

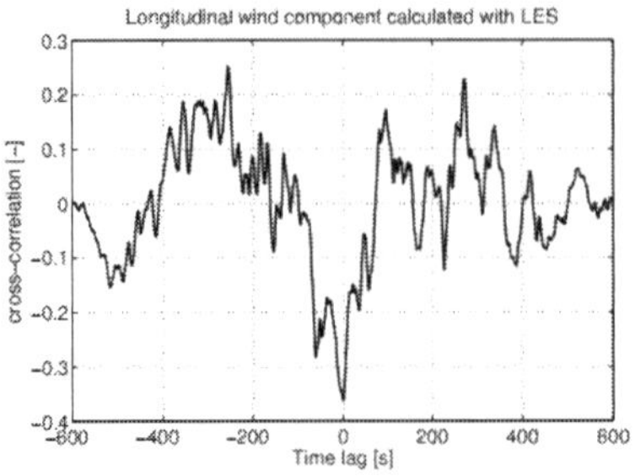

Figure 13. Cross-correlation of the longitudinal wind component sampled from the simulated wake of section 2.1 at hub height, at the opposite edges of the wake deficit, about 3 D downstream.

with no downstream offset to the longitudinal wind component of the simulated wind field described in section 2.1. Two points at hub height, about 3 D_T downstream and at opposite sides of the wake were considered. The related graph shown in figure 13 presents a clear evolution of the cross-correlation in which a negative peak and no time lag can be identified. This suggests that the results of the cross-correlation analysis from the full field lidar measurements could be improved if two points at the same downstream section were considered. Moreover, it indicates that the negative correlation found for the LOS measurements in figure 12 could be a real feature of the wind field and not just the outcome of an uncorrelated time series. For this reasons, a connection between the higher turbulence energy content at the edge of the wake and the wake meandering can be supposed.

5. Conclusion

In the first part of this work, simulations of lidar measurements at a fixed position in the wake of a wind turbine have been analysed to understand the influence of the misalignment between the LOS direction and the mean wind direction on the spectral analysis of the lidar measurements. It was demonstrated that LOS measurements at a fixed point can well represent the dynamic fluctuations of the main wind component at the edge of a wake deficit when they are oriented with a moderate offset with respect to the mean wind direction. The results suggest also that this constraint could be relaxed if only low-frequency fluctuations are considered.

To investigate the suitability of fixed point lidar measurements to the study of wake meandering, the normalized PSD of the lateral displacement of the wake center was compared with the one calculated for the LOS measurements simulated with no deviation from the wind direction. In both cases, the energy content mainly distributed within the low frequency range and then decays following a power law. No corresponding spectral peaks could be identified.

In the last part of the paper, full field data were considered. The measurement approach implemented in the simulations was reproduced in the offshore wind farm »alpha ventus« during an experimental campaign. In relation to other positions in the wake, a high level of spectral energy characterizes a well defined region around the lateral edges of the deficit in the transition between near and far wake. The meandering of the wake was suggested as possible explanation and a negative correlation between LOS measurements or simulations at opposite sides of the wakes supports this argument.

Fixed point lidar measurements showed to be a potential application for the study of the wake dynamics by means of spectral analysis. This preliminary study indicates a possible relation between the wake meandering and the spectral characteristics of appropriate lidar measurements at the lateral boundaries of the wake. However, this hypothesis has to be confirmed by further research.

Acknowledgments
The authors thank the whole »Lidar and Wakes« group at ForWind - University of Oldenburg which made the measurement campaign possible.

The research in the offshore wind farm »alpha ventus" was carried out in the frame of the RAVE (Research at alpha ventus) research project »GW Wakes - Part A«, funded by the German Federal Ministry for Economic Affairs and Energy (BMWi) based on a decision of the Parliament of the Federal Republic of Germany (grant number 0325397A).

References

[1] Frandsen S T and Thøgersen M 1999 *Wind Engineering* **23** 327–339 ISSN 0309-524X

[2] Larsen G C, Madsen H A, Thomsen K and Larsen T J 2008 *Wind Energy* **11** 377–395 ISSN 1099-1824 URL `http://dx.doi.org/10.1002/we.267`

[3] Trujillo J J and Kühn M 2009 *European Wind Energy Conference (EWEC)*

[4] Medici D and Alfredsson P H 2008 *Wind Energy* **11** 211–217 ISSN 1099-1824 URL `http://dx.doi.org/10.1002/we.247`

[5] España G, Aubrun S, Loyer S and Devinant P 2012 *Journal of Wind Engineering and Industrial Aerodynamics* **101** 24 – 33 ISSN 0167-6105 URL `http://www.sciencedirect.com/science/article/pii/S0167610511002157`

[6] de Mare M, Mann J, Churchfield Matthew J and Patton Edward G 2013 *EWEA Offshore*

[7] Bastine D, Björn W, Matthias W and Joachim P 2014 *Journal of Physics: Conference Series* **524**

[8] Trujillo J J, Bingöl F, Larsen G C, Mann J and Kühn M 2011 *Wind Energy* **14** 61–75 ISSN 1099-1824 URL `http://dx.doi.org/10.1002/we.402`

[9] Machefaux E, Larsen G C, Troldborg N, Gaunaa M and Rettenmeier A 2014 *Wind Energy* ISSN 1099-1824 URL `http://dx.doi.org/10.1002/we.1805`

[10] Trabucchi D, Trujillo J J, Bitter M and Kühn M 2014 *Meteorologische Zeitschrift* – URL `http://dx.doi.org/10.1127/metz/2014/0610`

[11] Raasch S and Schröter M 2001 *Meteorologische Zeitschrift* **10** 363–372 URL `http://dx.doi.org/10.1127/0941-2948/2001/0010-0363`

[12] Troldborg N, Sorensen J N and Mikkelsen R 2010 *Wind Energy* **13** 86–99 ISSN 1099-1824 URL `http://dx.doi.org/10.1002/we.345`

[13] Manolakis D G, Ingle V K and Kogon S M 2000 *Statistical and Adaptive Signal Processing* (McGraw-Hill) chap 5

[14] Bendat J and Piersol A 2000 *Random data analysis and measurement procedures* (Wiley Interscience)

[15] Beeken A, Neumann T and Westerhellweg A 2008 *DEWEK*

[16] Taylor John R 1997 *An introduction to error analysis* (University Science Books)

[17] España G, Aubrun S, Loyer S and Devinant P 2011 *Wind Energy* **14** 923–937 ISSN 1099-1824 URL `http://dx.doi.org/10.1002/we.515`

Wind turbine wake detection with a single Doppler wind lidar

H Wang and R J Barthelmie
Sibley School of Mechanical and Aerospace Engineering, Cornell University, USA

E-mail: hw524@cornell.edu

Abstract. Using scanning lidar wind turbine wakes can be probed in three dimensions to produce a wealth of temporally and spatially irregular data that can be used to characterize the wakes. Unlike data from a meteorological mast or upward pointing lidar, the spatial coordinates of the measurements are not fixed and the location of the wake also varies in three dimensions. Therefore the challenge is to provide automated detection algorithms to identify wakes and quantify wake characteristics from this type of dataset. Here an algorithm is developed and evaluated on data from a large wind farm in the Midwest. A scanning coherent Doppler wind lidar was configured to measure wind speed in the wake of a continuously yawing wind turbine for two days during the experiment and wake profiles were retrieved with input of wind direction information from the nearby meteorological mast. Additional challenges to the analysis include incomplete coverage of the entire wake due to the limited scanning domain, and large wind shear that can contaminate the wake estimate because of the height variation along the line-of-sight. However, the algorithm developed in this paper is able to automatically capture wakes in lidar data from Plan Position Indicator (PPI) scans and the resultant wake statistics are consistent with previous experiment's results.

1 Introduction

Wind turbine wakes are characterized by low wind speed and high turbulence intensity relative to the freestream. As a result, wind turbines in the wake of operating wind turbines can experience reduced power output and enhanced fatigue load [1]. To reduce the adverse effects of wind turbine wakes, accurate wake modelling plays a critical role in wind farm layout and operation optimization, and the accuracy of the wake models needs to be verified with observations [2]. While verification can be performed with in-situ measurements and downscaled experiments in wind tunnels, remote sensing technologies have shown potential in studying wake characteristics and testing wake models in full-scale wind farms. For example, a single coherent Doppler wind lidar (hereafter called lidar) can be used to study the wake recovery and expansion along the downwind distance from the wind turbine under various atmospheric conditions [3]. With two lidars it has been shown that wake has a fast recovery rate in convective atmospheric conditions [4]. Measurements from dual Doppler radars have been used to characterize the complex flow features such as wake merging in a large wind farm [5]. To further explore the potential of lidar for wind turbine wake characterization, this paper presents an experiment designed to test the performance of a single lidar for measuring the wake of a continuously yawing wind turbine. This paper is organized as follows: A brief description of lidar operation principle is given in Section 2, and then followed by the description of the experiment setup in Section 3. The wake profile estimation algorithm is described in Section 4. Section 5 presents the wake characteristics, such as wake deficit and wake width, derived from the lidar measurements. Conclusions are drawn in Section 6.

2 Lidar basics

A lidar only measures the radial velocity at each range gate. The radial velocity (v_R) is the projection of the wind velocity ($\boldsymbol{v} = [u, v, w]$) on the line-of-sight (LOS) which is the path of the light wave emitted by the lidar (Fig. 1). Mathematically, the radial velocity can be expressed as

$$v_R = u \cos\beta \sin\alpha + v \cos\beta \cos\alpha + w \sin\beta \tag{1}$$

where α and β are the azimuth angle and the elevation angle of the LOS, respectively, and u, v and w are the west-east, south-north and vertical components of the wind velocity. The radial velocity can also be defined in terms of the horizontal wind speed (V) and wind direction (θ_w):

$$v_R = -V \cos\beta \cos(\alpha - \theta_w) + w \sin\beta \tag{2}$$

where $V = \sqrt{u^2 + v^2}$ and $\theta_w = \pi + \arctan(u/v)$. The range gate is the area over which the echo samples are used to estimate the radial velocity, and therefore the measured v_R represents the mean wind condition over the range gate.

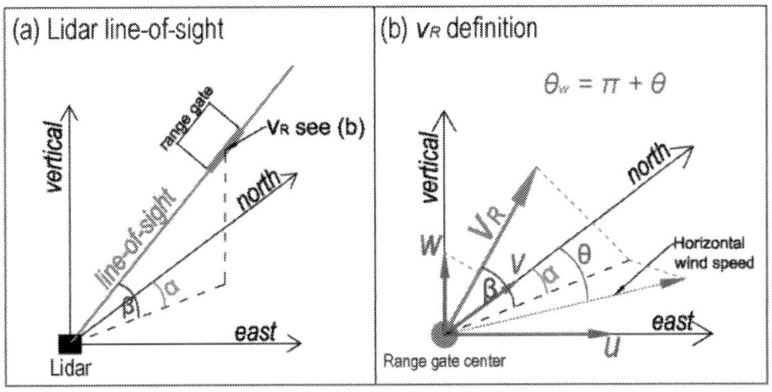

Figure 1. Schematics of (a) the orientation of the lidar line-of-sight in terms of the elevation angle (β) and the azimuth angle (α), and (b) the relationship between the radial velocity (v_R) and the wind velocity ($\boldsymbol{v} = [u, v, w]$), and the relationship between the radial velocity and the horizontal wind speed and wind direction (θ_w).

3 Experiment setup

This experiment was conducted in a large wind farm in Indiana from 7-20 May 2012 [6]. Results are presented here from a case study on May 16 and 17, 2012 when the wind direction was from the northeast and southeast. A Galion lidar (denoted as Galion hereafter) was deployed on the southwest edge of the wind farm. To probe the wake of the wind turbine denoted as WT2 in Fig. 2, the Galion was configured to perform 3-stack Plan Position Indicator (PPI) scans. Each stack of PPI scans has a fixed elevation angle and varying azimuth angles. The elevation angles were set to 8.0°, 8.5° and 9.0° for the three stacks, respectively. For each PPI stack, the azimuth angle changed from 60° to 90° with 1° increment. The range gate size was 30 m. The wind turbine WT2 has a hub height of 80 m and a rotor diameter of 77 m, and hence it was located between the range gate 17 and 18 of the three PPI stacks (Fig. 2). The Galion's measurement height decreases with decreasing range gate, but it can still detect the wake of WT2 at the range gate 10 which is about 44 m above the ground and higher than the bottom tip of the rotor (41.5 m). An 80-m tall meteorological mast was collocated with the Galion, and the wind direction measured at 77 m height from the meteorological mast are used to estimate the wake profiles from the Galion measurements. In addition, two ZephIR lidars were located about 10 m north of the Galion lidar measuring wind speeds at 40, 80, 120, 160 and 200 m height [6].

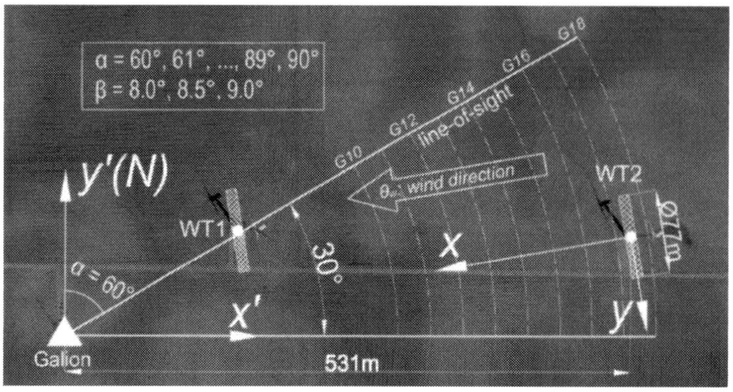

Figure 2. Top view of the locations of the Galion and the two wind turbines (WT1 and WT2), the scanning geometry used to probe the wake of WT2, and the two coordinate systems denoted as *x'*-*y'* and *x*-*y* (details see 4.2). The azimuth angles (α) and elevation angles (β) used are listed on the upper-left corner. The range gate locations are denoted by the dashed white lines. The underlying wind farm map is from the Google Earth.

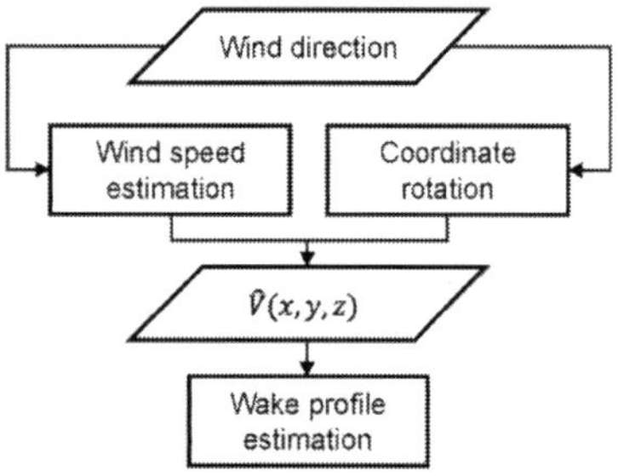

Figure 3. Wake profile estimation algorithm flow chart. The term $\hat{V}$ is the wind speed estimated using the lidar data.

4 Wake profile estimation algorithm

Wake characteristics are quantified by interpolating the wind speed estimated from the measured v_R to the locations of interest in the wake of WT2 using the Gaussian kernel smoother. Wake profiles are estimated for lines parallel to the rotor plane of WT2 at every 0.2D distance downwind of WT2, where D is the rotor diameter. The first line is located at 1.0D downwind of WT2. The conical surface formed by each PPI stack is treated as a slanted flat surface by ignoring the small curvature on the conical surface. Because of the non-zero elevation angle, the height associated with the estimated wind speed varies across one individual line and among multiple lines for one PPI stack. Wake profiles are estimated from the measured v_R of each PPI stack according to the flow chart shown in Fig. 3. The estimation method requires the wind direction as an input and uses the Gaussian kernel smoother to estimate the wake profiles of WT2. The details of the algorithm are given below.

4.1 Wind direction

The wake profile estimation algorithm requires *a priori* knowledge of the wind direction, which can be obtained from the wind direction measured at the nearby meteorological mast, or estimated from the Galion measurements in the freestream. For this project, the input wind direction is from the wind vane at 77 m height on the meteorological mast. Though wind directions are estimated from the measured radial velocities between the range gate 24 and 28, the estimated wind directions are not reliable because of the possible wind speed gradient at the edge of the wind farm. Thus, the wind direction estimated from the Galion measurements is not used.

4.2 *Coordinate rotation*

Two Cartesian coordinate systems are used in this algorithm. The first one is the conventional meteorological coordinate system (x', y') with its origin at the Galion location and x-coordinate and y-coordinate increasing eastward and northward, respectively (Fig. 2). The second one is the streamwise coordinate system (x, y) with its origin at the location of WT2 (Fig. 2). The relationship between the two coordinate systems is a function of the wind direction θ_w:

$$x = x' \cos(1.5\pi - \theta_w) + y' \sin(1.5\pi - \theta_w) \quad (3)$$

$$y = y' \cos(1.5\pi - \theta_w) - x' \sin(1.5\pi - \theta_w) \quad (4)$$

Noting here we assume the rotor axis is parallel to the wind direction (i.e. no yaw error).

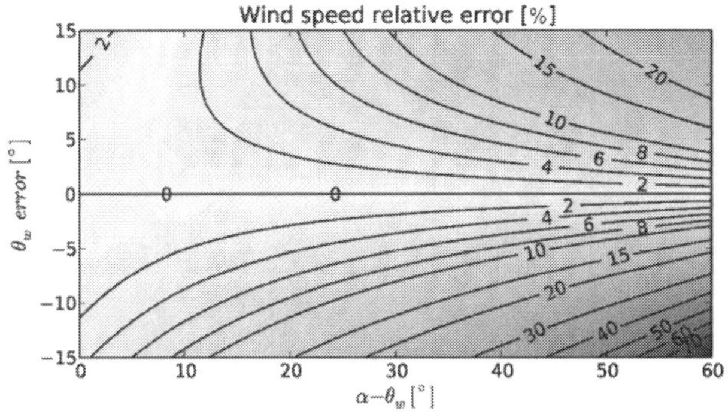

Figure 4. Contours of the wind speed relative error as a function of the wind direction (θ_w) error and the angle between the LOS and the wind direction ($\alpha - \theta_w$), when Eq. (5) is used to estimate the wind speed.

4.3 *Wind speed estimation*

When the vertical wind speed w is close to zero or the elevation angle β is small, the term $w \sin\beta$ in Eq. (2) can be dropped. Then the horizontal wind speed (V) can be calculated by rearranging Eq. (2) as

$$V = -\frac{v_R}{\cos\beta \cos(\alpha - \theta_w)} \quad (5)$$

An incorrect θ_w can introduce an error in the estimate of V, and the error is a function of both the wind direction error and the angle between the LOS and the wind direction , namely, $\alpha - \theta_w$. The error is small and insensitive to the wind direction error when $(\alpha - \theta_w)$ is small (Fig. 4). When $(\alpha - \theta_w) = 60°$, the wind direction needs to be accurate within ±5° in order to ensure that the relative error of the estimated wind speed < 10%.

4.4 *Wake profile estimation*

For each PPI stack, wake profiles are estimated for lines that are parallel to the rotor plane of WT2 at every 0.2D downwind of WT2. The estimation starts at 1.0D and ends when the *x*-axis in Fig. 2 is out of the scanning domain. The wind speed at a point of interest is estimated as the weighted average of all available wind speeds estimated with Eq. (5) from one PPI stack. The weight function is given by

$$w_{gt}(x, y) = \exp(-\frac{l^2}{2L^2}) \quad (6)$$

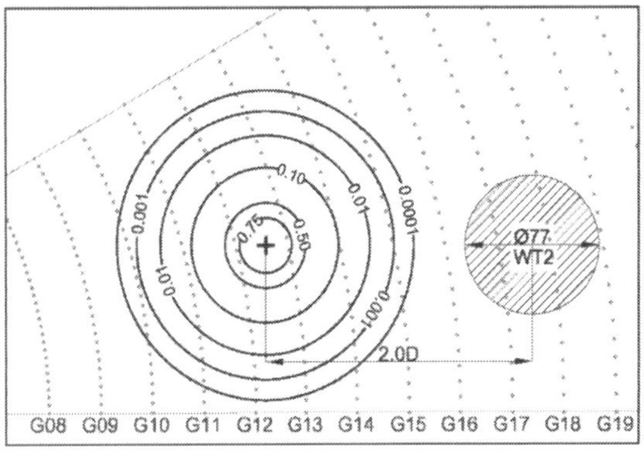

Figure 5. Contours of the weight used to calculate the wind speed at a point (dark cross) two rotor diameters (2.0D) downwind of WT2 using Eq. (6). The gray circles are the Galion range gate locations, and the range gates are indexed by the numbers at the bottom.

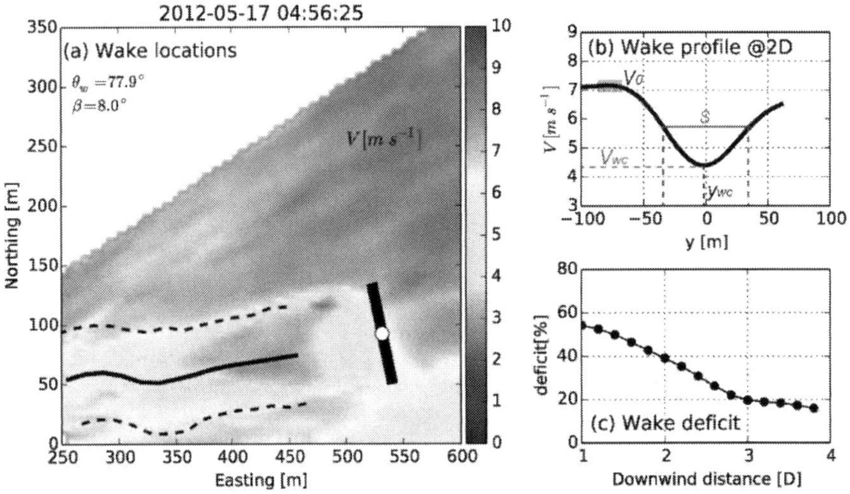

Figure 6. An example of wind speed retrieved for one stack of PPI scan. The wind speed over the entire scanned area is shown in (a). The wind turbine rotor orientation is denoted by the dark bar with a white circle indicating the location of WT2. The dark line is the wake center and the dashed lines denote the wake width. The estimated wind speed profile across the line that is parallel to the rotor plane at 2.0D downwind of WT2 is shown in (b). The definitions of wake center (y_{wc}), the FWHM (S), wake center wind speed (V_{wc}) and freestream wind speed (V_0) are also shown in (b). The wake deficit (δ in Eq. (7)) as a function of the downwind distance normalized by the rotor diameter (D) is shown in (c).

where w_{gt} is the weight assigned to the wind speed at a range gate located at (x, y), l is the distance between the range gate and the point of interest [7]. The term L is a constant and it is set to 20 m considering that the Galion's spatial resolution is about 40 m. The distribution of the weight for a point at 2.0D downwind of WT2 is shown in Fig. 5. An example of a wake profile estimated at 2.0D downwind of WT2 is illustrated in Fig. 6b. The wake center (y_{wc}) is the location of the minimum wind speed or the wake center wind speed (V_{wc}) on the line. The wake deficit (δ) is defined as

$$\delta = \frac{V_0 - V_{wc}}{V_0} \tag{7}$$

where V_0 is the freestream wind speed which is estimated as the mean of the five highest wind speeds from the wake profile. The Full Width at Half Maximum (FWHM, and denoted as S) is used to characterize the wake width. The FWHM is estimated as the distance between the wake center and the location at which wind speed recovers by 50% (i.e. the deficit is 0.5δ). As shown in Fig. 6, the algorithm described above can detect the wake center and width up to 3.8D downwind of WT2. The estimated wake deficit decreases with the downwind distance from ~60% at 1.0D to ~20% at 3.0D (Fig. 6c). This is consistent with other observations [3].

5 Results and discussions

Wakes of WT2 are observed during two periods marked as wake1 and wake2 in Fig. 7, respectively, and their characteristics such as wake deficit and wake width will be presented hereafter. The wake1 period lasted for about 3 hours. WT2 was in the wake of the whole wind farm as the wind direction gradually changed from 60° to 90°. Measurements from both the cup anemometers on the meteorological mast and the ZephIR lidars were affected by the wakes of WT1 and WT2 (Fig. 7). During the wake2 period, the wind veered from 100° to 125° as the wind speed ramped up to 11 m s^{-1}. Then the wind backed to 115° and the wind speed dropped to 8 m s^{-1}. Large wind shear was also observed during this period (Fig. 7). Wake profiles derived during the wake2 period represent single wakes from WT2 because WT2 is located in the first row of the wind farm. The freestream wind speed estimated from the wake profiles matches closely with measurements from the cup anemometer and the ZephIR lidar during the wake2 period (Fig. 7).

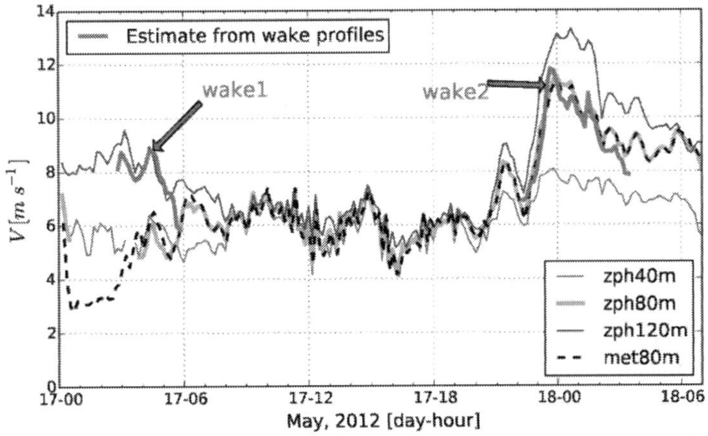

Figure 7. Time series for wind speed observed at 80 m on the meteorological mast and at 40 m, 80 m and 120 m heights by the ZephIR lidar. The two red lines are the time series of the freestream wind speed (V_0 in Eq. (7)) derived from the estimated wake profiles.

5.1 *Wake deficit*

Wake deficits are presented as a function of the downwind distance normalized by the rotor diameter (D) from WT2 in Fig. 8 and 9 for the wake1 and wake2 period, respectively. During the wake1 period, the lidar scanning geometry allows wake measurement up to 3.8D downwind of WT2. The wake deficit decreases with increasing downwind distance from 56%±4% at 1.0D to 12%±6% at 3.8D (Fig. 8). Note that the height at which the wake deficit is evaluated also decreases with increasing downwind distance because of the non-zero elevation angle (Fig. 8). Though the height is below the bottom tip of the rotor beyond 3.6D downwind of WT2, the wake deficit (~12%) still exists as the result of wake expansion. During the wake2 period, the lidar scanning geometry can only measure wake profiles up to 2.6D downwind of WT2. The height of the detected wake center remains almost the same along the downwind distance. The wake deficit has little variation between 1.0 D and 1.6D downwind of WT2 and gradually decreases with increasing downwind distance beyond 1.6D. Large variations in wake deficit are observed because WT2 was operating in a wide wind speed range (7.5−11.5 m s^{-1}) during this period. The thrust coefficient of WT2 is low at high wind speeds, causing the low wake deficit for cases with wind speed > 10.5 m s^{-1}. However, the low wake deficits observed at high wind speeds are overestimated (i.e. the wake deficit is not low enough) because of the large height variation along each wake profile and the large vertical wind shear associated with these cases (Fig. 10). For example, in the case shown in Fig. 10, the wake center is located at 65 m height with wind speed 7.1 m s^{-1}. The wake deficit is 35% based on the freestream wind speed (11 m s^{-1}) estimated as the mean of the five highest wind speeds on the wake profile. However, the estimated freestream wind speed is from 80 m height. The actual freestream wind speed at 65 m height is about 10 m s^{-1} according to the ZephIR measurement. Therefore, the actual wake deficit should be 29%, which is overestimated by 6% using the method presented in Sec. 4.4.

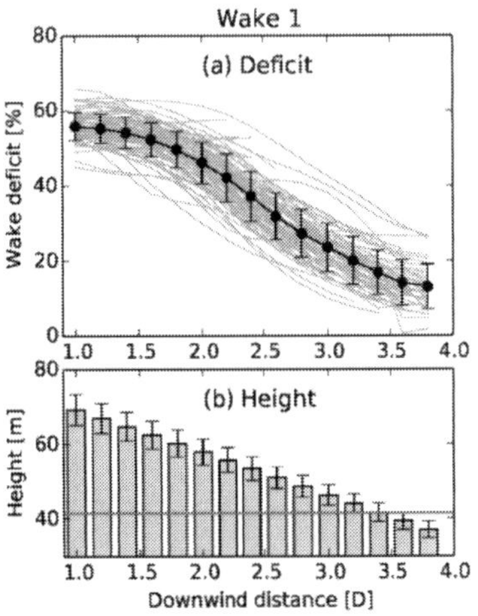

Figure 8. Wake deficit observed during the wake1 period as a function of the downwind distance in (a). The detected wake center height is given in (b) and the error bars denote the range. The red line in (b) denotes the rotor bottom tip.

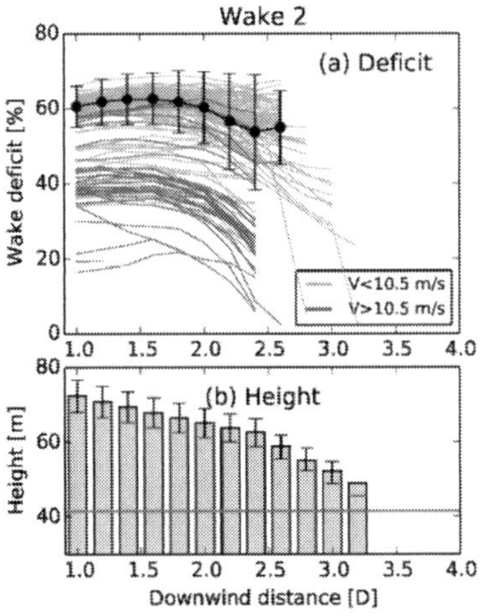

Figure 9. Wake deficit observed during the wake2 period as a function of the downwind distance in (a). The detected wake center height is given in (b) and the error bars denote the range. The red line in (b) denotes the rotor bottom tip.

5.2 Wake width

The wake width derived from the wake profile is the FWHM (S) at a height that varies with the wind direction. Assuming that the wake deficit on the cross section parallel to the rotor plane of WT2 is axisymmetric about the rotor axis, the FWHM at the hub height (S_{hub}) can be estimated by the following equation:

$$S_{hub} = \sqrt{S^2 + 4\Delta z^2} \tag{8}$$

where Δz is the height difference between the hub height and the detected wake center. By further assuming the wake deficit follows a Gaussian profile, the hub height equivalent wake width, defined as four times the standard deviation of the Gaussian profile, is calculated as $1.7S_{hub}$ and presented in Fig. 11.

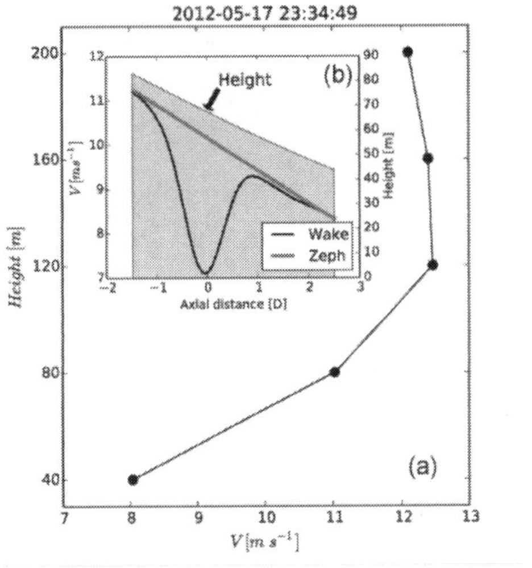

Figure 10. Wind shear measured by the ZephIR lidar in (a) and the wake profile derived from a PPI stack at 2D downwind of WT2 in (b) during the same 10-minute period. In (b), the dark line is the wake profile and the red line is the linearly interpolated from the ZephIR measurements at 40 m and 80 m in (a). The shaped area shows height variation across the wake profile.

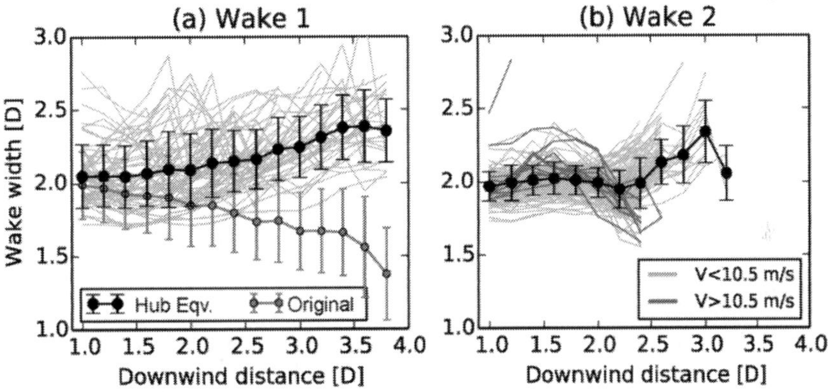

Figure 11. The equivalent hub height wake width as a function of the downwind distance normalized by the rotor diameter (D) for (a) the wake1 period and (b) the wake2 period. The gray and purple lines represent wake width estimates from individual PPI stacks for low and high wind speed cases, respectively. Error bars show the mean with plus/minus one standard deviation. The dark error bars are based on the equivalent hub height wake width and the red error bars are derived from the original wake width estimated directly from the Galion measurements. The purple lines the wake2 period are not included in calculating the error bars.

During the wake1 period, the hub height equivalent wake width on average increases from 2.0D to 2.3D along the downwind distance of WT2. However, the wake widths from some individual PPI stacks show large variability with the downwind distance, and the variability is mainly caused by the inhomogeneous wind field on the edge of the wind farm (e.g. wind speed gradient seen in Fig. 6a). The original wake width decreases with increasing downwind distance because the height of Galion measurements decreases as the wake moves away from WT2. During the wake2 period, the average wake width remains constant (about 2.0D) between 1.0D and 2.0D downwind of WT2. Note that wake widths during the wake2 period are derived from wake profiles on slanted lines (Fig. 10). The wake profile on a slanted line is not symmetric when the vertical wind shear is high (Fig. 10), causing difficulty in defining and detecting the wake width. The unexpected relationship between the wake width and the downwind distance for high wind speed cases (the purple lines in Fig. 11) could be the result of asymmetric wake profiles on the slanted line, and hence, should not be used to characterize the wake expansion downwind of WT2.

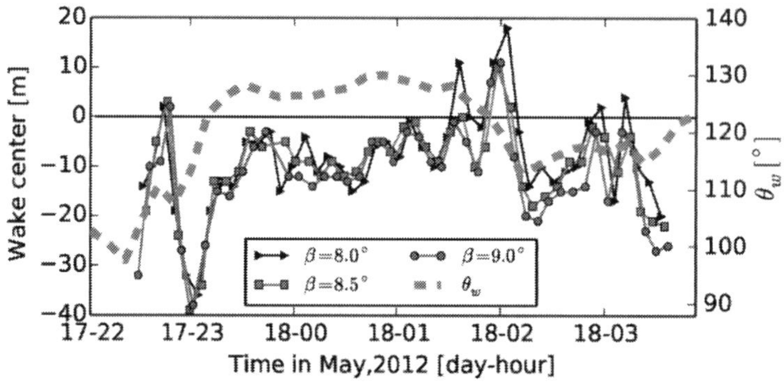

Figure 12. Time series of the estimated wake center during the wake2 period from PPI stacks with elevation angle 8.0°, 8.5° and 9.5°, respectively. The dashed red line is the time series of wind direction.

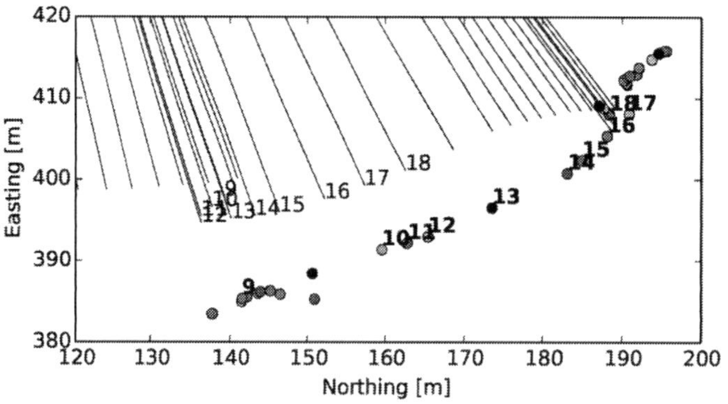

Figure 13. Locations of wake centers (circles) and wind directions (lines) observed between 22:40 and 23:59 on May 17, 2012 local time. The numbers denotes the time sequence of the wake centers and wind directions

5.3 Wake center

The time series of the wake center at 2.0D downwind of WT2 is shown in Fig.12 for the wake2 period. The wake center deviates from the rotor axis by about 40 m or 0.5D at the beginning of this period when wind veers from 95° to 125°. Meanwhile the temporal variation of wake center is large. When the wind direction becomes more constant, the wake center becomes stationary with a constant 10 m deviation from the rotor axis for about 2 hours. Then the wake center starts to fluctuate again as a result of the changing wind direction. It is worth noting that the wake center locations shown in Fig. 12 are relative to the rotor axis of WT2 that is assumed to be parallel to the wind direction. If the wind direction is incorrect or there is a yaw error, the orientation of the rotor axis will be wrongly determined, and consequently the deviation of the wake center from the rotor axis shown in Fig. 12 will be erroneous. The absolute wake center locations in the Galion-based coordinates system (x', y') are plotted in Fig. 13 for the first two hours of the wake2 period. The wake center moves as the wind direction changes (shifting from left to right in Fig. 13) and the wake center only varies within 10 m when the wind direction stabilizes (see upper right corner in Fig. 13). There is always an offset between the wake center and the wind direction or the rotor axis, indicating possible wind direction errors or wind turbine yaw errors. However, neither of these errors can be verified without the operational data of WT2.

6 Conclusions

This paper presents the results from an experiment designed to measure wakes of a continuously yawing wind turbine with a ground-based coherent Doppler wind lidar in a large wind farm. Three-stack PPI scans are used. Radial velocities from the lidar are converted to wind speeds using the concurrent wind direction measurements from a meteorological mast. By assuming no yaw error of the wind turbine, wake profiles are estimated for every 0.2D downwind of the wind turbine using a weighted averaging method. Wake statistics such as wake deficit, wake width and wake center are then automatically detected from the wake profiles. Consistent with previous observations, the wake deficit decreases and the wake width increases with increasing downwind distance. The location of the wake center shows no variation with downwind distance, but its movement with changing wind direction is observed.

While the lidar measurements from PPI scans can be used to automatically detect wakes from multiple directions, care must be taken with regards to the height variation over the conical surface. Depending on the wind direction or wind turbine orientation, the derived wake statistics can be contaminated by wind shear and become difficult to interpret. Hence, future work needs to seek the possibility of accounting for the height variation by incorporating three dimensional wake properties in wake profile estimate and provide practical guidelines for designing PPI scans for lidar wake measurements.

7 Acknowledgments

This project is funded by the National Science Foundation #1464383 and Department of Energy (DE EE0005379). Meanwhile, the authors gratefully acknowledge the owner of the wind farm and help from SgurrEnergy.

8 References

[1] Barthelmie, R.J., et al., *Modelling and measuring flow and wind turbine wakes in large wind farms offshore.* Wind Energy, 2009. **12**(5): p. 431-444. DOI: 10.1002/we.348.
[2] Barthelmie, R.J., et al., *Comparison of wake model simulations with offshore wind turbine wake profiles measured by sodar.* Journal of Atmospheric and Oceanic Technology, 2006. **23(7)**: p. 888-901.
[3] Aitken, M.L., et al., *Quantifying Wind Turbine Wake Characteristics from Scanning Remote Sensor Data.* Journal of Atmospheric and Oceanic Technology, 2014. **31**(4): p. 765-787.
[4] Iungo, G.V. and F. Porté-Agel, *Volumetric Lidar Scanning of Wind Turbine Wakes under Convective and Neutral Atmospheric Stability Regimes.* Journal of Atmospheric and Oceanic Technology, 2014. **31**(10): p. 2035-2048.
[5] Hirth, B.D., et al., *Coupling Doppler radar-derived wind maps with operational turbine data to document wind farm complex flows.* Wind Energy, 2015. **18**(3): p. 529-540.
[6] Barthelmie, R.J., et al., *3D wind and turbulence characteristics of the atmospheric boundary-layer.* Bulletin of the American Meteorological Society, 2014. **95**: p. 743–756.
[7] Daley, R., *Atmospheric data analysis.* 1993: Cambridge university press.

Comparison between experiments and Large-Eddy Simulations of tip spiral structure and geometry

S. Ivanell[1,2], T. Leweke[3], S. Sarmast[1], H.U. Quaranta[3], R.F. Mikkelsen[4], J.N. Sørensen[1,4]

[1]Uppsala University, Wind Energy Section, Campus Gotland, 621 67 Visby, Sweden
[2]KTH Mechanics, 100 44 Stockholm, Sweden
[3]IRPHE, CNRS, Aix-Marseille Université, Centrale Marseille, 13384 Marseille, France
[4]Wind Energy Department, Technical University of Denmark, 2800 Lyngby, Denmark

E-mail: `stefan.ivanell@geo.uu.se`

Abstract.
Results from Large-Eddy Simulations using the actuator line technique have been validated against experimental results. The experimental rotor wake, which forms the basis for the comparison, was studied in a recirculating free-surface water channel, where a helical vortex was generated by a single-bladed rotor mounted on a shaft. An investigation of how the experimental blade geometry and aerofoil characteristics affect the results was performed. Based on this, an adjustment of the pitch setting was introduced, which is still well within the limits of the experimental uncertainty. Excellent agreement between the experimental and the numerical results was achieved concerning the circulation, wake expansion and pitch of the helical tip vortex. A disagreement was found regarding the root vortex position and the axial velocity along the centre line of the tip vortex. This work establishes a good base for further studies of more fundamental stability parameters of helical rotor wakes.

1. Introduction

Studies focusing on understanding the stability properties of the wakes generated by wind turbine rotors have been intensified during the last decade. The reason is that wind farms are becoming larger and larger and therefore the importance of a better understanding of the wake physics is becoming more and more important.

The stability of tip vortex systems has been studied by Widnall [1] and Leishman *et al.* [2], with the main emphasis on helicopter rotors. In 2010, Ivanell *et al.* [3] performed a numerical investigation of the tip spiral instability of a wind turbine wake. This work has been followed up by Sarmast *et al.* [4] and by Sørensen *et al.* [5].

Experimental studies of wind turbine tip spiral instabilities have been carried out by Leweke *et al.* [6] and Bolnot [7]. In general, the stability results by Leweke *et al.* [6] agree well with the results of Ivanell *et al.* [3] and Sarmast *et al.* [4]. However, no study known to the authors have been performed to make a detailed comparison between experiments and simulations considering the same configuration, with focus on tip spiral instability.

With this article we take the first step towards a thorough comparison between an experimental study and a numerical method. When studying instabilities of tip vortices of a wind turbine using experimental methods, the detailed turbine characteristics are usually not of importance, since the aim in that case is to establish a realistic helical vortex system. It is of course important that the blades are identical and that there is no pitch offset in order not to introduce disturbances to the spiral system. But the design of the blades, the lift and drag force characteristics, are not an essential ingredient, as long as realistic distributions are employed. However, when comparing experimental studies with numerical methods, the detailed knowledge of the blade design is needed. In this study, the rotor is simulated using the actuator line method, in which the detailed geometry of the blades are not needed. Instead, aerofoil characteristics, i.e., lift and drag distributions, together with twist and chord length distributions, are essential. Considering the scale of the turbine, which in this specific case has a radius $R_0 = 80$ mm, the aerofoil characteristics, and especially the drag component, is difficult to establish, because of the low Reynolds number.

Before being able to make a comprehensive numerical/experimental comparison of the breakdown process of the tip spiral system, as it was performed separately by e.g. Sarmast *et al.* [4] and Leweke *et al.* [6], there are many challenges to ensure an adequate comparison of the basic wake configuration. The aim of this article is to establish a comparison of the main flow properties, to conclude on how different parameters match and to quantify the differences due to the setup of the experiment and the numerical method. Studies of the stability of the helical system will therefore not be included in this study, but will be treated in future work.

2. Experiments

The experimental rotor wake, which is the basis for the present comparison, was studied in the recirculating free-surface water channel at IRPHE Marseille, having a test section of dimensions 38 cm (height) × 50 cm (width) × 150 cm (length). A helical vortex was generated near the test section entry by a single-bladed rotor mounted on an ogive-tipped shaft (diameter 15 mm) and driven by a computer-controlled stepper motor outside the channel using a belt (figure 1a). The rotor blade geometry, shown schematically in figure 1(b), is based on the low-Reynolds number airfoil A18 by Selig *et al.* [8]. It is designed to operate in the wind turbine regime and produce a constant radial circulation distribution (Joukowsky rotor) over the outer 75% of the span, in order to generate a highly concentrated tip vortex. The (measured) chord and twist distributions of the blade are shown in figure 3, they correspond to the design distributions. The rotor has a radius $R_0 = 80$ mm and a tip chord $c_{tip} = 10$ mm. Details of the design procedure, as well as the experimental set-up and procedure, can be found in Bolnot [7] and Quaranta *et al.* [9]. For the present experiments, the blade is rotated at a frequency $f_0 = 6$ Hz, and placed in a uniform incoming flow of velocity $U_0 = 36$ cm/s. This results in a tip chord-based Reynolds number of about 30000 and a tip speed ratio $2\pi f_0 R_0/U_0 = 8.4$, representing the design conditions for the blade geometry. No cavitation occurred for these parameter values.

For the comparison, it is of utmost importance that the geometry of the rotor used in the comparison is identical to the one used in the experiments. The experimental set-up was initially not designed for an accurate analysis of rotor blade aerodynamics, but for the study of helical vortex dynamics in the rotor wake, for which a precise blade alignment was not needed. The relatively simple mounting of the rotor object (figure 1) on the shaft resulted in a limited precision of the alignment of the blade with respect to the axis of rotation. Although the mounting was always done in the same way, this could result in a constant offset of the order of 1° in the twist distribution of figure 3(b). Since it was not possible to measure this offset experimentally, the relevant value was found from comparison with numerical simulations. In figure 3 we show, as an example, the experimental evolution of the wake radius and various corresponding numerical results obtained for different twist offsets β. It is found that an offset angle of $\beta = 1.5°$ gives the best agreement between experiment and simulation. Additional comparisons (not shown) were also made for the other wake parameters discussed in section 4, resulting in the same optimal offset. Hence, this value is the one used in the following; the corresponding twist distribution is

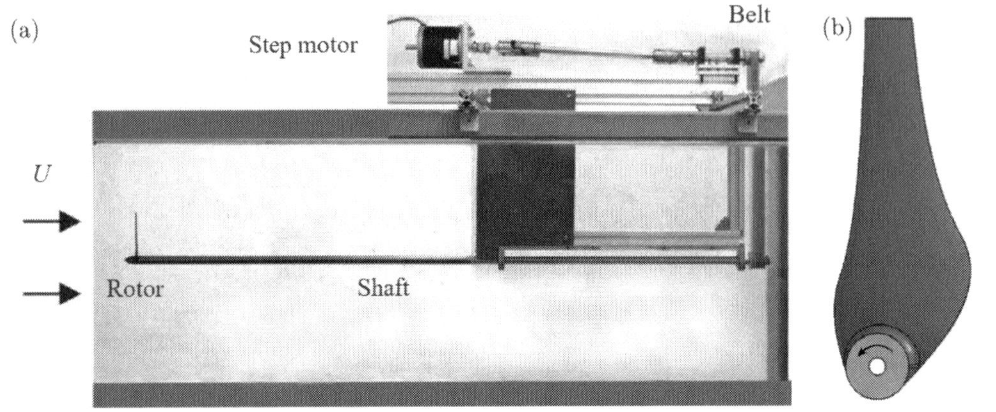

Figure 1. (a) Experimental set-up (side view of the water channel test section) and (b) schematic of the rotor blade geometry.

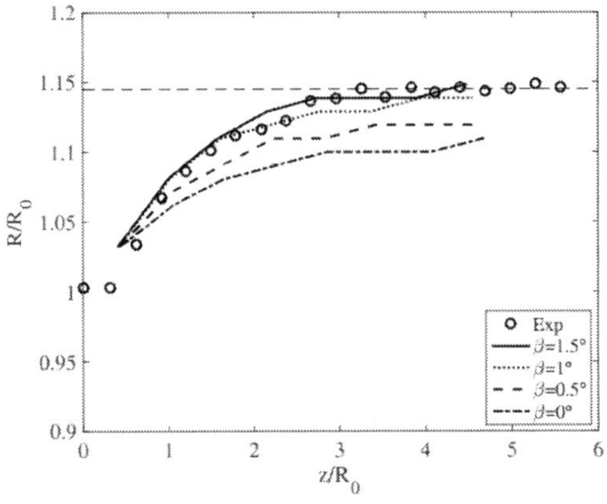

Figure 2. Downstream evolution of helix radius. β is the blade pitch offset.

shown as a line in figure 3.

The vortex structures are visualized using fluorescent dye illuminated by the light of an argon ion laser. The dye is either washed off the blade tip or injected at a fixed location near the blade tip trajectory. Quantitative measurements were carried out using two-component Particle-Image Velocimetry [11] in planes containing the helix axis. Further details can be found in [7] and [9].

3. Numerical method and setup

The actuator line (ACL) method was introduced by Sørensen & Shen [10], as an aerodynamic model for simulating rotor wakes. In this approach, the flow around the rotor is governed by the three-dimensional incompressible Navier–Stokes equations, while the influence of the blades on the flow field is approximated by body forces. These forces, which are determined using the local flow field combined with tabulated airfoil data, and are distributed radially along a line representing the blade of the wind turbine. At each point of the line, the force is smeared among neighboring nodes with a three-dimensional Gaussian smearing function [11]. The ACL method is implemented into the EllipSys3D code, developed by Michelsen [12] and Sørensen

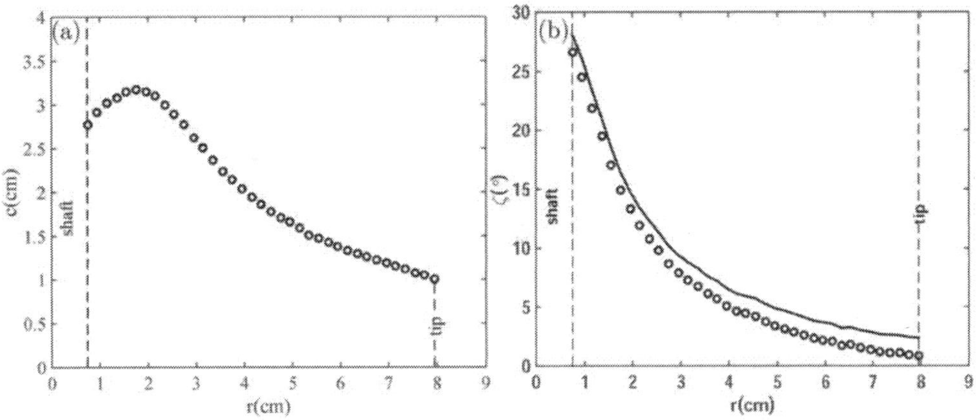

Figure 3. Radial distributions of (a) chord c and (b) twist angle ζ of the blade. Measurements (∘) and twist distribution used in the computations (—).

[13]. Large-Eddy Simulation (LES) is employed, in which the large scales are resolved and the small scales are modeled by the eddy viscosity based sub-grid scale model by Ta Phouc [14]. The computations are conducted using an axisymmetric 360° polar grid with 49 million grid points, in which each blade is represented by an actuator line of 105 grid points. The simulations are performed in a rotating frame of reference, where the actuator lines are standing still. In order to replicate experimental blockage effects, a cross section area corresponding to the one of the water tunnel is considered.

The current simulations are based on the operating conditions of the experiments, where the rotor operates at its design condition. Previous studies of Hansen *et al.* [15] and Sarmast *et al.* [16] show that the accuracy of the actuator line approach is directly related to the quality of the airfoil characteristics. Figure 4 shows the airfoil characteristics used in rotor computations. Prior measured 2D characteristics of an A18 airfoil has been reported for aerodynamic coefficients above $Re_c = 40000$, together with the lift coefficient distributions at $Re_c = 30000$. Xfoil computations are performed to obtain the missing drag coefficients at $Re_c = 30000$.

4. Results

4.1. Helical vortices

Figures 5(a) and 5(b) show the helical vortices produced by the one-bladed rotor operating at its optimum condition $\lambda = 8.4$ for the experiments and the actuator line computations, respectively. In the experiments, fluorescent dye is washed off from the blade tip, while the helical vortices are visualised using an iso-surface of the vorticity magnitude. This visual comparison indicates a very good qualitative agreement between the experiment and the simulation.

Figure 6 illustrates the downstream evolution of the helix radius, i.e. the wake expansion, and the helix pitch. The agreement is again very good.

Figure 7 shows a comparison between experimental and numerical results of the local vorticity in the vortex core. We here consider a helical vortex filament in a cylindrical reference frame (r, θ, z) linked to the rotor geometry, while an additional cylindrical reference frame (ρ, ϕ, ξ), linked locally to the tip vortex, is introduced.

The phase-averaged azimuthal (with respect to the rotor) vorticity field in the near-wake centre plane of the rotor, obtained from 100 instantaneous PIV fields, is shown in figure 7(a). Figure 7(b) shows the azimuthally (with respect to the vortex) averaged velocity profile

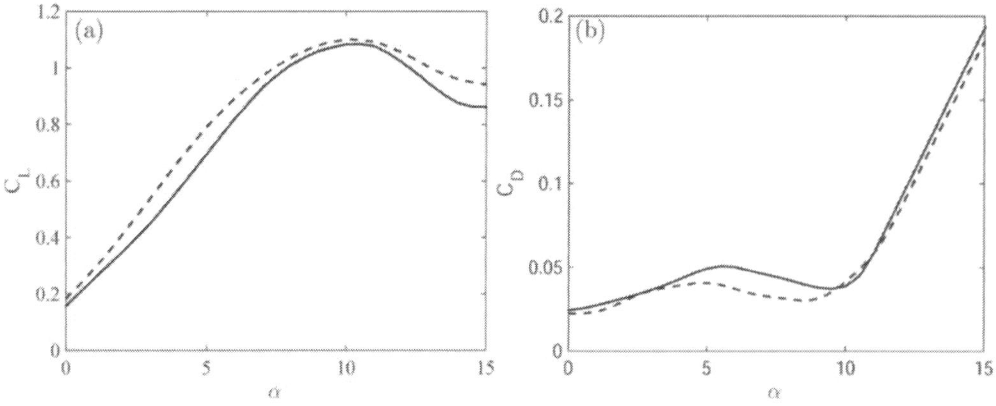

Figure 4. (a) Lift and (b) drag coefficient distributions of A18 airfoil. Re_c=30,000 (—) and Re_c=40,000 (- -). The drag coefficients for Re_c = 30000 are obtained using XFOIL computations while the rest of aerodynamic coefficients are measurements reported by Selig *et al.* [8].

of the blade tip vortex, from which the vortex core size parameter $a_{\max}$, i.e., the radius of maximum azimuthal velocity can be found. The comparison shows some differences, but overall the experimental and numerical results are very close.

Figure 8 shows the downstream evolution of the tip vortex circulation, calculated inside a circle of diameter $0.2R_0$ around the vortex centre (circle in figure 7a), and of the core radius $a_{\max}$. The circulation values agree well, even if it was not possible to extract experimental data further downstream. Although, some discrepancies in the trend can be seen the level is in general correct. The core radius is of the same order of magnitude in the near wake. In the simulations, a slight dependence on the grid resolution was found, and the core size increases with downstream distance. Part of this increase can be explained by the viscous diffusion of vorticity. The core growth associated with this effect can be estimated by considering a two-dimensional Lamb-Oseen vortex, with Gaussian vorticity distribution of the form $\omega_\theta(\rho) = (\Gamma/\pi a^2)\exp(-\rho^2/a^2)$. The core radius a, which is linked to the radius $a_{\max}$ of maximum swirl velocity by $a_{\max} \simeq 1.12a$, increases in time according to $a^2 = 4\nu(t - t_0)$, where t_0 is a virtual origin of time. Since the downstream convection velocity of the tip vortices is given by f_0h, time t can be related to the

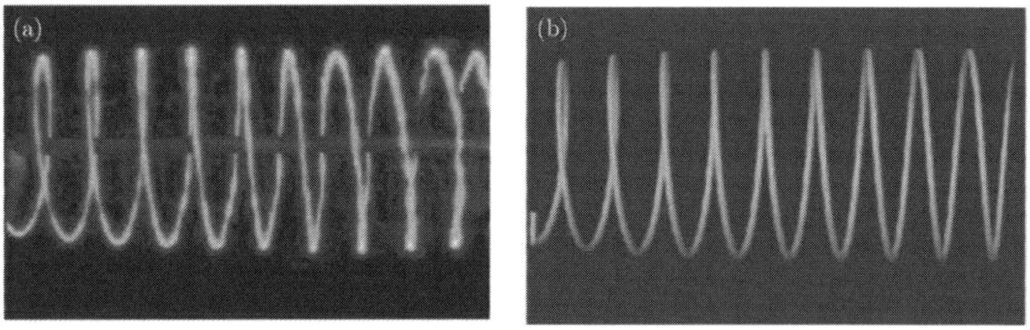

Figure 5. Visualisation of the helical vortices: (a) experiment and (b) numerical computation.

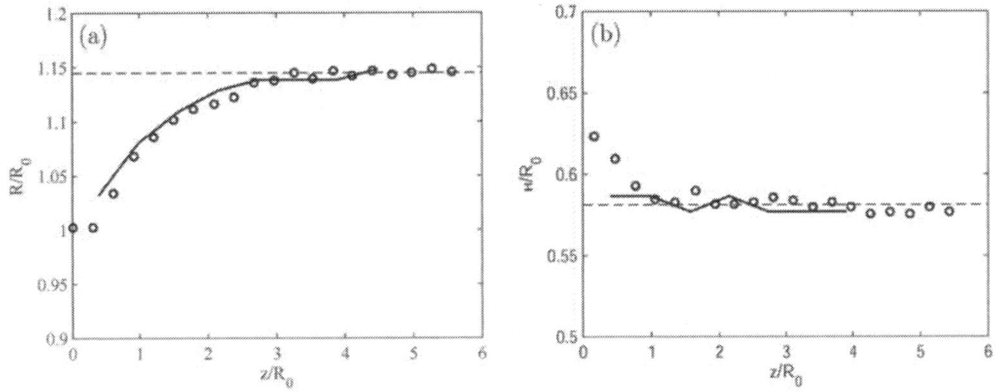

Figure 6. Downstream evolution of helix parameters: (a) radius and (b) pitch. Experiment (○) and numerical computations (—).

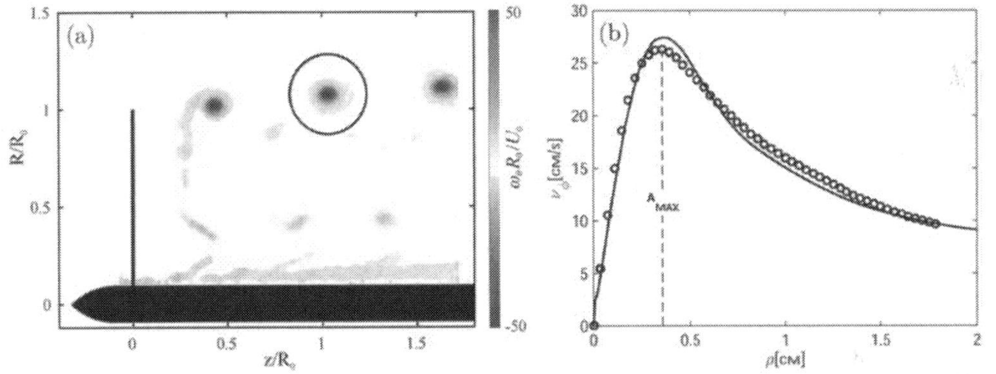

Figure 7. (a) Measured azimuthal (θ) vorticity in the central plane of the rotor. (b) Azimuthal (ϕ) velocity of the vortex inside the circle in (a): experiment (○) and numerical computation (—).

axial position z by $t = z/(f_0 h)$. The dashed line in figure 7(b) represents a fit of the experimental data to the theoretical growth given by $a_{\max} = 1.12\{4\nu[(z/R_0)f_0/(h/R_0) - t_0]\}^{1/2}$, with t_0 as fitting parameter. The faster growth observed in the simulations is possibly linked to an increased diffusion of vorticity due to the numerical scheme.

4.2. Near wake comparison

Figure 9 shows the wake development behind a rotor using the time-averaged azimuthal vorticity.

Figure 10 clearly shows the axial velocity deficit behind the rotor. From figures 9 and 10, it is clear that there is a disagreement in the position of the root vortices. The reason for this will be further discussed in the next section.

Considering the velocity field development, figure 11 shows profiles of axial velocity at three different streamwise positions. The outer wake region shows an increase of velocity due to blockage effects, which is well captured by the numerical computations. The main differences are observed in the region $0 < r/R_0 < 0.5$, which is related to the respective positions of the root vortices.

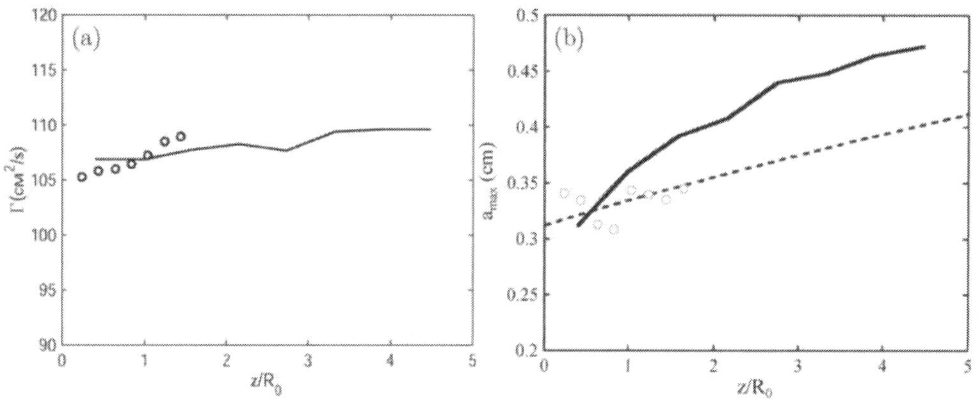

Figure 8. Vortex parameters, (a) circulation and (b) core size as determined from figure 7: experiment (○) and numerical computations (—). The dashed line in (b) represents the theoretical core growth of a Gaussian vortex.

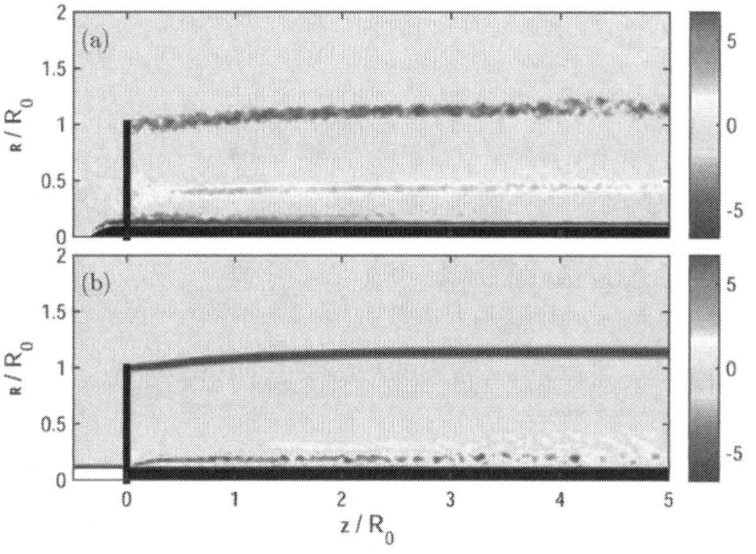

Figure 9. Azimuthal vorticity $\omega_\theta R_0/U_0$: (a) experiment and (b) numerical computation.

Figure 12 shows the axial velocity in the direction along the vortex centre line inside the tip vortices. Figure 12(a) contains the simulation results, extracted from the first four vortex cores in a plane orthogonal to the vortices. The averaged values from these curves are then depicted in figure 12(b), where it is compared to the experimentally observed profile. The profiles show a velocity deficit in the vortex core in the frame of reference of the blade, presumably due to losses in the blade boundary layers, represented by the blade drag. It is observed that the maximum axial velocity is significantly larger, and the profile narrower in the experiments. The total momentum loss (area under the curves) is of the same order in both cases, but it is about 30% higher in the experiments. The shape of the axial velocity profile in the simulations may

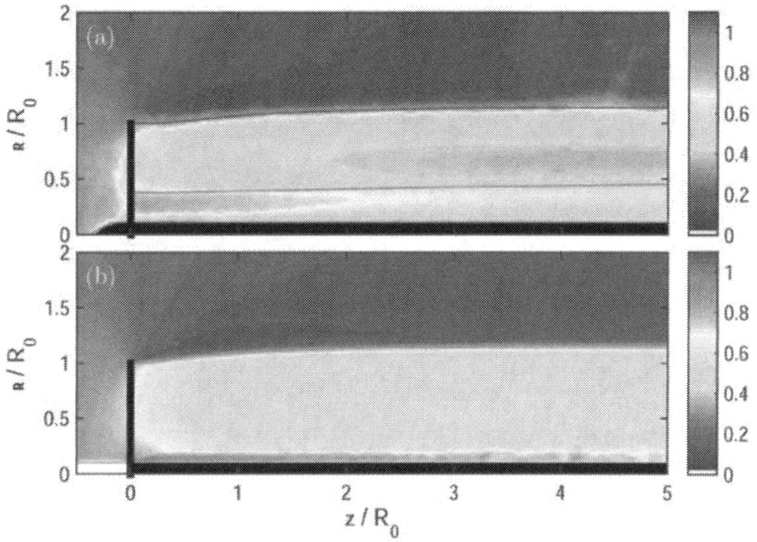

Figure 10. Axial velocity u_z/U_0: (a) experiment and (b) numerical computation.

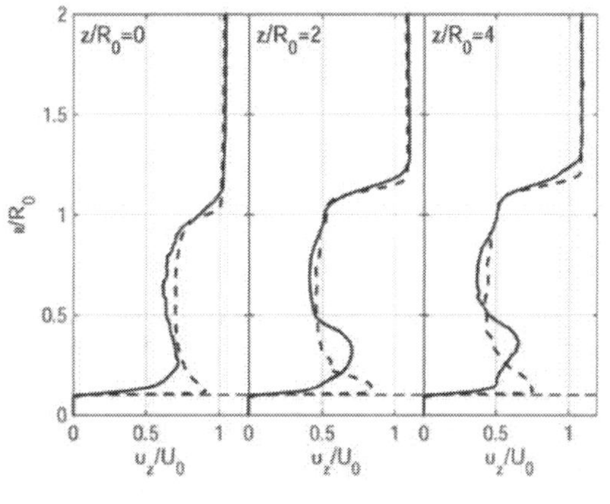

Figure 11. Axial velocity profiles computed at different streamwise positions along the wake. experiment (- -) and numerical computations (—).

depend on the resolution, and some qualitative difference may arise from the fact that the boundary layers are actually not represented in the actuator line approach. This aspect needs to be investigated in more detail.

5. Discussion and Conclusions

From the comparisons in the preceding section, it is clear that, when assuming a pitch misalignment of 1.5°, the experimental and simulated results of wake expansion, circulation, helical pitch, tip core properties, as shown in figures 6 to 8, are in very good agreement.

However, when considering the azimuthal vorticity in figure 9 and the axial velocity in figure 10, there is evidence that the root vortex radial position is closer to the rotor axis for the

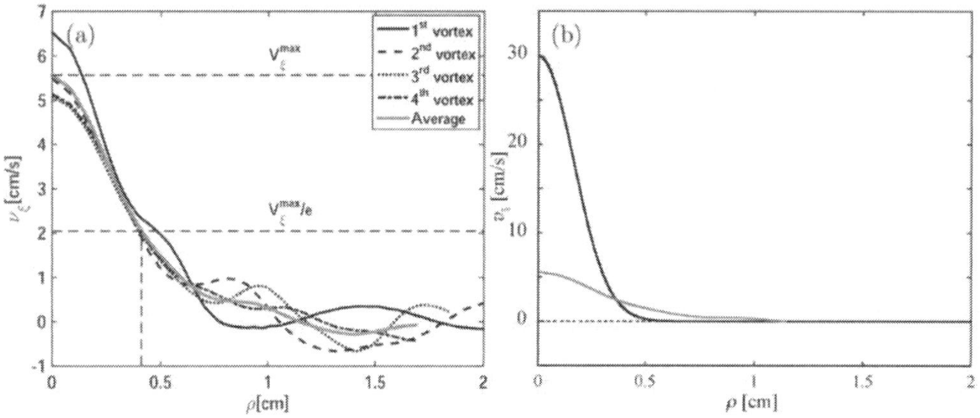

Figure 12. Axial velocity ξ of the vortex: (a) numerical computations profiles of first 4 vortices and (b) experiments (—) and numerical computations (—).

simulations compared to the experiments. This is further illustrated in figure 11, where the flow behaviour is shown at three axial positions. There can be two explanations for this. First, the centre shaft (see figure 1a), used to support the experimental rotor, is rotating at the same angular velocity as the rotor. This has been taken into account in the simulations by applying no-slip conditions in the rotation frame of reference at a radius corresponding to the shaft position in the experimental setup. There are some discrepancies with the actual experimental setup concerning the tip of the shaft extending in front of the rotor (compare figures 9a and 9b). However, the effect of this is found to be small. The simulation setup is based on an LES simulation method, combined with a body force approach. The original reason for this approach is to avoid resolving the boundary layers with the LES method. The mesh has been refined in this area but not to the extent of resolving all scales in the boundary layer. This is therefore considered to be one probable explanation for the different position of the root vortex.

The second explanation could be the body fore approach itself. The actuator line method distributes the body forces along a line representing the blades. The model turbine used in the experiment does, however, have a rather large solidity compared to a modern wind turbine. The blade chord extension is, in the inner part, significantly larger than the number of node points that the loading is applied on. This may result in an unphysical vortex roll-up process, causing a root vortex position that does not correspond to the experimental result.

The present investigation has resulted in good agreement between the experiment and numerical modelling, except for the position of the root vortex and the axial velocity along the centre line of the tip vortex. This study sets a good base for further work on more fundamental stability parameters of the tip vortex. The influence of the root vortex position needs to be further investigated. This study has indicated two possible limitations in the comparison and the numerical modelling. This also results in a good base for further studies of limitations of the actuator line method, as well as of the influence from the boundary layers in LES simulations.

Acknowledgments
This work is supported by the French *Agence Nationale de la Recherche* through the A*MIDEX grant (ANR-11-IDEX-0001-02), project LABEX MEC (ANR-11-LABX-0092) and project HELIX (ANR-12-BS09-0023-01). The simulations were performed on resources provided by the Swedish National Infrastructure for Computing (SNIC) within the project SNIC 2014/8-5.

References

[1] Widnall S E 1972 *J. Fluid Mech.* **4** 641–663

[2] Leishman G, Bhagwat M J and Ananthan S 2004 *J. Am. Helicopter Soc.* **49** 160–175

[3] Ivanell S, Mikkelsen R, Sørensen J N and Henningson D S 2010 *Wind Energy* **13** 705–715

[4] Sarmast S, Dadfar R, Mikkelsen R, Schlatter P, Ivanell S, Sørensen J and Henningson D 2014 *J. Fluid Mech.* **755** 705–731

[5] Sørensen J N, Mikkelsen R F, Henningson D S, Ivanell S, Sarmast S and Andersen S J 2015 *Philosophical Transactions of the Royal Society A: Mathematical, Physical and Engineering Sciences* **373** 20140071

[6] Leweke T, Bolnot H, Quaranta U and Dizès S L 2013 (Lyngby, Denmark: Int. Conf. on Aerodynamics of Offshore Wind Energy Systems and Wakes)

[7] Bolnot H 2012 *Instabilités des tourbillons hélicoïdaux: application au sillage des rotors* Ph.D. thesis Aix-Marseille Université Marseille, France

[8] Selig M S, Guglielmo J J, Broeren A P and Giguere P 1995 *Summary of Low-Speed Airfoil Data* (SoarTech)

[9] Quaranta H U, Bolnot H and Leweke T 2015 *J. Fluid Mech.* (submitted)

[10] Sørensen J N and Shen W Z 2002 *J. Fluid Eng.* **124** 393–399

[11] Mikkelsen R 2003 *Actuator Disc Methods Applied to Wind Turbines* Ph.D. thesis Dept. of Fluid Mech., Technical University of Denmark, DTU

[12] Michelsen J A 1994 Block structured multigrid solution of 2D and 3D elliptic PDE's Tech. Rep. AFM 94-06 Dept. of Fluid Mech., Technical University of Denmark, DTU

[13] Sørensen N N 1995 *General purpose flow solver applied to flow over hills* Ph.D. thesis Risø National Laboratory Roskilde

[14] Ta Phuoc L 1994 *Proc. of the DRET conference: 'Aérodynamique Instationnaire Turbulents - Aspects Numériques et Expérimentaux*

[15] Hansen M O L, Sørensen J N, Voutsinas S, Sørensen N and Madsen H A 2006 *Progress in aerospace sciences* **42** 285–330

[16] Sarmast S 2014 *Numerical study on instability and interaction of wind turbine wakes* Ph.D. thesis KTH, Stability, Transition and Control

Journal of Physics: Conference Series **625** (2015) 012019 doi:10.1088/1742-6596/625/1/012019

Aeroelastic large eddy simulations using vortex methods: unfrozen turbulent and sheared inflow

E. Branlard[1], G. Papadakis[2], M. Gaunaa[1], G. Winckelmans[3], T. J. Larsen[1]

[1] Department of Wind Energy, Technical University of Denmark, Frederiksborgvej 399, 4000 Roskilde, Denmark

[2] NTUA, School of Mechanical engineering, Greece

[3] Institute of Mechanics, Materials and Civil Engineering (iMMC), Université catholique de Louvain, Belgium

E-mail: ebra@dtu.dk

Abstract. Vortex particles methods are applied to the aeroelastic simulation of a wind turbine in sheared and turbulent inflow. The possibility to perform large-eddy simulations of turbulence with the effect of the shear vorticity is demonstrated for the first time in vortex methods simulations. Most vortex methods formulation of shear, including segment formulations, assume a frozen shear. It is here shown that these formulations omit two source terms in the vorticity equation. The current paper also present unfrozen simulation of shear. The infinite support of the shear vorticity is accounted for using a novel approach relying on a Neumann to Dirichlet map. The interaction of the sheared vorticity with the wind turbine is shown to have an important impact on the wake shape. The obtained wake shape are closer to the one obtained using traditional computational fluid dynamics: Results with unfrozen shear do not have the severe upward motion of the wake observed in vortex methods simulation with frozen shear. The interaction of the shear and turbulence vorticity is shown to reduce the turbulence decay otherwise observed. The vortex code implemented is coupled to an aeroelastic code and examples of aeroelastic simulations under sheared and turbulent inflow are presented.

1. Introduction

Conventional aeroelastic codes use the Blade Element Momentum method (BEM) to determine the induced velocities and the aerodynamic loads at the blade. A precomputed synthetic velocity component is added to the induced velocities to include the effect of turbulence [1]. The turbulent field, referred to as a "turbulent box", convects with the free-stream velocity under the assumption of Taylor's hypothesis of frozen turbulence [2].

Aeroelastic simulations have also been performed using higher order aerodynamic tools such as vortex methods [3] and traditional Computational Fluid Dynamics (CFD) [4], at a successively increasing computational cost. The methods have been applied to steady inflow conditions. Vortex methods have been recently used to perform non-aeroelastic Large Eddy Simulations (LES) of wind turbines by Chatelain et al. [5]. The current study is conducted by performing aeroelastic LES of wind turbines using vortex methods and by further examining the inclusion of shear in the method.

The paper is structured as follows. The theoretical and computational models are presented in a first section. The different components of the implementation are then studied in more detail. The final part presents combined analyses of the different components.

2. Model and computational setup

2.1. Problem definition

A wind turbine operating under sheared and turbulent inflow is considered. The shear and the turbulence are assumed to be given at a distance upstream of the turbine, while the process responsible for their generation is not modelled. The turbulence is assumed to be generated by an upstream mask of finite cross section upstream. The layout is sketched in Figure 1. The

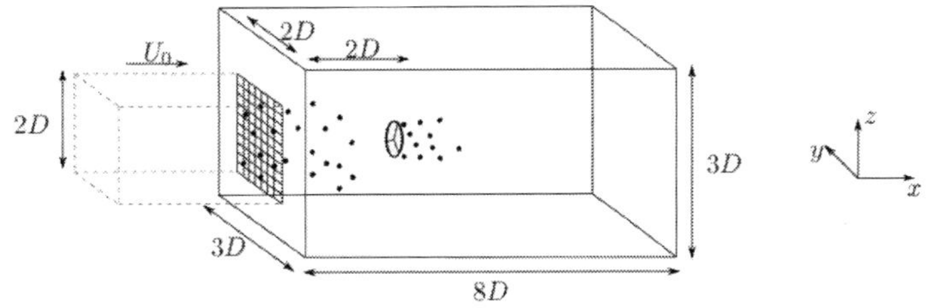

Figure 1: Computational setup. Turbulent vortex particles generated by a mask (represented by a grid) are inserted at each time step. The wind turbine wake is modeled with vortex particles. The domain is filled with particles (not shown) representing the shear vorticity, also inserted at each time step.

incompressible Navier-Stokes equation in vorticity formulation and in Lagrangian coordinates $\boldsymbol{X}(\boldsymbol{\xi},t)$, with $\boldsymbol{X}(\boldsymbol{\xi},0)=\boldsymbol{\xi}$, takes the form:

$$\frac{d\boldsymbol{\omega}}{dt}(\boldsymbol{X},t) = [\boldsymbol{\nabla u}(\boldsymbol{X},t)]\cdot\boldsymbol{\omega}(\boldsymbol{X},t) + \nu\nabla^2\boldsymbol{\omega}(\boldsymbol{X},t), \qquad \frac{d\boldsymbol{X}}{dt}(\boldsymbol{\xi},t) = \boldsymbol{u}(\boldsymbol{X},t) \tag{1}$$

where $\boldsymbol{u}$ is the flow velocity, ν is the fluid viscosity, and $\boldsymbol{\omega} = \nabla\times\boldsymbol{u}$ is the vorticity. Of the wind turbine, only the blades are modelled and a lifting line approach is used: The blades are represented by a line of bound vorticity which shed and trailed vorticity in the wake (see e.g. [6]). The entire system of the fluid and the turbine is thus modelled in a unified vorticity formulation.

2.2. Vorticity inversion and Helmholtz decomposition

For most physical applications, a velocity field $\boldsymbol{u}$ can be written according to the Helmholtz decomposition [7]: $\boldsymbol{u} = \boldsymbol{U}_0 + \boldsymbol{u}_\omega + \boldsymbol{u}_\phi$ where $\nabla\times\boldsymbol{U}_0 = 0$, $\nabla\cdot\boldsymbol{U}_0 = 0$, $\nabla\cdot\boldsymbol{u}_\omega = 0$ and $\nabla\times\boldsymbol{u}_\phi = 0$. The divergent free component is written as function of a vector potential $\boldsymbol{\psi}$ as $\boldsymbol{u}_\omega = \nabla\times\boldsymbol{\psi}$. Choosing the gauge $\nabla\cdot\boldsymbol{\psi} = 0$, the definition $\boldsymbol{\omega} = \nabla\times\boldsymbol{u}$ leads to $\Delta\boldsymbol{\psi} = -\boldsymbol{\omega}$. For incompressible flows $\boldsymbol{u}_\phi$ accounts for boundary conditions. In absence of boundaries and for a flow occupying the entire space $\boldsymbol{u}_\phi = 0$. In an unbounded space and in the absence of boundaries, the velocity can be retrieved from the vorticity field using the Biot-Savart law:

$$\boldsymbol{u}_\omega(\boldsymbol{x},t) = \int_D \boldsymbol{K}(\boldsymbol{x}-\boldsymbol{x}')\times\boldsymbol{\omega}(\boldsymbol{x}',t)\,d\boldsymbol{x}' , \qquad \boldsymbol{K}(\boldsymbol{r}) = -\frac{\boldsymbol{r}}{4\pi\, r^3} \tag{2}$$

where the integral is taken over $D = D_\omega$, with D_ω the support of vorticity, possibly infinite. If one restricts the integral to a smaller domain $D = D_{\text{in}}$, the contribution from the vorticity outside of the domain is accounted for by means of a Neumann to Dirichlet map (or generalized Helmholtz decomposition) which ensures the continuity of tangential and normal velocity at the domain interface ∂D_{in} [8]. The velocity field is then written $\boldsymbol{u} = \boldsymbol{U}_0 + \boldsymbol{u}_\omega + \boldsymbol{u}_{\text{ext}}$ with

$$\boldsymbol{u}_{\text{ext}}(\boldsymbol{x}) = \int_{\partial D_{\text{in}}} \left[-\boldsymbol{K}(\boldsymbol{x}-\boldsymbol{x}')\,u_n(\boldsymbol{x}') + \boldsymbol{K}(\boldsymbol{x}-\boldsymbol{x}')\times\boldsymbol{u}_\tau\right]\,d\boldsymbol{x}' \tag{3}$$

where, u_n is the component of the input field normal to ∂D_{in} such that $u_n = \boldsymbol{u} \cdot \boldsymbol{n}$, with $\boldsymbol{n}$ pointing towards the interior of the domain, and $\boldsymbol{u}_\tau = \boldsymbol{n} \times \boldsymbol{u}$. The gradient $\nabla \boldsymbol{u}_{\text{ext}}$ is directly obtained from the expression of $\boldsymbol{u}_{\text{ext}}$. In the current study D_{in} represents the computational domain and it is assumed for simplicity that the contribution from $\boldsymbol{u}_{\text{ext}}$ is time invariant.

2.3. *The numerical tool*

Equation 1 is solved using a vortex particle approximation [9]. The vorticity field is written as a sum of particle strengths $\boldsymbol{\alpha}_p = \int_{V_p} \boldsymbol{\omega}\, dV$ weighted by a local function ζ, as

$$\boldsymbol{\omega}(\boldsymbol{x}, t) \approx \boldsymbol{\omega}_\alpha(\boldsymbol{x}, t) = \sum_p \zeta(|\boldsymbol{x} - \boldsymbol{x}_p|)\, \boldsymbol{\alpha}_p(t) \tag{4}$$

where $\boldsymbol{x}_p$ is the position of the particle and V_p a volume attributed to the particle. In the present study a second order exponential function is used for ζ.

The velocity and its gradient in the domain can be determined in two ways. Both approaches are tested in this study. Inserting Equation 4 into Equation 2, the particle positions are used as quadrature points and the Biot-Savart integral becomes a summation over the particles. A second order multipole tree algorithm is used for a fast evaluation of the summation [10]. This will be here referred to as the "mesh-free" method and was added to the in-house solver *Omnivor* since its previous description [11]. Alternatively, the vorticity carried by the particles can be projected onto a grid and the velocity is then retrieved using a Poisson solver (Particle Mesh Methods): solving $\nabla^2 \boldsymbol{\psi} = -\boldsymbol{\omega}$ for $\boldsymbol{\psi}$ and getting $\boldsymbol{u} = \nabla \times \boldsymbol{\psi}$, or solving $\nabla^2 \boldsymbol{u}_\omega = -\nabla \times \boldsymbol{\omega}$ for $\boldsymbol{u}$ directly. The second order Poisson solver, *Yaps*, developed at NTUA is used [12] and is coupled to the *Omnivor* library. In order to facilitate performance of the Poisson solver, the method of local corrections is applied [13] together with a domain decomposition method described in [14]. Regardless of the approach used, particles are convected using a first-order forward Euler scheme. Redistribution (re-meshing) of the particles on a regular lattice every few time-steps is performed to ensure the coverage of the space and retain accuracy [15]. Even in the "mesh-free" formulation, an underlying grid and computational domain is thus defined.

The vortex code *Omnivor* was coupled [11] to the aero-servo-elastic code *HAWC2* [16] using a strong coupling: At each sub-iteration of the structural solver new aerodynamic loads on the lifting-line are computed by the vortex code based on the new blade position using an iteration procedure on the bound and near-wake circulation [6] and using a form of the Beddoes-Leishman dynamic stall model [17]; at the end of a time step Equation 1 is solved for the free vorticity.

2.4. *Viscosity and LES*

For the high Reynolds number considered here (Re $\approx [10^5 - 10^7]$), the omission of the diffusion associated with the molecular viscosity can be justified. Yet, the inclusion of turbulence calls for an account of viscous effects. In the current study, the effects of the smaller scales is modelled by limiting the amount of stretching and by using a low-order kernel in the re-meshing scheme to act as a subgrid-scale (SGS) filter [18, p. 140]. The λ_1 kernel is here used [15]. This approximation was used in a previous study by the authors [19]. The approach is convenient when the mesh-free method is used. Indeed, the computation of viscous terms [20] or SGS models [21] is possible by integral approximation of the differential operators but the method can be computationally expensive for large amount of particles. Using the Poisson-solver approach though, finite difference tools are directly available on the grid and can be used for computing the diffusion and different sub-grid scale models [5, 15]. Considering the coupling with the *Yaps* Poisson solver the finite-difference sub-grid scale models will be tried in the near future.

2.5. *The simulation setup*

The computational domain used in the present study has dimensions $8D \times 3D \times 3D$ as shown in Figure 1. The three-bladed Nordtank 500kW stall regulated wind turbine located at one of DTU Wind Energy's test sites is modelled. The turbine diameter is $D = 2R = 41$m and it rotates at the constant speed $\Omega = \dot{\theta} = 27.1$ RPM. The free-stream velocity is chosen as $U_0 = 7.5\text{m.s}^{-1}$ and the tip-speed ratio is then $\lambda = 7.7$. The shear is assumed linear of slope $du_s/dz = 0.079\text{s}^{-1}$ so that the free-stream velocity at the blade tip varies between 6 and 9m.s^{-1}. The domain consists of a grid of $256 \times 96 \times 96$ cells resulting in 2.4M particles. Re-meshing is performed every 10 time steps. The time step chosen is $dt = 0.086$s ($d\theta = 14°$), while the total simulation time is 300s. Particles exiting the domain are discarded but a hybrid wake approach could be used [22].

3. Separate analysis of the different models involved

3.1. *Accuracy of the numerical tools*

The accuracy of vortex particle methods depends to a large extent on the determination of the velocity $\boldsymbol{u}$ and deformation $[\boldsymbol{\nabla u}].\boldsymbol{\omega}$ from the distribution of vorticity. The accuracy of both the mesh-free Biot-Savart law and the Poisson solver approaches are studied by comparison with the analytical formulae of the Hill's spherical vortex [23, 24]. The particles are initialized using the analytical vorticity value on a regular grid of spacing h/a where a is the radius of the Hill's sphere. The domain extends from $-1.2a$ to $1.2a$ is all directions. The velocity is then computed by the two methods on two control volumes: one entirely located inside the sphere and one crossing the sphere surface. Both volumes are made of 1000 control points that remain unchanged for the different grid spacings tried. The second control volume is more challenging since the vorticity experiences a discontinuity at the sphere boundary. The mean relative error between the numerical and analytical velocity and deformation on the two control volumes are shown in Figure 2. The Poisson solver uses a grid of the same spacing as the particle spacing. Different grid spacing h were used, the lowest grid resolution corresponding to 2000 particles and the highest to 14M. Resolutions up to 628M particles were also tried. Both methods are seen to be second order as expected. Better performance of the Poisson solver was observed in the determination of the velocity. For the mesh-free approach, the accuracy can easily be improved by using higher order 3D regularization kernels ζ [25, 26] as long as the vorticity distribution has sufficient continuous derivatives [27]. For the Poisson solver, both the order of the solver and the order of the differentiation scheme are important. Since the solution of the Poisson problem highly depends on the quality of the information projected on the grid, it is usually recommended to use more than one particle per cell to increase accuracy and limit diffusion during the re-meshing procedure. These will be considered in future studies.

3.2. *Representation of shear*

Inviscid flows are used in this section but the results can be extended to viscous flows.

Shear only A vertical shear profile $\boldsymbol{u}_s = U_s(z)\boldsymbol{e}_x$ is associated to a vorticity profile $\boldsymbol{\omega}_s = \frac{dU_s}{dz}\boldsymbol{e}_z$ and the gradient of the velocity field is $[\boldsymbol{\nabla U}_s] = \frac{dU_s}{dz}\,\boldsymbol{e}_y \otimes \boldsymbol{e}_z$. The vorticity equation for a shear flow is directly obtained from Equation 1 as:

$$\frac{\partial \boldsymbol{\omega}_s}{\partial t} + [\boldsymbol{\nabla \omega}_s] \cdot \boldsymbol{U}_s = [\boldsymbol{\nabla U}_s] \cdot \boldsymbol{\omega}_s \qquad \text{(Unfrozen Shear - Shear only)} \qquad (5)$$

All three terms are zero for a uniform and steady shear flow.

Shear and additional vorticity Another source of vorticity is here assumed to be present in the flow such that the total vorticity is $\boldsymbol{\omega} = \boldsymbol{\omega}_s + \tilde{\boldsymbol{\omega}}$ and the total velocity is $\boldsymbol{u} = \boldsymbol{U}_s + \tilde{\boldsymbol{u}}$. The

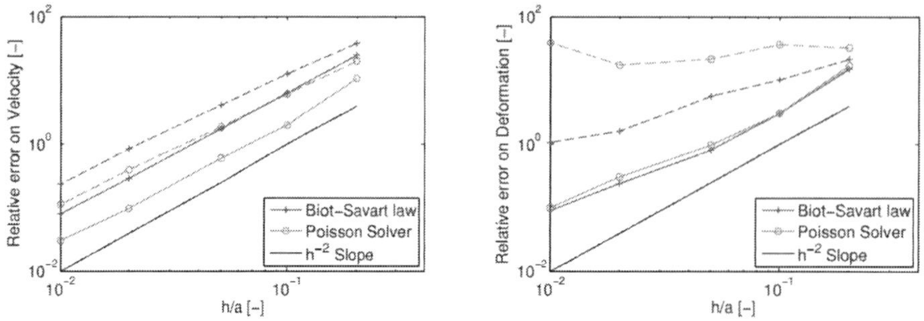

Figure 2: Accuracy of the current particle method using the Biot-Savart law or the Poisson solver by comparison with the analytical solution from the Hill's vortex. Relative error in velocity (left) and deformation (right). The errors were evaluated in a domain inside the Hill's vortex (plain lines), and in a domain surrounding part of the boundary of the vortex (dashed lines). The latter present more challenges due to the discontinuity of the vorticity across the boundary.

evolution of the total vorticity is given in Equation 1, written again below:

$$\frac{\partial \boldsymbol{\omega}}{\partial t} + [\boldsymbol{\nabla}\boldsymbol{\omega}]\cdot\boldsymbol{u} = [\boldsymbol{\nabla}\boldsymbol{u}]\cdot\boldsymbol{\omega} \qquad \text{(Unfrozen Shear - Full interaction)} \qquad (6)$$

Frozen shear The term "frozen shear" will be used to denote the case where the additional vorticity does not affect the shear vorticity. Developing the expressions of $\boldsymbol{u}$ and $\boldsymbol{\omega}$ in Equation 6, and using the fact that Equation 5 holds if the shear is frozen, leads to:

$$\frac{\partial \tilde{\boldsymbol{\omega}}}{\partial t} + [\boldsymbol{\nabla}\tilde{\boldsymbol{\omega}}]\cdot(\boldsymbol{U}_s+\tilde{\boldsymbol{u}}) = ([\boldsymbol{\nabla}\boldsymbol{U}_s]+[\boldsymbol{\nabla}\tilde{\boldsymbol{u}}])\cdot\boldsymbol{\omega} - [\boldsymbol{\nabla}\boldsymbol{\omega}_s]\cdot\tilde{\boldsymbol{u}} + [\boldsymbol{\nabla}\tilde{\boldsymbol{u}}]\cdot\boldsymbol{\omega}_s \qquad \text{(Frozen Shear)} \quad (7)$$

Frozen shear - Erroneous approach In all vortex methods wind turbine simulations known to the authors, the wind shear is assumed frozen and included as part of the "free-stream" [28, 29, 22]. In vortex particle simulations, the term $[\boldsymbol{\nabla}\boldsymbol{U}_s]$ is added to the gradient $[\boldsymbol{\nabla}\tilde{\boldsymbol{u}}]$ in order to include the deformations due to the shear. In vortex segments simulations this term is automatically included since the vortex segments are convected as material elements. For such particle or segment simulation the vorticity equation solved for is:

$$\frac{\partial \tilde{\boldsymbol{\omega}}}{\partial t} + [\boldsymbol{\nabla}\tilde{\boldsymbol{\omega}}]\cdot(\boldsymbol{U}_s+\tilde{\boldsymbol{u}}) = ([\boldsymbol{\nabla}\boldsymbol{U}_s]+[\boldsymbol{\nabla}\tilde{\boldsymbol{u}}])\cdot\tilde{\boldsymbol{\omega}} \qquad \text{(Frozen Shear - Erroneous)} \qquad (8)$$

Comparison of Equation 7 and 8 reveals that most vortex code implementations with frozen shear omit two terms. The two additional terms are easily implemented in a vortex particle simulation. The term $[\boldsymbol{\nabla}\boldsymbol{\omega}_s]\cdot\boldsymbol{u}$ is zero for a linear shear but the term $[\boldsymbol{\nabla}\tilde{\boldsymbol{u}}]\cdot\boldsymbol{\omega}_s$ is non negligible. The consequence of this omission for wind turbine wakes will be discussed in subsection 4.2.

Numerical implementation for Unfrozen Shear A novel approach to account for the shear vorticity and its possible interaction with the additional vorticity is described in the following. The shear vorticity contained within the computational domain is interpolated onto vortex particles. Since the shear vorticity has an infinite support, the external map from Equation 3 is used to account for the velocity $\boldsymbol{u}_{\text{ext}}$ and gradient $[\boldsymbol{\nabla}\boldsymbol{u}_{\text{ext}}]$ due to the vorticity outside of the computational domain. In a time-stepping simulation, the "shear" vortex particles convecting outside of the domain at the outlet are discarded whereas new particles are inserted at the inlet. The particles are inserted on a regular grid of spacing h in the $y-z$ plane of the inlet. The validation of the whole procedure, with and without time-stepping, is shown in Figure 3 for a step shear and a linear shear. The contribution of the external map from Equation 3 is crucial to reconstruct the exact shear and allow long term simulations with the proper deformation of the vortex particles. The method can be applied to any kind of shear with the same accuracy since the developments of subsection 2.2 are exact. The only approximation comes in the evaluation of the Biot-Savart and surface map integrals. In the current study, the surface map integral is computed using point sources and point vortices which result in loss of accuracy close to the boundary D_{in}. Small instabilities were observed at the end of the computational domain due to this point-wise approximation. The rest of the domain is not seen to be significantly affected by these instabilities as seen in Figure 3. The use of piecewise constant surface panels and the account of the principal value of the integral at the surface itself will increase the accuracy.

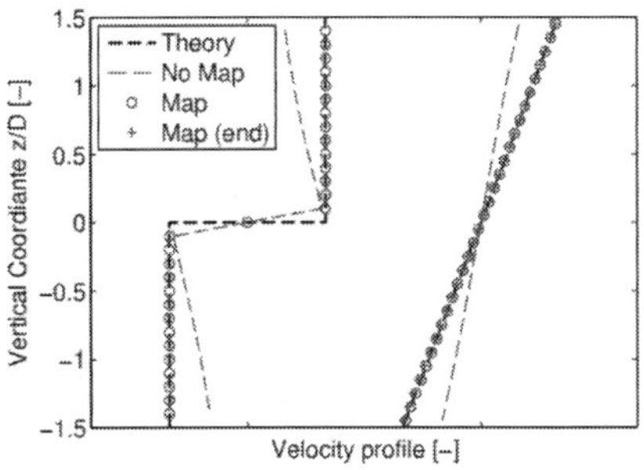

Figure 3: Modelling of shear using vortex particles. Velocity profiles obtained at the center of the domain for a step shear and a linear shear. The inclusion of the surface map from Equation 3 is crucial. The curved labelled "(end)" is extracted at the end of the simulation. The shear is seen to be conserved with time despite the numerical error that could be introduced by the re-meshing, numerical stretching and convection.

3.3. Representation of turbulence using vortex particles

The approach used by Chatelain et al. [5] is followed: A synthetic isotropic turbulent velocity field is computed prior to the simulation using the Mann model [30]; the velocity field is smoothly clipped on the side, by multiplication with a function that smoothly drops to zero, to avoid strong vortex lines at these locations; the vorticity of the field is computed and interpolated onto vortex particles; at the initial time step the domain is filled with these "turbulent" vortex particles while more particles are inserted at each time-step at the inlet of the domain. The method was previously used in the literature [31, 32, 33] and recently applied by the authors to study the impact of a wind turbine on turbulence [19]. More details can be found in these references. Instead of clipping the turbulent vorticity field, an external map could also be used but the surface integral would need to be evaluated at each time step since the flow on the surface is unsteady.

The vorticity obtained from the particle approximation is represented in the right of Figure 4 for the initial and final time steps. The spectra of the velocity fields are shown in the left of Figure 4. The evolution of the turbulence intensity can be seen in Figure 6. The current particle approximation was estimated to have a cut-off frequency of about $f_{\text{cut}} \approx U_0/10h$ [19]. This explains the differences between the input Mann spectrum and the "Frozen" spectrum obtained by purely convecting the particle turbulent field using the free-stream velocity and without allowing it to deform. When the particle field is allowed to deform ("Unfrozen Turbulence"), the turbulent kinetic energy k_t decays with the distance from the insertion point. The same behavior was found in CFD simulations where turbulence is inserted as body forces [34]. In the presence of viscous forces the decay of k_t behind a mask is expected if there is no driving force to sustain turbulence [35]. Batchelor and Townsend [36] observed a linear decay of the kinetic energy independently of the viscosity value. This is consistent with the theory from Kármán and Howarth [37] that obtained the decay as $\frac{dk_t}{dt} = -10\nu\frac{k_t}{\lambda^2}$, where λ denotes the representative length scale, when λ satisfies $\lambda^2 \propto \nu/\sqrt{k_t}$. More advanced turbulence decay models can be found e.g. in the work of Thormann and Meneveau [38]. The reproduction of the early linear decay seen in the curve labelled "No Shear" in Figure 6 confirms the validity of the current model. The other curves of the figure are discussed in subsection 4.1. More analyses on the turbulence modelling were performed by the authors in [19].

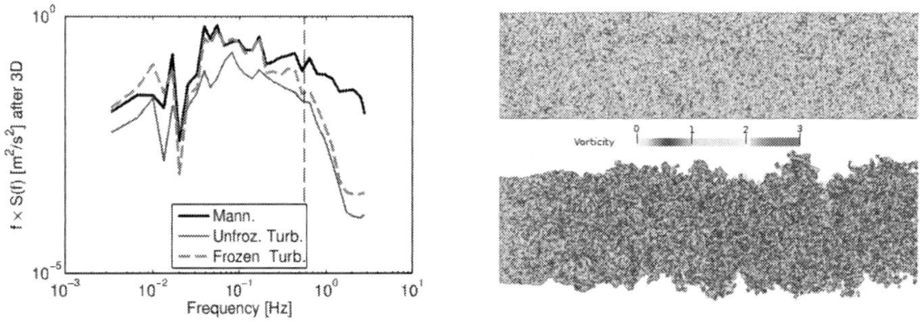

Figure 4: Statistics of the turbulent velocity fields represented using vortex particles. Left: comparison of the original velocity spectrum with the ones obtained from the particle approximation. The unfrozen spectrum is shown at $x = 0$, after the turbulence has evolved over 2 diameters. Right: Vorticity for frozen turbulence (top) and unfrozen decaying turbulence (bottom). $x \in [-2D, 6D]$

3.4. Wind turbine wake simulation using particles

The *Omnivor* library in its segments wake formulation was previously validated against BEM codes, actuator-line and actuator-disk (AD) simulations results and measurements [11, 39]. The principle of the vortex segment wake formulation used is similar to the one described for the *AWSM* code [6]. The particle formulation simply consists in converting the wake segments into vortex particles. The bound segments may also be temporarily converted to particles. The particle wake formulation is here validated against the segment wake formulation.

Vortex particles can be seen as lower order vortex elements compared to vortex segments. As a first step, the difference between the velocity induced by a vortex segment and a vortex particle is investigated. A given vortex segment of length l and intensity $\mathbf{\Gamma}$ is modeled using a distribution of n equally spaced vortex particles of intensity $\boldsymbol{\alpha} = \mathbf{\Gamma}l/n$. The velocity field from the segment or the collection of particles is evaluated on a surface described by h = constant, where h is the minimum distance to the segment. The maximum relative error on this surface for different values of h and n is shown in the left of Figure 5. The mean error was found to be about 2.5 lower than the maximum error.

The circulation along the span of the wind turbine blade for a simulation under uniform inflow with a vortex segment wake and with a particle wake is shown in the right of Figure 5. The agreement between the two cases validates the implementation of the particle wake method.

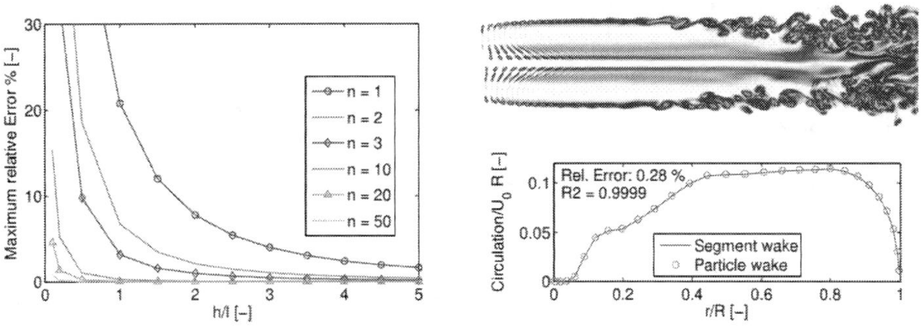

Figure 5: Comparison of segment and particle formulations. Left: error in the velocity field surrounding a vortex filament when represented using a distribution of n particles. Bottom-Right: Circulation along the blade at the end of the simulation using a vortex segment wake or a particle wake. Top-Right: vorticity values obtained from the particle simulation ($x \in [0D; 6D]$).

4. Combined analyses

4.1. Combination of shear and turbulence

It was seen in subsection 3.3 that the "unfrozen" turbulence tends to loose energy progressively as the eddies evolve away from the mask where they had been generated. The inclusion of shear is expected to sustain the turbulence. Shear and turbulence particles are here inserted as described individually in subsection 3.2 and subsection 3.3. When combining shear and turbulence, the velocities on the boundary ∂D_{in} will not be exactly the ones from the undisturbed shear. This is clear at the inlet and outlet of the turbulence but also on all other boundaries due to the induced velocities of the turbulence vortex particles on ∂D_{in} and due to the interaction between the shear and turbulence particles which may affect the strength of the shear particles. The difference between the unsteady velocity at the boundary and the steady state shear velocities are expected to be of second order and they are here omitted for simplicity. As seen in Equation 3, the errors involved would decrease in $1/r^2$ away from the boundaries, and an extension of the computational domain would thus reduce the error involved. No visible differences was observed using a lateral extent of 5D instead of 3D. Since meandering occurs, some shear particles might exit the domain on one side of the domain and holes may be created on the other side. Shear vortex particles are also inserted on the sides of the domain when such holes are detected. At the inlet, the shear particles are inserted as described in subsection 3.2: More shear vorticity is inserted where the wind velocity is higher. This approach is not followed for the turbulence as the turbulent particles are inserted with the free-stream. Further, the turbulence box used to generate the vortex particles is isotropic and it does not account for an inherent shear. The influence of these two limitations will be investigated in the future.

The evolution of the turbulence intensity across the domain is shown in Figure 6 for different cases. All simulations corresponds to "unfrozen turbulence". It is observed in the figure that the frozen shear does not significantly affect the turbulence kinetic energy which is seen to decay in the same proportion as the non-sheared case. On the other hand, the account for the full interaction is seen to reduce the decay and a trend towards a stabilisation of the turbulent kinetic energy can be foreseen. Analysis of the velocity profiles revealed that the shear slope was reduced when the full interaction was allowed. The turbulence extracts energy from the shear and tends to homogenize the profile. Troldborg et al. performed CFD simulations with prescribed linear shear and insertion of turbulence in an upstream plane of the domain [40]. In one of their test cases, the boundary condition consisted in artificially prescribing the velocity on the lower and higher boundary. This situation can be linked to the one used here where the surface map is assumed constant. The authors observed that the inclusion of shear was sustaining turbulence and constant statistical properties of turbulence were obtained [40]. The current results are not as pronounced but the difference between the uniform and sheared inflow cases follow the same trends as their results. The fact that a decay occurs in the unfrozen shear case reveals inconsistencies between the numerical model and the input field. Future work will address this issue in order to obtain a sustained turbulence field throughout the domain in the presence of shear. This can be done by recirculating the turbulence field in a fashion similar to the study of Troldborg et al. [40] or by adjusting the parameters of the input turbulence box.

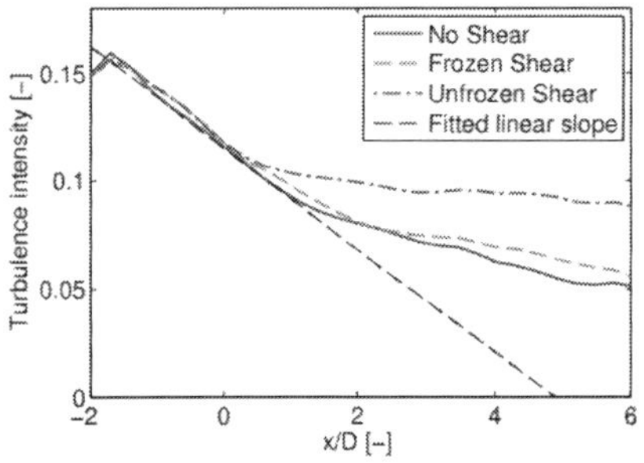

Figure 6: Evolution of the turbulence intensity with the distance from the turbulence insertion point. Accounting for the interaction between the shear and turbulence ("unfrozen shear") is seen to increase the turbulence intensity.

4.2. Combination of shear and wind turbine

Simulations of the Nordtank wind turbine are here presented for three cases: "Frozen Shear - Erroneous", "Frozen Shear", and "Unfrozen Shear". Vorticity contours for the "Frozen Shear - Erroneous" and "Unfrozen Shear" cases are shown in Figure 7. The wake shape for the case "Frozen Shear" is also shown on the figure. Results from the "Frozen Shear - Erroneous" case, in the top of Figure 7, are consistent with vortex segments simulations performed without the inclusion of the shear vorticity [29]. In the current study, it is seen that the interaction between the shear vorticity and the turbine has an important impact on the wake shape. The wake does not have such a significant upward motion when the full interaction is included. This result is consistent with CFD simulations [41]. The cancellation of the longitudinal vorticity of the wake induced by the shear and the longitudinal vorticity of the shear induced by the wake is likely to be the source of this result. It can be proven using a simple vortex model [42]. For the current simulation, the omission of this interaction leads to an overestimation of 4% of the power and 1% of the thrust. The comparison between the two frozen shear cases reveals the importance of the additional terms of Equation 7: Less upward motion of the wake is found and the wake shape compare better with the full-interaction case.

4.3. Aeroelastic LES of wind turbine with shear and turbulence

A proof of concept of the applicability of the different models to aeroelastic wind turbine simulations is presented here using the *HAWC2* and *Omnivor* tools. Turbulent simulations with different shear modelling are run. The simulation time for the full unfrozen case, was 7h30 with 160 CPUs using the tree-code algorithm. A total of 1h was spent in the iterative interaction with the structural code. Many future optimizations of the code are yet possible

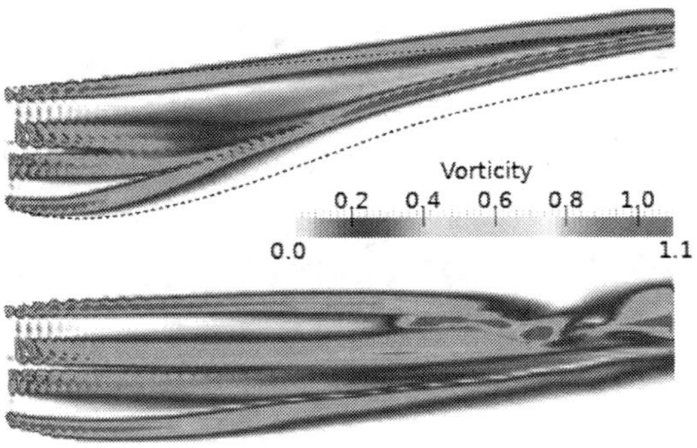

Figure 7: Vorticity contours in the wake of the turbine ($x \in [0D, 6D]$) for the case "Frozen Shear - Erroneous" (top) and "Unfrozen Shear - Full interaction" (bottom). The black dashed lines represent the wake shape for the correct "Frozen Shear" simulation (Equation 7). The interaction between the shear vorticity and the turbine has an important impact on the wake shape. Implementing the correct frozen shear equation improves the results but not enough to reach the results of the full interaction.

since new bottlenecks appeared in this study. Further speed up will be obtained by using the Poisson solver and by optimizing the re-meshing procedure (accounting for 30min here). The simulation time is roughly half the time of a typical CFD LES simulation with similar resolution at the rotor. Results are shown in Figure 8. The vertical lines corresponds to values obtained using the BEM aerodynamic module of *HAWC2* without turbulence. Both the BEM and vortex code predicts a reduction of loads and power when shear is present for the simulated operational conditions. Unfreezing the shear is seen to reduce the power mean value but increases its standard deviation. Little impact is seen on the flap-wise moment. Further investigations will be the topic of future work.

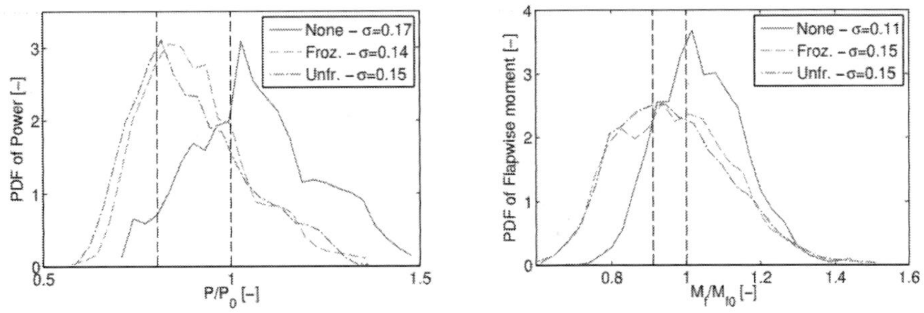

Figure 8: Probability density functions of power and flap-wise moments for turbulent aeroelastic simulations perform using the vortex code coupled with *HAWC2* under different shear conditions: no shear, frozen shear, unfrozen shear. The turbulence intensity is TI = 12% (see Figure 6 at $x = 0$). Vertical lines represent BEM value for TI = 0%, $P_0 = 148$kW, $M_{f,0} = 1.36$kNm. Unfreezing the shear is seen to reduce the power but has little influence on the flapwise moment.

5. Conclusions

Vortex particle methods were applied to turbulence, shear, a wind turbine and the combinations of these cases. For the first time, a Neumann to Dirichlet map was used to account for the vorticity outside of the domain in the modelling of shear. The interaction of the shear and turbulence vorticity was seen to avoid the decay of turbulent kinetic energy. If shear is present but the interaction is not a two-way interaction, the decay occurs. It was shown that most vortex methods implementation of frozen shear were lacking two terms in the vorticity equation. The inclusion of these two terms reduces the upward motion of the wake in a sheared flow. The full interaction of the shear and the wind turbine vorticity further reduces this upward motion. Solving for the full interaction appeared of importance both for the wake shape and the loads. The possibility to perform aeroelastic simulations of wind turbine under sheared and turbulent conditions was demonstrated. It is the first time, known to the authors, that fully coupled lifting-line aeroelastic simulations including shear and turbulence have been performed. Future work should address the possibility to include unsteady boundary conditions in the method, more consistent SGS models, and further investigate the different interactions of vorticity.

Acknowledgments

The work was funded by the Danish Council for Strategic Research (DSF), under contract 2104-09-0026, Center for Computational Wind Turbine Aerodynamics and Atmospheric Turbulence. The computational cluster was provided by DCSC and the DTU central computing facility.

References

[1] G. C. Larsen, H. A. Madsen, K. Thomsen, and T. J. Larsen. Wake meandering. *Wind Energy*, 11(4):377–395, 2008.

[2] G. I. Taylor. The spectrum of turbulence. *Proceedings of The Royal Society of London Series A-Mathematical And Physical Sciences*, 164(A919):0476–0490, 1938.

[3] V. A. Riziotis, D. I. Manolas, and S. G. Voutsinas. Free-wake aeroelastic modelling of swept rotor blades. *European Wind Energy Conference and Exhibition 2011, EWEC 2011*, 2011.

[4] J. C. Heinz. *Partitioned Fluid-Structure Interaction for Full rotor computations using CFD*. PhD thesis, Technical University of Denmark, 2013.

[5] P. Chatelain, S. Backaert, G. Winckelmans, and S. Kern. Large eddy simulation of wind turbine wakes. *Flow Turbulence And Combustion*, 91(3):587–605, 2013.

[6] A. van Garrel. Development of a wind turbine aerodynamics simulation module. Technical Report ECN-C–03-079, ECN, 2003.

[7] S. M. Richardson and A. R. H. Cornish. Solution of three dimensional incompressible flow problems. *Journal of Fluid Mechanics*, 82:pp. 309–319, 1977.

[8] G. Papadakis and S. G. Voutsinas. In view of accelerating CFD simulations through coupling with vortex particle approximations. *Journal of Physics: Conference Series*, 524(1):012126, 2014.

[9] G. S. Winckelmans and A. Leonard. Contributions to vortex particle methods for the computation of 3-dimensional incompressible unsteady flows. *Journal Of Computational Physics*, 109(2):247–273, 1993.

[10] J. K. Salmon, M.S. Warren, and G.S. Winckelmans. Fast parallel tree codes for gravitational and fluid dynamical n-body problems. *Intl. J. Supercomput. Appl. High Perf. Comp.*, 8(2):192–142, 1994.

[11] E. Branlard, E. Machefaux, M. Gaunaa, H.H. Brandenborg Sørensen, and N. Troldborg. Validation of vortex code viscous models using lidar wake measurements and CFD. In *Proceedings*. EWEA - The European Wind Energy Association, 2014.

[12] Giorgos Papadakis. *Development of a hybrid compressible vortex particle method and application to external problems including helicopter flows*. PhD thesis, National Technical University of Athens, 2014.

[13] C. R. Anderson. A method of local corrections for computing the velocity field due to a distribution of vortex blobs. *Journal of Computational Physics*, 62(1):111–123, 1986.

[14] GT Balls and P. Colella. A finite difference domain decomposition method using local corrections for the solution of poisson's equation. *Journal Of Computational Physics*, 180(1):25–53, 2002.

[15] G.-H. Cottet and P. Koumoutsakos. *Vortex methods: theory and practice*. Cambridge University Press, 2000.

[16] T. J. Larsen and A. M. Hansen. *HAWC2 - User manual*. DTU-Risø-R-1597, 2007.

[17] J. G. Leishman and T.S. Beddoes. A semi-empirical model for dynamic stall. *Journal of the American Helicopter Society*, 34(3):p3–17, 1989.

[18] G. S. Winckelmans. *Encyclopedia of computational mechanics - Volume 1. Chapter 5: Vortex Methods*, chapter 5. J. Wiley & Sons, New-York, N.Y., 2004.

[19] E. Branlard and M. Gaunaa. Impact of a wind turbine on turbulence: un-freezing the turbulence by means of a simple vortex particle approach, 2015. (Submitted to the Journal of Wind Engineering and Industrial Aerodynamics, currently under review).
[20] P. Degond and S. Mas-Gallic. The weighted particle method for convection-diffusion equations. part 1: The case of an isotropic viscosity. *Mathematics of Computation*, 53(188):485–507, 1989.
[21] G. Daeninck. *Developments in hybrid approaches: Vortex method with known separation location Vortex method with near-wall Eulerian solver RANS-LES coupling*. PhD thesis, Université catholique de Louvain, Faculté des sciences appliiquées, 2006.
[22] S. G. Voutsinas. Vortex methods in aeronautics: how to make things work. *International Journal of Computational Fluid Dynamics*, 2006.
[23] M. J. M. Hill. *On a spherical vortex*. the Royal Society, 1894.
[24] D. G. Akhmetov. *Vortex rings*. Springer, 2009.
[25] J. T. Beale and A. Majda. High Order Accurate Vortex Methods with Explicit Velocity Kernels. *Journal of Computational Physics*, 208:188–208, 1985.
[26] M. N. Hejlesen, J. T. Rasmussen, P. Chatelain, and J-H. Walther. A high order solver for the unbounded poisson equation. *Journal of Computational Physics*, 252:458–467, 2013.
[27] J. T. Beale and A. Majda. Rates of convergence for viscous splitting of the navier-stokes equations. *Mathematics of Computation*, 37:243–259, 1981.
[28] H.Aa. Madsen, V. Riziotis, F. , Zahle, M.O.L. Hansen, H. Snel, F. Grasso, T.J. Larsen, E. Politis, and F. Rasmussen. Blade element momentum modeling of inflow with shear in comparison with advanced model results. *Wind Energy*, 15:63–81, 2012.
[29] F. Grasso. Ground and Wind Shear Effects in Aerodynamic Calculations. Technical Report ECN-E–10-016, Energy research center of the Netherlands (ECN), 2010.
[30] J. Mann. The spatial structure of neutral atmospheric surface-layer turbulence. *Journal of Fluid Mechanics*, 273:p141–168, 1994.
[31] J. T. Rasmussen, M. M. Hejlesen, A. Larsen, and J-H. Walther. Discrete vortex method simulations of the aerodynamic admittance in bridge aerodynamics. *Journal of Wind Engineering and Industrial Aerodynamics*, 98(12):754–766, 2010.
[32] S. Backaert, P. Chatelain, G. Winckelmans, and I. De Visscher. Vortex particle-mesh simulations of atmospheric turbulence effects on wind turbine blade loading and wake dynamics. In M. Hölling, J. Peinke, and S. Ivanell, editors, *Wind Energy - Impact of Turbulence*, volume 2 of *Research Topics in Wind Energy*, pages 135–140. Springer Berlin Heidelberg, 2014.
[33] D. Sale, A. Aliseda, and Y. Lic. Simulation of hydrokinetic turbines in turbulent flow using vortex particle methods. In *Proceedings of the 2nd Marine Energy Technology Symposium: METS2014-April 15-18, 2014, Seattle, WA*, 2014.
[34] N. Troldborg, J. N. Sorensen, and R. Mikkelsen. Actuator line simulation of wake of wind turbine operating in turbulent inflow. *Science Of Making Torque From Wind*, 75(1):–, 2007.
[35] A. S. Monin and A. M. Yaglom. *Statistical fluid mechanics mechanics of turbulence. vol. 1*. MIT press, Cambridge, 1971.
[36] G. K. Batchelor and A. A. Townsend. Decay of vorticity in isotropic turbulence. *Proceedings of the Royal Society of London A: Mathematical, Physical and Engineering Sciences*, 190(1023):534–550, 1947.
[37] T. de Kármán and L. Howarth. On the statistical theory of isotropic turbulence. *Proceedings of The Royal Society of London Series A-Mathematical And Physical Sciences*, 164(917):192–215, 1937.
[38] A. Thormann and C. Meneveau. Decay of homogeneous, nearly isotropic turbulence behind active fractal grids. *Physics of Fluids*, 26(2):1, 2014.
[39] E. Branlard, M. Gaunaa, and E. Machefaux. Investigation of a new model accounting for rotors of finite tip-speed ratio in yaw or tilt. *Journal of Physics: Conference Series (Online)*, 524(1), 2014.
[40] N. Troldborg, J. N. Sørensen, R. F. Mikkelsen, and N. N. Sørensen. A simple atmospheric boundary layer model applied to large eddy simulations of wind turbine wakes. *Wind Energy*, 17(4):657–669, 2014.
[41] Frederik Zahle and Niels Sørensen. Rotor aerodynamics in atmospheric shear flow. *EWEC 2008 Conference Proceedings*, pages 120–124, 2008.
[42] M. Gaunaa, N. Troldborg, and E. Branlard. Modelling the influence of wind shear using a simple vortex rotor model, 2015. (To appear in *Wind Energy*).

Journal of Physics: Conference Series **625** (2015) 012020 doi:10.1088/1742-6596/625/1/012020

Evaluation of wind farm effects on fatigue loads of an individual wind turbine at the EnBW Baltic 1 offshore wind farm

A Bustamante, L Vera-Tudela, M Kühn
ForWind – University of Oldenburg, Institute of Physics, Ammerländer Heerstr. 136, 26129, Oldenburg, Germany.

Email: augusto.bustamante@uni-oldenburg.de

Abstract. Turbulence in wake has special interest due to its strong connection with increment of fatigue loads. The aim of this paper is to evaluate the wind farm effects on fatigue loads of an individual wind turbine at the EnBW Baltic 1 offshore wind farm and compare the results with the statements suggested by IEC 61400-1 ed. 3 [1].
From measurements, the study provides strong evidence that considerable wake effects up to 15 rotor diameters (D) downstream are related to the increase of fatigue loads in the analyzed turbine. The influence of the behavior of the upstream wind turbine's thrust coefficient (C_T) can be observed in the analyzed curves for turbulence intensity and damage equivalent loads (DELs) in blade root flapwise and edgewise direction.

1 Introduction

Wind turbines extract energy from the turbulent wind and while doing so create additional turbulence [2], increasing the accumulated damage in the components of downstream wind turbines. The turbulence model based on wind velocity developed by Frandsen and suggested by the IEC 61400-1 ed. 3 in its Annex D is widely used to account for extra turbulence generated by neighboring wind turbines. In addition, Frandsen [3] suggests a model based on thrust coefficient (C_T). In this study, both models are briefly described and one year measured data from EnBW Baltic 1 offshore wind farm is analyzed. The evaluation of measurements includes the analysis of fatigue loads and turbulence intensity in function of velocity and direction of the incoming wind. The analysis is limited to blade root bending moments in flapwise and edgewise direction in free stream and wake flow conditions.

2 EnBW Baltic 1 offshore wind farm

The wind farm is situated in the Baltic Sea, 16 km north of the coast of the island Fischland Darß, and it is arranged in a triangular layout with an area of approximately 7 km^2, shown in Figure 1.

EnBW Baltic 1 comprises 21 wind turbines Siemens SWT 2.3-93 on monopile foundations in 16 to 19 m depth. This is a three-bladed model, upwind rotor with a diameter of 93 meters and a hub height of 67 m. The cut-in, rated, cut-out wind speeds are 4 m/s, 13-14 m/s and 25 m/s, respectively. The variable speed turbines have asynchronous generators with 2.3 MW capacity and are pitch-regulated, more details can be found in [4].

The meteorological information for the design-basis was obtained from three sources [5]: FINO-2 mast and two meteorological stations of the German Federal Maritime and Hydrographic Agency (BSH) i.e. Darßer Schwelle and Arkona Becken. The wind is mainly coming from southwest with a low turbulence intensity of 5% at 15 m/s [6], which is comparable to other offshore sites. Along this analysis, the used data was taken from the anemometers located on the turbines. The data was corrected by the developer of the project and validated with Lidar information on the site.

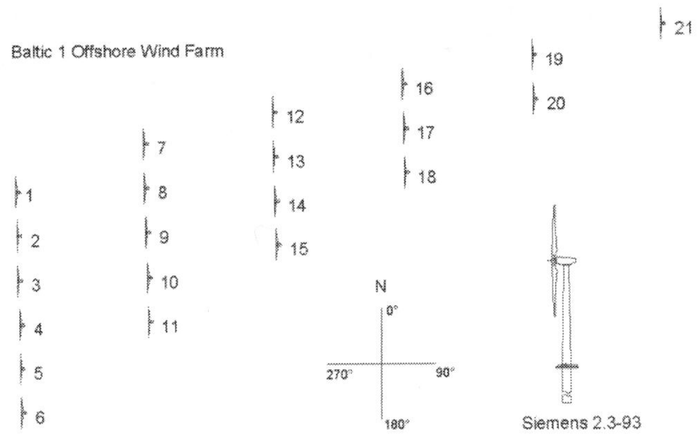

Figure 1. Layout of EnBW Baltic 1.

3 Turbulence intensity

Turbulence intensity is conceived as the instantaneous, random deviation of the mean wind speed [7]. According to Frandsen, turbulence in wake condition has two main components: ambient turbulence and turbulence added by the wake. The total turbulence modelled by Frandsen [3] is:

$$I_T = \sqrt{\frac{1}{\left(1.5 + \frac{0.8 \times D}{C_T}\right)^2} + I_0^2} \tag{1}$$

Where I_0 is the ambient turbulence modeled as σ/V_{hub}; σ is the standard deviation of the wind speed at V_{hub}; V_{hub} is the average of 10-minutes wind speed at hub height; D is the normalized distance between turbines in rotor diameters; C_T is the specific wind turbine thrust coefficient.

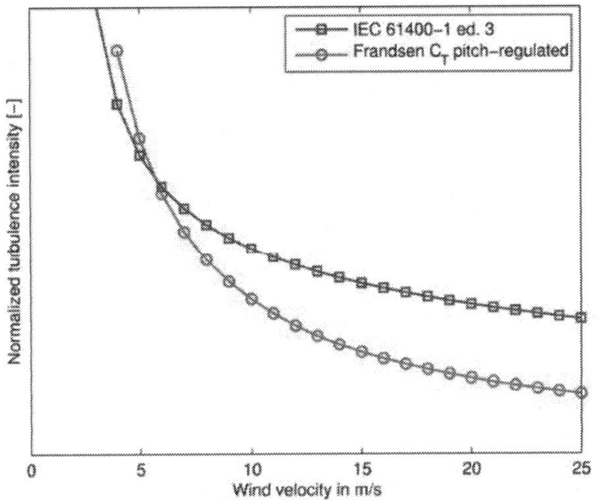

Figure 2. Normalized turbulence intensity for IEC and Frandsen model for pitch controlled turbine (measured), assuming 10 D and Normal Turbulence Model (NTM) turbulence reference (Iref) 0.16.

Meanwhile, the IEC 61400-1 ed. 3 [1] formulates I_T as in equation (2) , with c being a constant value equal to 1 m/s:

$$I_T = \sqrt{\frac{0.9 \times V_{hub}^2}{\left(1.5 + 0.3 \times D \times \sqrt{V_{hub} / c}\right)^2} + I_0^2} \tag{2}$$

Moreover, both models propose that the influence of neighboring turbines, with respect to the turbulence in wake, should be considered when they are located within 10 D [1].

4 Fatigue loads in wind turbines

As the wind speed is varying in time, the loads acting on the wind turbine blade are also varying and ultimately the stresses acting on the wind turbine blade are also changing. This leads to the fatigue of the different components in the wind turbine [8]. Turbulence is used as primary wind turbine design parameter, substantially describing the fatigue damage. In blades apart from the weight, turbulence is a main cause of material fatigue [7]. A large turbine nowadays can experience between 10^8 and 10^9 cycles over its lifetime [9]. Therefore, wind turbines suffer a huge number of cyclic loads and that is why they are known as the perfect "Fatigue Machine". A general methodology to calculate fatigue loads in wind turbines can be found in [7].

For this research, it is assumed a generalized S-N fatigue slope $m = 10$ for blades, which is a widely used value for good performance on a basic unidirectional laminate material as glass fiber [10].

The number of load cycles that a component suffer is known as fractional damage and it is modelled as $d=n/N$ [9], where n is the number of constant amplitude cycles and N is the number of cycles to failure.

A convenient format to represent the variable spectrum loads that a wind turbine withstand is the damage equivalent load (DEL). The rainflow counting method is used to count and bin the load cycle time histories, converting the variable load spectra into constant range load spectra and it is modelled as follows:

$$DEL = \left[\sum_i \frac{S_i^m \times N_i}{N_{eq}}\right]^{1/m} \tag{3}$$

Where S is the stress range; an equivalent number of load cycles of $N_{eq}=10^7$ is normally used or the number of load cycles corresponding to 1 Hz. Nevertheless, for this study it will be referred to the number of load cycles corresponding to 50 Hz. In this study the damage equivalent load concept is used to compare different types of measured loads. In order to obtain an accurate basis for comparison the S-N slopes m must have the same numerical values.

5 Load measurement program in EnBW Baltic 1

Load measurements were performed at turbine 1 and 8. Several sensors installed in the two turbines, following the Standard IEC 61400-13, collected the data at various components e.g. blades and tower. In this study power production state is examined as Measure Load Case (MLC) which corresponds to Design Load Case (DLC) 1.2.

5.1 Fatigue analysis as function of wind velocity

The 10-minutes measured data acquired from turbine 1 is sampled at 50 Hz and is sorted in 1 m/s wind speed bins, from 4 to 17 ± 0.5 m/s. Three scenarios are analyzed in this section: free stream for directions going from 210° - 300° and wake condition at 90° and 180° directions are evaluated. Due to the different amount of 10-minutes filtered data at each wind speed bin and at each scenario, the mean DEL is computed. It has to be mentioned that over the 10-minute period, the variation in wake exposure due to wind direction changes, will contribute to the creation of few but large load cycles, due to low frequency turbulence added by those inflow variations.

Figure 3 shows that the flapwise damage equivalent load curves are mainly influenced by turbulence. It can be observed that for the upstream turbines referred to turbine 1 e.g. turbine 2 and 8, from rated wind speed on, the thrust coefficient varies due to the entry into operation of the pitch control, influencing the downstream turbulence. This action leads to a change in the full load behavior in turbine 1, tending the loads to decrease to the range of the mean DELs generated in free stream.

The analysis of the DELs in the edgewise direction is shown in Figure 4. The mean DELs in both wake conditions show similar values. On the other hand, compared with the analysis made in flapwise direction, the deviation between DELs in free stream and wake condition is lower since edgewise bending moment is driven mainly by gravity and torque. From rated speed the DELs tend to decrease based on what is explained in the previous paragraph.

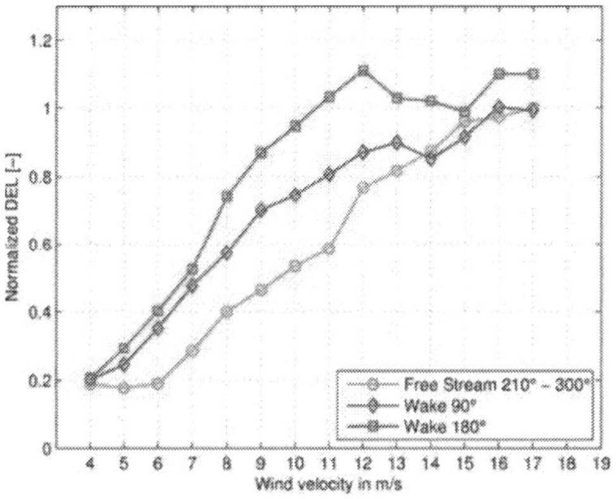

Figure 3. Mean flapwise DELs for free stream vs wake conditions.

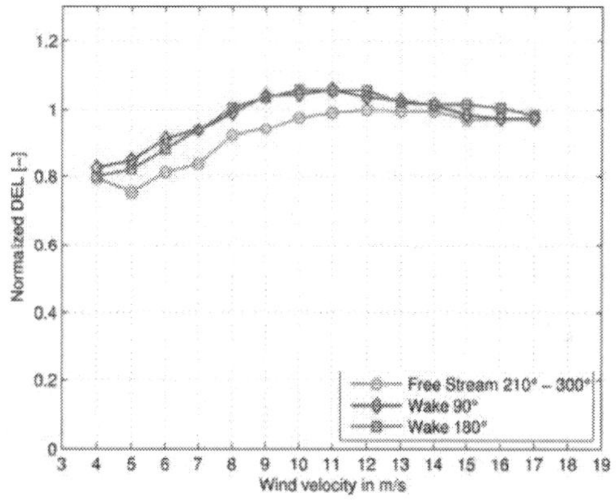

Figure 4. Mean edgewise DELs for free stream vs wake conditions.

5.2 *Fatigue analysis as function of sectors*

In this section, the wind directions are derived on the basis of turbine orientation. The polar wind direction 0°- 350° is covered with a resolution of 10°. The data analyzed are referred to wind speed values going from 4 to 20 m/s. Due to the different amount of 10-minutes filtered data at each sector, the mean values were computed. The values of turbulence intensity and DELs in flapwise direction are normalized based on the free stream data. To simplify the analysis, it is assumed that for every wind direction, turbine 1 is affected by only one other turbine of the wind farm, even though for some wind angle configurations turbine 1 is partially loaded by more than one wake.

According to the standard IEC-61400-1 ed. 3, the effects from "hidden" turbines do not need to be considered in wake analysis. Figure 5 and Figure 6 depict the distances and the angles between the analyzed turbine and its neighbors, respectively.

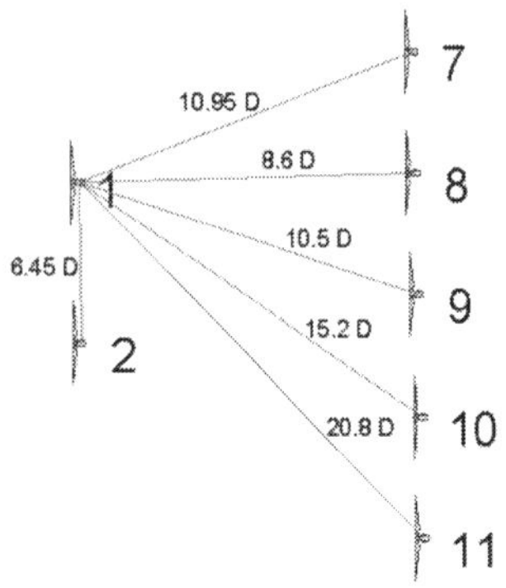

Figure 5. Normalized distances between turbine 1 and its neighbors.

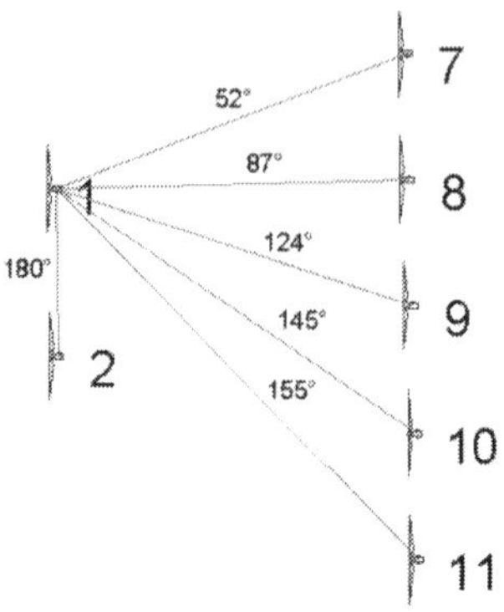

Figure 6. Angles between turbine 1 and its neighbors.

Figure 7 shows the mean turbulence intensities based on nacelle anemometer. In free stream, the turbulence intensity shows the least values. The highest sectorial turbulence intensity value is located at 180°, due to turbine 2, which is located 6.45 D from turbine 1, as it is depicted in Figure 5.

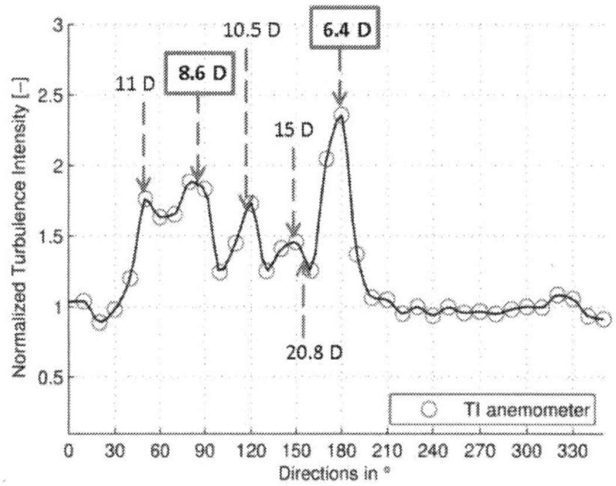

Figure 7. Normalized sectorial mean turbulence intensity in turbine 1. Values in rectangles show which sectors should be included to account for additional turbulence sources according to IEC 61400-1 ed. 3 Annex D.

Figure 7 shows five turbulence peaks affecting turbine 1 detailed in Table 1. The peaks located at 90° and 180°, corresponding to the turbines 8 and 2 respectively, are within the range to be considered as source of additional turbulence, as it is suggested by IEC 61400-1 ed. 3 – Annex D. It is also important to remark that even the turbulence coming from turbines installed at 20.8 D, does not decrease to the ambient turbulence level. This is probably due to at these positions, turbine 1 is partially loaded by more than one turbine.

Table 1. Sectorial turbulence intensity analysis of turbine 1

Location of turbulence intensity peaks	Location of the neighbor turbines	Turbine number	Distance respecting the Turbine 1	Radio to free stream turbulence
50°	52°	Turbine 07	10.9 D	1.75
90°	87°	Turbine 08	8.6 D	1.80
120°	124°	Turbine 09	10.5 D	1.65
140°	145°	Turbine 10	15.2 D	1.45
150°	155°	Turbine 11	20.8 D	1.30
180°	180°	Turbine 02	6.45 D	2.30

Figure 8 shows the sectorial mean turbulence intensity measured at turbine 1 and the NTM for an Iref of 0.16 correspondent to class A according to [1]. It can be observed that the thrust coefficient of the upstream turbine influences the turbulence curves seen by turbine 1. In partial load the turbulence intensity shows larger values for directions where the neighboring turbines are located closer, decreasing when the turbine upstream starts to pitch the blades from rated wind speed.

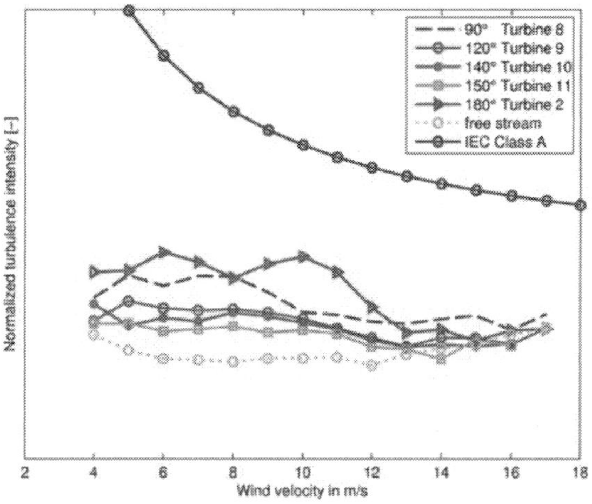

Figure 8. Sectorial measured mean turbulence intensities at turbine 1 according to the location of neighbor turbines and the NTM with Iref 0.16.

A partial load scenario at 8 ± 0.5 m/s in flapwise direction is depicted in Figure 9. With this figure, it can be seen the close connection between turbulence intensity and DELs in flapwise direction. Furthermore, as it was shown similarly in the previous turbulence analysis, there are DELs peaks related with turbines that are located at distances larger than 10 D e.g. 50°, 145°. At 155° where turbine 11 is located, the mean DELs are up to 1.3 times compared to the loads in the free stream conditions.

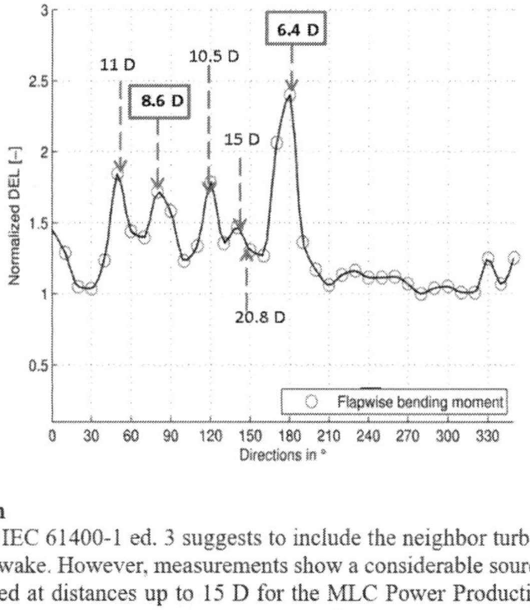

Figure 9. Mean DELs in flapwise direction at 8±0.5 m/s. Values in rectangles show which sectors should be included to account for additional turbulence sources according to IEC 61400-1 ed. 3 Annex D.

6 Discussion

The Standard IEC 61400-1 ed. 3 suggests to include the neighbor turbines that are located within 10 D for turbulence in wake. However, measurements show a considerable source of turbulence and loads caused by turbines located at distances up to 15 D for the MLC Power Production. From sectors, where a neighbor turbine is located within 20.8 D, mean DELs in flapwise direction and mean turbulence intensity do not decrease to the ambient conditions values. Nevertheless at this position, turbine 1 could be partially loaded by more than one turbine.

It was found in the measurements that the turbine's thrust coefficient C_T influences the behavior of the turbulence intensity and the mean DELs in partial load of the blade root in flapwise direction of the downstream turbine. Although the measured turbulence intensities are covered within the IEC 61400-1 ed. 3 model, the turbulence curves in wake conditions seem to be not well represented. In load analysis turbulence intensity plays an important role, therefore to achieve higher load accuracy, a model which can replicates the behavior of the turbulence intensity in wake conditions is necessary.

It is important to point that in blade design, other criteria are also considered e.g. buckling, ultimate stress, tower clearance. Therefore, in some cases, additional fatigue loading in the blades might not be critical, but indeed has to be considered.

Acknowledgments

This work would not have been completed without the support from EnBW Erneuerbare Energie GmbH and Bundesministerium für Wirtschaft und Energie BMWi.

References

[1] International Electrotechnical Commission, IEC 61400-1:2005 Wind Turbines - Part 1: Design Requeriments, Geneva, 2005.

[2] J. Peinke, A. Brand and J. Mann, Turbulence and wind turbines, 13th European Turbulence Conference, 2013.

[3] S. Frandsen, Turbulence and turbulence-generated structural loading in wind turbine clusters, Roskilde, 2007.

[4] SIEMENS, "Wind Turbine SWT-2.3-93," [Online]. Available: http://www.energy.siemens.com. [Accessed 12 August 2014].

[5] S. Jankowski, Validierung des Farm Layout Programms (FLaP) durch Analyse der Windparkeffekte im Offshore-Windpark EnBW Baltic 1, Oldenburg, 2013.

[6] A. Westerhellweg, B. Canadillas and T. Neumann, Direction dependency of offshore turbulence intensity in the German Bight, 2010.

[7] R. Gasch and Jochen Twele, Wind Power Plants, Berlin: Springer, 2012.

[8] S. Tawade, T. Sachin and H. Ashwinikumar, "Fatigue life optimization of wind turbine blade," International Journal of Research in Engineering and Technology, May 2014.

[9] J. F. Manwell, J. G. McGowan and A. L. Rogers, Wind Energy Explained, John Wiley & Sons, 2002.

[10] G. Freebury and W. Musial, Determining Equivalent Damage Loading for Full-Scale Wind Turbine Blade Fatigue Tests, 2000.

[11] T. Burton, N. Jenkins, D. Sharpe and E. Bossanyi, Wind Energy Handbook, John Wiley & Sons, 2011.

[12] S. Frandsen and M. Thøgersen, "Integrated fatigue loading for wind turbines in wind farms by combining ambient turbulence and wakes," Wind Energy, 1999.

Wake Conference 2015 IOP Publishing
Journal of Physics: Conference Series **625** (2015) 012021 doi:10.1088/1742-6596/625/1/012021

Evaluation of the Actuator Line Model with coarse resolutions

M Draper, G Usera

Universidad de la República, Facultad de Ingeniería, Julio Herrera y Reissig 565 11300, Montevideo, Uruguay

E-mail: mdraper@fing.edu.uy

Abstract. The aim of the present paper is to evaluate the Actuator Line Model (ALM) in spatial resolutions coarser than what is generally recommended, also using larger time steps. To accomplish this, the ALM has been implemented in the open source code caffa3d.MBRi and validated against experimental measurements of two wind tunnel campaigns (stand alone wind turbine and two wind turbines in line, case A and B respectively), taking into account two spatial resolutions: R/8 and R/15 (R is the rotor radius). A sensitivity analysis in case A was performed in order to get some insight into the influence of the smearing factor (3D Gaussian distribution) and time step size in power and thrust, as well as in the wake, without applying a tip loss correction factor (TLCF), for one tip speed ratio (TSR). It is concluded that as the smearing factor is larger or time step size is smaller the power is increased, but the velocity deficit is not as much affected. From this analysis, a smearing factor was obtained in order to calculate precisely the power coefficient for that TSR without applying TLCF. Results with this approach were compared with another simulation choosing a larger smearing factor and applying Prandtl's TLCF, for three values of TSR. It is found that applying the TLCF improves the power estimation and weakens the influence of the smearing factor. Finally, these 2 alternatives were tested in case B, confirming that conclusion.

1. Introduction

Wind energy has evolved significantly during the last decades, becoming one of the fastest-growing energy sources. Despite this amazing development, it is still a hard task to simulate accurately the wind flow over stand alone wind turbines and wind farms [1].

To overcome the computational cost required to simulate the blade boundary layers, in the past few years actuator models have been proposed: Actuator Disk Model (without and with rotation, ADM-NR and ADM-R respectively) [3][4], Actuator Sector Model (ASeM) [5], Actuator Line Model (ALM) [6][7] and Actuator Surface Model (ASM) [8]. For a comprehensive review see [1].

The ALM [5] has been widely validated, taking into account different inflow conditions, from uniform to shear flow with and without turbulent fluctuations, with special focus on the near wake or the far wake and the interaction between wind turbines. For example in [9] the ALM has been implemented in a finite volume code and results were presented of a Large Eddy Simulation (LES) of a wind farm that consists of 48 wind turbines with a rated power of 2.3MW, obtaining good results for the time-averaged power production, except for the sixth turbine and beyond in each wind-aligned. In [10] results obtained with the ADM-NR, ADM-R and ALM were compared with wind tunnel measurements of the wake of a model wind turbine subject to turbulent-shear inflow conditions, finding in general that the characteristics of the simulated wake with the ADM-R and ALM were in good agreement with the measurements. A similar comparison between the ADM-R and the ALM against a full rotor simulation was performed in [11], obtaining a fairly good agreement of the mean axial velocity between them up to 2 rotor diameters downstream in the wake for laminar inflow, but further downstream the full rotor simulation predicted a faster recovery, related to a higher turbulence level respect to the ADM-R and ALM. However, for turbulent inflow conditions, the differences in axial velocity and turbulent kinetic energy were reduced significantly. As a conclusion the authors suggested that when simulating wind turbines in the atmospheric boundary layer (ABL), it is sufficient to use the ADM-R or ALM, and that there are no big differences between them in the predicted wake. In [12] the ALM is used to simulate the MEXICO rotor under yawed conditions, computing loads as well as wake characteristics. Finally, in [13] the capability of the ALM to compute vortex structures in the near wake was evaluated taking into account the MEXICO experimental data, obtaining satisfactory results of the wake expansion, circulation and velocity distributions.

These references are just a few of many possible examples showing the capabilities of the ALM to predict the wake characteristics as well as loads and power performance. In general, it is recommended to use a spatial resolution of at least R/30 and to set the time step size in order to limit the movement of the rotor tip to a grid cell length in each time step [14]. For example, in [9] the spatial resolution was R/26 and it took one million processor-hours to simulate ten minutes of wind plant flow. In [11] and [12] the spatial resolution used in the simulations with the ALM was R/30. However, in [13], taking into account the main objective of that work, the rotor is discretized using 80 grid points, distributed so that the resolution near the tip region is almost R/122.

Taking into account those recommendations and the good performance of the ADM-R, it is uncertain whether the ALM could be used with coarser resolutions and larger time steps. Regarding the latter, as the computed aerodynamic forces are projected into the domain through a smearing function, with spatial resolutions where the chord length is close to or less than the grid length (blades almost immersed in a grid cell), the restriction of limiting the tip movement one grid cell length may not be strictly required, as the forces are smeared to neighbor cells despite the chord is immersed in just one grid cell. In that sense, it is like the ALM approaches the ASeM.

The aim of the present paper is to evaluate the ALM in spatial resolutions coarser than what is generally recommended (R/30-R/60), also using larger time steps. The paper is organized as follows: section 2 presents the solver and the ALM implementation, section 3 describes the experimental campaigns used for the evaluation and the numerical setup, section 4 presents the main results and summary is given in section 5.

2. Numerical method

In this section a brief description of the main features of the Navier-Stokes solver is provided and then the ALM implementation is presented shortly.

2.1. Solver

caffa3d.MBRi [15][16] is an open source, finite volume (FV) code, second order accurate in space and time, parallelized with MPI, in which the domain is divided in unstructured blocks of structured grids. Representation of complex geometries can be handled through a combination of body fitted grids and the immersed boundary method over both Cartesian and body fitted grid blocks. Geometrical properties and flow properties, which are expressed in primitive variables, are always expressed in a Cartesian coordinate system, using a collocated arrangement.

Recently, the ALM and the ADM-R have been implemented in the code.

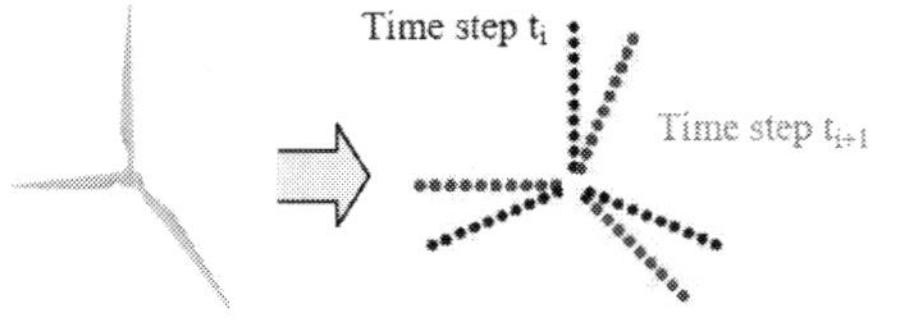

Figure 1. ALM rotor representation.

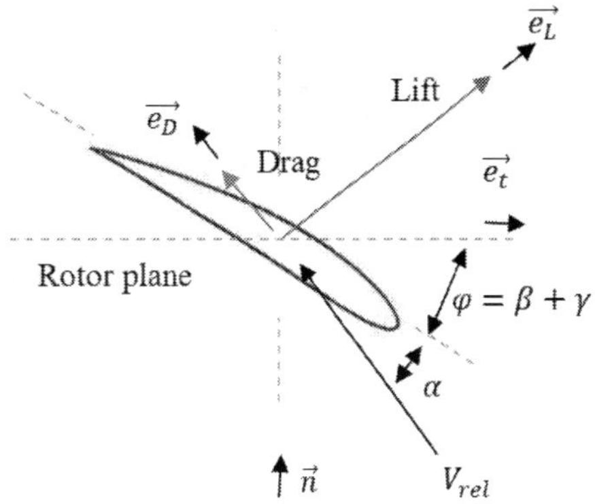

Figure 2. A cross-sectional airfoil radial section.

2.2. ALM

In the ALM, the wind turbine rotor is represented as a body force field, instead of resolving the blade boundary layers each blade is represented as a line that moves with the rotational speed of the rotor and it is discretized in radial sections where the aerodynamic forces are computed (Figure 1). To compute the force in each radial section, geometrical properties (chord length and twist angle) as well as aerodynamic properties (lift and drag coefficients) are found. The latter are obtained from tabulated airfoil data.

At each radial section the aerodynamic force is computed as

$$\vec{f} = -\frac{1}{2}\rho V_{rel}^2 c(C_L\vec{e_L} + C_D\vec{e_D})dr \quad (1)$$

where ρ is the air density, V_{rel} is the relative velocity, c is the chord length, C_L is the lift coefficient, C_D is the drag coefficient, $\vec{e_L}$ is a unit vector in the direction of the lift force, $\vec{e_D}$ is a unit vector in the direction of the drag force (Figure 2) and dr is the length of the radial section.

After computing the aerodynamic forces, it is required to project them from the representation of the rotor onto the computational domain. To accomplish this, a smearing Gaussian function (smearing factor E) is often used, taking into account the distance between each grid cell and radial section (d), in order to compute the additional source term.

$$f(d) = \frac{1}{(E\pi^{0.5})^3} e^{-\left(\frac{d}{E}\right)^2} \quad (2)$$

3. Validation cases

3.1. Experimental campaigns

Two experiments, performed at the wind tunnel of the Department of Energy and Process Engineering at The Norwegian University of Science and Technology, are selected for validation. The wind tunnel has a test section of 1.80m (height) x 2.71m (width) x 11.14m (length).

The first experiment, hereafter case A, consists of one model wind turbine subject to uniform inflow condition almost without turbulent fluctuations (turbulence intensity at the inlet was 0.3%). The rotor diameter is 0.894m and the hub height is 0.817m. The blades use the NREL S826 airfoil along the entire span. For further information please see [17].

The second experiment, hereafter case B, consists of two model wind turbines subject to uniform inflow condition almost without turbulent fluctuations (turbulence intensity at the inlet was 0.3%). The rotor diameter of the upstream wind turbine is 0.944m and the rotor diameter of the downstream wind turbine is 0.894m, while the hub height of both is 0.817m. The blades use the NREL S826 airfoil along the entire span of both wind turbine models. For further information please see [18].

3.2. *Numerical setup*

Two spatial resolutions are tested, R/15 and R/8. Details of the numerical setup are presented in Table 1. The inflow conditions are of uniform non-turbulent velocity, while zero velocity gradient is imposed at the outlet. No slip condition is used at walls in order to represent the wind tunnel.

Table 1. Numerical setup

Resolution	Lx(m)	Ly(m)	Lz(m)	Nx	Ny	Nz	Δx (m)	Δy (m)	Δz (m)
R/15	11.14	2.71	1.82	384	96	64	0.029	0.028	0.028
R/8	11.14	2.71	1.82	192	48	36	0.058	0.056	0.051

We use the second order backward Euler scheme to advance in time and the scale dependent Smagornsky dynamic model to compute the subgrid scale stress. The convective term is discretized using a variable blending factor (varying between 0.95 and 1.00), depending on the distance and position of a grid cell relative to the wind turbine rotor (in a similar way as presented in [14]). To accomplish this, three regions are defined taking into account the distance from the rotor, the blending factor close to the rotor is 0.95 and far away is 1.00, while it varies linearly between those regions. To represent the wind turbine rotor the ALM is used with 27 radial sections in each line, distributing the forces with a three dimensional Gaussian function defined by a smearing factor E. The presence of the tower and nacelle are taking into account in a similar approach as presented in [4], through drag coefficients. The airfoil data is taken from [19].

First a sensitivity analysis for case A is performed in order to check the influence of the smearing factor E and the size of the time step dt in the wake as well as in the power and thrust coefficients (Cp and Ct respectively) without applying a tip loss correction factor and for design conditions (tip speed ratio, TSR=6). The following smearing factors and time step sizes are considered:

Table 2. Sensitivity analysis - Smearing factor

Smearing factor (m)	Line type/colour[a]	Ratio $Ei/\Delta x$	
		R/15	R/8
0.044 (E1)	Blue	1.51	0.76
0.065 (E2)	Red	2.25	1.12
0.116 (E3)	Green	4.00	2.00

[a]Used in Figure 3

Table 3. Sensitivity analysis - time step size

Time step size (s)	Line type/colour[a]	CFL-tip[b]	
		R/15	R/8
0.0002 (T1)	Solid	0.4	0.2
0.0010 (T2)	Dotted	2.1	1.1
0.0015 (T3)	Dashed	3.2	1.6

[a]Used in Figure 3
[b]CFL computed with tip speed

From the sensitivity analysis, a value for the smearing factor is defined, Es, for the larger dt and spatial resolution R/8 in order to calculate precisely the power coefficient at TSR=6. The results with this smearing factor Es are compared with another simulation where Prandtl's tip loss correction factor (TLCF) is applied and thus a larger smearing factor Ep is calibrated to match the Cp at TSR=6, for TSR=3, 6 and 10 in case A and TSR=2.5 and 4 in case B. Those values in the latter case are related to the downstream wind turbine, while the upstream wind turbine operates at design condition (TSR=6).

Results

3.3. *Sensitivity analysis*

A sensitivity analysis is performed in order to get some insight into the influence of the smearing factor and time step size in the wake as well as the power and thrust coefficients. In the legend of the following figure, type and colour of each line is selected as defined in Table 2 and Table 3. Figure 3 depicts the velocity deficit (left) and the turbulent kinetic energy (*TKE*, right), both normalized with the inlet velocity, at 3D from the rotor plane (where D is the rotor diameter) in case A for a *TSR* value of 6 (design condition), using grid R/8. Comparing this results with the corresponding results using grid R/15 (not shown here), it seems that the tower model is highly affected by the spatial resolution.

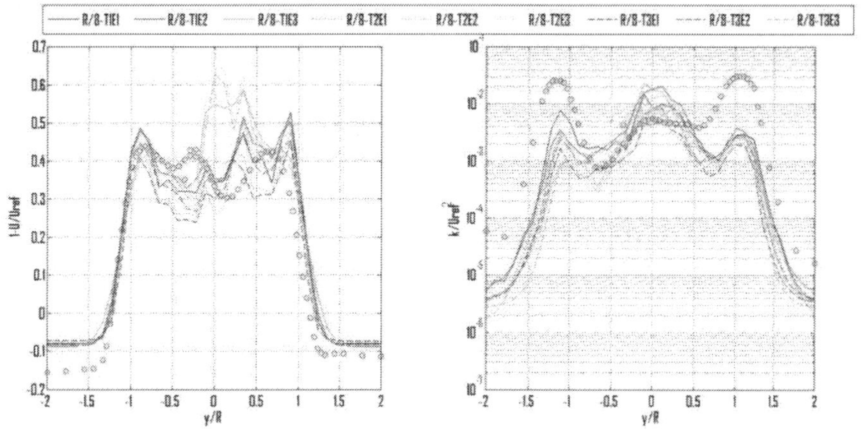

Figure 3. Normalized mean velocity deficit (left) and normalized turbulent kinetic energy (right), without applying a TLCF. Distance: 3D. TSR: 6. Case A. Spatial resolution: R/8.

Figure 4 presents the error in the computed *Cp* for the 18 simulation runs, combining different settings. The error is plotted against the smearing factor, normalized by the rotor radius (left) and by the spatial resolution (right). From this figure, we can concluded that the computed *Cp* depends strongly on the smearing factor and also on the time step size, but the influence of the latter is related to the value of the smearing factor. For higher values of the smearing factor, the dependency on the time step size is less significant. In addition to this, there is not a clear relationship between the spatial resolution, the smearing factor, the time step size and the error in the computed *Cp*. A set of criteria could not be obtained to ensure the convergence of the error to a small value.

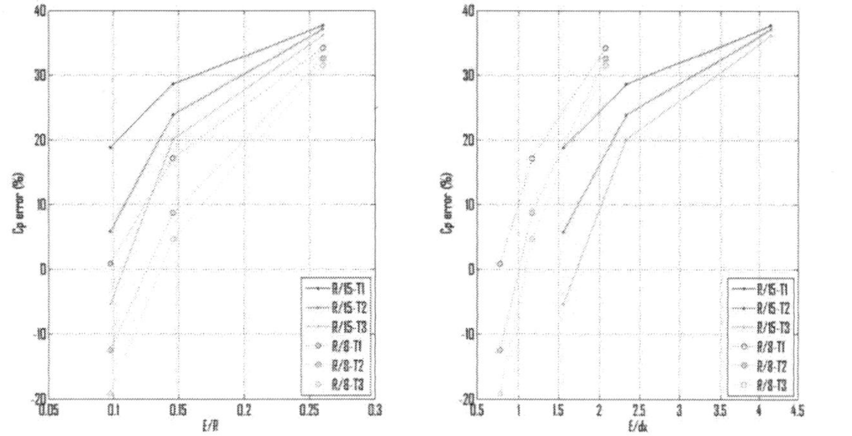

Figure 4. Error in *Cp*. TSR: 6. Case A.

Taking into account the above results, it is concluded that as the smearing factor is larger/the time step size is smaller the power is increased, but the velocity deficit is not as much affected.

3.4. Influence of applying a tip loss correction factor

From the above analysis, a value of 0.0608m (*E/R*=0.136) for the smearing factor is selected, using the larger time step (*T3*) and the coarser spatial resolution (R/8), in order to estimate the *Cp* value precisely (please see Figure 4). In addition to this, another option is considered that takes a value of 0.0715m and applies Prandtl's TLCF. The latter should be larger in order to achieve a better estimation of the *Cp* value (with the same smearing factor, *E/R*=0.136, a lower *Cp* would be obtained).

First, the wake characteristics are presented, reporting the mean velocity deficit and turbulence statistics. Figure 6, Figure 5 and Figure 7 depicts the mean velocity deficit (left) and *TKE* (right), both normalized with the inlet velocity, at 3D from the rotor plane in case A for a *TSR* value of 6, 3 and 10 respectively. In general, there is a good agreement in the mean velocity deficit and the ALM is able to capture the main characteristics, despite finding some discrepancies close to the rotor centre probably related to the rough models used to represent the tower and the nacelle. Regarding turbulent characteristics, the *TKE* profile for design condition is almost one order of magnitude smaller than the experimental data in the tip region, similar results were presented in [17]. In [20] it is argued that small fluctuations in the upcoming flow help to trigger instabilities in the vortices, finding better results with a low-intensity free-stream turbulence than with laminar inflow.

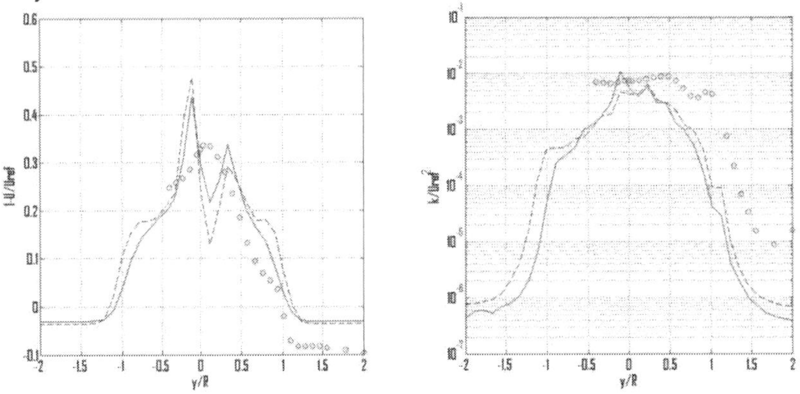

Figure 5. Normalized mean velocity deficit (left) and normalized turbulent kinetic energy (right), without applying a TLCF (dashed black line) and applying a TLCF (solid blue line). Experimental data: red circles. Distance: 3D. *TSR*: 3. Case A.

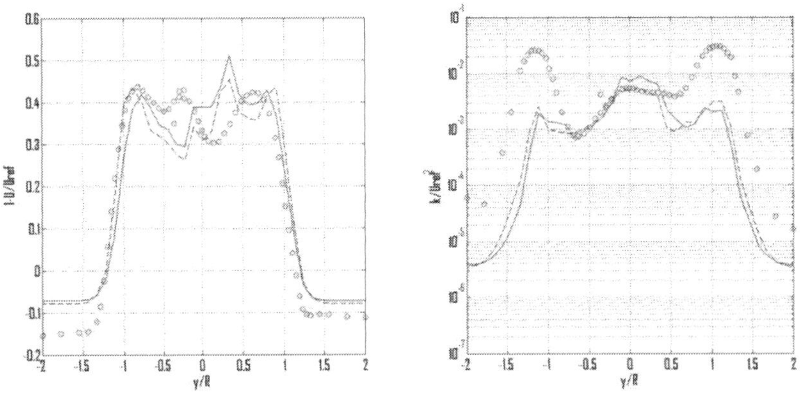

Figure 6. Normalized mean velocity deficit (left) and normalized turbulent kinetic energy (right), without applying a TLCF (dashed black line) and applying a TLCF (solid blue line). Experimental data: red circles. Distance: 3D. *TSR*: 6. Case A.

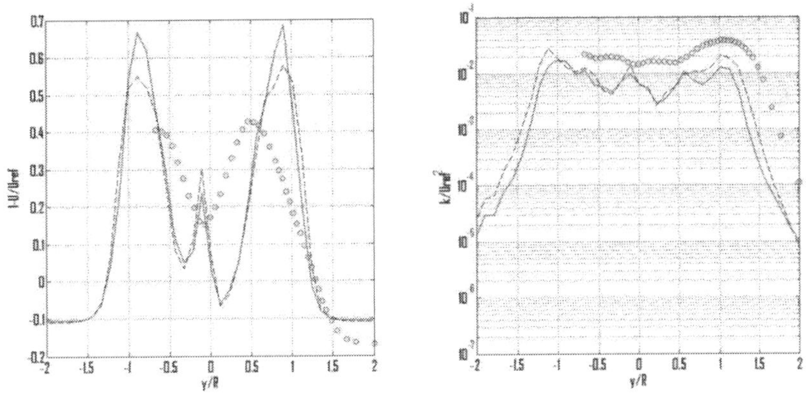

Figure 7. Normalized mean velocity deficit (left) and normalized turbulent kinetic energy (right), without applying a TLCF (dashed black line) and applying a TLCF (solid blue line). Experimental data: red circles. Distance: 3D. *TSR*: 10. Case A.

Regarding case B, Figure 8 and Figure 9 depict the mean velocity normalized with the inlet velocity (left) and the turbulence intensity (right) at 4D from the rotor plane in case B for a *TSR* value of 4 (design condition) and 2.5 respectively of the downstream wind turbine while the upstream wind turbine operates at design condition (*TSR*=6). The turbulent statistics have improved, compared to case A, but there is a lower recovery in the mean velocity compared to the experimental data. This could be related to the inlet condition applied in this study, that generates a lower turbulence level in the wake.

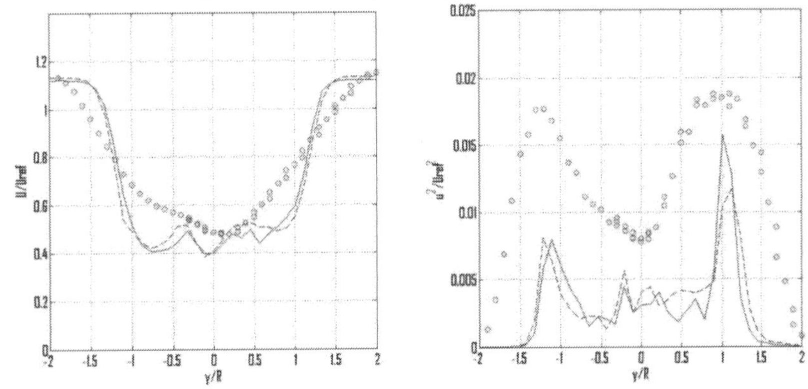

Figure 8. Normalized mean velocity (left) and normalized turbulent kinetic energy (right), without applying a TLCF (dashed black line) and applying a TLCF (solid blue line). Experimental data: red circles. Distance: 4D. *TSR*: 6 and 4 (upstream and downstream wind turbine respectively). Case B.

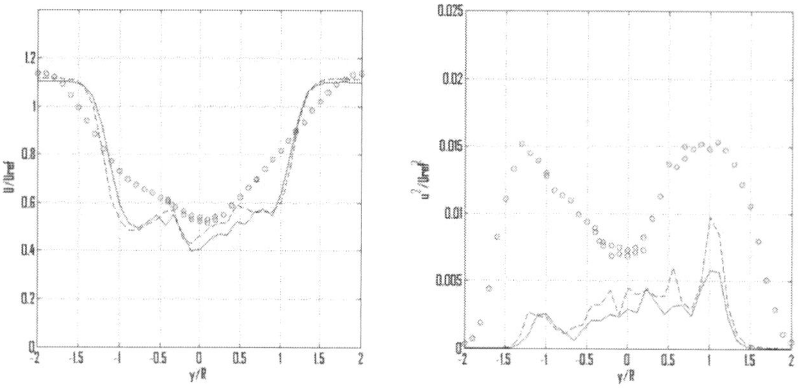

Figure 9. Normalized mean velocity (left) and normalized turbulent kinetic energy (right), without applying a TLCF (dashed black line) and applying a TLCF (solid blue line). Experimental data: red circles. Distance: 4D. *TSR*: 6 and 2.5 (upstream and downstream wind turbine respectively). Case B.

From those figures, it can be concluded that the ALM with a coarse resolution captures the main characteristics of the wake and that the TLCF has not a significant influence on the wake. Regarding the performance coefficients, Figure 10 presents the *Cp* (left) and *Ct* (right) computed in case A (top) and in case B (bottom). There is a better agreement when applying a TLCF for all the *TSR* values considered. Moreover, in case A, despite adjusting the smearing factor to compute precisely the *Cp* at design condition without applying a TLCF, the *Cp* obtained when the wind turbine is operating under different conditions is overestimated by 40% (*TSR*=3) or underestimated by 37% (*TSR*=10), while those errors are reduced significantly when applying a TLCF (8% and 1% respectively). In case B, the *Cp* is overestimated with both approaches, but the error is lower when applying a TLCF (26% compared to 36% for *TSR*=4 and 25% compared to 46% for *TSR*=2.5). Nevertheless, despite using the same model and parameters for both wind turbines, the error in *Cp* for the downstream wind turbine is quite higher than the error for the upstream wind turbine.

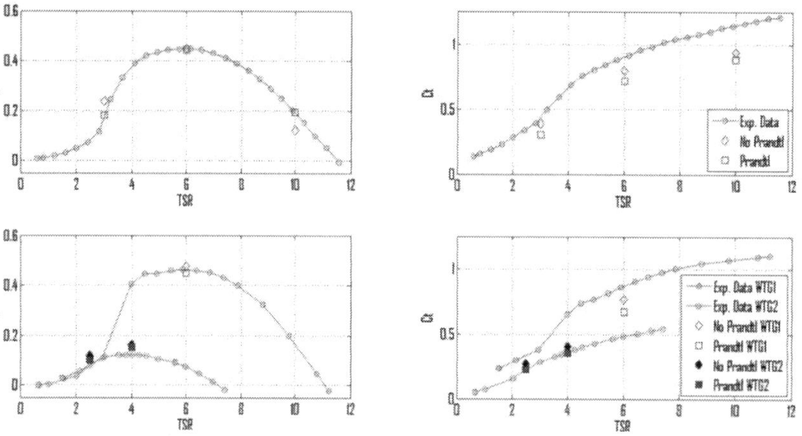

Figure 10. Power coefficient (left) and thrust coefficient (right), without applying a TLCF (diamond black marker) and applying a TLCF (square blue marker). Experimental data: red circles. Case A: top, case B: bottom (full markers correspond to the downstream wind turbine).

Tests with Blind Test 3 were performed (not shown here) and better results of *Cp* were obtained, which suggests that the required smearing factor is affected by the inflow condition at each wind turbine. Regarding the *Ct*, its value is under-predicted regardless of the operating condition. A similar under-prediction was obtained in [20][21] with the ALM and in [22] with a blade element momentum method. As reported in [21] following [22], the accuracy of the ALM is based on, among other parameters, the airfoil data used, so our results may be affected by this.

4. Conclusions

The ALM has been implemented in the open source code caffa3d.MBRi and validated against wind tunnel experimental campaigns using coarser spatial resolutions and larger time steps. Regarding the latter, generally it is recommended to set the time step size in order to limit the movement of the rotor tip to a grid cell length in each time step. In the present work it is argued that as the computed aerodynamic forces are projected into the domain through a smearing function, with spatial resolutions where the chord length is close to or less than the grid length, the restriction of limiting the tip movement one grid cell length may not be strictly required, as the forces are smeared to neighbor cells despite the chord is immersed in just one grid cell, so the time step size should be related to the region of the domain where the forces are projected.

A sensitivity analysis, taking into account two spatial resolutions (R/8 and R/15), was performed in order to assess the influence of the smearing factor and time step size on power and thrust as well as the wake characteristics. Power and thrust are highly dependent on the projection function as well as spatial resolution, but the mean velocity deficit in the wake is not as much affected. The influence of the time step size is diminished with larger smearing factors. Nevertheless, further investigation is needed in order to determine a procedure to adjust the evaluated parameters. Moreover, more complex models to represent the tower and the nacelle should be tested in order to improve the results in the wake centre.

From the above analysis, a value for the smearing factor was defined for the larger time step size and coarser resolution in order to calculate precisely the power coefficient at *TSR*=6 for the validation case with one model wind turbine. The results with this smearing factor were compared with another simulation choosing a larger smearing factor and applying Prandtl's tip loss correction factor, for *TSR*=3/6/10. It is found that applying the tip loss correction factor improves the power estimation in all operational conditions. This two alternatives were tested in a validation case with two model wind turbines, confirming that conclusion, although with higher errors in power production.

Overall, we have shown that the ALM is able to capture the main characteristics of the wake in coarser spatial resolutions and using larger time steps. In addition to this, applying a tip loss correction factor improves the power estimation, besides obtaining the expected behavior of the aerodynamic forces along the blades.

References

[1] Vermeer L J, Sorensen J N and Crespo A 2003 Wind turbine wake aerodynamics *Progress in Aerospace Sciences* **39** 467-510.

[2] Sanderse B, van del Pijl S P and Koren B 2011 Review of computational fluid dynamics for wind turbine wake aerodynamics *Wind Energy* **14** 799-819.

[3] Mikkelsen R 2003 Actuator Disc Methods Applied to Wind Turbines PhD thesis, Technical University of Denmark.

[4] Wu Y T and Porté-Agel F 2011 Large-eddy simulation of wind-turbine wakes: evaluation of turbine parametrisations *Boundary Layer Meteorology* **138** 345-66.

[5] Storey R C, Norris S E and Cater J E 2014 An actuator sector method for efficient transient wind turbine simulation *Wind Energy* doi: 10.1002/we.1722.

[6] Sørensen J N, Shen W Z 2002 Numerical modelling of wind turbine wakes *Journal of Fluids Engineering* **124** 2 393-99.

[7] Troldborg N 2008 Actuator Line Modeling of Wind Turbine Wakes. PhD thesis, Technical University of Denmark.

[8] Shen W Z, Sorensen J N and Zhang J H 2007 Actuator surface model for wind turbine flow computations 2007 *Proc. of European Wind Energy Conference and Exhibition* (Milan).

[9] Churchfield M J, Lee S, Moriarty P, Martínez L A, Leonardi S, Vijayakumar G and Brasseur J G 2012 A large-eddy simulation of wind-plant aerodynamics *50th AIAA Aerospace Sciences Meeting* (Nashville, Tennessee).

[10] Porté-Agel F, Lu H and Wu Y T 2010 A large-eddy simulation framework for wind energy applications *The Fifth International Symposium on Computational Wind Engineering* (North Carolina).

[11] Troldborg N, Zahle F, Réthoré P E and Sorensen N N 2014 Comparison of wind turbine wake properties in non-sheared inflow predicted by different computational fluid dynamics rotor models *Wind Energy* doi: 10.1002/we.1757.

[12] Shen W Z, Sorensen J N and Yang H 2011 Actuator Line/Navier stokes computations for flows past the yawed MEXICO rotor *Wake Conference* (Visby).

[13] Nilsson K, Shen W Z, Sorensen J N, Breton S P and Ivanell S 2014 Validation of the actuator line method using near wake measurements of the MEXICO rotor *Wind Energy* **18** 3 499-514.

[14] Pankjak K J, Churchfield M J, Moriarty P J and Schmitz S 2014 Guidelines for volume force distributions within actuator line modeling of wind turbines on large-eddy-simulation type grids *Journal of Solar Energy Engineering* **136** 031003-1-11.

[15] G. Usera, A. Vernet, J.A. Ferré. “A parallel block-structured finite volume method for flows in complex geometry with sliding interfaces”. Flow, Turbulence and Combustion, 2008, 81, 471-495.

[16] Mendina M, Draper M, Kelm Soares A P, Narancio G and Usera G 2014 A general purpose parallel block structured open source incompressible flow solver *Cluster Computing* **17** 2 231-41.

[17] Krogstad P A and Eriksen P E 2013 "Blind test" calculations of the performance and wake development for a model wind turbine *Renewable Energy* **50** 325-33.

[18] Pierella F, Krogstad P A and Saetran L 2014 Blind Test 2 calculations for in-line model wind turbines where the downstream turbine operates at various rotational speeds *Renewable Energy* **70** 62-77.

[19] Krogstad P A, Saetran L 2013 Invitation to the 2013 "Blind test 3" workshop - two in-line wind turbines with spanwise offset (Norwegian University of Science and Technology - NTNU).

[20] Sarlak H 2014 Large Eddy Simulation of Turbulent Flows in Wind Energy PhD thesis, Technical University of Denmark.

[21] Martínez-Tossas L A, Churchfield M J, Leonardi S 2014 Large eddy simulation of the flow past wind turbines: actuator line and disk modeling *Wind Energy* **18** 6 1047-1060.

[22] Krogstad P A, Lund J A 2012 An experimental and numerical study of the performance of a model turbine Wind Energy **15** 3 443-457.

Wake Conference 2015 IOP Publishing
Journal of Physics: Conference Series **625** (2015) 012022 doi:10.1088/1742-6596/625/1/012022

Large-eddy simulations of wind farm production and long distance wakes

O Eriksson[1], K Nilsson[1,2], S-P Breton[1], S Ivanell[1,2]
[1]Uppsala University, Department of Earth Sciences, Wind Energy Campus Gotland, 621 67 Visby, Sweden
[2]KTH , Department of Mechanics, Linné Flow Centre, Stockholm, Sweden

E-mail: ola.eriksson@geo.uu.se

Abstract. The future development of offshore wind power will include many wind farms built in the same areas. It is known that wind farms produce long distance wakes, which means that we will see more occasions of farm to farm interaction, namely one wind farm operating in the wake of another wind farm. This study investigates how to perform accurate power predictions on large wind farms and how to assess the long distance wakes generated by these farms. The focus of this paper is the production's and wake's sensitivity to the extension of the grid as well as the turbulence when using Large-eddy simulations (LES) with pregenerated Mann turbulence. The aim is to determine an optimal grid which minimizes blockage effects and ensures constant resolution in the entire wake region at the lowest computational cost.

The simulations are first performed in the absence of wind turbines in order to assess how the atmospheric turbulence and wind profile are evolving downstream (up to 12,000 m behind the position where the turbulence is imposed).

In the second step, 10 turbines are added in the domain (using an actuator disc method) and their production is analyzed alongside the mean velocities in the domain. The blockage effects are tested using grids with different vertical extents. An equidistant region is used in order to ensure high resolution in the wake region. The importance of covering the entire wake structure inside the equidistant region is analyzed by decreasing the size of this region. In this step, the importance of the lateral size of the Mann turbulence box is also analyzed.

In the results it can be seen that the flow is acceptably preserved through the empty domain if a larger turbulence box is used. The relative production is increased (due to blockage effects) for the last turbines using a smaller vertical domain, increased for a lower or narrower equidistant region (due to the smearing of the wake in the stretched area) and decreased when using a smaller turbulence box (due to decreased inmixing). The long distance wake behind the row is most impacted by the use of a smaller turbulence box, while the other simulation setups have less influence on these results.

In summary, the results show the importance of having relatively large extensions of the domain, large extensions of the equidistant region and especially large extensions of the turbulence box.

1. Introduction

The future development of wind power offshore will include many wind farms built in the same areas. It is known that these wind farms will produce long distance wakes [1]. This means that we will see more instances of farm to farm interaction where one wind farm is in the wake of another wind farm. A step towards a better understanding of this interaction is to study the long distance wake behind single wind farms using numerical models.

The scale of the long distance wakes is in the order of 10 km [2]. Different types of wake models and meso scale models can be used to study the long distance wakes [1][3]. Large eddy simulations are, although more computational demanding, another alternative to these models considering the continuously increasing computational resources available.

Large eddy simulations (LES) have been used for simulations of wakes inside wind farms in a wide range of studies, see e.g. [4][5][6][7][8][9][10][11]. LES in combination with the actuator disc (ACD) methodology was recently used to study the relative production of the Lillgrund wind farm with good comparison to measurements [11].

Modeling of larger wind farms with long extension and the long distance wakes downstream is, however, associated with increased uncertainties. Large eddy simulations in combination with ACD have been used by Eriksson et al. [12] to study the wake behind Horns Rev. The results showed an increased production at the end of the farm and a faster velocity recovery in the long distance wake compared to the measurements. A parameter study on a row of ten turbines has been performed to determine the impact of grid resolution, Reynolds number, turbulence intensity and internal spacing on production as well as on the velocity recovery in the long distance wake [13]. The results were most sensitive to the turbulence level. The increase in the production in the end of the row could not be explained.

The present study uses LES with ACD on a row of ten turbines and investigates the impact of the grid dimensions, the dimensions of the equidistant region and the extension of the imposed turbulence planes. The study is presented in Section 3 and the simulation setups are presented further in Section 4. First the numerical model is presented in Section 2. The simulations are analysed with respect to production and the development of the wind shear, the velocity as well as the turbulence throughout the domain, see Section 5.

The aim of the present study is to acquire additional knowledge on how to perform accurate power predictions of long wind farms and how to assess the long distance wakes generated by these farms. The focus is on the sensitivity to the extension of the grid and the turbulence when using LES with pregenerated Mann turbulence.

2. Numerical model

The computations are carried out as large-eddy simulations using the EllipSys3D code. The wind turbines are parameterized using an actuator disc model [14]. A controller is used to maintain the optimal rotational speed for each turbine[15]. The sub-grid-scale model from Ta-Phouc[16] is used for the smallest (less than 2 times the grid resolution) non-resolved eddies in the LES.

The atmospheric turbulence is pregenerated using the Mann model [17]. The turbulence and the wind shear are imposed in the computational domain using body forces[10]. The simulations are performed assuming a neutral boundary layer.

The domain is stretched in the inlet, the outlet, towards the sides and towards the top. In the region of interest the domain is equidistant in all directions (uniform). The boundary conditions are fixed values for the inlet (according to the shear), periodic for the sides, convective for the outlet, far field for the top and for the ground. The far field correspond for the ground to a noslip condition and for the top to a fixed velocity (according to the shear).

3. Study of one row of turbines

The study is performed on a row of 10 turbines and includes the long distance wake up to over 6000 m (150 R) downstream from the last turbine, see Figure 1. The simulations are first performed in the absence of wind turbines in order to study the preservation of the flow characteristics throughout the domain (up to 12,000 m behind the position where the turbulence is imposed).

As a second step, 10 turbines are added in the domain and their production is analyzed alongside the mean velocities in the domain.The aim of this step is to determine an optimal grid which minimizes blockage effects and ensures a constant resolution in the entire wake region at the lowest computational cost. The blockage effects (acceleration dependent of the cross section of the domain versus rotor size) are investigated using grids with different vertical extents. In order to have a high resolution in the wake region an equidistant region is used. The importance of covering the entire wake structure inside the equidistant region is analyzed by decreasing this region. In this step, the importance of the size of the Mann turbulence box is also analyzed.

4. Simulation setup

The 10 turbines used in the simulation have a radius of 80 m, their hub height is 70m (1.75 R). This turbine is based on a NREL 5 MW [18] and scaled down, to correspond to a Vestas V80 regarding power and thrust coefficient [11].

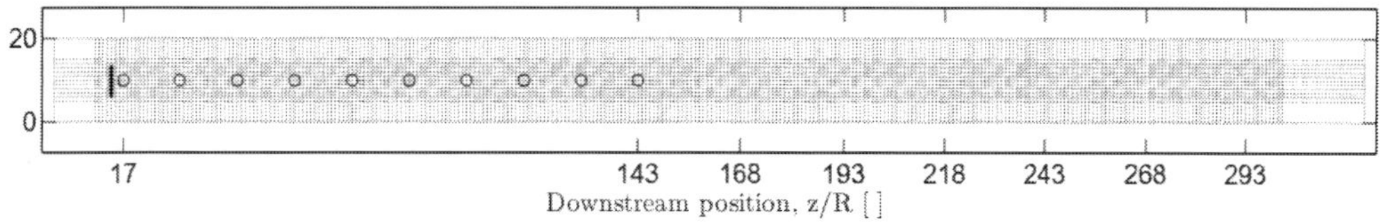

Figure 1. Row of 10 turbines.

These are put into a domain, see Figure 1, that has an equidistant region with the length of 302 R (in the streamwise direction z). The extension in width (x-direction) and height (y-direction) varies. The turbulence is inserted at 13 R, which corresponds to 3 R into the equidistant region, and the first turbine is inserted at 17 R. The internal spacing in the row is 14 R. The resolution in the equidistant region is chosen to 0.1 R (in accordance with earlier studies [11]) and a non dimensionalized (with U_0 and R) timestep of 0.025 is used for all cases. Full convergence could however not be shown for the downstream turbines with this resolution in an earlier parameter study [13]. The chosen grid resolution also has an impact on to what extent the sub grid model is used. The flow passes through the domain for 26,000 time steps (50 min) before the averaging is started. The averaging is performed over a physical time period of 50 min.

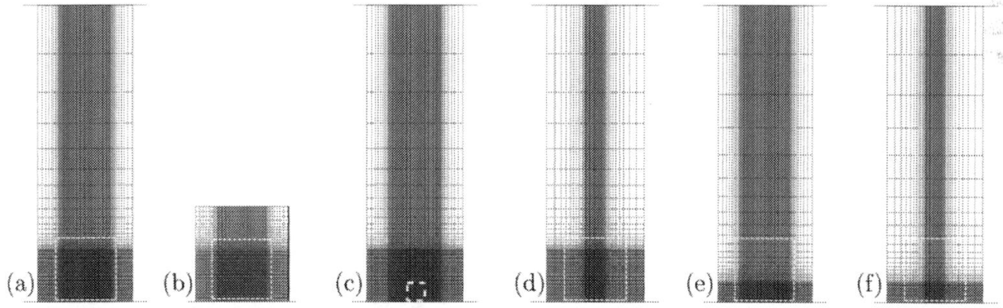

Figure 2. The grid (in the x-y plane) and extension of the turbulence box (marked with dashed white lines) for (a) Reference case, (b) Lower domain, (c) Turbulence box -small, (d) Equidistant region: High/narrow, (e) Equidistant region: Wide/low, (f) Equidistant region: Small.

The boundary layer in the simulations is described with respect to wind shear and turbulence level. The wind shear follows the logarithmic profile in Equation 1, with a roughness length value (z_0) of 0.0001 and constant giving a wind speed of 8 m/s at the hub heigth 70 m.

$$U(y) = \frac{u_*}{\kappa} * ln\left(\frac{y}{z_0}\right) = 0.5944 * ln\left(\frac{y}{0.0001}\right) \quad (1)$$

The Mann turbulence is created using the same roughness length (z_0) of 0.0001 and the same wind speed of 8 m/s at the hub height 70 m. This gives $\alpha\epsilon^{2/3}$ $[m^{4/3}s^{-2}]$ =0.009040, the eddy lifetime constant of γ = 3.886 and the length scale L = 41.3327 $[m]$. For a detailed description the reader is referred to Mann [19]. The length of the Mann box corresponds to a physical time of 10 min (using the Taylor frozen hypothesis). The resolution is slightly larger in the Mann box compared to the grid with dz = 0.117 R and dx = dy = 0.125-0.195 R (depending on case).

4.1. Studied cases

The simulations are performed for different setups of the domain and sizes of the turbulence box. All modifications are compared to a reference case. The reference case, see Figure 2a, is chosen with rather large extensions and can be seen as the "best affordable". The other setups study the impact of decreasing the extension to save computational power. The setup for the different cases along with used grid size can be seen in Table 1. The grid and the turbulence box (marked with dashed white lines) are also visually presented in Figures 2.

Table 1. The setup used for the reference and modified cases along with the grid size. The bold text indicate the difference to the reference case.

Case description	**Short name**	**Domain (x*y) [R*R]**	**Equidistant (x*y) [R*R]**	**Turbulence (x*y) [R*R]**	**Grid size [M cells]**
Reference case	Ref	20*60	10.75*10.75	12.5*12.5	50.3
Lower domain	Low	**20*20**	10.75*10.75	12.5*12.5	$\leq$ **50.3**
Turbulence box -Small	Turb s	20*60	10.75*10.75	**4*4**	50.3
Equidistant region High (and narrow)	Equ h	20*60	**4*10.75**	12.5*12.5	**25.2**
Equidistant region -Wide (and low)	Equ w	20*60	**10.75*4**	12.5*12.5	**25.2**
Equidistant region -Small	Equ s	20*60	**4*4**	12.5*12.5	**12.6**
Small turbulence, Small equidistant, Lower domain	All s	20*20	**4*4**	**4*4**	**12.6**

5. Results and discussion

The study is performed with and without turbines. The simulations performed in the absence of wind turbines shows the flow adjusting to the domain and evolving downstream. In the simulations where 10 turbines are added in the domain their production is analyzed alongside the flow in the domain.

The studied parameters are the streamwise flow and a streamwise turbulence measure, (here the streamwise turbulence is defined as the standard deviation of axial wind divided by the mean free stream velocity). The results include the vertical profiles and the values at hub height (both taken at positions along the centerline of the row of turbines).

The legends in Figures 3- 24 refer to the short name for the different cases given in Table 1. Note that all results are presented non dimensionalized with the rotor radius (R) and the incoming wind speed at hub height (U).

5.1. Downstream development of velocity, turbulence and wind shear - without turbines

The downstream development of the shear for the different cases can be seen in Figure 3 (for the domain without turbines). The profiles follow the initial profile to a relatively high extent also further downstream in the domain. An acceleration of the flow can be seen for the case with a smaller turbulence box. In Figures 4-11 below the development of wind speed and turbulence at hub height can be seen for the different cases.

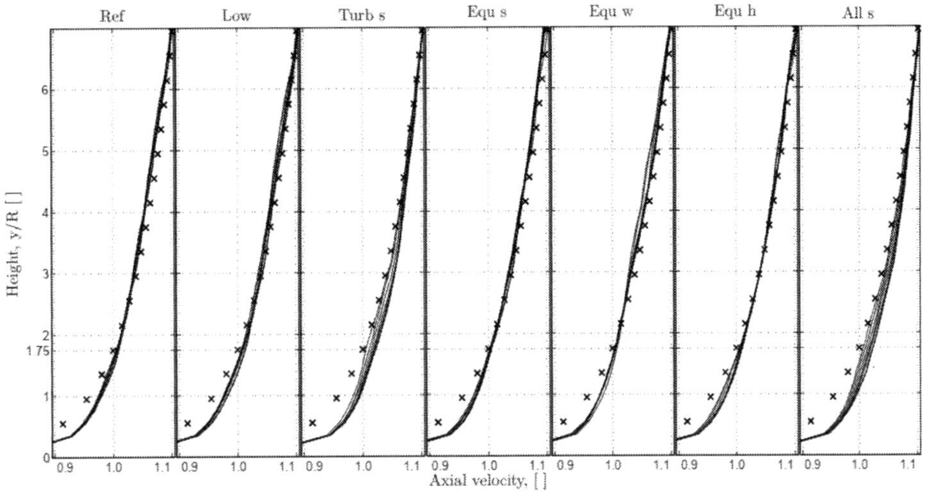

Figure 3. The downstream wind shear (without turbines) compared to the inlet profile (marked with x's). For each case, 6 profiles between 168 R - 293 R (every 25 R / 1 km behind the row) are plotted with increasing line width further downstream. The hub height at 1.75 R is marked in the grid.

The wind speed at hub height for both the reference case and the case with lower domain stays close to 1 throughout the empty domain, as can be seen in Figure 4. However a slight increase of the wind speed can be seen, especially for the lower domain. The turbulence shows in both cases the same trend, namely an increase in the beginning followed by a decrease towards a stable value further downstream, see Figure 5.

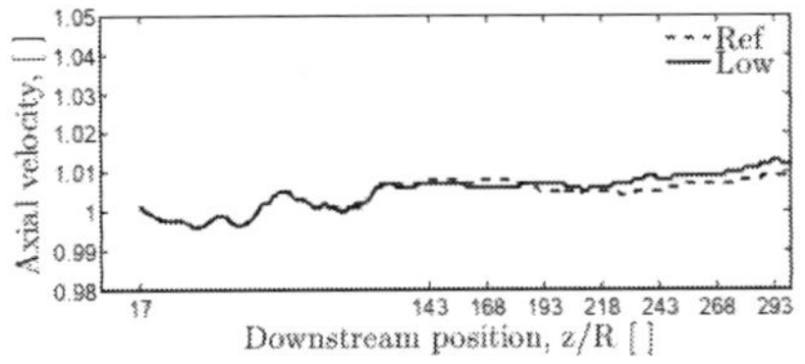

Figure 4. Impact of height of domain on velocity at hub height, no turbines.

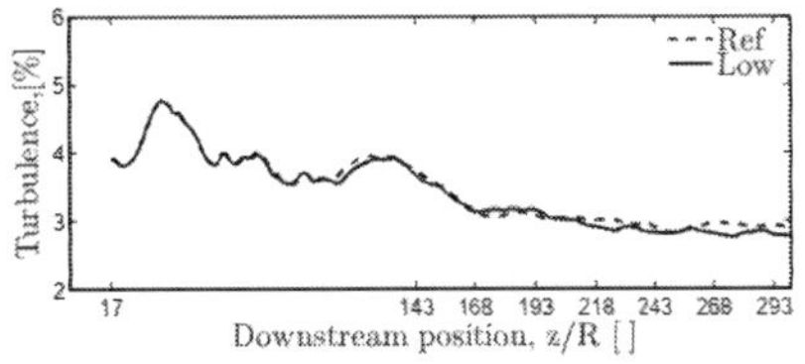

Figure 5. Impact of height of domain on turbulence at hub height, no turbines.

The wind speed stays around 1 and follows the same trend for all the different sizes of equidistant region throughout the empty domain, see Figure 6. A slightly higher wind speed can, however, be seen downstream in the case with the higher equidistant region. The development of the turbulence shows the same trend for all cases, see Figure 7. Only the reference case and the wide case give slightly lower values (at the end of the domain)

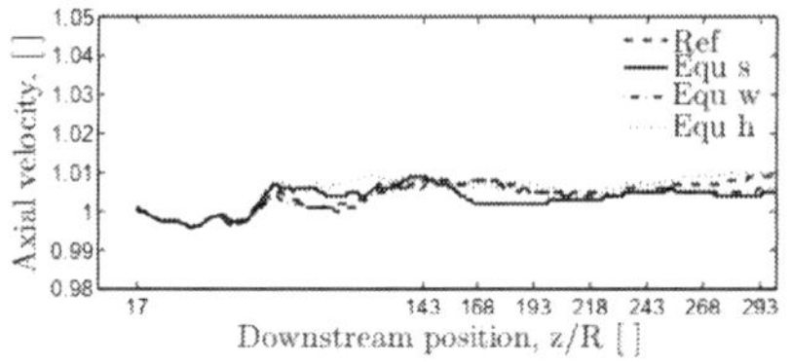

Figure 6. Impact of size of equidistant region on velocity at hub height, no turbines.

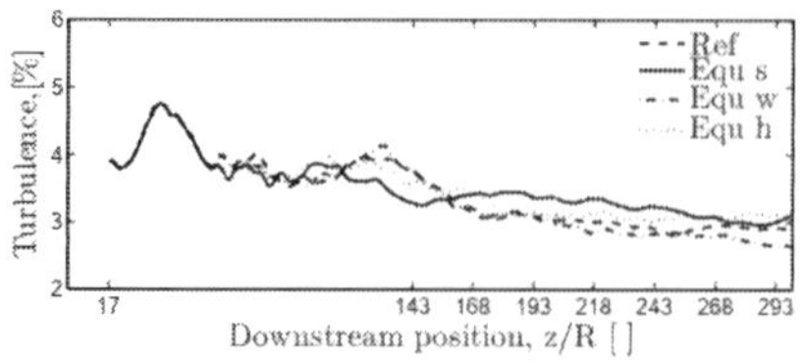

Figure 7. Impact of size of equidistant region on turbulence at hub height, no turbines.

The clearest impact on the downstream development is caused by the size of the turbulence box. In the empty domain the smaller turbulence box gives a downstream increase of the wind speed, see Figure 8. The turbulence peaks in the beginning of the domain and decreases towards a clearly lower value in case of a smaller turbulence box, see Figure 9.

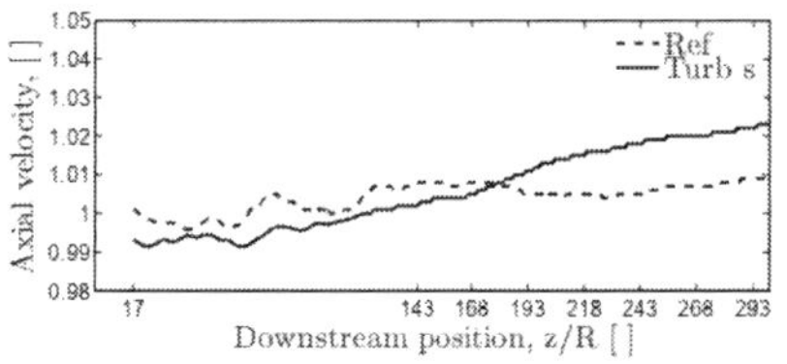

Figure 8. Impact of size of turbulence box on velocity at hub height, no turbines.

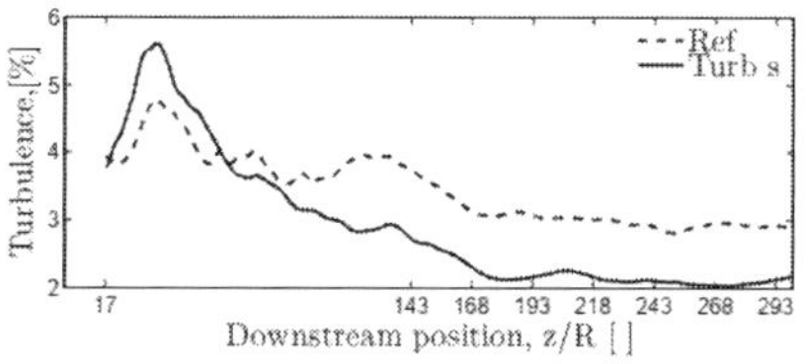

Figure 9. Impact of size of turbulence box on turbulence at hub height, no turbines.

In the extreme cases (that using a small equidistant region, a small turbulence box and a low domain) the trends mostly follow the case with the small turbulence box, see Figures 10, 11. Some smaller impact can, however, also be seen downstream from the other parameters with a slightly higher wind speed from the small equidistant region and a slightly higher turbulence from the small equidistant region.

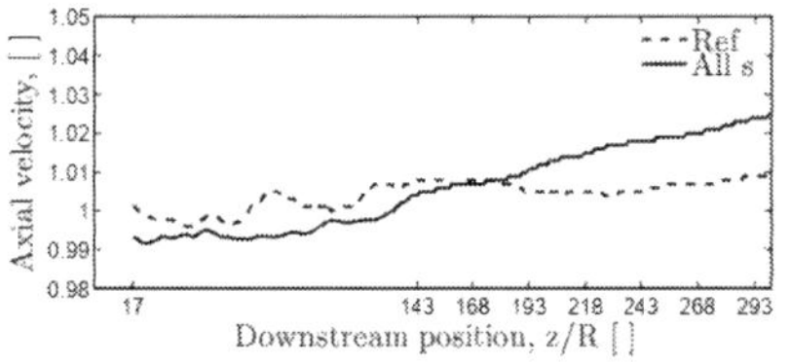

Figure 10. Impact of the "all small case" on velocity at hub height, no turbines.

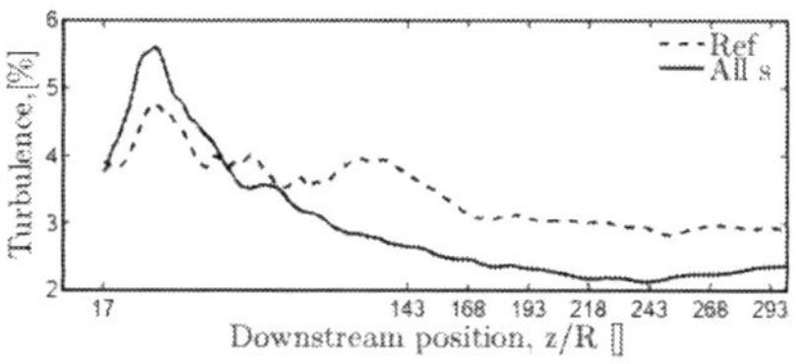

Figure 11. Impact of the "all small case" on turbulence at hub height, no turbines.

5.2. Production

The relative production shows a clearer dependency on the simulation setup in the different cases. In Figure 12, it can be seen that the relative production for the reference case and the case with the lower domain are similar for the first turbines. For the downstream portion of this long row of turbines a higher production can be seen when using a lower domain.

The size of the equidistant region, see Figure 13, has less impact. The production follows the same trend for all cases. However, the Equ w case shows a slightly higher relative production from turbine 2 and the Eqw h case shows a slightly higher relative production further downstream. This could be explained by the smearing of the wake then reaching outside the equidistant region and that the turbines are closer to the stretched area upwards.

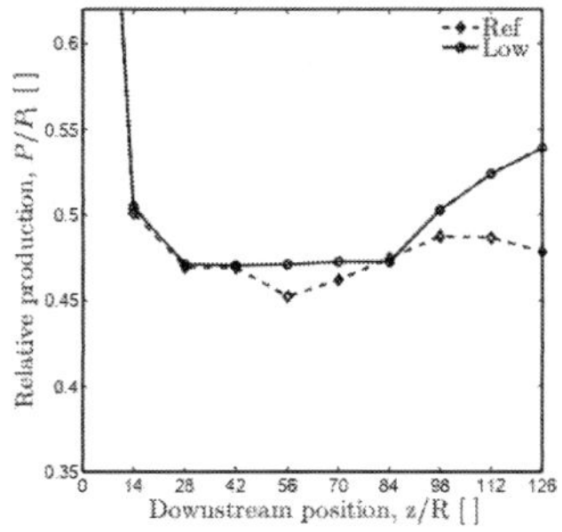

Figure 12. Impact of height of domain on production.

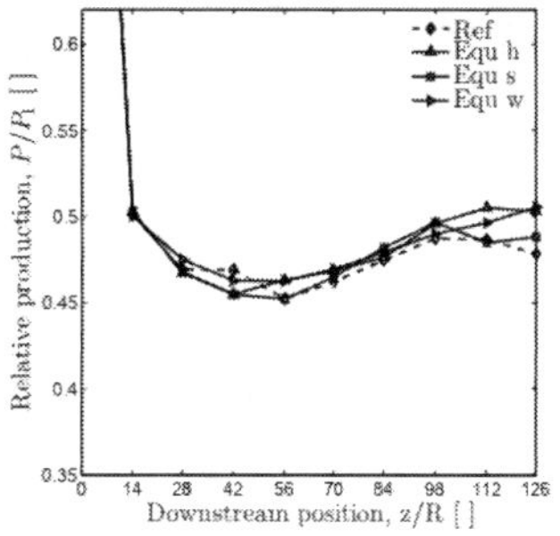

Figure 13. Impact of size of equidistant region on production.

The size of the turbulence box, see Figure 14, shows a high impact. The production decreases gradually when using a smaller box to a level clearly lower than the reference case. This could be explained by the decreased mixing of the flow from above and the sides.

The results of the extreme case (a small equidistant region, a small turbulence box and a low domain) compared to the reference case can be seen in Figure 15. The gradually lower relative production downstream follows the same trend as in Figure 14 with the small turbulence box and the increase in relative production for the most downstream turbines follows the same trend as in Figure 12 with the lower domain.

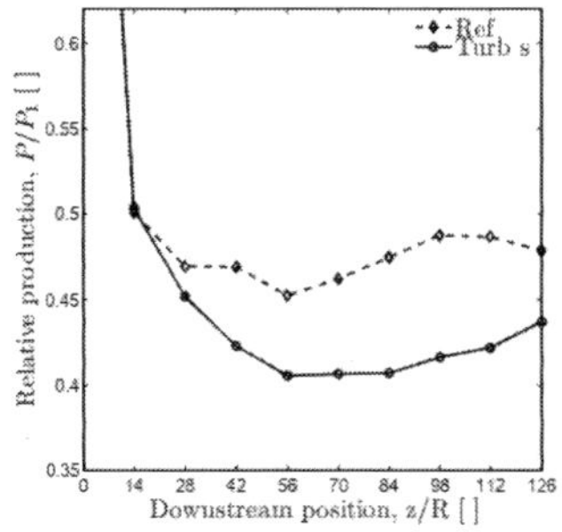

Figure 14. Impact of size of turbulence box on production.

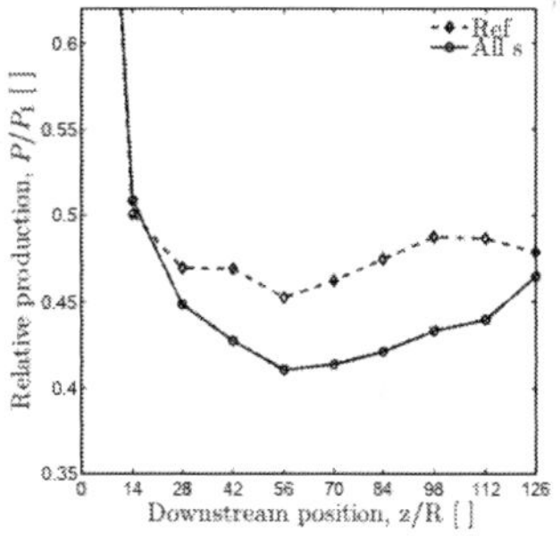

Figure 15. Impact of the "all small case" on production.

5.3. Downstream development of velocity, turbulence and wind shear - with turbines

The downstream development of the shear behind the wind farm (each km / 21.5 R) for the different cases is shown in Figure 16. It is noticeable that the profile for the lower domain shows a faster velocity recovery for all heights indicating some blockage effect. For the smaller turbulence box a slower velocity recovery can be seen as well as a higher vertical extension of the wake.

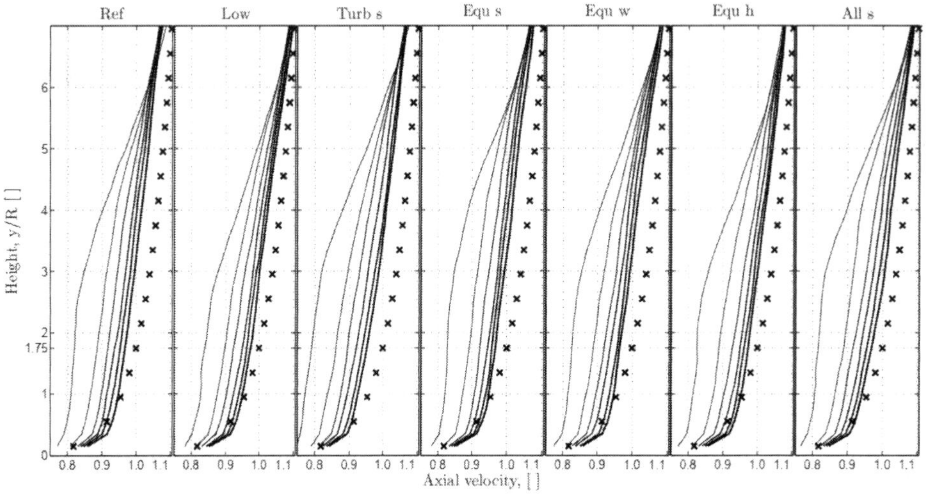

Figure 16. The downstream wind shear (with turbines) compared to the inlet profile (marked with x's). The cases are separated 0.1 in the x-axis. For each case, 6 profiles between 168 R - 293 R (every 25 R / 1 km behind the row) are plotted with increasing line width further downstream. The hub height at 1.75 R is marked in the grid.

Figures 17-24 show more details of the downstream trends for wind speed and turbulence at hub height for the different cases.

The long distance wake behind the row shows relatively low dependency on the domain height, see Figures 17, 18. The lower domain shows a slightly faster velocity recovery and a higher turbulence level.

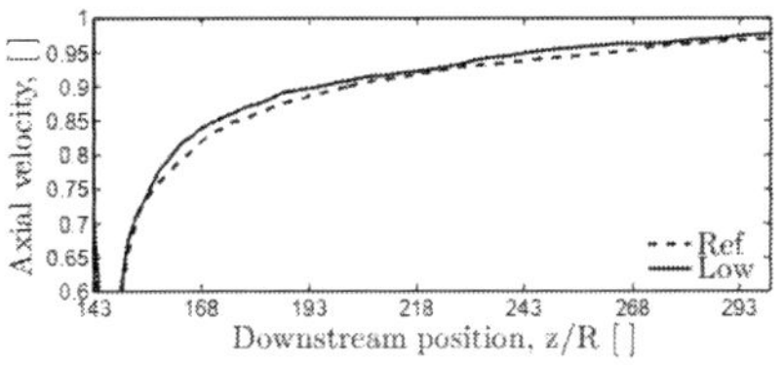

Figure 17. Impact of height of domain on velocity at hub height.

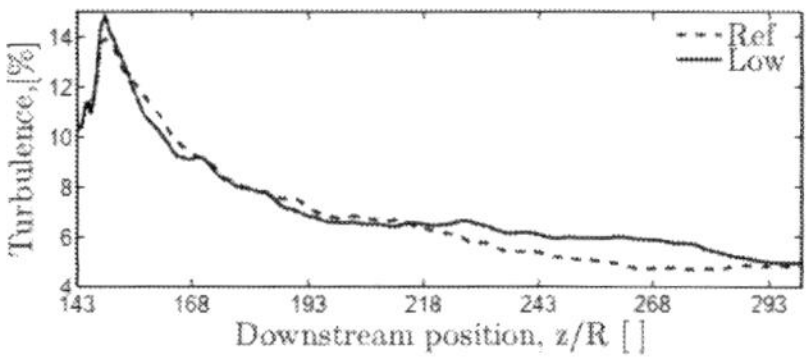

Figure 18. Impact of height of domain on turbulence at hub heights.

The size of the equidistant region is of less importance for the long distance wake, see Figures 19, 20.

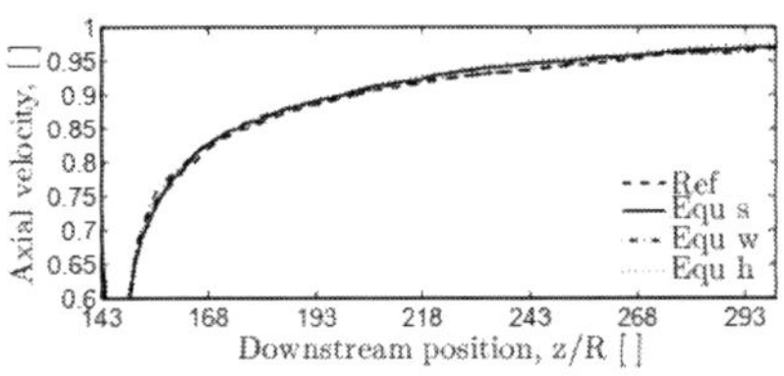

Figure 19. Impact of size of equidistant region on velocity at hub height.

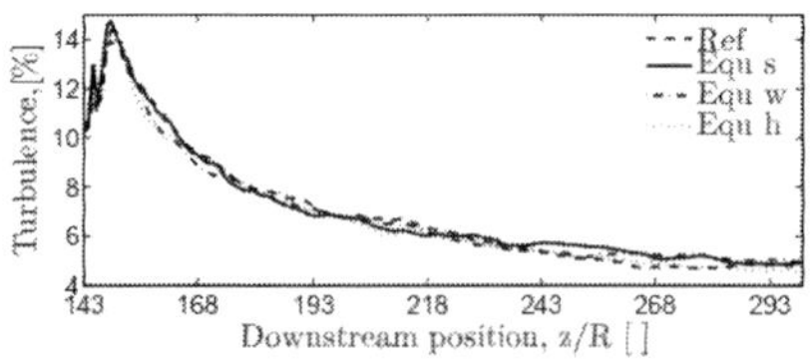

Figure 20. Impact of size of equidistant region on turbulence at hub height.

The size of the turbulence box is the most important parameter for the long distance wake, see Figures 21, 22. The wake recovery follows the same trend in both cases. However the wake recovers faster in the reference case. The turbulence level is slightly higher for the reference case in the beginning of the farm wake but levels out to the same value further downstream.

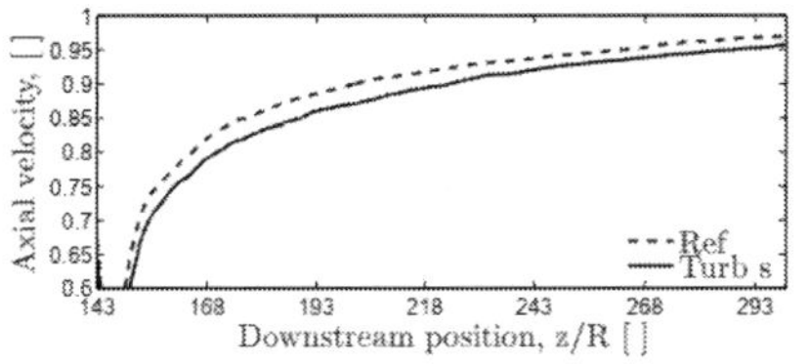

Figure 21. Impact of size of turbulence box on velocity at hub height.

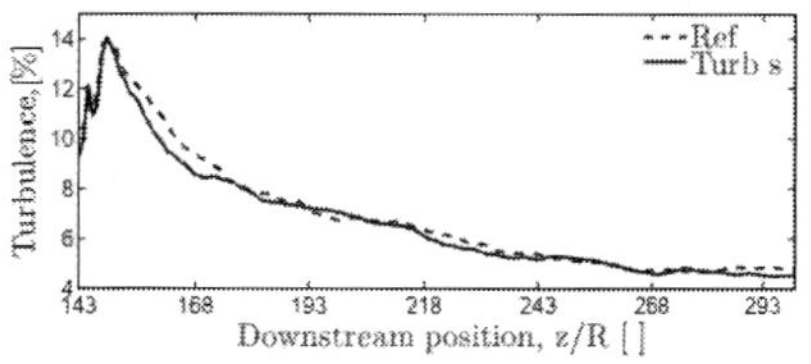

Figure 22. Impact of size of turbulence box on turbulence at hub height.

In the extreme case (using a small equidistant region, a small turbulence box and a low domain) the accumulated impact results in very small differences for the farm wake, see Figures 23, 24. As in the low domain case the turbulence level is slightly higher compared to the reference case at longer distances behind the row of turbines.

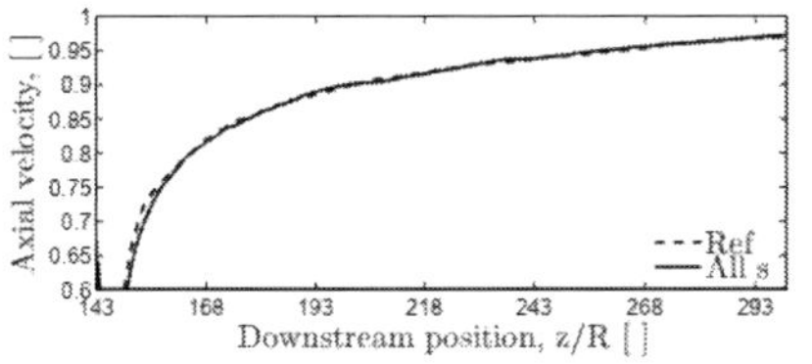

Figure 23. Impact of the "all small case" on velocity at hub height.

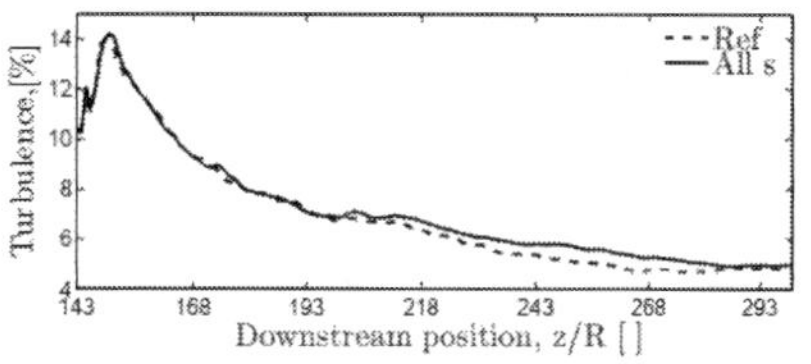

Figure 24. Impact of the "all small case" on turbulence at hub height.

6. Conclusion

Large eddy simulations using the actuator disc method have been performed, both with and without turbines, in order to study the production of a row of turbines and how the atmospheric turbulence and wind profile are evolving downstream in the domain (up to 12,000 m). The sensitivity of the results to varying sizes of domains, different extensions of the imposed turbulence and different extensions of the equidistant region has been studied.

The results show that in the empty domain the conservation of the wind shear and wind speed at hub height are acceptable throughout the domain. The turbulence increases slightly in the beginning of the domain and decreases to a stable but slightly lower value compared to the imposed value. The downstream preservation of both wind speed and turbulence is however impacted negatively by the usage of a smaller turbulence box.

The relative production shows that for a long row of turbines the downstream part is impacted by blockage effects when using a too low domain. The extensions of the equidistant region has some (but less) impact on the relative production. The equidistant wide (Equ w) case shows an increased relative production from turbine 2 and the equidistant high (Equ h) case shows an increased relative production further downstream . The parameter with the greatest impact on the relative production was the size of the turbulence box. The relative production is clearly lower for the small turbulence box.

The size of the turbulence box also caused the largest differences in the long distance wake. Although the farm production was higher in the reference case the velocity recovery in the farm's wake was faster for the reference case compared to the case with the smaller turbulence box.

Concluding from the results the domain size impacts the relative production downstream, due to blockage effects, and is it therefore important for studies of large and long wind farms to simulate large enough domains. The equidistant region needs to be extended to cover the entire wake to avoid a smearing of the wake in the stretched region. The turbulence box needs to cover an area bigger than the crossection of the wake to allow the mixing in of energy from the surrounding flow. For all cases without turbines a downstream adjustment of the inlet flow and turbulence to the domain can be seen. The characteristic of this adjustment and impact of this on the production needs to be studied in further detail [20].

It can be summarized that it is important to have relatively large extensions of the domain, large extensions of the equidistant region and especially large extension of the turbulence box.

Acknowledgments
The simulations were performed on resources provided by the Swedish National Infrastructure for Computing (SNIC) at the National Supercomputer Centre in Sweden (NSC).

References
[1] Frandsen S, Barthelmie R, Rathmann O, Jørgensen H, Badger J, Hansen K, Ott S, Rethore P, Larsen S and Jensen L 2007 *Summary report: The shadow effect of large wind farms: measurements,data analysis and modelling.* (Risø-R-1615(EN), Denmark)
[2] Eriksson O and Ivanell S 2012 *A survey of available data and studies of farm-farm interaction.* (EAWE PhD Seminar on Wind Energy in Europe, 2012)
[3] Brand A 2009 *Wind Power Plant North Sea – Wind farm interaction, The effect of wind farming on mesoscale flow.* (ECN, the Neatherlands)
[4] Ivanell S 2009 *Numerical Computations of Wind Turbine Wakes.* (PhD thesis, ISBN 978-91-7415-216, KTH Engineering Sciences, Sweden)
[5] Lu H and Porté-Agel F 2011 *Large eddy simulation of a very large wind farm in a stable atmospheric boundary layer.* (Physics of Fluids 2011; 23: 065101)
[6] Wu Y T and Porté-Agel F 2012 *Atmospheric turbulence effects on wind-turbine wakes: An LES study.* (Energies 2012, 5, 5340-5362)
[7] Troldborg N, Sørensen J N and Mikkelsen R 2010 *Numerical simulations of wake characteristics of a wind turbine in uniform inflow.* (Wind Energy 2010; 13: 86–99)
[8] Troldborg N, Larsen G C, Hansen K S, Sørensen J N and Mikkelsen R 2011 *Numerical simulations of wake interaction between two wind turbines at various inflow conditions.* (Wind Energy 2011; 14: 859–876.)
[9] Keck R E, Mikkelsen R, Troldborg N, de Maré M and Hansen K S 2013 *Synthetic atmospheric turbulence and wind shear in large eddy simulations of wind turbine wakes.* (Wind Energy 2013; DOI: 10.1002/we.1631)
[10] Troldborg N, Sørensen J N, Mikkelsen R and Sørensen N N 2014 *A simple atmospheric boundary layer model applied to large eddy simulations of wind turbine wakes.* (Wind Energy 2014; 17: 657-669.)
[11] Nilsson K, Ivanell S, Hansen K S, Mikkelsen R, Breton S P and Henningson D 2014 *Large-eddy simulations of the Lillgrund wind farm.* (Wind Energy 2014; DOI: 10.1002/we.1707)
[12] Eriksson O, Mikkelsen R, Nilsson K and Ivanell S 2012 *Analysis of long distance wakes of Horns rev 1 using actuator disc approach.* (J. Phys.: Conf. Ser. 555 012032)
[13] Eriksson O, Nilsson K, Breton S P and Ivanell S 2014 *Analysis of long distance wakes behind a row of turbines - a parameter study* (J. Phys.: Conf. Ser. 555 012032)
[14] Mikkelsen R 2003 *Actuator Disc Methods Applied to Wind Turbines.* (DTU, Denmark)
[15] Breton S P, Nilsson K, Ivanell S, Olivares-Espinosa H, Masson C and Dufresne L 2012 *Study of the effect of the presence of downstream turbines on upstream ones and use of a controller in CFD wind turbine simulation models.* (J. Phys.: Conf. Ser. 555 012014)
[16] TaPhouc L 1994 *Modéles de sous maille appliqués aux écoulmements instationnaires décollés.* (Tech. rep. LIMSI 93074 LIMSI France.)
[17] Mann J 1998 *Wind field simulation.* (Risø, Denmark)
[18] Jonkman J, Butterfield S, Musial W and Scott G 2009 *Wind Turbine for Offshore System Development.* (NREL, the United states of America)
[19] Jakob M, Ott S, Jørgensen B H and Frank H P 2013 *WAsP Engineering 2000.* (Risø–R–1356(EN))
[20] Nilsson K, Eriksson O, Svensson N, Breton S P and Ivanell S 2015 *Large-eddy simulations of the evolution of imposed turbulence in prescribed boundary layers in a very long domain.* (To be submitted to Renewable Energy)

Coalescing Wind Turbine Wakes

S Lee[1a], M. Churchfield[a], S. Sirnivas[a], P. Moriarty[a], F. G. Nielsen[b,c], B. Skaare[b], and E. Byklum[b]
[a]National Renewable Energy Laboratory, Golden, CO, U.S.A.
[b]Statoil, Norway
[c]University of Bergen, Geophysical Institute

E-mail: Sang.Lee@nrel.gov

Abstract. A team of researchers from the National Renewable Energy Laboratory and Statoil used large-eddy simulations to numerically investigate the merging wakes from upstream offshore wind turbines. Merging wakes are typical phenomena in wind farm flows in which neighboring turbine wakes consolidate to form complex flow patterns that are as yet not well understood. In the present study, three 6-MW turbines in a row were subjected to a neutrally stable atmospheric boundary layer flow. As a result, the wake from the farthest upstream turbine conjoined the downstream wake, which significantly altered the subsequent velocity deficit structures, turbulence intensity, and the global meandering behavior. The complexity increased even more when the combined wakes from the two upstream turbines mixed with the wake generated by the last turbine, thereby forming a "triplet" structure. Although the influence of the wake generated by the first turbine decayed with downstream distance, the mutated wakes from the second turbine continued to influence the downstream wake. Two mirror-image angles of wind directions that yielded partial wakes impinging on the downstream turbines yielded asymmetric wake profiles that could be attributed to the changing flow directions in the rotor plane induced by the Coriolis force. The turbine wakes persisted for extended distances in the present study, which is a result of low aerodynamic surface roughness typically found in offshore conditions.

1. Introduction

Turbine wakes interacting with neighboring wakes is a complex physical process that significantly impacts not only the global energy capture of an operational wind farm but the longevity of the individual turbines, which leads to uncertainty in the cost of energy during a 20-year design life. Although empirically formulated turbulence intensity models exist to predict the fatigue life of turbines placed within an array, inaccuracies are thought to have led to stringent designs that ultimately sacrifice the turbines' overall economic competiveness. Even more perplexing is the lack of understanding of the influence on the power and loads of merging turbine wakes within wind farms. Therefore, in the present study, we aimed to characterize the mean wake deficits, turbulent kinetic energy produced by merged wakes, and their altered configurations caused by varying wind directions.

2. Numerical Methodology

During this study, a team of researchers from the National Renewable Energy Laboratory (NREL) and Statoil performed large-eddy simulations (LES) of the atmospheric boundary layer with three 6-MW, three-bladed, horizontal-axis wind turbines placed in a row. To accomplish this, we used an

[1] To whom any correspondence should be addressed.

incompressible formulation of the continuity equation (Eq. 1) and the momentum equation (Eq. 2), which includes the Coriolis force, buoyancy force, and aerodynamic force at the actuator point, and the potential temperature transport equations (Eq. 3). These equations are shown in the following:

$$\frac{\partial u_i}{\partial x_i} = 0 \tag{1}$$

$$\frac{\partial u_i}{\partial t} + \frac{\partial}{\partial x_j}(u_i u_j) = -\frac{\partial \hat{p}}{\partial x_i} - \frac{\partial \tau_{ij}^D}{\partial x_j} - \frac{1}{\rho_0}\frac{\partial p_0(x,y)}{\partial x_i} - \varepsilon_{i3k} f_3 u_k + g\left(\frac{\theta-\theta_0}{\theta_0}\right)\delta_{i3} + \frac{1}{\rho_0}F_i \tag{2}$$

$$\frac{\partial \theta}{\partial t} + \frac{\partial}{\partial x_j}(u_j\theta) = \frac{\upsilon_{SGS}}{Pr_{SGS}}\frac{\partial^2\theta}{\partial x_j^2} \tag{3}$$

Note that all of the primitive variables are spatially resolved components. The modified pressure is defined as $\hat{p} = p'/\rho_0 + \frac{\tau_{kk}}{3}$, where $p'(x,y,z,t) = p(x,y,z,t) - p_0(x,y) + \rho_0 gz$. The filtered static pressure term is p, and ρ_o is the constant density. The mean pressure term is $p_0(x,y)$, in which the spatial gradient acts to drive the flow convection. The last term, $\rho_0 gz$, represents the hydrostatic pressure. The deviatoric part of the fluid stress tensor is $\tau_{ij}^D = \tau_{ij} - \tau_{kk}\delta_{ij}/3$, where δ_{ij} is the Kronecker delta. The subgrid-scale (SGS) stresses are included in the τ_{ij} term. The SGS flux is computed using the dynamic Lagrangian scale-independent model [1]. The ε_{ijk} is the alternating unit tensor. The Coriolis parameter is defined as $f = 2\omega[0, \cos(\varphi), \sin(\varphi)]$, where ω is the planetary rotation rate ($7.27\text{x}10^{-5}$ rad/s) and φ (43.2°) is the latitude. Coriolis forces are important because they cause a mean wind direction change with height described by Ekman spiral/transport theory. Including this effect is essential because the mean wind direction change over the rotor disk can be a significant contributor to cyclic blade loading that causes fatigue, especially during stable atmospheric conditions. The buoyancy effect is calculated using the Boussinesq approximation, where g is the gravity, θ is the resolved potential temperature, θ_0 is the reference temperature taken to be 300 K. The last term, F_i, is the force field generated by the actuator line model, which is discussed in the following section. The transport equation for the potential temperature is shown in (3) where υ_{SGS} and Pr_{SGS} are the SGS viscosity and turbulent Prandtl number following Moeng's formulation [2].

The turbine model consists of an actuator line representation of the turbine blades coupled with NREL's computer-aided engineering tool FAST [3]. FAST employs a combined modal and multibody dynamics formulation. Within FAST, each turbine blade is represented by a set of discrete elements along the blade that move in space as the turbine blades rotate and bend. The tool then uses the local flow field at each blade element to compute aerodynamic forces, via the dynamic airfoil coefficient lookup tables in the AeroDyn module [4]. Using the updated blade-element positions, the aerodynamic forces are projected onto the flow field as a body force term in the LES equation. Therefore, this actuator line representation consists of two-way coupling, in which the flow information is used as an input to compute the aerodynamic forces that are pressed on to the flow with new positions.

The preceding equations addressing the incompressible Navier-Stokes with the Coriolis force and the potential temperature flux are solved using Simulator fOr Wind Farm Applications, which is the NREL LES code [5] that was developed using OpenFOAM libraries. The equations are discretized using an unstructured collocated finite-volume formulation. The second-order central-differencing scheme is employed and uses the Rhie and Chow interpolation method [6] to avoid checkerboard pressure-velocity decoupling. For the time advancement, pressure implicit with splitting operation [7] is used, with three substep corrections to maintain a second-order temporal accuracy. FAST's aeroelastic calculation employs the fourth-order Adams-Bashforth predictor and Adams-Moulton corrector time integration scheme.

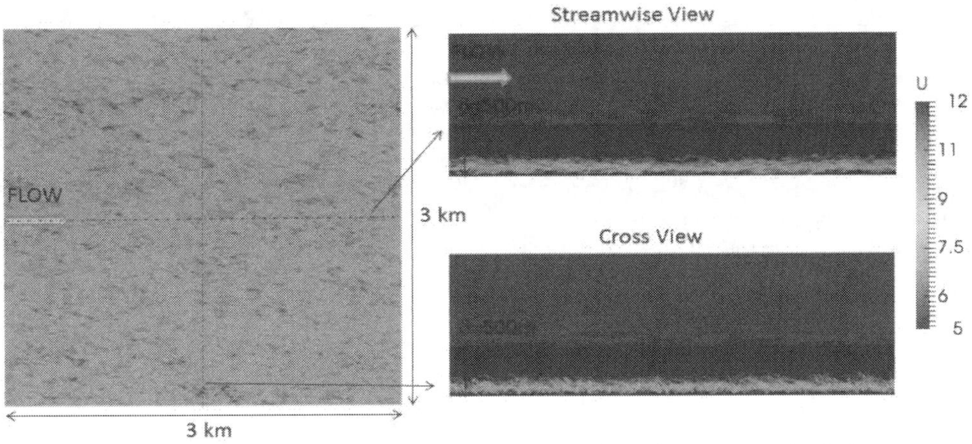

Figure 1. Instantaneous streamwise velocity contours from the precursor simulation.

The turbulent inflow boundary condition is generated via a precursor simulation. In this study, the size of the precursor computational domain was set at 3 km x 3 km x 1 km to accommodate a wide range of length scales of the turbulent structures. The grid resolution was fixed with a 10-m uniform mesh and the driver for the wind was based on the pressure gradient to obtain a mean wind speed of 10.4 m/s at the turbine hub height (z = 98.2-m). The turbulence intensity was adjusted to 4.2% of the freestream flow with an aerodynamic surface roughness of 0.0002 m. The same pressure gradient was used for the turbine simulations because the changes in the surface roughness were moderately small. The atmospheric stability was neutral by fixing the potential temperature flux at the surface to be zero. Samples of the instantaneous streamwise velocity contours at various locations are provided in Figure 1, which shows the boundary layer height, δ, at approximately 500 m.

3. Computational Domain

The LES study was performed using a domain extended in the general wind direction to accommodate

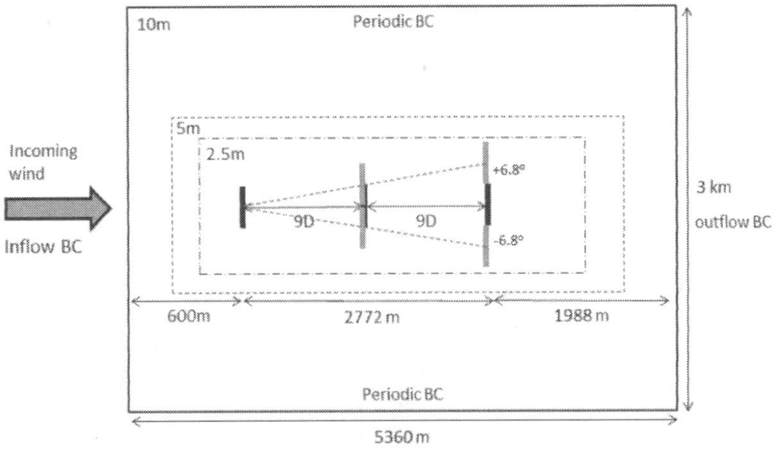

Figure 2. Computational domain schematic.

three turbines that are separated by 9 rotor diameters (D). As shown in Figure 2, two layers of grid nesting were created within the existing 10-m resolution grid to better capture the turbine wake structures that yielded a 2.5-m resolution in the subdomain where the turbines are embedded. The total number of cells was approximately 80 million. The turbulent inflow boundary condition was imposed at the west boundary with the precursor database. An outflow boundary condition on the east boundary allowed the turbulent structures to smoothly exit the domain and enforce zero-net-mass conservation per numerical time step. The north and south boundaries are connected via periodic boundary conditions, which yielded a minimal effect on the turbine wake propagation as a result of the far distance. The generic, 6-MW offshore turbine has a rotor diameter of 154 m and its structural and aerodynamic properties were modeled in FAST. Change in wind direction is mimicked by rotating the turbine row by ±6.8° relative to the first turbine.

4. Results

4.1. Wind direction at 270°

Figure 3 shows three wind turbines in a row subjected to a neutrally stable atmospheric boundary layer flow, in which both the instantaneous velocity contours of the streamwise and vertical components (Figure 3a and 3c, respectively) exhibit significant wake meandering motions where the wake structure randomly traverses in lateral directions [8, 9]. As shown in the instantaneous velocity contours, the meandering of the wake becomes worse at each passing of downstream turbines, which is attributed to increased turbulent mixing incurred by the interaction between the waked inflow and rotors. Increased wake diffusion is clearly observed in the transverse velocity contour (Figure 3c), where the wake width spreads significantly along the downstream distance. In the time-averaged streamwise velocity contour shown in Figure 3b, the wake deficit regions still persist, even at a 9-D

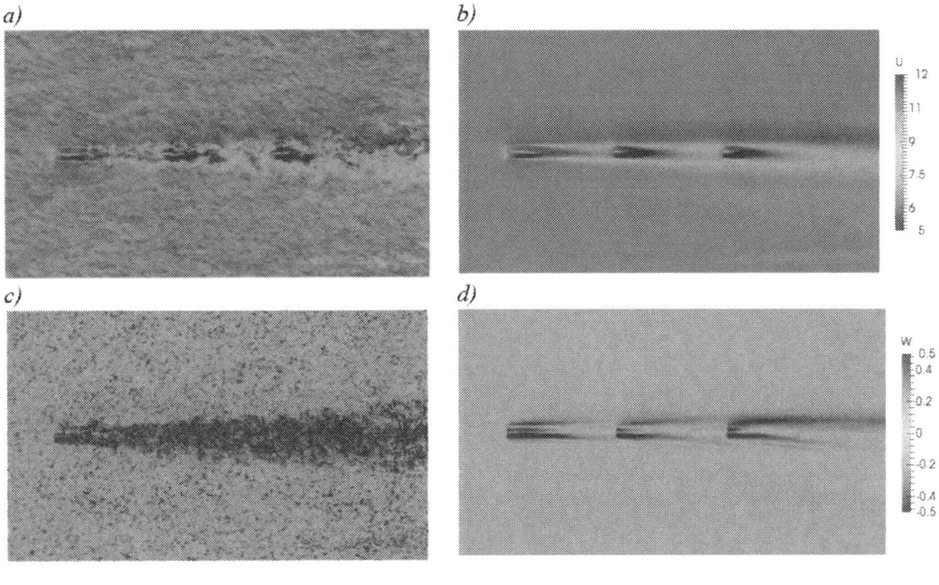

Figure 3. Instantaneous (left column) and time-averaged (right column) velocity contours of streamwise (U: a – b) and vertical (W: c – d) components. The mean freestream velocity at hub height is 10.4 m/s.

distance, where the adjacent turbines are located; however, the recovery of the velocity deficit is improved for turbine 2 and 3, even though the velocity deficits are stronger in the near wake regions. The wake structure, in general, rotates counter-clockwise in the opposite direction to the blade rotation. As shown in Figure 3c and 3d, consistent updrafts are present on the right region when looking downstream and, conversely, down drafts are prevalent on the left region.

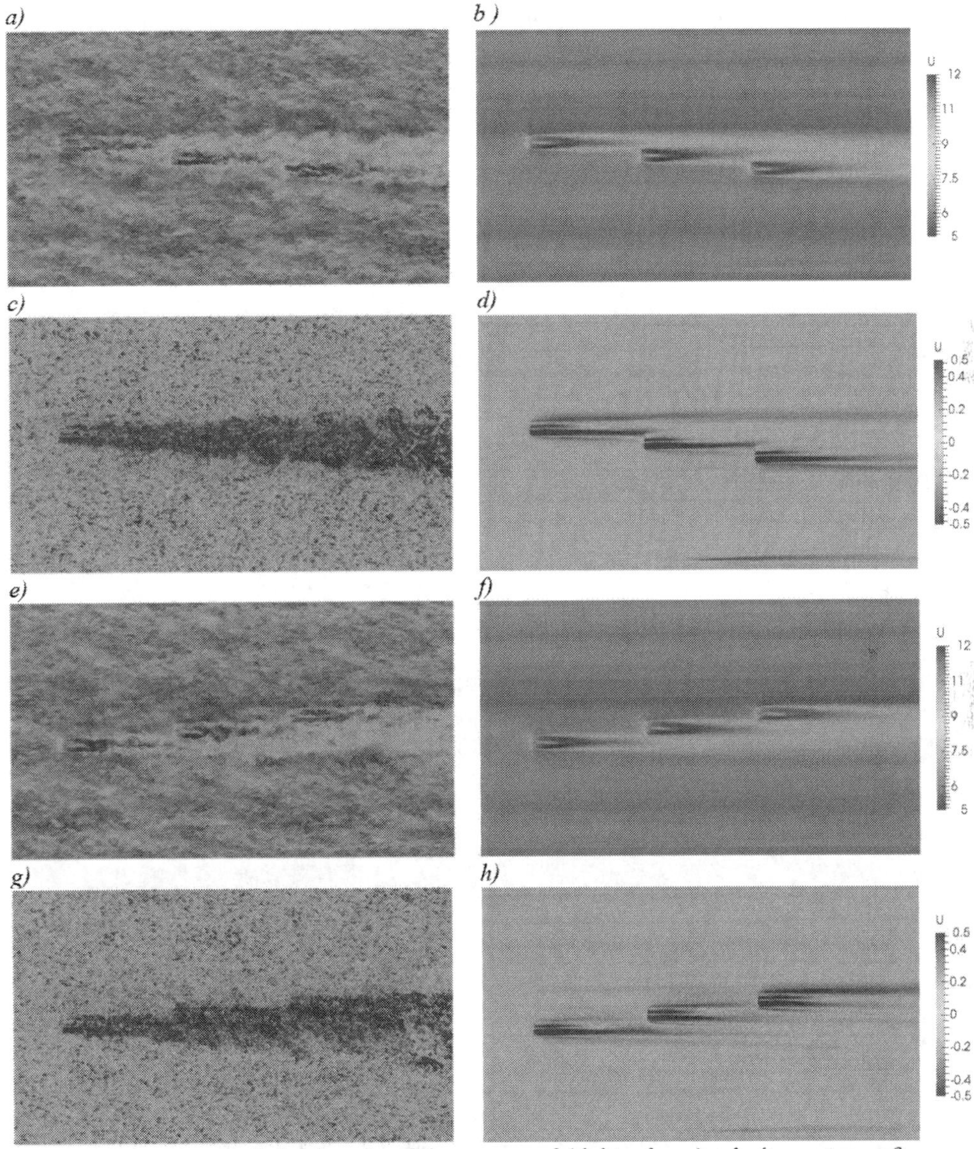

Figure 4. Instantaneous (left column) and time-averaged (right column) velocity contours of streamwise (U) and vertical (W) components for the 263.2° wind direction: a) – d) and 276.8° wind direction: e) – h).

4.2. Wind directions at 263.2° and 276.8°

A 6.8° shift in the wind direction from the previous case is mimicked by rotating the turbine row by ±6.8° relative to the freestream wind at 270°, which are shown in Figure 4. The skewed wind direction resulted in a partial waking for the downstream turbines, in which approximately half of the upstream turbine wake impinged on the downstream rotor plane. Parcels of low-speed fluid that avoided contact with the downstream rotor plane can be seen to convect past the downstream turbines as shown in Figure 4a and 4e. Note that the meandering wake is mitigated for the second turbine compared to the fully waked case shown in Figure 3a, which is attributed to less turbulent mixing.

Figure 5 shows cross-cut planes of time-averaged streamwise velocity contours looking upstream of the flow for the 263.2° case and the 276.8° case. The relative positions, with respect to the immediate upstream turbine locations, were taken at 3, 5 and 7 rotor diameters downstream. The primary wake structures exhibit ellipsoidal shapes in which the blue regions shown at 3 D are seen to reduce in size at 7 D. A small footprint of the wake from the upstream turbine can be found in the right region of the primary wake for the 263.2° case as shown in Figure 5a through Figure 5f. Conversely, the secondary

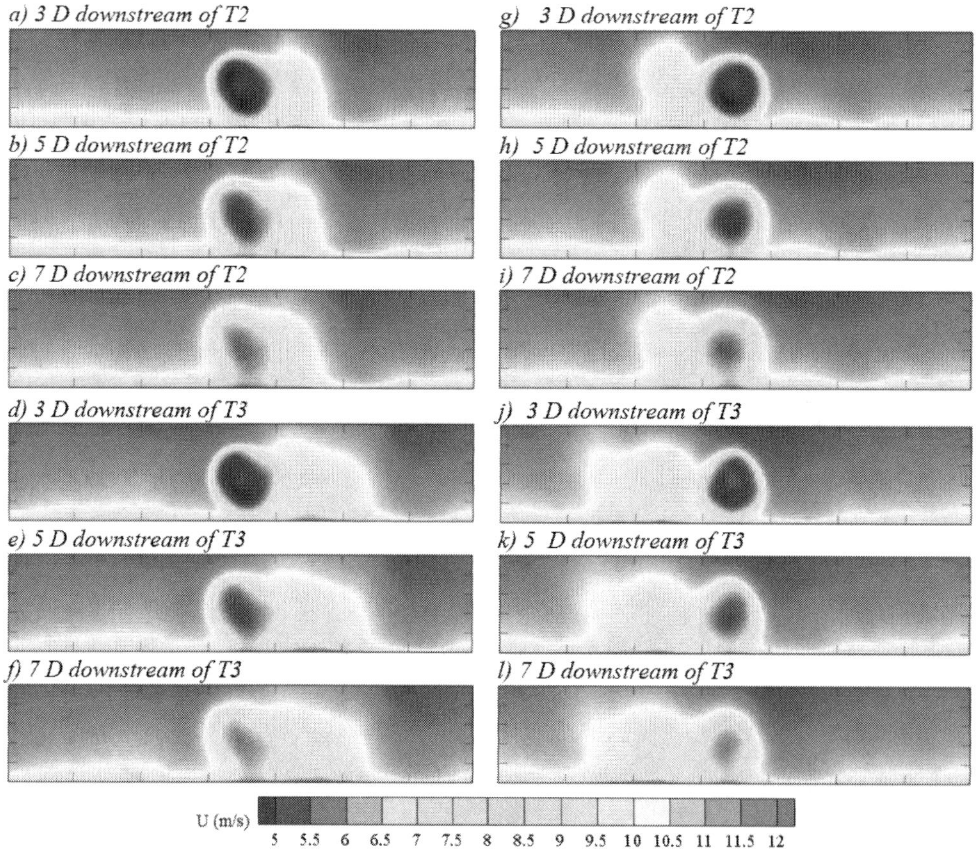

Figure 5. Cross-cut view of time-averaged streamwise velocity contours for 263.2° (left column) and 276.8° (right column). The relative positions, with respect to the immediate upstream turbine locations, are described in the figure subindex.

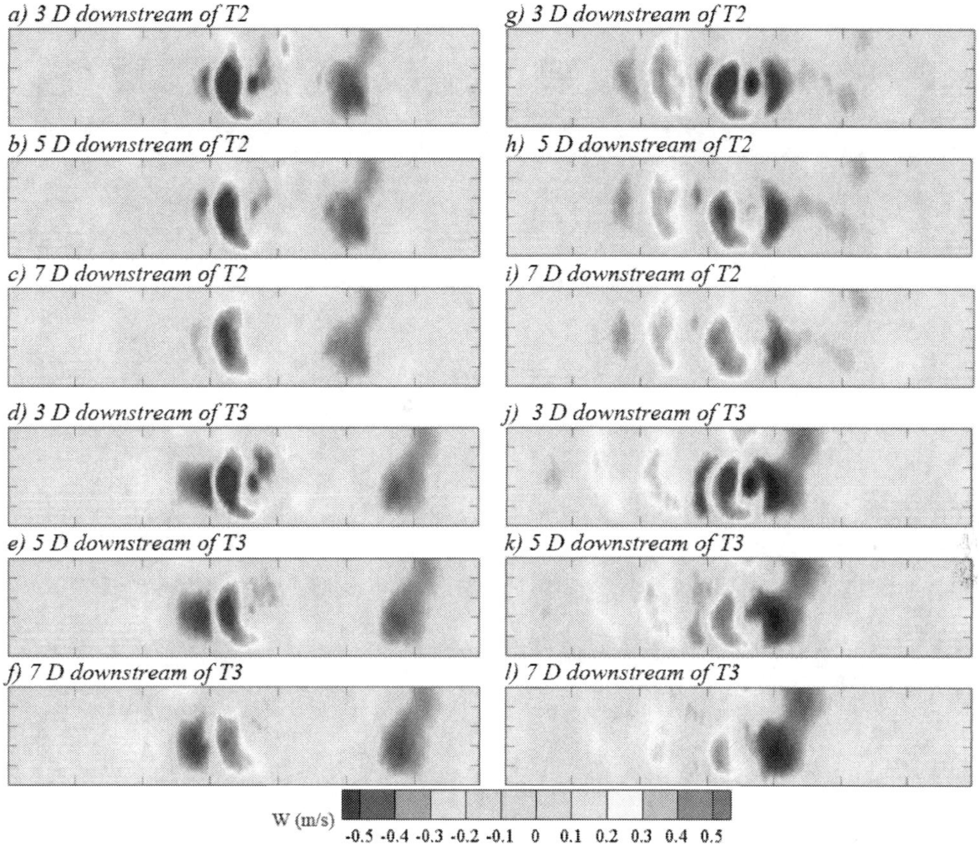

Figure 6. Cross-cut view of time-averaged vertical velocity contours for 263.2° (left column) and 276.8° (right column). The relative positions, with respect to the immediate upstream turbine locations, are described in the figure subindex.

wake structures present on the left side of the primary wake in Figure 5g through Figure 5l can be seen for the 276.8° case and they are more circular than the opposite wind direction. We speculate that the asymmetry in the wake deficit regions is a result of the Coriolis effect, which skews the wake structure by flow directional change that is combined with the merging wakes that rotate. The persistence of the multiple wake structure can be detrimental to the downwind turbines' energy production and fatigue damage.

The time-averaged, vertical velocity contours at the same locations are shown in Figure 6. The updraft and downdraft regions denoted in red and blue are clearly shown, suggesting that the wake structure continues to slowly rotate further downstream. Despite the viscous effect that causes the wake to dissipate along the downstream distance, the rotating wake structure from the previous upstream turbines persist which are shown as light blue and yellow structure pairs. The extended survivability of the upstream wake structures can be attributed to the relatively low ambient turbulence intensity. Note that in the 276.8° case, the upstream wake structures appear more visible than in the 263.2° case.

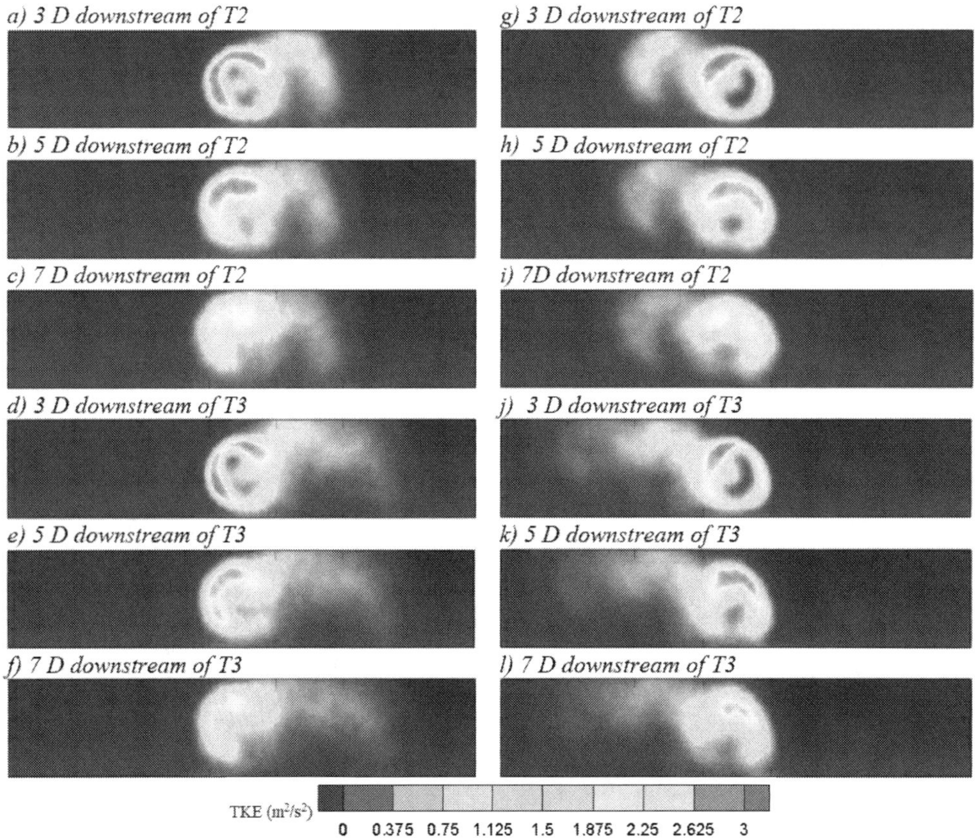

Figure 7. Cross-cut view of time-averaged turbulent kinetic energy contours for 263.2° (left column) and 276.8° (right column). The relative positions, with respect to the immediate upstream turbine locations, are described in the figure subindex.

The turbulent kinetic energy contours reveal that high intensity regions are mostly found at the circumferential edge of the primary wake structure because of the high shear in the flow. The lack of a nacelle model in the actuator line method attributed to the low turbulent kinetic energy at the centroid of the primary wake region. As expected, the merging wakes from the far upstream turbine have lower turbulent kinetic energy. It is worth noting that the wake signatures from the two upstream turbines are still present, yet they have diffused and decayed in intensity.

5. Conclusions

Large-eddy simulations of three 6-MW, three-bladed, horizontal-axis turbines subjected to a neutral atmospheric boundary layer flow were performed with three different wind directions. By imparting the turbine arrays to 263.2°, 270°, and 276.8° wind direction, the resulting flow yielded different wake longevity and merging scenarios. One extreme case was the 270° wind direction simulation, in which the persisting wake was fully merged with the wakes generated by the downstream turbines. The meandering of the wake's tail significantly increased with each passing rotor plane as a result of higher levels of turbulent mixing from each subsequent turbine. By shifting the wind direction by

±6.8°, we found that the footprints of the wake continued to propagate further past the turbines that were spaced at 9 D; however, the persisting lengths of the wake structure were shown to have asymmetric results in which the 263.2° case exhibited wakes from the previous upstream turbine to partially mix with the adjacent primary wake structure, causing the wake rotation to decay quickly. Conversely, the 276.8° case was able to capture triplet wake structures at 7 D downstream from the third turbine. Despite the symmetry in the two wind directions, the varying flow direction with respect to the height in the boundary layer caused by the Coriolis force appeared to dominantly influence the asymmetry in the wake development. Various conditions including atmospheric stability, aerodynamic surface roughness, and mean wind speed, which drive the turbulent mixing in the boundary layer, dominantly influence the level of asymmetric wake propagation and should be better quantified in future research. This asymmetric behavior of wake propagation can have significant implications with respect to the energy capture and fatigue damage that accumulates over a wind turbine's service lifetime, which should be accounted for when designing wind farms.

6. Acknowledgments

The authors gratefully acknowledge the funding support provided by Statoil and the Department of Energy. All the computations were performed on the Peregrine of the National Renewable Energy Laboratory and on the Hexagon of the University of Bergen's high-performance computing system. The authors also thank the project team members for their insightful discussions: Jim Green, Zachary Parker, Aaron Smith, Andrew Platt, George Scott, Tyler Stehly, Donna Heimiller, Pieter Gebraad and Paul Fleming from the National Renewable Energy Laboratory and Andrea Nina Eugster from Statoil and Michelle Burns and Nick Wimer from the University of Colorado at Boulder.

7. References

[1] Meneveau C, Lund T and Cabot W 1996 *A lagrangian dynamic subgrid-scale model of turbulence*, J. Fluid Mech. **319**, pp. 353-385

[2] Moeng C H 1984 *A large eddy simulation model for the study of planetary boundary layer turbulence*, J. Atmospheric Sciences, **41**, pp. 2052-2062

[3] Jonkman J M and Buhl M L 2005 *FAST User's Guide*, NREL/EL-500-3823

[4] Moriarty P J and Hansen A C 2005 *AeroDyn Theory Manual*, NREL/EL-500-36881

[5] *Simulator fOr Wind Farm Applications*, http://wind.nrel.gov/designcodes/simulators/sowfa/

[6] Rhie C M and Chow W L 1983 *Numerical Study of the Turbulent Flow Past an Airfoil with Trailing Edge Separation*, AIAA J. **21**, pp. 1525-1532

[7] Issa R 1985 *Solution of the implicitly discretized fluid flow equations by operator-splitting*, J. Comput. Phys., **62**, pp. 40–65

[8] Larsen G C, Madsen H A, Bing¨ol F, Mann J, Ott S, Sørensen J N, Okulov V, Troldborg N, Nielsen M, Thomsen K, Larsen T J and Mikkelsen R 2007 *Dynamic Wake Meandering Modeling*, Tech. Rep. R-1607(EN), Risø National Laboratory, Technical University of Denmark, Roskilde, Denmark

[9] Larsen G C, Madsen H A, Thomsen K and Larsen T J 2008 *Wake Meandering – A Pragmatic Approach*, Wind Energy, Vol. 11, No. 4, pp. 377–395

Large Eddy Simulation of wind turbine wakes: detailed comparisons of two codes focusing on effects of numerics and subgrid modeling

Luis A. Martínez-Tossas[1], Matthew J. Churchfield[2] and Charles Meneveau[1]

1. Johns Hopkins University, Baltimore, MD, USA
2. National Renewable Energy Laboratory, Golden, CO

E-mail: tony.mtos@gmail.com

Abstract. In this work we report on results from a detailed comparative numerical study from two Large Eddy Simulation (LES) codes using the Actuator Line Model (ALM). The study focuses on prediction of wind turbine wakes and their breakdown when subject to uniform inflow. Previous studies have shown relative insensitivity to subgrid modeling in the context of a finite-volume code. The present study uses the low dissipation pseudo-spectral LES code from Johns Hopkins University (LESGO) and the second-order, finite-volume OpenFOAM code (SOWFA) from the National Renewable Energy Laboratory. When subject to uniform inflow, the loads on the blades are found to be unaffected by subgrid models or numerics, as expected. The turbulence in the wake and the location of transition to a turbulent state are affected by the subgrid-scale model and the numerics.

1. Introduction

During recent years large-eddy simulation (LES) has become one of the main tools for performing highly resolved numerical studies of wind turbine wakes and wind farms [1, 2, 3]. The main models for representing wind turbines in LES are the actuator disk and, for more highly resolved LES the Actuator Line Model (ALM). In ALM the turbines are represented as forces along the blades which enter the filtered Navier-Stokes equation as body forces [4, 5]. In this work we study the effects of subgrid-scale modeling in LES of a wind turbine wake using the ALM. Simulations of uniform inflow past a wind turbine are performed with two different codes, a finite-volume based solver, and a low dissipation pseudo-spectral solver. Different subgrid-scale (SGS) models of the Smagorinsky type are compared. Previous studies have shown insensitivity to the SGS model when the turbine is subject to turbulent inflow [6]. In this study we consider the more simplified case and compare the effect of SGS modeling and numerical method on simulations with uniform inflow.

2. Numerics

The present study compares two codes, the Simulator for Wind Farm Applications (SOWFA) package [7] from the National Renewable Energy Laboratory (NREL) and LESGO, the Johns Hopkins University (JHU) Turbulence Research Group's code [7, 8]. The codes have different

JHU LESGO	NREL SOWFA
Smagorinsky $C_S = 0.16$	Smagorinsky $C_S = 0.16$
Smagorinsky $C_S = 0.08$	Smagorinsky $C_S = 0.08$
Scale Dependent Lagrangian	Lagrangian Dynamic

Table 1: Turbulence models for the JHU and NREL codes.

numerical implementations in order to solve the filtered Navier-Stokes equations with an SGS turbulence model. The effect of the numerics is addressed by comparing both codes with the same SGS model. Moreover, different SGS models are compared for each code, specifically a detailed comparison is presented between standard Smagorinsky SGS models and Dynamic type models. The table below describes the cases simulated for each code. All models represent the deviatoric part of the SGS stress tensor, τ_{ij}, using a Smaroginsky type SGS eddy viscosity according to

$$\tau_{ij} = -2\nu_{SGS}\widetilde{S}_{ij} = -2(C_S\Delta)^2|\widetilde{S}|\widetilde{S}_{ij}, \tag{1}$$

where C_S is the Smagorinsky coefficient and $\widetilde{S}_{ij}$ is the symmetric part of the resolved velocity gradient tensor [9]. Dynamic models calculate the C_S coefficient as a function of space and time making use of the Germano identity [10, 11, 12].

The turbine used in the present study is the NREL 5MW Reference [13]. The turbine diameter is $D = 126m$ and the dimensions of all the simulations are as follows:

- $L_x = 24$ D
- $L_y = 6$ D
- $L_z = 6$ D
- $N_x = 768$
- $N_y = 192$
- $N_z = 2N_y = 384$
- turbine location: 3 D
- Inflow: $U_\infty = 8$ m/s
- ρ= 1.0 kg/m^3

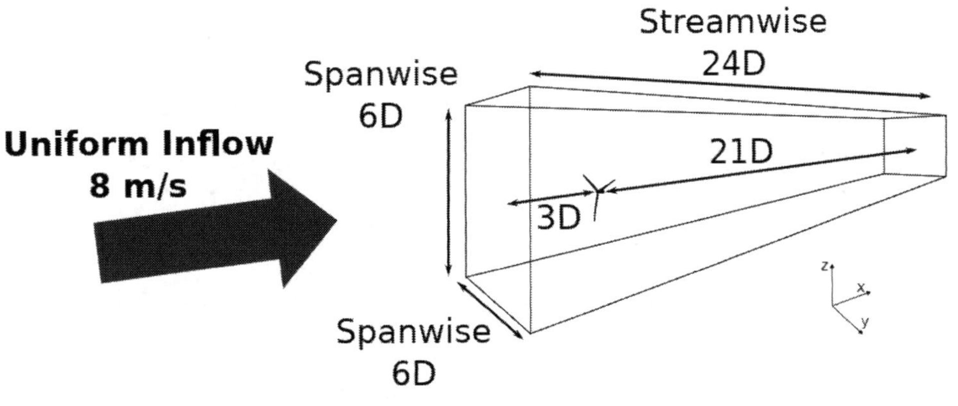

Figure 1: Computational domain.

When smearing the actuator line forces onto the computational grid, a smoothing parameter ϵ is introduced as the smearing length-scale [4, 14]. This value is related to the geometry of the blades, but there is not a general guideline based on the physics of the problem to establish its size [5]. From prior experience using a finite volume implementation the known numerical limit of the value under uniform inflow is related to the grid size as $\epsilon = 2\Delta x$ [5]. For the current case a value of $\epsilon = 10m$ was used, which is on the safe side with $\epsilon = 2.5\Delta x$.

2.1. Simulator for Wind Farm Applications (SOWFA)
SOWFA is an LES solver meant for wind farm simulations implemented under the OpenFOAM framework. It is a finite-volume code with second-order numerics in both space and time. An ALM implementation is used to represent the turbines. The SGS models are of the standard Smagorinsky type with fixed C_S coefficient and a dynamic model with Lagrangian averaging [11].

2.2. LESGO
LESGO is the pseudo-spectral LES code from the Turbulence Research Group at Johns Hopkins University (JHU) [8]. It is based on the early work of Moeng's [15] and Albertson and Parlange [16]. It is pseudo-spectral in two directions with the third direction using second-order centered finite differencing. Time advancement is done using a second order Adams-Bashforth scheme. The SGS models are of the standard Smagorinsky type with fixed C_S coefficient and a scale dependent dynamic model (see Bou-Zeid [12]).

For the JHU LESGO code, the x and y directions are spectral while the z direction uses centered finite differencing. It is not known a priori which resolution to use for the spectral and finite difference directions. Initial simulations were carried out at the same resolutions for both and it was observed that the Reynolds stresses differed in both directions. Based on the modified wave number, it seems like a good choice to have the finite difference resolution to be twice the spectral resolution, since the modified wavenumber up to 1/2 of the cutoff wavenumber limits the attenuation to about 33% only. With this value the Reynolds stress profiles look now very similar for the two different directions.

3. Results
3.1. Comparisons to Blade Element Momentum
In this section simulation results are compared to Blade Element Momentum Theory (BEM), which assumes inviscid flow and a zero velocity component in the radial direction [17, 18]. When non-turbulent uniform inflow passes through the rotor, the flow does not break down until several diameters downstream. This can be seen in Figure 2, where the lines for all turbulence models lie exactly on top of each other. This means that the turbulence model has negligible impact in this portion of the flow domain. There are differences between the LES simulations and the predictions of BEM. There are several reasons for these differences. The first being that there is a blockage effect in the computational box. The current domain is 6D by 6D in the spanwise directions. In order to eliminate the blockage effect the domain would have to be greater than 10D by 10D which is constrained by the computational resources available. The other difference, which is more noticeable is the induction. It is known that the induction created by the turbine is affected by the way the forces are smeared onto the domain [5]. In our case, the width of the Gaussian kernel is quite large compared to the chord length scale. This causes a difference in the axial velocity and thus, affects the angle of attack and hence the lift and drag. The differences in angle of attack seem to be small, but these differences amplify when calculating the lift and drag coefficients. These differences could be further reduced, but it requires many more grid points, exceeding our computational resources. Another important difference between the BEM and LES-ALM is that the former assumes locally 2D pressure distributions around the blade. The radius-dependent foil sections induce radial changes that the real pressure field will smooth out in the radial direction, while the BEM assumes locally constant 2D conditions to prevail with sudden jumps in axial velocity and forces which cannot happen in reality. Next, we study the behavior of the wakes, where we look into the numerics and the SGS modeling.

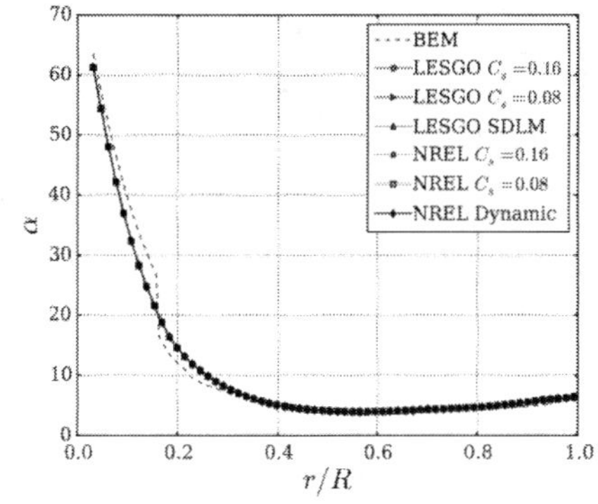

(a) Angle of Attack

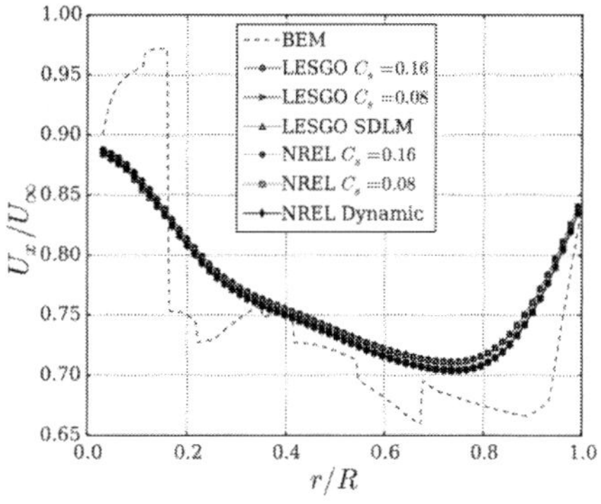

(b) Axial velocity

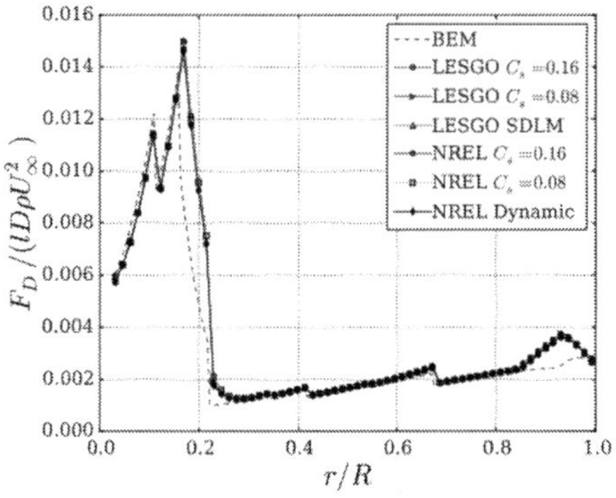

(c) Drag per unit length

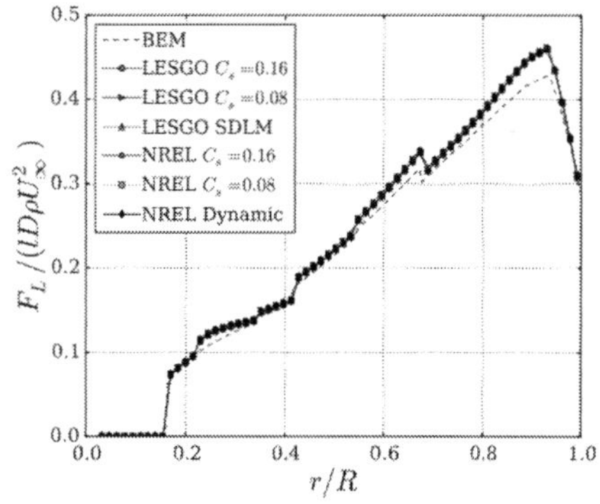

(d) Lift per unit length

Figure 2: Comparison with Blade Element Momentum Theory. Lift and Drag have been non-dimensionalized using the diameter D, inflow velocity U_∞, density ρ and width l.

3.2. Wake Profiles

The wake of a turbine in uniform inflow is characterized by two main regions. A mostly non-turbulent 'smooth' region, the near wake, and a turbulent region where the breakdown happens and the components of the Reynolds stress tensor have much higher values. Figure 3 shows a volume rendering of the instantaneous streamwise velocity component, where the two regions of the wake are clearly shown. The contour plot shows the streamwise velocity component on a plane across the wake. Volume rendering shows the streamwise velocity component in the wake.

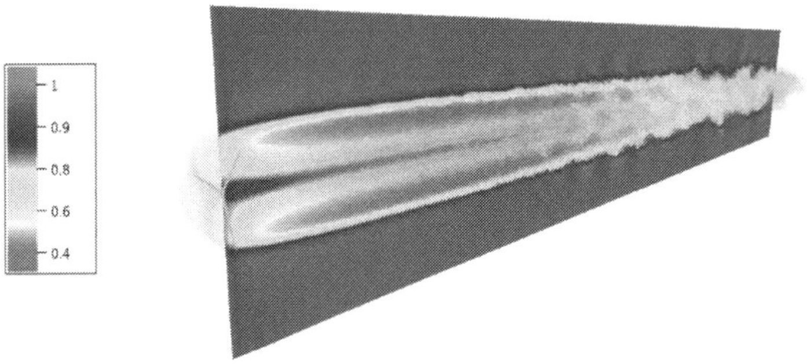

Figure 3: Snapshot of the streamwise velocity component u/U_∞ in the wake of a turbine under uniform inflow.

In order to study the differences in the wakes, contours of streamwise velocity component are shown for all cases in Figure 4. The wake distributions are very similar, specially in the near wake. This suggests that the SGS model has very little effect in this region of the flow. Greater differences are observed as we move downstream, where the breakdown of the wake occurs. The cases for the same SGS model but different codes show the effects of the different numerics of each code. The contours of the dynamic models and the Smagorinsky case with $C_S = 0.08$ are very similar. These similarities are because the C_S values obtained from the Dynamic type models are closer on average to $C_S = 0.08$.

It can be observed that the Smagorinsky model case with $C_S = 0.16$ from the SOWFA package differs from the rest. There are streamwise inhomogeneities that look like structures in the center not present in the other cases. Figure 5 shows the velocity component normal to the plane for the Smagorinsky model case with $C_S = 0.16$. The flow on the outside part of the wake keeps separating from the center. To conserve angular momentum the fluid in the inner region of the wake starts counter rotating. The reason for this behavior in these simulations is still unclear. The most noticeable aspect for this case is that it has the highest dissipation from the turbulent model and numerical diffusion as a result of the second-order finite volume discretization.

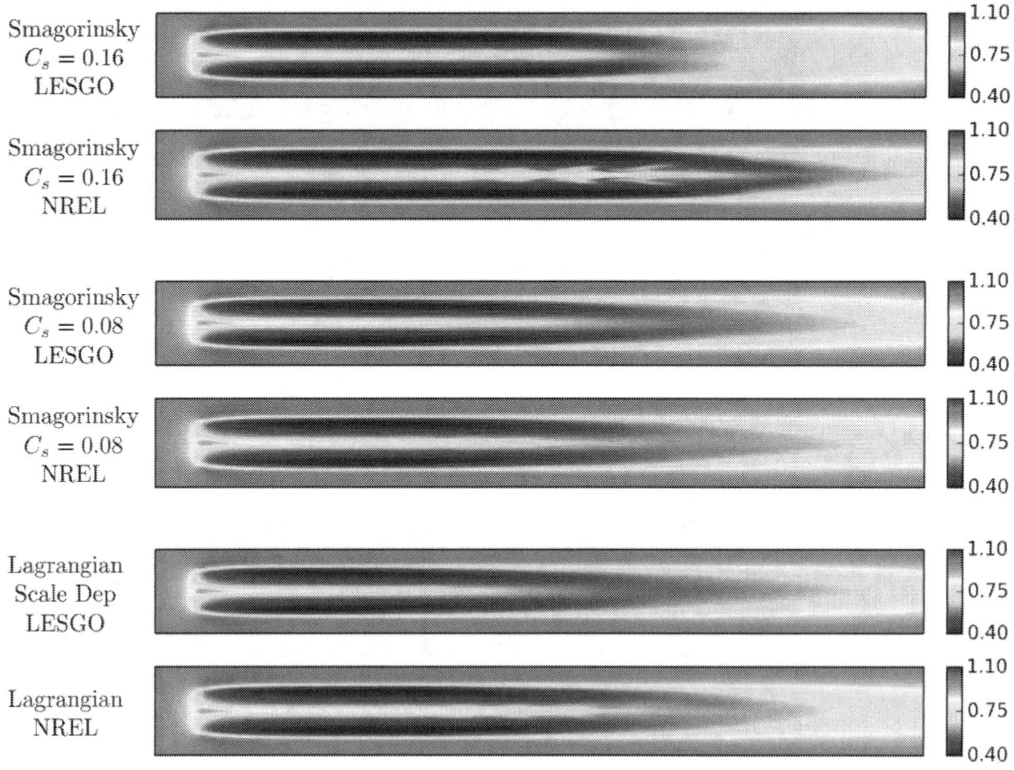

Figure 4: Streamwise velocity $\overline{u}/U_\infty$ contours of the wake with flow going from left to right.

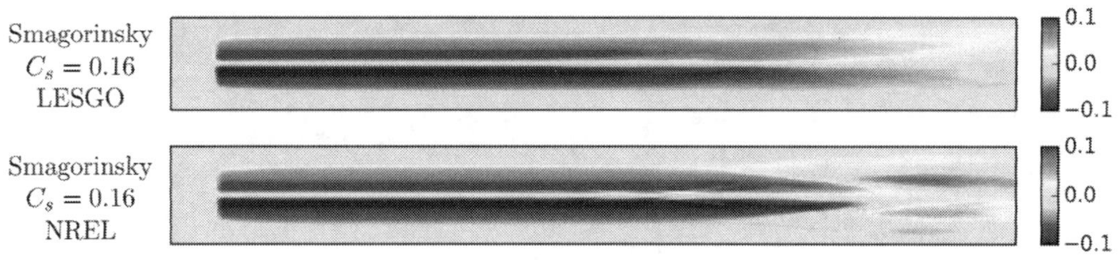

Figure 5: Velocity component normal to the plane $\overline{w}/U_\infty$ contours of the wake with flow going from left to right.

Contours of the streamwise normal $\overline{u'u'}/U_\infty^2$ component of the Reynolds stress tensor in Figure 6 show these differences more clearly. It can be concluded quite clearly that the Smagorinsky model with a coefficient of $C_s = 0.16$ (derived value for homogeneous isotropic turbulence [9]) predicts a delayed transition of the wake. This difference is caused by a higher SGS viscosity (ν_{SGS}) that damps the flow structures in the shear layer.

Journal of Physics: Conference Series **625** (2015) 012024 doi:10.1088/1742-6596/625/1/012024

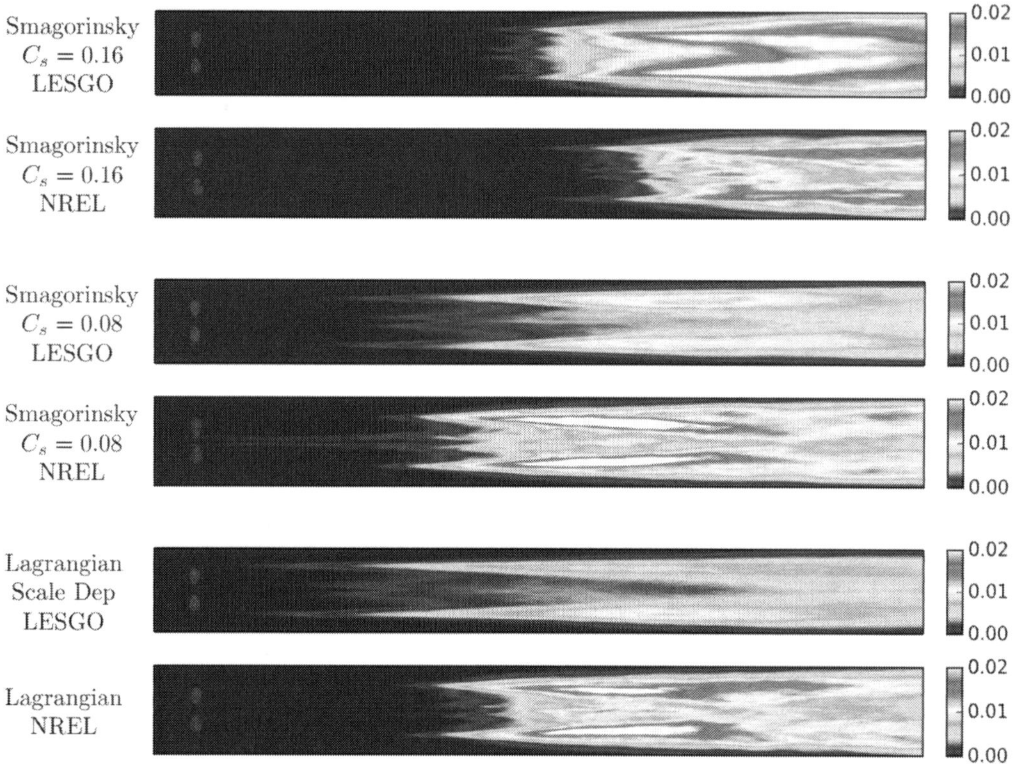

Figure 6: Streamwise normal component of the Reynols stress $\overline{u'u'}/U_\infty^2$ contours of the wake with flow going from left to right.

Differences can also be observed when the same SGS model is used in both codes. The biggest difference is a delayed transition to turbulence in the Finite-Volume SOWFA framework. This is expected since the second order numerics will introduce more dissipation. This will delay the breakdown of the shear layer in the wake. There are also some differences in the Scale Dependent Lagrangian Dynamic Model simulation (LESGO) and the Dynamic Model without scale dependence (NREL). The comparison shows a delayed transition of the wake in the Dynamic (NREL) case. These differences are caused by the fundamental differences in the models, but as in the other cases it is due mostly to the numerics which add more dissipation.

Figures 7 and 8 show mean velocity profiles of streamwise velocity u/U_∞ and Reynolds stress $\overline{u'u'}/U_\infty^2$ profiles along lines across the wake 1, 4, 7, and 9 diameters downstream. This clearly shows how the velocity field is unaffected by the SGS model for the first 4 diameters downstream and noticeable differences are observed past 4 diameters downstream. More noticeable differences can be seen for the Reynolds stress component. The main feature to note is that for profiles downstream, the Reynolds stress components for Smagorinsky $C_s = 0.16$ in both codes differ from the rest of the cases. The Dynamic models and Smagorinsky $C_s = 0.08$ cases show very similar profiles. This suggests that adjusting the Smagorinsky coefficient will give results very close to what a dynamic model will predict as long as one knows what correct value to use for C_s.

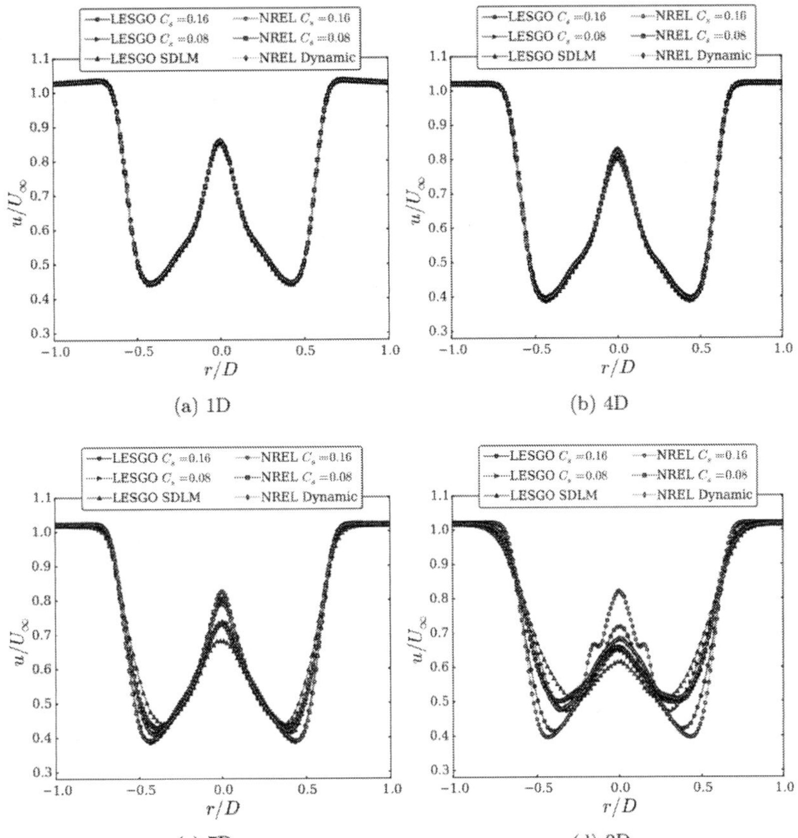

(a) 1D

(b) 4D

(c) 7D

(d) 9D

Figure 7: Streamwise velocity $\overline{u}/U_\infty$ across the wake 1, 4, 7 and 9 diameters downstream.

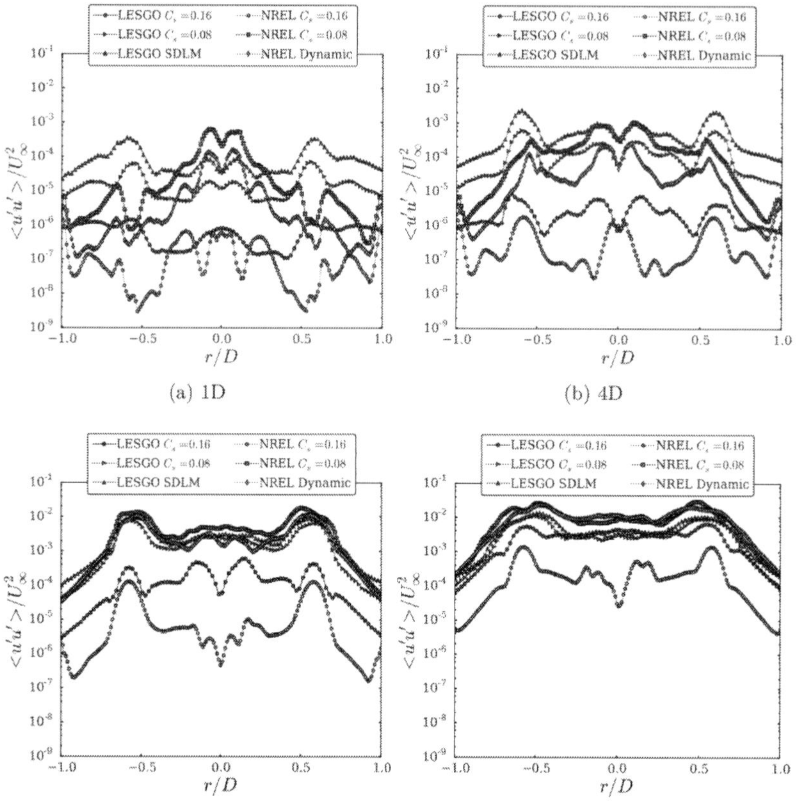

(a) 1D (b) 4D

(c) 7D (d) 9D

Figure 8: Streamwise normal component of the Reynols stress $\overline{u'u'}/U_\infty^2$ component of the Reynolds Stress tensor across the wake 1, 4, 7 and 9 diameters downstream. Note the logarithmic vertical scale.

4. Conclusions

An ALM implementation has been tested using detailed comparisons of results obtained using two LES codes with different numerical schemes and BEM. Subgrid scale modeling makes no significant difference in the loads and flow field near the turbine when running simulations under uniform inflow. The SGS model makes a difference further downstream where there is a breakdown of the wake. The numerics in LES of wind turbine wakes under uniform inflow play a very important role. The dissipation added by second order finite difference numerics damps the turbulence which delays the breakdown of the wake. A properly chosen value of the Smagorinsky coefficient C_S can provide very similar results as a Dynamic model, but with a reduced computational cost. It is hard to predict an exact C_S value for each case, in our cases similar results were obtained with $C_S = 0.08$ (i.e. about half of the derived value for homogeneous isotropic turbulence) to those using Lagrangian type models.

Acknowledgements

We acknowledge helpful discussions with Jens Sørensen and Robert Mikkelsen from Danish Technical University (DTU) which motivated initial comparisons with BEM during a visit to

DTU. Richard Stevens (JHU) provided valuable help and guidance on using and implementing the ALM into LESGO. LAMT was supported by NSF (IGERT), while international travel and CM were supported by NSF (IIA 124382, the WINDINSPIRE project). Simulations were performed on XSEDE supercomputers. Post-processing was done using the open-source python library matplotlib, and Vapor (NCAR).

References

[1] Churchfield M J, Lee S, Moriarty P J, Martinez L A, Leonardi S, Vijayakumar G and Brasseur J G 2012 *AIAA paper*
[2] Calaf M, Meneveau C and Meyers J 2010 *Physics of Fluids* **22** 015110
[3] Wu Y T and Port-Agel F 2012 *Boundary-Layer Meteorology* **146** 181–205
[4] Sorensen J N and Shen W Z 2002 *Journal of Fluids Engineering* **124** 393–399
[5] Martnez-Tossas L A, Churchfield M J and Leonardi S 2014 *Wind Energy*
[6] Sarlak H, Meneveau C and Srensen J N 2015 *Renewable Energy* **77** 386–399
[7] Churchfield M J and Lee S SOWFA | NWTC Information Portal URL `https://nwtc.nrel.gov/SOWFA`
[8] Graham J and Meneveau C 2012 *Physics of Fluids (1994-present)* **24** 125105
[9] Pope S B 2000 *Turbulent Flows* 1st ed (Cambridge ; New York: Cambridge University Press)
[10] Germano M 1992 *Journal of Fluid Mechanics* **238** 325–336
[11] Meneveau C, Lund T S and Cabot W H 1996 *Journal of Fluid Mechanics* **319** 353–385
[12] Bou-Zeid E, Meneveau C and Parlange M 2005 *Physics of Fluids (1994-present)* **17** 025105
[13] Jonkman J M and National Renewable Energy Laboratory (US) 2009 *Definition of a 5-MW reference wind turbine for offshore system development* (Golden, Colo.: National Renewable Energy Laboratory)
[14] Troldborg N, Sorensen J N and Mikkelsen R 2010 *Wind Energy* **13** 86–99
[15] Moeng C H 1984 *Journal of the Atmospheric Sciences* **41** 2052–2062
[16] Albertson J D and Parlange M B 1999 *Water Resources Research* **35** 2121–2132
[17] Burton T 2001 *Wind energy: handbook* (John Wiley and Sons)
[18] Hansen M O L 2007 *Aerodynamics of Wind Turbines, 2nd edition* 2nd ed (London; Sterling, VA: Routledge)

Data-driven RANS for simulations of large wind farms

G.V. Iungo[1], F. Viola[2], U. Ciri[1], M.A. Rotea[1] and S. Leonardi[1]

[1] The University of Texas at Dallas, Mechanical Engineering Department, 75080 Richardson, TX

[2] Ecole Polytechnique Fédérale de Lausanne (EPFL), Laboratory of Fluid Mechanics and Instabilities (LFMI), Lausanne, Switzerland

E-mail: valerio.iungo@utdallas.edu

Abstract. In the wind energy industry there is a growing need for real-time predictions of wind turbine wake flows in order to optimize power plant control and inhibit detrimental wake interactions. To this aim, a data-driven RANS approach is proposed in order to achieve very low computational costs and adequate accuracy through the data assimilation procedure. The RANS simulations are implemented with a classical Boussinesq hypothesis and a mixing length turbulence closure model, which is calibrated through the available data. High-fidelity LES simulations of a utility-scale wind turbine operating with different tip speed ratios are used as database. It is shown that the mixing length model for the RANS simulations can be calibrated accurately through the Reynolds stress of the axial and radial velocity components, and the gradient of the axial velocity in the radial direction. It is found that the mixing length is roughly invariant in the very near wake, then it increases linearly with the downstream distance in the diffusive region. The variation rate of the mixing length in the downstream direction is proposed as a criterion to detect the transition between near wake and transition region of a wind turbine wake. Finally, RANS simulations were performed with the calibrated mixing length model, and a good agreement with the LES simulations is observed.

1. Introduction

As recently published on the Wall Street Journal (http://blogs.wsj.com/numbers/what-is-the-most-efficient-source-of-electricity-1754/), wind energy is the most efficient source of electricity on a US national average. However, in 2013 the US Department of Energy estimated that typical power losses for a wind power plant are about 20% of its annual production (http://energy.gov/eere/wind/atmosphere-electrons), which are mainly due to wind turbine wake effects, such as chaotic wake interactions and shadowing due to upstream wind turbines. Moreover, wake-related phenomena affect not only power production, but also the overall life-cycle of wind turbines, thus their reliability. Indeed, wind turbines operating in the wake produced by upstream turbines experience highly heterogeneous, unsteady, and turbulent wind flows, which generate larger fatigue loads compared to the typical loads connected to the atmospheric wind.

Today there is an increasing need for tools for wind turbine wake prediction in real-time, which can be executed within a relatively short time frame and coupled with wind turbine control systems in order to maximize power production from a power plant and increase durability of the turbines.

Computational fluid dynamics (CFD) techniques with different degrees of complexity and accuracy have been applied to simulate wind turbine wake flows. More recently, large eddy simulation (LES) frameworks have been developed for wind energy applications. LES computes explicitly large energy-containing non-universal scales of the flow, while the small scales are modeled [1, 2, 3, 4, 5]. The wind induced loads, namely thrust and torque, are parameterized using different types of wind turbine models [6], such as the actuator disc model, which distributes uniformly the loads over the rotor disk, and the actuator line model, which distributes the forces along a lifting line that rotates mimicking the blade motion. Although accuracy on the evaluation of wind turbine wake flows has been improved over the last decade, LES still remains far too costly, which makes it inapplicable for operations.

In practice simpler and less accurate analytical models are extensively uses for wind farm siting and prediction of wind turbine wakes. Indeed, even though their fidelity is very limited, these built-to-purpose wind resource methodologies can easily be implemented thanks to their low computational cost [7], such as the commercial software WAsP [8], WindPRO [9], WindSim [10], WindFarmer [11] and OpenWind [12]. One of the pioneering analytical wake model, which is available since the late 1980s, is the one proposed by Jensen [13], which assumes a top-hat shape for the wake velocity deficit. This simple wake model is obtained by only applying the mass conservation. Similar models are derived from the mass and momentum conservation, such as for the Frandsen model [14]. More recently an improved analytical wake model has been proposed in [15].

Summarizing, LES simulations of wind turbine wakes can achieve very accurate results; however, they still imply high computational costs, which makes them not suitable for real wind turbine operations. On the other hand, analytical models have very low computational costs, but they can be very inaccurate due to their simplistic assumptions. In addition, both LES and analytical models may only apply in the diffusive region of the wake, i.e. downstream to the rollup and formation of the wake vorticity structures.

Other CFD tools have been developed for the wind energy field by adopting the Reynolds-Averaged Navier-Stokes (RANS) approach, in which the mean flow is computed and the effects of turbulence are parameterized using classical closure models [16, 17, 18, 19, 20]. The RANS approach present the undeniable advantage of requiring very low computational costs compared to LES simulations, thus to be suitable for real-time applications. In [21], the RANS approach is used with a semi-empirical eddy-viscosity model. In that work, the eddy-viscosity model is formulated as a function of the wake velocity deficit evaluated at the downstream location of two rotor diameters, d, and of several empirical parameters representing the effects of wind shear, turbulence intensity, and atmospheric thermal stability. Moreover, a filter function is applied in order to take the turbulence mechanical production occurring in the near wake into account. This filter function is empirically limited for downstream locations with $d < 5.5$. The main limitation of this approach clearly consists in the extensive use of empirical parameters, which are difficult to tune and, more importantly, to dynamically update for dynamic and operational conditions.

The aim of this paper is to propose a CFD tool that requires low computational costs, as typical for a 2D RANS algorithm, in order to provide real-time predictions of wind turbine wakes. At this same time, this CFD tool ensures an adequate accuracy for real-time wind turbine operations, which is achieved the data-driven calibration. This numerical tool is mainly meant for a framework in which a measuring system monitors wind turbine operations, such as a wind LiDAR network [22, 23, 24, 25], and the available data are embedded in a numerical algorithm in order to improve accuracy of the numerical model. Therefore, this approach can be classified as a data-driven algorithm.

The low computational cost is achieved by using a RANS approach , thus by predicting the time-averaged velocity field and by modeling the wind turbulence effects through turbulence closure models [26]. The accuracy of the wake predictions is mainly dependent on the turbulence model, which is dynamically tuned through data assimilation. Specifically, a mixing length model is calibrated by measuring the Reynolds stress between the axial and radial velocity component, and the gradient of the axial velocity in the radial direction. Since the turbulence closure model is calibrated, by measuring a velocity profile at a certain downstream location, it is possible to predict accurately the downstream evolution of the wake velocity field, thus the wind field impacting on downstream waked turbines.

The paper is organized as follows. The LES database used for the present work is described in section 2, while the the RANS equations and the formulation of the mixing length model are briefly summarized in section 3. The statistical characterization of the turbulent wake flows obtained for a turbine operating at different tip speed ratios follows in section 4. In this section, the calibration of the mixing length model and the RANS simulations are also presented. Finally, the conclusions are given in section 5.

2. Dataset obtained from LES simulations

In this work, high fidelity LES simulations of a utility scale wind turbine are used for the tuning of the turbulence closure model adopted for the RANS simulations. The LES simulations were performed by the CFD Lab at UT Dallas with the in-house developed code UTD Wind Farm (UTDWF). The flow around the wind turbine blades is modeled using the actuator line model. The lift and drag coefficients are obtained from documented airfoil data using the calculated local angle of attack. The latter is corrected taking into account the tip and root vortex by means of a modified Prandtl correction factor. The applied aerodynamic forces are distributed smoothly among the points corresponding to the section of the airfoil in order to avoid a singular point force that may causes numerical instabilities. The forces are distributed in a two-dimensional manner by applying a regularization Gaussian distribution kernel. The tower and nacelle are treated by the efficient immersed boundary technique described in detail by [27]. This method allows the flexibility of imposing complex geometries inside the domain without the need of body fitted grids, thus reducing the computational time.

The code UTDWF has been validated with experiments performed at Norwegian University of Science and Technology using a wind turbine placed inside a wind tunnel. The computational box is $9d \times 10d \times 3d$ (d is the rotor diameter of 126 m) in the streamwise, wall normal and spanwise direction, respectively. The grid is $512 \times 256 \times 384$ in streamwise, spanwise and wall normal direction, respectively. The grid is stretched in vertical direction, and it is uniform at the rotor heights. Along the blade span, 40 grid points are present.

A uniform velocity profile is used as inflow condition. Radiative boundary conditions are imposed at the outlet, a no slip condition is imposed on the lower boundary (the ground) and on the tower and nacelle, while free slip is assumed at the upper boundary. Periodicity is imposed in the spanwise direction to mimic infinite columns of turbines.

3. RANS formulation and mixing length turbulence model

Data-driven and low computational cost simulations of the wind turbine wake were performed with the Reynolds-Averaged Navier-Stokes (RANS) equations formulated in cylindrical coordinates. The equations are made non-dimensional by taking as reference length the rotor diameter d and the oncoming velocity at hub-height U_{hub}. The pressure is made non-dimensional by the fluid density and U_{hub}. Consequently, the Reynolds number is defined as $Re = U_{hub}\, d/\nu$. The RANS equations are the following:

$$\begin{aligned} \mathbf{U} \cdot \nabla \mathbf{U} &= -\nabla p + \frac{1}{Re} \Delta \mathbf{U} - \nabla \cdot (\overline{\mathbf{u}'\mathbf{u}'}); \\ \nabla \cdot \mathbf{U} &= 0; \end{aligned} \tag{1}$$

where $\mathbf{U}$ are p are the time-averaged velocity vector and pressure, respectively. The time-derivatives in the Navier-Stokes equations are omitted because only steady state is simulated. The last term in the right-hand-side of the momentum equation represents the Reynolds stresses, and $\mathbf{u}'$ is the turbulent fluctuating velocity vector. The operator $\overline{(\cdot)}$ denotes time averaging.

In order to solve the RANS equations, the Reynolds stresses term has to be modeled [26]. Through the Boussinesq hypothesis, the deviatoric part of the Reynolds stress tensor is expressed as a function of the strain rate tensor $S = \frac{1}{2}[\nabla + \nabla^T]\mathbf{U}$, while the isotropic part is absorbed in the the pressure term:

$$-\overline{\mathbf{u}'\mathbf{u}'} + \frac{2}{3}\overline{q}I \simeq 2\nu_t S, \tag{2}$$

where q is the turbulent kinetic energy (TKE) and I is the 3x3 identity matrix. Each component of the Reynolds stress tensor evaluated through the Boussinesq hypothesis is reported in the following:

$$\begin{aligned}
-\overline{u'_x u'_x} &= \nu_t\left(2\frac{\partial U_x}{\partial x}\right) - \frac{2}{3}\overline{q};\\
-\overline{u'_\theta u'_\theta} &= \nu_t\left(\frac{2}{r}\frac{\partial U_\theta}{\partial \theta} + 2\frac{U_r}{r}\right) - \frac{2}{3}\overline{q};\\
-\overline{u'_r u'_r} &= \nu_t\left(2\frac{\partial U_r}{\partial r}\right) - \frac{2}{3}\overline{q};\\
-\overline{u'_x u'_r} &= \nu_t\left(\frac{\partial U_x}{\partial r} + \frac{\partial U_r}{\partial x}\right);\\
-\overline{u'_x u'_\theta} &= \nu_t\left(\frac{1}{r}\frac{\partial U_x}{\partial \theta} + \frac{\partial U_\theta}{\partial x}\right);\\
-\overline{u'_r u'_\theta} &= \nu_t\left(\frac{\partial U_\theta}{\partial r} + \frac{1}{r}\frac{\partial U_r}{\partial \theta} - \frac{U_\theta}{r}\right);
\end{aligned} \tag{3}$$

The turbulent eddy-viscosity, ν_t, is modeled by using the generalized mixing-length algebraic model suitable for shear-flows [26]. Specifically ν_t is expressed as a function of the strain rate tensor and the mixing-length l_m, as

$$\nu_t(r,x) = l_m^2(x)(2\overline{S} : \overline{S})^{1/2}, \tag{4}$$

where l_m is only a function of the streamwise position. By considering an axisymmetric wake flow, the term $(2S : S)^{1/2}$ reduces to:

$$(2S:S)^{1/2} = \left[2\left(\frac{\partial U_r}{\partial r}\right)^2 + 2\left(\frac{U_r}{r}\right)^2 + 2\left(\frac{\partial U_x}{\partial x}\right)^2 + \left(\frac{\partial U_\theta}{\partial r} - \frac{U_\theta}{r}\right)^2 + \left(\frac{\partial U_x}{\partial r} + \frac{\partial U_r}{\partial x}\right)^2 + \left(\frac{\partial U_\theta}{\partial x}\right)^2\right]^{1/2} \tag{5}$$

Therefore, the full set of RANS equations for an axisymmetric flow results:

$$\begin{aligned}
&\Gamma U_r - \frac{U_\theta^2}{r} = -\frac{\partial p}{\partial r} + \left(\frac{1}{Re} + \nu_t\right)\left[\frac{1}{r}\frac{\partial}{\partial r}\left(r\frac{\partial U_r}{\partial r}\right) + \frac{\partial^2 U_r}{\partial x^2} - \frac{U_r}{r^2}\right] + 2\frac{\partial U_r}{\partial r}\frac{\partial \nu_t}{\partial r} + \left(\frac{\partial U_x}{\partial r} + \frac{\partial U_r}{\partial x}\right)\frac{\partial \nu_t}{\partial x}\\
&\Gamma U_\theta + \frac{U_r U_\theta}{r} = \left(\frac{1}{Re} + \nu_t\right)\left[\frac{1}{r}\frac{\partial}{\partial r}\left(r\frac{\partial U_\theta}{\partial r}\right) + \frac{\partial^2 U_\theta}{\partial x^2} - \frac{U_\theta}{r^2}\right] + r\frac{\partial U_\theta/r}{\partial r}\frac{\partial \nu_t}{\partial r} + \frac{\partial U_\theta}{\partial x}\frac{\partial \nu_t}{\partial x}\\
&\Gamma U_x = -\frac{\partial p}{\partial x} + \left(\frac{1}{Re} + \nu_t\right)\left[\frac{1}{r}\frac{\partial}{\partial r}\left(r\frac{\partial U_x}{\partial r}\right) + \frac{\partial^2 U_x}{\partial x^2}\right] + \left(\frac{\partial U_x}{\partial r} + \frac{\partial U_r}{\partial x}\right)\frac{\partial \nu_t}{\partial r} + 2\frac{\partial U_x}{\partial x}\frac{\partial \nu_t}{\partial x}\\
&\frac{1}{r}\frac{\partial r U_r}{\partial r} + \frac{\partial U_x}{\partial x} = 0
\end{aligned} \tag{6}$$

where the advection operator Γ is

$$\Gamma = U_r \frac{\partial}{\partial r} + U_x \frac{\partial}{\partial x}. \tag{7}$$

At the inlet a specific velocity profile is given. Free-stresses conditions are imposed at the outlet and at the lateral boundary, while symmetry conditions are set on the axis.
Eq. 6 together with the boundary conditions are discretized using a staggered pseudospectral Chebyshev-Chebyshev collocation method implemented in Matlab. The three velocity components are collocated at the Gauss-Lobatto-Chebyshev (GLC) nodes, whereas the pressure is defined on the Gauss-Chebyshev nodes (GC). Specifically, the momentum equation is collocated at the GLC nodes, and the pressure is interpolated from GC points to GLC points. Conversely the continuity equation is defined on the GC grid and the velocity components are interpolated from the GLC grid. Consequently, the two grids are mapped in the physical domain through a linear mapping.

Being the time-averaged Navier-Stokes equations steady, the solution is obtained by Newton-Raphson method. The iterative procedure is arrested when the L^2-norm of the difference between two consecutive solutions is less than 10^{-8}.

4. Turbulence statistics of the wind turbine wake and data-driven RANS

The turbulent wake velocity field produced by the wind turbine is investigated in order to perform an efficient calibration of the RANS turblence closure model from the available data. To this aim, LES simulations are considered, which were performed for a turbine operating with different tip speed ratios (TSR), i.e. the ratio between the rotational velocity at the tip of the turbine blade and the velocity of the incoming wind at hub height. In Fig. 1, the time-average of the axial velocity evaluated over the horizontal plane at hub height is shown for the wind turbine operating with TSR equal to 5.8, 6.5, 7 and 7.5. The cross-width of the wake seems to be barely affected by the the TSR variation; however, an enhanced velocity deficit is observed by increasing TSR, which is related to a larger power harvesting. Indeed, TSR=7.5 is to a good approximation the optimal set-point for the tested turbine.

The different terms in Eq. 5 are now investigated in order to evaluate their contribution to the eddy-viscosity formulated in Eq. 4. In Fig. 2, the results related to the simulation performed with TRS=5.8 are presented. It is evident that the main contribution to $(2\overline{S} : \overline{S})^{1/2}$ (Fig. 2(a)) is represented by the term $(\partial U_x/\partial r + \partial U_r/\partial x)$ reported in Fig. 2(f). Moreover, within this term

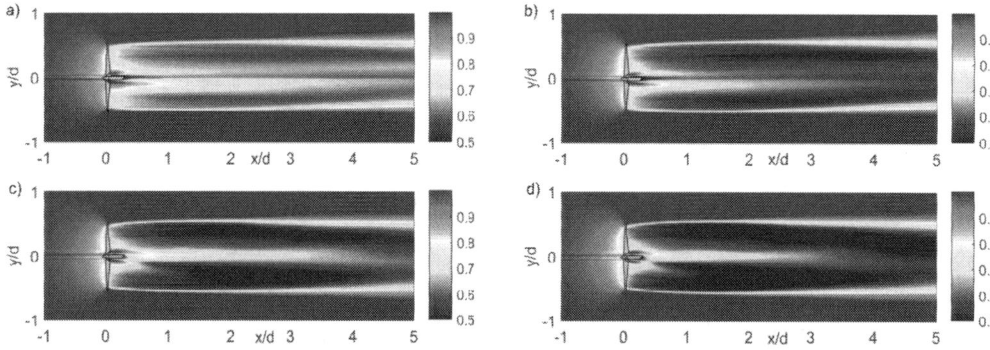

Figure 1. Time-averaged axial velocity over the horizontal plane at hub height for different tip speed ratio: a) TSR=5.8; b) TSR=6.5; c) TSR=7; d) TSR=7.5.

the most significant contribution is provided only by the term $(\partial U_x/\partial r)$ in Fig. 2(g), which is the gradient of the axial velocity in the radial direction.

By considering the Boussinesq hypothesis used for modeling the Reynolds stress tensor, in Eq. 3 the only component affected by $(\partial U_x/\partial r)$ is $\overline{u'_x u'_r}$. The latter represents the turbulent interaction between the wake flow and the surrounding unperturbed wind. Therefore, the higher $\overline{u'_x u'_r}$, more flow entrainment is promoted while the wake moves downstream. In Fig. 3, the Reynolds stress component $\overline{u'_x u'_r}$ is evaluated over the horizontal plane at hub height for the turbine operating with different TSR values. Generally, it increases by proceeding downstream with the maximum intensity located at a radial distance roughly equal to the rotor radius. Moreover, the region characterized by a significant value of $\overline{u'_x u'_r}$ becomes wider by moving downstream, which represents the enhanced diffusive phenomena in the far wake. With increasing TSR, the intensity of $\overline{u'_x u'_r}$ becomes more significant, which is connected with the larger mechanically produced turbulence, thus to the higher gradient of the axial velocity in the radial direction, $(\partial U_x/\partial r)$. Therefore, it is concluded that, up to the optimal TSR the larger the TSR, more power is generated, thus producing a larger velocity deficit. Consequently, the larger gradient of the axial velocity in the radial direction leads to an enhanced mechanically produced turbulence, which promotes flow entrainment and wake recovery.

Accordingly to the discussion above, it derives that the most meaningful procedure to evaluate the mixing length through Eqs. 3 and 4, is to fit the Reynolds stress component $\overline{u'_x u'_r}$ through

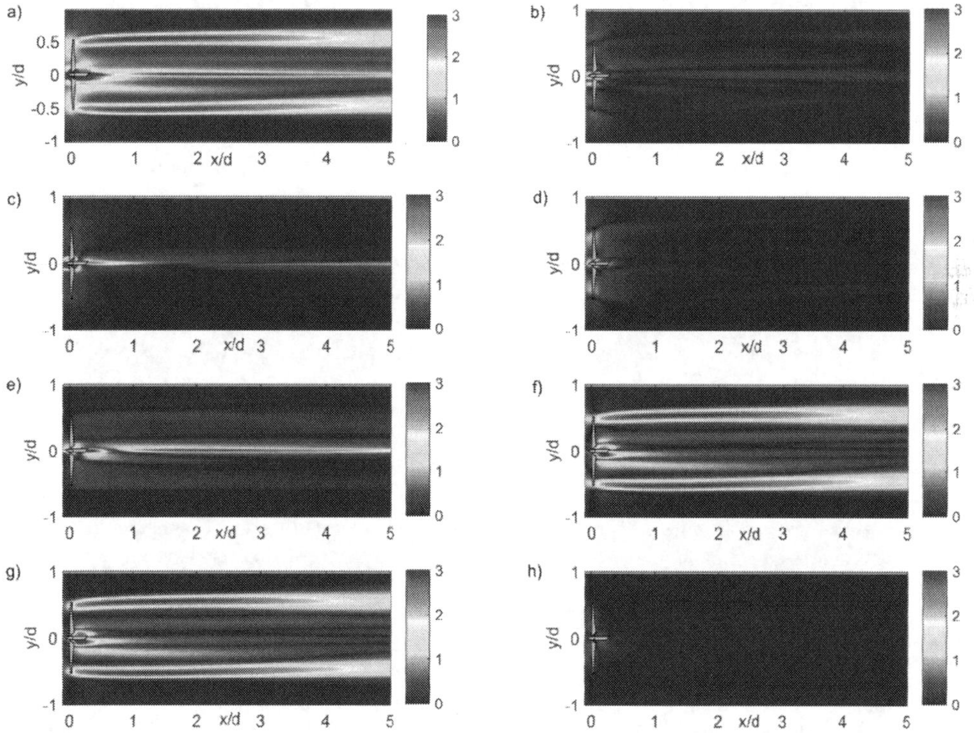

Figure 2. Split of term $(2\overline{S}:\overline{S})^{1/2}$ of the eddy-viscosity in Eq. 4 for the case TSR=5.8: a) $(2\overline{S}:\overline{S})^{1/2}$; b) $\sqrt{2}\partial U_r/\partial r$; c) $\sqrt{2}U_r/r$; d) $\sqrt{2}\partial U_x/\partial x$; e) $\partial U_\theta/\partial r - U_\theta/r$; f) $\partial U_x/\partial r + \partial U_r/\partial x$; g) $\partial U_x/\partial r$; h) $\partial U_\theta/\partial x$.

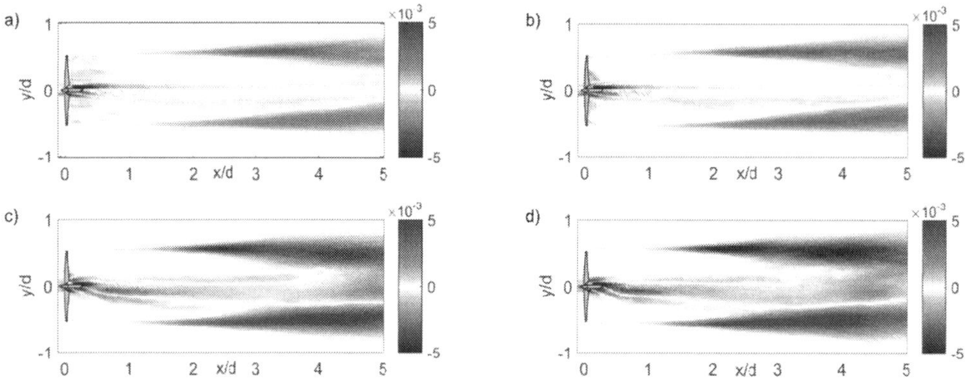

Figure 3. Reynolds stress component $\overline{u'_x u'_r}$ between the streamwise and transversal velocities evaluated over the horizontal plan at hub height: a) TSR=5.8; b) TSR=6.5; c) TSR=7; d) TSR=7.5.

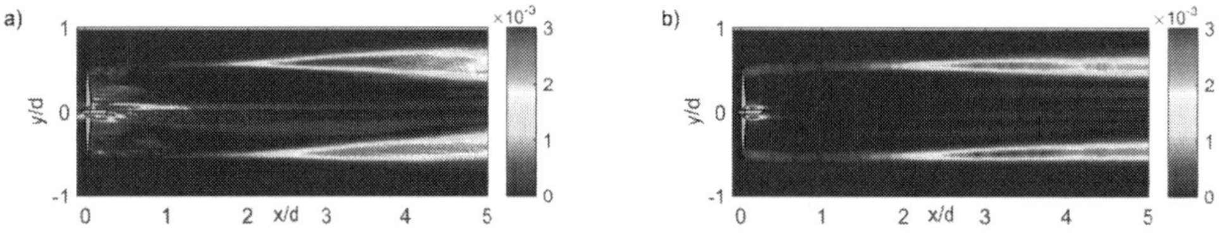

Figure 4. Best fitting procedure for the evaluation of the mixing length as a function of the downstream position. TSR=5.8 : a) absolute value of $\overline{u'_x u'_r}$; b) $l_m^2\,(\partial U_x/\partial r)^2$.

the time-averaged velocity gradient $(\partial U_x/\partial r)$ as follows:

$$-\overline{u'_x u'_r} = l_m^2\,(\partial U_x/\partial r)^2\,. \tag{8}$$

The results of the best fit procedure are reported in Fig. 4 for the case with TSR=5.8. In Fig. 4(a), the absolute value of the Reynolds stress component $\overline{u'_x u'_r}$ is compared to $l_m^2\,(\partial U_x/\partial r)^2$ in 4(b), where the mixing length is obtained at each downstream location through a best fit procedure.

The mixing length evaluated for the different TSR values over the horizontal plane at hub height and over the vertical symmetry plane of the wind turbine wake is reported in Fig. 5. For all the analyzed cases, the mixing length is roughly invariant for different downstream locations in the very near wake, i.e. for $x/d < 1$, then it increases for larger downstream distances. This wake feature can be very useful for a sharp detection of the near wake from the transition region, denoted also as diffusive region [28]. Indeed, in the near wake the rollup of the wake vorticity structures and the wake formation occur, whereas further downstream diffusive phenomena dominate leading to the wake recovery.

It is also interesting to observe that at the beginning of the diffusive region, i.e. at $x/d \approx 1$, the mixing length has roughly the same value for the different TSR values, which is expectable due to the same incoming wind condition simulated for the different cases. Another general feature is the roughly linear increase of the mixing length with the downstream distance, which is in agreement with previous wind tunnel experiments presented in [29, 30]. Moreover, the slope of the mixing length as a function of the downstream location increases with increasing TSR

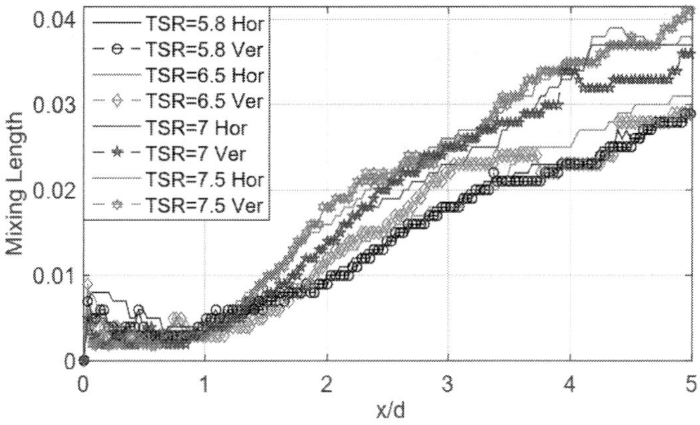

Figure 5. Mixing length as a function of the downstream location evaluated for different tip speed ratio values over the horizontal plane at hub height (Hor), and over the vertical symmetry plane (Ver).

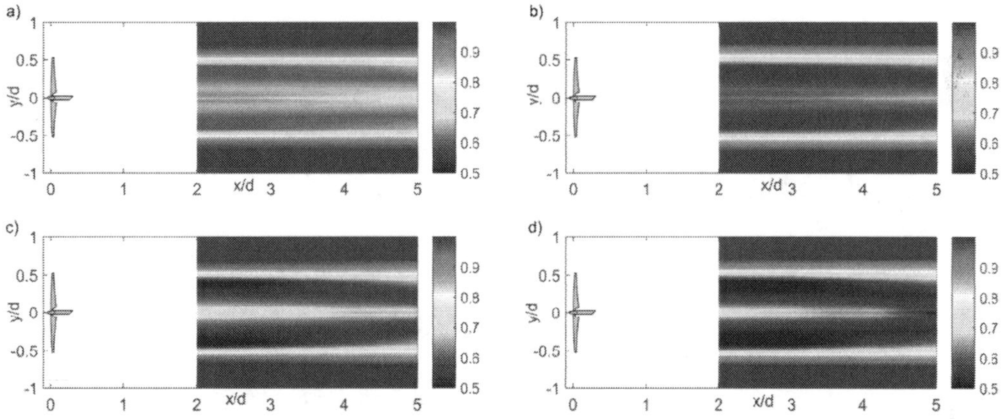

Figure 6. Time-averaged axial velocity over the horizontal plane at hub height obtained through the data-driven RANS simulations: a) TSR=5.8; b) TSR=6.5; c) TSR=7; d) TSR=7.5.

value. Therefore, wind turbines operating with larger TSR capture more power and produce a larger velocity deficit, but their wakes are characterized by a higher recovery rate. Indeed, the enhanced diffusive phenomena are related to the enhanced mechanically produced turbulence that, in turn, are related to the high axial velocity shear in the radial direction.

Since the mixing length model is calibrated through the available data as explained in the previous section, RANS simulations for the evaluation of the downstream evolution of the wind turbine wake flow can be easily executed within a very limited amount of time. For this work, the velocity profile at a downstream location of $x/d = 2$ is provided as boundary condition at the inlet of the computational domain, and the downstream evolution is calculated till a downstream distance of $x/d = 5$. The velocity field over the horizontal plane at hub height is shown in Fig. 6 for different TSR values. A general good agreement is observed with respect to the time-averaged velocity field obtained through the LES simulations reported in Fig. 1.

5. Final discussion and conclusions

In this paper a data-driven RANS framework for low computational cost and accurate simulations of wind turbine wake flows is proposed. The limited computational requirements are obtained through the RANS formulation, and an adequate accuracy is achieved through a data assimilation procedure for the calibration of the turbulence closure model. Specifically, a mixing length model for swirling flow is adopted for which the eddy-viscosity is only function of the mixing length and the gradient of the axial velocity along the radial direction. Moreover, the mixing length is only a function of the downstream location.

High fidelity LES simulations of a wind turbine operating with different tip speed ratios are used as test case for the present work. It has been shown that the main contribution of the time-averaged velocity field to the evaluation of the eddy-viscosity through the Boussinesq hypothesis is represented by the gradient of the axial velocity component along the radial direction. Consequently, the most meaningful turbulence parameter for the evaluation of the mixing length results to be the Reynolds stress component between the axial velocity and the radial velocity. Indeed, the mixing length is evaluated at each downstream location as best fit between the above-mentioned Reynolds stress component and the gradient of the axial velocity along the radial direction.

It has been found that the mixing length is roughly invariant in the very near wake, and its value is not affected by the operational conditions of the wind turbine. This is an important feature that allows in the near wake to evaluate the mixing length only through a characterization of the incoming wind field, and by neglecting the wind turbine characteristics. At a fixed location, which is one rotor diameter downstream for this work, the mixing length starts increasing linearly with the downstream location. This feature, i.e. the slope change of the mixing length as a function of the downstream location, is considered an effective criterion to sharply detect the transition from the near wake region and the transition region. The former is the wake region where the rollup of the wake vorticity structure occurs, whereas the latter is the area where the diffusive phenomena dominate and the wake gradually recovers to the incoming wind conditions. In the transition region, the slope of the mixing length in the downstream direction raises with increasing tip speed ratio. This suggests that a turbine operating with a higher tip speed ratio is characterized by a faster wake recovery. This feature is a consequence of the enhanced flow entrainment and mechanically produced turbulence generated by the higher shear of the axial velocity in the radial direction. Since the mixing length is calibrated through available wake measurements, the downstream evolution of a wind turbine wake can be quickly evaluated through RANS simulations by providing the velocity profile at a certain downstream location.

This data-driven RANS approach will be tested to simulate multiple turbine wakes, where the near wake velocity profile can be predicted through different model, such as actuator line or actuator disc. The main objective of the proposed tool is to perform quick predictions of wind turbine wakes by performing a limited number of wake measurements.

Acknowledgments
S. Leonardi was supported by NSF PIRE Award IIA 1243482. TACC is acknowledged for providing computational time.

References

[1] Calaf M, Meneveau C and Meyers J 2010 *Phys. Fluids* **22** 015110
[2] Lu H and Porté-Agel F 2011 *Phys. Fluids* **23** 065101
[3] Chatelain P 2013 *Flow, Turbulence and Combustion.* **91** (3) 587-605
[4] Martinez-Tossas L A, Churchfield M J, Leonardi S 2014 *Wind Energy.* doi:10.1001/we.1747
[5] Sarlak H, Meneveau C and Sorensen J N 2015 *Renewable Energy* **77** 386-399

[6] Porté-Agel F, Wu Y-T, Lu H and Conzemius R 2011 *J. Wind Eng. Ind. Aerodyn.* **99** 154-168
[7] Crespo A, Hernandez J and Frandsen S T 1999 *Wind Energy* **2** 124
[8] Barthelmie R J, Folkerts L, Larsen G C, Rados K, Pryor S C, Frandsen S T at al. 2005 *J. Atmos. Ocean. Tech.* **23** 881-901
[9] Thogersen M L 2005 *Tech. Rep. Niels Jernes Vej 10*
[10] Crasto G, Gravdahl A, Castellani F and Piccioni E 2012 *Energy Procedia* **24** 385-392
[11] Garrad Hassan and Partners Ltd 2009
[12] Openwind ver. 1.3 *AWS Truepower*
[13] Jensen N 1983 *Tech. report* Ris-M-2411
[14] Frandsen S T, Barthelmie R J, Pryor S, Rathmann O, Larsen S, Hojstrup J et al. 2006 *Wind Energy* **9** 39-53
[15] Bastankhah M and Porté-Agel F 2014 *Renewable Energy* **70** 116-123
[16] Alinot C and Masson C 2002 *Proc. ASME Wind Energy Symp. WIND2002* **42** 206-215
[17] Sorensen J N, Michelsen J A and Schreck S 2002 *Wind Energy* **5** 151-169
[18] Gomez-Elvira R, Crespo A, Migoya E, Manuel F and Hernandez J 2005 *J.Wind Eng. Ind. Aerodyn.* **93** 797-814
[19] Sezer-Uzol N and Long L N 2006 *Proc. 44th AIAA Aerospace Sciences Meeting and Exhibit* 0934
[20] El Kasmi A and Masson C 2008 *J.Wind Eng. Ind. Aerodyn.* **96** 103-122
[21] Ainslie J F 1988 *J. Wind Eng. Ind. Aerodyn.* **27** 213-224
[22] Iungo G V, Wu Y-T and Porté-Agel F 2013 *J. Atmos. Ocean. Tech.* **30** (2) 274-287
[23] Iungo G V and Porté-Agel F 2013 *Adv. Sci. Res.* **10** 71-75
[24] Iungo G V and Porté-Agel F 2014 *J. Atmos. Ocean. Tech.* **31** (10) 2035-2048
[25] Carbajo Fuertes F, Iungo G V and Porté-Agel F 2014 *J. Atmos. Ocean. Tech.* **31** (7) 1549-1556
[26] Pope S B 2000 Turbulent flows *Cambridge Univ. Press*
[27] Orlandi P and Leonardi S 2006 *J. of Turbulence* **7** 53
[28] Sorensen J N, Mikkellsen R, Sarmast S, Ivanell S and Henningson D 2014 *J. Phys.: Conf. Ser.* **524** 012155
[29] Iungo G V, Viola F, Camarri S, Porté-Agel F and Gallaire F 2013 *J. Fluid Mech.* **737** 499-526
[30] Viola F, Iungo G V, Camarri S, Porté-Agel F and Gallaire F 2014 *J. Fluid Mech.* **750** R1

Predicting wind farm wake interaction with RANS: an investigation of the Coriolis force

M. P. van der Laan[1], K. S. Hansen[2], N. N. Sørensen[1], P.-E. Réthoré[1]

[1]Technical University of Denmark, DTU Wind Energy, Risø Campus, DK-4000 Roskilde, Denmark

[2]Technical University of Denmark, DTU Wind Energy, Lyngby Campus, DK-2800 Kgs. Lyngby, Denmark

E-mail: plaa@dtu.dk

Abstract. A Reynolds-averaged Navier-Stokes code is used to simulate the interaction of two neighboring wind farms. The influence of the Coriolis force is investigated by modeling the atmospheric surface/boundary layer with three different methodologies. The results show that the Coriolis force is negligible for a single wind turbine, small for a single wind farm, but important for simulations of wind farm wake interaction.

1. Introduction

A large number of off-shore wind farms are planned to be build in the North Sea with a relatively close wind farm spacing. Nygaard [1] used the simple NO Jensen engineering wake model [2] to show that wind farm wakes can cause annual energy losses to neighboring wind farms. [1]. While wind turbine wake interaction within wind farms have been studied extensively using engineering wake models [3], Computational Fluid Dynamics (CFD) [4, 5, 6, 7] and mesoscale models [8, 9], the mutual interaction between wind farm wakes is a relatively new research area.

In the EERA-DTOC project, power measurements of a Danish wind farm cluster consisting of two off-shore wind farms, Rødsand II and Nysted, are investigated. In this paper, the measurements are compared with a Reynolds-averaged Navier-Stokes (RANS) code using a modified k-ε turbulence model, which is labeled as the k-ε-f_P turbulence model [10]. While the standard k-ε model is unable to predict the near wake deficit, the k-ε-f_P model shows comparable results with measurements and large-eddy simulations (LES). In previous work, the RANS setup including the k-ε-f_P has been successfully tested for single wind turbine wakes [10] and complete wind farms [7]. RANS does not require large computational resources compared with LES, because RANS is steady state and it allows a relative coarse grid (e.g. a cell spacing of 1/8 of the rotor diameter), while LES is transient and a fine grid is necessary to resolve the important turbulent scales. This enables us to use the RANS setup for large domains, e.g. a wind farm cluster covering an area of 100 km^2, which requires a grid size in the order of 100 million cells.

It is common to assume a simple shear flow in wind turbine wake simulations using RANS. In other words, only the atmospheric surface layer (ASL) is modeled, i.e. a logarithmic velocity profile in neutral stratification. However, the scale of the wind farm cluster is so large that the interaction of the Coriolis force with wind farm wakes can become important. In the present work, the effect of the Coriolis force on wind turbine wake and wind farm wake interaction is investigated by modeling the entire atmospheric boundary layer (ABL). We use a simplified version of the k-ε model of Koblitz et al.[11] to model an ABL, and it is coupled with the k-ε-f_P model. Only neutral stratification is investigated in order to isolate the effect of the Coriolis force from buoyancy forces. The simplified k-ε model of Koblitz et al. is based on two components: a turbulent length scale limiter to limit the boundary layer height, and the addition of the Coriolis force in the momentum equations. The effect of these components are individually tested and compared with the traditional ASL modeling by defining three methodologies:

- Method I: ASL.
- Method II: ABL (turbulent length scale limiter).
- Method III: ABL Coriolis (turbulent length scale limiter and Coriolis force).

The three methodologies are explained in sections 3-5, and they are applied to three test cases, which are presented in section 2. The results are discussed in section 6.

2. Test cases and measurements

Table 1. Summary of cases and corresponding input parameters for numerical computations.

Case	Description	Measurement data	$I_{H,\infty}$ [%]	$U_{H,\infty}$ [m/s]	D [m]	z_H [m]
1	ASL/ABL profiles	-	7	8	-	-
2	Single wake	-	7	8	82.4	69.0
3	Rødsand II &	SCADA data	7	8	92.6	68.5
	Nysted				82.4	69.0

Three test cases are defined in table 1, and they are used to compare the three methodologies from sections 3-5. All test cases are simulated with a turbulence intensity $I_{H,\infty}$ and velocity $U_{H,\infty}$ at hub height z_H of 7% and 8 m/s, respectively. Note that D corresponds to the rotor diameter. In the first test case, the capability of simulating and sustaining ASL and ABL profiles is investigated. Subsequently, the effect of the ASL/ABL modeling is tested for a single wind turbine wake, that is based on the Bonus 2.3 MW/82 wind turbine. The third test case is a Danish wind farm cluster consisting of the Rødsand II wind farm (90 SWT-2.3-93 wind turbines) and the Nysted wind farm (72 Bonus 2.3 MW/82 wind turbines). The layout is shown in figure 1. The wind farm spacing is around 3 km.

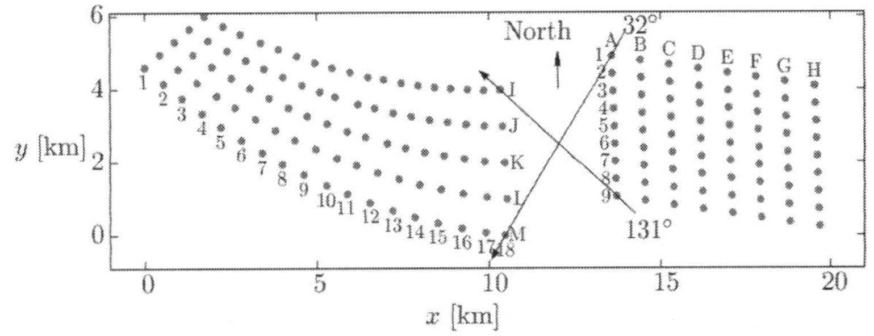

Figure 1. Wind farm layout of the Rødsand II (left) and Nysted (right) wind farms.

Power measurements (SCADA data) of all 162 wind turbines are available; however, the data set of each wind farm cannot be synchronized because the time stamps are not given, which makes it difficult to determine the free-stream conditions when Rødsand II is operating in the wake of Nysted. This limits the comparison of the measured wind farm interaction with the simulations. In addition, it is not possible to filter the data set from non-neutral atmospheric conditions. We choose to use the measurements for trend comparison only. More information about the measurements can be found in Hansen et al. [12]

3. Method I: ASL

3.1. General setup

EllipSys3D is used as a flow solver, which is originally developed by Sørensen [13] and Michelsen [14]. The RANS equations are solved with a SIMPLE algorithm [15], and a QUICK scheme [16] is used to discretize convective terms. Decoupling of the pressure, velocity and body forces are avoided with a modified Rhie-Chow algorithm [17, 18].

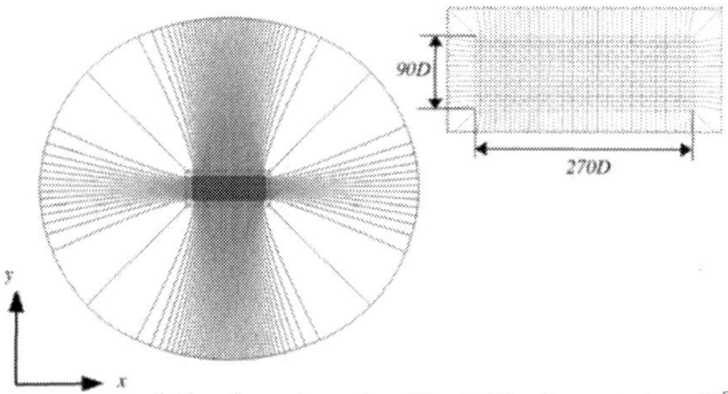

Figure 2. Blocking structure of the flow domain. Each block contains 48^3 cells. Wake domain is shown in red dotted box. Normalization based on $D = 82.4$ m.

The multi-block structure of the flow domain for the wind farm cluster is shown in figure 2, where the dimensions are normalized with $D = 82.4$ m. A region with a uniform spacing of D/8 in all direction is defined, which is labeled as the wake domain. The cell spacing is based on a grid dependency study of single wind turbine simulations [10]. The wake domain has horizontal dimensions of $270{\times}90\text{D}^2$, large enough to cover both wind farms including a margin of 1 km. The flow and wake domain extend 10D and 3D in the vertical direction, respectively. The wake domain is surrounded by an O-grid with a radius of 50 km. The total grid consist of 795 blocks of 48^3 cells, i.e. 88 million cells. Near the wall, at $z = 0$, the cells are refined in the z-direction, towards a first cell height of 0.5 m. The cells are stretched outside the wake domain. Except for the rough wall at $z = 0$, the flow solver decides if a boundary is an inlet or an outlet, based on the local flow direction. A fully developed flow is assumed at the outlet, while a neutral logarithmic solution is set at the inlet:

$$U(z) = \frac{u_*}{\kappa} \ln\left(\frac{z}{z_0}\right), \qquad k = \frac{u_*^2}{\sqrt{C_\mu}}, \qquad \varepsilon = \frac{u_*^3}{\kappa z}, \tag{1}$$

where U is the stream-wise velocity, u_* is the friction velocity, $\kappa = 0.4$ is the von Kármán constant, z_0 is the roughness height, k is the turbulent kinetic energy, ε is the turbulent dissipation, and C_μ is the eddy-viscosity coefficient. The logarithmic solution is retained through out the domain by setting a rough wall condition at $z = 0$. At the rough wall, the boundary conditions for U and ε are based on an equilibrium of turbulent production and dissipation, while a zero gradient condition is used for the turbulent kinetic energy [19].

The ambient turbulence intensity at hub height $I_{H,\infty}$ is set by the roughness height z_0 in:

$$I_{H,\infty} \equiv \frac{\sqrt{\frac{2}{3}k}}{U_{H,\infty}} = \frac{\kappa\sqrt{\frac{2}{3}}}{\ln\left(\frac{z_H}{z_0}\right)\sqrt[4]{C_\mu}}. \quad (2)$$

Note that C_μ is often used to set $I_{H,\infty}$; however, the behavior of the f_P function in the k-ε-f_P model, as defined in equation (6), changes unphysically with C_μ [20]. The friction velocity u_* is set to obtain the desired hub height velocity. The resulting velocity profile deviates less than 2% in rotor area from a profile that is set with a z_0 based on the site (e.g. 10^{-4} m).

3.2. *Wind turbine modeling*

The thrust and power curves of the Bonus 2.3 MW/82 and SWT-2.3-93 wind turbines are available, while detailed information about the rotor geometry or wind turbine control is unknown. Therefore, the wind turbines are represented as actuator disks (ADs) [21, 22, 23], that use a simple variable force method [24]. In previous work, it has been shown that the rotational force has a negligible effect on the power deficit [7]. Therefore, it is acceptable to only consider a variable normal force F_N [24]:

$$F_N = \frac{1}{2}\rho C_T^* A\langle U_{AD}^2\rangle, \qquad C_T^* = C_T\frac{U_{H,\infty}^2}{\langle U_{AD}^2\rangle} \quad (3)$$

where $\rho = 1.225$ kg/m^3 is the air density, C_T^* is the thrust coefficient that is based on the disk-averaged velocity $\langle U_{AD}\rangle$, while C_T is the thrust coefficient that is based on the free-stream $U_{H,\infty}$, and A is the rotor area. The normal force is distributed over the AD by multiplying F_N with a normalized force distribution that is obtained from a Detached Eddy Simulation of the NREL-5MW reference wind turbine [23, 25]. During the wind farm simulation, C_T^* is updated per iteration from a C_T^*-$\langle U_{AD}\rangle$ curve, until a satisfactory convergence is reached. The wind turbine power is estimated from a P-$\langle U_{AD}\rangle$ curve. The C_T^*-$\langle U_{AD}\rangle$ and P-$\langle U_{AD}\rangle$ curves are determined from 6 single wind turbine simulations with free-stream velocities between 4 and 9 m/s. In these simulations, a constant thrust force is applied, where F_N is based on a known C_T and a free-stream velocity $U_{H,\infty}$. The procedure is carried out for both wind turbine types.

3.3. *Turbulence modeling*

The k-ε-f_P turbulence model from van der Laan et al. [10] is applied to the wind farm simulations. The model is based on the standard k-ε model from Launder and Spalding [26]. Both turbulence models can only predict isotropic Reynolds-stresses $\overline{u_i'u_j'}$ because the Boussinesq approximation [27] is used:

$$\overline{u_i'u_j'} = \frac{2}{3}k\delta_{ij} - \nu_T\left(U_{i,j} + U_{j,i}\right), \quad (4)$$

where δ_{ij} is the Kronecker delta, $U_{i,j}$ are the mean velocity gradients and ν_T is the turbulent eddy viscosity:

$$\nu_T = C_\mu f_P\frac{k^2}{\varepsilon}, \quad (5)$$

with C_μ as a constant and ε as the turbulent dissipation. The effective eddy-viscosity coefficient $C_\mu f_P$ is a constant in the standard k-ε model because $f_P = 1$, while f_P is a scalar function in the k-ε-f_P model that depends on the local shear parameter: $\sigma \equiv \frac{k}{\varepsilon}\sqrt{(U_{i,j})^2}$. The effective eddy-viscosity coefficient $C_\mu f_P$ is variable, instead of a constant, which is the only difference with the standard k-ε model. The scalar function f_P in the k-ε-f_P model is defined as [28]:

$$f_P\left(\sigma/\tilde{\sigma}\right)=\frac{2f_0}{1+\sqrt{1+4f_0\left(f_0-1\right)\left(\frac{\sigma}{\tilde{\sigma}}\right)^2}},\qquad f_0=\frac{C_R}{C_R-1},\tag{6}$$

with $\tilde{\sigma}=1/\sqrt{C_\mu}$ as the shear parameter in an idealized (logarithmic) neutral atmospheric surface layer and C_R is a calibration parameter. In the neutral logarithmic solution, $f_P=1$ because $\sigma=\tilde{\sigma}$. In regions with a high shear parameter, i.e. $\sigma>\tilde{\sigma}$, $f_P<1$, and the turbulent eddy viscosity from equation (5) is decreased. The near wind turbine wake is characterized by high velocity gradients, where $\sigma\gg\tilde{\sigma}$. As a result, the k-ε-f_P delays the wake recovery compared to the standard k-ε model. The chosen value of C_R is based on a calibration against LES using eight different single wind turbine cases, as performed in previous work [10].

The standard transport equations for k and ε are used:

$$\frac{Dk}{Dt}=\nabla\cdot\left[\left(\nu+\frac{\nu_T}{\sigma_k}\right)\nabla k\right]+\mathcal{P}-\varepsilon,\qquad \frac{D\varepsilon}{Dt}=\nabla\cdot\left[\left(\nu+\frac{\nu_T}{\sigma_\epsilon}\right)\nabla\varepsilon\right]+\left(C_{\varepsilon,1}\mathcal{P}-C_{\varepsilon,2}\varepsilon\right)\frac{\varepsilon}{k},\tag{7}$$

where $\mathcal{P}$ is the turbulent production, ν is the kinematic molecular viscosity and $C_{\varepsilon,1}, C_{\varepsilon,2}$, σ_k, σ_ε are constants. The values of the constants are listed in table 2.

The turbulent eddy viscosity from equation (5) can be written as a turbulent velocity scale, i.e. the friction velocity u_*, and a turbulent length scale ℓ_t:

$$\nu_T=C_\mu^{\frac{1}{4}}k^{\frac{1}{2}}C_\mu^{\frac{3}{4}}f_P\frac{k^{\frac{3}{2}}}{\varepsilon}=u_*\ell_t,\qquad\Rightarrow\ell_t=C_\mu^{\frac{3}{4}}f_P\frac{k^{\frac{3}{2}}}{\varepsilon},\tag{8}$$

hence f_P can be seen as a local turbulent length scale limiter, that limits the turbulent length scales in the near wind turbine wake.

Table 2. Model constants.

C_R	C_μ	$C_{\varepsilon,1}$	$C_{\varepsilon,2}$	σ_k	σ_ε	κ
4.5	0.03	1.21	1.92	1.00	1.30	0.40

4. Method II: ABL

4.1. General setup

The wind farm cluster ABL simulations are carried out with the same flow solver and flow domain as discussed in section 3.1. Instead of prescribing the inlet boundaries with the logarithmic solution of equation (1), the inflow is based on a precursor simulation. In the precursor, the flow is driven by a prescribed mass flux, that translates in to a pressure gradient. The flow domain of the precursor simulation is a box that is 6 km high and 1×1 km^2 wide, and it consists of 192×32×32 cells. The lateral boundaries in the inflow direction of the domain are specified as periodic, the bottom boundary is a rough wall as described in section 3.1 and the top and sides are symmetry boundaries where all gradients are set to zero. It is found that a mass flux with a mean velocity of $U=30.0$ m/s and a roughness of $z_0=0.002$ m gives the desired hub height conditions.

4.2. Wind turbine modeling

The wind turbines are modeled in the same way as discussed in section 3.2. It should be noted that the calibration of the thrust coefficient C_T^* is not re-done for the ABL simulations. Hence, it is assumed that the C_T^*-$\langle U_{AD}\rangle$ curve is not affected.

4.3. Turbulence modeling

The turbulence modeling of the ABL simulations is based on the k-ε ABL model of Koblitz et al. [11], where the boundary layer height is finite by limiting the global turbulent length scale ℓ_t:

$$\ell_t = C_\mu^{\frac{3}{4}} \frac{k^{\frac{3}{2}}}{\varepsilon}, \tag{9}$$

using a limiter on ε:

$$\varepsilon = \max \left(C_\mu^{\frac{3}{4}} \frac{k^{\frac{3}{2}}}{l_{t,\max}}, \varepsilon \right), \tag{10}$$

where $l_{t,\max}$ is a chosen maximum turbulent length scale. $l_{t,\max}$ is used to enforce an equilibrium in the transport equation of ε outside the logarithmic region of the ABL, through a height dependent $C_{\varepsilon,1}$, labeled as $C^*_{\varepsilon,1}$, as proposed by Apsley and Castro [29]:

$$C^*_{\varepsilon,1} = C_{\varepsilon,1} + (C_{\varepsilon,2} - C_{\varepsilon,1}) \frac{\ell_t}{l_{t,\max}}. \tag{11}$$

The global turbulent length scale limiter also affects wind turbine wakes similarly to the local turbulent length scale limiter f_P from equation (6). In order to avoid double counting of length scale limitation in the wind farm simulations, the global length scale limiter is switched off in the wake region by multiplying $C^*_{\varepsilon,1}$ with the blending function f_1:

$$f_1 = 0.5 \tanh\left(50\left[f_P - 0.9\right]\right) + 0.5, \tag{12}$$

In addition, equation (10) is not used in the wake domain of the wind farm simulations. The same transport equations for k and ε are used, as defined in equation (7), where the constant $C_{\varepsilon,1}$ is replaced by $f_1 C^*_{\varepsilon,1}$. Ambient values of k and ε are forced by adding the source terms S_k and S_ε to the k and ε equations, respectively, as proposed by Spalart and Rumsey [30]:

$$S_k = \varepsilon_{\text{amb}}, \qquad S_\varepsilon = C_{\varepsilon,2} k^2_{\text{amb}} / \varepsilon_{\text{amb}}, \tag{13}$$

where k_{amb} and ε_{amb} are chosen ambient values: $k_{\text{amb}} = 10^{-4}$ and $\varepsilon_{\text{amb}} = 7.208 \times 10^{-8}$ such that $\ell_t = 1$ above the ABL height.

5. Method III: ABL including Coriolis force

5.1. General setup

The wind farm cluster ABL simulations including the Coriolis force are carried out in the same way as the ABL simulations without Coriolis, as described in section 4; however, the flow is now driven by the Coriolis force and a counter balancing pressure gradient, instead of a prescribed mass flux. A precursor simulation is used to calculate the inflow profiles for the wind farm simulations, as described in section 5. Note that all lateral boundaries are periodic to account for wind veer. It is found that a geostrophic wind of 10.2 m/s and a roughness height of 0.007 m gives the desired conditions at hub height.

The Coriolis force is added to the momentum equation, and it is balanced by a specified constant pressure gradient that is a function of the geostrophic wind G. The resulting volume source $S_{v,i}$ becomes:

$$S_{v,i} = \rho f_c \epsilon_{ijk} e_k \left(U_j - G_j\right), \tag{14}$$

where ϵ_{ijk} is the Levi-Civita symbol, e_k is the unit vector in the normal direction, G_j is a component of the geostrophic wind G, and f_c is the Coriolis parameter defined as:

$$f_c = 2\omega \sin(\phi), \tag{15}$$

where $\omega = 7.292115 \times 10^{-5}$ rad/s is the rotation of the earth, and ϕ is the latitude. The latitude of the wind cluster is 55.6°, and is assumed to be constant in the simulation, which means that f_c is also a constant. In other words, the global rotation that is induced by the Coriolis force is not modeled. The assumption is valid for the current domain size of 10^2 km, because it is an order smaller than the Rossby radius of 10^3 km. Local flow deflections are still possible, because the Coriolis force is a function of U_j. Hence, wind veer and the interaction of the Coriolis force with wind turbine wakes are modeled. This local flow deflection can be explained using a simple analysis of a flow crossing a roughness change, e.g. a coastal site where the flow moves from sea to land or vice versa. The x and y components of the Coriolis force from equation (14) become:

$$S_{v,x} = \rho f_c \left(V - G_y\right), \quad S_{v,y} = -\rho f_c \left(U - G_x\right), \tag{16}$$

where U and V are the velocity components in x-direction and y-direction, respectively. For simplicity, we consider an incoming flow that is aligned with the x-direction at a certain height, hence $V = 0$ and $S_{v,x}$ is a constant, and only $S_{v,y}$ is responsible for the flow deflection. The local behavior of the Coriolis force is described in table 5.1. Two different flow cases are listed, applicable to the Northern Hemisphere where $f_c > 0$. When the flow moves from sea to land, the flow decelerates and $S_{v,y}$ increases using equation (16), which turns the flow to the left, as also shown by Orr et al. [31]. The opposite occurs when the flow moves from land to sea. The same principle applies to the interaction of the Coriolis force with wind turbines wakes. In regions where the flow is slowed down, the Coriolis force turns the wake to the left, while opposite occurs in regions of wake recovery. The analysis changes if the coastal area or wind farm is located in the Southern Hemisphere, where $f_c < 0$, hence the Coriolis force turns the flow to the right and left for flow deceleration and acceleration, respectively.

Table 3. Local behavior of the Coriolis force in a coastal area in the Northern Hemisphere.

Flow deceleration			Flow acceleration		
sea	$\rightarrow$	land	land	$\rightarrow$	sea
$z_{0,\text{sea}}$	$<$	$z_{0,\text{land}}$	$z_{0,\text{land}}$	$>$	$z_{0,\text{sea}}$
U_{sea}	$>$	U_{land}	U_{land}	$<$	U_{sea}
$S_{v,y,\text{sea}}$	$<$	$S_{v,y,\text{land}}$	$S_{v,y,\text{land}}$	$>$	$S_{v,y,\text{sea}}$
$\rightarrow$ Turning to the left			$\rightarrow$ Turning to the right		

5.2. Wind turbine modeling

The wind turbines are modeled in the same way as discussed in sections 3.2 and 4.2. It should be noted that the wind turbines are not dynamically yawed during the simulations. This means that the local wind direction changes that are caused by the interaction of the Coriolis force and the velocity deficit, can lead to small yaw misalignments. However, similar errors are made in half wake conditions (with or without Coriolis), where a real wind turbine would yaw because the rotor experiences a difference in blade loading, as shown by Schepers et al. [32]

5.3. Turbulence modeling

The global length scale limitation, as discussed in section 4.3, is also necessary in the ABL simulations including Coriolis. Following the k-ε ABL model of Koblitz et al. [11], $l_{\max}$ is estimated from measurements, as described by Blackader [33]:

$$l_{t,\max} = 0.00027\frac{G}{f_c}, \tag{17}$$

where G is the geostrophic wind and f_c is the Coriolis parameter, as defined in equation (15).

6. Results and Discussion

6.1. Test case I: ASL/ABL profiles

The simulated ASL/ABL profiles are shown in figure 3. The dashed lines are ABL results from the precursor simulations or represent the analytical ASL profiles, as defined in equation (1). The solid lines are profiles extracted from empty wind farm domain simulations. All three methods are capable of sustaining the inlet profiles, where only the k-profiles shows some deviation, which is related to the known problem of sustaining k near the wall.

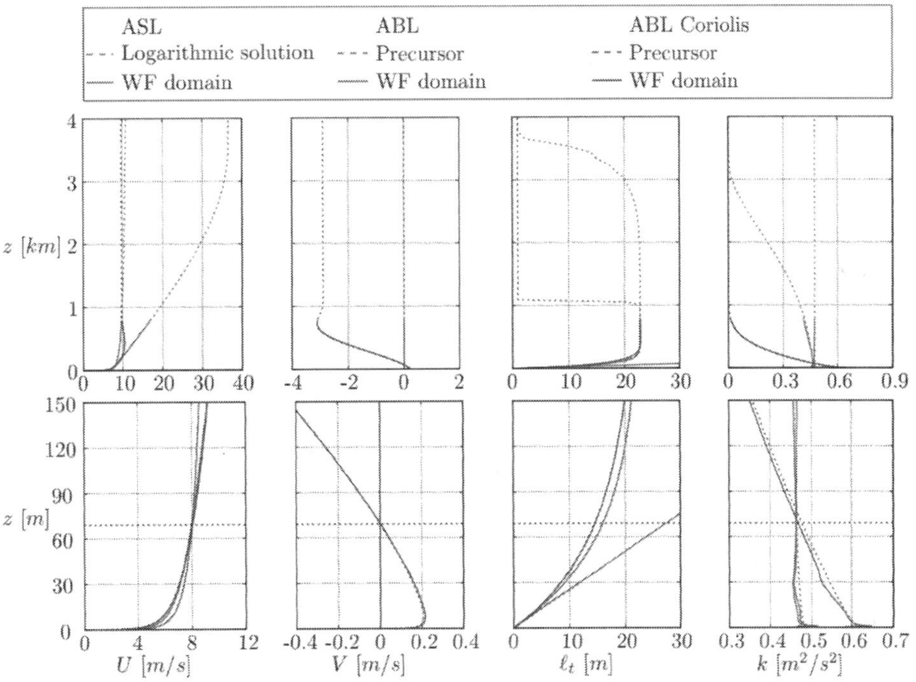

Figure 3. Test case I: Simulated profiles of the precursor and corresponding wind farm (WF) simulations (without wind turbines), extracted at the center of the domain. Bottom plots are a zoomed view of the top plots. Black dotted line represents the hub height (69 m).

As expected, the ABL Coriolis method predicts a wind veer, which is about 0.5 m/s over the rotor area. In addition, the ABL Coriolis method predicts a much lower boundary layer height than ABL method. This is caused by the high mean velocity in the precursor simulation of the ABL method that needs to be set in order to obtain the desired conditions at hub height.

The turbulent length scale ℓ_t predicted by the ASL method grows linearly with z, while the two ABL methods limit ℓ_t. As a results, the ASL-predicted ℓ_t is twice as large at hub height as predicted by the ABL methods.

6.2. Test case II: Single wake

In figure 4, the velocity deficit at hub height of a single wind turbine is plotted at 2D, 7D and 12D downstream. The ABL methods predict similar deficits, while they differ from the ASL method (5% and 3% difference of velocity at the wake center at 7D and 12D, respectively). This is caused by the large difference in the inflow turbulent length scale ℓ_t at hub height, as discussed in section 6.1. A lower ℓ_t reduces the turbulent eddy viscosity, which delays the wake recovery. The wake of the ABL Coriolis method is also slightly skewed because of the Coriolis force, but the impact is negligible.

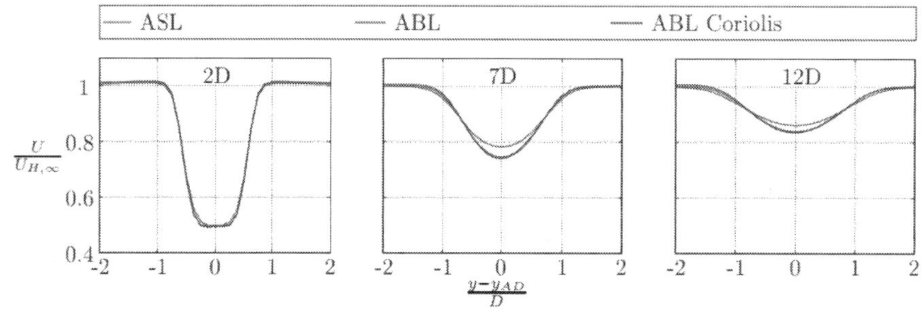

Figure 4. Test case II: Single wake velocity deficit at hub height of the Bonus 2.3 MW/82 wind turbine.

6.3. Test case III: Rødsand II-Nysted

In figure 5, contours of velocity at hub height are plotted in the wind farm cluster simulation. The magnitude of the horizontal velocity components $\sqrt{U^2+V^2}$, and the cross-flow velocity V are shown for a wind direction of 97°. The ABL methods predict larger velocity deficits compared to the ASL method because the free-stream turbulent length scale at hub height is lower, as discussed in section 6.2. The ABL Coriolis method shows similar deficits as the ABL method, although the Coriolis force deflects the wind farm wakes towards the right. This is best visible in the bottom plot of figure 5, where the difference in cross-flow velocity between the ABL Coriolis and the ABL methods is plotted. The difference in cross-flow velocity in the wind farm wake of Nysted is in the order of 0.1 m/s. As a result, the merged wind turbine wakes that originate from rows 1-9 of Nysted move around 0.5D north when they reach the first wind turbines of Rødsand II, 3 km downstream. The lateral movement changes the inflow condition of the wind turbines I18, J18 and K18 (figure 1), from half wake to full wake, as shown in the velocity magnitude plots of the ABL and the ABL Coriolis methods from figure 5. The Coriolis force turns the wind farm wake to the right because the regions of acceleration (wake recovery) are much larger than the regions of deceleration that mainly occur in near vicinity of a wind turbine. Note that the theoretical interaction of the Coriolis force with wind turbines wakes is discussed in section 5.1.

The wind farm efficiency of Rødsand II for a wind direction range of 32-132° is depicted in figure 6. Solid and dashed lines are results with and without Nysted, respectively, while the three colors represent the three different methods from sections 3-5. As expected, all methods predict a lower efficiency of Rødsand II when it is operating in the wake of Nysted. The ASL

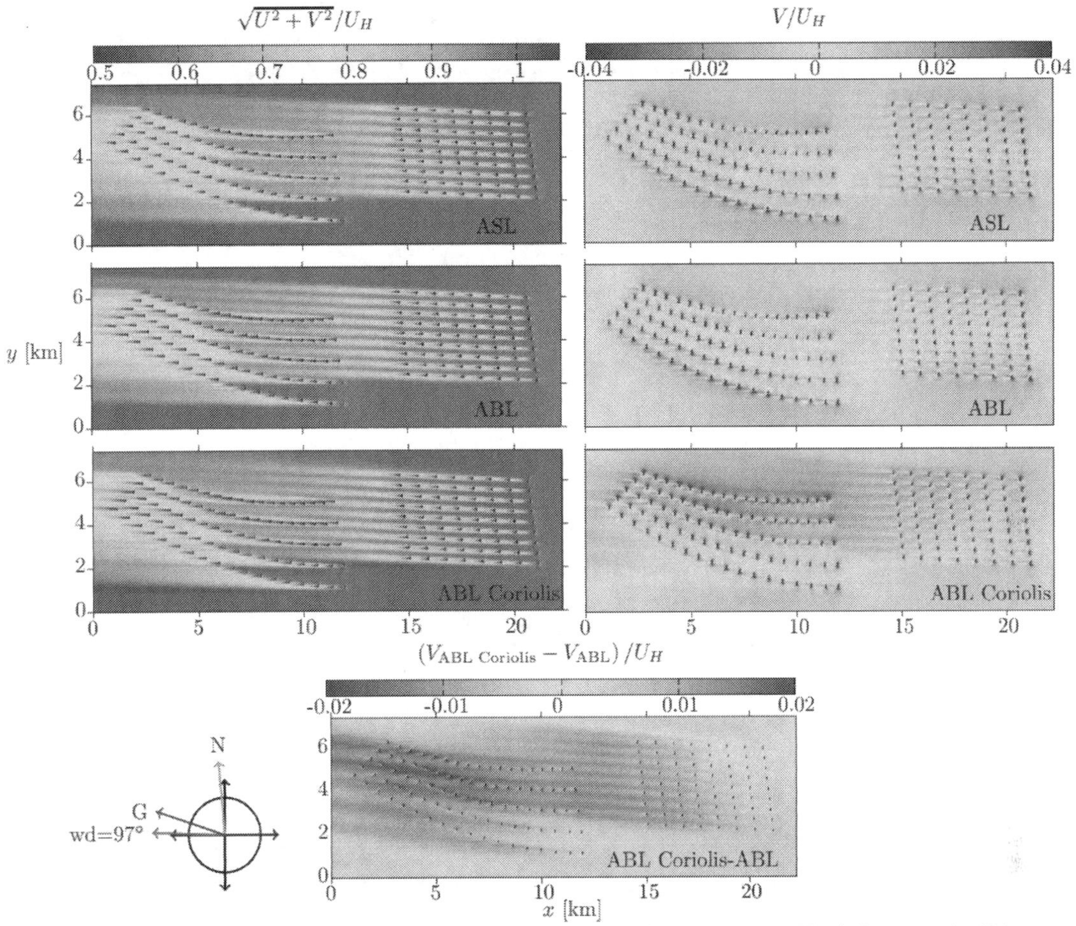

Figure 5. Test case III: Velocity contours in the wind farm cluster. Wind direction is 97° and the north is directed towards a wind direction (wd) of 175°. Left column: magnitude of horizontal velocity, right column: cross-flow velocity, bottom plot: difference in cross-flow velocity between ABL Coriolis and ABL methods.

method predicts a lower impact of Nysted compared to the other the ABL methods, because of the faster wake recovery due to the high free-stream turbulent length scale at hub height in the ASL simulations, as discussed previously. Although not shown in figure 6, it has been found that the influence of Rødsand II to Nysted is small for the Eastern wind sector (0.6% averaged loss in Nysted's wind farm efficiency for wind directions between 32-132°).

The effect of the Coriolis force is best seen when the ABL simulations are compared, such that the effect of the length scale limiter is disregarded. While the Coriolis force has a negligible impact on the single wind turbine simulations (as discussed in section 6.2) and a minor influence on single wind farm simulations (dashed lines), it has a large effect on the wind farm cluster simulations (solid lines). This can be explained as follows. The Coriolis force is a function of the local velocity. When the local velocity is changed by a wind turbine, the local Coriolis force also changes. In the Northern Hemisphere, the Coriolis force increases when the flow decelerates, and the flow is turned to the left, while the opposite occurs for flow acceleration, as discussed in section 5.1. The effect is not visible for a single wind turbine simulation because the region of modified velocity is too small. In other words, a flow particle does not have enough time to be deflected in such a small space. However, the flow deflection gets stronger when the length scale of the regions with a modified velocity increases, e.g. a wind farm wake. Hence, the flow is most deflected in the wind farm cluster simulations, which in this case increases the alignment of the flow with the arcs in the Rødsand II wind farm. Therefore, the orientation of the curved rows in Rødsand II is unfavorable.

Figure 6 also indicates that Rødsand II loses in order of 10-15% in wind farm efficiency for wind directions between 70-100° because of the presence of Nysted. This eastern wind sector has a probability of 10%, which results into an overall loss of 1-2% for a wind speed of 8 m/s.

The measurements in figure 6 show a similar trend in wind farm efficiency as seen in the wind farm cluster simulations, except for the 107° data point, which is not understood. Unfortunately, the lack of free-stream measurements makes it difficult to make stronger statements.

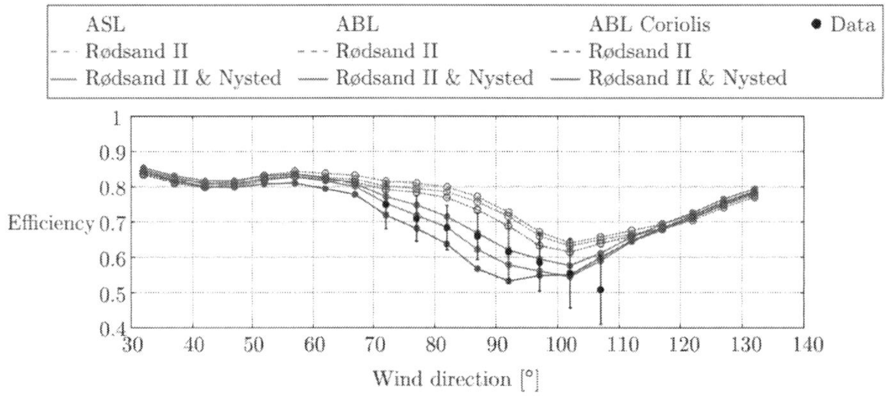

Figure 6. Test case III: Wind farm efficiency of Rødsand II. Solid and dashed lines: with and without Nysted, respectively. Limited data shown for trend comparison only.

7. Conclusions

The effect of the Coriolis force on wind farm wake interaction is investigated in RANS. Three different methodologies of modeling the atmospheric surface/boundary layer are used: ASL method (only modeling the neutral logarithmic region), ABL method (modeling the neutral ABL with a turbulent length scale limiter to limit the boundary height) and the ABL Coriolis method (modeling the neutral ABL with a turbulent length scale limiter including Coriolis forces). All methods can sustain their corresponding inflow profiles in a large wind farm cluster domain. The ABL methods predict a turbulent length scale at hub height that is half of the one simulated by the ASL method, which delays the wake recovery. Therefore, the effect of the Coriolis can only be investigated by comparing the ABL method and the ABL Coriolis method. While the Coriolis force has a negligible impact on a single wind turbine wake and a small effect on a single wind farm, it has a large influence on the wind farm cluster simulation, where the Rødsand II wind farm is operating in the wake of the Nysted wind farm. The Coriolis force deflects the recovering wind farm wake towards the right, which increases the alignment of the Nysted wakes with the curved rows in the Rødsand II wind farm. Therefore, the orientation of the curved rows is predicted to be unfavorable with respect to the sign of the Coriolis force that is present in the Northern Hemisphere. Since the effect of the Coriolis force is significant in a

RANS simulation of a wind farm cluster, it is recommended to include it when wind farm wake interaction is studied.

Acknowledgments

This work is supported by the Center for Computational Wind Turbine Aerodynamics and Atmospheric Turbulence funded by the Danish Council for Strategic Research, grant number: 09-067216, and the EERA-DTOC project funded by FP7-ENERGY-2011-1, grant number: 282797. Computational resources were provided by DCSC and the DTU central computing facility.

References

[1] Nygaard N G 2014 *J. Phys.: Conf. Series* **524** 1
[2] Jensen N O 1983 A note on wind generator interaction Tech. Rep. Risø-M-2411 Risø National Laboratory Roskilde, Denmark
[3] Gaumond M, Réthoré P E, Ott S, Peña A, Bechmann A and Hansen K S 2013 *Wind Energy* **17** 1169
[4] Churchfield M J, Lee S, Moriarty P J, Martinez L A, Leonardi S, Vijayakumar G and Brasseur J G 2012 *AIAA Conference, Nashville, USA* p 1
[5] Port´e-Agel F, Wu Y T and Chen C H 2013 *Energies* **6** 5297
[6] Nilsson K, Ivanell S, Hansen K S, Mikkelsen R, Sørensen J N, Breton S P and Henningson D 2014 *Wind Energy* Published online: DOI: 10.1002/we.1707
[7] van der Laan M P, Sørensen N N, Réthoré P E, Mann J, Kelly M C, Troldborg N, Hansen K S and Murcia J P 2014 *Wind Energy* Published online: DOI: 10.1002/we.1804
[8] Fitch A C, Olson J B, Lundquist J K, Dudhia J, Gupta A K, Michalakes J and Barstad I 2012 *Mon. Wea. Rev.* **140** 3017
[9] Volker P J H, Badger J, Hahmann A N and S O 2015 *Geosci. Model Dev. Discuss.* **8** 3481
[10] van der Laan M P, Sørensen N N, Réthoré P E, Mann J, Kelly M C, Troldborg N, Schepers J G and Machefaux E 2015 *Wind Energy* **18** 889
[11] Koblitz T, Bechmann A, Sogachev A, Sørensen N and Réthoré P E 2015 *Wind Energy* **18** 75
[12] Hansen K S, Réthoré P E, Palma J, Hevia B, Prospathopoulos J, Penã A, Ott S, Schepers G, Palomares A, van der Laan M P and Volker P 2015 *Submitted to J. Phys.: Conf. Series*
[13] Sørensen N N 1994 Ph.D. thesis Risø National Laboratory Roskilde, Denmark
[14] Michelsen J A 1992 Basis3d - a platform for development of multiblock PDE solvers. Tech. Rep. AFM 92-05 Technical University of Denmark, Lyngby, Denmark
[15] Patankar S V and Spalding D B 1972 *Int. J. Heat. Mass. Trans.* **15** 1787
[16] Leonard B P 1979 *Comput. Methods in Appl. Mech. Eng.* **19** 59
[17] Réthoré P E and Sørensen N N 2012 *Wind Energy* **15** 915
[18] Rhie C M and Chow W L 1983 *AIAA Journal* **21** 1525
[19] Sørensen N N, Bechmann A, Johansen J, Myllerup L, Botha P, Vinther S and Nielsen B S 2007 *J. Phys.: Conf. Series* **75** 1
[20] van der Laan M P, Sørensen N N, Réthoré P E, Mann J, Kelly M C and Schepers J G 2013 *Proc. ICOWES2013, Copenhagen, Denmark* p 514
[21] Mikkelsen R 2003 Ph.D. thesis Technical University of Denmark, Mek dept Lyngby, Denmark
[22] Réthoré P E 2009 Ph.D. thesis Aalborg University, Risø DTU Roskilde, Denmark
[23] Réthoré P E, van der Laan M P, Troldborg N, Zahle F and Sørensen N N 2014 *Wind Energy* **17** 919
[24] van der Laan M P, Sørensen N N, Réthoré P E, Mann J, Kelly M C and Troldborg N 2014 *Wind Energy* Published online: DOI: 10.1002/we.1816
[25] Jonkman J, Butterfield S, Musial W and Scott G 2009 Definition of a 5-MW Reference Wind Turbine for Offshore System Development Tech. rep. National Renewable Energy Laboratory
[26] Launder B E and Spalding D B 1972 *Mathematical models of turbulence* (Academic Press, London, UK)
[27] Boussinesq M J 1897 *Théorie de l'écoulement tourbillonnant et tumultueux des liquides* (Gauthier-Villars et fils, Paris, France)
[28] Apsley D D and Leschziner M A 1998 *Int. J. Heat. Fluid. Fl.* **19** 209
[29] Apsley D D and Castro I P 1997 *Bound-Lay. Meteorol.* **83** 75
[30] Spalart P and Rumsey C 2007 *AIAA J.* **45** 2544
[31] Orr A, Hunt J, Capon R, Sommeria J, Cresswell D and Owinoh A 2005 *Weather* 291
[32] Schepers J G, Obdam T S and Prospathopoulos J 2012 *Wind Energy* **15** 575
[33] Blackader A K 1962 *J. Geophys. Res.* **67** 3095

Quantifying variability of Large Eddy Simulations of very large wind farms.

S J Andersen[1a], B Witha[2], S-P Breton[3], J N Sørensen[1,3], R F Mikkelsen[1], and S Ivanell[3]

[1] DTU Wind Energy, Technical University of Denmark, 2800 Lyngby, Denmark.
[2] ForWind - Center for Wind Energy Research, Carl von Ossietzky University Oldenburg, Ammerländer Heerstr. 136, 26131 Oldenburg, Germany.
[3] Uppsala University, Dep. of Earth Sciences, Campus Gotland, Visby, Sweden.

E-mail: [a] sjan@dtu.dk

Abstract. Large Eddy Simulations are inherently dynamic as the largest scales are resolved and the smallest scales are modeled temporally. This raises challenges for simulations including very large scales such as atmospheric flows, which require very long simulation times. Simple averages fail at capturing these dynamics and potentially yield misleading interpretations concerning the capabilities of different models when tested in blind tests or in benchmarking exercises such as Wakebench, where results from different flow models are compared. This article will present results from very large wind farm simulations using Actuator Disc (AD) and Line (AL) models for two different turbine spacings with turbulent inflow. The results of each numerical flow model include a certain variability, and it will be examined if different models result in comparable probability distributions.

1. Introduction

The state-of-the-art in numerical simulations of wind turbines and wind farms consists of Large Eddy Simulations (LES) combined with Actuator Disc (AD) or Actuator Line (AL) methods to model the flow behind and between wind turbines. LES are inherently dynamic as the largest scales are resolved and the smallest scales are modeled temporally. This yields important advantages but also raises challenges in comparing results from otherwise comparable simulations. Simple averages do not capture the dynamics. Particularly, too short averaging times which do not filter out large turbulent structures, a common feature in the atmospheric boundary layer, may yield misleading interpretations concerning the capabilities of different models. This provides a challenge when models are tested and compared with measurements, benchmarking exercises or in blind tests such as the Bolund Experiment [1]. LES require long simulation times before statistical convergence is achieved, see e.g. Andersen [2] who examined the statistical convergence of the turbulent quantities in infinitely large wind farms. This is resource consuming and the required convergence time might differ for different models and setups. Yet, it is also standard procedure in atmospheric and wind energy related measurements to report 10 min averages.

The present article aims at examining the variability from different model simulations performed by different users in order to explore how the variability can be presented and compared best. Furthermore, the analysis will investigate if the variability can be exploited to provide valuable insights in terms of overall confidence in model results as well as in potentially building a bridge to the natural variability found in measurements, where results and measurements are often filtered to compare well.

The results presented have been derived from a benchmarking exercise of the IEA Wind Task 31: Wakebench[1]. In that benchmark very large wind farms are to be simulated to determine if and when an asymptotic wake state as in an infinite wind farm is reached. A range of different methods to simulate the turbulent flow and model the effect of the wind turbines are used by the participants. Hence, the aim of this article is to present graphical methods for comparing and evaluating the temporal and spatial variability resulting from the different approaches.

This article will present results from very large wind farm simulations using AL and AD models in EllipSys3D [3], [4] from Technical University of Denmark (DTU) and Uppsala University (UU) and PALM [5], [6] from ForWind (FW) for different turbine spacings. Each numerical flow model produces a spatially and temporally variable flow, and it will be examined if different models result in comparable probability distributions.

2. Methodology

This section gives a brief introduction to the difference in methodology applied by the different participants but refers to other papers for detailed information on the theoretical background.

2.1. Numerical Solvers

Results from two different CFD codes are used.

2.1.1. EllipSys3D EllipSys3D is a 3D flow solver developed as a collaboration between DTU (Michelsen [3]) and the former Risø(Sørensen [4]), and has been applied for wind energy research for more than two decades. EllipSys3D solves the discretized incompressible Navier-Stokes equations in general curvilinear coordinates using a block structured finite volume approach. EllipSys3D is formulated in primitive variables (pressure-velocity) in a collocated grid arrangement. Additional details can for instance be found in Mikkelsen [7] and Troldborg [8].

2.1.2. PALM PALM (Parallelized LES Model) is an open source highly parallelized LES model which has been applied for the simulation of a variety of atmospheric and oceanic boundary layers for more than 15 years. PALM solves the filtered, incompressible, non-hydrostatic Navier-Stokes equations under the Boussinesq approximation on an equidistant Cartesian grid. The sub-grid scale turbulence is parameterized by a 1.5th order closure after Deardorff [9]. Further details can be found in Maronga et al. [6]. Recently, it has been adapted by FW for the simulation of wind turbine wakes, see Witha et al. [10].

2.2. Turbine Modeling

2.2.1. EllipSys3D The wind turbines are modeled using either the AL (by DTU) or AD (by UU) method where body forces are distributed along rotating lines or on a disc, respectively, see Mikkelsen [7] and Troldborg et al. [11] as well as Sørensen and Shen [12] for details on the EllipSys3D implementation of the AL and AD. Tabularized airfoil data is used to determine the local blade forces based on the local inflow conditions. In the DTU-AL model, the 2D airfoil data is corrected to account for 3D effects, see e.g. Hansen et al. [13]. Furthermore, the body forces in the DTU implementation are calculated through a full coupling with Flex5, a full aeroelastic code for calculating deflections and loads on wind turbines, see Øye [14] for details on Flex5 and Sørensen et al. [15] for a description of the coupling.

[1] `http://www.ieawind.org/task_31.html`

2.2.2. PALM The PALM implementation contains two different turbine models. The standard Actuator Disc model with uniform force distribution (FW-AD-U) uses an aerodynamic thrust curve and the velocity averaged over the rotor disc together with an axial induction factor to determine the thrust force which is the same all over the rotor plane. The second model is an enhanced AD model with rotation (FW-AD-R) in which local body forces are calculated by means of BEM theory. In contrast to the AL method, the forces are distributed across the rotor plane. Both FW-AD-U and FW-AD-R include tower and nacelle effects modeled by a drag force approach. See Dörenkämper et al. [16] for details of the PALM implementation.

2.2.3. Turbine Controller Three of the four models (DTU-AL, UU-AD and FW-AD-R) include a turbine controller which effectively means that the applied body forces are governed by the inflow conditions, i.e. the turbines are not constantly loaded, but operate as "real turbines". A general description of such controllers can be found in Larsen and Hanson [17] or Hansen et al. [18].

2.2.4. Turbine Data Two different turbines have been used: the NM80 and the NREL 5MW. Both are three bladed horizontal axis wind turbines. The NM80 turbine is proprietary to Vestas Wind Systems A/S, but reference is made to the DAN-AERO MW Experiments, see Madsen et al. [19]. NM80 has a radius $R = 40$ m and is rated to 2.75 MW at a hub height velocity of $U_{hub} = 14$ ms^{-1}. The NREL 5MW has a radius of $R = 63$ m, a hub height of $z_{hub} = 90$ m and is rated to 5 MW at $U_{hub} = 11.4$ ms^{-1}, see Jonkman et al. [20].

Table 1 summarizes the different methods used by the participants.

3. Simulations Setup

The coordinate system used in this work is defined such that x, y and z correspond to the streamwise, crosswise and vertical directions, respectively. The grids used the simulation are equidistant in the horizontal directions. The grid used in the DTU simulation is also vertically uniform whereas the grids used in the FW and UU simulations are stretched in the vertical direction from a height far above the wind farm (FW: 9.5 R, UU: 8 R). Further details of the setup are given in Table 1 and in the text below.

3.1. Atmospheric Boundary Layer and Turbulence

All participants simulated a neutrally stratified atmospheric boundary layer and aimed at obtaining an average streamwise inflow velocity of $U_0 = 8$ ms^{-1} and an average inflow turbulence intensity (TI, streamwise component) of $\sigma_u/U_0 = 15\%$, both at hub height.

3.1.1. EllipSys3D EllipSys3D uses the prescribed boundary layer method (PBL) which utilizes body forces to achieve any arbitrary vertical wind shear profile, see Mikkelsen et al. [21] and Troldborg et al. [22]. A comparison of PBL with a wall model approach was performed by Sarlak et al. [23] which showed that the two approaches yield very comparable vertical profiles of mean streamwise velocity, shear stress, and streamwise velocity fluctuations in the rotor region when modeling large wind farms. The modeling of the ambient turbulence is performed by introducing pregenerated synthetic ambient turbulence. The ambient turbulence is generated using the Mann model, [24]. Turbulence planes are imposed at an axial position of 6 R and 13 R for the DTU and UU simulations, while the first simulated turbine is located at 10 R and 30 R, respectively.

3.1.2. PALM The atmospheric boundary layer is modeled in PALM using a no-slip bottom boundary condition and Monin-Obukhov similarity theory between the surface and the first grid level. Initially, random perturbations are imposed on the velocity fields until atmospheric turbulence has developed in a precursor simulation, which is performed on a smaller domain and with periodic boundary conditions in streamwise and lateral directions. The precursor results are used to initialize the full simulations with non-periodic boundary conditions in the streamwise direction and additionally turbulence recycling was applied, see Maronga et al. [6] for details.

3.2. Wind Farm
A very large wind farm has been simulated, containing 16 turbines in the streamwise direction and an infinite number of turbines in the lateral direction (by applying periodic boundary conditions). Two different streamwise turbine spacings of $S_x = 12\ R$ and $S_x = 20\ R$ have been simulated by all participants (see Table 2). The lateral spacing is $S_y = 20\ R$ for both simulations.

Table 1. Summary of methods and simulations by DTU and FW.

Method	*DTU-AL*	*FW-AD-U*	*FW-AD-R*	*UU-AD*
CFD Solver	EllipSys3D	PALM	PALM	EllipSys3D
Domain Size [R]	200/320 × 20 × 20	343/533 × 38 × 35	343/533 × 38 × 35	374 × 20 × 40
Equiv. roughness [m]	N/A	1 m	1 m	0.5 m
Turbine	NM80	NREL 5MW	NREL 5MW	NREL 5MW
R	40 m	63 m	63 m	63 m
z_{hub}	80 m	90 m	90 m	90 m
Controller	Yes	No	Yes	Yes
3D effects correction	Yes	No	No	No
Resolution [R]	0.0625	0.127	0.127	0.0781

4. Results and Discussion
All datasets have been interpolated in time to have a sampling rate of 1 Hz and all raw data were post-processed in the same way. The data have been sampled after running the models the equivalent time of at least two flow-throughs in the domain in order to avoid any transients from the simulation startup. Different types of data will be presented: instantaneous data, averages of the entire sampling period (as given in Table 2), as well as 10 min averages. The 10 min averages are computed within the total sampling period by shifting the averaging window by 1 min to get a larger number of samples[2].

4.1. Inflow Conditions
Despite aiming at setting up and simulating exactly the same case, the different methodologies and codes result in different inflows. The temporally averaged streamwise velocity and streamwise TI profiles extracted 1 R upstream the first turbine are shown in Figure 1. TI is here defined as σ_u/U_0. The vertical profiles have been shifted to be aligned at hub height, which is marked by dashed lines. The velocity for the FW simulations is slightly less than for the DTU simulation. At the same time, the vertical shear is larger, which could be attributed

[2] *i.e.* the 10 min averages are not independent in a statistical sense

Table 2. Overview of simulations by DTU and FW with the time periods (from the end of the simulation) used for evaluation.

Method	*DTU-AL*	*FW-AD-U*	*FW-AD-R*	*UU-AD*
$S_x = 12\ R$	60 min	60 min	60 min	30 min
$S_x = 20\ R$	30 min	60 min	60 min	30 min

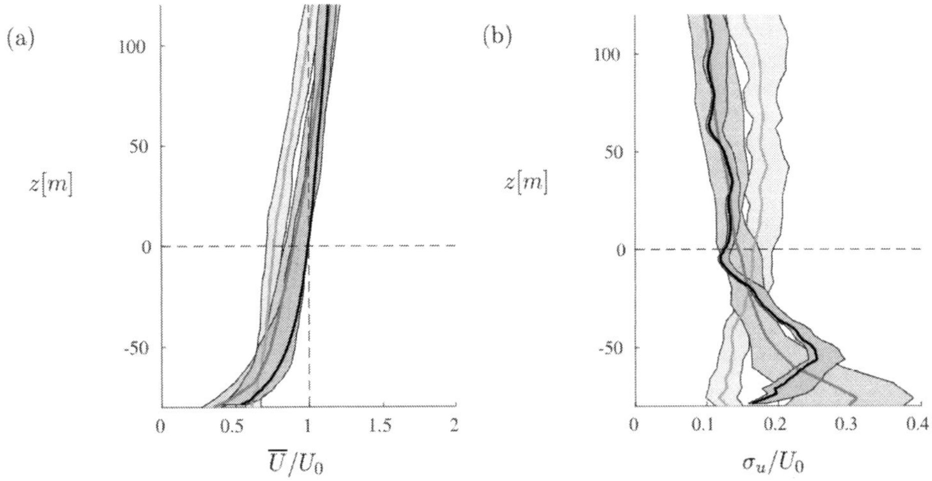

Figure 1. Vertical profiles of a) average streamwise velocity and b) streamwise fluctuations extracted 1 R upstream the first turbine. The profiles are shifted to have hub height at $y = 0$ to enable a direct comparison. Black: DTU-AL. Red: FW-AD-R. Green: UU-AD. The thick lines denote the temporal average over all 10 min averages within the evaluation period given in Table 2. The shaded areas give the maximum and minimum of the 10 min averages.

to the large effective roughness which had to be used in the FW simulations to reach the desired TI. The UU simulation shows a lower velocity than both the FW and DTU simulations, except close to the ground, where it finds itself somewhere between FW and DTU. The lower inflow velocities in the FW and UU simulations can be attributed to a longer induction zone of the first turbine compared to the DTU simulation, so that 1 R upstream, the FW and UU flow is already affected by the first turbine. See section 4.2 for further elaboration. Above hub height, the FW and DTU TI profiles are very similar whereas they differ to some extent below hub height. The UU profile is considerably different from the two others and does not follow the typical shape of atmospheric boundary layer turbulence with increasing turbulence towards the ground. Whereas the TI is higher around and above hub height, it is lower near the ground, compared to the other simulations. The observed differences could be explained by the different approaches used to generate turbulence, which highlights how a direct comparison can be very difficult, when different methods are compared. The turbulence in the DTU simulations is scaled to fit measured turbulence close to the ground, which makes it quite similar to the FW profiles. Such a scaling was not performed in the UU simulations, explaining the differences obtained close to the ground.

4.2. Average Velocities and Turbulent Fluctuations.

The average streamwise velocity at hub height through the wind farm is compared in Figure 2. The data has been averaged over the entire time period given in Table 2. Despite the different inflow conditions, all simulations follow a very comparable pattern, almost sinusoidal, and it is evident how the average streamwise velocity very quickly converges to a constant level between the turbines[3], at least for the DTU and FW simulations. The UU-AD results are less constant and appear to peak around the 11th turbine for $S_x = 12\ R$ and the 8th turbine for $S_x = 20\ R$. Generally, the average streamwise velocity has reached a converged level already at the fourth turbine, after which the streamwise velocity trend does not vary significantly. The dashed lines in Figure 2 indicates the mean of the average streamwise velocity 1 R upstream the turbines after the fourth turbine. The EllipSys3D (DTU and UU) results systematically show higher velocities compared to the results provided by FW for both spacings. This holds especially true for the wake of the first two turbines. The two FW simulations provide very similar results, although there is a tendency of higher amplitude between the "wave crest" and "trough" between each set of turbines with the standard AD-U.

The converged velocity or asymptotic wake deficit is often denoted as the "infinite wind farm" scenario. Several low fidelity models estimate the converged velocity ratio in an "infinite wind farm", e.g. N. O. Jensen [25] and Frandsen et al. [26]. These simple engineering models are independent of rotor details (except the thrust coefficient), the vertical shear and TI of the inflow. Hence, the present simulations could be used to scale the model parameters, similar to what was done by Andersen [27] for infinite wind farm simulations. The converged streamwise velocity extracted 1 R upstream the turbines has been averaged from the fourth to the 16th turbine and plotted with a dashed line in Figure 2 and summarized in Table 3. The converged velocity ratios can be compared to velocity ratios obtained by Andersen [2], who reported $\overline{U}/U_0 = 0.8$ for turbine spacings of $S_x = 20\ R$, where the "infinite wind farm" was achieved by applying cyclic boundary conditions. The simulations used an inflow velocity of 15 ms^{-1}, no added turbulence and a smaller vertical shear, but as seen the converged result is the same. It has to be noted, however, that the internal wind farm boundary layer (IWBL) above the turbines is continuously growing until the last of the 16 simulated turbines. So, although the velocity deficit converges very quickly after three or four turbines (for most cases), a true "infinite wind farm" scenario would require many more turbines. Tests by FW with 50 turbine rows have shown that the IWBL is still (slowly) increasing until the last row.

It is also interesting to note the difference in inflow to the first turbine as previously shown in Figure 1. The difference in inflow velocity at 1 R is here seen to be related to the induction upstream the first turbine. The apparent large difference in Figure 1 merely stems from the steep slope around 1 R. Further upstream the inflow is comparable. The flow at 1 R upstream each turbine is taken as representative of the inflow to each turbine inside the farm. The length of the induction zone in front of each turbine can also be estimated based on the distance from the local maximum velocity in front of each turbine to the turbine itself, i.e. using the fact that the wake velocity is no longer recovering from the latter position due to the adverse pressure gradient from the wind turbines. This simple analysis yields very similar estimates of the induction zone from the different simulations (see Table 3), although it is noteworthy how the length of the induction zone increases from approximately 2 R to 3 R as the spacing between the turbines increases from $S_x = 12\ R$ to $S_x = 20\ R$. This is attributed to the stronger wake recovery for the larger spacing, i.e. the wake is less dominant and hence more sensitive to the influence of the following turbine. A more detailed investigation of the wind evolution within the induction zone has previously been performed by Simley et al. [28].

The streamwise turbulent fluctuations, as shown in Figure 3 are initially higher in the DTU

[3] When speaking of convergence we mean the velocity at the same relative position between each set of turbines, e.g. the values of the relative maxima and minima behind each turbine

(a)

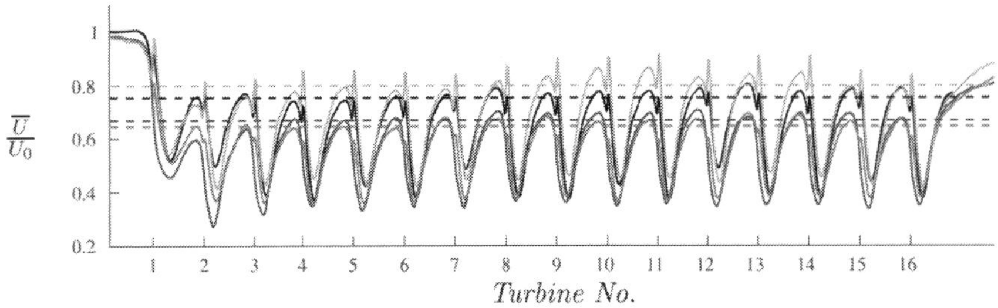

(b)

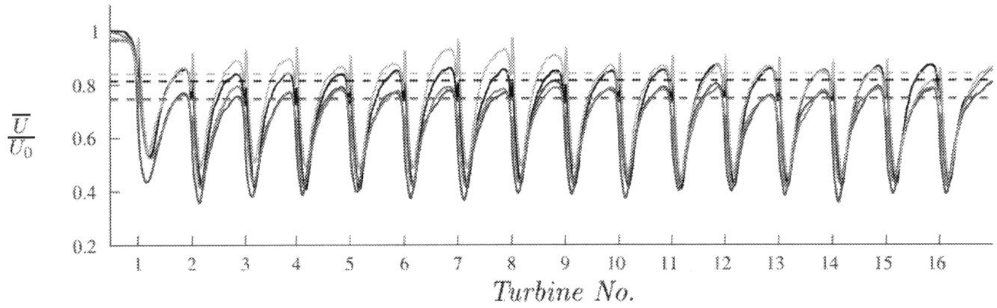

Figure 2. Mean streamwise velocity through the wind farms at hub height for different spacings: a) $S_x = 12\ R$. b) $S_x = 20\ R$. Black: DTU-AL. Blue: FW-AD-U. Red: FW-AD-R. Green: UU-AD.

Table 3. Converged velocity ratios $\overline{U}/U_0$ and length of induction zone Δ_I for very large wind farms for all simulations.

		DTU-AL	*FW-AD-U*	*FW-AD-R*	*UU-AD*
$S_x = 12\ R$	$\overline{U}/U_0$	0.71	0.67	0.65	0.80
	Δ_I	-2.01 R	-2.06 R	-2.20 R	-2.06 R
$S_x = 20\ R$	$\overline{U}/U_0$	0.80	0.75	0.75	0.84
	Δ_I	-3.16 R	-3.10 R	-3.04 R	-3.23 R

and UU simulations, leading to a significantly higher velocity behind the first two turbines, but generally and consistently the streamwise turbulent fluctuations of the FW and UU simulations exceed those found in the DTU simulations after the third turbine. This is particularly the case for the FW-AD-R with higher TI between the turbines, but lower in the rotor plane compared with the same simulations with AD-U. The higher TI in the FW simulations could possibly be explained by the substantially higher TI in the lower rotor plane (see Figure 1) which is lifted up by mixing and rotation of the wake. This explanation is however not supported by the fact that the UU simulation, which features a very low TI below hub height, shows a similarly high TI as the FW-AD-R simulation in Fig. 3. Furthermore, a consistently higher turbulence intensity should intuitively result in greater wake recovery, hence increasing streamwise velocities further into the farm. However, this is not the case as the velocity maxima between the turbines are nearly constant, see Figure 2.

So far only long-time averages have been considered. For a better assessment of the dynamics and variability of the different LES, distributions of instantaneous velocity and wind direction

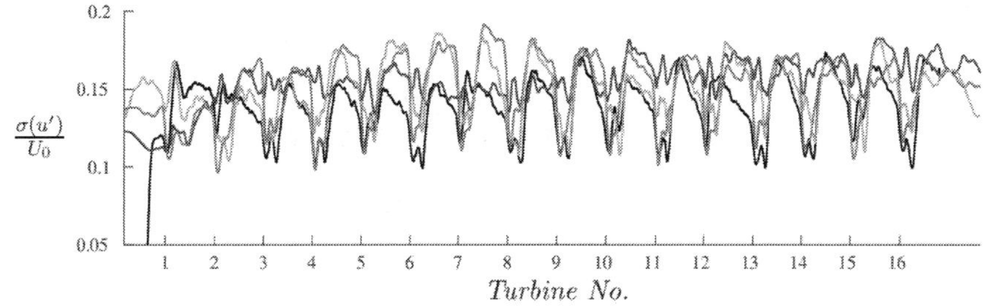

Figure 3. Mean streamwise TI through the wind farms at hub height for $S_x = 12\ R$. Black: DTU-AL. Blue: FW-AD-U. Red: FW-AD-R. Green: UU-AD.

will be investigated in the following section.

4.3. Velocity Distributions

Figure 4 shows box plots of the instantaneous (1 Hz) streamwise velocities extracted 1 R upstream of each of the turbines. A box plot is a simple way of depicting the variability of a data and is drawn by horizontal lines defining the first quartile q_1, median, and third quartile q_3, as well as vertical lines. The line extends to $q_3 \pm w(q_3 - q_1)$, where $w = 1.5$. This would correspond to $\pm 2.7\sigma$ equivalent to 99.3% coverage if the data is normally distributed. Finally, any outliers outside this extent are marked by crosses.

As expected, the median level follows the same trend as the long term averages. Generally, the variability is comparable, though slightly smaller for the DTU results, seen by comparing the box extent between the first and third quartile. The first and third quartile span a maximum variability of 21.5%, 24.4%, 25.0%, and 27.2% on the instantaneous axial velocity extracted from the simulations by DTU-AL, FW-AD-U, FW-AD-R, and UU-AD, respectively. For the larger turbine spacing, the variability is slightly reduced to 18.7%, 22.2%, 24.7%, and 23.1% for the four datasets. There are also more outliers in the UU results, which again indicates a larger variability. The outliers constitute approximately 0.33%, 0.85%, 1.45%, and 2.92% of the instantaneous axial velocities for $S_x = 12\ R$ from DTU-AL, FW-AD-U, FW-AD-R and UU-AD, respectively. Similar values are obtained for $S_x = 20\ R$.

The averages shown for the streamwise velocity and fluctuations are averaged over long periods (30-60 min). Within wind energy and atmospheric studies, it is standard to report 10 min averages, which is also often set as criteria for blind test and benchmarks. Hence, Figure 5 depicts box plots based on 10 min averages of the streamwise velocity. Averaging reduces the average span between the first and third quartile to 6.2%, 9.4%, 9.1%, and 5.0% for the four datasets for $S_x = 12\ R$. Interestingly, the variability is reduced relatively more in the UU-AD results compared to the other simulations. Similar variability is seen for $S_x = 20\ R$, but left out for brevity.

4.4. Directional Variation

The variability is not only temporal, but also spatial or directional. Figure 6 shows the distributions of wind direction based on the instantaneous horizontal velocity components for the first and the ninth turbine with spacing $S_x = 12\ R$. The ninth turbine is shown, because the box plots are most similar for this particular turbine. The directional variability of the inflow to the first turbine is very similar and very symmetric about 270° for the DTU and FW-AD-U results. The case FW-AD-R is not as symmetric. The wind is from $270° \pm 2.5°$ in 30% of the time in all these three simulations, and from the adjacent sectors of $265 \pm 2.5°$ and $275 \pm 2.5°$

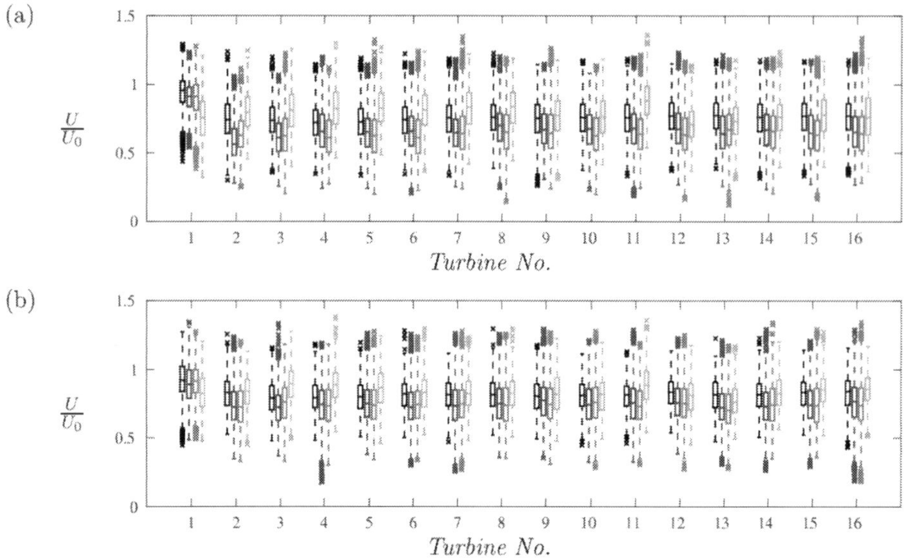

Figure 4. Box plots comparing instantaneous (1 Hz) streamwise velocity distributions extracted 1 R upstream of each of the turbines. a) S_x = 12 R. b) S_x = 20 R. Black: DTU-AL. Blue: FW-AD-U. Red: FW-AD-R. Green: UU-AD.

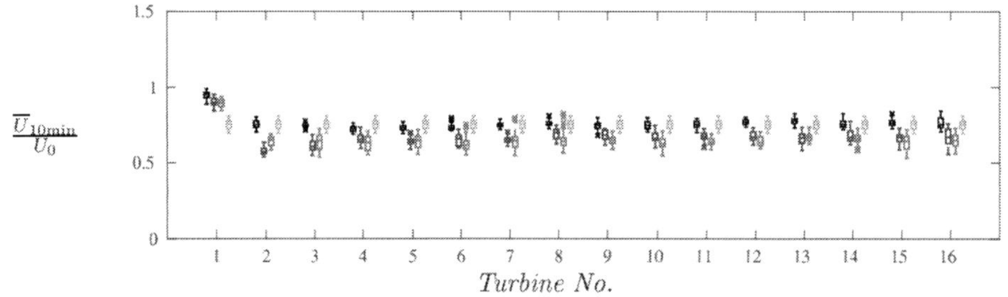

Figure 5. Box plots distributions of 10 min averages of streamwise velocities extracted 1 R upstream of each of the turbines for S_x = 12 R. Black: DTU-AL. Blue: FW-AD-U. Red: FW-AD-R. Green: UU-AD.

approximately 10% of the time, except for FW-AD-R with slight bias to higher degrees. In this case, the inflow is not exactly parallel to the streamwise direction, because the FW simulations include the Coriolis force, thus the wind is turning with height. It is not trivial to find the suitable prescribed wind speed to obtain a flow exactly aligned with the x-axis at hub height. Methods have been proposed to adjust the geostrophic wind during the simulation to obtain an aligned flow (e.g. by Sescu and Meneveau [29]) but not been implemented in the PALM code, so far. Apart from the very slight differences in wind direction, not modeling the Coriolis force in the DTU and UU simulations is not expected to affect the presented results. As previously seen, the DTU results show a slightly higher velocity. The UU-AD results are seen to be quite different from the three other cases, as the directional variability around the streamwise direction is much larger upstream of the first turbine. This is stems from the direct application of the unscaled Mann turbulence. The DTU simulations also apply Mann turbulence, but this has been scaled

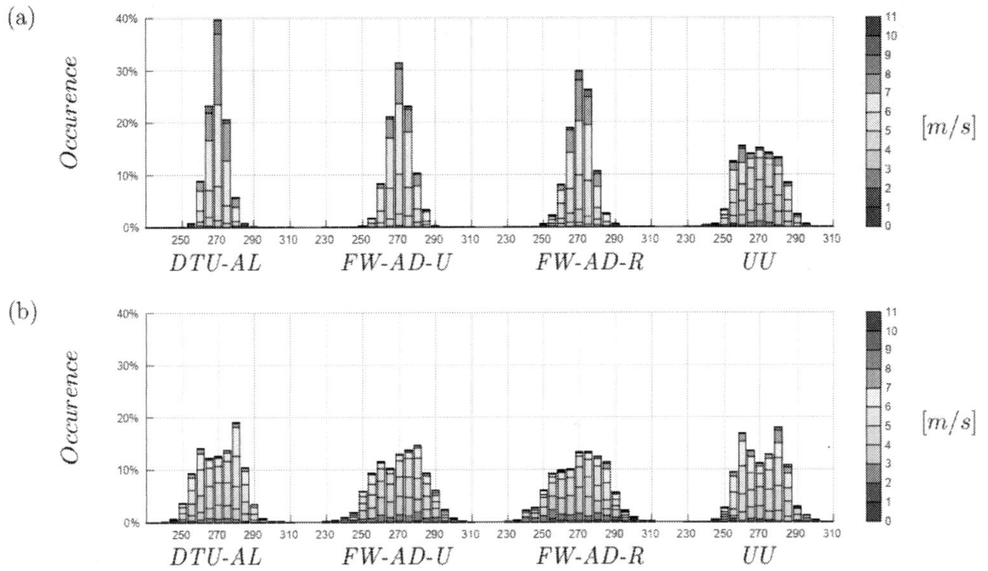

Figure 6. Directional variability of horizontal velocities (U and V components) extracted 1 R upstream the first and ninth turbine for $S_x = 12\ R$. 270 degrees corresponds to completely aligned inflow.

to fit measure turbulence statistics as mentioned in 4.1, which yields a decrease of the spanwise fluctuations close to the ground.

The directional variability for the ninth turbine is quite comparable between the four simulations. However, the distribution is considerably wider than for the first turbine ($\approx 270^\circ \pm 30^\circ$) and it is noteworthy how the dominant direction is shifted towards $280^\circ \pm 2.5^\circ$ for the DTU and FW simulations. The UU simulation also has a peak here, but is more symmetric with a second minor peak at $260^\circ \pm 2.5^\circ$. Such a minor peak is also present in the DTU results, although the shifted distribution is more comparable to the FW results.

This change in dominant direction is not as clear for the larger spacing $S_x = 20\ R$ (not shown).

5. Conclusions

A total of eight datasets derived from LES simulations of very large wind farms with two different turbine spacings have been compared using four different model setups. Despite differences in model setup and inflow conditions, the simulations yielded a number of comparable results. At least three of the four simulations showed that the "infinitely large wind farm" in terms of average streamwise velocity at hub height (at the same relative position between each set of turbines) was achieved by the fourth turbine, although the simulations resulted in slightly different velocity levels. The streamwise fluctuations were comparable, though one simulation showed slightly lower values compared to the other simulations. The temporal variability of streamwise velocity, both instantaneous and 10 min averages, was also very comparable. However, the variability analysis showed that there could still be significant differences between individual 10 min averages from the same simulation despite averaging. Hence, longer simulation times and longer averages are required to enable reporting such variability for comparing in e.g. benchmarks. Finally, the directional variability was examined through wind distribution plots, which were also equivalent in most cases despite the model differences.

Examining the variability of LES simulations could bridge the gap between measurements and simulations within very large wind farms. As the results show both temporal and directional variability, the variability could potentially be compared with the variability found in measurements, whereas the variability is usually filtered out in order to compare measurements and simulations.

Acknowledgments

The work has been carried out with the support of the Danish Council for Strategic Research for the project *Center for Computational Wind Turbine Aerodynamics and Atmospheric Turbulence* (grant 2104-09-067216/DSF) (COMWIND: http://www.comwind.org), in the framework of the project *Parallelrechner-Cluster für CFD und WEA-Modellierung* funded by the German Federal Ministry for Economic Affairs and Energy (FKZ: 0325220), as part of the EUROTECH Wind project, and with support of the Nordic Consortium on Optimization and Control of Wind Farms funded by the Swedish Energy Agency. Computer time was granted by the Swedish National Infrastructure for Computing (SNIC) and by the North German Supercomputing Alliance (HLRN). Furthermore, the proprietary data for Vestas' NM80 turbine has been used.

References

[1] Bechmann A, Sørensen N, Berg J, Mann J and Réthoré P 2011 The Bolund Experiment. Part II: Blind Comparison of Microscale Flow Models *Boundary-Layer Meteorol.* **141** 245–271 ISSN 0006-8314

[2] Andersen S J 2013 *Simulation and Prediction of Wakes and Wake Interaction in Wind Farms* Ph.D. thesis Technical University of Denmark, Wind Energy.

[3] Michelsen J A 1992 Basis3D – a Platform for Development of Multiblock PDE Solvers Tech. rep. Danmarks Tekniske Universitet

[4] Sørensen N N 1995 *General Purpose Flow Solver Applied to Flow over Hills* Ph.D. thesis Technical University of Denmark

[5] Raasch S and Schröter M 2001 PALM - A large-eddy simulation model performing on massively parallel computers *Meteorol. Z.* **10** 363–372

[6] Maronga B, Gryschka M, Heinze R, Hoffmann F, Kanani-Sühring F, Keck M, Ketelsen K, Letzel M O, Sühring M and Raasch S 2015 The Parallelized Large-Eddy Simulation Model (PALM) version 4.0 for atmospheric and oceanic flows: model formulation, recent developments, and future perspectives *Geosci. Model Dev. Discuss.* **8** 1539–1637

[7] Mikkelsen R 2003 *Actuator Disc Methods Applied to Wind Turbines* Ph.D. thesis Technical University of Denmark, Mek dept

[8] Troldborg N 2008 *Actuator Line Modeling of Wind Turbine Wakes* Ph.D. thesis DTU Mechanical Engineering, Technical University of Denmark, DTU, DK-2800 Kgs. Lyngby, Denmark Denmark

[9] Deardorff J 1980 Stratocumulus-capped mixed layers derived from a three-dimensional model *Boundary-Layer Meteorol.* **18** 495–527

[10] Witha B, Steinfeld G and Heinemann D 2014 *Wind Energy - Impact of Turbulence* Research Topics in Wind Energy 2 ed Hölling M, Peinke J and Ivanell S (Springer Berlin Heidelberg) pp 175–181

[11] Troldborg N, Gaunaa M and Mikkelsen R 2010 Actuator Disk Simulations on Influence of Wind Shear on Power Production of Wind Turbines *Proceedings of the V European Conference on Computational Fluid Dynamics, 14-17 June 2010, Lisbon, Portugal*

[12] Sørensen J N and Shen W Z 2002 Numerical modelling of Wind Turbine Wakes *J. Fluids Eng.* **124** 393–399

[13] Hansen M O L, Sørensen J N, Voutsinas S, Sørensen N and Madsen 2006 State of the art in wind turbine aerodynamics and aeroelasticity *Prog. Aerosp. Sci.* **42** 285–330

[14] Øye S 11-12 April 1996 Flex4 simulation of wind turbine dynamics (Lyngby, Denmark: Danmarks Tekniske Universitet) pp 71–76

[15] Sørensen J, Mikkelsen R, Henningson D, Ivanell S, Sarmast S and Andersen S 2015 Simulation of wind turbine wakes using the actuator line technique *Phil. Trans. R. Soc. A* **373** ISSN 1364-503X

[16] Dörenkämper M, Witha B, Steinfeld G, Heinemann D and Kühn M 2015 The impact of stable atmospheric boundary layers on wind-turbine wakes within offshore wind farms *J. Wind Eng. Ind. Aerodyn.* Accepted for publication

[17] Larsen T and Hanson T 2007 A method to avoid negative damped low frequent tower vibrations for a floating, pitch controlled wind turbine *J. Phys.: Conf. Ser.* **75** 11

[18] Hansen M H, Hansen A, Larsen T J, Oye S, Sorensen P and Fuglsang P 2005 *Control design for a pitch-regulated, variable speed wind turbine* Denmark. Forskningscenter Risœ. Risœ-R (Danmarks Tekniske Universitet, Risø Nationallaboratoriet for Bæredygtig Energi)

[19] Aagaard Madsen H, Bak C, Schmidt Paulsen U, Gaunaa M, Fuglsang P, Romblad J, Olesen N, Enevoldsen P, Laursen J and Jensen L 2010 *The DAN-AERO MW Experiments* Denmark. Forskningscenter Risø. Risø-R (Danmarks Tekniske Universitet, Risø Nationallaboratoriet for Bæredygtig Energi)

[20] Jonkman J, Butterfield S, Musial W and Scott G 2009 Definition of a 5-MW Reference Wind Turbine for Offshore System Development URL `http://www.nrel.gov/wind/pdfs/38060.pdf`

[21] Mikkelsen R, Sørensen J and Troldborg N 2007 Prescribed wind shear modelling with the actuator line technique *EWEC*

[22] Troldborg N, Sørensen J N, Mikkelsen R and Sørensen N N 2014 A simple atmospheric boundary layer model applied to large eddy simulations of wind turbine wakes *Wind Energy* **17**

[23] Sarlak H, Mikkelsen R and Sørensen J 2015 Comparing wall modeled LES and prescribed boundary layer approach in infinite wind farm simulations *Proc. of 33rd Wind Energy Symposium* (AIAA)

[24] Mann J 1998 Wind field simulation *Probabilistic Engineering Mechanics* **13** 269–282 ISSN 0266-8920

[25] Jensen N O 1983 A note on wind generator interaction

[26] Frandsen S, Barthelmie R, Pryor S, Rathmann O, Larsen S, Højstrup J and Thøgersen M 2006 Analytical modelling of wind speed deficit in large offshore wind farms *Wind Energy* **9** 39–53

[27] Andersen S J, Sørensen J N, Ivanell S and Mikkelsen R F 2014 Comparison of Engineering Wake Models with CFD Simulations *J. Phys.: Conf. Ser.* **524** 012161

[28] Simley E, Pao L Y, Gebraad P and Churchfield M 2014 Investigation of the Impact of the Upstream Induction Zone on LIDAR Measurement Accuracy for Wind Turbine Control Applications using Large-Eddy Simulation *J. Phys.: Conf. Ser.* **524** 012003

[29] Sescu A and Meneveau C 2014 A control algorithm for statistically stationary large-eddy simulations of thermally stratified boundary layers *Quart. J. R. Met. Soc.* **140** 2017–2022

Wake downstream of the Lillgrund wind farm - A Comparison between LES using the actuator disc method and a Wind farm Parametrization in WRF

O Eriksson[1], J Lindvall[2], S-P Breton[1], S Ivanell[1]

[1]Uppsala University, Department of Earth Sciences, Wind Energy Campus Gotland, 621 67 Visby, Sweden

[2]Kjeller Vindteknikk, Luntmakargatan 22, 111 37 Stockholm, Sweden

E-mail: ola.eriksson@geo.uu.se

Abstract.

Simulations of the Lillgrund wind farm (located between Malmö and Copenhagen) are performed using both Large Eddy Simulation (LES) and mesoscale simulations in WRF. The aim is to obtain a better understanding of wakes generated by entire wind farms in order to improve the understanding of farm to farm interactions. The study compares the results from the two used models for the energy production and the wake characteristics downstream of the wind farm. A comparison is also performed with regards to the production data from the Lillgrund wind farm which has been filtered to be comparable to the case used in the simulations.The studied case, based on a prerun in WRF without any wind farm, has an inflow angle of 222 ± 2.5 deg, a wind speed at hub height of 9.8 m/s and a near neutral atmosphere. A logarithmic wind shear is used in LES and the turbulence intensity is 5.9%.

The WRF simulations use a parameterization for wind farms. The wind farm is treated by the model as a sink of the resolved atmospheric momentum. The total energy extraction and the electrical power are respectively proportional to specified thrust and power coefficients. The generated turbulent kinetic energy are the difference between the total and the electrical power.

The LES are performed using the EllipSys3D code applying the actuator disc methodology for representing the presence of the rotors. Synthetic atmospheric turbulence is generated with the Mann model. Both the atmospheric turbulence and the wind shear are introduced using body forces.

The production was found to be better estimated in LES. WRF show a slightly higher recovery behind the farm. The internal boundary layer is for the compared simulation setups higher in LES while the wake expansion is about the same in both models. The results from the WRF parameterization could potentially be improved by increasing the grid resolution. For farm to farm interaction a combination of the two methods is found to be of interest.

1. Introduction

More and larger wind farms are planned offshore in Europe. One of the main reasons for this development is the good wind conditions offshore. The number of highly suitable sites for offshore wind farms is however limited by factors such as water depth and distance from shore. As more offshore wind farms are built there will be more occasions when the wake from one wind farm will interact with other close-standing wind farms. This expected development requires an increased understanding of the so-called farm to farm interaction in order to ensure a correct estimation of production and loads for future wind farms. It is therefore necessary to not only study the near and far wakes behind single turbines and the interaction inside farms but also the long distance wakes impacting the wind conditions at neighboring sites.

It is known that wind farms produce long distance wakes [1]. The scale of the long distance wakes is in the order of 10 km [2]. Different types of wake and mesoscale models can be applied to simulate wind farm wakes [1][3]. Even Large Eddy Simulations (LES) are becoming feasible due to the increasing computational resources. LES have been used for the prediction of wakes inside wind farms in a wide range of studies [4][5][6][7][8][9][10][11]. LES with an actuator disc method was used to study the velocity recovery behind an infinitely wide wind farm [12].

Mesoscale models are used for atmospheric simulations and can by including parameterization of wind turbines [13] also be used to study the wakes behind farms.

LES and mesoscale models can be combined in two ways. The mesoscale output can be used as an input for the LES (1-way nesting) or the results obtained from this procedure can then even be fed back into the mesoscale simulation (2-way nesting) [14]. By combining the models one can take advantage of the relatively coarse grid in the mesoscale simulations which gives a good description of the atmosphere and the finer grid in LES which can resolve the wake flow.

In the current study the same case is studied both in the mesoscale model WRF (Weather Research and Forecasting Model) and in LES using an actuator disc method. This enables a comparison between the two models regarding the calculated energy production (also compared to wind farm data), the recovery of the flow behind the farm, the impact of the boundary layer and the expansion of the wake. The aim is to obtain a better understanding of wakes generated by entire wind farms in order to improve the understanding of farm to farm interactions.

2. Study of wakes in and behind the Lillgrund wind farm

The object of the simulations is the Swedish wind farm Lillgrund which is located between Malmö and Copenhagen. The wind farm consists of 48 Siemens wind turbines with a hub height of 65 m, a rotor radius (R) of 93 m and a rated power of 2.3 MW. A description of the Lillgrund wind farm is given by Nilsson [11] and in other earlier studies [15],[16]. Figure 1 displays the layout of the farm. The wake behind the farm is studied up to 7 km downstream.

One suitable atmospheric case with near neutral atmosphere and a relatively stable wind (regarding wind speed and direction) is found from presimulations without the wind farm in WRF. The studied wind direction is 222 deg +-2.5 deg where the wind is aligned with the rows. The wind shear determined from WRF along with the shear used in LES are shown in Figure 2. The turbulence intensity from WRF is 5.9% and the wind speed at hub height (U_0) is 9.82 m/s.

This case is then run in LES using an actuator disc method (see Section 2.1) and respectively in WRF with the wind farm parameterization (see Section 2.2). The production data from the site is filtered to be comparable to the simulated conditions (see Section 2.3). The comparison of the simulation results and measured production is presented in Section 3.

2.1. LES

Large Eddy Simulations (LES) resolve the largest most energetic eddies, while the smallest eddies are modeled using a subgrid-scale model. They can therefore be used to get a good description of the wake flow behind the wind turbines.

2.1.1. Numerical model The EllipSys3D code, non dimensionalized with R and U is used to carry out LES where the turbines are modeled according to an actuator disc method based on airfoil data [17]. The rotational speed of the turbines is individually controlled by a generator-torque algorithm in order to ensure a realistic and production optimized operation of every turbine throughout the simulation [18]. The subgrid-scale model from Ta-Phouc[19] is used for the small modeled eddies. The Mann model is used to pregenerate a realistic atmospheric

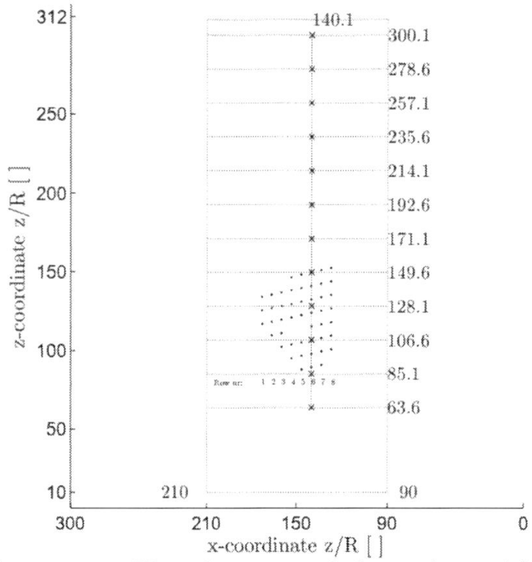

Figure 1. The placement of the turbines (•) in the domain covering 300 R * 322 R with the marked equidistant region of 120 R * 300 R. The flow is studied along the marked lines and for vertical profiles at the x's.

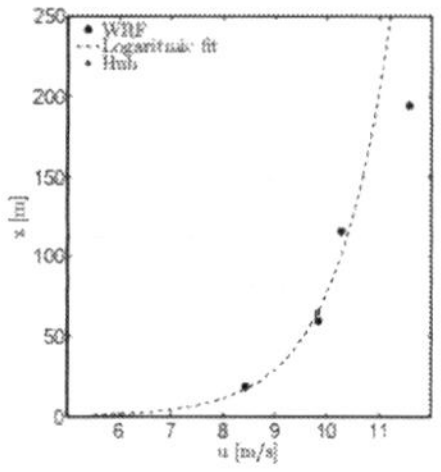

Figure 2. Logarithmic fit to the 3 lowest WRF-levels.

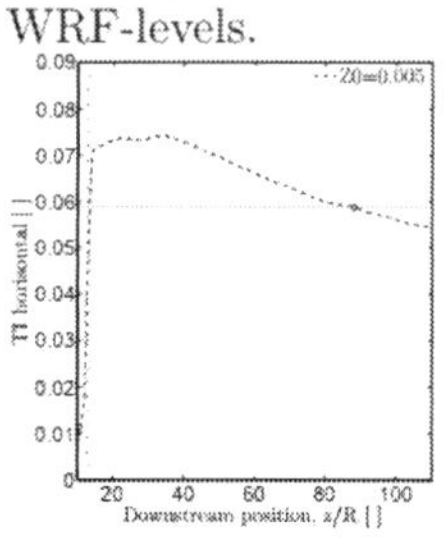

Figure 3. Downstream development of horizontal turbulence intensity (TI) at hub height (spanwise mean over 10 R).

turbulence [20][21]. The wind shear as well as the turbulence are imposed on the computational domain using body forces[10]. The model setup is considering a neutral boundary layer.

2.1.2. Simulation setup The simulations are performed for an area of 300 R in the spanwise/lateral direction (x) and 322 R in the streamwise/axial direction (z). This allows the wake to be studied to an extension of about 7 km behind the farm. In the region of interest the domain is equidistant. In Figure 1 the grid extensions, the equidistant area and the placement of the turbines can be seen. The farm is turned in such a way that a flow parallel to the z-axis corresponds to a wind direction of 222 degrees. The height (y-direction) of the equidistant region is 7.5 R (350m). The grid is stretched in the inlet (10 R), the outlet (10 R), towards the respectively side (90 R) and towards the top (up to the height of 50 R). This gives an equidistant region with 120 R width, 7.5 R height and 300 R length. The boundary conditions are fixed values for the inlet (according the wanted wind shear), cyclic for the sides, convective for the outlet, far field for the top and for the ground. The far field for the ground is setup to function as a noslip condition. The resolution (dx) in the equidistant region is 0.1 R (4.65 m) and in order to fulfill the CFL condition in a conservative way a non dimensionalized (with U_0 and R) timestep of 0.025 is used. The grid consists of 400 M cells.

The wind shear used is calculated using a logarithmic fit to the three lowest grid points in WRF, to get the most comparable shear at the turbine height, see Figure 2. The profile that has a wind speed of 9.82 m/s at hub height can be obtained from Equation 1.

$$U(z) = \frac{u_*}{\kappa} * ln\left(\frac{z}{z_0}\right) = 1.0366 * ln\left(\frac{z}{0.005}\right) \quad (1)$$

The Mann turbulence is created using the same roughness length (z_0) of 0.005 m and the wind speed 9.82 m/s at the hub height of 65 m as in the logarithmic wind shear used. In a simulation without turbines (in a similar but shorter and narrower grid) the downstream development of the turbulence has been studied at hub height, see Figure 3. It can be seen that the turbulence intensity first increases and goes towards a value just below 6% further downstream. The first turbine is therefore placed on the domain at z=88 R where the horizontal turbulence intensity is 5.9% (the same level as in WRF). The turbulence planes are imposed to the domain at z=13 R and are convected downstream by the flow. The Mann box has an axial length of 10 min using the Taylor frozen hypothesis and it covers the equidistant region in x-y (120 R * 8 R) well with a resolution of 1.17 dx /1.25 dx/1.237 dx (width/height/length).

In two of the flow cases the flow is turned ±2.5 deg from 222 deg. The grid, the placement of the turbulence fields and the placement of the turbines remain unchanged. The inlet wind shear is however turned 2.5 deg and the turbines are yawed. To be normal respectively parallel to the incoming flow the fluctuations in the Mann box are therefore turned (but the Taylor frozen hypothesis is still used based on the assumption that the flow is aligned with the z-axis).

The airfoil data used in this study is based on the NREL 5 MW turbine [22] and scaled down to correspond to a Siemens SWT93-2.3MW turbine regarding both power and thrust coefficients [11] . The turbines are controlled by a generator-torque algorithm which is applied up to the point where the rated power is reached [18]. A pitch controller would be needed in order to handle higher wind speeds. However, the impact on the results is limited since there are only a few occasions in which the first turbines of the rows experience wind speeds above the rated wind speed. In these cases the rotational speed of these turbines only show a minor increase over a short period before it adjusts itself back to a value under the critical level.

The simulations first run for 20,000 timesteps, allowing the flow to establish and to pass through the entire domain length. The flow is thereafter studied for 40 min to get mean values of the flow variables. The result for the flow sector 222 deg ± 2.5 deg is finally calculated as the mean of the simulations for the three directions 219.5 deg, 222 deg and 224.5 deg.

2.2. Mesoscale simulations

The Weather Research and Forecasting (WRF) Model is a mesoscale model used in atmospheric research and numerical weather predictions. The model was created and is maintained by the National Center for Atmospheric Research (NCAR). Version v3.5.0 is used in this study and a further description of the model can be found in the technical note [23].

2.2.1. Numerical model WRF provides a parameterization for wind farms that uses a turbine drag coefficient [13], [24]. This parameterization is here used to simulate the wind farm for the chosen case. The wind farm is treated by the model as a sink of the resolved atmospheric momentum. The total energy extraction is a function of the wind speed and proportional to a specified generic thrust coefficient. The electrical power is also a function of the wind speed but proportional to a specified generic power coefficient. The generated turbulent kinetic energy is the difference between the total energy extraction and the electrical power. The fraction of the resolved atmospheric momentum that is extracted is given by specified thrust coefficients.

2.2.2. Simulation setup The first simulation is performed without any wind turbines. A second simulation is performed with the same setup as the first one but with the wind farm implemented according to the parameterization described above [13], [24]. The model is run with data from ERA Interim reanalysis [25] as input on the boundaries. WRF uses a nested grid with a horizontal resolution of 333m * 333m in the region of interest. The lowest vertical grid points (18 m, 59 m, 116 m, 195 m and 295 m) are given in Figure 2.

The studied result is the difference between the case run with and without the wind farm parameterization. The chosen case is characterized by a wind direction in the interval of 222 ± 2.5 deg lasting for more than 10 min. The stability is near neutral on the stable side (the potential temperature gradient is 0.6 K/km from sea surface up to 115m).

2.3. Production data

The production data is based on SCADA data collected over 5 years [15]. The data is filtered to include the inflow angles from 219.5 to 224.5 degrees and the wind speed interval of 9.82 m/s +-0.5 m/s. The stability classes that are included are near unstable, neutral and near stable, based on the Monin Obukhov length in the range of $L <$ -500 or $L >$500.

3. Results

The following sections provide a comparison between WRF and LES regarding relative production, velocity recovery, boundary layer development and wake expansion. Figure 1 shows the turbines and the position of the data outtakes. The data has been interpolated from the LES and WRF results. The wake is visualized based on the LES results in Figures 20 and 21.

The presented turbulence intensity is defined as the root mean square of the fluctuations (in z for streamwise and x for horizontal) divided by U_0 giving the real turbulence intensity for a wind speed equal to U_0 and a relative measure of the turbulent energy in all other cases. The generated turbulent kinetic energy from the farm is called Excess TKE and is defined as the the difference between the TKE in the wake and before the farm. The velocity deficit is in a similar way defined as the velocity reduction compared to the undisturbed wind before the farm. The results are non dimensionalized with R and U_0.

3.1. Production

The relative production (normalized with the mean of the measured production for the front turbines of rows 5-7) is shown in Figures 4-11.

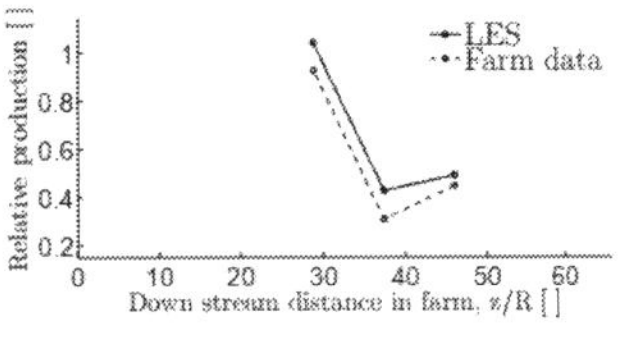

Figure 4. Row 1.

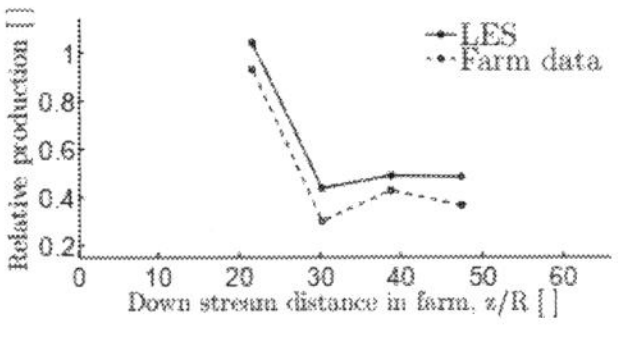

Figure 5. Row 2.

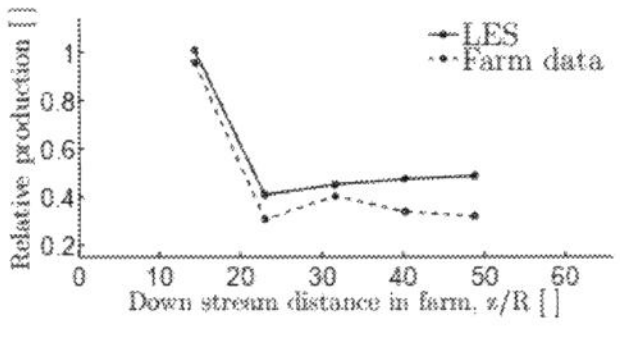

Figure 6. Row 3.

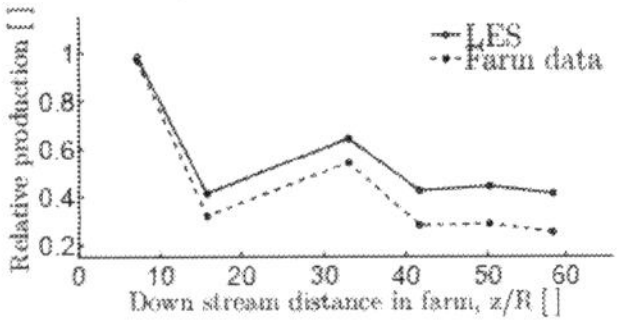

Figure 7. Row 4.

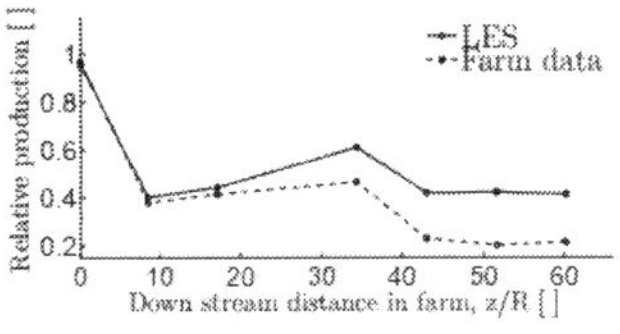

Figure 8. Row 5.

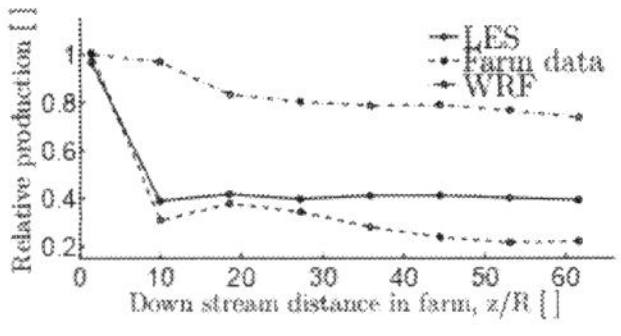

Figure 9. Row 6.

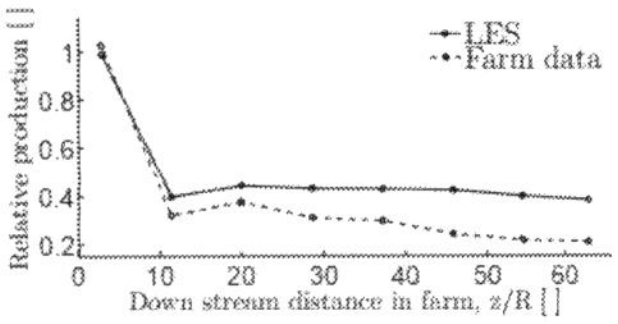

Figure 10. Row 7.

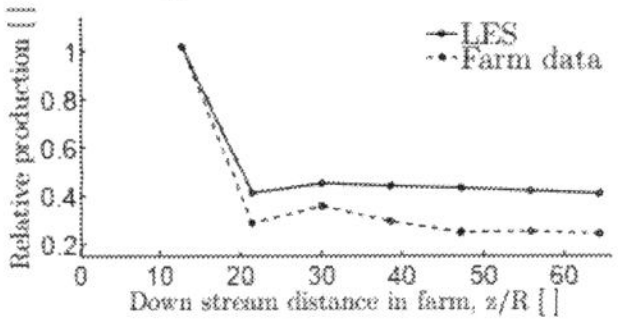

Figure 11. Row 8.

Figures 4-11 The relative production normalized with the mean of the production (SCADA-data) of the first turbines in rows 5-7. (In WRF normalized with the first turbine in the row.)

The distances are relative to the first turbine at z=88 R. The LES results follow the trend of the farm data relatively well. The production is however slightly overpredicted in the LES results, which is more pronounced towards the end of the farm. For rows 1-3 the front turbines also show higher values in the LES case. This can partly be explained by the fact that the farm data also includes partially stable atmospheric conditions which have lower turbulence intensity levels and a higher power deficit compared to the neutral conditions [15]. The averaging used over the sector in LES might also give too much weight to the $\pm$2.5 deg inflow. The LES results also assume the same turbine efficiency (electrical/aero dynamic power) for all wind speeds, which might be lower in reality for partial loads. For row 6 the results from WRF can also be seen. The WRF parameterization used gives too high production. An explanation for this could be the low (333m) resolution of the grid.

3.2. Velocity recovery

Figures 12-15 show the downstream flow development at hub height along row 6 (the dotted lines indicate the turbine positions).

The LES results in Figure 12 show that inside the farm the downstream development of the wind speed (oscillating between 0.4 and 0.6) is relatively stable. Behind the farm the velocity in LES recovers to 0.81 at 2km, 0.86 at 4km and 0.907 at 6 km (for comparison the measured values behind the Horns Rev wind farm are around 0.85 at 2 km and 0.9 at 6 km [12]). The velocity deficit in Figure 13 shows clearly lower values inside the farm for WRF compared to LES. This (likewise the higher production in WRF) can be partially explained by the low grid resolution. The velocity reduction determined by the parametrization is here distributed over 1-2 cells of 333m in width and 3 cells/ 0-159 m in height. The downstream turbines experience consequently a slightly lower mean wind speed, but no full wake situation. WRF gives, as expected considering the behavior inside the farm, also higher velocities behind the wind farm (around 0.82 at 2 km and 0.95 at 6 km), but the differences are not as significant as in the farm.

The streamwise turbulence intensity in LES is 6.5% before the farm (at z=63.8 R), see Figure 14. Inside the farm the turbulence intensity increases for the 4 first turbines until a stable maximum value of around 14 % is reached. Behind the farm the turbulence intensity decreases over a few km down to a value of 7 %. The Excess TKE can be seen in Figure 15. It is clear that the value in WRF is much higher than LES's. This is also seen in other studies [26]. This could be explained by the fact that in WRF all the energy that is taken out from the flow either becomes electrical power or TKE, neglecting for example the efficiency of the turbine drive train [13]. Also the differences between the used C_T-curves could have an impact. The higher TKE level gives higher mixing and a faster velocity recovery.

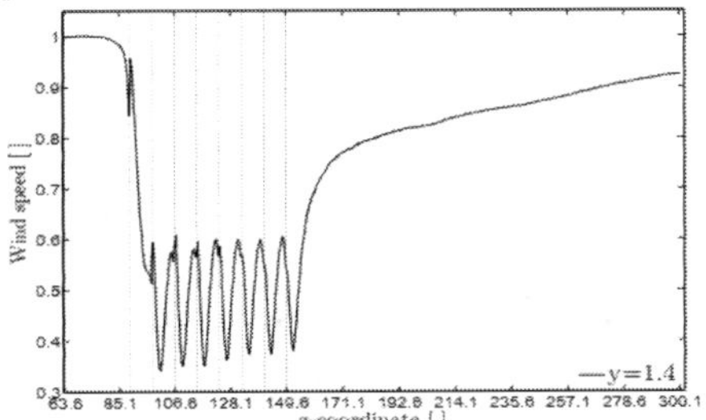

Figure 12. Streamwise velocity LES.

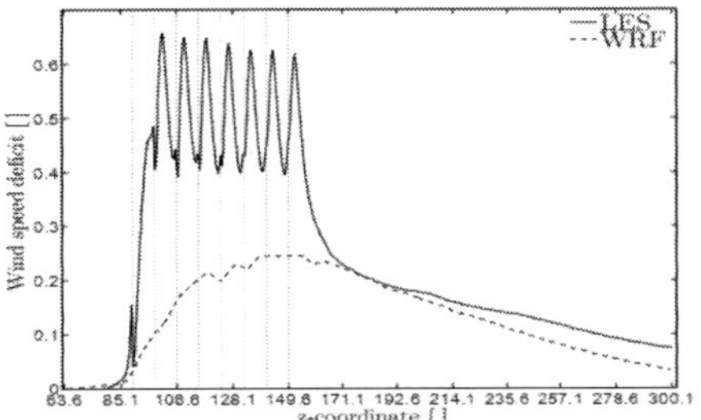

Figure 13. Streamwise velocity deficit (compared to z=63.8 R) , LES and WRF.

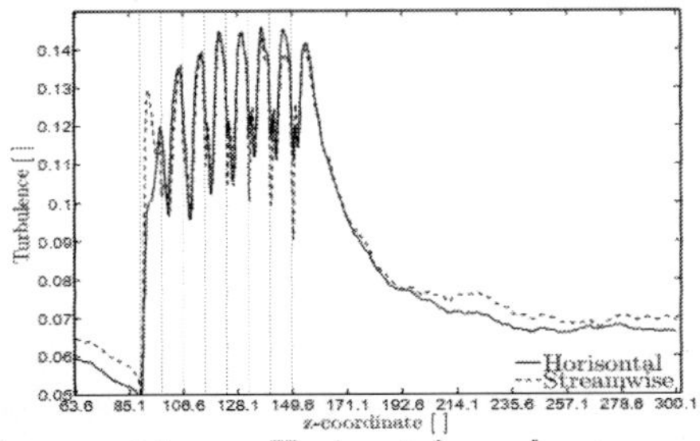

Figure 14. Horizontal and streamwise turbulence intensity LES.

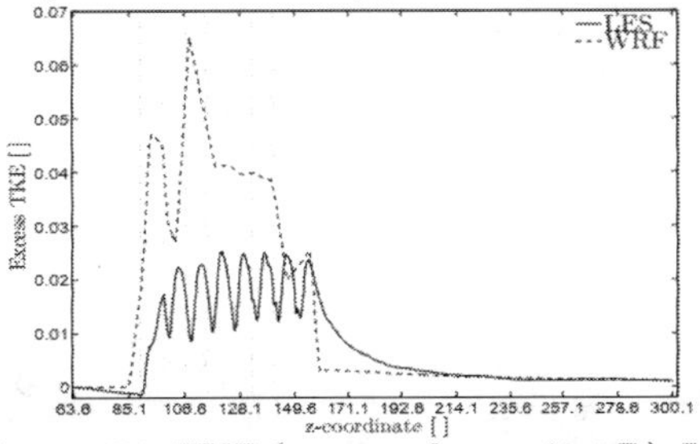

Figure 15. TKE (compared to z=63.8 R), LES and WRF.

3.3. Boundary layer

Figures 16-19 show the vertical profiles taken at each km along row 6 (before, in and behind the farm).

The two first wind shear profiles (from LES), representing the flow upstream of the farm, follow the logarithmic inlet wind shear relatively well, see Figure 16. The three next profiles are inside the farm and show the development of the internal boundary layer. The impact of the farm can be seen up to the height of around 4.2 R at 106.6 R , 5.7 R at 128.1R and 6.25 R over the last turbine at 149.6 R. Also after the farm an increased impact at greater heights can be seen. At the height of the rotor plane (0.4-2.4 R) the wind shear is almost vertical for the first profiles after the wind farm. The slope may increase slightly, but even after 7 km the shear is less sharp compared to the inlet. The turbulence intensity shows a similar vertical development resulting in increased turbulence intensity at greater heights downstream, see Figure 18.

A comparison between the wind shear profiles in LES and WRF is made in Figure 17. A lower velocity deficit at the height of the rotor plane can be noticed in WRF, especially in but also behind the farm. The LES method gives also a larger decrease in wind speed at greater heights. The resulting shear after the wind farm (estimated from the velocity deficit) is sharper in WRF compared to LES. The sharper shear in WRF compared to LES could (due to increased momentum transport) explain the faster velocity recovery in the farm wake. The TKE levels are higher in WRF, both inside the farm and at greater heights, Figure 19.

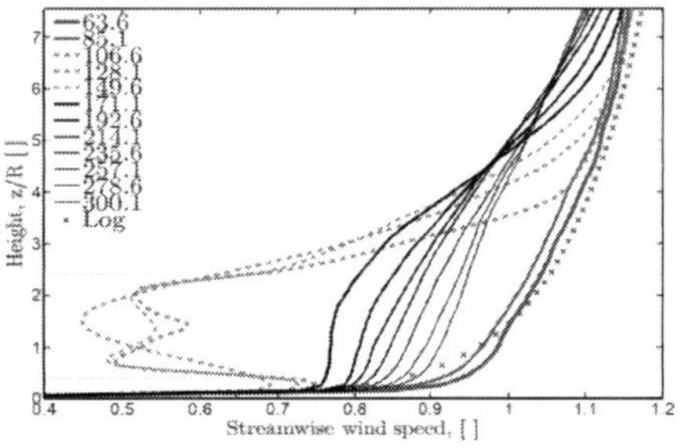

Figure 16. Streamwise velocity LES.

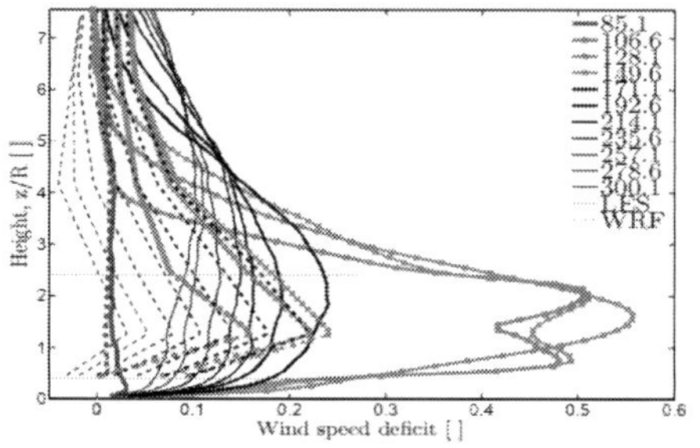

Figure 17. Streamwise velocity deficit (compared to z=63.8 R), LES and WRF.

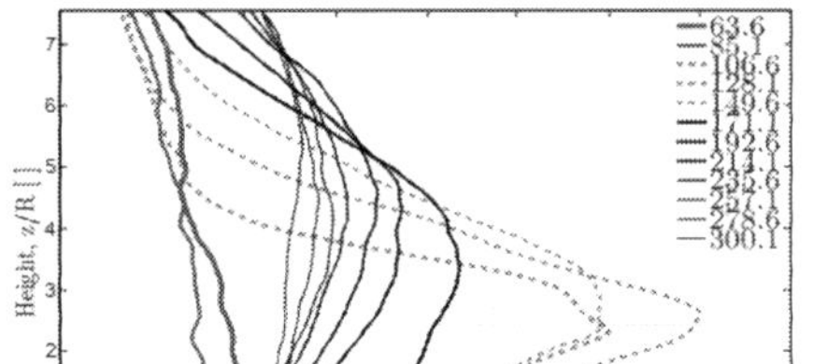

Figure 18. Horizontal turbulence intensity LES.

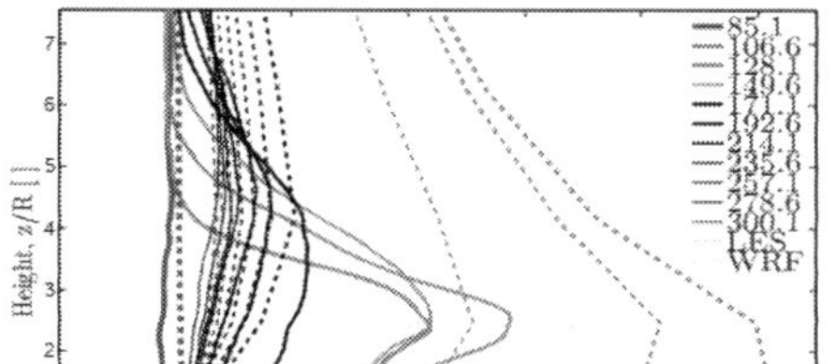

Figure 19. TKE (compared to z=63.8 R), LES and WRF.

3.4. Wake expansion

Figure 22 shows the wind speed along a spanwise line at hub height and Figure 24 the respective turbulence intensities in the LES case. The wind speed at the two first lines before the farm varies around 1. The following lines show the positions 1-7 km behind the farm. The values for the different lines follow the expected trend considering the number of turbines in each row (dotted lines) . The wake expansion downstream can be seen and outside the wake a slight

Journal of Physics: Conference Series **625** (2015) 012028 doi:10.1088/1742-6596/625/1/012028

acceleration of the flow can be found. As expected, the expansion is slightly larger towards the right as the wake flow is more decelerated on that side due to the longer rows. The extensions of the wake can also clearly be seen in Figures 20 and 21 which covers the wake flow up to 150 R / 6 km behind the farm.

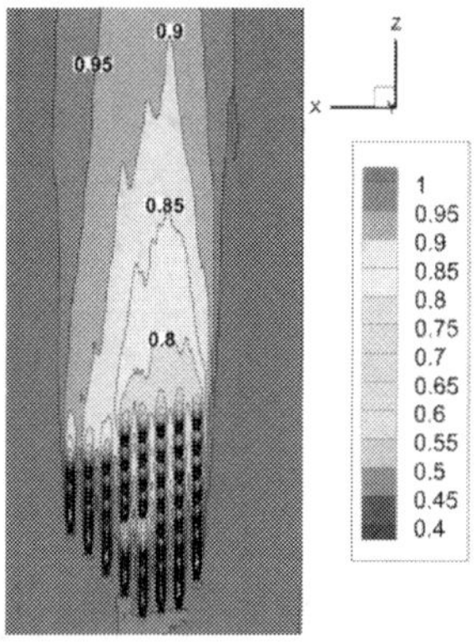

Figure 20. Streamwise velocity at hub-height from the equidistant region in LES.

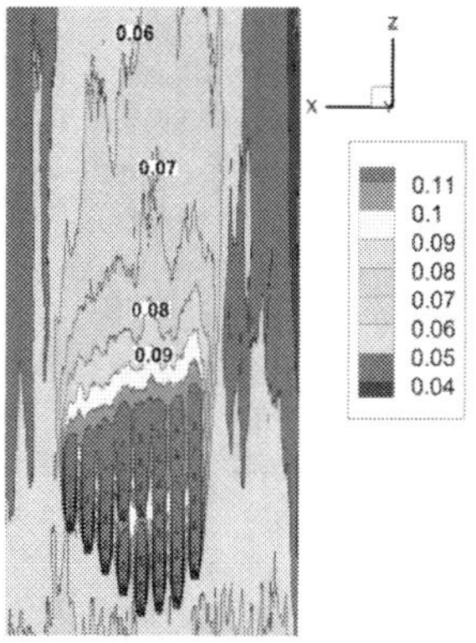

Figure 21. Horizontal turbulence intensity at hub height in LES.

The wake expansion is in the same order in WRF and LES, see Figure 23. In the WRF result an unexpected large acceleration can be seen for the left side. The excess TKE is also in the same order, except closest to the farm where the levels are higher in LES, see Figure 25. In LES the expansion of TKE is slightly larger. The general level of TKE outside the wake and downstream of the farm is lower in the LES.

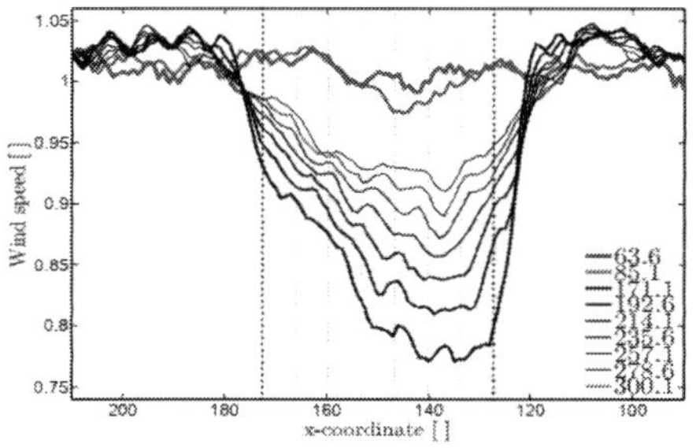

Figure 22. Streamwise velocity LES.

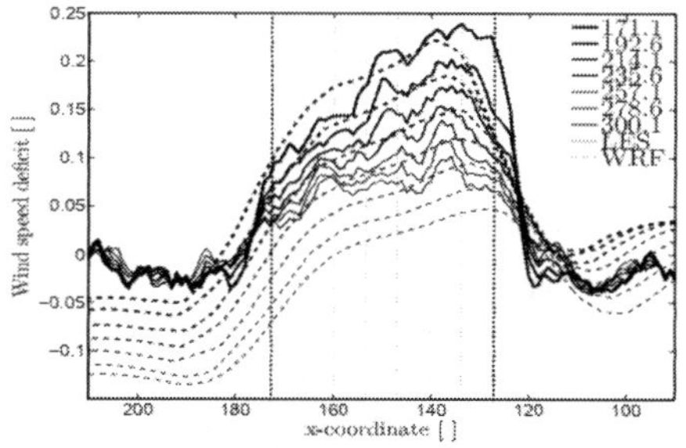

Figure 23. Streamwise velocity deficit (compared to z=63.8 R) , LES and WRF.

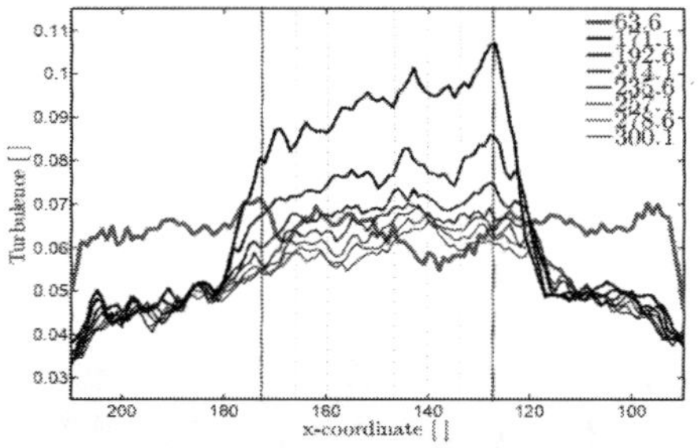

Figure 24. Horizontal turbulence intensity LES.

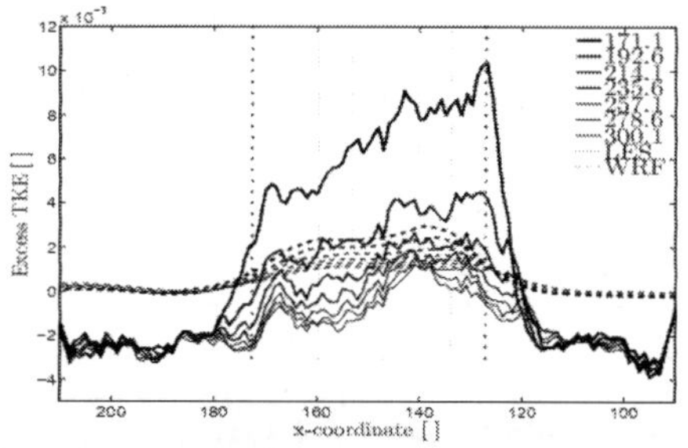

Figure 25. TKE (compared to z=63.8 R), LES and WRF.

4. Conclusion

One neutral flow case for the Lillgrund wind farm has been simulated both in the MESO-scale model WRF and as an LES with an actuator disc method. The input wind profile and turbulence intensity for the LES has been taken from WRF. The results have been compared with respect to the calculated energy production (also compared to wind farm data), the recovery of the flow behind the farm, the boundary layer development and the expansion of the wake.

The LES was shown to slightly overpredict production compared to the farm data, while WRF clearly overestimated the production.

The velocity reduction inside the farm is significantly larger for LES compared to WRF. For the recovery of the flow behind the farm a slightly faster recovery is seen in WRF.

Regarding the impact of the farm on the development of the boundary layer, a reduction of velocity could be found at greater heights in the LES as compared to WRF. For the turbulence intensity the increase was found to be larger in the WRF results. The estimated slopes of the shear profiles are higher in WRF indicating a larger downward momentum transport. The expansion of the wake was in the same order in both simulation models.

The main differences between the results could be related to the lower grid resolution in WRF and the higher TKE levels added from the parameterization. The low grid resolution in WRF causes a smearing of the velocity deceleration from the turbine parametrization over a large cross section. This results in an increased production for the downstream turbines. Although the outtake of energy from the flow by the turbines in WRF is larger, a lower wake deficit behind the wind farm was seen in this case. This could be due to the sharper resulting shear and the related downward momentum transport. It has to be pointed out that the comparison is performed for one specific flow occasion in WRF. The introduced uncertainties could have been reduced by averaging over a number of similar flow cases.

The production results from LES show relatively good agreement with production results from Lillgrund. For the farm wake flow no site data is available but comparing to measurements on Horns rev the velocity recovery is in the same order at 2 km and 6 km behind the farm.

The differences between the results indicate that the results from LES are in better agreement with the farm data. The needed computational resources are however much larger. WRF could potentially give better results with a higher resolution over the farm. The results from LES could possibly be used to adopt the WRF parameterization for the downstream turbine.

To get better understanding of farm to farm interaction a first step is to study the wake behind entire wind farms. Some differences in the results from WRF and LES were here pointed out. The production was especially shown to be closer to the farm production in the LES case. The velocity recovery was reasonable in LES. The WRF case showed a slightly faster velocity recovery. For farm to farm interaction studies it can be of interest to combine the two models to take advantage from both the relatively coarse grid in the mesoscale simulations which gives a good description of the atmosphere and the finer grid in LES which can resolve the wake flow. One alternative is to nest LES into the WRF simulations, this would however be more computational demanding. A potential configuration for future farm to farm interaction studies could be to use WRF for the first wind farm and then use the farm wake flow conditions as an input (similar to what was done in this study for the inlet profile) for a LES of the second wind farm.

Acknowledgments

The LES were performed on resources provided by the Swedish National Infrastructure for Computing (SNIC) at the National Supercomputer Centre in Sweden (NSC). The WRF-simulations were performed on the Abel Cluster, owned by the University of Oslo and the Norwegian meta center for High Performance Computing (NOTUR), and operated by the Department for Research Computing at USIT, the University of Oslo IT-department. http://www.hpc.uio.no/. Jan-Åke Dahlberg at Vattenfall AB is acknowledged for providing measurement data from the Lillgrund wind farm as well as Kurt Hansen at DTU Wind for post processing of these data. This work was supported financially by the Top-Level Research Initiative (TFI) project, Improved Forecast of Wind, Waves and Icing (IceWind) and Vindforsk IV.

References

[1] Frandsen S, Barthelmie R, Rathmann O, Jørgensen H, Badger J, Hansen K, Ott S, Rethore P, Larsen S and Jensen L 2007 *Summary report: The shadow effect of large wind farms: measurements,data analysis and modelling.* (Risø-R-1615(EN), Denmark)

[2] Eriksson O and Ivanell S 2012 *A survey of available data and studies of farm-farm interaction.* (EAWE PhD Seminar on Wind Energy in Europe, 2012)

[3] Brand A 2009 *Wind Power Plant North Sea – Wind farm interaction, The effect of wind farming on mesoscale flow.* (ECN, the Neatherlands)

[4] Ivanell S 2009 *Numerical Computations of Wind Turbine Wakes.* (PhD thesis, ISBN 978-91-7415-216, KTH Engineering Sciences, Sweden)

[5] Lu H and Porté-Agel F 2011 *Large eddy simulation of a very large wind farm in a stable atmospheric boundary layer.* (Physics of Fluids 2011; 23: 065101)

[6] Wu Y T and Porté-Agel F 2012 *Atmospheric turbulence effects on wind-turbine wakes: An LES study.* (Energies 2012, 5, 5340-5362)

[7] Troldborg N, Sørensen J N and Mikkelsen R 2010 *Numerical simulations of wake characteristics of a wind turbine in uniform inflow.* (Wind Energy 2010; 13: 86–99)

[8] Troldborg N, Larsen G C, Hansen K S, Sørensen J N and Mikkelsen R 2011 *Numerical simulations of wake interaction between two wind turbines at various inflow conditions.* (Wind Energy 2011; 14: 859–876.)

[9] Keck R E, Mikkelsen R, Troldborg N, de Maré M and Hansen K S 2013 *Synthetic atmospheric turbulence and wind shear in large eddy simulations of wind turbine wakes.* (Wind Energy 2013; DOI: 10.1002/we.1631)

[10] Troldborg N, Sørensen J N, Mikkelsen R and Sørensen N N 2014 *A simple atmospheric boundary layer model applied to large eddy simulations of wind turbine wakes.* (Wind Energy 2014; 17: 657-669.))

[11] Nilsson K, Ivanell S, Hansen K S, Mikkelsen R, Breton S P and Henningson D 2014 *Large-eddy simulations of the Lillgrund wind farm.* (Wind Energy 2014; DOI: 10.1002/we.1707)

[12] Eriksson O, Mikkelsen R, Nilsson K and Ivanell S 2012 *Analysis of long distance wakes of Horns rev 1 using actuator disc approach.* (J. Phys.: Conf. Ser. 555 012032)

[13] Fitch A C, Olson J B, Lundquist J K, Dudhia J, Gupta A K, Michalakes J and Barstad I 2012 *Local and Mesoscale Impacts of Wind Farms as Parameterized in a Mesoscale NWP Model.* (Monthly Weather Review, Volume 140, Issue 9 .)

[14] Lundquist J K, Mirocha J D and Kosovic B 2010 *Nesting large-eddy simulations within mesoscale simulations in WRF for wind energy applications.* (Proceedings of the Fifth International Symposium on Computational Wind Engineering, Chapel Hill, NC)

[15] Hansen K S 2013 *Preliminary benchmark for Lillgrund performed in EERA-DTOC.* (EERA-DTOC. Presentation DTU 20th June 2013.)

[16] Churchfield M, Lee S, Moriarity P J, Martinez L A, Leonardi S, Vijayakumar G and Brasseur J G 2012 *A Large-Eddy Simulation of Wind-Plant Aerodynamics.* (NREL, the United states of America)

[17] Mikkelsen R 2003 *Actuator Disc Methods Applied to Wind Turbines.* (DTU, Denmark)

[18] Breton S P, Nilsson K, Ivanell S, Olivares-Espinosa H, Masson C and Dufresne L 2012 *Study of the effect of the presence of downstream turbines on upstream ones and use of a controller in CFD wind turbine simulation models.* (J. Phys.: Conf. Ser. 555 012014)

[19] TaPhouc L 1994 *Modéles de sous maille appliqués aux écoulmements instationnaires décollés.* (Tech. rep. LIMSI 93074 LIMSI France.)

[20] Mann J 1998 *Wind field simulation.* (Risø, Denmark)

[21] Jakob M, Ott S, Jørgensen B H and Frank H P 2013 *WAsP Engineering 2000.* (Risø–R–1356(EN))

[22] Jonkman J, Butterfield S, Musial W and Scott G 2009 *Wind Turbine for Offshore System Development.* (NREL, the United states of America)

[23] Skamarock W C, Klemp J B, Dudhia J, Gill D O, Barker D M, Duda M G, Huang X, W W and Powers J G 2008 *A Description of the Advanced Research WRF Version 3.* (Technical Note NCAR/TN-475+STR)

[24] Fitch A C, Olson J B, Lundquist J K, Dudhia J, Gupta A K, Michalakes J, Barstad I and Archer C L 2013 *Corrigendum.* (Monthly Weather Review, 141, 1395–1395.)

[25] Dee D P and et al 2011 *The ERA-Interim reanalysis: configuration and performance of the data assimilation system.* (Q.J.R. Meteorol. Soc., 137: 553–597)

[26] Abkar M and Porté-Agel F 2015 *A new wind-farm parameterization for large-scale atmospheric models.* (Journal of Renewable and Sustainable Energy, 7.1: 013121)

The effect of blockage on power production for laterally aligned wind turbines

A R Meyer Forsting[1], **N Troldborg**[1]

[1] DTU Wind Energy, Department of Wind Energy, Technical University of Denmark, Risø Campus, DK-4000 Roskilde, Denmark

E-mail: alrf@dtu.dk

Abstract. This paper studies the change in the individual power coefficients for a laterally aligned row of wind turbines over a single, free turbine in the context of varying inflow directions via numerical simulations. All turbines were rotating in-line with the main flow direction. The problem definition is similar to that of many wind turbine testing sites and wind farms. Hence any changes in the individual turbine power production could have implications regarding power curve validation procedures.These changes are relatively small and therefore the size of the computational domain was identified to be detrimental in avoiding any domain-inflicted blockage. Increasing the misalignment of the main flow direction with the row of turbines led to significant variations in the power production across turbines. At the largest inflow angle of 45° it varied from -1.1 % to 2 %. As a whole, the power production increased by about 0.5 %, almost independent of the inflow direction.

1. Introduction

The lateral spacing between wind turbines is an important parameter in wind farm development, as it limits the number of wind turbines for a given site. Recent technological advances in the wind turbine industry have left many prime wind farm sites underperforming to their power production potential, due to their older, sub-MW turbines. Many wind parks have to undergo extensive modernisation programs in the years to come. These include upgrading to taller towers and larger rotors. In most cases, however the tower positions are fixed under the initial planning permits. As a consequence the lateral distance between blade tips is further reduced. Moving rotors ever more closely together means that the individual blockage effects, that are caused by the pressure jump across turbines, start to interact. Therefore the velocity field upstream is not only altered locally for a single turbine but, also on the global, wind farm scale. One would hence expect significant patterns to emerge in the individual power coefficients. Performance analyses of wind farms, though mainly focusing on wake effects, reveal significant variations along the first row of wind turbines, even if their causation has not so far been identified [1].

Whereas the effect might be negligible over an entire wind farm they can be significant in turbine power curve verification. Wind turbine testing sites normally align several turbines so that they lie on an axis perpendicular to the prevailing wind direction. Representative of this are the DTU sites Høvsøre and Østerild in Denmark. The widely used IEC standard for power curve validation does not include any specifications regarding lateral spacing, instead focusing on the avoidance of wake interaction from neighbouring objects and other turbines. Depending on the layout of the testing site valid wind sectors might include wind directions of up to 45° to the alignment axis. The underlying assumption is of course that the neighbouring wind turbines are not influencing each other significantly, even at large inflow angles. Evaluating and quantifying the power production under the effect of interacting blockages, thus may proof important for future testing procedures and standards, as these assumptions are tested. In particular the influence on short-range remote sensing technology could be significant as it relies heavily on quantifying the local velocity field upstream.

Blockage has attracted most attention in marine research, as it has been shown experimentally [2] and numerically [3, 4] that it significantly increases the power output of tidal wave turbines. This conclusion is not transferable to wind turbines, which operate in a free domain unrestricted by water depth. In wind energy the focus has been largely on longitudinal spacing and wind farm staggering, as these parameters significantly influence wake interaction. Its impact on overall wind farm power production and turbine lifetime explain the proliferation of papers published on this subject. Previous attempts to quantify upstream blockage in a wind engineering context heavily relied on low Reynolds number experiments and a limited number of flow conditions. This includes work of Medici *et al.*, who used hot-wire anemometry and particle image velocimetry on a single turbine and concluded that blockage was noticeable more than three turbine diameters (D) upstream [5]. McTavish *et al.* studied the effect with similar limitations for three turbines by attempting to reach power maximization through an optimal spatial arrangement [6]. A power increase of 8 % was observed for a lateral spacing of 1.5D during experiments, which was validated by vortex particle simulations.

None of the current literature quantifies the blockage effect to the extent that any conclusions could be drawn for wind turbine testing. All existing experiments' flows were operating at small Reynolds numbers and the flow orientation was at all times perpendicular to the line of turbines. Furthermore the impact of tunnel blockage on results was hastily rejected and not properly assessed. The approach of this paper is more systematic. It employs computational fluid dynamics (CFD) to evaluate any kind of blockage and more specifically its possible effects on power curve measurements, where special attention is given to uncertainty reduction.

1.1. Problem definition

To allow for the conclusions of this paper to be relevant for turbine testing procedures, the arrangement adopted is loosely related to the DTU Høvsøre testing site. As shown in figure 1.1 five turbines are aligned on an axis with a 3D lateral spacing. Taking current measuring procedures as a reference, the angle between the line of wind turbines and the main wind direction (θ) was varied from $0°$ to $45°$ in $15°$ steps and the wind speed set to 8 ms^{-1}. Note that θ will sometimes be referred to as 'yaw' in the following passages. The inflow was set to be uniform and the solution assumed to be time invariant.

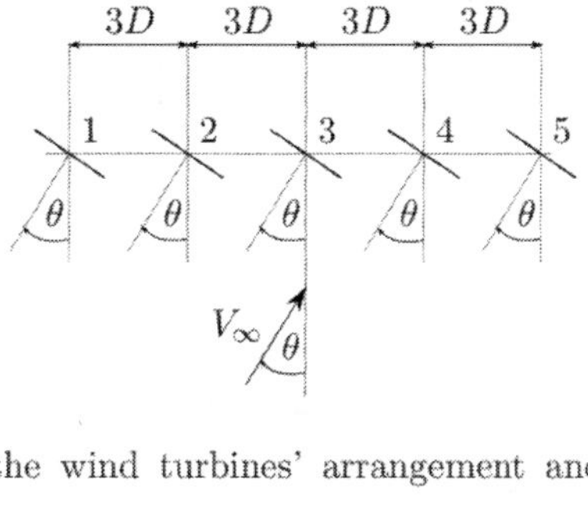

Figure 1. Schematic of the wind turbines' arrangement and reaction to changing inflow directions.

Journal of Physics: Conference Series **625** (2015) 012029 doi:10.1088/1742-6596/625/1/012029

2. Computational method

2.1. Flow solver

The finite volume code EllipSys3D solves the incompressible Reynolds Averaged Navier-Stokes (RANS) equations over a discretised block-structured domain [7–9]. The SIMPLE algorithm [10] solves the pressure-linked terms of the Navier-Stokes equations, while the QUICK scheme [11] is applied to the convective terms. To allow the implementation of discrete body forces without jeopardising the coupling between velocity and pressure a modified Rhie-Chow algorithm has been implemented [12]. Further the Menter k–ω shear-stress transport modelled the turbulence [13].

2.2. Turbine model

Allowing the computations to be easily reproducible and representative of wind turbines currently employed, the well known NREL 5-MW [14] turbine with a diameter of 126 m was used. At the set wind speed of 8 ms^{-1}, the rotor performs 9.21 rotations per minute (RPM).

In the computational domain the rotor is represented by permeable body forces, which are derived from local velocity and 2-D airfoil data. In this particular simulation the actuator shape model [15] distributed these forces inside the domain by calculating the intersection between the rotor and flow grid.

2.3. Numerical domain

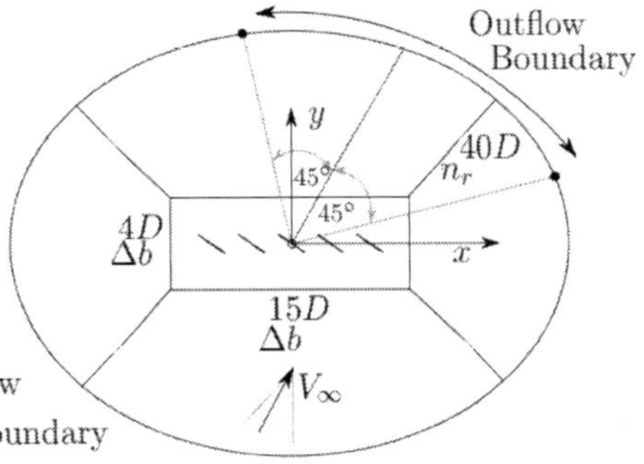

Figure 2. Top view of the numerical domain. All dimensions are scaled by the turbine diameter D. Note that the boundaries are a function of the inflow angle.

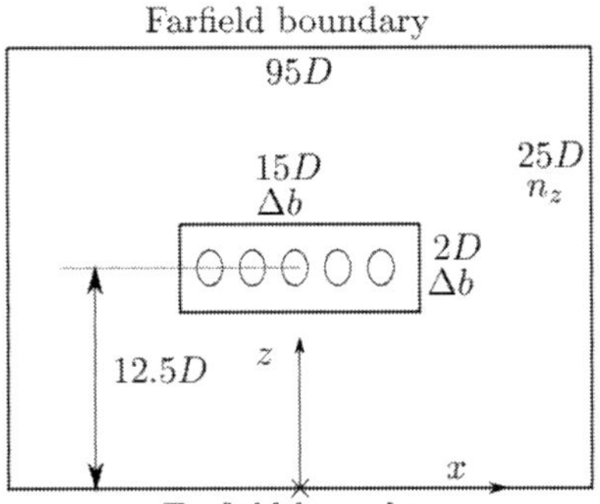

Figure 3. Frontal view of the numerical domain.

2.3.1. Layout The numerical domain was discretised by a structured O-type meshing methodology, containing a centrally located equispaced box mesh. The latter encompassed the row of turbines. Farfield boundary conditions were applied to both, top and bottom of the domain. The inflow and outflow boundaries were applied dynamically to the sides of the O-mesh by the preprocessor, depending on the inflow angle. More specifically the outflow sector is $\pm$ 45° centred around the inflow direction. A schematic layout of the domain and its coordinate system is shown in figures 2 and 3. All dimensions were scaled by D. The spacing inside the central box was fully defined by the inner grid spacing Δb, that resulted from the number of grid points specified per diameter (n_{rot}). From there the mesh grew hyperbolically outwards in the radial and z directions, where the dimension of the first cell connecting to the inner box mesh was equal to Δb. The number of grid spaces along the radial and z directions were defined by n_r and n_z respectively.

2.3.2. Grid resolution and domain size To allow the conclusions drawn from this paper to be generally applicable, the grid resolution had to be sufficiently fine and the domain large enough to ensure the simulation results to be independent of these two factors.
Preliminary investigations showed that a relatively large domain was needed with respect to other types of investigations, to avoid 'tunnel' blockage influencing the results. 'Tunnel' blockage refers to constriction of the cross-sectional area through an object, leading to flow acceleration, which will hence influence the C_P of each turbine. In fact the equivalent 'tunnel' blockage, if defined as the rotor swept area over the cross-sectional area of the numerical domain, was 0.165 %, much lower than was seen as acceptable in previous investigations [6]. To further reduce the uncertainty deriving from the domain size, an outflow correction can be implemented at the domain boundaries to remove some of the excess mass flow. Koning [16] worked with potential flow assumptions to show that an analytical solution can be obtained to estimate this outflow. Here the actuator disc is represented by a doublet pressure source, which determines the flowfield everywhere. The final outflow velocity induced by a single turbine in the local disc coordinate system, denoted by $\bullet'$, is given as:

$$u'_i = \frac{1}{2} a V_\infty \frac{R^2 x'_i}{\left(\sum_{i=1}^{3} x'^2_i\right)^{3/2}} \quad \text{with } x'_i = \{x'y'z'\} \text{ and } u'_i = \{u'v'w'\} \tag{1}$$

,where a is the axial induction factor and R the disc radius. The former is estimated by momentum theory ($C_T = 4a(1-a)$) [17], where the thrust coefficient (C_T) is taken from a non-corrected simulation. Note that applying this correction is to a certain extent equivalent to increasing the domain size. No correction is needed for unrestricted domain methods.

The following paragraph addresses the second important factor regarding the discritisation of the domain and its influence on the results. As previously mentioned the changes in the C_P to be investigated were assumed to be relatively small, thus already at the first grid iteration, the resolution was aimed to be high. For wake investigations setting the number of points per diameter (n_{rot}) to 8, results in an acceptably low error [18]. Here it was initially set to 32, instead. In table 1 a summary of all grid parameters is presented. A higher grid level derived from halving the inner box spacing Δb.

Table 1. Mesh parameters.

Mesh	n_{rot} $[D^{-1}]$	Δb $[D]$	n_r	n_z	Cells $\times 10^6$
coarse	32	0.0313	32	128	2.05
fine	64	0.0156	64	256	8.21

Since the intention of this paper is to investigate the variation of C_P along a row of turbines, its sensitivity to changes in the domain size and grid level was analysed to evaluate the uncertainty in the simulations. The minimum and maximum yaw (θ) settings were chosen as evaluation points, as these were expected to yield the most extreme simulation solutions, making grid and domain errors more noticeable.

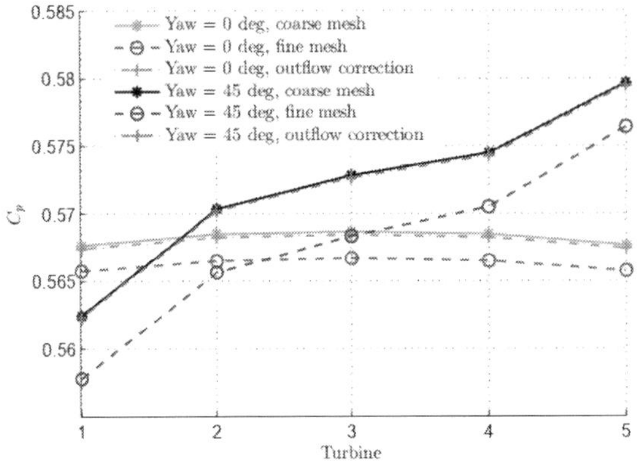

Figure 4. Evolution of C_P along a row of turbines for a coarse, fine and coarse mesh with outflow correction.

In figure 4 the evolution of C_P with turbine number is presented for the fine, coarse and coarse mesh with the outflow correction at the two extreme yaw positions. All curves agree in terms of their overall trend, though there is a visible offset between the solutions of the coarse and fine meshes. A numerical summary of the grid and domain errors is presented in table 2. The errors are calculated by taking the non-corrected coarse mesh solution as base and then calculating the resulting percentage difference in C_P at each turbine. As the aim of this study is to access relative changes in C_P across the row of turbines any absolute offsets are acceptable as long as they stay constant across the row. This implies that its standard deviation ($\sigma(\Delta C_P)$) has to be small. Between the fine and coarse meshes the largest error in predicting the overall trend was 0.1 %. This is acceptable for proving a significant change in C_P and therefore the coarse mesh was sufficiently resolved. The outflow correction had an even smaller impact, thus showing that the domain was acceptably large.

Table 2. Grid and domain errors calculated with the coarse mesh solution for C_P as base.

	θ [°]	$\langle \Delta C_P \rangle$ [%]	$\sigma(\Delta C_P)$ [%]
Grid	0	0.344	0.00636
	45	0.738	0.11200
Outflow	0	0.0433	0.000928
	45	0.0272	0.005070

2.4. Turbine orientation and location correction

The power coefficient for each turbine is not independent of its grid location. Therefore single turbine calculations are needed at each turbine location. Furthermore yawing the turbines inside the numerical domain (see figure 2) changes the way the body forces from the actuator disc are implemented inside the flow domain, thus making single turbine calculations at each yaw setting necessary. The resulting C_P values are then used to normalise the simulations:

$$\Delta C_P|_{i,j} = \frac{C_P|_{i,j} - C_{P_{single}}|_{i,j}}{C_{P_{single}}|_{i,j}} \times 100 \qquad (2)$$
$$,\text{with } i = \{1, \ldots, 5\}, j = \{0, 15, \ldots, 45\}$$

3. Results

3.1. Power evolution

Figure 5 shows the variation in the global power coefficient across the row of turbines for the four inflow settings. These results are summarised in figure 6 in terms of their respective mean, minimum and maximum at each inflow angle. For a yaw of 0° the result's overall trend agrees well with those of McTavish [6]. McTavish also found an increased mean power production and a rise towards the centre of the row. A direct comparison is not possible as his lateral spacing was only 1.5D, half of what is used in this simulation. Yawing the inflow changes the behaviour markedly. Whereas the middle turbine (3) shows an increase in the power production close to the mean value, the upstream turbines (1-2) are losing their gain in power in line with an increase in yaw. For the first turbine losses are observed from a yaw of 15° onwards. The losses upstream are counteracted by almost equally sized increases on the downstream side (4-5). The existing balance between the losses upstream and gains downstream is underlined by the mean of all turbines being practically independent of yaw (see figure 6).

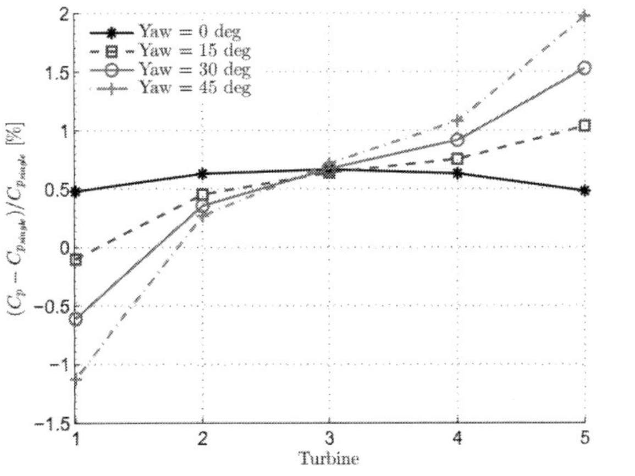

Figure 5. Percentage change in the power coefficient for different yaw angles.

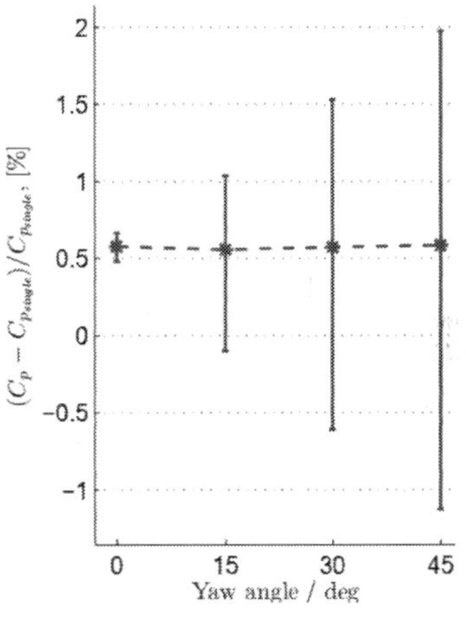

Figure 6. Summary of mean, minimum and maximum values of the results in figure 5 as a function of yaw angle.

3.2. The flowfield

The changes in the power coefficient along a row of turbines presented in the previous section can be directly linked to the emerging flow phenomena.

The main factors that are interacting and causing these flow alterations are the blockage, both on a local and global scale, and the acceleration of the flow between wind turbines and neighbouring wakes. The overall flow alterations are best explained via the extreme example of

$\theta = 45°$. Contours of the free-stream normalised axial velocity component (v_a) for this case are shown in figure 7. All wind turbines taken together act as one large flow obstruction and thus cause the emergence of a 'global' blockage effect or, equivalently, induction zone. Its presence is clearly visible in figure 7, where the flow is decelerating over the entire length of the wind farm. Notably at a yaw of 45° this deceleration is visibly reduced along the row of turbines.

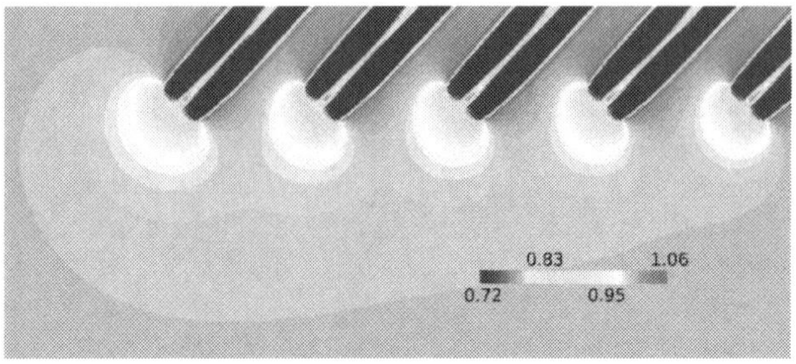

Figure 7. Normalised axial velocity contours (v_a) at a yaw angle (θ) of 45°.

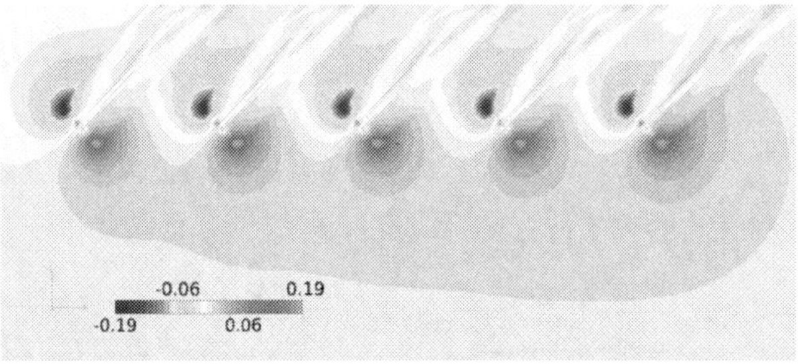

Figure 8. Normalised cross flow velocity contours (u) at a yaw angle (θ) of 45°.

This can be partly attributed to the local blockage from each turbine. As the flow approaches the turbine furthest upstream (1) the local induction reduces the mass flow forcing some of the flow to be diverted around it and thus towards turbine 2. There some of the flow is channelled through the space between the neighbouring wake and the turbine, hence causing the flow to accelerate, reducing the size of the local induction zone of turbine 2. This is also reflected by the axial velocity v_a distributions along the disc for turbines 3 and 5 shown in figure 9. This flow diversion continues all along the row of turbines leading to a continues build up of mass flow, as not all the excess flow escapes in-between turbines and wakes. This build up is a clear feature in the cross flow velocity contours in figure 8. It is this effect that leads to a decrease in the global induction zone along the row of turbines, as the flow accelerates. Note that the axial forcing from the turbines, defined by the thrust coefficient C_T, is actually increasing towards turbine 5. This shows that it is indeed the acceleration causing the decrease in the global induction zone. The process described in the paragraph above is aggravated by an increase in θ, as the distance between the turbines and neighbouring wakes decreases. This reduces the mass flow in-between turbines and wakes, causing an increase in the cross flow. Linked to this decrease in distance is a stronger induction felt by turbine 1 (see figure 9). The increased induction at turbine 1 and the subsequent increase in v_a along the row can be better understood by conceiving the row of turbines as a partially permeable flat plate.

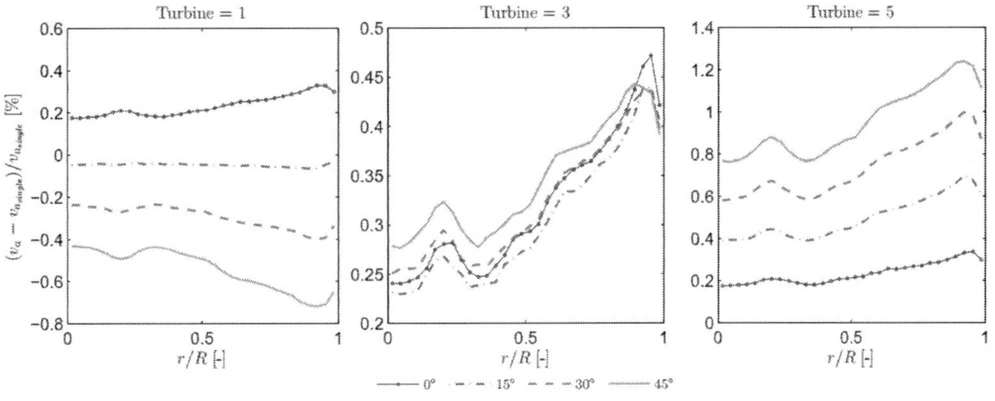

Figure 9. Azimuthally averaged axial velocity v_a distributions at the rotor disc at each yaw configuration.

4. Conclusion

The blockage effect and its impact on power production were assessed for a scenario similar to that of a wind turbines testing site using CFD. Extensive attention was paid to uncertainty reduction by assessing the influence of the numerical methodology on the results. Detrimental to a reliable investigation of this rather small effect, was selecting a sufficiently large domain. As an outcome the blockage ratio was significantly lower than deemed usually acceptable at 0.165 %. It is generally recommended to use similar ratios for domain restricted methods or alternatively applying corrections.

The blockage effect for laterally aligned wind turbines significantly alters the upstream flow, leading to marked changes in individual power coefficients. An inflow orthogonal to the row of turbines only had limited impact, causing a mean increase in the power coefficient across all turbines of 0.57 %. The different turbines' C_P values encompassed a range of 0.18 %. Yawing the inflow in respect to the line of alignment did not greatly affect the mean, in contrast to this the range of C_P extended in line with turbine rotation up to 3.1 % at 45°.

This was attributed to local and global blockage effects. Global blockage increased with turbine rotation, making the row of turbines adopt flow characteristics similar to those of a flat plate, diverting increasing amounts of mass flow down the row of turbines. Furthermore a stronger induction zone appeared at the upstream turbine causing a decrease in power, whereas downstream reacceleration lead to increases, peaking at the last turbine (2.0 % at 45°).

For wind turbine testing sites the accuracy with which a power performance evaluation can be carried out is of utmost importance. Even relatively small changes in the individual power coefficients, registered in this study, might prove significant. Nevertheless to make specific recommendations for real world applications, a more realistic flow, including an atmospheric boundary layer should be implemented and the actual measurement procedure be included.

Acknowledgments

This work was performed inside the UniTTe project (unitte.dk), which is financed by The Innovation Fund Denmark, grant number 1305-00024B. Further to the authors R. Wagner (DTU Wind Energy) contributed to this work through her valuable insights into power curve validation procedures. Computational resources were provided by the Risø DTU central computing facility.

References

[1] Mitraszewski K, Hansen K, Nygaard N and Rethoré P E Wall effects in offshore wind farms *Torque 2012*
[2] Bahaj A, Molland A, Chaplin J and Batten W 2007 *Renewable Energy*
[3] Turnock S R, Phillips A B, Bank J and Nicholls-Lee R 2011 *Ocean Engineering*
[4] Nishino T, Richard and Wilden H 2012 *J. Fluid Mech.*
[5] Medici D, Ivanell S, Dahlberg J A and Alfredsson P 2011 *Wind Energy*
[6] McTavish S, Rodrigue S, Feszty D and Nitzsche F 2014 *Wind Energy* DOI: 10.1002/we.1806
[7] Sørensen N 1995 *General purpose flow solver applied to flow over hills* Ph.D. thesis RisøNational Laboratory
[8] Michelsen J 1994 Basis3d - a platform for development of multiblock pde solvers Tech. rep. Dept. of Fluid Mechanics, Technical University of Denmark, DTU
[9] Michelsen J 1994 Block structured multigrid solution of 2d and 3d elliptic pdes Tech. rep. Dept. of Fluid Mechanics, Technical University of Denmark, DTU
[10] Patanker S and Spalding D 1972 *International Journal of Heat and Mass Transfer*
[11] Leonard B 1979 *Computer Methods in Applied Mechanics and Engineering*
[12] Rethoré P E and Sørensen N 2012 *Wind Energy*
[13] Menter F R 1993 *AIAA Journal*
[14] Jonkerman J, Butterfield S, Musial W and Scott G 2009 Definition of a 5-mw reference wind turbine for offshore system development Tech. rep. NREL
[15] Rethoré P E, van der Laan P, Troldborg N, Zahle F and Sørensen N 2014 *Wind Energy*
[16] 1976 *Aerodynamic Theory: A General Review of Progress* (Peter Smith) chap Devision M: Influence of the propeller on other parts of the airplane structure
[17] 2008 *Aerodynamics of Wind Turbines 2nd Ed.* (Earthscan)
[18] van der Laan P, Sørensen N N, Rethoré P E, Mann J, Kelly M C, Troldborg N, Schepers J G and Machefaux E 2014 *Wind Energy*

Wake Conference 2015 IOP Publishing
Journal of Physics: Conference Series **625** (2015) 012030 doi:10.1088/1742-6596/625/1/012030

How Many Model Evaluations Are Required To Predict The AEP Of A Wind Power Plant?

J P Murcia[1], P E Réthoré[1], A Natarajan[1] and J D Sørensen[2]
[1]DTU Wind Energy, Technical University of Denmark, Risø Campus, Frederiksborgvej 399, 4000 Roskilde, Denmark
[2]Department of Civil Engineering, Aalborg University
E-mail: jumu@dtu.dk

Abstract. Wind farm flow models have advanced considerably with the use of large eddy simulations (LES) and Reynolds averaged Navier-Stokes (RANS) computations. The main limitation of these techniques is their high computational time requirements; which makes their use for wind farm annual energy production (AEP) predictions expensive. The objective of the present paper is to minimize the number of model evaluations required to capture the wind power plant's AEP using stationary wind farm flow models. Polynomial chaos techniques are proposed based on arbitrary Weibull distributed wind speed and Von Misses distributed wind direction. The correlation between wind direction and wind speed are captured by defining Weibull-parameters as functions of wind direction. In order to evaluate the accuracy of these methods the expectation and variance of the wind farm power distributions are compared against the traditional binning method with trapezoidal and Simpson's integration rules.

The wind farm flow model used in this study is the semi-empirical wake model developed by Larsen [1]. Three test cases are studied: a single turbine, a simple and a real offshore wind power plant. A reduced number of model evaluations for a general wind power plant is proposed based on the convergence of the present method for each case.

1 Introduction

The evaluation of the performance of a wind power plant requires to calculate the expected energy production over the years. The most common measure of wind farm performance used in the planning or evaluation stages is the annual energy production. AEP is proportional to the expected mean power over all the possible atmospheric conditions: $AEP = 8760\,\mathbb{E}(P)$ in [W.h]. The expected power is defined in eq. 1.1. The likelihood of occurrence of the different atmospheric conditions is represented by the joint probability density function, $\mathbb{PDF}(u, \theta)$, where u and θ are the Reynolds-averaged wind speed and wind direction.

$$\mathbb{E}(P) = \int_0^{2\pi}\int_0^{\infty} P(u,\theta)\,\mathbb{PDF}(u,\theta)\,du\,d\theta \tag{1.1}$$

The common practice in the wind energy industry is to consider the Reynolds-averaged wind speed variations over the years at a given location to follow a Weibull distribution. On the other hand, the mean wind direction is modeled by defining sectors or bins. Additionally, the correlation between wind direction and wind speed is captured by defining different wind speed Weibull-parameters for each sector. As a result the wind speed and wind direction are independent in each of the R_θ wind direction sectors: $[\theta_j, \theta_{j+1}]$, see eq. 1.2. In this equation, $\mathbb{P}_j$ is the the probability of occurrence of each sector.

$$\mathbb{PDF}(u,\theta) \approx \mathbb{PDF}(u|\theta \in [\theta_j, \theta_{j+1}])\,\mathbb{PDF}(\theta|\theta \in [\theta_j, \theta_{j+1}])\,\mathbb{P}_j \tag{1.2}$$

This assumptions implies that the mean power can be computed in each sector independently and then weighted averaged with respect $\mathbb{P}_j$.

$$\mathbb{E}(P) \approx \sum_{j=1}^{R_\theta} \mathbb{P}_j \left(\int_{\theta_j}^{\theta_{j+1}} \int_0^\infty P(u,\theta)\,\mathbb{PDF}(u|\theta \in [\theta_j, \theta_{j+1}])\,\mathbb{PDF}(\theta|\theta \in [\theta_j, \theta_{j+1}])\,du\,d\theta \right) \tag{1.3}$$

There are different approaches to numerically approximate eq. 1.3: trapezoidal and Simpson's integration rules consist in building linear or quadratic interpolation lines between a number of wind speed and direction evaluation points, $P(u_k, \theta_k)$. Such techniques are the common practice in the industry and they can be interpreted as a weighted average of the evaluation points, see eq. 1.4. The weights shown in this equation, w_k, depend on the integration technique and on the $\mathbb{PDF}(u,\theta)$.

$$\mathbb{E}(P) \approx \sum_{j=1}^{R_\theta} \mathbb{P}_j \left(\sum_{k=1}^{N} w_k\, P(u_k, \theta_k) \right) \tag{1.4}$$

Polynomial chaos with semi-spectral collocation techniques can be used to have a higher accuracy prediction to integrals in the same way as presented in eq. 1.4 with a reduced number of model evaluations. The idea originally introduced in [2], presents quadrature rules based on a normal distribution by the construction of an orthonormal polynomial basis with respect the normal probability density function. The generalized polynomial chaos (gPC) techniques presented in [3] expand this technique to uniform, beta and gamma probability distributions. In the present work, data driven polynomial chaos (aPC) techniques introduced in [4] are applied to the Weibull distribution. Furthermore, the multi-element polynomial chaos presented in [5] is implemented in this paper to deal with integrals over multiple regions of wind speed and wind direction.

2 Polynomial Chaos for AEP

2.1 Multi-element Data-driven Polynomial Chaos (MEPC)

PC techniques enables the user to find the statistical properties of a model with a reduce number of simulations. This is achieved by defining a set of polynomials that are used to interpolate the model response. The purpose of using multiple element PC is to separate the integration region into small sub-regions in which the integration variables can be assumed to be independent. In the present section the theory for building quadrature rules is presented. The theory is presented as a tool to integrate the function $f(x)$ under a random variable, x, characterized by its probability density function $\mathbb{PDF}(x)$. The integral is separated over R_x regions, defined by its end points, x_i.

$$\int_{-\infty}^{\infty} f(x)\,\mathbb{PDF}(x)\,dx = \sum_{i=1}^{R_x} \mathbb{P}_i \int_{x_i}^{x_{i+1}} f(x)\,\mathbb{PDF}(x|x \in [x_i, x_{i+1}])\,dx \tag{2.1}$$

An inner product is defined based on the conditional probability density function such that:

$$\mathbb{E}_i(f) = \langle f \rangle_i = \int_{x_i}^{x_{i+1}} f(x)\, \mathbb{PDF}(x|x \in [x_i, x_{i+1}])\, dx \tag{2.2}$$

$$\mathbb{E}_i(f\,g) = \langle f, g \rangle_i = \int_{x_i}^{x_{i+1}} f(x)\, g(x)\, \mathbb{PDF}(x|x \in [x_i, x_{i+1}])\, dx \tag{2.3}$$

A polynomial of order n is defined as:

$$\pi_n(x) = a_{0n} + a_{1n}\, x + \dots + a_{nn}\, x^n = \sum_{m=0}^{n} a_{mn}\, x^m \tag{2.4}$$

An orthonormal polynomial basis based on the probability distribution consists in a group of polynomials that are orthogonal and that have unitary norm with respect the inner product defined in eq. 2.3. These conditions are summarized in eq. 2.5-2.6. These equations are recursive as they use the coefficients from the lower order polynomials to find the next one. The first polynomial is assumed to be: $\pi_0(x) = 1$. Furthermore the statistical moments of the truncated distribution, $\mathbb{E}_i(x^m)$, up to the $(2j-1)$-th order are required.

$$\langle \pi_n, \pi_k \rangle_i = 0 = \sum_{m=0}^{n} a_{mn}\, \langle x^m, \pi_k \rangle_i = \sum_{m=0}^{n} a_{mn} \left(\sum_{l=0}^{k} a_{lk}\, \mathbb{E}_i(x^{m+l}) \right) \quad \forall k < n \tag{2.5}$$

To close this system of equations orthogonality is solved first. This is done by setting the last coefficient of each polynomial to be: $a_{n\,n} = 1$. Finally the polynomials are normalized:

$$\pi_n = \frac{\pi_n}{\sqrt{\langle \pi_n, \pi_n \rangle}} \tag{2.6}$$

Note that the polynomial basis will be different for each region, since each one has a different conditional $\mathbb{PDF}$. The polynomial basis for the i-th region will be denoted as π_{in}.

The fundamental theorem of Gaussian quadrature states that integrals with respect the $\mathbb{PDF}(x)$ can be approximated using only N evaluation points, x_k. The quadrature rules calculate exactly the integral of functions, $f(x)$, that are a polynomial of order equal or smaller than $2N-1$.

$$\mathbb{E}_i(f) = \langle f \rangle_i = \int_{x_i}^{x_{i+1}} f(x)\, \mathbb{PDF}(x|x \in [x_i, x_{i+1}])\, dx \approx \sum_{k=1}^{N} w_k\, f(x_k) \tag{2.7}$$

where the evaluation points, x_k, are the roots of the N-th polynomial in the basis, $\pi_{i\,N}$, and the weights, w_k, are computed as:

$$w_k = \frac{1}{\sum\limits_{n=0}^{N} (\pi_{in}(x_k))^2} \tag{2.8}$$

By projecting any function, f, into the orthonormal polynomial basis one can approximate its response as a polynomial. Note that since the method can handle different truncation orders in each region in general there are going to be discontinuities in the polynomial response between the regions; this does not produce any problem as the integration in each region is done independently.

$$f(x) = \sum_{n=0}^{\infty} \langle f, \pi_{in} \rangle_i \, \pi_{in}(x) \approx \sum_{n=0}^{N} \langle f, \pi_{in} \rangle_i \, \pi_{in}(x) = \sum_{n=0}^{N} c_{in} \, \pi_{in}(x) \quad \forall x \in [x_i, x_{i+1}] \tag{2.9}$$

The coefficients of the function in the polynomial basis, c_{ij}, can be computed numerically using the Gaussian quadrature rules presented in eq. 2.8; this method is also known as spectral collocation:

$$c_{in} = \langle f, \pi_{in} \rangle_i \approx \sum_{k=1}^{N} w_k \, f(x_k) \, \pi_{in}(x_k) \quad \forall n < N \tag{2.10}$$

Finally the statistical moments of the function can be obtained from its coefficients in the polynomial basis:

$$\mathbb{E}(f) = \sum_{i=1}^{R_x} \mathbb{P}_i \, \mathbb{E}_i(f) = \sum_{i=1}^{R_x} \mathbb{P}_i \, c_{i\,0} \tag{2.11}$$

$$\mathbb{V}(f) = \sum_{i=1}^{R_x} \sum_{k=0}^{N} \mathbb{P}_i \, c_{ik}^2 - \mathbb{E}(f)^2 \tag{2.12}$$

It is very important to notice that the polynomial surrogate function is not used for computing neither the $\mathbb{E}(f)$ nor $\mathbb{V}(f)$. In fact, the $\mathbb{E}(f)$ can be computed using the quadrature rule directly as presented in eq. 2.7; which is equivalent to computing only the 0-th order coefficients; and then using eq. 2.11. If the variance is required then all the coefficients in the polynomial basis are computed and used in eq. 2.12.

2.2 MEPC for wind speed and wind direction:
Due to the independence between wind direction and speed inside a direction sector, a polynomial basis can be constructed independently for wind speed and for wind direction. The 2D polynomial basis is then the product between the one dimensional polynomial basis. Similarly the 1D quadrature weights can be used to define the 2D quadrature rules for each sector. A polynomial basis of orders N_u for wind speed and N_θ for wind direction are built by solving the equations 2.5 and 2.6. To build the orthonormal polynomials the statistical moments of the truncated Weibull are required, as well as the truncated moments for the wind direction distribution. Expressions for the truncated statistical moments of the Weibull distribution are presented in [6].

$$\mathbb{E}(P) \approx \sum_{i=1}^{R_u} \sum_{j=1}^{R_\theta} \langle P \rangle_{ij} \, \mathbb{P}_j \, \mathbb{P}_i \tag{2.13}$$

$$\langle P \rangle_{ij} = \int_{\theta_j}^{\theta_{j+1}} \int_{u_i}^{u_{i+1}} P(u,\theta) \, \mathbb{PDF}(\theta | \theta \in [\theta_j, \theta_{j+1}]) \, \mathbb{PDF}(u | u \in [u_i, u_{i+1}]) \, du \, d\theta$$

$$\approx \sum_{k=1}^{N_u} \sum_{l=1}^{N_\theta} w_{jl} \, \nu_{ik} \, P(u_k, \theta_l) \tag{2.14}$$

The notation used is: w_{jl} represents the weight associated to the l-th root of the N_θ-th order polynomial for the wind direction inside the j-th sector. ν_{ik} represents the weight associated to the k-th root of the N_u-th order polynomial for the wind speed inside the i-th sector. The roots and weights are computed independently as in the single variable case from eq. 2.8. Once the polynomial families have been obtained for each sector, one can capture the statistical properties of the power as produced by the random wind speed and wind direction by projecting the power curve function into the 2D polynomial basis:

$$P(u,\theta) \approx \sum_{k=0}^{N_u} \sum_{l=0}^{N_\theta} c_{ijkl}\, \pi_{ik}(u)\, \pi_{jl}(\theta) \quad \forall \theta \in [\theta_j, \theta_{j+1}] \quad \text{and} \quad \forall u \in [u_i, u_{i+1}] \tag{2.15}$$

The coefficients of the power in the polynomial basis can be computed numerically using the quadrature rules:

$$c_{ijkl} = \langle P, \pi_{ik}\pi_{jl} \rangle_{ij} \approx \sum_{k=1}^{N_u} \sum_{l=1}^{N_\theta} w_{jl}\, \nu_{ik}\, P(u_k, \theta_l)\; \pi_{ik}(u_k)\, \pi_{jl}(\theta_l) \tag{2.16}$$

Finally the statistical moments of the power distribution can be obtained from the polynomial basis coefficients:

$$\mathbb{E}(P) = \sum_{i=1}^{R_u} \sum_{j=1}^{R_\theta} \mathbb{P}_i\, \mathbb{P}_j \left(c_{ij00} \right) \tag{2.17}$$

$$\mathbb{V}(P) = \sum_{i=1}^{R_u} \sum_{j=1}^{R_\theta} \left(\mathbb{P}_i\, \mathbb{P}_j \sum_{k=0}^{N_u} \sum_{l=0}^{N_\theta} c_{ijkl}^2 \right) - \mathbb{E}(P)^2 \tag{2.18}$$

3 Single wind turbine case

A single wind turbine case is considered in order to verify the concept of using MEPC techniques to estimate the AEP. For this case the power is only a function of the Weibull distributed wind speed. The wind speed operational range of the wind turbine and optionally the region with rated power are used to define the regions of integration. Some examples of the orthonormal polynomial basis for the truncated Weibull distribution are shown in figure 1, as well as the obtained quadrature points and the polynomial surrogates. In this figure it can be observed that the polynomial response passes through the evaluation points. And it can be expected that oscillations might appear for high polynomial orders, especially for regions with constant power.

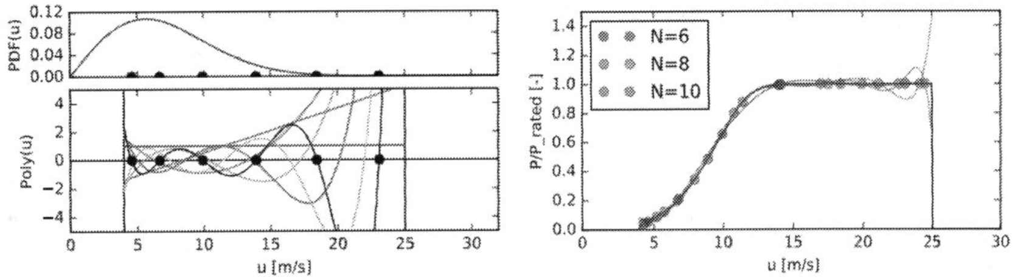

Figure 1. (Left) 6-th order polynomial basis for an individual region Weibull distributed wind speed: $u \in [4, 25]$ [m/s] (Right) Power curve (black line), quadrature evaluation points and polynomial surrogate of the power curve.

Figure 6 shows the convergence for different integration schemes for the relative error of $\mathbb{E}(P)$ and of $\mathbb{V}(P)$ vs. the number of model evaluations used in the integration N. The relative error is calculated with respect the trapezoidal integration rule with 10^5 points. The instabilities in the MEPC quadrature results are produced by the difference in the location of the evaluation points. The location of the evaluation points and the accuracy of the quadrature rules depend on the Weibull parameters because the method depends on the $\mathbb{PDF}(u)$. The benefit obtained for using MEPC is considerable; as it reduced the number of model evaluations from 21 (current common practice) down to 6 model evaluations for similar relatives errors of 0.1% in expected power and variance. Note that the method becomes numerically unstable at high polynomial orders (larger than 10) even with careful handling of number precision because of the large differences in the values of the high order statistical moments, $\mathbb{E}(x^m)$. The results presented in this article use the following python libraries: the polynomial handling class on numpy (http://www.numpy.org), scipy (http://www.scipy.org) for its implementation of the trapezoidal/Simpson's methods and mpmath (http://mpmath.org/doc/current/index.html) for handling the polynomial construction and root finding algorithms with an arbitrary high precision.

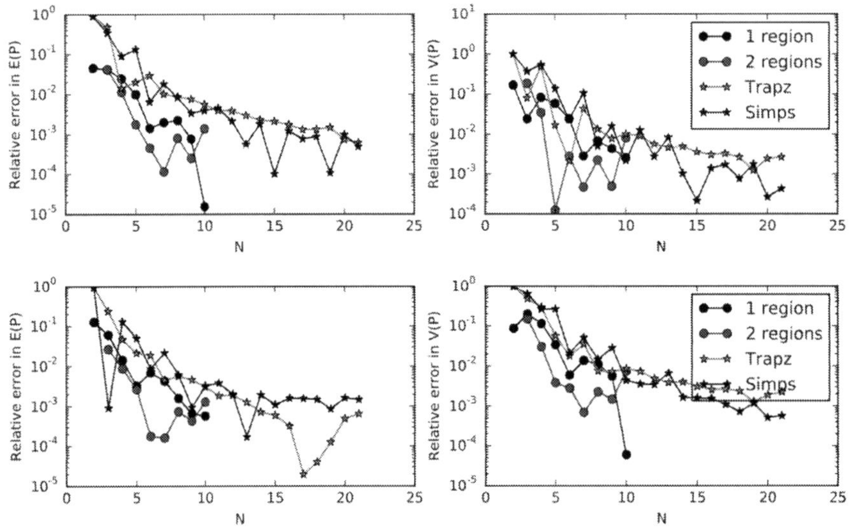

Figure 2. Convergence for different integration schemes vs. number of model evaluations N for the relative error of (left) $\mathbb{E}(P)$ and (right) $\mathbb{V}(P)$. Weibull distribution with shape parameter $k_W = 2$ and scale parameter (Top) $A = 8.0$ [m/s] (Bottom) $A = 10.0$ [m/s].

4 Simple wind power plant case

In this section the multi-element polynomial chaos technique is applied to a simple wind power plant. The power is computed using G. C. Larsen's semi-empirical wake model [1]. The local weather is characterized by a Weibull distributed wind speed and an uniformly distributed wind direction inside each wind direction sector. Note that this type of assumption is the standard in most wind energy flow models such as WAsP. On the other hand, the method proposed in this paper could be applied for an arbitrary distribution of wind direction inside each sector. The only requirement is that the statistical moments are known or computable (some examples of marginal distributions for wind direction are: Multiple mixture of Von Mises, or kernel density estimated $\mathbb{PDF}$).

Under the assumptions of the present model, the correlation between wind speed and wind direction is modeled by having different parameters of the Weibull distribution as a function of the wind direction sector. The 1D polynomial basis for the wind direction are the Legendre polynomials since the conditional probability density function for the wind direction inside each sector is assumed to be uniform. The final number of evaluation points is the tensorial product (meshgrid combination) of the wind speed and the wind direction quadrature points.

The layout of the simple offshore power plant is shown in figure 3, as well as the power and thrust coefficient curves for the Vestas V80-offshore turbines. The wind rose and Weibull parameters for the site are shown in figure 4.

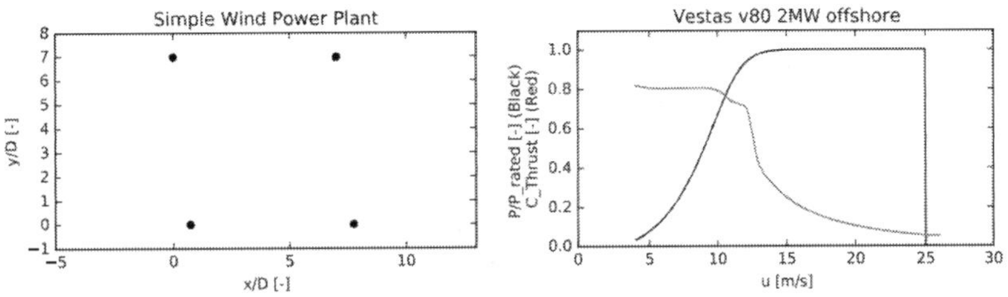

Figure 3. Simple wind power plant: (Left) layout and (Right) power and thrust coefficient curves for the turbines.

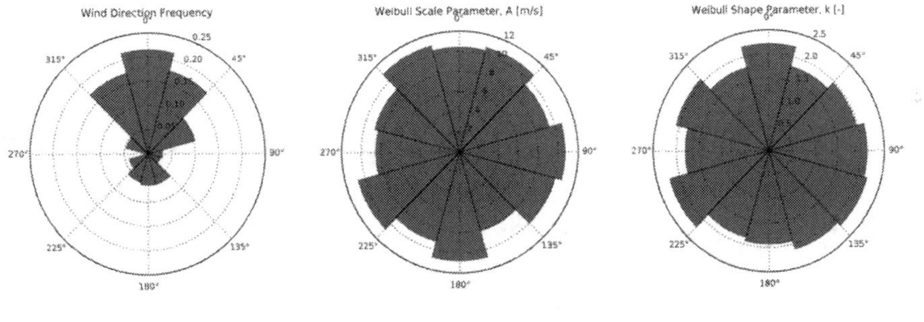

Figure 4. (Left) Wind rose of the site. (Middle) Weibull scale parameter for each sector. (Right) Weibull shape parameter for each sector.

The wind park's power curve , $P(u, \theta)$ is shown in figure 5 (Left). This power curve is computed using 30960 model evaluations: every 0.5 [deg.] and 0.5 [m/s]. This figure additionally shows an example of the resulting polynomial surrogate power curve. It can be observed that the power surrogate captures the rough behavior of power. As expected the polynomial surrogates present oscillations in the region of rated power as well as discontinuities between the regions; this does not cause any problem due to the fact that the integration in each region is done independently and since the oscillations around the *real* power curve cancel out when computing the $\mathbb{E}(P)$ or $\mathbb{V}(P)$. It might be of interest to visualize the polynomial surrogate because if the surrogate presents to many oscillations it means that the order of the polynomial is to high, and a further division of integration region can be executed.

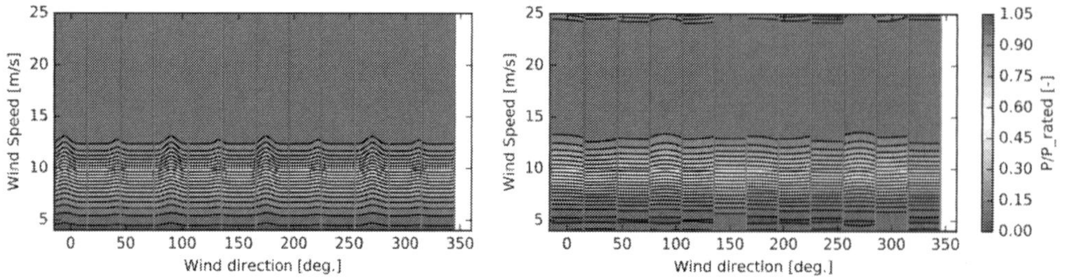

Figure 5. Power plant power curve: (Left) computed from $720 \times 43 = 30960$ model evaluations. (Right) Polynomial surrogate of the power curve order computed in $R_D = 12$ sectors, with $N_u = 8$-th order in wind speed and $N_\theta = 5$-th order in wind direction for a total of $12 \times 5 \times 9 = 540$ simulations.

The convergence of the expected power and of the standard deviation of power normalized by the rated wind farm power are shown in figure 6. The same numerical instabilities produced by the shifting of root locations can be observed. Furthermore it can be observed that the MEPC algorithm converges very fast to the desired values with relatives error of order of 1%. This relative error is computed with respect the predictions with 30960 model simulations and it is show in the figure as the shaded purple region. It is important to remark that surrogates with a low polynomial degree, 2-nd in wind direction per sector and 4-th in wind speed, capture the expected power and its variance within 1% accuracy with around 10 times less model evaluations than the trapezoidal integration scheme used in the traditional binning method. Note that when $R_u = 2$, the regions are defined as: $[4, 14]$ and $[14, 25]$ [m/s], and the order on the rated power region is always $N_u[2] = 1$.

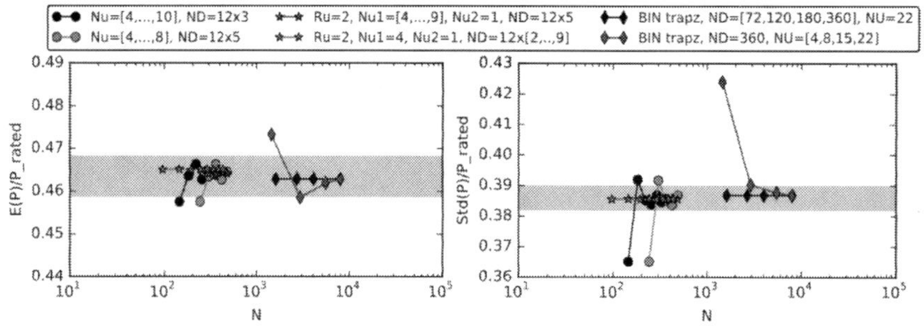

Figure 6. Convergence for different integration schemes vs. number of point evaluation N for the (left) normalized expected power $\mathbb{E}(P)$ and (right) normalized standard deviation of power, $\mathbb{S}(P)$ for number of model evaluations.

5 Real wind power plant case - Horns Rev 1

In this section the multi-element polynomial chaos technique is applied to the Danish offshore wind power plant Horns Rev 1 (HR1) co-owned by Vattenfall and DONG Energy. This wind farm is located in the Western coast of Denmark, at a distance of 9 to 12 [km] from the coast. This plant has has been studied in several articles and it is one of the benchmark cases for wind turbine wake models. The layout of HR1 is shown in figure 7, along with the power and thrust coefficient curves for the Vestas V80-offshore turbines. The wind rose and Weibull parameters for the site are shown in figure 8.

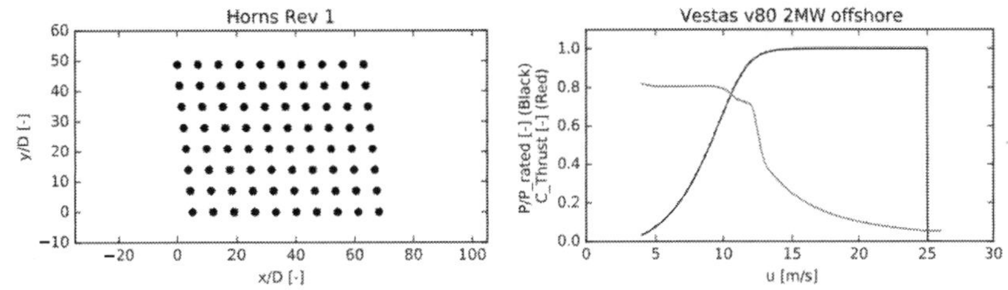

Figure 7. Horns Rev 1 power plant: (Left) layout and (Right) power and thrust coefficient curves for the turbines.

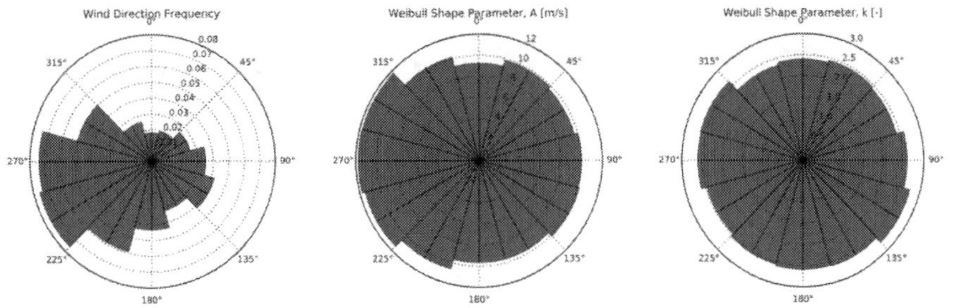

Figure 8. Horns Rev 1 plant: (Left) Wind rose of the site. (Middle) Weibull scale parameter for each sector. (Right) Weibull shape parameter for each sector.

The power curve of Horns Rev 1, $P(u, \theta)$ is shown in figure 9. As in the previous section the power curve is computed using 30960 simulations of G. C. Larsen's model [1]. An example of a polynomial surrogate for the power curve is also presented in this figure. As expected, the polynomial surrogate presents discontinuities between the sectors and it contains oscillations in the region of rated power. As discussed in the previous section, the discontinuities do not cause problems in the computation of AEP because each region is treated individually. Note that the 12 wind direction sectors were divided into 2 regions in order to reduce the degree of the polynomial basis required to capture the dependency on wind direction.

Convergence of power expectation and standard deviation can be observed in figure 10. It can be observed that the MEPC algorithm converges very fast to the desired values with relatives error of order of 1%. This relative error is computed with respect the predictions done using 30960 model simulations and it is show in the figure as the shaded purple region. It is important to remark that surrogates with a low polynomial degree, 2-th in wind direction per sector and 4-th in wind speed, capture the expected power and its variance within 1% accuracy using 10 times less model evaluations than the trapezoidal integration scheme used in the traditional binning method. Note that when $R_u = 2$, the regions are defined as: $[4, 16]$ and $[16, 25]$ [m/s], and the order on the rated power region is always $N_u[2] = 1$.

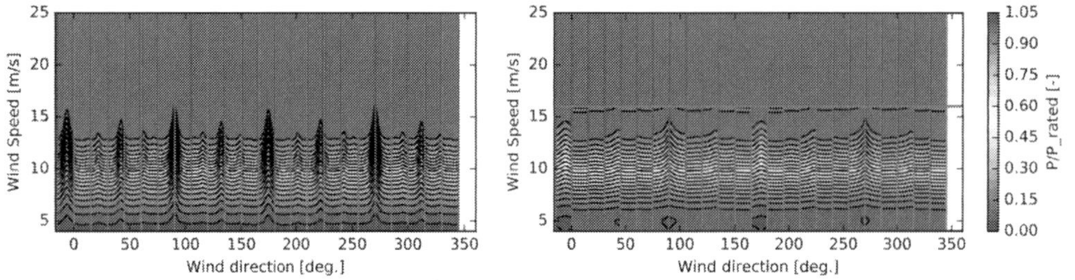

Figure 9. Horns Rev 1 power curve: (Left) computed from $720 \times 43 = 30960$ model evaluations. (Right) Polynomial surrogate of the power curve order computed in $R_D = 24$ sectors, with 4-th order in wind speed and 3-th order in wind direction for a total of $24 \times 3 \times 6 = 288$ simulations.

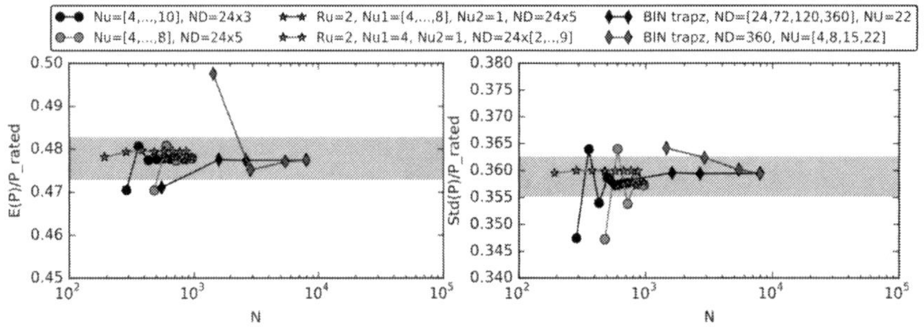

Figure 10. Convergence for different integration schemes vs. number of point evaluation N for the (left) normalized expected power $\mathbb{E}(P)$ and (right) normalized standard deviation of power, $\mathbb{S}(P)$ for number of model evaluations.

6 Discussion

The conceptual methodology of MEPC has been proven as having the potential to handle arbitrary local wind resources probability distributions. The methodology presented in the current paper has been implemented to handle user defined number/machine precision. Despise these efforts the polynomial basis building algorithm presents instabilities for large order of polynomial basis order. The quadrature rules are a fundamental step to find the statistical parameters of the power. Further research is still required to assure stability in the convergence rates of polynomial chaos for arbitrary Weibull distributions.

AEP production calculations with high fidelity flow models are restricted by their large computational requirements, in particular due to the large number of model simulations required to estimate the AEP with a certain level of accuracy. There is a need for methods to reduce the number of model evaluations required to compute power statistics. Furthermore analysis of uncertainty in power production have shown that the uncertainty in AEP can go up to 10% depending on the uncertainty in the input parameters [7]. The minimum requirement of AEP accuracy should be between $0.1 - 1\%$. MEPC method can achieve a relative error in the calculation of AEP smaller than 1% for a reduced number of evaluations. A reduction by a factor of 10 in comparison to current techniques can be achieved using MEPC with $192 = 24 \times 4 \times 2$ model simulations: this accounts for 2 regions of wind speed polynomial basis of orders $N_U = [4, 1]$ and a 2-nd order basis for wind direction.

The presented method is designed to be applied in optimization under uncertainty in which AEP calculations are going to be calculated on every optimization step (possible thousands of times). Each optimization step will propose a new wind power plant layout which makes the use of previous simulations impossible. Building the MEPC surrogate is a fast process, specially if the polynomial basis in wind direction and wind speed are precomputed and stored before the optimization technique. Furthermore, in optimization algorithms the accuracy of the AEP is gradually refined as the objective function increment diminishes, this can be achieved with the present method by increasing the order of the polynomial basis. Consequently a good estimation of AEP with few model simulations can speed up the process of optimization significantly and a more accurate AEP estimation will only be computed for the final steps.

Further investigation of the use of other types of functional basis are planed to be developed such as: (1) the use of polynomial basis for correlated wind velocity vector and turbulence intensity, (2) the use of wavelets for wind speed to deal with the discontinuities in the power curve and (3) a mixtures of polynomials and Fourier series for wind speed, wind direction and turbulence intensity. These techniques promise to have a better estimation of AEP but at the cost of increasing the number of model evaluations in comparison to the method presented in this article.

Acknowledgments

This work was supported by the International Collaborative Energy Technology R&D Program of the Korea Institute of Energy Technology Evaluation and Planning (KETEP), granted financial resource from the Ministry of Trade, Industry & Energy, Republic of Korea. (No. 20138520021140).

References

[1] Larsen G C 2009 A simple stationary semi-analytical wake model Technical report, risø-r-1713(en) Risø-DTU

[2] Wiener N 1938 *American Journal of Mathematics* 897–936

[3] Xiu D and Karniadakis G E 2002 *SIAM Journal on Scientific Computing* **24** 619–644

[4] Oladyshkin S and Nowak W 2012 *Reliability Engineering & System Safety* **106** 179–190

[5] Wan X and Karniadakis G E 2006 *SIAM Journal on Scientific Computing* **28** 901–928

[6] McEwen R P and Parresol B R 1991 *Communications in Statistics-Theory and Methods* **20** 1361–1372

[7] Clerc A, Anderson M, Stuart P and Habenicht G 2012 *Journal of Wind Engineering and Industrial Aerodynamics* **111** 85–94

Large-eddy simulation of the diurnal variation of wake flows in a finite-size wind farm

Mahdi Abkar, Ahmad Sharifi, Fernando Porté-Agel

École Polytechnique Fédérale de Lausanne (EPFL), Wind Engineering and Renewable Energy Laboratory (WiRE), EPFL-ENAC-IIE-WIRE, CH-1015 Lausanne

E-mail: fernando.porte-agel@epfl.ch

Abstract. In this study, large-eddy simulation (LES) is used to study the evolution of the wind-turbine wakes and their effects on power losses inside an idealized finite-size wind farm in the course of a full diurnal cycle. In the LES, turbulent subgrid-scale stresses are modeled using tuning-free Lagrangian scale-dependent dynamic models, while the turbine-induced forces are parameterized using a dynamic actuator disk model with rotation. The simulation results show a strong effect of atmospheric stability on the wind farm wakes and associated power losses. During the night, the relatively low turbulence intensity of the ambient ABL flow results in a relatively slow rate of entrainment of momentum into the wake and, consequently, a slow wake recovery. In contrast, during the day the positive buoyancy flux and associated turbulence production lead to a relatively high turbulence level in the background ABL flow, which enhances turbulent mixing and wake recovery. As a result, the averaged power deficit in the wind farm is found to increase with increasing thermal stability. In particular for that day, the averaged power deficit increased from 28% under the most convective condition to about 57% under the most stable condition.

1. Introduction

In a wind farm, with a relatively small spacing between turbines, the wind speed in the turbine wakes does not recover to its upstream unperturbed value. Hence, the wakes from upstream turbines impact the performance of downstream turbines. Any mechanism that changes the recovery rate of wake flow can increase or decrease the significance of this impact as the greater loss of power production in the downstream turbines is associated with slower rate of wake recovery. In several studies, it has been shown that besides land or sea surface characteristics such as the roughness and topography, thermal stratification has an important effect on the wind-turbine wakes and the performance of wind farms (e.g. [1, 2, 3, 4, 5]). For instance, Zhang et al. [6] carried out wind-tunnel experiments of the down-scaled wind-turbine models to investigate the effect of thermal stability on wind-turbine wakes under the convective and neutral conditions. They found a smaller velocity deficit (about 15% at the wake center) in the CBL compared with the wake of the same wind turbine in the neutral boundary layer. Barthelmie and Jensen [7] analyzed power output data from Nysted wind farm and found reduced power efficiency in the wind farm under stable conditions. Churchfield et al. [8] used LES (Open-FOAM) to study the effects of unstable stratification on the power deficit as well as fatigue loads on a wind turbine placed downstream of another turbine. They concluded that under unstable stratification, the power deficit was 15-20% lower in comparison with neutral condition; also fatigue loads were greater. Despite the valuable insights provided by the above-mentioned studies, a number of key questions remain regarding the impact of thermal stability on the dynamics of turbine wakes and its effects on the power deficit inside finite-size wind farms.

Here, an in-house LES code (the WiRE LES) is used to investigate the effects of atmospheric stability changes through a full diurnal cycle on the performance of a finite size wind farm and the wake flow structure, dynamics and associated power losses. In the WiRE LES, the SGS turbulent fluxes are parameterized using tuning-free Lagrangian scale-dependent dynamic models [9] and the turbine-induced forces are modelled using the rotational dynamic actuator disk model [10, 11]. A short introduction of the LES framework is presented in Section 2. Section 3 presents the the LES result for precursor simulation of the ABL flow in through the wind turbines. Finally, a summary and conclusions are provided in Section 4

2. Large-Eddy Simulation Framework

2.1. LES Governing Equations

LES solves the filtered continuity equation, the filtered Navier-Stokes equations (written here using the Boussinesq approximation), and the filtered transport equation for potential temperature:

$$\frac{\partial \widetilde{u}_i}{\partial x_i} = 0,$$

$$\frac{\partial \widetilde{u}_i}{\partial t} + \widetilde{u}_j \frac{\partial \widetilde{u}_i}{\partial x_j} = -\frac{\partial \widetilde{p}^*}{\partial x_i} - \frac{\partial \tau_{ij}^d}{\partial x_j} + \delta_{i3} g \frac{\widetilde{\theta} - \langle \widetilde{\theta} \rangle}{\theta_0} + f_c \varepsilon_{ij3}(\widetilde{u}_j - G_j) - \frac{f_i}{\rho}, \tag{1}$$

$$\frac{\partial \widetilde{\theta}}{\partial t} + \widetilde{u}_j \frac{\partial \widetilde{\theta}}{\partial x_j} = -\frac{\partial q_j}{\partial x_j},$$

where the tilde represents a spatial filtering at scale $\widetilde{\Delta}$, t is time; $\widetilde{u}_i$ is the instantaneous resolved velocity in the $i-$direction (with $i = 1, 2, 3$ corresponding to the streamwise (x), spanwise (y) and vertical (z) directions, respectively); $\widetilde{\theta}$ denotes the resolved potential temperature; θ_0 is the reference temperature, the angle brackets represent a horizontal average; G is the geostrophic wind; g refers to the gravitational acceleration; δ_{ij} is the Kronecker delta; $\widetilde{p}^* = \widetilde{p}/\rho + \frac{1}{3}\tau_{kk}$ is the modified kinematic pressure; f_i is a body force (per unit volume) used to model the effect of the turbine on the flow, ρ is the fluid density; q_j denotes the SGS heat flux; τ_{ij} represents the kinematic SGS stress; and τ_{ij}^d is its deviatoric part. Note that τ_{ij}^d and q_j are unknown and need to be parameterized as a function of filtered (resolved) fields. Regarding the SGS turbulent fluxes, the SGS stress and the SGS heat flux are computed with eddy-viscosity and eddy-diffusivity models, respectively, as follows [12]:

$$\tau_{ij}^d = -2\widetilde{\Delta}^2 C_S^2 |\widetilde{S}| \widetilde{S}_{ij}, \qquad q_j = -2\widetilde{\Delta}^2 C_S^2 Pr_{sgs}^{-1} |\widetilde{S}| \frac{\partial \widetilde{\theta}}{\partial x_j}, \tag{2}$$

where $\widetilde{S}_{ij} = (\partial \tilde{u}_i/\partial x_j + \partial \tilde{u}_j/\partial x_i)/2$ is the resolved strain-rate tensor whose magnitude is $|\widetilde{S}|$. C_S and $C_S^2 Pr_{sgs}^{-1}$ are the Smagorinsky and the lumped coefficients, respectively, where Pr_{sgs} is the SGS Prandtl number. In this study, we use Lagrangian scale-dependent dynamic models [9] to compute the local optimized value of the model coefficients without any *ad hoc* tuning. More details can be found in Porté-Agel et al. [13], Porté-Agel [14], and Stoll and Porté-Agel [9].

To parameterize the turbine-induced forces, the actuator-disk model with rotation [15] is used. Through this model, the aerodynamic forces are determined using the lift and drag characteristics of the airfoil type as the well as the local flow conditions. Figure 1 shows a cross-sectional element of radius r in the (θ, x) plane, where x is the axial direction. Different forces, velocities and angles are shown in this figure. $V_x = V_x(r, \theta)$ and $V_\theta = V_\theta(r, \theta)$ are axial and tangential velocities of the incident flow at the blades, respectively, in the inertial frame of reference. The local velocity relative to the rotating blade is defined as $V_{rel} = (V_x, \Omega r - V_\theta)$, where Ω is the turbine angular velocity. The angle of attack is defined as $\alpha = \varphi - \gamma$, where $\varphi = \tan^{-1}(V_x/(\Omega r - V_\theta))$ is the angle between V_{rel} and the rotor plane and γ is the local pitch angle. To determine the forces acting on the rotor disk, we consider an annular area of differential size $dA = 2\pi r dr$. The resulting force per unit rotor area is given by:

$$\mathbf{f}_{disk} = \frac{d\mathbf{F}}{dA} = \frac{1}{2}\rho V_{rel}^2 \frac{Bc}{2\pi r}(C_L\mathbf{e}_L + C_d\mathbf{e}_D), \tag{3}$$

where B is the number of blades, $C_L = C_L(\alpha, Re_c)$ and $C_D = C_D(\alpha, Re_c)$ are the lift and drag coefficients, respectively, Re_c is the Reynolds number based on relative velocity and chord length, c, and $\mathbf{e}_L$ and $\mathbf{e}_D$ denote the unit vectors in the directions of the lift and the drag, respectively. Through the simulation, the tangential (V_θ) and the axial (V_x) velocities at the rotor plane are known. Hence, the local velocity relative to the rotating blade (V_{rel}) and the angle of attack $\alpha = \varphi - \gamma$ can be computed. Then, the resulting force is obtained using Equation 3.

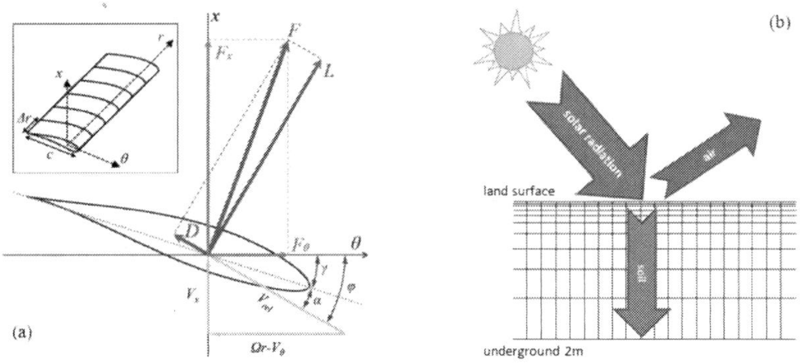

Figure 1. (a) A cross-sectional airfoil element. (b) schematic of energy balance, and structured grid for soil showing a depth of 2[m] using a logarithmic spacing in vertical direction.

2.2. Numerical Setup

The LES code used in this study is a modified version of the one described by Albertson and Parlange [16], Porté-Agel et al. [13], Stoll and Porté-Agel [9], Wu and Porté-Agel [15], and Ablar and Porté-Agel [17]. In the code, the computational domain is broken uniformly into $N_x \times N_y \times N_z$ grid points with a spatial resolution of $\Delta x \times \Delta y \times \Delta z$ in the streamwise, spanwise and wall-normal directions, respectively.The grid planes are staggered in the $z-$direction, with the first vertical velocity plane on the surface, and the first horizontal velocity plane $\Delta_z/2$ from the surface. The vertical derivatives are approximated with second-order central differences and the horizontal directions are discretized pseudo-spectrally which requires periodic boundary conditions. Full dealiasing of the nonlinear terms is obtained by padding and truncation according to the 3/2 rule [18]. The time advancement is carried out using a second-order-accurate Adams-Bashforth scheme [19]. The upper boundary condition consists of a stress/flux-free condition. At the bottom surface, the standard formulation based on local application of Monin-Obukhov similarity theory [20] is used. Although this theory was developed for mean quantities, it is a common practice in LES of atmospheric flows to compute the instantaneous filtered surface momentum flux [21, 22] as follows, $\tau_{i3}|_w = -u_*^2 \frac{\tilde{u}_i}{\tilde{u}_r} = -(\frac{\tilde{u}_r \kappa}{\ln(z/z_o) - \Psi_M})^2 \frac{\tilde{u}_i}{\tilde{u}_r}$, where

$\tau_{i3}|_w$ is the instantaneous local wall stress; u_* is the friction velocity; z_o is the aerodynamic surface roughness; κ is the von Kármán constant; and $\tilde{u}_r$ is the local filtered horizontal velocity at the first level $z = \Delta z/2$. In a similar manner, the surface heat flux is computed as, $q_3|_w = \frac{u_*\kappa(\theta_s - \tilde{\theta})}{\ln(z/z_{ot}) - \Psi_H}$, where θ_s is the surface (ground level) potential temperature, z_{ot} is the surface roughness for potential temperature and is set to 10% of the aerodynamic surface roughness;and $\tilde{\theta}$ denotes the resolved potential temperature at the first vertical level. Ψ_M and Ψ_H are the stability corrections for momentum and heat, respectively, and are defined as follows [23, 24]:

$$\Psi_M = \begin{cases} -4.7\frac{z}{L} & \text{for stable conditions,} \\ 2\ln[\frac{1}{2}(1+x)] + \ln[\frac{1}{2}(1+x^2)] - 2\tan^{-1}[x] + \frac{\pi}{2} & \text{for unstable conditions,} \end{cases} \tag{4}$$

and

$$\Psi_H = \begin{cases} -7.8\frac{z}{L} & \text{for stable conditions,} \\ 2\ln[\frac{1}{2}(1+x^2)] & \text{for unstable conditions,} \end{cases} \tag{5}$$

where $L = -(u_*^3\theta_0)/(\kappa g q_3|_w)$ is the local Obukhov length, θ_0 is the reference temperature and is set to $\theta_0 = 293K$, and $x = (1 - 15\ z/L)^{1/4}$.

The LES of the diurnal cycle of ABL is performed over a domain of size 4.8 km $\times$2.4 km $\times$1.6 km discretized into 120 $\times$ 160 $\times$ 160 grid-points. Hence, the spatial resolution is $\Delta_x = 40$m, $\Delta_y = 15$m, and $\Delta_z = 10$m. It is worth mentioning that based on previous grid-resolution sensitivity results [15, 25], the grid resolution used in this study is well suited for the LES framework to account for the most significant characteristics of wind-turbine wakes. The boundary layer is driven by an imposed, uniform geostrophic wind (G) of 10 m/s. The Coriolis parameter is $f_c = 1 \times 10^{-4}$ rad/s corresponding to a latitude of about 43.3^o. The aerodynamic surface roughness z_o is set to 0.05 m, and the reference potential temperature is $\theta_0 = 293$ K. In this study, in order to determine the land-surface temperature and surface heat flux, we adopt a surface thermal energy balance approach along with 10 levels of soil temperature to a depth of 2 m using a logarithmic spacing, as shown in Fig. 1b. A similar method has been used by Deardorff [26], but in this study we release the assumption of horizontally homogeneous, and solve a three-dimensional heat transfer equation for the soil as: $\frac{\partial \tilde{\theta}}{\partial t} = \alpha_{soil}\nabla^2\tilde{\theta}$, where α_{soil} is the soil thermal diffusivity. The simulations are initialized with a constant streamwise velocity $U_g = 10$ m/s and zero velocity for the spanwise and vertical components. The potential temperature is initialized with a uniform temperature of 293 in the lowest 100 m. In this layer, a very small random perturbations are added to the initial velocity and potential temperature fields. Above this layer, the ABL is set to be laminar with a constant lapse rate of 10 K/km. The initial soil temperature is also uniform and equal to 293 K. The soil is assumed to be dry ground, and its diffusivity is 5.0×10^{-7} m^2/s (averaged value of desert and concrete [26, 23]). The heat capacity of the soil is 1.3×10^6 J/m^{3}K [23]. For the surface thermal energy balance, we use the typical profile for the net radiative flux throughout the day [23] as shown in Fig. 2. The simulations are carried out for two full diurnal cycles. To minimize the effects of the initial conditions on the results, our analysis is focused on the second diurnal cycle.

In the simulation of the diurnal ABL flow through the wind farm, 36 Vestas V-80 2MW wind turbines arranged in 6 rows and 6 columns are immersed in the flow. The wind farm is infinite in the spanwise direction as a result of the implicit periodic boundary condition at the spanwise boundaries. Each turbine has a hub height of $h_{hub} = 70$ m and rotor diameter of $D = 80$ m. More details on the turbine characteristics can be found in Wu and Porté-Agel [11]. $S_x = 7D$ and $S_y = 5D$ denote the streamwise and spanwise spacings between turbines, respectively. To avoid the downwind flow affecting the flow upwind of the wind turbine due to the periodic boundary conditions, a buffer zone upstream the turbine is employed to adjust the flow from the downwind condition to that of a fully turbulent boundary-layer inflow condition. The inflow

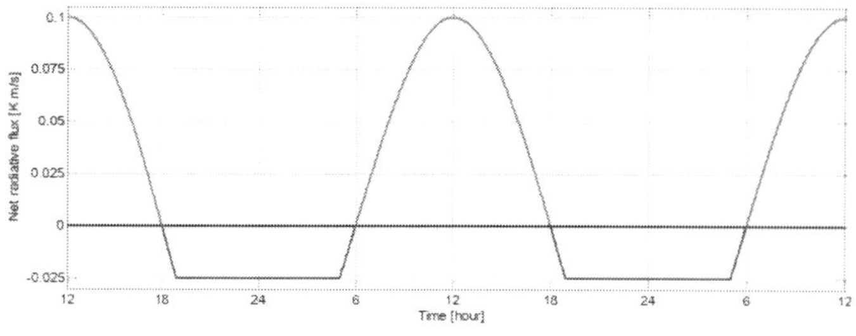

Figure 2. Variation of the net radiative flux throughout two full days.

condition is obtained from separate (without turbine) simulations of fully-developed ABL flows over horizontally-homogeneous flat surfaces [27, 15, 28]. In order to ensure that the mean velocity direction is always aligned with the turbine axes at the hub-height level, a control algorithm similar to Sescu and Meneveau [29] is used. In this method a gradual adjustment of the geostrophic wind is applied by adding a source term in the momentum equation. This method allows us to study the effect of wind-farm layout on the ABL flow as well as the power output from wind turbines [30].

3. Results

Figure 3 shows the the temporal and spatial evolution of the turbulent kinetic energy (TKE) during the two diurnal cycles. The growth of the boundary layer during the day, due to the entrainment from the overlying inversion layer, is clear in this figure. An strong contrast in the magnitude and vertical extent of the TKE between the daytime and nighttime profiles is also observed. As shown in this figure, the turbulence associated with the convective heating during the day is almost disappeared after the sun set.

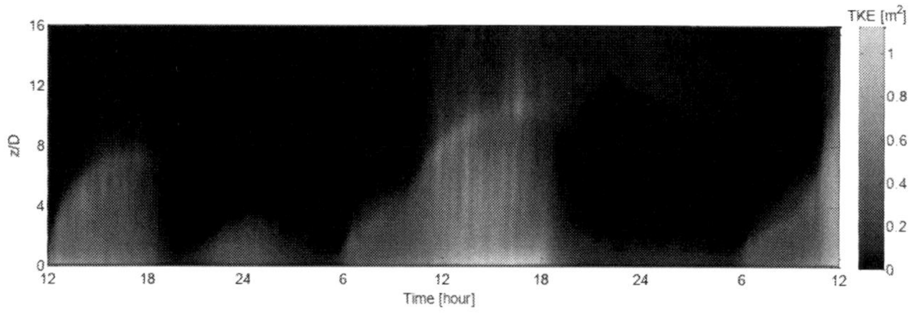

Figure 3. Time and space variation of turbulent kinetic energy (TKE) during the 2 full diurnal cycles.

Figure 4 illustrates temporal evolution of the Obukhov length throughout the two full diurnal cycles. The Obukhov length is a common measure to determine the state and strength of the thermal stability in the ABL [31]. In this study, in order to study the wake flow variability in more detail, two time spans (from the second day), based on the magnitude of the Obukhov length, are selected in which the ABL is strongly stable, and strongly unstable.During the the 2nd day, the first hour from 12 : 00 to 13 : 00 (the time span confined between the two red lines in Fig. 4), with the Obukhov length around $L \cong -95$ m, is selected to study the turbulent ABL under the condition. Beside that, in the morning of the 2nd day, from 04 : 00 to 05 : 00 (the time span confined between the two blue lines in Fig. 4), the Obukhov length is about $L \cong 76$ m which indicates a strong stability condition in the ABL. As a result, this time span is also selected to study the wake flow under the stable stratification.

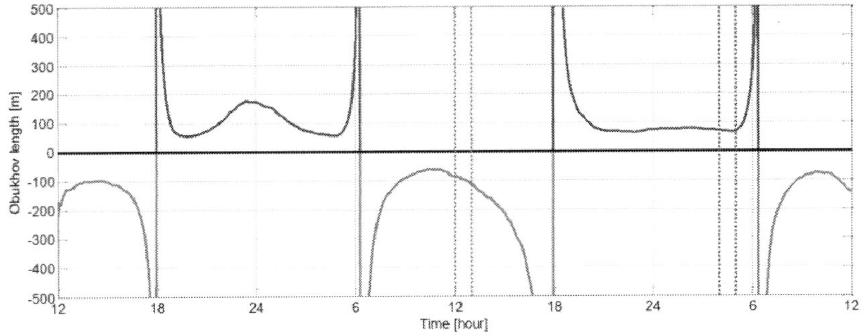

Figure 4. Time variation of the Obukhov length in the simulations.

Figures 5a and 5b display vertical profiles of the horizontally-averaged wind velocity magnitude, and wind direction (obtained from the precursor run), respectively, for each case. The blue lines denote the stable case, and the red lines denote the unstable case. As expected, under the stable condition, the mean wind shear is stronger near the surface compared with the convective case (Fig. 5a). It is also clear that, as a result of the Coriolis effects, the wind direction changes with height, resulting in a lateral shear in addition to the vertical shear (Fig. 5b). Especially for the stable condition, the boundary layer is very shallow, which leads to a strong lateral shear. Figure 5c displays vertical profiles of the streamwise turbulence intensity for each case. As seen in this figure, in the case of very stable stratification, the strong negative buoyancy flux acts against the shear production from the surface and decreases the ambient turbulence intensity in comparison with the unstable case. In contrast, during the day, the strong positive buoyancy flux stirs the air above the surface and increases the turbulence intensity significantly. Similarly, the total shear stress in the stable condition is lower compared with the unstable one (Fig. 5d). Figure 5e illustrates vertical profiles of the horizontally-averaged potential temperature for each case. The vertical distribution of horizontally-averaged soil temperature for the two cases is also plotted in Fig. 5f. As shown in this figure, and consistent with the analytical solution (see e.g., [23]), during the day with positive heat flux from the surface soil temperature decays exponentially with depth, and that changes mostly occur within a shallow layer near the surface (0 − 50cm).

Figure 6 shows two-dimensional fields of the mean simulated wind velocity in a horizontal $x - y$ plane through the turbine hubs. The normalized velocity through the center of the turbines (at hub height), averaged over all 6 rows of turbines, is also plotted with downwind distance in Fig. 7. In all these figures, the reduction in velocity inside the wakes is clearly visible. Nonetheless, obvious differences are observed between the structure and strength of the wakes in the two stability regimes. Specifically, in the convective boundary layer, with a higher turbulence level, the wake recovers faster. This is due to the fact that, in the unstable condition, the positive buoyancy flux at the surface creates thermal instabilities which enhance the TKE and the turbulent mixing. In contrast, in the stable condition, the negative buoyancy flux at the surface damps the turbulence and, consequently, reduces the turbulent mixing. As a result, it takes a longer distance for the wake to recover. This is consistent with recent experimental and

Journal of Physics: Conference Series **625** (2015) 012031 doi:10.1088/1742-6596/625/1/012031

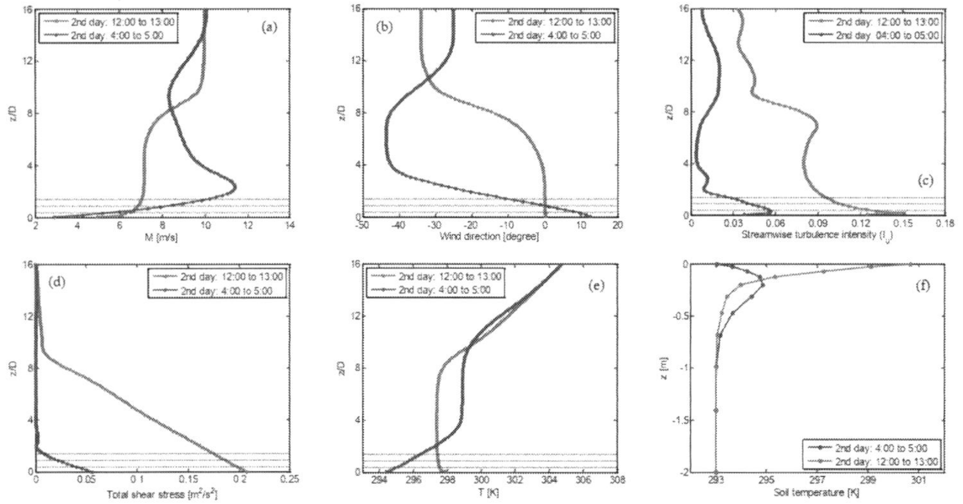

Figure 5. Vertical profile of the horizontally-averaged a) wind velocity magnitude, b) wind direction, c) streamwise turbulence intensity, d) total shear stress, e) air potential temperature, and f) soil temperature. The horizontal dotted lines show the top-tip, hub and bottom-tip heights.

numerical studies that show that the recovery rate of the wake from a stand-alone wind turbine recovers faster with decreasing level of stratification [6, 1, 32, 5].

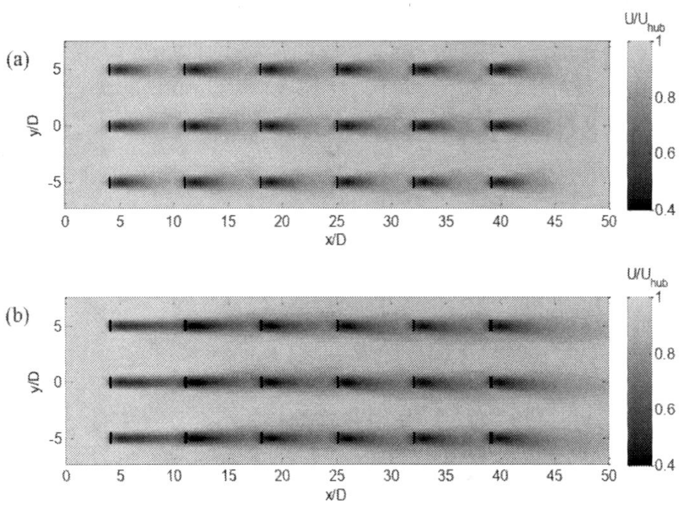

Figure 6. Contour plots of the normalized time-averaged streamwise velocity on a horizontal plane at the hub level for: a) the very unstable case (2nd day from 12:00 to 13:00); b) the very stable case (2nd day from 04:00 to 05:00). Only a section of the domain is shown.

Journal of Physics: Conference Series **625** (2015) 012031 doi:10.1088/1742-6596/625/1/012031

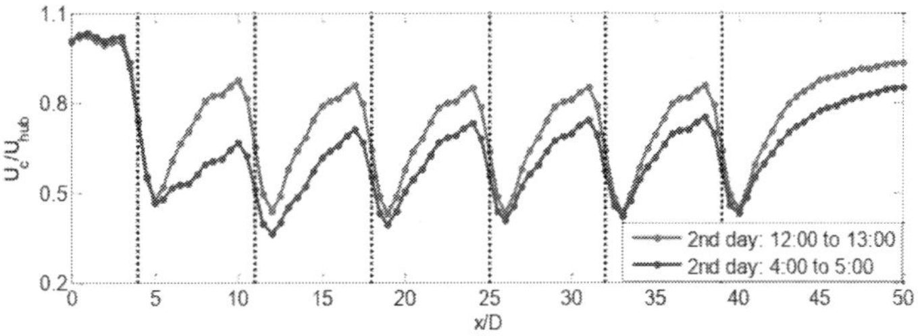

Figure 7. Normalized velocity at the center of the turbine with downwind distance. The vertical dotted lines show the location of turbines.

Figure 8 shows contours of the mean streamwise turbulence intensity in the horizontal $x - y$ plane at the hub-height level. As can be seen in the contours, for both stability conditions, the turbulence intensity behind each turbine exhibits a dual-peak pattern with larger values near the two side-tip positions, which is related to the intense production of turbulent kinetic energy associated with the strong shear at those locations. In addition, inside the wind farm, the turbulence intensity is higher in the unstable case in comparison with the stable one. This is mainly related to the fact that, as demonstrated in Fig. 5c, the turbulence level of the incoming wind is higher in the convective condition, which leads to also a larger wake turbulence level compared with the stable counterpart. The higher turbulence intensity at the inflow, and the higher turbulence intensity inside the farm, contribute to the faster recovery of the wakes in the unstable case observed in Figs. 6 and 7. This, in turn, leads to a reduction in the power losses due to wake effects in the wind farm comparing to the stable case, as shown in Fig. 9.

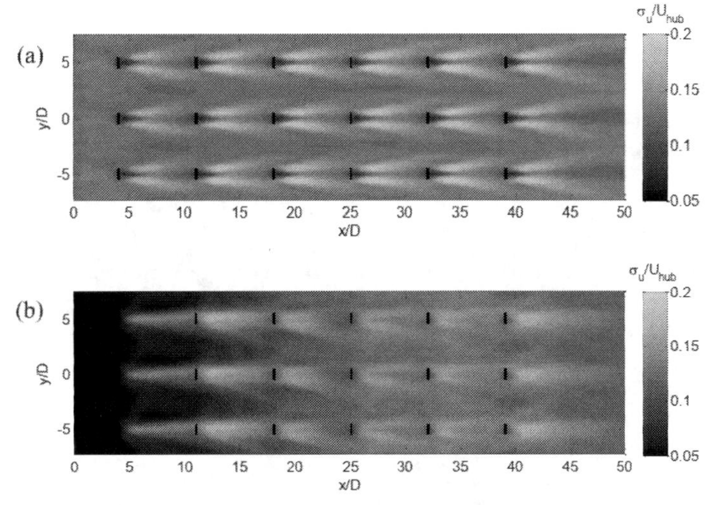

Figure 8. Contour plots of the streamwise turbulence intensity on a horizontal plane at the hub level for: a) the very unstable case (2nd day from 12:00 to 13:00); b) the very stable case (2nd day from 04:00 to 05:00). Only a section of the domain is shown.

Figure 10 shows the averaged power deficit throughout the diurnal cycle. As seen in this

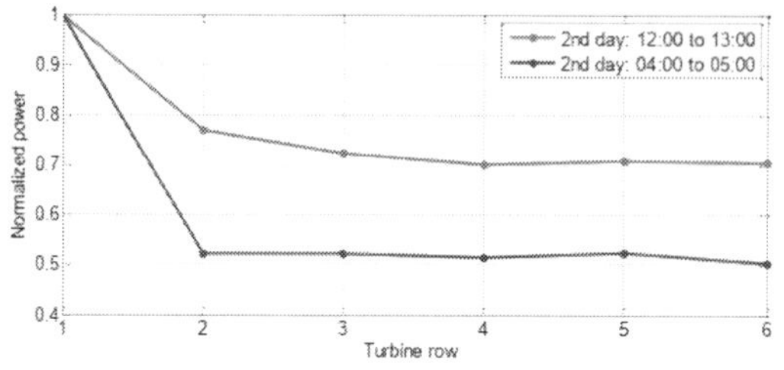

Figure 9. Normalized power output from the wind farm as a function of turbine row for the two different stability conditions.

figure, on average, the power deficit is larger during the night in the stable regime compared to the daytime in the convective condition. As mentioned before, this is mainly related to the fact that the turbulence level of the incoming wind to the turbines is larger under the convective condition compared to the stable one. One the other hand, during the night, the turbulence level decreases due to the negative buoyancy fluxes at the surface, the power deficit increases. In particular, for the specific day discussed here, the averaged power deficit increases from 28% under the most unstable condition to about 57% under the most stable case.

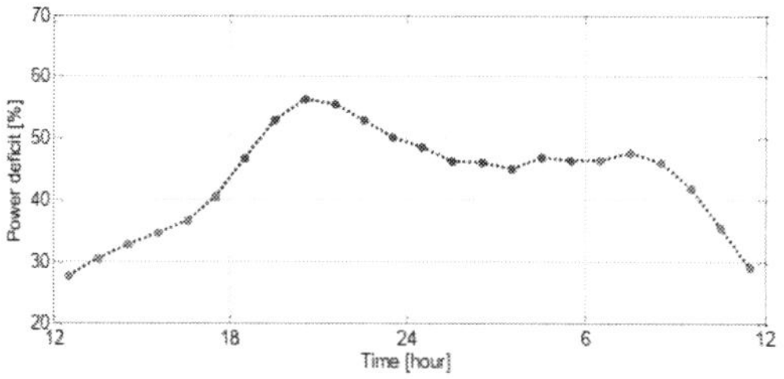

Figure 10. Time variation of the averaged power deficit throughout a full diurnal cycle. Each point shows the averaged power deficit over one hour.

4. Conclusion

In this study, LES combined with the actuator-disk model is used to study the turbine wakes and their effects on the power loses inside an idealized finite-size wind farm throughout a full diurnal cycle. In the simulations, the tuning-free Lagrangian scale-dependent dynamic models are used to model the SGS fluxes, and the turbine-induced forces (e.g. lift and drag) are parameterized with the dynamic actuator disk model with rotation.

The simulation results demonstrate that, in the absence of turbines, transient nature of the forcing terms such as surface fluxes of momentum and heat significantly affect the structure and dynamics of the ABL in the course of a day. In particular, during the day, when the net radiative flux is positive, the boundary layer grows due to the entrainment from the overlying inversion layer. In addition, the positive buoyancy flux at the surface creates thermal instabilities which enhance the TKE and the turbulent mixing. This enhanced turbulence level diminishes very quickly as the sun sets, and is very small during the nighttime. That is because, the negative surface buoyancy fluxes in the stable condition damps the turbulence, which leads to lower turbulence levels in all three directions. The simulated velocity and temperature fields were then used as inflows to study the the wake flow variability throughout a full day.

The results obtained from the LES of the ABL over a finite-size wind farm show that the atmospheric stability has significant effect of the wind-farm wakes and associated power losses. Specifically, during the day, under the convective condition, the wake recovery is faster compared to the one during the night. This enhancement in the wake recovery rate is mainly related to the higher turbulence level of the incoming wind in the unstable condition, due to positive buoyancy flux and associated turbulence production, which leads to a higher turbulent entrainment flux into the wake. In contrast, during the night, the relatively low turbulence intensity of the ambient ABL flow results in a relatively slow rate of entrainment of momentum into the wake and, consequently, a slow wake recovery. As a result, the averaged power deficit in the wind farm is found to increase with increasing thermal stability. In the particular case considered in this study, the averaged power deficit increased from 28% under the most convective condition to about 57% under the most stable condition.

Future research will address the effect of wind-farm configuration as well as wind direction on the structure and characteristics of wind-turbine wakes and their interaction with the ABL throughout the diurnal cycle.

Acknowledgements. This research was supported by the Swiss National Science Foundation (grants 200021-132122 and IZERZ0-142168) and the Swiss Innovation and Technology Committee (CTI) within the context of the Swiss Competence Center for Energy Research FURIES: Future Swiss Electrical Infrastructure. Computing resources were provided by the Swiss National Supercomputing Center (CSCS) under project ID s306.

References

[1] Hancock P E and Pascheke F 2014 *Boundary-Layer Meteorol.* **151.1** 23–37
[2] Jensen L 2007 *Proc. 2007 EWEC Conf*
[3] Lu H and Porté-Agel F 2011 *Phys. Fluids* **23** 065101
[4] Abkar M and Porté-Agel F 2014 *Renewable Energy* **70** 142–152
[5] Abkar M and Porté-Agel F 2015 *Phys. Fluids* **27** 035104
[6] Zhang W, Markfort C D and Porté-Agel F 2013 *Boundary-Layer Meteorol.* **146** 161–179
[7] Barthelmie R J and Jensen L E 2010 *Wind Energy* **13** 573–586
[8] Churchfield M J, Lee S, Michalakes J and Moriarty P J 2012 *J. Turbu.* **13** N14
[9] Stoll R and Porté-Agel F 2006 *Water Resour. Res.* **42.1** W01409
[10] Porté-Agel F, Wu Y T and Chen C H 2013 *Energies* **6(10)** 5297–5313
[11] Wu Y T and Porté-Agel F 2015 *Renewable Energy* **75** 945–955
[12] Smagorinsky J 1963 *Mon. Weather Rev.* **91** 99–164
[13] Porté-Agel F, Meneveau C and Parlange M B 2000 *J. Fluid Mech.* **415** 261–284
[14] Porté-Agel F 2004 *Boundary-Layer Meteorol.* **112** 81–105
[15] Wu Y T and Porté-Agel F 2011 *Boundary-Layer Meteorol.* **138** 345–366
[16] Albertson J D and Parlange M B 1999 *Water Resour. Res.* **35** 2121–2132
[17] Abkar M and Porté-Agel F 2013 *Energies* **6** 2338–2361
[18] Orszag S A 1970 *J. Atmos. Sci.* **27** 890–895
[19] Canuto C, Hussaini M, Quarteroni A and Zang T 1988 *Spectral Methods in Fluid Dynamics* (Springer)
[20] Monin A and Obukhov M 1954 *Tr. Akad. Nauk SSSR Geophiz. Inst.* **24** 163–187
[21] Moeng C 1984 *J. Atmos. Sci.* **46** 2311–2330
[22] Stoll R and Porté-Agel F 2009 *J. Atmos. Sci.* **66(2)** 412–431
[23] Stull R 1988 *An introduction to boundary-layer meteorology* (Kluwer Academic Publishers, Dordrecht)
[24] Businger J A, Wyngaard J C, Izumi Y and Bradley E F 1971 *J. Atmos. Sci.* **28** 181–189
[25] Wu Y T and Porté-Agel F 2013 *Boundary-Layer Meteorol.* **146** 181–205
[26] Deardorff J W 1974 *Boundary-Layer Meteorol.* **7** 81–106
[27] Tseng Y H, Meneveau C and Parlange M B 2006 *Environ. Sci. Technol.* **40** 2653–2662
[28] Abkar M and Porté-Agel F 2012 *J. Turbul.* **13** 1–18
[29] Sescu A and Meneveau C 2014 *Q. J. R. Meteorol. Soc.*
[30] Abkar M and Porté-Agel F 2015 *J. Renewable Sustanable Energy* **7** 013121
[31] Gryning S E, Batchvarova E, Brümmer B, Jørgensen H and Larsen S 2007 *Boundary-Layer Meteorol.* **124** 251–268
[32] Iungo G V and Porté-Agel F 2014 *J. Atmos. Oceanic Technol.* **31** 2035–2048

LES and RANS simulation of onshore Bessaker wind farm: analysing terrain and wake effects on wind farm performance

Mandar Tabib[1], **Adil Rasheed**[1] **and Trond Kvamsdal**[1,2]
[1]Applied Mathematics, SINTEF ICT, Strindveien 4, 7035, Trondheim, Norway
[2] Mathematical Sciences, NTNU, Alfred Getz vei 1, 7491, Trondheim, Norway

E-mail: mandar.tabib@sintef.no, adil.rasheed@sintef.no, trond.kvamsdal@sintef.no

Abstract. This work compares the predictive performance of RANS and LES solver in capturing the effect of terrain and wakes on the performance of the Bessaker wind farm. This 25 turbine wind farm is located in a highly complex terrain and is exposed to predominantly high westerly and south easterly winds. A one-equation sub-grid scale LES turbulence model has been used to help capture the wake dynamics: particularly the effects of wake meandering and wake-turbine interactions. A comparison between RANS and LES models highlights the influence of turbulence model on wake decay and its subsequent effect on prediction of power production. The LES model predicts delayed decay of the wake and more pronounced wake interference leading to a lower power production in wind farm than the RANS case. The RANS model overpredicts turbulence, which cause faster turbulent momentum diffusivity and faster wake recovery. This study has given some insights regarding the power production at Bessaker wind farm for neutral conditions and westerly flow.

1. Introduction

Accurate simulation of turbulent conditions (like, terrain induced turbulence and wakes interference effects) can help in strategizing a wind farm operation and to reduce its cost of energy. The potential of wind energy is immense provided the cost of wind-energy (COE) is competitive. Significant improvements in turbine technology and operating strategies have helped to reduce the COE in recent decades. Research devoted towards accurate estimate of wind resource can reduce the COE by enabling optimized power production. Wind resource within a particular farm site is influenced by surrounding terrain conditions, atmospheric stratification and wake effects (wherein upstream turbine generated wake influences downstream turbines) [1]-[4]. A model that resolves the large scale dynamics (like LES) can help to more accurately capture the turbulence than the RANS model (which models all the turbulent scales rather than resolving it). However, most of the numerical wind models used to predict wind conditions/power production over simplify the actual physics for the sake of faster computational time, in at least one of these three ways (a) by using less advanced turbulence models, like RANS model in place of the LES model, (b) by not accounting for a complex terrain and over simplifying its effect by using a roughness factor and a flat terrain, (c) by assuming only neutral atmospheric conditions and ignoring stratified conditions and accompanying buoyancy effects. Simulations involving these physics needs large computational resources, while with simplified assumptions

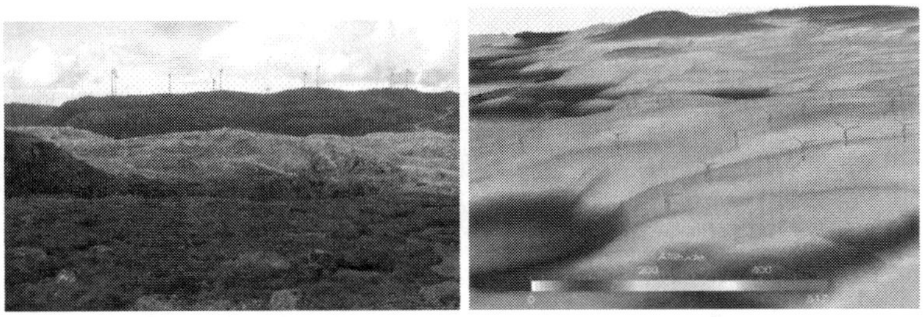

(a) Bessaker wind farm (b) Digital model

Figure 1. Bessaker wind farm

faster results are obtained for enabling engineering solutions which makes them more popular. However, such simplifications are bound to cause a loss of information and accuracy. Many notable contributions have been made to study the gain in accuracy and information using LES turbulence models ([2],[5]-[14]), in case of both the single-turbine and multi-turbine set-ups. These works have helped to highlight the importance of LES, but they have mostly been applied to very simplistic terrain or flatter terrains or off-shore conditions. In this work, we compare the effect of using an LES turbulence model vis-a-vis RANS turbulence model for an industrial scale wind farm located on a highly complex terrain. This model has been applied on a realistic industrial wind farm called Bessaker Wind Farm, operated by Trønder-Energi. The technical aspects about the Bessaker Wind Farm is given in the next section, followed by details of this model, solution methodology, results and discussion and conclusions.

2. Terrain and Bessaker wind farm details

Figure 1 shows the Bessaker wind farm, which is located on a complex terrain in mid-Norway region. This farm is operated by one of the largest energy providers in Norway: Trønder Energy AS. The terrain has an altitude difference of about $500m$, starting from sea level to the highest peak. This terrain houses 25 turbines of Enercon E70 turbine make. The lowest turbine base is located at $240m$ above sea level and the highest turbine is located at about $385m$ above sea level. The rest of the turbines are within this altitude range. All turbines have a rated power of $2.3MW$ with rotor diameter of $71m$ and a hub height of $64m$. The cut-off speed is about $28m/s$ wind. Generally, most dominant wind directions recorded at this site at $35m$ above the ground is westerly and south easterly and the average wind speed is recorded in the range of $10 - 15m/s$ implying that the turbines are rarely idle. However, a closer look at the power production of individual turbines reveals that a few turbines are almost always underperforming and overall power production is low. Computational Fluid Dynamics simulations were used to understand why this could be occurring. Our previous RANS based studies suggested that this drop in power production at wind farm could be owing to flow channeling and delayed wake decay [15]-[16] due to thermal stratification. This current work is conducted for neutral conditions and helps to understand the effect of turbulence modelling on wakes and power production in wind farm.

3. CFD Model and Equations solved

3.1. Governing Equations

A transient 3D CFD model incorporating the actuator line model is developed. The model computes the flow field (velocity, pressure, temperature) and the turbulence field arising out of shear induced turbulence, buoyancy induced turbulence and turbine induced wake effects in the wind farm. The turbulence is modeled using two different turbulence models: One-equation sub-grid scale turbulent kinetic energy LES model and the RANS k-ϵ model. The equations for LES are derived by applying filtering operator to the NavierStokes equations, while equations for RANS are derived by applying the Reynolds averaging to the Navier-Stokes equations. The equation set for both cases can be represented by Equations 1-??, where velocity $\mathbf{u}$ represents the filtered (or resolved) velocity in-case of LES and the ensemble averaged velocity in case of RANS. Thus, the three equations below represents a filtered (or averaged) mass continuity equation (Equation 1), filtered (or averaged) and momentum transport equation (Equation 2).

$$\nabla \cdot (\rho \mathbf{u}) = 0 \tag{1}$$

where ρ is the density.

$$\frac{D\mathbf{u}}{Dt} = -\nabla \left(\frac{p}{\rho}\right) + \frac{1}{\rho}\nabla \cdot \mathbf{R} + \mathbf{f} \tag{2}$$

where, p is pressure, t is time, $\mathbf{f}$ refers to external forces arising from actuator line model, $R_{ij} = \nu_T \left(\frac{\partial u_i}{\partial x_j} + \frac{\partial u_j}{\partial x_i}\right) - \frac{2}{3}k\delta_{ij}$ arises owing to averaging procedure and is referred to turbulent stresses. The computation of eddy viscosity needed for the closure of above equation set is dependent upon the turbulence model (as described below). When $k - \epsilon$ model is used, the turbulent eddy viscosity is formulated in terms of turbulent kinetic energy (k) and eddy dissipation rate (ϵ). These two are obtained by solving their transport equation as in Equation 4 and 5.

$$\mu_t = C_\mu \frac{k^2}{\epsilon} \tag{3}$$

where C_μ is a constant.

$$\frac{DK}{Dt} = \nabla \cdot \left(\frac{\nu_T}{\sigma_k}\nabla K\right) + P_k - \epsilon \tag{4}$$

$$\frac{D\epsilon}{Dt} = \nabla \cdot \left(\frac{\nu_T}{\sigma_e}\nabla \epsilon\right) + (C_1 P_k)\frac{\epsilon}{k} - C_2\frac{\epsilon^2}{k} \tag{5}$$

where ν_T is turbulent diffusivity, σ_e and σ_e are turbulent prandtl numbers, P_k is production of turbulent kinetic energy due to shear. The production due to buoyancy term is neglected here for neutral flows.

$$P_k = \nu_T \left(\frac{\partial u_i}{\partial x_j} + \frac{\partial u_j}{\partial x_i}\right)\frac{\partial u_i}{\partial x_j} \tag{6}$$

When LES model is used, the turbulent eddy viscosity is formulated as follows:

$$\mu_t = \rho(C_k \Delta)k_{sgs}^{1/2} \tag{7}$$

where C_k is a constant and k_{sgs} is obtained by solving a transport equation for sub-grid scale turbulent kinetic energy. The value of C_k used in this study is 0.094.

3.2. Actuator Line Model

The turbines are modeled using actuator line model (ALM) approach, which was first developed by Sørensen and Shen [17] . The actuator line model (Equations 8, 9 and 10) uses the velocity field input from the CFD model and outputs body force, which are used as the sink term in the momentum equation. The ALM approach resolves each blade of the turbine as a rotating line (made of N actuator segments), over which the forces are computed. The force at each segment comprises of lift force and drag forces (Equations 8, 9), which are computed from the local relative velocity (V_{rel}) , local twist angle, blade chord (c), local actuator width (w) and local angle of attack (α) at a given actuator segment. The local angle of attack is computed from the tangential and normal component of relative velocity at the segment. The lift coefficient (C_l) and drag coefficient (C_d) at each segment (in Equations 8, 9 and 10) are a function of local angle of attack, and this dependency is provided as an input (blade airfoil data) to the ALM model. The force at an actuator segment i (f_i) is a point force and it is translated on to the fluid domain as a volumetric body force (F_i) using Gaussian projection (Equation 10). The regularization parameter (ε) in Equation 10 represents the width of the Gaussian and determines the concentration of the force. Larger the parameter, more smoothed out the force is on the flow field. The negative sign in Equation 10 accounts for the fact that the force exerted by turbine on the flow field is equal and opposite to the force experienced by it due to the flow. At the location (x, y, z) of the fluid domain, the overall body force is summation of force over all N actuator segments of the turbine, where (x_j, y_j, z_j) is the location of the jth segment and r_j is the distance between segment j and the fluid domain location. In the present work, the aerodynamic data for Enercon E70 turbine blades was not available. So, we modified the aerodynamic data for NREL 5MW Reference blade such that a power-curve similar to an Enercon E70 blade is obtained.

$$L = \frac{1}{2} C_l(\alpha) \rho V_{rel}^2 cw \tag{8}$$

$$D = \frac{1}{2} C_d(\alpha) \rho V_{rel}^2 cw \tag{9}$$

$$F_i^T(x, y, z) = -\sum_{j=1}^{N} f_i^T(x_j, y_j, z_j, t) \frac{1}{\varepsilon^3 \pi^{3/2}} exp\left[-\frac{r_j^2}{\varepsilon}\right] \tag{10}$$

4. Simulation methodology : Solver details and set-up parameters

4.1. Solver details

The solver has been created in OpenFOAM-2.3.0 (OF). To ensure continuity, an elliptic equation for the modified pressure is created by combining continuity equation with divergence of momentum equation. This elliptic equation along with the momentum equation, energy equation and turbulence equation are solved in a segregated manner using the PISO-SIMPLE algorithm (PIMPLE algorithm). The PIMPLE algorithm ensures use of a higher time step for transient simulations. The OF uses a finite volume discretization technique, wherein all the equations are integrated over control volumes (CV) using Green Gauss divergence theorem. The gauss divergence theorem converts the volume integral of divergence of a variable into a surface integral of the variable over faces comprising the CV. Thus, the divergence term defining the convection terms can simply be computed using the face values of variables in the CV. The face values of variables are obtained from their neighboring cell centered values by using convective scheme. In this work, all the equations (except k and turbulence equations) use second order linear discretization scheme, while the turbulent equations use upwind convection schemes. Similarly, the diffusion term involving Laplacian operator (the divergence of the gradient) is simplified to computing the gradient of the variable at the face. The gradient term can be split into contribution from the orthogonal part and the non-orthogonal parts, and both these contributions have been accounted for. The mesh details, CFD domain set up, boundary and initial conditions are defined below.

4.2. CFD set up

The Bessaker wind farm CFD set up involves a domain size of $6.98km \times 4.7km \times 1.3km$ (Figure 2). This domain size ensures that the entire wind/farm area is considered and the location of domain boundary will not affect the flow profile within wind farm. A westerly wind flow is

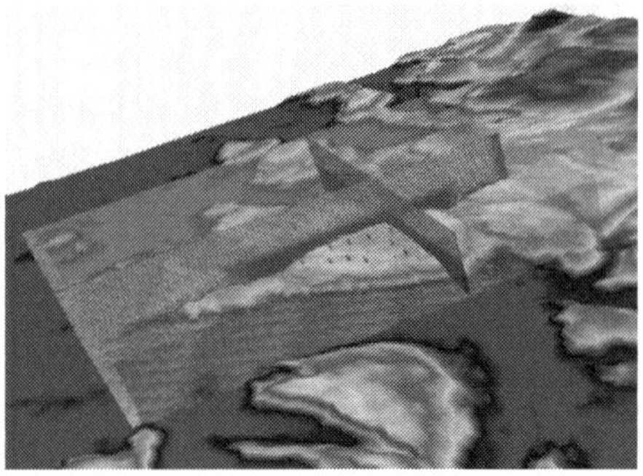

Figure 2. CFD domain and internal mesh enclosing the terrain. Westerly wind flows from inlet at the left and the outlet on right. Top and side walls are slip boundary.

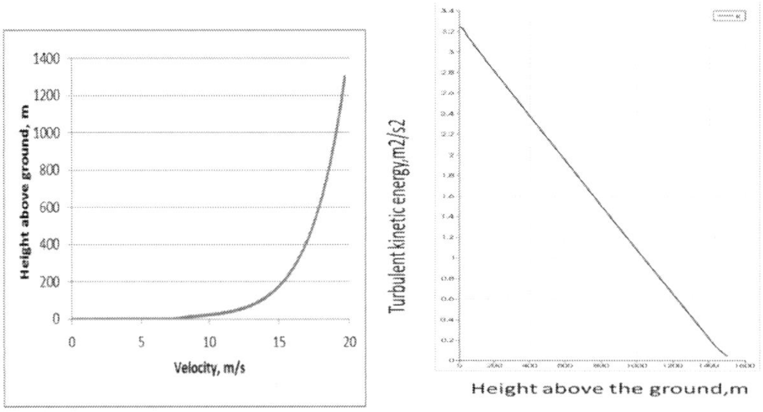

(a) Inlet Velocity (b) Inlet turbulent kinetic energy

Figure 3. Inlet boundary profiles

considered for this simulation as this is one of the dominant wind directions at the site. The terrain geometry is included in the domain as an stereo lithography file (STL) and the domain and mesh is built using it. The mesh size for this domain is about 13 million cells. The grid is finer in the wind farm region and near the terrain, where the velocity gradient is high and effect of wake has to be captured. The finest grid size is $6m$ in the wind farm location. This ensures that there are at least 10 numerical finite volume cells within the turbine area, as the rotor blade is of $60m$ diameter. The time step used for simulation is more stringent than normal CFL criteria of 1. This time step was selected based on the tip speed and grid cell size in the actuator line model domain. The time step was such that the blade (rotating line) does not move more than one grid cell within a time step. To achieve this, a time step of $0.015s$ was used which gave a CFL of 0.12. An Euler discretization scheme is chosen for temporal discretization. The solver was run on 216 parallel processors. The simulation has been carried out for two cases: (a) RANS neutral atmospheric condition and (b) LES neutral atmospheric condition. Figure 3 shows the inlet conditions specifed as a function of vertical distance from ground : velocity profile is used in both LES and RANS case and the inlet turbulent kinetic energy profile is used only for RANS study. No potential temperature equation is required to be solved for the neutral

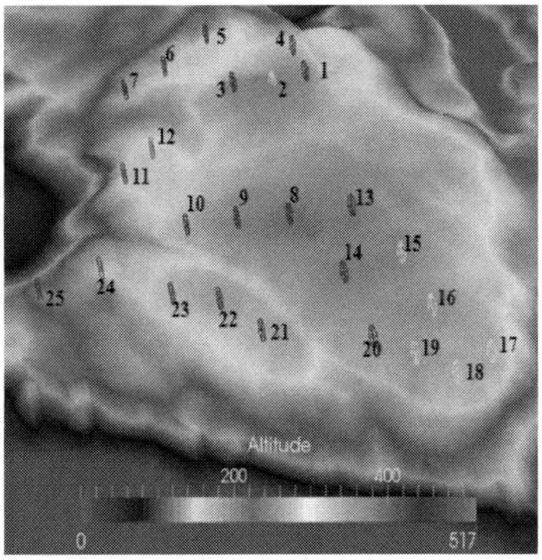

Figure 4. Location and performance of 25 Turbines. Green coloured turbines producing highest power, blue coloured turbine producing lowest power and rest black coloured are in between

boundary condition. The side and top boundary conditions of the domain are slip walls. The outlet has fixed pressure boundary conditions. The bottom terrain geometry is no-slip boundary condition. A continuouse wall function for rough wall as suggested by Spalding[18] has been used to simulate the near wall profiles. Figure 4 shows the location of the 25 turbines on the terrain. The effect of turbines is accounted using the actuator line method as explained in section 3. For the actuator line method, about 40 actuator segments (i.e parameter N in equation 10) in were used in the simulation. The regularization parameter (ε in equation 10) in ALM was chosen to be about two times the cube root of grid volume size in that region. This was selected so as to ensure that the force is not overly concentrated to cause numerical oscillations/solver instability, and neither does the force becomes too smoothed so as to cause no resistance to the wind flowing through the turbine. The LES runs were initialized from the RANS runs. The LES results were run for 6500 s of simulation time (which corresponds to approximately 15 times flow passage in the domain at wind free stream velocity,nearly 18m/s) and averaging was done for last 3000 s. The simulation took around 3 weeks of computational time when run parallely on 216 processors (each processor having 2.6 GHz processing speed). The next section discusses the results obtained.

5. Results and discussions

5.1. Prediction of power produced by RANS and LES

For neutral atmospheric conditions and westerly wind direction, the overall power output predicted in the wind farm by RANS Simulation ($30 - 31MW$) is around $8 - 10\%$ more than that predicted by LES simulation (around $27.528MW$). Most of the turbines produce higher power with RANS simulation than for the LES simulation. The reasons for this is the way turbine induced wake effects and terrain induced effects are captured by the two turbulence models (explained in sections below).

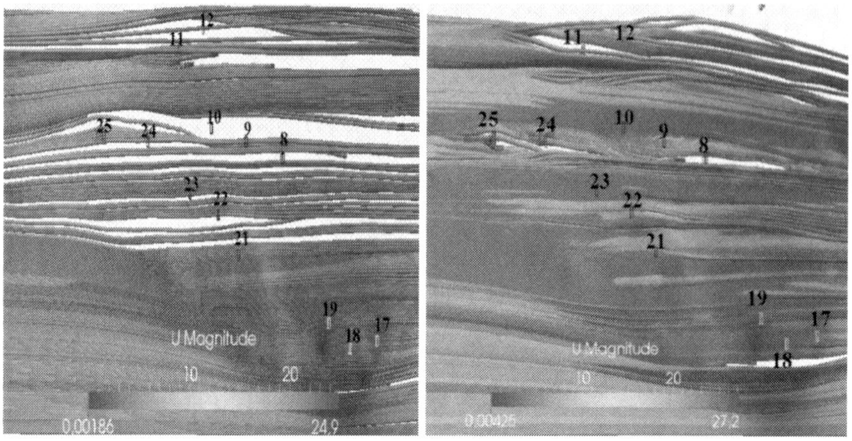

(a) RANS streamline (b) LES instantaneous streamline

Figure 5. Turbine performance : green coloured turbines (number 15-19) are exposed to higher velocity wind and perform well while blue coloured turbines (number : 25,24,11,12 in streamline) perform the worst owing to lower wind-velocity.

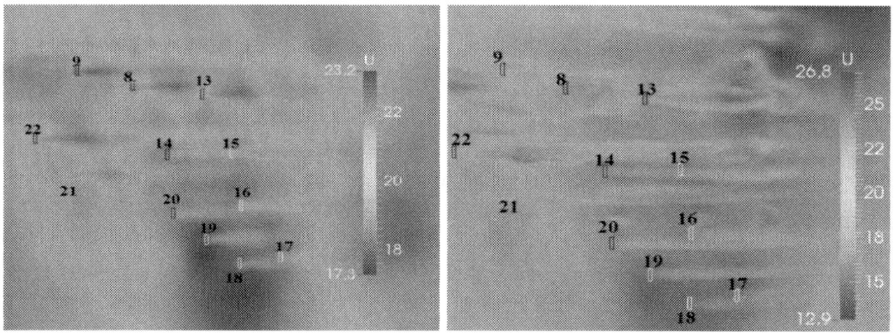

(a) RANS velocity at a given instant (b) LES instantaneous velocity

Figure 6. Comparision of RANS and LES velocity at a instant and relation to turbine performance

5.2. Terrain effects and wake effects captured by the two models

The atmospheric wind when exposed to any terrain related obstructions undergoes a change in its flow pattern. This involves wind speed-up over the hill, boundary layer separations, channeling, mountain waves, etc. Turbulence generated by rough terrain plays an important role in diffusing momentum and it affects the flow pattern. Figure 5-6 compares results at a given instant for streamline and velocity, while figures 7-9 compares the mean-fields of velocity,vorticity and turbulent knietic energy. In all the figures, the turbine numbers are marked to enable easy comprehension. The figures 7-9 are plotted by merging two planes on a common color-map : the plane involving turbine 8,9 and 13 at top is located at a higher altitude of 435 m (to capture the wakes at their hub height) and the plane at bottom involving rest of turbines is located at altitude of 405 m. Figure 5 shows that for both the turbulence models, the turbines shown in

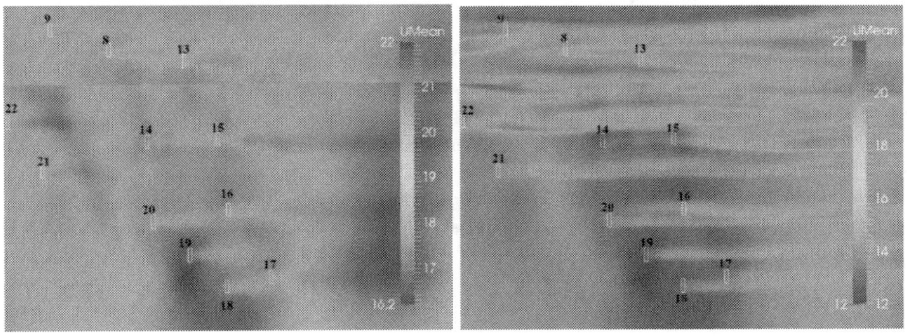

(a) RANS predicted mean velocity (b) LES predicted mean velocity

Figure 7. Mean velocity contours

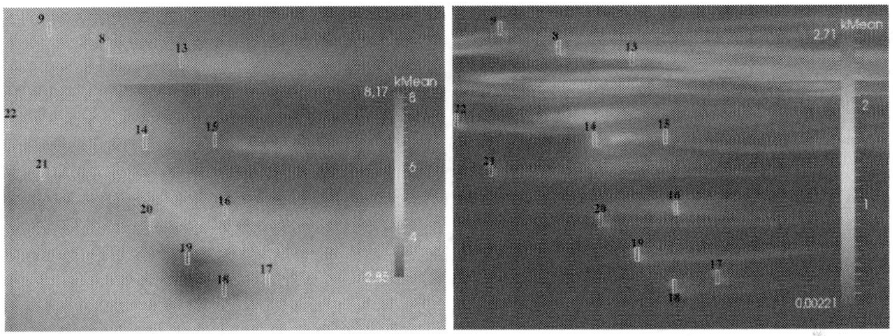

(a) RANS (b) LES

Figure 8. Mean turbulent kinetic energy contours

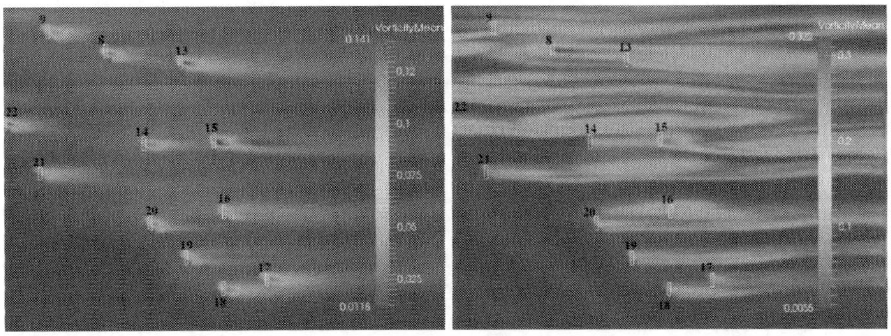

(a) RANS (b) LES

Figure 9. Mean vorticity contours

green color in Figure 5 (Turbine number 15, 16, 17, 18, 19) are producing higher power output, while the turbines shown in blue color (Turbine number 11, 12, 24, 25) are producing the least power output. The reason for this behavior can be attributed to the wind flow experienced by these turbines. The instantaneous streamlines obtained from LES (in Figure 5(b)) shows higher velocity wind region (wind speed above $20m/s$, up-to $27m/s$) at location of green colored turbines. These turbines are located at higher altitude and experience good wind speeds owing to speed-up over altitude. While, the turbines in blue are at lower altitude and experience comparatively lower wind speed between $5-10m/s$ and hence, provides lower power output. Similar trends were observed by streamlines from RANS model [16] as shown in Figure 5(a) , except that the obtained velocity range was different than LES. RANS revealed wind-speeds of about $15-22m/s$ in the high altitude regions where the better performing green colored turbines lie, and wind speeds of $5-15m/s$ at lower altitude regions where low performing blue colored turbines lie. The instantaneous LES velocity contour (Figure 6(b)) also shows a wider velocity range (difference between the minimum and maximum velocity values) than the RANS predicted field (Figure 6(a)) at a given instant. Figure 7 compares the time-averaged velocity predicted by LES and RANS. The comparision shows that LES predicts a more pronounced wake region than RANS and thus more downstream turbines are exposed to wakes with lower velocity. For example, compare the wake behind turbines 8-13, 14-15, 16-20 and 17-21 (in Figure 7-9) for the two cases of RANS and LES. These figures shows that LES predicts higher wake interference effect, which results in a lower prediction of power production as compared to RANS. The differences in prediction of wake effect by the two models could be because of the way the shear induced turbulence (owing to terrain and velocity gradient in atmospheric flow) are captured. RANS models predict higher values of time-averaged turbulent kinetic energy than the LES (as seen in Figure 8) , which causes larger turbulent momentum diffusivity. This results in faster wake decay. The overall effect is that for RANS simulations the turbines are exposed to a wind profile that causes higher overall power production. This is seen in vorticity plots too. Figure 9 compares the time-averaged vorticity plot obtained from LES and RANS. The LES mean vorticity plot reveals that wakes induced by turbine 8 reaches out further till turbine 13, and similar wake interference can be seen in turbine 14-15 and turbine 17-18. In contrast, in the RANS vorticity prediction, the wake vortices seems to decay faster. Also, a higher degree of meandering of wakes is observed in LES case than in RANS. Thus, for the current study involving westerly wind direction, the current inter-turbine distance (around four times the rotor diameter between turbines 14-15) leads to an inefficient windfarm operation perspective. However, in-order to draw a general criteria about optimum inter-turbine spacings, more studies at different wind conditions and inter-turbine spacings needs to be done.

6. Conclusion and future work

A one-equation sub-grid scale turbulent kinetic energy LES model and a RANS turbulence model have been used to observe the wake dynamics and the power production in an industrial multi-turbine wind farm. Both the models incorporate the actual terrain effects along with an actuator line model for understanding wake behaviour. The RANS model predicts a higher power production in wind farm than the LES as it predicts faster decay of the wake. This is owing to higher terrain induced and wake turbulence predicted by RANS model, which cause faster turbulent momentum diffusivity and faster wake recovery. The LES model reveals that for this particular wind farm layout, terrain and wind direction condition, the placement of downstream turbines at an inter-turbine distance of four rotor turbine diameter is not good from wind farm operation perspective owing to wake interference. The RANS shows the wake effects but the turbine-turbine wake interactions is not as pronounced as in LES. To draw a general criteria about optimum turbine spacings, more studies at different wind conditions and inter-turbine spacings needs to be done. Future work may also involve inclusion of stratification effects in LES model for wake-plume interaction and nesting the meso-scale CFD model with the LES model for better inlet boundary conditions.

Acknowledgments

The authors would like to acknowledge the financial support from the Norwegian Research Council and the industrial partners of the FSI-WT (216465/E20) (http://www.fsi-wt.no) project (Kjeller Vindteknikk, Statoil, Trønder Energi AS and WindSim). We are also thankful to Ingrid Vik and Magne Røen from TrønderEnergie for providing information on the locations of turbines in the Bessaker wind farm and the information on turbines.

References

[1] Churchfield MJ, Lee S, Michalakes J, Moriarty PJ, A numerical study of the effects of atmospheric and wake turbulence on wind turbine dynamics,*Journal of Turbulence* 2012;N14.
[2] Churchfield MJ, Lee S, Moriarty PJ, Martnez LA, Leonardi S, Vijayakumar G, Brasseur JG, A large-eddy simulation of wind-plant aerodynamics, 50th AIAA Aerospace Sciences Meeting Including the New Horizons Forum and Aerospace Exposition, 2012
[3] Keck RE, Maré MDe, Churchfield MJ, Lee S, Larsen G, Madsen HA, On Atmospheric Stability in the Dynamic Wake Meandering Model, *Wind Energy* , 2014;17:1689-1710.
[4] Lee S, Churchfield MJ, Moriarty PJ, Jonkman J, Michalakes J, A Numerical Study of Atmospheric and Wake Turbulence Impacts on Wind Turbine Fatigue Loadings, *Journal of Solar Energy Engineering*, 2013;135(3):31001-10
[5] Calaf M, Meneveau C, Meyers J. (2010). Large eddy simulation study of fully developed wind-turbine array boundary layers. *Physics of Fluids*, 22 (1), art.nr. 015110, 1-16.
[6] Jimenez A, Crespo A, Migoya E, Garcia J. 2008, Large-eddy simulation of spectral coherence in a wind turbine wake, *Environmental Research Letters*, 3(1),
[7] Martnez-Tossas LA, Churchfield MJ and Leonardi S. 2014,Large eddy simulations of the flow past wind turbines: actuator line and disk modeling, *Wind Energy* 2014,
[8] Mehta D, Zuijlen AH, Koren B, Holierhoek JG, Bijl H. 2014 Large Eddy Simulation of wind farm aerodynamics: A review, *Journal of Wind Engineering and Industrial Aerodynamics* 2014, 133, 1-17.
[9] Porté-Agel F, Wu YT, Lu H, Conzemius RJ. 2011. Large-eddy simulation of atmospheric boundary layer flow through wind turbines and wind farms. *Journal of Wind Engineering and Industrial Aerodynamics* 99(4): 154-168.
[10] VerHulst C and Meneveau C. 2014 Large eddy simulation study of the kinetic energy entrainment by energetic turbulent flow structures in large wind farms, *Physics of Fluids* 26, 025113
[11] Basu S, Porté-Agel F. 2006. Large-eddy simulation of stably stratified atmospheric boundary layer turbulence: A scale-dependent dynamic modeling approach. *Journal of the Atmospheric Sciences* 63(8): 2074-2091.
[12] Lu H, Porté-Agel F. 2011. Large-eddy simulation of a very large wind farm in a stable atmospheric boundary layer. *Physics of Fluids* 23(6).
[13] Wan F, Porté-Agel F. 2011b. Large-Eddy Simulation of Stably-Stratified Flow Over a Steep Hill. *Boundary-Layer Meteorology* 138(3): 367-384.
[14] Wu YT, Porté-Agel F. 2011. Large-Eddy Simulation of Wind-Turbine Wakes: Evaluation of Turbine Parametrisations. *Boundary-Layer Meteorology* 138(3): 345-366.
[15] Rasheed A, Sørli K, Holdahl R, Kvamsdal T, A Multiscale Approach to Micrositing of Wind Turbines, Energy Procedia 2012;14:1458-1463.
[16] Tabib M, Rasheed A, and Kvamsdal T, Investigation of the impact of wakes and stratification on the performance of an onshore wind farm, Submitted to *Energy Procedia*
[17] Sørensen J, Shen W. 2002. Numerical modelling of Wind Turbine Wakes. *Journal of Fluids Engineering* 124 (2): 393-399.
[18] Spalding D. 1961.A single formula for the law of the wall. *Transactions ofth ASIDE, Series E: Journal of Applied Mechanics* 28,455-458.

Effects of incoming wind condition and wind turbine aerodynamics on the hub vortex instability

R. Ashton[1], F. Viola[2], F. Gallaire[2] and G.V. Iungo[1]

[1] The University of Texas at Dallas, Mechanical Engineering Department, Wind Fluids and Experiments Lab (WindFluX), 75080 Richardson, TX

[2] Ecole Polytechnique Fédérale de Lausanne (EPFL), Laboratory of Fluid Mechanics and Instabilities (LFMI), Lausanne, Switzerland

E-mail: valerio.iungo@utdallas.edu

Abstract. Dynamics and instabilities occurring in the near-wake of wind turbines have a crucial role for the wake downstream evolution, and for the onset of far-wake instabilities. Furthermore, wake dynamics significantly affect the intra-wind farm wake flow, wake interactions and potential power losses. Therefore, the physical understanding and predictability of wind turbine wake instabilities become a nodal point for prediction of wind power harvesting and optimization of wind farm layout. This study is focused on the prediction of the hub vortex instability encountered within wind turbine wakes under different operational conditions of the wind turbine. Linear stability analysis of the wake flow is performed by means of a novel approach that enables to take effects of turbulence on wake instabilities into account. Stability analysis is performed by using as base flow the time-averaged wake velocity field at a specific downstream location. The latter is modeled through Carton-McWilliams velocity profiles by mimicking the presence of the hub vortex and helicoidal tip vortices, and matching the wind turbine thrust coefficient predicted through the actuator disc model. The results show that hub vortex instability is promoted by increasing the turbine thrust coefficient. Indeed, a larger aerodynamic load produces an enhanced wake velocity deficit and axial shear, which are considered the main sources for the wake instability. Nonetheless, wake swirl also promotes hub vortex instability, and it can also affect the azimuthal wavenumber of the most unstable mode.

1. Introduction

Prediction of wind turbine wakes for different conditions of the incoming wind and different aerodynamic settings of the turbines, such as yaw and pitch angles, still remains a big challenge for the wind energy community. Accurate descriptions of the wake flow produced by a turbine array can be obtained through LES simulations by mimicking the forcing of the turbines on the atmospheric boundary layer flow with different actuator models (see e.g. [1, 2]). However, the high computational cost required for those simulations make them a prohibitive tool for real applications, for which a large set of configurations typically needs to be investigated in a relatively short time.

These predictions are also of interest to the fluid mechanics community, because the mean operational properties of the turbine are intimately linked to the complex flow features in the wind turbine wake, which involves vorticity structures with completely different characteristic scales, the variability of the atmospheric boundary layer flow, and heterogeneity of the environment.

As extensively explained in [3], wind turbine wakes are characterized by a significant velocity deficit consequent to the power extracted from the turbine, which gradually recovers to the incoming wind conditions by proceeding downstream. Moreover, turbine rotation adds azimuthal momentum to the wake flow, which acts in the opposite direction than for the turbine rotation. Besides the global swirling motion of the turbine wake, highly coherent vorticity structures are present in the near-wake, which enhance the complexity of this swirling wake flow. Specifically, a vorticity structure mainly oriented along the wind turbine axis is typically observed at the wake core, which is denoted as hub vortex. Furthermore, vortices detach from the tip of each blade, which are then advected downstream forming a set of helicoidal tip vortices. All those vorticity structures evolve and interact by proceeding downstream, encounter different instabilities, and finally dissipate producing a smooth swirling wake with a Gaussian-like profile of the axial velocity field (see e.g. [4, 5, 6]).

In addition to this, the prediction of a wind turbine wake flow is a challenging task due to the variability introduced by the broad range of atmospheric conditions that can be experienced during real turbine operations. Indeed, different levels of turbulent intensity, shear and veer of the incoming wind can be encountered. Those different conditions of the incoming wind lead to completely different power performance and, in turn, wind turbine wake flows. Recently, wind LiDAR measurements of utility-scale wind turbines performed under different stability conditions of the atmospheric boundary layer showed that wind turbine wakes recover faster under convective regimes than for neutral ones [7]. The variability of wind turbine wakes can be enhanced even more in presence of heterogeneous terrains due to the topography, urban environment, and surrounding infrastructures.

The vorticity structures constituting the wake undergo several instability processes as they evolve by moving downstream. The coherent motion induced by the wake instabilities can significantly modify the instantaneous wake velocity field, thus the shear stresses. The latter represents the main contribution to the entrainment of momentum and energy from the surrounding boundary layer flow into the wind turbine wake. Therefore, wake instabilities can noticeably affect downstream wake recovery, wake interactions, and the performance of a whole wind farm. Moreover, the presence of wake dynamics and instabilities can lead to wind velocity fluctuations persisting further downstream, thus producing a potential hazard for added fatigue loads on downstream waked turbines.

Different types of instabilities of the helicoidal tip vortices has been observed in [8, 9, 10], such as long and short waves instabilities. The downstream persistence of the helicoidal tip vortices plays a significant role on the whole wake recovery. Indeed, a faster diffusion of the tip vortices can lead to an enhancement of the shear stresses over the wake boundaries, then promoting flow entrainment. In [11] it was observed that the far-wake meandering motion of the turbine wake was triggered by the diffusion of the tip vortices, which in turn was promoted by interactions with the hub vortex.

The hub vortex also undergoes to instability by proceeding downstream. A single-helix counter-winding instability was observed in [3] through wind tunnel experiments, then confirmed by means of linear stability analysis. The same phenomenon was confirmed by LES simulations of a hydro-kinetic wind turbine in [12], and through water tunnel experiments of a down-scaled wind turbine model in [13]. In [14] effects of turbulence are included in a linear stability analysis formulation in order to predict the hub vortex instability by using as base flow the time averaged velocity field. This approach is based on the Boussinesq hypothesis and by using three different mixing length models, which are tuned on the turbulence statistics of the base flow.

The main objective of this paper is the prediction of the hub vortex instability under different operational conditions of the wind turbine. This survey is performed by mimicking the

Journal of Physics: Conference Series **625** (2015) 012033 doi:10.1088/1742-6596/625/1/012033

radial distribution of the axial and azimuthal velocity field with multiple Carton-McWilliams vortices [15]. Accordingly to previous LES simulations [2, 5], LiDAR data [6] and wind tunnel experiments [3], this vortex model can accurately represent the typical velocity field connected with turbine wakes, which is affected by the presence of hub and tip vortices. For each condition of the incoming wind and wind turbine loading, the radial distribution of the velocity field is obtained through an iterative method in order to match the thrust coefficient of the wind turbine predicted via actuator disc model. The different wake velocity profiles are then analyzed via linear stability analysis, following the formulation proposed in [14], in order to take the turbulence effects on wake instability into account. Specifically, in this paper the hub vortex instability is predicted for different induction factor, i.e. power coefficient of the turbine, and tip speed ratio.

The paper is organized as follows. The formulation of the linear stability analysis, including turbulence effects with a mixing length model presented in [14], is briefly summarized in section 2. Then, the iterative method adopted to evaluate the wake velocity profiles for different conditions of the incoming wind and turbine loading is described in section 3. Results of the stability analysis obtained by varying the wind turbine induction factor and tip speed ratio are discussed in details in section 4. Finally, the survey and main results are summarized in the conclusions, section 5.

2. Linear stability analysis with eddy-viscosity models

Following the approach proposed in [16] the unsteady flow, $\mathbf{U}(\mathbf{x}, t)$, is decomposed in the time-averaged base-flow, $\overline{\mathbf{U}}(\mathbf{x})$, the coherent fluctuation, $\tilde{\mathbf{u}}(\mathbf{x}, t)$, and the turbulent motion, $\mathbf{u}'(\mathbf{x}, t)$:

$$\mathbf{U} = \overline{\mathbf{U}} + \tilde{\mathbf{u}} + \mathbf{u}', \tag{1}$$

where the sum of the time-averaged flow and the coherent fluctuation coincides with the ensemble averaged flow $< \mathbf{U} >= \overline{\mathbf{U}} + \tilde{\mathbf{u}}$ (see e.g. [17]).

The stability of the flow field is given by the tendency of $\tilde{\mathbf{u}}$ to grow (unstable) or decay (stable) in time and space. Thus, stability can be verified by a modal analysis of the linearized dynamics of $\tilde{\mathbf{u}}$. To this purpose the nonlinear evolution of the coherent perturbation for an incompressible flow can be written as (see e.g. [16]):

$$\begin{gathered}\nabla \cdot \tilde{\mathbf{u}} = 0 \\ \frac{\partial \tilde{\mathbf{u}}}{\partial t} + \nabla \tilde{\mathbf{u}} \cdot \overline{\mathbf{U}} + \nabla \overline{\mathbf{U}} \cdot \tilde{\mathbf{u}} = -\nabla \tilde{p} + \frac{1}{\mathrm{Re}} \Delta \tilde{\mathbf{u}} - \nabla \cdot [\tilde{\mathbf{u}}\tilde{\mathbf{u}} - \overline{\tilde{\mathbf{u}}\tilde{\mathbf{u}}}] - \nabla \cdot [< \mathbf{u}'\mathbf{u}' > - \overline{\mathbf{u}'\mathbf{u}'}]\end{gathered} \tag{2}$$

where variables t and p represent time and pressure, respectively. In the framework of a linear analysis with respect to the coherent fluctuations $\tilde{\mathbf{u}}$, the third term of the rhs is neglected. However, the system of equations is not closed and the last term of the rhs, related to the turbulent diffusion, has to be modeled. In [14] this term was modeled via Boussinesq hypothesis, in analogy to [16] and [18], where the Reynolds stresses are linearly proportional to the strain rate yielding:

$$\overline{R} = \overline{\mathbf{u}'\mathbf{u}'} \simeq -2\nu_t \overline{S} + \frac{2}{3}\overline{q} I, \tag{3}$$

where R is the Reynolds stress tensor, S is the strain rate tensor, q is the turbulent kinetic energy (TKE), and I is the 3x3 identity matrix.

As well documented by several wind tunnel and LES experiments, see e.g. [2, 3, 4, 5], the wind turbine wake flow is practically axisymmetric and with a negligible radial velocity. Therefore, in the framework of a local stability analysis of a wind turbine wake flow, we have $\overline{U}_r \approx 0$,

$\partial\overline{\mathbf{U}}/\partial\theta \approx 0$, and $\partial\overline{\mathbf{U}}/\partial x << \partial\overline{\mathbf{U}}/\partial r$, which implies that the term $\overline{R}_{x\theta} = \overline{u'_\theta u'_x}$ is null, and the model leads to null normal stresses [19]. Consequently, the only non-null components of the modeled tensor $\overline{R}$ are $\overline{R}_{r\theta}$, $\overline{R}_{rx}$, and their symmetric counterparts.

In this work a generalized mixing-length model for swirling-flows is applied (see [20] for more details):

$$\nu_t(r) = l_m^2(2\overline{S}:\overline{S})^{1/2} = l_m^2\left[\left(r\frac{\partial}{\partial r}\left(\frac{\overline{U}_\theta}{r}\right)\right)^2 + \left(\frac{\partial\overline{U}_x}{\partial r}\right)^2\right]^{1/2}, \tag{4}$$

where ν_t is the eddy-viscosity, which is a function of the downstream and radial position, while the mixing length, l_m, is only a function of the downstream location. By linearizing the eddy viscosity model at first order with respect to $\tilde{\mathbf{u}}$, and further manipulating Eq. (2), it is obtained (see [14] for details):

$$\frac{\partial\tilde{\mathbf{u}}}{\partial t} + \nabla\tilde{\mathbf{u}}\cdot\overline{\mathbf{U}} + \nabla\overline{\mathbf{U}}\cdot\tilde{\mathbf{u}} = -\nabla\tilde{p} + \frac{1}{\mathrm{Re}}\Delta\tilde{\mathbf{u}} + \nabla\cdot(\nu_t(\overline{\mathbf{U}})[\nabla + \nabla^T]\tilde{\mathbf{u}}) + \nabla\cdot((\nabla_{\mathbf{U}}\nu_t(\overline{\mathbf{U}})\cdot\tilde{\mathbf{u}})[\nabla + \nabla^T]\overline{\mathbf{U}}) \tag{5}$$

where the term $\nu_t(\overline{\mathbf{U}})$ in Eq. (5) can be evaluated from the statistics of the experimental data, as in [19], while the term $\nabla_{\mathbf{U}}\nu_t(\overline{\mathbf{U}})\cdot\tilde{\mathbf{u}}$ is obtained by the linearization of the turbulence model used to close the equations. Eq. (5) is analogous to the one used in [16, 18, 21], but the last term in the rhs is included to take the linearization of the turbulence model into account.

In the framework of weakly-non-parallel stability analysis, Eq. (5) is now applied to a parallel flow $\overline{\mathbf{U}} = (\overline{U}_x, \overline{U}_\theta, 0)$ evaluated at a given streamwise location. This allows for a modal expansion of the coherent fluctuation in the following form:

$$\tilde{\mathbf{u}}(x,\theta,r,t) = \hat{\mathbf{u}}(r)\exp(ikx + im\theta - i\omega t) \tag{6}$$

where k and m are the axial and azimuthal wavenumber, respectively, and ω is the frequency. When this modal form is substituted in Eq. (5), an eigenvalue problem is obtained. In the temporal stability analysis k is real and assigned, while ω is the complex eigenvalue of the problem. The opposite choice is done for the spatial stability analysis. In both cases, m is a free integer parameter. In this work only temporal stability analysis was performed.

For the stability analysis, Eq. (5) together with continuity equation are discretized using a code based on a Chebyshev spectral collocation method. In the present analysis the number of collocation points is $N = 120$ and the size of the domain in the radial direction is $r_{max}/d = 50$. This choice provides the convergence of the most unstable eigenvalue with a five-digit accuracy, which is sufficient for the present purposes.

3. Evaluation of the wake velocity profiles for different loading conditions

Linear stability analysis was performed by using as base flow the time-averaged velocity field evaluated at a reference downstream location. The axial velocity profile, U, is modeled by representing the velocity deficit produced by the wake vorticity structures through Gaussian functions, which is a model already assessed by previous experimental and numerical works [3, 4, 14]:

$$\frac{U(r)}{U_\infty} = 1 - U_{hub}\exp\left[-\left(\frac{r}{\sigma_{hub}}\right)^2\right] - U_{tip}\exp\left[-\left(\frac{r - r_{tip}}{\sigma_{tip}}\right)^2\right] \tag{7}$$

U_∞ is the freestream velocity, U_{hub} is the maximum velocity deficit connected with the hub vortex, and U_{tip} is the one related to the tip vortex. The radial position from the wake center is r, while r_{tip} is the radial position of the tip vortex. The parameters σ_{hub} and σ_{tip} are proportional to the cross-dimension of the hub and tip vortex, respectively.

Journal of Physics: Conference Series **625** (2015) 012033 doi:10.1088/1742-6596/625/1/012033

From the momentum budget formulated for the streamwise direction, the thrust coefficient of the wind turbine, C_T, can be evaluated as follows:

$$C_T = \frac{8}{R^2}\int_0^R \frac{U}{U_\infty}\left(1-\frac{U}{U_\infty}\right) r\,\mathrm{d}r. \tag{8}$$

By injecting Eq. (7) into Eq. (8), it is obtained:

$$\begin{aligned} C_T = \frac{8\,U_{hub}}{R^2}\int_0^R &\left(1 - U_{hub}\left\{\exp\left[-\left(\frac{r}{\sigma_{hub}}\right)^2\right] - k_U\exp\left[-\left(\frac{r-r_{tip}}{\sigma_{tip}}\right)^2\right]\right\}\right) \\ &\left\{\exp\left[-\left(\frac{r}{\sigma_{hub}}\right)^2\right] + k_U\exp\left[-\left(\frac{r-r_{tip}}{\sigma_{tip}}\right)^2\right]\right\} r\,\mathrm{d}r, \end{aligned} \tag{9}$$

where $k_U = U_{tip}/U_{hub}$.

The azimuthal velocity profile, V_θ, is modeled through Carton-McWilliams vortices [15], which is a model previously used to investigate instabilities of swirling jets in [22], and to describe the plane instabilities of the so-called isolated or screened vortices, i.e. vortices with circulation decreasing away from the core, in [23] and [24]. The azimuthal velocity is expressed as follows:

$$\frac{V_\theta(r)}{U_\infty} = \Omega_{hub}\, r\,\exp\left[-\left(\frac{r}{\sigma_{\theta,hub}}\right)^\alpha\right] - \Omega_{tip}(r-r_{tip})\,\exp\left[-\left(\frac{r-r_{tip}}{\sigma_{\theta,tip}}\right)^\alpha\right] \tag{10}$$

where Ω_{hub} is the axial vorticity peak of the hub vortex, while Ω_{tip} is the one for the tip vortex. The parameters $\sigma_{\theta,hub}$ and $\sigma_{\theta,tip}$ are proportional to the core radius of the hub and tip vortex, respectively. The parameter α can modify the radial distribution of the vorticity field and for this study is considered equal to 2.6 accordingly to previous wind tunnel experiments [3, 14].

The thrust coefficient of a wind turbine is evaluated now from the momentum budget in the azimuthal direction:

$$C_T = \frac{4\,TSR}{R^3}\int_0^R \frac{V_\theta}{U_\infty}\left(1+\frac{1}{2\,TSR}\frac{R}{r}\frac{V_\theta}{U_\infty}\right) r^2\,\mathrm{d}r, \tag{11}$$

where TSR is the tip speed ratio of the wind turbine, i.e. the ratio between the azimuthal velocity of the blade tip and the incoming wind velocity at hub height. By injecting Eq. (10) in Eq. (11), it is obtained

$$\begin{aligned} C_T = \frac{4\,TSR\,\Omega_{hub}}{R^3}\int_0^R &\left\{r\,\exp\left[-\left(\frac{r}{\sigma_{\theta,hub}}\right)^\alpha\right] - k_\Omega(r-r_{tip})\exp\left[-\left(\frac{r-r_{tip}}{\sigma_{\theta,tip}}\right)^\alpha\right]\right\} \\ \left(1+\frac{\Omega_{hub}}{2\,TSR}\frac{r}{R}\right.&\left.\left\{r\,\exp\left[-\left(\frac{r}{\sigma_{\theta,hub}}\right)^\alpha\right] - k_\Omega\,(r-r_{tip})\exp\left[-\left(\frac{r-r_{tip}}{\sigma_{\theta,tip}}\right)^\alpha\right]\right\}\right) r^2\,\mathrm{d}r, \end{aligned} \tag{12}$$

where $k_\Omega = \Omega_{tip}/\Omega_{hub}$.

Accordingly to the actuator disc model, the loading conditions of a wind turbine can be represented through the induction factor, a, which is a direct measurement of the velocity deficit produced through the turbine rotation [25]:

$$a = 1 - \frac{U_R}{U_\infty}. \tag{13}$$

U_R is the averaged axial velocity at the rotor disc. Consequently, the thrust coefficient, C_T, can be expressed as a function of the induction factor

$$C_T = 4a(1-a), \tag{14}$$

and the power coefficient, C_P, can be evaluated as:

$$C_P = 4a(1-a)^2. \tag{15}$$

For a given loading condition, i.e. induction factor and tip speed ratio, the wake velocity profiles of a wind turbine are evaluated through an iterative method. First, several parameters in Eqs. (7) and (10) were fixed in order to mimic a certain aerodynamic design of the turbine blades. In this work, their respective values were selected accordingly to previous wind tunnel experiments and are reported in Tab. 1 [3, 14]. The only free parameters are U_{hub}, which is the maximum axial velocity deficit located at center of the wake, and Ω_{hub}, which is the peak of the axial vorticity at the wake center and represents the wake swirl.

For a given induction factor, a, the thrust coefficient, C_T, is evaluated through the actuator disc model as in Eq. (14). Then, the parameter U_{hub} is iteratively varied until the C_T evaluated through Eq. (9) matches the value predicted through the actuator disc model in Eq. (14). Therefore, the axial velocity profile is univocally estimated by fixing the induction factor, a.

A similar iterative method is then applied for the evaluation of the azimuthal velocity, V_θ. However, in Eqs. (11) and (12) it is observed that C_T is a function of the induction factor a and of the tip speed ratio TSR. Therefore, in order to obtain univocally V_θ, they both need to be fixed for a certain loading condition. Then, Ω_{hub} is varied iteratively until the C_T evaluated through Eq. (12) matches the value obtained through the actuator disc model in Eq. (14).

Table 1. Parameters of the velocity profiles adopted to mimic a fixed aerodynamic design of the turbine blades.

Parameter	Value
$k_U = U_{tip}/U_{hub}$	0.33
σ_{hub}	0.26
σ_{tip}	0.085
r_{tip}	0.45
$k_\Omega = \Omega_{tip}/\Omega_{hub}$	0.25
$\sigma_{\theta,hub}$	0.26
$\sigma_{\theta,tip}$	0.065
α	2.6

4. Effects of the turbine induction factor and tip speed ratio on the hub vortex instability

Different loading conditions, which produce different wake velocity profiles, were considered in order to evaluate the effects of wind turbine power and thrust on the hub vortex instability. Indeed, the axial induction factor, a, was varied between 0.04 and 0.44, which leads to C_T from 0.15 to 0.99, and C_P from 0.15 to 0.55. For each case, the iterative method described in the previous section was used to calculate the velocity profiles U and V_θ, as shown in Fig. 1. For this test case, the tip speed ratio of the turbine is $TSR = 7$. As the induction factor increases, the aerodynamic load and thrust coefficient, C_T, become larger, which lead to a larger axial velocity deficit and a greater axial shear, i.e. the radial gradient of the axial velocity, see Fig. 1(a). On the other hand, the increased wind turbine thrust causes an enhanced wake swirl, thus to a larger azimuthal velocity as seen in Fig. 1(b).

Temporal linear stability analysis was performed for the wake velocity profiles obtained by varying the induction factor, a. In Fig. 2, the growth rate ω_i of each unstable mode is plotted as

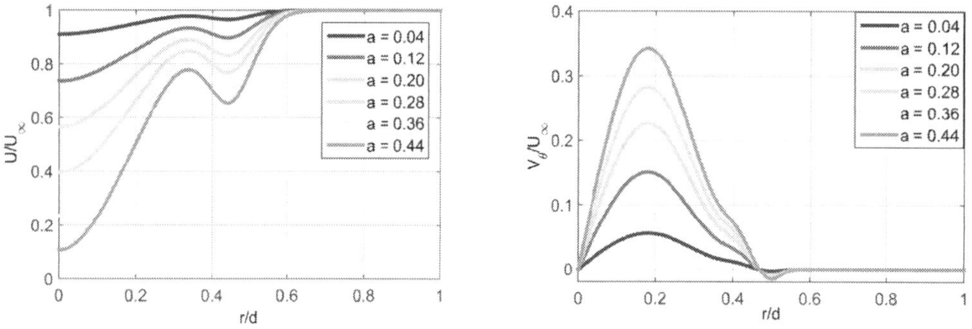

Figure 1. Velocity profiles evaluated for different induction factors and TSR=7: a) axial velocity component, U/U_∞; b) azimuthal velocity, V_θ/U_∞.

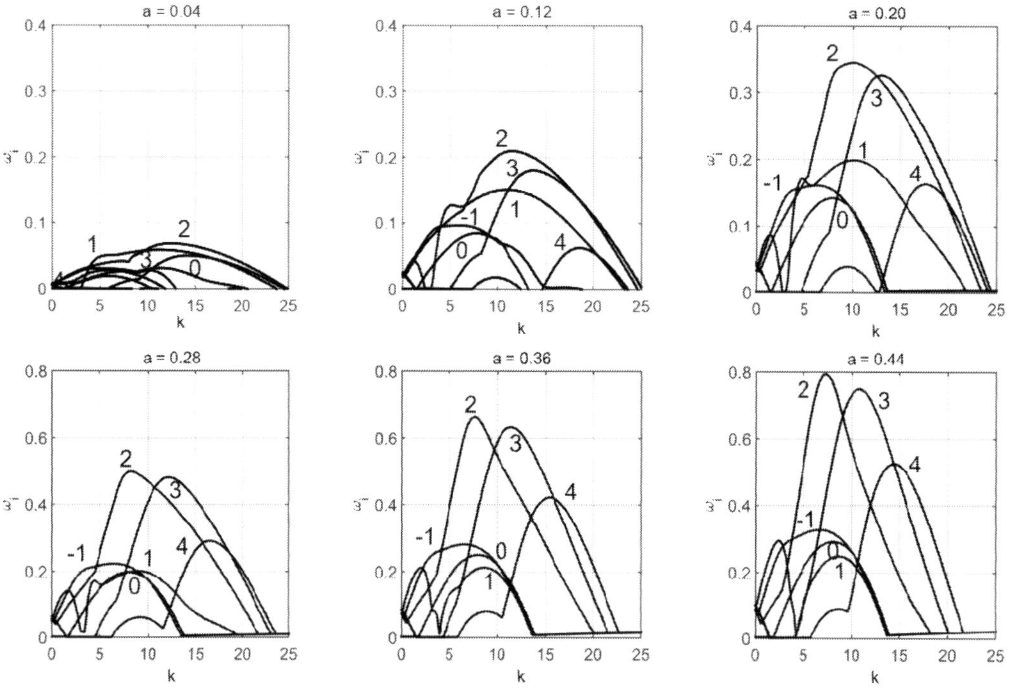

Figure 2. Growth rates, ω_i, of the unstable modes as a function of axial wavenumber, k, for various induction factors and TSR=7. The induction factor increases from left to right, top to bottom. N.B. the vertical axis limit is 0.4 in the top row, while it is 0.8 in the bottom row.

a function of the axial wavenumber, k. It is evident that the growth rate of each mode increases as the induction factor increases. However, for each case the dominant unstable mode is always the one with azimuthal number $m = 2$. Therefore, the hub vortex instability is promoted by larger aerodynamic load and thrust coefficient, thus by a higher axial shear.

In order to investigate effects of turbine rotation on the hub vortex instability, the tip speed

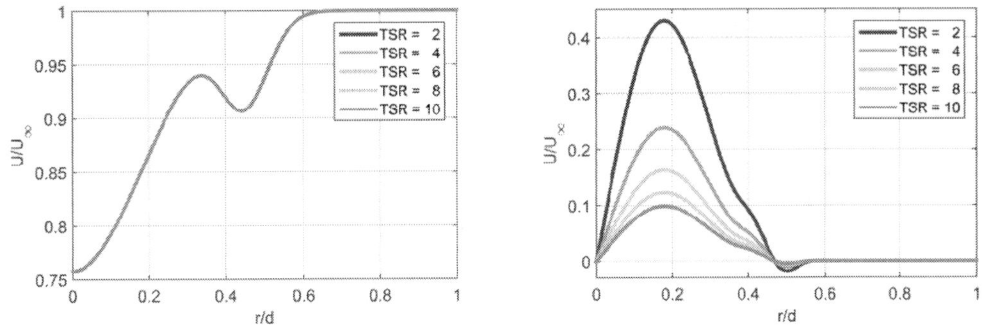

Figure 3. Velocity profiles of the wind turbine wake evaluated for different tip speed ratio, TSR, and with axial induction factor $a = 0.1106$: a) axial velocity component, U/U_∞; b) azimuthal velocity, V_θ/U_∞.

ratio, TSR, was varied between 2 and 10. For this study, the induction factor was fixed to $a = 0.1106$, which corresponds to a power coefficient $C_P = 0.35$. The axial velocity profile is not a function of TSR for a fixed C_T, as shown in Eqs. (8) and (9). Therefore, the profile of the axial velocity is the same for all the tested cases as reported in Fig. 3(a). The profiles of the azimuthal velocity for different tip speed ratios are reported in Fig. 3(b). For a fixed C_T, as TSR increases the swirl in the wake decreases. Indeed, the same power can be extracted with a low tip speed ratio and high aerodynamic load, which produces in turn higher wake swirl, or viceversa with a high tip speed ratio and low aerodynamic load.

The spectra obtained through the temporal linear stability analysis are reported in Fig. 4. For a lower TSR, thus for higher azimuthal velocity, the instability growth rate is increased. Therefore, the wake swirl promotes the hub vortex instability. Moreover, a different TSR leads to a different azimuthal wavenumber of the most unstable mode, m. Indeed, unstable modes with higher values of m are selected for increasing TSR.

5. Final remarks and conclusions

In this paper the hub vortex instability encountered in wind turbine wakes was investigated for different turbine loading conditions. This survey was performed via temporal stability analysis of the wake velocity field estimated at a reference downstream location. The linear stability analysis is formulated in order to take turbulence effects on wake instability into account. To this end, the Reynolds stresses are embedded in the linearized Navier-Stokes equations, and the turbulence closure model is formulated by using Boussinesq hypothesis and a mixing length model adapted for swirling flows.

The radial profiles of the axial and azimuthal velocity fields connected with the wind turbine wake are modeled with multiple Carton-McWilliams vortices. This technique resulted to be an efficient method to mimic the presence of the hub vortex and tip vortices within the wind turbine wake. An iterative procedure has been applied to vary the maximum axial velocity deficit and axial vorticity of the Carton-McWilliams vortices in order to match the thrust coefficient of the wind turbine predicted through the actuator disc model for each considered condition.

Stability analysis has been performed by varying the induction factor of the turbine, i.e. thrust and power coefficient, and tip speed ratio. It has been shown that an increased turbine thrust coefficient leads to an enhanced maximum velocity deficit and, in turn, to a larger axial shear. This wake flow feature is considered to be one of the dominant mechanisms underpinning the increase of the growth rate of the hub vortex instability with increasing induction factor.

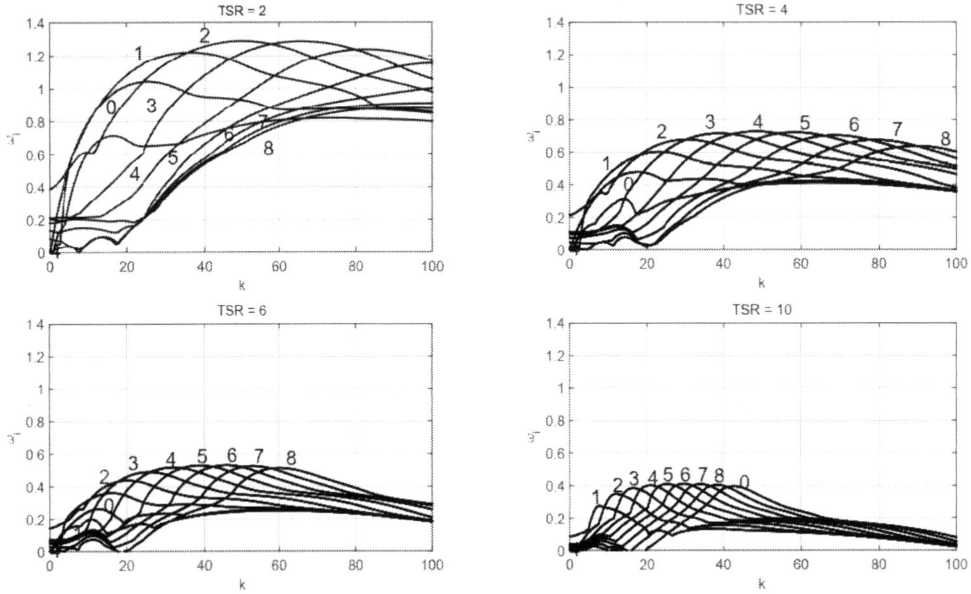

Figure 4. Growth rates, ω_i, of the unstable modes as a function of axial wavenumber, k, for various tip speed ratios and with axial induction factor $a = 0.1106$.

However, different values of the induction factor do not affect the selected azimuthal wavenumber of the most unstable mode.

The tip speed ratio of the turbine also affects the hub vortex instability. Indeed, wake swirl induced by the turbine rotation promotes the hub vortex instability. Moreover, the variation of the tip speed ratio can also modify the azimuthal wavenumber of the dominant unstable mode. This feature can affect significantly the downstream wake recovery due to a different spatial variability of the velocity field. A larger azimuthal wavenumber of the most unstable mode enhances the mechanically produced wake turbulence, which can lead to larger shear stresses and an earlier wake recovery.

Summarizing, in this paper it has been shown that a large variability of the hub vortex instability in wind turbine wakes, specifically growth rate and azimuthal wavenumber, can be observed by varying the operational conditions of the wind turbine. As mentioned above, a different mode of the hub vortex instability implies a modification of the downstream evolution of the wind turbine wake. This variability in the downstream evolution of wind turbine wakes affects wake interactions within a wind farm, thus performance for both power harvesting and fatigue loads on waked turbines. Therefore, these results strengthen the value of this methodology for prediction of wind turbine wake instabilities, consisting of modeled wake velocity profiles for a certain loading condition. This procedure enables the analysis of a large set of realistic operational conditions of wind turbines, which could be otherwise prohibiting through CFD simulations or wind tunnel experiments.

References

[1] Porté-Agel F, Wu Y-T, Lu H and Conzemius R J 2011 *J. Wind Eng. Ind. Aerodyn.* **99** 154-168
[2] Martinez-Tossas L A, Churchfield M J, Leonardi S 2014 *Wind Energy.* doi:10.1001/we.1747
[3] Iungo G V, Viola F, Camarri S, Porté-Agel F and Gallaire F 2013 *J. Fluid Mech.* **737** 499-526

[4] Chamorro L P and Porté-Agel F 2009 *Boundary-Layer Meteorol.* **132** 129-149
[5] Wu Y-T and Porté-Agel F 2011 *Boundary-Layer Meteorol.* **138** 345-366
[6] Iungo G V, Wu Y-T and Porté-Agel F 2013 *J. Atmos. Ocean. Technol.* **30** 274-287
[7] Iungo G V and Porté-Agel F 2014 *J. Atmos. Ocean. Technol.* **31** 2035-2048
[8] Felli M, Camussi R and Di Felice F 2011 *J. Fluid Mech.* **682** 5-53
[9] Ivanell S, Mikkelsen R, Sorensen J N and Henningson D 2010 *Wind Energy.* **13** 705-715
[10] Sarmast S, Dadfar R, Mikkelsen R F, Schlatter P, Ivanell S, Sorensen J N and Henningson D S 2014 *J. Fluid Mech.* **755** 705-731
[11] Chamorro L P, Hill C, Morton S, Ellis C, Arndt R E A and Sotiropoulos F 2013 *J. Fluid Mech.* **716** 658-670
[12] Kang S, Yang X and Sotiropoulos F 2014 *J. Fluid Mech.* **744** 376-403
[13] Okulov V L, Naumov I V, Mikkelsen R F, Kabardin I K and Sorensen J N 2014 *J. Fluid Mech.* **747** 369-380
[14] Viola F, Iungo G V, Camarri S, Porté-Agel F and Gallaire F 2014 *J. Fluid Mech.* **750** R1
[15] Carton X and McWilliams J 1989 *Mesosc./Synop. Coher. Struct. Geoph. Turbul.* 225-244
[16] Reynolds W C and Hussain K M 1972 *J. Fluid Mech.* **54** 263-288
[17] Reau N and Tumin A 2002 *Eur. J. Mech (B/Fluids).* **21** 143-155
[18] Cossu C, Pujals G and Depardon S 2009 *J. Fluid Mech.* **619** 79-94
[19] Kitsios V, Cordier L, Bonnet J-P, Ooi A and Soria J 2010 *J. Fluid Mech.* **664** 74-107
[20] Pope S B 2000 Turbulent flows *Cambridge Univ. Press.*
[21] Del Alamo J C and Jimenez J 2006 *J. Fluid Mech.* **559** 205-213
[22] Gallaire F and Chomaz J 2003 *J. Fluid Mech.* **494** 223-253
[23] Carnevale G and Kloosterziel R 1994 *J. Fluid Mech.* **259** 305-331
[24] Orlandi P and Carnevale G 1999 *J. Fluid Mech.* **381** 239-269
[25] Burton T, Sharpe D, Jenkins N and Bossanyi E 2002 *Wiley*

Instability of wind turbine wakes immersed in the atmospheric boundary layer

Francesco Viola[1], Giacomo Valerio Iungo[2], Simone Camarri[3], Fernando Porté-Agel[4] and François Gallaire[1]

[1] Ecole Polytechnique Fédérale de Lausanne (EPFL), Laboratory of Fluid Mechanics and Instabilities (LFMI), Lausanne, Switzerland

[2] The University of Texas at Dallas, Mechanical Engineering Department, Wind Fluids and Experiments Lab (WindFluX), 75080 Richardson, TX

[3]Department of Civil and Industrial Engineering, University of Pisa, Pisa 56122, Italy

[4] Ecole Polytechnique Fédérale de Lausanne (EPFL), Wind Engineering and Renewable Energy (WIRE) Lab, Lausanne, Switzerland

E-mail: `Valerio.Iungo@utdallas.edu`

Abstract. In this work a technique capable to investigate the near-wake stability properties of a wind turbine immersed in the atmospheric boundary layer is presented. Specifically, a 2D local spatial stability analysis is developed in order to take into account typical flow features of real operating wind turbines, such as the presence of the atmospheric boundary layer and the turbulence heterogeneity of the oncoming wind. This stability analysis can be generally applied on either experimental measurements or numerical data. In this paper it was carried out on wind tunnel experiments, for which a downscaled wind turbine is immersed in a turbulent boundary layer. Through spatial stability analysis, the dominant mode in the near wake, i.e. the most amplified one, is characterized and its frequency matches the hub-vortex instability frequency measured in the wind tunnel. As in the case of [10], where an axisymmetric wake condition was investigated, the hub-vortex instability results in a single-helical mode.

1. Introduction

In the wake of wind turbines two main vorticity structures are typically observed: the helicoidal tip vortices and the hub vortex. The tip vortices are shed from the tip of the blades and are rapidly convected downstream because located at the wake periphery where high streamwise velocity is present. The hub vortex is a streamwise-oriented vorticity structure located approximately at the wake centre.

Wind tunnel measurements in the wake produced by a down-scaled wind turbine immersed in a uniform flow showed that these vorticity structures undergo to instabilities in the near-wake and are diffused proceeding downstream [8, 10]. The tip-vortices instability is mainly driven by the mutual inductance between adjacent spirals, in cooperation with short-wave and long-wave instabilities ([1, 2]) which favorite the tip-vortices diffusion within one diameter past the turbine ([3],[4]). Conversely, the hub vortex is characterized by oscillations of a frequency equal to roughly 0.34 times the rotational frequency of the wind turbine rotor which are detected up to two diameters downstream, see [5] and [8]. In previous works these flow fluctuations were ascribed to wake wandering or meandering, which consists in transversal oscillations of the wind turbine wake. Wake meandering was investigated in detail by [5] where it is stated that this phenomenon is mainly excited by the shedding of vorticity structures from the rotor disc acting as a bluff body. In [8], in the case of uniform oncoming flow, this phenomenon was explained as the appearance of a counter-winding and co-rotating single-helix unstable mode amplified in the wake of the wind turbine. More specifically temporal and spatial linear stability analyses were performed on the time-averaged wind tunnel velocity measurements in order to identify the most amplified vortical structures originating at the hub-vortex instability. Furthermore, simultaneous hot-wire measurements confirmed the presence of a helicoidal unstable mode of the hub vortex with a streamwise wavenumber roughly equal to that predicted from the linear stability analysis. This results have been confirmed in the experimental work of [6] where, through laser doppler anemometry (LDA) and particle image velocimetry (PIV) visualization, these authors found that the wake dynamics is associated with a precession (rotation) of the helical vortex core at a constant Strouhal number. Moreover, [7] carried out LES simulations which included all geometrical details of the wind turbine by a curvilinear immersed-boundary method with evidence of large-scale meandering motions.

In [10] a stability analysis is performed by taking into account the effects of the Reynolds stresses by means of eddy-viscosity models. Other studies in the literature have considered eddy-viscosity models to close the linearized equations for the coherent velocity field in a turbulent flow following the seminal work of [9]. In [10] three eddy-viscosity models are considered, where the model parameters are calibrated on the wind tunnel data. One model is based on the assumption of a uniform eddy viscosity for each streamwise location, whereas for the other two models a mixing length is estimated. It was shown that with the proposed improved formulation, stability analysis allows not only the unambiguous identification of the hub vortex instability, but also the accurate frequency prediction.

In the present work the analysis is extended to the case of wind turbines immersed in a turbulent atmospheric boundary layer (ABL). The presence of the ABL in the streamwise direction breaks the axisymmetry of the mean flow, thus the harmonic modal expansion in the azimuthal direction is not valid anymore and a 2D local stability analysis has to be carried out. With this formulation the effect of the condition of the ABL such as shear veer or turbulence heterogeneity on the near wake-dynamics can be included within the present analysis. This formulation is here applied to wind tunnel data acquired in the wake of a downscaled wind turbine subjected to a turbulent oncoming boundary layer flow.

This paper is organized as follows: the wind tunnel data are described in 2, then the formulation of the stability analysis, which takes the modeled Reynolds stresses into account, and the numerical method are presented in 3. The results of the spatial stability analysis of the mean wake flow are then described in 4. Finally, conclusions are drawn in 5.

2. Experimental data

In this work a data-set acquired in the boundary layer wind tunnel of the Wind Engineering and Renewable Energy Laboratory (WIRE) of the Ecole Polytechnique Federale de Lausanne (EPFL) has been considered and investigated by 2D local stability analysis. The experimental facility is a closed loop wind tunnel with an inlet providing a contraction with 5:1 area ratio. Several turbulence devices consisting of coarse meshes and honeycomb flow straighteners are used to increase flow quality. The wind turbine down-scaled model used is a three-bladed GWS/EP-6030x3 anticlockwise. The rotor, with a diameter, d, of 152 mm, is connected to a DC motor with a diameter of 10 mm. The wind turbine model is mounted with a stem of height 127 mm and is directly installed on the wind tunnel floor. By considering its frontal area, the blockage ratio is less than 0.5%, which is very small to neglect possible effects due to the size of the wind tunnel. More details about a similar experimental setup can be found in [8]. For the tests considered here, the tip speed ratio of the turbine (TSR), which is the ratio between

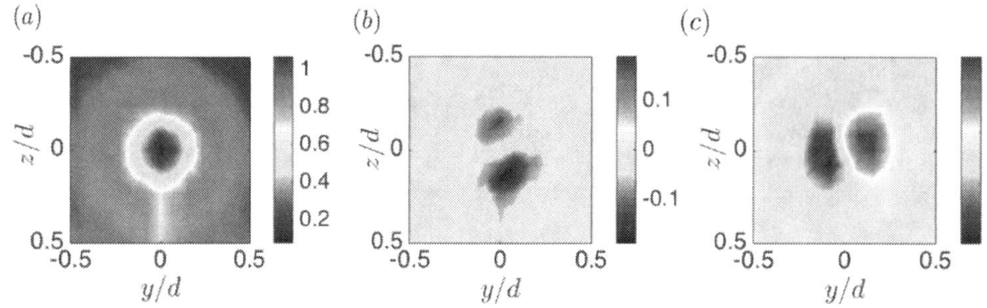

Figure 1. Non-dimensional time averaged mean flow at the streamwise position $x/d = 0.375$. (a) U_x/U_{hub}, (b) U_y/U_{hub} and (c) U_z/U_{hub}.

the speed of the blade tip and the oncoming velocity at hub height, U_{hub}, is 4.7. For these tests the mean rotational frequency, f_{hub}, was 62 Hz.The velocity at hub height is 6.3 m/s and the turbulence level is equal to 7%. A Reynolds number of 70000 is obtained by considering U_{hub} as reference velocity, and the rotor diameter, d, as reference length.

Measurements of three components of the velocity were performed with a customized Cobra probe produced by Turbulent Flow Instrumentation, which is a four-hole pressure probe. This probe, with an external diameter of 1.5 mm, can measure velocity fluctuations characterized by frequencies lower than 300Hz.

The reference frame used has its origin placed just behind the hub of the wind turbine, with the x-axis corresponding to the streamwise direction, positive pointing downstream. The z-axis is along the vertical direction, positive from the bottom towards the top, while the y-axis is along the spanwise direction oriented so as to produce a right-handed Cartesian coordinate system. Velocity measurements were performed from a downstream distance of $0.375d$ up to $4d$. In figure 1 the three velocity components of the time averaged mean flow field at the streamwise position in the wake $x/d = 0.375$ are shown. In figure 1(a) the streamwise component, U_x is characterized by a strong velocity deficit which is progressively recovered further downstream. The streamwise velocity is higher at the top periphery of the wake than in the lower due to the presence of the ABL and of the sting supporting the turbine. The significant peaks of the spanwise velocity components U_y and U_z (fig. 1(b) and (c) respectively) are detected for radial positions $r/d \approx 0.15$, which are connected to the rotational velocity induced by the hub vortex. In other words the spanwise components correspond to the azimuthal velocity of the hub vortex projected in the cartesian reference frame. The corresponding tangential components of the Reynolds stress tensor are shown in figure 2.

3. Stability analysis, turbulent model and numerical method

Following the approach proposed originally by [9] the unsteady flow, $\mathbf{U}(\mathbf{x}, t)$, can be decomposed in three contributions: the time-averaged base-flow, $\overline{\mathbf{U}}(\mathbf{x})$, the coherent fluctuation, $\tilde{\mathbf{u}}(\mathbf{x}, t)$, and the turbulent motion, $\mathbf{u}'(\mathbf{x}, t)$

$$\mathbf{U} = \overline{\mathbf{U}} + \tilde{\mathbf{u}} + \mathbf{u}'$$

Specifically the sum of the time-averaged flow and the coherent fluctuation coincides with the ensemble averaged flow $< \mathbf{U} >= \overline{\mathbf{U}} + \tilde{\mathbf{u}}$. Recalling the formulation detailed in [10] the non-linear evolution of the coherent perturbation, $\tilde{\mathbf{u}}$, on top of the time averaged mean flow, $\overline{\mathbf{U}}$, is given

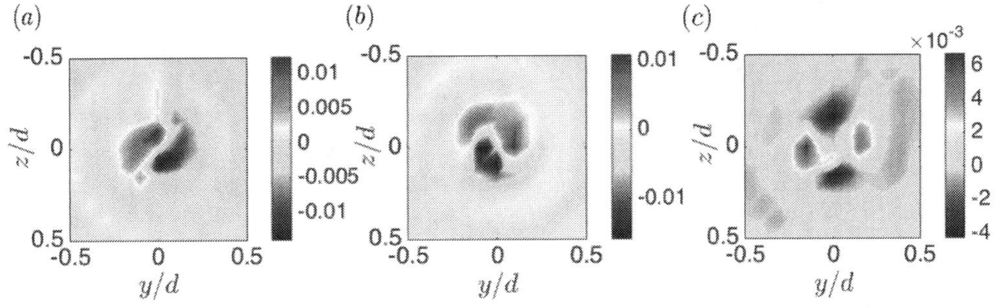

Figure 2. Non-dimensional tangential components of Reynolds stress tensor at the streamwise position $x/d = 0.375$. (a) $\overline{u'_x u'_y}/U^2_{hub}$, (b) $\overline{u'_x u'_z}/U^2_{hub}$ and (c) $\overline{u'_y u'_z}/U^2_{hub}$.

by the equation (1):

$$\begin{aligned} &\nabla \cdot \tilde{\mathbf{u}} = 0 \\ &\frac{\partial \tilde{\mathbf{u}}}{\partial t} + \nabla \tilde{\mathbf{u}} \cdot \overline{\mathbf{U}} + \nabla \overline{\mathbf{U}} \cdot \tilde{\mathbf{u}} = -\nabla \tilde{p} + \frac{1}{\mathrm{Re}} \Delta \tilde{\mathbf{u}} - \nabla \cdot [\tilde{\mathbf{u}}\tilde{\mathbf{u}} - \overline{\tilde{\mathbf{u}}\tilde{\mathbf{u}}}] - \nabla \cdot [< \mathbf{u}'\mathbf{u}' > - \overline{\mathbf{u}'\mathbf{u}'}] \end{aligned} \tag{1}$$

where the third and last terms of the rhs are respectively the non-linear term and the difference among the Reynolds stresses obtained by ensamble and time averages. From now on, the equations are made non-dimensional with the turbine diameter d and the velocity at the hub U_{hub}. In the linear framework the coherent perturbation is assumed to be small and by using the Boussinesq approximation for the Reynolds stresses, eq. (1) is rewritten as follows:

$$\frac{\partial \tilde{\mathbf{u}}}{\partial t} + \nabla \tilde{\mathbf{u}} \cdot \overline{\mathbf{U}} + \nabla \overline{\mathbf{U}} \cdot \tilde{\mathbf{u}} = -\nabla \tilde{p} + \frac{1}{\mathrm{Re}} \Delta \tilde{\mathbf{u}} + \nabla \cdot (\nu_t(\overline{\mathbf{U}})[\nabla + \nabla^T]\tilde{\mathbf{u}}) + \nabla \cdot ((\nabla_{\mathbf{U}} \nu_t(\overline{\mathbf{U}}) \cdot \tilde{\mathbf{u}})[\nabla + \nabla^T]\overline{\mathbf{U}}) \tag{2}$$

Equation(2) is the linear evolution of the coherent perturbation on a turbulent time-averaged mean flow with enforced Boussinesq hypothesis, where the eddy-viscosity terms need to be modeled. As in [10] we adopt here the generalized mixing-length model, hence ν_t is given by:

$$\nu_t = l_m^2 (2\overline{S} : \overline{S})^{1/2} \tag{3}$$

where S is the strain rate tensor. The mixing-length l_m is tuned on the experimental data through a best fit procedure on the Boussinesq equation, in a similar way to [10]. This is done separately at each streamwise section considered for the local stability analysis, so that the value of l_m varies with x. The term $\nabla_{\mathbf{U}} \nu_t \cdot \tilde{\mathbf{u}}$ represents the eddy-viscosity variation due to the coherent fluctuation and it is given by the linearization of the model around the mean flow:

$$\nabla_{\mathbf{U}} \nu_t(\overline{\mathbf{U}}) \cdot \tilde{\mathbf{u}} = l_m^2 \frac{2\overline{S} : \tilde{s}}{(2\overline{S} : \overline{S})^{1/2}} \tag{4}$$

In the framework of weakly-non-parallel stability analysis, eq. (2) is now applied to a parallel flow $\overline{\mathbf{U}} = (\overline{U}_x(y, z), \overline{U}_y(y, z), \overline{U}_z(y, z))$ extracted at a given streamwise location. Note that, the three components of the time averaged mean flow vary in both spanwise directions y and z, consequently the local analysis is bidimensional. In contrast, the x-dependence of the flow is considered by carrying out different stability analysis at different sections in the wake, as typical in local analysis assuming weak variations of the flow in that direction. This allows for a modal expansion of the coherent fluctuation in the following form:

$$\tilde{\mathbf{u}}(x,y,z,t) = \hat{\mathbf{u}}(y,z)\exp(i(kx-\omega t)) \tag{5}$$

where k is axial wavenumber and ω is the frequency. When this modal form is substituted in Eq. (2), an eigenvalue problem is obtained and eigenfunction solutions exist only if k and ω satisfy a dispersion relation of the form:

$$D[k,\omega] = 0. \tag{6}$$

The spatial branches, which describe the amplification of perturbation in space, are obtained by solving the dispersion relation with complex wavenumbers k, and real frequency ω, have been considered here (see [8]). Given $k = k_r + ik_i$, i being the imaginary unit, the term $-k_i$ is the spatial amplification rate, while k_r corresponds to the streamwise wavenumber of the traveling pertubation with frequency ω. Due to the weakly non-parallelism of the meanflow, $\overline{\mathbf{U}}$, the derivatives in the streamwise direction are of higher order and the term $2\overline{S}:\overline{S}$ appearing in equation (3) at leading order reduces to:

$$(2\overline{S}:\overline{S})^{1/2} \approx \left[\left(\frac{\partial \overline{U}_x}{\partial y}\right)^2 + \left(\frac{\partial \overline{U}_x}{\partial z}\right)^2 + \left(\frac{\partial \overline{U}_y}{\partial z} + \frac{\partial \overline{U}_z}{\partial y}\right)^2 + 2\left(\frac{\partial \overline{U}_y}{\partial y}\right)^2 + 2\left(\frac{\partial \overline{U}_z}{\partial z}\right)^2\right]^{1/2} \tag{7}$$

Similarly the term $2\overline{S}:\tilde{s}$ in the expression of the eddy-viscosity variation due to the coherent fluctuation (4) is given by:

$$2\overline{S}:\tilde{s} \approx \left(\frac{\partial \overline{U}_x}{\partial y}\right)\left(\frac{\partial \tilde{u}_x}{\partial y} + \frac{\partial \tilde{u}_y}{\partial x}\right) + \left(\frac{\partial \overline{U}_x}{\partial z}\right)\left(\frac{\partial \tilde{u}_x}{\partial z} + \frac{\partial \tilde{u}_z}{\partial x}\right) + \left(\frac{\partial \overline{U}_y}{\partial z} + \frac{\partial \overline{U}_z}{\partial y}\right)\left(\frac{\partial \tilde{u}_y}{\partial z} + \frac{\partial \tilde{u}_z}{\partial y}\right)$$
$$+ 2\left(\frac{\partial \overline{U}_y}{\partial y}\right)\left(\frac{\partial \tilde{u}_y}{\partial y}\right) + 2\left(\frac{\partial \overline{U}_z}{\partial z}\right)\left(\frac{\partial \tilde{u}_z}{\partial z}\right) \tag{8}$$

Equations (2) together with modal expansion (5) are discretized using a staggered pseudospectral Chebyshev-Chebyshev collocation method. The three velocity components are defined at the Gauss-Lobatto-Chebyshev (GLC) nodes, whereas the pressure is staggered on a different grid, which is generated with Gauss-Chebyshev nodes (GC). In both y and z directions the algebraic mapping with domain truncation is used, in order to cluster the nodes at the wake center. In the present work $N_y = 40$ and $N_z = 40$ points are used in the axial and radial direction respectively, this resolution having shown to provide the desired convergence of the amplification factors.

4. Hub vortex instability

The local spatial analysis has been carried out at several streamwise positions in the wake of the wind turbine where the experimental data were acquired. In figure 3 the results of the spatial analysis at the first available section, $x/d = 0.375$, are reported. In fig. 3(a) the spatial growth rates $-k_i$ are depicted as a function of the frequency of the coherent fluctuation ω, while in (b) the associated axial wavenumbers are shown. Three modes, labeled as modes 1, 2 and 3, result to be amplified in the frequency band $0 \leq \omega \leq 7$. In particular, in the frequency range $2.5 \leq \omega \leq 4$ the three modes coexist and three unstable eigenvalues are found when carrying out the stability analysis. Each mode is characterized by a *preferred* frequency, defined as the frequency associated with the higher spatial growth-rate, which is equal approximately to 2.7 for mode 1, to 5 for mode 2 and to 2.8 for mode 3. In figure 4 the axial velocity of the eigenmodes 1, 2 and 3 at the respective preferred frequency, are reported, showing that: the detected unstable modes and related branches present different spatial structures. Specifically, in fig. 4(a-d) the complex eigenfunction clearly exhibit a dominant single-helical shape, which is in agreement with the stability analysis results in the wake of a wind turbine invested by a uniform oncoming flow ([8, 10]) even though in this analysis no modal expansion is imposed in the azimuthal direction and eigenfunctions can have a generic shape in the $y-z$ plane. The eigenfunction of mode 2 in fig. 4(b-e) represents a double-helical mode. Lastly, the mode 3 in (c-f) is dominated by a single-helical symmetry, but more diffused respect to mode 1.

By observing the spatial growth rates in fig. 3(a), the most amplified perturbation at $x/d = 0.375$ is the single-helical mode 1 at the dominant frequency $\omega = 2.7$. The same analysis has been carried out in the other streamwise sections in the wind turbine wake and a similar scenario is obtained, with the single helix mode 1 more amplified at the low frequency range. However proceeding downstream the spatial growth rates reduce due to the recovery of the wake deficit and the increase of the turbulence intensity which favorites the diffusion of the vortex structures.

In order to determine the most amplified coherent fluctuation through the wake and its frequency, the local spatial growth rates are integrated in the streamwise direction according to:

$$G(\omega) = \exp \int_{x_i}^{x_s} -k_i(\omega, x)\mathrm{d}x. \tag{9}$$

The global gain represents the integral measure of the amplification in the wake between the first section available, $x_i = 0.375d$, and x_s which is the streamwise position where the mode becomes stable. $G(\omega)$ is reported in figure 5 as a function of the frequency ω, where the three modes amplified in the wake are labeled as before. Therefore mode 1 results to be the most amplified mode overall and its dominant frequency is in good agreement with the hub-vortex instability frequency measured in the wind tunnel, whose non-dimensional value is $\omega \approx 3$ and is depicted by the red line in fig. 5. The limited deviation in the frequency prediction is believed to be related to a non-linear modulation which can not be captured by this linear stability approach. The velocity field of the coherent perturbation at a given frequency ω is given by equation (10)

$$\tilde{\mathbf{u}}(x, y, z, t) = Re\left\{\hat{\mathbf{u}}(x; y, z)e^{i\left(\int_{x_i}^{x} k(x',\omega)\mathrm{d}x' - \omega t\right)}\right\}. \tag{10}$$

Hence, at a given time the term $e^{i\int_{x_i}^{x} k_r(x')\mathrm{d}x'}$, where $k_r(x)$ is the local wavenumber, is setting the phase of the local complex eigenfunction $\hat{\mathbf{u}}(x; y, z)$. Since the $k_r(x)$ are positive quantities, the phase increases proceeding downstream, providing different orientations to the local velocity field yielding the typical helical shape of the hub-vortex instability, see [8]. Consequently, the unsteady term $e^{-i\omega t}$ consists in a uniform change of phase in time, resulting in a constant rate rotation of the perturbation around its axis. The axial velocity of mode 1 at frequency $\omega = 2.7$, which is the most amplified coherent perturbation according to spatial analysis, is reported in figure 6 at different streamwise positions in the wind turbine wake. In all the considered sections the single-helical symmetry is the dominant one. Moreover, by moving downstream, the eigenfunction is less concentrated around the hub-vortex core due to turbulent diffusion, as known by experimental evidence. Hence, also in this case of wind turbine immersed in the atmospheric boundary layer the hub-vortex instability is associated with an helical precession of the vortex-core. Moreover, it is interesting to notice that while proceeding downstream, the helicoidal structure observed is slightly displaced downwards. This feature can be related to the presence of the boundary layer. This general lowering of the wake vorticity structures is also typically observed for wakes produced by bluff bodies immersed in a boundary layer, as for the case for a triangular prism in [11].

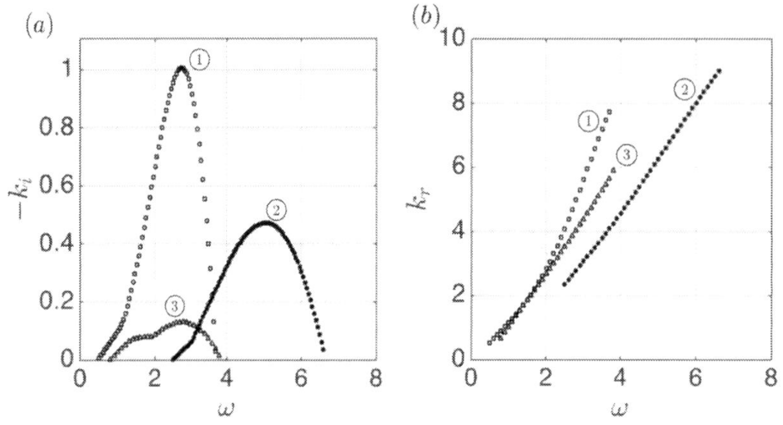

Figure 3. Spatial analysis results at the section $x/d = 0.375$. In (a) the growth rates, $-k_i$, are reported as a function of the frequency ω, while in (b) the corresponding axial wavenumbers, k_r, are shown. Each branch corresponds to a different amplified mode.

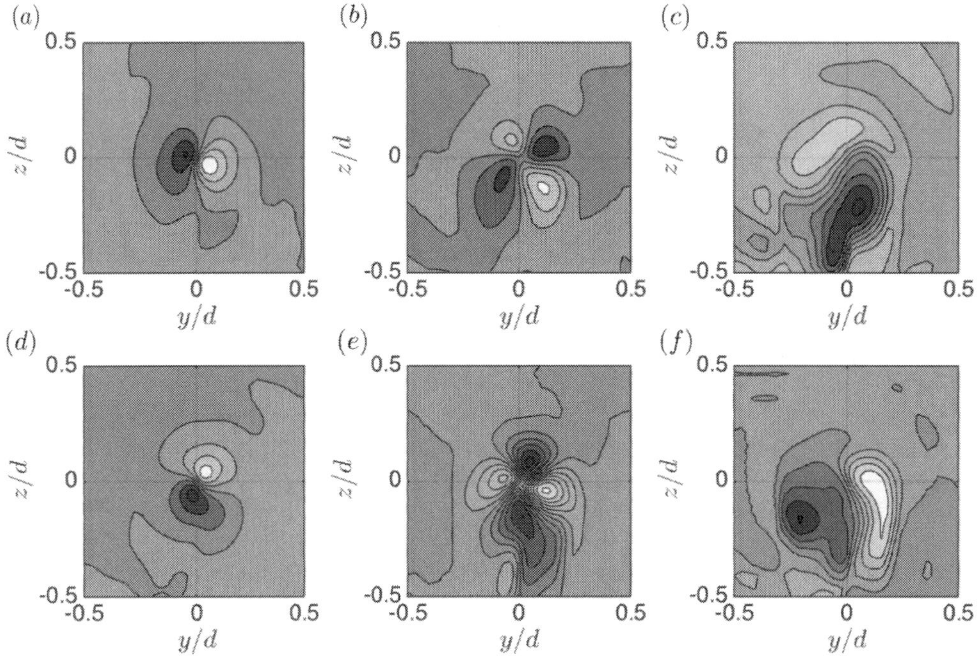

Figure 4. Spatial analysis results at the section $x/d = 0.375$. Real (a) and imaginary (d) part of the axial flow of the eigenfunction associated to mode 1 at frequency $\omega = 2.7$. In (b) and (e) the same quantities are shown for mode 2 at frequency $\omega = 5$ and in (c) and (f) for mode 3 at frequency $\omega = 2.8$.

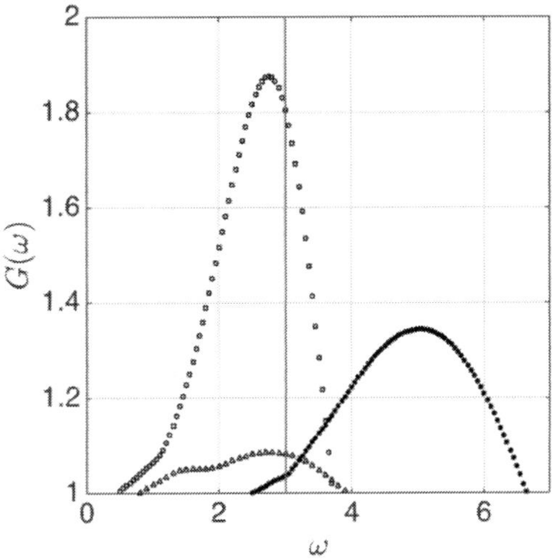

Figure 5. Integral amplification factor of coherent perturbations in the wake as a function of the frequency ω. Three modes, labeled as mode 1, mode 2 and mode 3 are amplified through the domain.The vertical red line represents the frequency of the hub vortex instability measured experimentally, corresponding to a non-dimensional pulsation $\omega_{hubvortex} = 3$.

5. Conclusions

A method suitable for local stability analysis in the near wake of wind turbine wakes has been developed and here presented. The method is based on the triple decomposition of the flow field, where an equation describing the dynamics of the coherent perturbations on the time-averaged flow and taking into account the turbulent diffusion is provided. Due to the weakly non parallelism of the flow in the case of wind turbine wakes, a local stability approach has been adopted. More specifically, at a given streamwise position in the wake, x, both y and z directions are discretized by using Chebyshev polynomials and the spatial amplification of perturbations as a function of the frequency are computed. Thus, in this method no modal expansion is imposed to the geometry of the perturbation in the azimuthal direction and the modes investigated are allowed to have a generic shape in the $y-z$ plane. Conversely, a complex wavenumber k is given, where its real and imaginary part define respectively the axial wavelength of the perturbation and its spatial amplification.

In this framework the effect on the near-wake dynamics of typical flow features of the incoming wind, such as the atmospheric boundary layer, turbulence heterogeneity, veer and shear, can be included in the stability analysis. Hence, real operating wind turbines conditions can be investigated with the present tool.

The analysis has been here applied on wind tunnel data, where the mean wind direction is aligned with the wind turbine axis. In particular, the mean wake flow acquired in the wind tunnel is used directly in the stability analysis and the measured Reynolds stresses are used to tune the mixing-length model, which is adopted to take into account turbulent diffusion in the dynamics of the hub-vortex instability. The local spatial growth rates are computed at every streamwise position in the wake as a function of the frequency of the perturbation. Three modes, which manifest a dominant helical shape, are found to be amplified in the wake. Specifically, the single helical mode 1 is more unstable in the low frequency range $0 \leq \omega \leq 4$, conversely the double helical mode 2 resonates in the higher frequency interval $4 \leq \omega \leq 6.5$. Finally mode 3 results to be less amplified than modes 1 and 2 in all the frequency range. The dominant mode

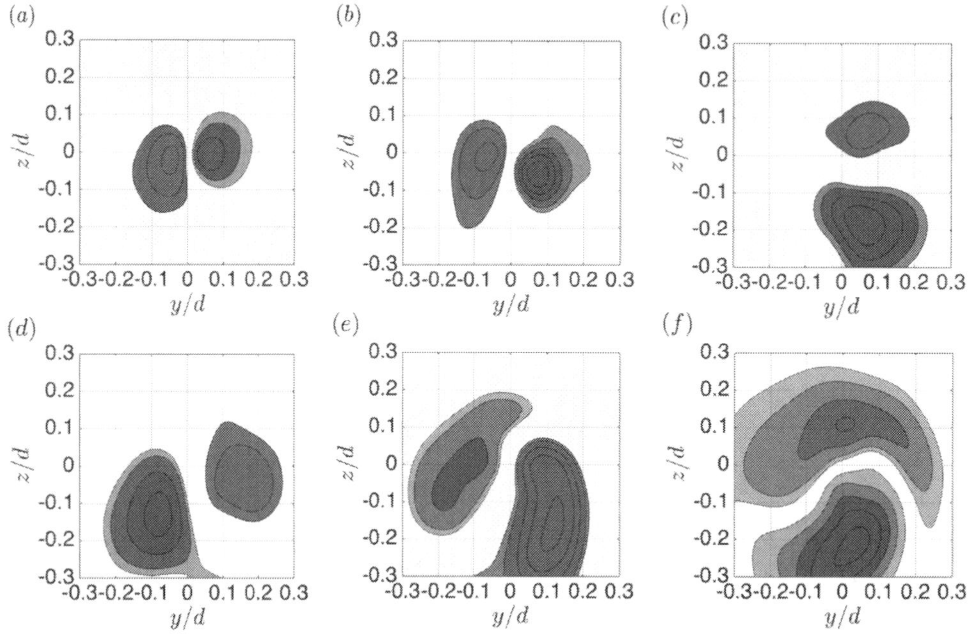

Figure 6. Axial velocity of the most amplified coherent perturbation (mode 1 with frequency $\omega = 2.7$) at different streamwise positions. (a) $x/d = 0.375$, (b) $x/d = 0.50$, (c) $x/d = 0.75$, (d) $x/d = 1.00$, (e) $x/d = 1.50$ and (f) $x/d = 2.00$.

in the wind turbine wake, and its frequency, are detected by means of the integral amplification factor, which is obtained by integrating the spatial growth rates for the different downstream locations. The spatial shape of this resonating vorticity structure results mode 1, i.e. a single-helix in agreement with previous works which is a remarkable result recalling that no azimuthal modal expansion has been applied to the perturbation field. Moreover, the frequency predicted for the most unstable mode is in good agreement with the one measured in the wind tunnel.

The applicability of this 2D spatial local stability analysis is very general since it is suitable to be directly used to investigate the wake stability of wind turbine flows experimentally measured in the wind tunnel, wind LiDARs, or simulated by RANS or LES approaches.

References

[1] Widnall S E, 1972, J. Fluid Mech. 54 (4), 641-663
[2] Felli M, Camussi R and DiFelice F, 2011, J. Fluid Mech. 682, 5-53.
[3] Sarmast S, Dadfar R, Mikkelsen R F, Schlatter P, Ivanell S, Sørensen J N and Henningson D S, 2014, J. Fluid Mech. 755, 705-731.
[4] Zhang W, Markfort C D and Porté-Agel F, 2012, Exp. in Fluids. 52-5, 1219-1235.
[5] Medici D, and Alfredsson P H, 2008 Wind Energy 11, 211-217.
[6] Okulov V L, Naumov I V, Mikkelsen R F, Kabardin I K and Sørensen J, 2014, J. Fluid Mech. 747, 369-380.
[7] Kang S, Yang X and Sotiropoulos F, 2014, J. Fluid Mech. 744, 376-403.
[8] Iungo G V, Viola F, Camarri S, Porté-Agel F and Gallaire F 2013 J. Fluid Mech. 737 499-526
[9] Reynolds W C and Hussain K M, 1972 J. Fluid Mech. 54, 263-288.
[10] Viola F, Iungo G V, Camarri S, Porté-Agel F and Gallaire F 2014 J. Fluid Mech. 750 R1
[11] Buresti G and Iungo G V 2010 J. Wind Eng. Ind. Aerodyn. 98 253262

Comparing satellite SAR and wind farm wake models

CB Hasager[1a], P Vincent[b], R Husson[b], A Mouche[c], M Badger[a], A Peña[a], P Volker[a], J Badger[a], A Di Bella[a], A Palomares[d], E Cantero[e], PMF Correia[e]
[a]DTU Wind Energy, Frederiksborgvej 399, 4000 Roskilde, Denmark
[b]CLS, Avenue La Pérouse, Bâtiment le Ponant, 29280 Plouzané, France
[c]IFREMER, Pointe du Diable, 29280 Plouzané, France
[d]CIEMAT, Avenida Complutense 40, 28040 Madrid, Spain
[e]CENER, Avda Ciudad de la Innovación 7, 31621 Sarrigurren (Navarra), Spain,

E-mail: cbha@dtu.dk; pvincent@cls.fr; romain.husson@cls.fr; alexis.mouche@ifremer.fr; mebc@dtu.dk; aldi@dtu.dk; pvol@dtu.dk; jaba@dtu.dk; adia@dtu.dk; ana.palomares@ciemat.es; ecantero@cener.com; pmfernandez@cener.com

Abstract. The aim of the paper is to present offshore wind farm wake observed from satellite Synthetic Aperture Radar (SAR) wind fields from RADARSAT-1/-2 and Envisat and to compare these wakes qualitatively to wind farm wake model results. From some satellite SAR wind maps very long wakes are observed. These extend several tens of kilometres downwind e.g. 70 km. Other SAR wind maps show near-field fine scale details of wake behind rows of turbines. The satellite SAR wind farm wake cases are modelled by different wind farm wake models including the PARK microscale model, the Weather Research and Forecasting (WRF) model in high resolution and WRF with coupled microscale parametrization.

1. Introduction

The wake of the offshore wind farms are observed from Synthetic Aperture Radar (SAR) satellite data [1, 2, 3]. In the present study SAR data are extracted from the Canadian RADARSAT-1/-2 satellites and the European Space Agency satellite Envisat. The basic measurement principle consists of emitted microwave radiation at C-band (wavelength around 5 cm) that is backscattered from the natural surface and objects herein. For natural ocean surface it is the wind field dominating the backscatter but also ocean current, natural slicks from algae and oil seepage may influence backscatter. Furthermore, ocean swell waves, tidal flow over shallow water areas and sea ice produce imprint. The ocean objects observed are ships, wind turbines, oil/gas platforms and other man-made structures at sea. For the ocean surface, the roughness of the sea appears darker for lower wind speed and brighter for higher wind speed because the backscatter from capillary and short-gravity waves at the surface of the ocean relates to surface wind speed. More wind produces more short waves. Small white dots seen in SAR images over the ocean are most likely ships or wind turbines. The backscatter is very high for hard targets [4].

[1] To whom any correspondence should be addressed.

Journal of Physics: Conference Series **625** (2015) 012035 doi:10.1088/1742-6596/625/1/012035

The investigation on wind farm wake is based on high-resolution SAR satellite data as this data source provides interesting overview of one or several wind farms at a specific time, i.e. the few seconds of observations while the satellite orbits above the area. The observed backscatter images may be processed into wind speed maps. The wind speed is retrieved from SAR using a geophysical model function (GMF) and with a priori input of wind direction information as well as information on the viewing angle geometry and azimuth angle [5].

This paper presents three cases where wakes are captured in SAR images. The wake modelling undertaken in the study includes the PARK microscale model [6] for the first case while the WRF mesoscale model [7] coupled to microscale parametrization is used for the second case. These results are from DTU. WRF is operated at high spatial resolution at CENER and CIEMAT and here used for the third case. The comparison of observed wind farm wake and model results is qualitative.

2. Wake case at Horns Rev 2 with fine-scale structure

The first wind farm wake case considered is from Horns Rev 2 in the Danish North Sea observed at 3 August 2011 at 09:49 UTC. The roughness image is shown in Figure 1.

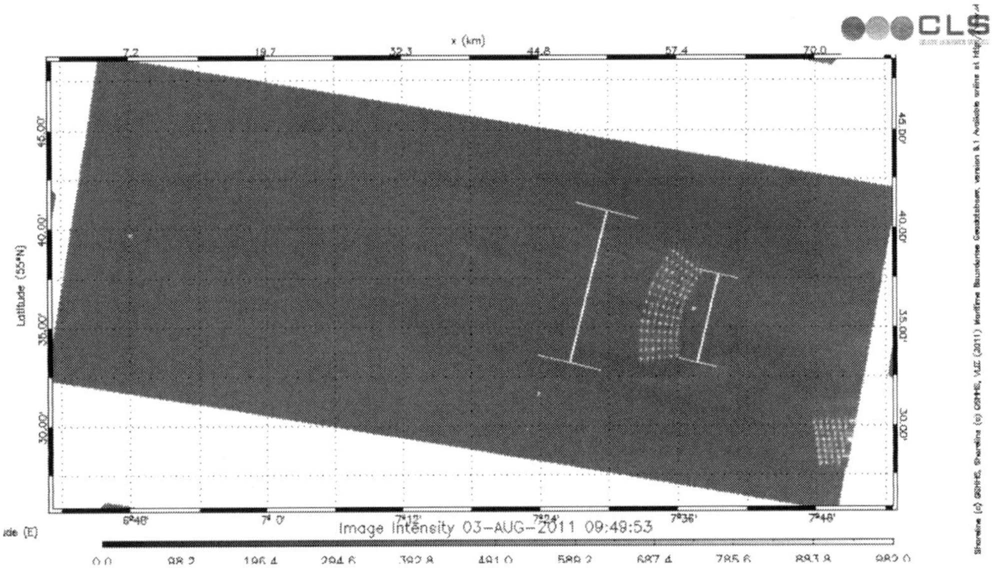

Figure 1. The roughness map is observed from Envisat ASAR IMG. The wind direction is from the east-southeast. Two horizontal transect lines are included upwind and downwind of Horns Rev 2.

In Figure 2 is shown the Normalized Radar Cross Section (NRCS) in dB along the horizontal transect west (downwind) of the wind farm while the NRCS along the upwind transect is shown in Figure 3. The polarization is horizontal transmit and receive (HH). From the SAR image the wake direction has been estimated at 111° and using this input for wind direction, wind speed has been retrieved using CMOD-IFR2 [8] and using polarization ratio 2 of [9]. The calculated equivalent neutral wind value is presented in Figure 2 and 3 for the downwind and upwind transects, respectively.

Looking at Figure 1 the wind farm wake is observed as fine-scale linear features aligned with the rows of wind turbines. The wake is most clearly visible in the northern half of the wind farm wake region.

In the downwind transect shown in Figure 2 the NRCS varies in a systematic pattern with up to 1 dB variation at very short distance. The retrieved wind speed shows variations from 4.8-6.0 ms^{-1}. Very interestingly the upwind horizontal transect shows wind speed values around 4.4 ms^{-1} with much smaller variation (around 0.3 ms^{-1}). The small variation in the upwind transect is most likely due to the natural uncertainty in the SAR data (so-called speckle noise) while the larger variation in the downwind transect most likely is due to influence from the wind turbines. It may also be noticed that there is speed up effect from the upwind to downwind transect. Speed up for winds blowing offshore is well-known [10].

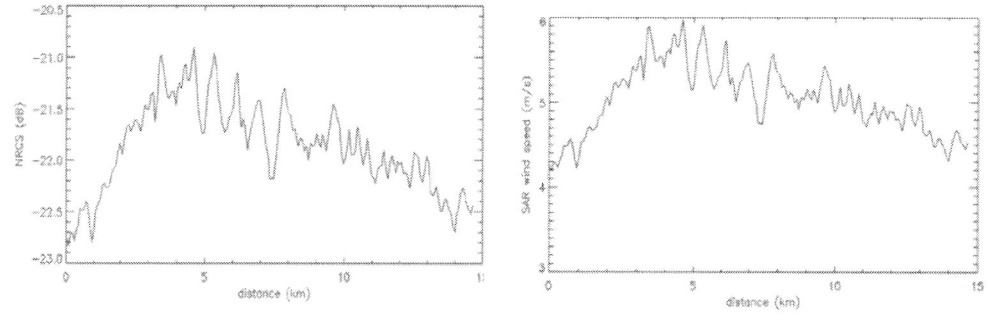

Figure 2. Horizontal transect from Figure 1. It is 15 km long and located downwind of the wind farm: (left) NRCS observed in the roughness map; (right) the wind speed retrieved from CMOD-IFR2.

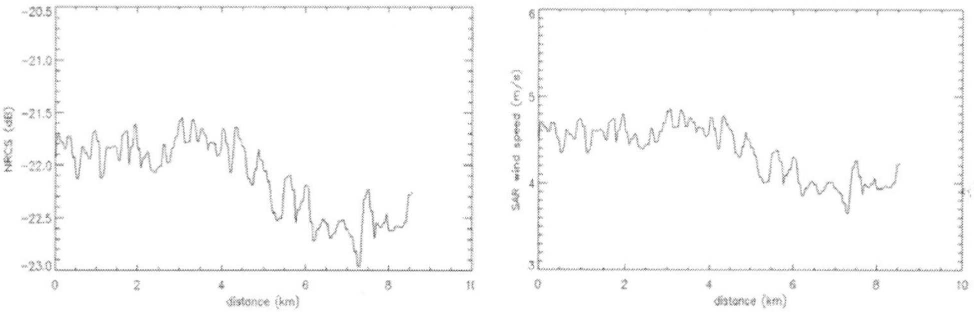

Figure 3. Horizontal transect from Figure 1. It is 8 km long and located upwind of the wind farm: (left) NRCS observed in the roughness map; (right) the wind speed retrieved from CMOD-IFR2.

To investigate the case further, the PARK microscale wake model has been used to assess the case. Figure 4 shows the map with position of wind turbines and results from PARK along the transect. The wind speed is near cut-in. The variation in wind speed behind the nearest turbines and between the rows is variable from 5.7-6.0 ms^{-1} in a regular pattern. This compare well to the observed wind speed variation in Figure 2. It is of course not obvious what 'free stream' wind speed value to use if the wind farm wake velocity deficit should be calculated, as the upwind wind speed is not constant along the wind direction due to coastal speed up effects. The 'free stream' wind speed observed north or south of the downwind transect vary from 4.2-4.5 ms^{-1}. In summary, the PARK model reproduces the fine-scale wind farm wake features observed in SAR. From the modelling point of view the challenges are that there is a clear coastal wind speed gradient and the flow is not homogeneous at the time of observation. Finally, it may be noted that in Figure 1 between Horns Rev 1 and 2 wind farms, a wavy large scale bright area is seen. This could be wake or it could be due to ocean current and bathymetry effects at the shallow areas of the Horns Rev. Thus care should be taken in the interpretation of this bright area to have higher wind speed that the local areas nearby with darker tones.

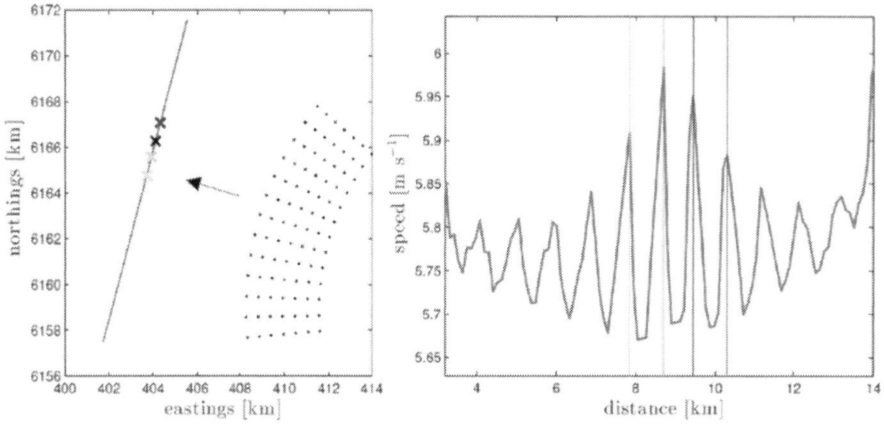

Figure 4. (left) The map of wind turbines at Horns Rev 2 and transect with four points highlighted. (right) the wake results from PARK along the transect with the four points indicated.

3. Wake case at Belwind1 and Thornton Bank with very long wakes

At the offshore wind farms Belwind1 and Thornton Bank 1+2+3 in the Belgium North Sea, two very long wind farm wakes, 79 and 65 km, respectively, were observed by an acquisition from RADARSAT-2 on the 1st of July of 2013 at 17:34 UTC. The flow is from the southwest. Figure 5 shows the intensity of the radar return signal and wind speed. The retrieved SAR wind map using CMOD-IFR2 is presented in Figure 6. This map clearly shows the wind deficit in the wind farms wake regions. The wind speed in the free stream region compared to the wake regions vary up to 1.5 ms^{-1} from 5.0-6.5 m s^{-1}. The wakes turns possible due to Coriolis effects, roughness changes and mesoscale meandering.

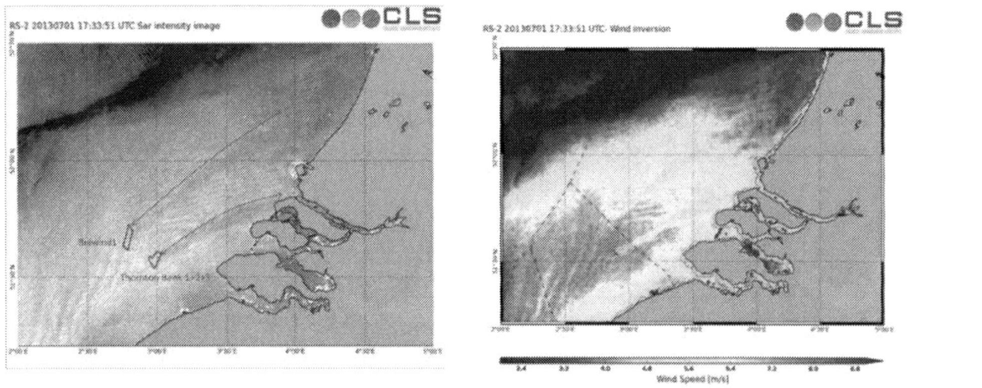

Figure 5. Long wind farm wakes observed behind Belwind1 and Thornton Bank from RADARSAT-2 intensity map (left) and wind speed map (right).

At DTU Wind Energy the WRF mesoscale model [7] has been used to model the divergence of the wakes induced by the Belwind1 and Thornton Bank wind farms. The challenge of wake modelling with mesoscale models is the limited model's horizontal resolution. To avoid resolving the completely parametrized turbulence, the horizontal grid spacing, should be much larger than the turbulence length scale [11]. Consequently, the local turbine induced wake remains completely unresolved by the model and interaction takes place only between grid-cells. For the parametrization of the wind farms we use the Explicit Wake Parametrization (EWP) [12]. It applies a grid-cell averaged deceleration to the model's flow equation and additional turbulence is produced by the Planetary Boundary Layer (PBL) scheme from the changed vertical shear in horizontal velocity. This renders the EWP parametrization PBL-scheme independent.

For the simulation we use WRF V3.4. The model was set-up with two nests inside the most outer domain I, which is driven by ERA-Interim reanalysis data [13]. The horizontal resolution is for the three domains 18, 6 and 2 km, respectively. The most inner (most expensive) nest is run twice, without and with the wind farm parametrization. The number of vertical layers is set to 60, with around 30 layers in the first 600 m. The second mass level is at around 12 m above sea level and is used for the comparison to the satellite images. The PBL-Scheme with the second order closure Mellor-Yamada-Nakanishi-Niino (MYNN) [14] was used. For completeness we summarized the most important physics parametrizations and nudging options used: Convection [15] (Domain I and II) and microphysics [16], long wave radiation [17], shortwave radiation [18], surface scheme [19] and nudging of U and V Domain I outside the PBL.

The most inner domain extends 262 by 208 grid-cells in the west-to-east and south-to-north direction, respectively. The simulation period is three days, starting at the 29th of June. Figure 6 shows the simulated normalized velocity and the velocity deficit at 10m above sea level (ASL) over the sea surface only.

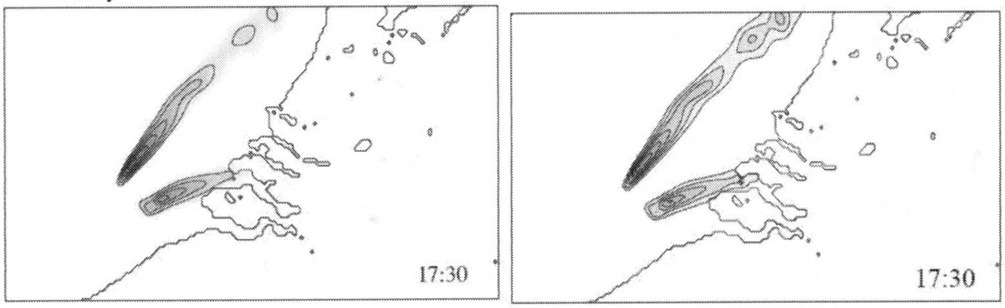

Figure 6. WRF model results on wind farm wakes behind Belwind1 and Thornton Bank: (left) normalized velocity at 10m ASL with velocity contours in ranges from 0.8 to 0.99; (right) velocity deficit at 10m ASL with contours in ranges from -1.25 to -0.1 ms^{-1} over the sea surface only.

From the satellite image (Figure 5) a clear wake of around 70 km long behind the Thornton Bank wind farm is found and the wake behind the Belwind1 wind farm appears to be even longer. Besides a bending, also a divergence of the wakes can be noticed. Similar features can be found in Figure 6 from WRF. Compared to the satellite image the wake's orientation is somewhat too far to the south in the model simulation. The wake interacts with the Dutch island Schouwen-Duivenland instead of with the most northerly delta island Goeree-Overakkee. Furthermore, the simulations show an even bigger divergence of the two wakes compared to that from the satellite image. Also the Belwind1 wind farm wake, when referring to the most outer contours of -0.1ms^{-1}, seems to extend further compared to the satellite image. However, the considered deficit is very small and perhaps not visible in the satellite image.

The wind farm induced wakes from the satellite images include mesoscale features, such as a divergence of wakes from two nearby wind farms and are also the result of unsteady flow conditions. In this section we compared the satellite images visually to a velocity deficit obtained from a WRF simulation without wind farm to one with wind farms parametrized with the EWP scheme. The results show that the mesoscale model can capture the mesoscale features. However, we cannot expect the "snap-shots" of the WRF simulations to match the observed velocity and wind directions in the SAR images perfectly. Therefore, the results for single events should be used qualitatively and a quantitative comparison could be made from statistics over longer periods.

4. Wake case at alpha ventus and Horns Rev-1/-2 with long wakes

The RADARSAT-1 image is acquired on 15th August 2012 at 05:49:27 UTC. There are long wind farm wakes observed at alpha ventus and Horns Rev-1/-2 and less visible wakes behind Bard Offshore, see Figure 7. The wind direction is from the southeast. The corresponding SAR wind retrieval maps covering the Horns Rev and alpha ventus wind farms are also shown in Figure 7. The free stream wind velocity is around 7 ms^{-1} while the wake region has winds 5 ms^{-1}.

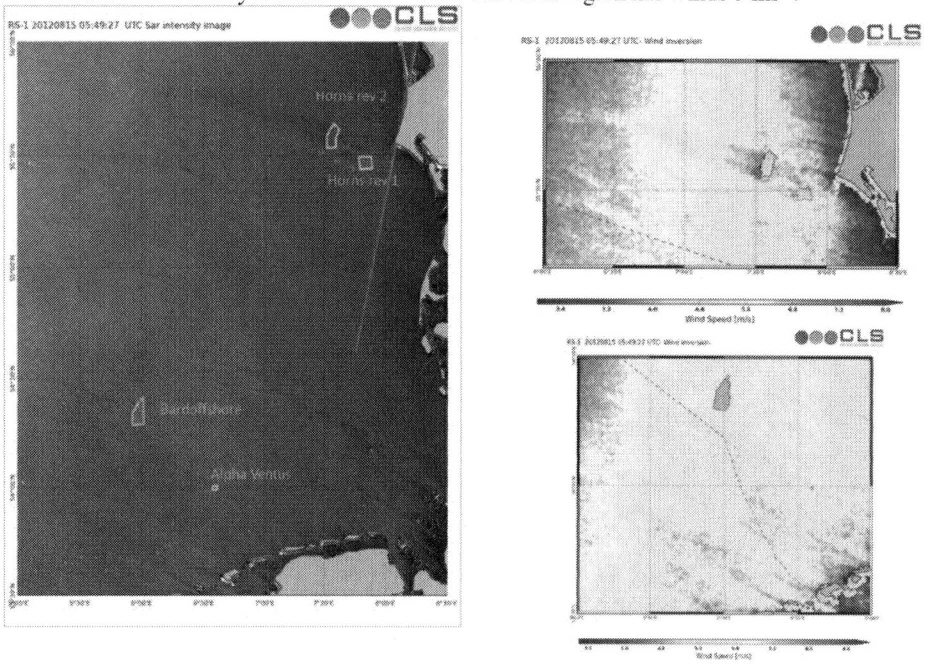

Figure 7. Long wind farm wakes observed behind alpha ventus, Horns Rev-1/-2 and too lesser extend BARD Offshore from RADARSAT-1 intensity map (left) and wind speed at Horns Rev-1/-2 (upper left) and alpha ventus (lower left).

At CENER the WRF model, version 3.4 [7] is configured using 4, two-way nested domains with resolutions of 10km, 3.33km, 1.11km and 370m with 35 vertical levels spreading from the surface until the top of the atmosphere. The domain is centred at 54.01°N, 6.606°E with sizes 250x250, 268x2688, 286x286 and 328x328. The simulation is driven by GFS 12UTC input data which have a horizontal resolution of 0.5° by 0.5°. The physics parameterizations used are: Microphysics (WRF Single-Moment (WSM) [20]); Long wave radiation (Rapid Radiative Transfer Model [17]); Shortwave radiation (Dudhia scheme [18]); Surface layer MYNN [14], Land surface (Noah [19]), PBL (MYNN 2.5 level TKE scheme [14]) and Cumulus (Kain-Fritsch (new Eta) [21]).

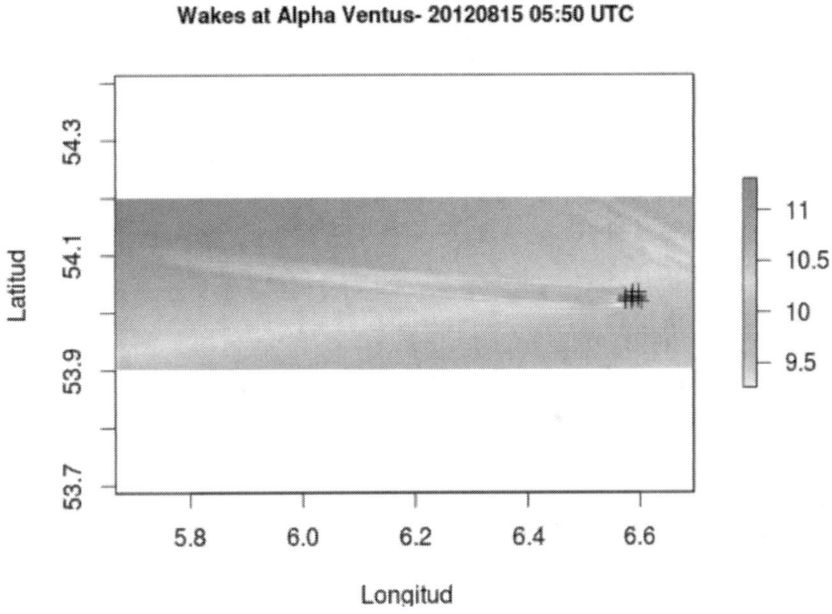

Fgure 8. Wind farm wake modelled by WRF at alpha ventus with wind speed scale in ms^{-1}.

A very clear wake is visible stretching for several kilometres behind the alpha ventus wind farm in Figure 8 as modelled in WRF. In this case, the simulated wake seems to be able to emulate both shape and direction of the observed wake. The output from the mesoscale model can only be visually compared with the satellite images that correspond to the same situation. The result is useful as a first approach to the development of a robust methodology to predict or simulate the effects on wind speed of neighbouring offshore wind farms in an area of interest.

The Horns Rev-1/-2 wind farm wake has been investigated by CIEMAT using WRF. The WRF model (version 3.4) [7] has been configured using a total of five two-way domains interacting with a three to one spatial refinement in order to reach a horizontal resolution of 333 m in the innermost domain that comprises the Horns Rev wind farm. We prescribed a total of 36 vertical levels, five of them within the first 200 m of the atmosphere. The five domains used in the WRF simulation with 27,000, 9000, 3000, 1000 and 333 m of horizontal resolution. The number of points (west-east by north-south) for each one of the five domains are 70x65, 49x49, 58x49, 43x43, and 43x43. Usually, each simulation consists of a sequence of short runs of the model. In each set of runs, the model is initialized at 0 UTC every day and run for 48 h storing the output every hour. The first day is discarded as a spin up of the model, whereas the output for the second day is retained as the simulation for that day. The initial and boundary conditions necessary to run the mesoscale model come from the ERA-Interim reanalysis project [22]. In this particular case, as the goal is to get the instantaneous WRF output to compare to the SAR image of the wake, the model is initialized the day before. A number of sub-grid scale processes are parameterized by WRF. The shortwave and longwave radiation schemes follow the works by [18] and [17] respectively. Microphysical processes are represented with the WRF single-moment six class scheme [20], whereas the cumulus effects are parameterized only in the outermost three domains [14, 23]. The air–sea momentum flux is parameterized following the work by Charnock. The effects of the turbulent vertical mixing within the planetary boundary layer are parameterized using a 1.5 order scheme that predicts the turbulent kinetic energy [14] and advects it with the wind. The scheme is based on the work by Mellor and Yamada [25] but includes a better formulation of buoyancy effects and a master length scale that depends on atmospheric stability. The effects of the wind turbines over the atmospheric flow are parameterized following the work by [26]. The wind turbines are represented as an elevated sink of momentum and a source of turbulent kinetic energy that responds to the wind speed according to a specified function approximating the effects of the turning rotors on the flow. The scheme is comprehensively documented in the work by [26]. This method has been shown to produce a realistic representation of wind farm wakes [20, 26] but a comprehensive evaluation of its performance has not been performed yet.

In order to obtain the wind speed deficit, that is plotted to compare with the satellite wake images, the WRF model is run twice, the first one without the wind farm (Horns Rev I) , and the second one with the wind farm inside. The result is shown in Figure 9. The wake shape is very similar to the one on the satellite image.

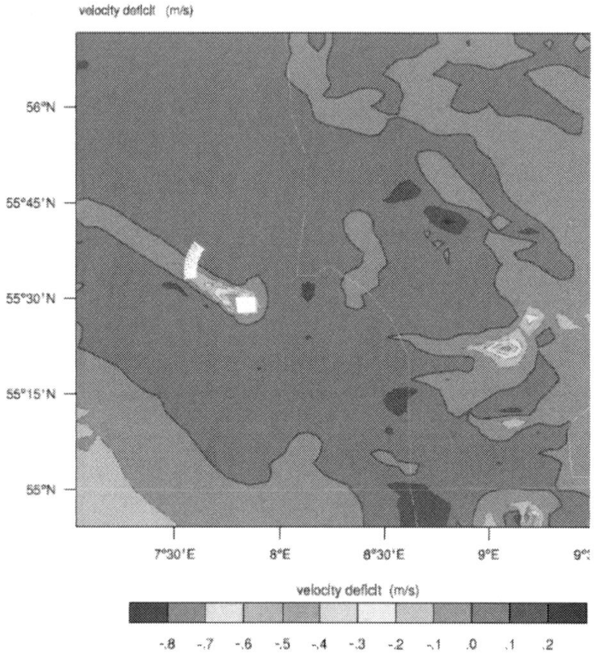

Figure 9. Wind farm wake modelled by WRF at Horns Rev-1/-2 showing velocity deficit in ms^{-1} near the time of satellite observation on 15 August 2012.

5. Summary and conclusion

The simulations performed by the micro- and meso-scale models tried to reproduce the real effects of several wind farms on the downstream wind observed through SAR satellite images. In some of the satellite observations, clearly wakes were visible stretching for dozens of kilometres, which by itself is a proof of the importance of this type of phenomena. Only qualitative comparison is possible.

The wind farm parameterization used by the WRF mesoscale model could provide a useful tool when studying the effects of a single or cluster of wind farms on the wind speed in a neighbouring area. The simulated wakes seem to emulate fairly well the shape and directions of the observed ones, but there is still room for improvement. The final goal is to obtain estimation of the expect wind speed deficit on an area of interest by the surrounding operating or planned wind farms. This estimation could constitute a decisive factor when deciding the layout or location of a new wind farm, or in the prediction of expected losses on wind energy production. In order to accomplish this final goal, measurements of wind speed or production in operating offshore wind farms would be crucial in order to properly calibrate the current state of the art mesoscale models.

Acknowledgments
Support from the EERA DTOC project FP7-ENERGY-2011-1/ n°282797 and satellite images from RADARSAT-1/-2 from Data and Products © MacDonald, Dettewiler and Associates Ltd are acknowledged.

References

[1] Christiansen M B and Hasager C B 2005 *Wake effects of large offshore wind farms identified from satellite SAR*. Remote Sens. Environ. v. 98 p. 251-268
[2] Christiansen M B and Hasager C B 2006 *Using airborne and satellite SAR for wake mapping offshore*. Wind Energy v. 9 p. 437-455
[3] Li X and Lehner S 2013 *Observation of TerraSAR-X for Studies on Offshore Wind Turbine Wake in Near and Far Fields*. IEEE Journal of Selected Topics in Applied Earth Observations and Remote Sensing 05/2013; 6(3).
[4] Longépé N. Hajduch G, Pelich R, Habonneau J and Lebras, J-Y 2013 *SAR-based Ship Monitoring: Advanced Methodologies with Medium Resolution Images (from WSM ASAR to EWS/IWS S1 Mission)*. ESA Living Planet Symposium, 9-13 September 2013, Edinburgh, United Kingdom.
[5] Dagestad K-F, Horstmann J, Mouche A, Perrie W, Shen H, Zhang B, Li X, Monaldo F, Pichel W, Lehner S, Badger M, Hasager C B, Furevik B, Foster R C, Falchetti S, Caruso M J and Vachon P 2013 *Wind retrieval from synthetic aperture radar - an overview*. Proceedings of SEASAR 2012, 4th SAR Oceanography Workshop (SEASAR 2012). Tromsø, Norway, 18-22 June 2013, ESA SP709, pp. 22.
[6] Katic I, Højstrup J and Jensen N O 1986 *A simple model for cluster efficiency*. Proceedings of the European Wind Energy Association Conference & Exhibition. Rome, Italy
[7] Skamarock W C, Klemp J B, Dudhia J, Gill D O, Barker D M, Duda M, Huang X Y, Wang W and Powers J G. 2008 *A description of the advanced research WRF version 3*. Technical Report TN-475+STR, NCAR, 2008. [WRF is a community model provided by the National Center for Atmospheric Research, Boulder CO, sponsored by the National Science Foundation].
[8] Quilfen Y, Chapron B, Elfouhaily T, Katsaros K and Tournadre J 1998 *Observation of tropical cyclones by high-resolution scatterometry*, Journal of Geophysisal Research, 103(C4), 7767–7786. doi:10.1029/97JC01911.
[9] Mouche A, Hauser D, Daloze J-F and Guerin C 2005 *Dual-polarization measurements at C-band over the ocean: results from airborne radar observations and comparison with Envisat ASAR data*. IEEE Transactions on Geoscience and Remote Sensing, 43(4), 753-769. 10.1109/TGRS.2005.843951

[10] Barthelmie R J, Badger J, Pryor S C, Hasager C B, Christiansen M B and Jørgensen B H 2007 *Offshore coastal wind speed gradients: Issues for the design and development of large offshore wind farms.* Wind Engineering, 31(6), 369-382.

[11] Wyngaard J C 2004 *Toward Numerical Modeling in the Terra Incognita.* Mon. Weather Rev., 61:1816- 1826, 2004.

[12] Volker P J H 2014 *Wake Effects of Large Offshore Wind Farms - a study of the Mesoscale Atmophere.* DTU Wind Energy, Ph.D. Thesis, p.132, 2014.

[13] Uppala S M, Kallberg P W, Simmons A J, Andrae U, da Costa, Bechtold V, Fiorino M, Gibson J K, Haseler J, Hernandez A, Kelly G A, Li X, Onogi K, Saarinen S, Sokka N, Allan R P, Andersson E, Arpe K, Balmaseda M A, Beljaars A C M, van de Berg L, Bidlot J, Bormann N, Caires S, Chevallier F, Dethof A, Dragosavac M, Fisher M, Fuentes M, Hagemann S, Holm E, Hoskins B J, Isaksen L, Janssen P A E M, Jenne R, McNally A P, Mahfouf J-F, Morcrette J-J, Rayner N A, Saunders RW, Simon P, Sterl A, Trenberth K E, Untch A,Vasiljevic D, Viterbo P and Woollen 2005 *The ERA-40 re-analysis.* Quart. J. R. Meteorol. Soc., 131, 2005

[14] Nakanishi M and Niino H 2009 *Development of an improved turbulence closure model for the atmospheric boundary layer.* J of the Meteorol Soc of Japan, 87:895-912, 2009.

[15] Kain J S 2004 *The Kain-Fritsch convective parameterization: An update.* J. Appl. Meteor., 43:170{181, 2004

[16] Thompson G, Field P R, Rasmussen R M and Hall W D 2008 *Explicit forecasts of winter precipitation using an improved bulk micro- physics scheme. Part II: Implementation of a new snow parameterization,* Mon. Weather Rev., 136, 5095–5115, 2008

[17] Mlaver E J, Taubman S J, Brown P D, Iacono M J and Clough S A 1997 *Radiative transfer for inhomogeneous atmosphere: RRTM, a validated corrected-k model for the long wave.* J. Geophys. Res., 102:16663{16682, 1997.

[18] Dudhia J 1089 *Numerical study of convection observed during the wind monsoon experiment using a mesoscale two-dimensional model.* J. Atmos. Sci., 46:3077{3107, 1989.

[19] Chen F and Dudhia J 2001 *Coupling an advanced land-surface/ hydrology model with the Penn State/ NCAR MM5 modeling system. Part I: Model description and implementation.* Mon. Wea. Rev.,129:569{585 , 2001

[20] Hong S Y, Noh Y and Dudhia J 2006 *A new vertical diffusion package with an explicit treatment of entrainment processes.* Monthly Weather Review 2006; 134: 2318–2341.

[21] Kain J S and Fritsch J M 1990 *A one-dimensional entraining/detraining plume model and its application in convective parameterization,* Journal of the Atmospheric Sciences, vol. 47, no. 23, pp. 2784âˇA ,S2802

[22] Dee D P, Uppala S M, Simmons A J, Berrisford P, Poli P, Kobayashi S, Andrae U, Balmaseda M A, Balsamo G, Bauer P, Bechtold P, Beljaars A C M, van de Berg L, Bidlot J, Bormann N, Delsol C, Dragani R, Fuentes M, Geer A J, Haimberger L, Healy S B, Hersbach H, Hólm E V, Isaksen L, Kållberg P, Köhler M, Matricardi M, McNally A P, Monge-Sanz B M, Morcrette J-J, Park B-K, Peubey C, de Rosnay P, Tavolato C, Thépaut J-N and Vitart F 2011 *The Wind Energ.* (2014) © 2014 John Wiley & Sons, Ltd. DOI: 10.1002/we

[23] Kain J S and Fritsch J M 1990 *A one-dimensional entraining/detraining plume model and its application in convective parameterization.* J of Atmospheric Sciences 1990; 47: 2784–2802.

[24] Kain J S and Fritsch J M 1993 *Convective Parameterization for Mesoscale Models: The Kain-Fritcsh Scheme. The Representation of Cumulus Convection in Numerical Models,* Meteor. Monogr., No. 24. American Meteorological Society: Boston, 1993. pp. 165–170.

[25] Mellor G L and Yamada T 1982 *Development of a turbulence closure model for geophysical fluid problems.* Reviews of Geophysics and Space Physics 1982; 20: 851–875.

[26] Fitch A C, Lundquist J K and Olson J B 2013 *Mesoscale influences of wind farms throughout a diurnal cycle.* Monthly Weather Review 2013; 141: 2173–2198.

Wake meandering under non-neutral atmospheric stability conditions – theory and facts

G C Larsen[1], E Machefaux and A Chougule
Department of Wind Energy, Technical University of Denmark
P.O. Box 49, 4999 Roskilde, Denmark

E-mail: gula@dtu.dk

Abstract. This paper deals with modelling of wake dynamics under influence of atmospheric stability conditions different from neutral. In particular, it is investigated how the basic split in turbulent scales, on which the Dynamic Wake Meandering model is based, can be utilized to include atmospheric stability effects in this model. This is done partly by analyzing a large number of turbulence spectra obtained from sonic measurements, partly by analyzing dedicated full-scale LiDAR measurements from which wake dynamics can be directly resolved. The theory behind generalizing the Dynamic Wake Meandering model to non-neutral conditions are summarized and linked to the results of the full-scale experimental results. It is concluded that there is a qualitative match between the conjecture behind the Dynamic Wake Meandering model and the dependence of turbulence structure on atmospheric stability conditions, and consequently that there is a potential for generalizing the Dynamic Wake Meandering model to include effects of atmospheric stability.

1. Introduction

Wakes cause significant power losses in wind farms. On average these losses amount to 10 - 20% of the wind farm power output. In addition to the power losses comes the increased (fatigue) loading of turbines operated under wake conditions, and wakes are therefore a major factor in wind farm economics. Wake effects relate to a multitude of aspects related to the interplay between wind farm topology and site wind field characteristics such as mean wind speed distribution, wind direction distribution, atmospheric turbulence and atmospheric stability [1], [2].

This paper deals with modelling of wake dynamics under influence of atmospheric stability conditions different from neutral and addresses thereby both power and load predictions of turbines located in wind farms under such conditions. Whereas friction is dictating the structure of turbulence in the atmospheric boundary layer (ABL) under neutral conditions, buoyancy effects adds to friction when it comes to the turbulence structure under ABL stability conditions different from neutral.

The major impact from buoyancy on the ABL turbulence structure is on the large turbulent scales, and being largely based on a convenient split in turbulence scales it may be straight forward to include atmospheric stability aspects into the framework of the Dynamic Wake Meandering (DWM) model [3], [4]. Crucial in this regard is if the specific scale split applied in the DWM model is such, that ABL stability affects primary the (lateral and vertical) turbulent scales, which drives the wake meandering. This is the basic conjecture on which the work in [3] and [4] is based.

[1] To whom any correspondence should be addressed.

For a variety of stability conditions, we will explore this conjecture partly by analyzing full-scale velocity measurements and partly by analyzing organized wake deficit flow structures as well as wake dynamics as based on full-scale LiDAR measurements. Finally, the performance of a newly developed spectral tensor, consistently including buoyancy effects in the ABL turbulence description [5], will be investigated.

2. Simulation of wake affected flow fields under non-neutral conditions

Buoyancy and wake affected flow fields can be modelled using advanced CFD models – e.g. CDF LES combined with actuator disc or actuator line representation of wind turbine rotors [6]. However, such simulations are computationally very demanding and consequently not yet suited for neither wind turbine design simulations nor for wind farm layout optimization. Both these disciplines call for simplified models describing the essential physics of the problem(s). One possible candidate is the DWM model ([7], [8]), which has recently been included in the IEC standard as a recommended practice, and which offers simulation of in-stationary wind farm flow fields with an acceptable computational effort.

2.1. The DWM conjecture and atmospheric stability implications

Based on a priori knowledge of the undisturbed ambient wind climate on a wind farm site, the DWM model provides detailed information of the inflow wind field to each individual wind turbine inside the wind farm. The DWM model describes the essential physics of the problem, and accounts both for the observed increased *turbulence level* of wake flow fields and for the modified *turbulence structure*. The model has successfully been verified against both full-scale measurements [9], [10] and against detailed CFD LES actuator line (ACL) computations [8].

The core of the model is a *split of scales* in the wake flow field, with large scales being responsible for stochastic *wake meandering*, and small scales being responsible for wake *attenuation* and *expansion* in the meandering frame of reference as caused by turbulent mixing. Thus, essentially the DWM model assumes that the transport of wakes in the atmospheric boundary layer (ABL) can be modeled by considering the wakes to act as passive tracers driven by a combination of large-scale turbulence structures and a mean downstream advection velocity, adopting the Taylor hypotheses. With the large turbulent structures being modeled using the Navier-Stokes (N-S) consistent Mann spectral tensor [12], and modeling of the organized wake deficits flow structures being based on the thin layer approximation of N-S in their rotationally symmetric form combined with an dedicated eddy viscosity turbulence closure, the DWM approach resembles the characteristics of a CFD Large Eddy Simulation (LES) approach and may thus be considered as the "poor man's LES".

While the original version of the DWM model was intended for simulation of neutrally stratified atmospheric boundary layers (ABL), the generalization to ABL stability conditions different from neutral is straight forward. This is because of the DWM spilt of scales matches nicely with the fact that the various ABL stability characteristics primarily affect the low frequency part of the turbulence spectra, and thus primarily the large-scale turbulent structures. Therefore a particular stability classification has a direct impact on the wake meandering part of the DWM formulation, whereas the small scale turbulence regime of DWM is less affected.

Within the DWM framework we consequently assume ABL stability *only* to impact the turbulent scales within the meandering regime. This was first conjectured in [11] and later further elaborated on in [4]. However, in these initial studies the approach was limited by the lack of a consistent kinematic model for turbulence modelling under non-neutral ABL stability conditions, and the Mann spectral tensor, although developed for neutral conditions, was used in approximate manner to cover also non-neutral conditions. This shortcoming has recently been overcome by the development of buoyancy dependent spectral tensor [5], whereby ABL stability effects can consistently be accounted by within the framework of the DWM model.

2.2. Buoyancy dependent spectral tensor

The generalized spectral tensor, $\Phi_{ij}(\boldsymbol{k})$, resulting from the governing Rapid Distortion Theory (RDT) equations including buoyancy effects [5], contains two extra parameters, in addition to those of the "classic" Mann spectral tensor model, where $\boldsymbol{k}(t) = (k_1, k_2, k_{30}\text{-}k_1(\mathrm{d}U/\mathrm{d}z)t)$ is a three dimensional wave vector, and t is time. These parameters are: 1) a stability parameter (the Richardson number); and 2) the rate of destruction of temperature variance.

The RDT equations, which include the linearized momentum and the temperature equation in Fourier space, evaluates in time under the influence of a constant wind shear ($\mathrm{d}U/\mathrm{d}z$) and a constant gradient of potential temperature ($\mathrm{d}\theta/\mathrm{d}z$) from an initial isotropic state of turbulence. In isotropic turbulence, the velocity-spectrum tensor is

$$\Phi_{ij}(\boldsymbol{k}_0) = \frac{E(k)}{4\pi k^2}\left(\delta_{ij} - \frac{k_i k_j}{k^2}\right),$$

where $\boldsymbol{k}_0 = \boldsymbol{k}(0)$ and k is the length of the vector $\boldsymbol{k}$. The energy spectrum, $E(k)$, is given by [13] as

$$E(k) = \alpha\varepsilon^{2/3}L^{5/3}\frac{(kL)^4}{(1+(kL)^2)^{17/6}},$$

where $\alpha \approx 1.7$ is the Kolmogorov constant, ε is the rate of viscous dissipation of specific turbulent kinetic energy (TKE), and L is a turbulence length scale.

In order to make the model stationary, the time dependency in the model is removed by incorporating the general concept of an eddy life time, $\tau(k)$. The parameterization of $\tau(k)$ is adapted from the "classic" Mann spectral tensor model [12]. In the inertial sub-range, the life time of eddies are proportional to $k^{-2/3}$, and the assumption in the Mann model, for scales larger than the inertial sub-range, is that the eddy life time is proportional to k^{-1} divided by their characteristic velocity given by

$$\left(\int_k^\infty E(p)dp\right)^{1/2},$$

thus resulting in eddy life times proportional to $k^{-2/3}$ for $k \to \infty$ and to k^{-1} for $k \to 0$.

For the temperature variable, the isotropic three-dimensional spectrum is given as

$$\Phi_{\theta\theta}(\boldsymbol{k}_0) = \frac{S(k)}{4\pi k^2},$$

where $S(k)$ is the potential temperature energy spectrum containing the form of the inertial sub-range [14] as

$$S(k) = \beta\varepsilon^{-1/3}\varepsilon_\theta L^{5/3}\frac{(kL)^2}{(1+(kL)^2)^{11/6}},$$

Here ε_θ is the dissipation rate for half the temperature variance, and $\beta = 0.8$ is a universal constant [15]. Based on the above formulations of isotropic velocity and temperature spectra combined with the Mann eddy life time formulation, RTD results in an anisotropic spectral tensor including buoyancy of the form $\Phi_{ij}(\boldsymbol{k}) = \Phi_{ij}(\boldsymbol{k};\, \alpha\varepsilon^{2/3},\, L,\, \Gamma,\, Ri,\, \eta_\theta)$, where Ri denotes the Richardson number [14] resulting from the temperature equation, and

$$\eta_\theta \equiv \frac{\varepsilon_\theta}{\varepsilon}\left[\frac{g}{\theta}\left(\frac{dU}{dz}\right)^{-1}\right]^2,$$

where the potential temperature, θ, as well as $\mathrm{d}U/\mathrm{d}z$ and $\mathrm{d}\theta/\mathrm{d}z$ are representative of the height of interest.

To summarize, the five adjustable model parameters, which are attainable from single-point measurements are: 1) $\varepsilon^{2/3}$; 2) L, which represents a representative size of the energy containing eddies; 3) Γ, which is a measure of the degree of turbulence isotropy; the Richardson number Ri [14]; and 5) η_θ as defined above. Note, that for $Ri = 0$ and $\eta_\theta = 0$, the generalized spectral tensor degenerates to the "classic" Mann spectral tensor.

Simulation of consistent 3D synthetic turbulence fields requires knowledge of cross-spectra between turbulence components. The spectral tensor model provides the cross-spectrum, χ_{ij}, between any two velocity components, or between any velocity component and temperature, as

$$\chi_{ij}(k_1, \Delta y, \Delta y; \alpha\varepsilon^{2/3}, L, \Gamma, Ri, \eta_\theta) \equiv \alpha\varepsilon^{2/3} L^{5/3} \int \Phi_{ij}(\boldsymbol{k}; \alpha\varepsilon^{2/3}, L, \Gamma, Ri, \eta_\theta)\, e^{i(k_2\Delta y + k_3 \Delta z)} dk_2 dk_3 ,$$

where Δy and Δz are transverse and vertical separations, respectively. Thus, based on such cross-spectra, the requested turbulence field is simulated using the approach described in [16], however, presently only implemented for the velocity components.

3. Examination and validation of basic DWM conjecture

As mentioned, the basic conjecture behind the DWM modelling of non-neutral wake affected flows is that ABL stability predominantly impacts only the turbulent scales within the meandering regime. In this section we will examine this conjecture using a 500kW example turbine. The analysis will focus on lateral turbulence characteristics, since this turbulence component is the most important regarding wake meandering. First a large number of full-scale sonic measurements are analyzed with the available data material binned with respect to both mean wind speed (1m/s bins) and ABS stability (7 stability classes). Subsequently dedicated full-scale LiDAR measurements are used to resolve and compare wake characteristics for three different stability conditions with mutually comparable inflow mean wind speeds.

3.1. Classification of ABL stability

Classification of atmospheric stability conditions is in this context based on the Monin-Obukhov length, which can be derived directly from sonic measurements. The Monin-Obukhov length, L, is defined as [14]

$$L = -\frac{u_*^3}{\kappa (g/T)\langle w'\Theta_v'\rangle} ,$$

where g and T are respectively acceleration of gravity and absolute temperature, u_* is the friction velocity, and $\langle w'\Theta_v'\rangle$ denotes the mean kinematic virtual heat flux with w and Θ_v being the vertical components of the velocity and the virtual potential temperature, respectively. A prime denotes fluctuations, and κ is the von Kármán constant (≈ 0.4).

The Monin-Obukhov length expresses the height where production of mechanical and convective turbulence is equal, and thus offers a natural way to quantify the degree of dominance of buoyancy over mechanical and shearing effects. Adopting the classification scheme proposed in [17], the following 7 stability classes are defined:

- Very stable: $10 \leq L < 50$
- Stable: $50 \leq L < 200$
- Near neutral-stable: $200 \leq L < 500$
- Neutral: $|L| \geq 500$
- Near neutral-unstable: $-500 < L \leq -200$
- Unstable: $-200 < L \leq -100$
- Very unstable: $-100 < L \leq -50$

3.2. Full-scale sonic measurements

The measurement campaign was conducted from June 2011 to early January 2012 at the DTU Risø Campus, and sonic data recorded 16.5m a.g.l. are used for this analysis. To ensure "homogeneous" inflow conditions, only data from the (prevailing) wind direction sector (120° - 150°) is used, resulting in 1122 available 10-minute time series covering the mean wind speed regime ranging from 4-10m/s.

These data are binned with respect to mean wind speed and ALB stability. The resulting bin matrix is shown in Table 1, where the numbers reflect the number of available 10-minute time series available within each particular bin.

Table 1. Bin matrix specifying number of available 10-minute time series.

Wind speed interval [m/s]	Very unstable	Unstable	Near neutral-unstable	Neutral	Near neutral-stable	Stable	Very stable
4-5	6	2	2	12	12	68	108
5-6	40	20	12	28	28	84	90
6-7	36	12	18	58	16	26	6
7-8	38	56	44	66	12	4	0
8-9	8	12	56	60	6	0	0
9-10	0	2	4	18	0	0	0

For each bin, the power spectrum of the *lateral* turbulence component is evaluated as based on the available 10-minute time series. Prior to the spectral analysis all data have normalized with their respective mean wind speeds and de-trended assuming a linear trend. This is done to assure un-weighted averaging (assuming constant turbulence intensity within stability-wind-speed bins) in the subsequent averaging of spectra belonging to the same bin, which is performed to improve statistical significance of the spectral estimates. Aiming at investigating the spectral characteristics wind speed bin-wise, the described bin-wise normalization of spectra does not provide any restriction for the present data analysis.

Examples of bin-normalized spectra, associated with mean wind speeds in the interval [6; 7]m/s, are shown in Figure 1.

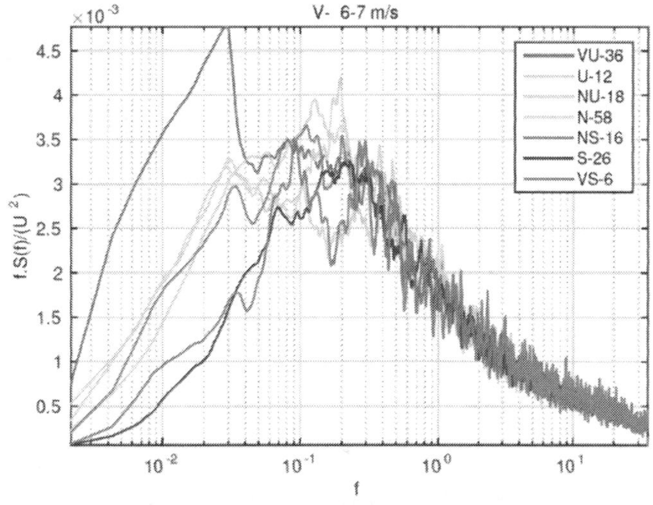

Figure 1. Normalized spectra for various stability conditions and associated with mean wind speeds in the interval [6; 7]m/s.

In neutral conditions the turbulence generation is dictated by mechanical friction. For non-neutral ABL conditions buoyancy inters the scene as an additional turbulence source term. The consequence is that *turbulence intensity* as well as *turbulence structure* varies with stability condition. As seen from Figure 1, ABL stability conditions hardly affect the spectral inertial subrange regime, but significantly alter the large scale energy-containing spectral regime [14], where the turbulent energy is produced. In the present analysis we will, as already indicated, explore non-neutral stability conditions mean wind speed bin-wise with focus on derived effects on wake meandering dynamics.

For this purpose, we will now focus on the energy balance, quantified in normalized numbers, between "large" and "small" scale turbulence as function of ABL stability condition. Analogues, we will investigate the dependence of the total spectral energy on stability conditions. To relate to the DWM split in scales [7], "large" scale turbulence is in this context defined as turbulence associated with frequencies below a frequency split given by $f_s = U/(2D)$, where U denotes the mean wind speed and D is the diameter of the rotor in question, and "small" scale turbulence in analogy defined as turbulence associated with frequencies above f_s. In the present analysis, U will refer to the *average* of 10-minute mean wind speeds belonging to a particular mean wind speed bin, and D is defined by the previously mentioned Nordtank 500kW example turbine with a rotor diameter of 41m. This turbine was selected as example turbine, because it is the turbine on which the subsequent LiDAR measurement analysis is based.

Table 2 shows the dependence of turbulence variance on ABL stability conditions for the analyzed mean wind speed bins, whereas Table 3 and Table 4 show the ABL stability dependence of "large" scale turbulence variance and of "small" scale turbulence variance, respectively.

Table 2. Bin specific variance normalized with respect to variance associated with neutral conditions.

Wind speed interval [m/s]	Very unstable	Unstable	Near neutral-unstable	Neutral	Near neutral-stable	Stable	Very stable
4-5	1,00	0,83	1,20	1,00	1,12	0,77	0,50
5-6	1,51	2,06	1,07	1,00	0,94	0,77	0,54
6-7	1,23	1,00	0,93	1,00	0,95	0,75	0,77
7-8	1,22	1,19	1,04	1,00	0,97	1,28	-
8-9	0,96	0,80	0,88	1,00	0,77	-	-
9-10	-	0,86	0,83	1,00	-	-	-

For all mean wind speed regimes Table 2 reflects, as expected, a clear trend with the turbulent energy increasing relatively for unstable ABL conditions and decreasing relatively for stable ABL conditions. As seen, this difference in energy level is up to a factor between 2 and 3 for low mean wind speeds which, however, is expected to be reduced for higher mean wind speeds, where the mechanically generated turbulence gradually increases relative to the buoyancy generated part. A few "outliers" are observed which is attributed to the limited number of statistical degrees of freedom available for spectral averaging in certain bins (cf. Table 1).

Table 3. Bin specific "large" scale variance normalized with respect to "large" scale variance associated with neutral conditions.

Wind speed interval [m/s]	Very unstable	Unstable	Near neutral-unstable	Neutral	Near neutral-stable	Stable	Very stable
4-5	1,78	1,37	2,93	1,00	0,89	0,76	0,52
5-6	2,23	3,63	1,47	1,00	0,90	0,63	0,37
6-7	1,71	1,07	1,09	1,00	0,96	0,50	0,65
7-8	1,51	1,41	1,16	1,00	0,95	1,33	-
8-9	0,99	0,72	0,88	1,00	0,59	-	-
9-10	-	0,60	0,77	1,00	-	-	-

The results in Table 3 reflects the same clear trend as identified for the total spectral energy, namely that the "large" scale spectral energy increases significantly with increasing buoyancy related turbulence production and vice versa.

Table 4. Bin specific "small" scale variance normalized with respect to "small" scale variance associated with neutral conditions.

Wind speed interval [m/s]	Very unstable	Unstable	Near neutral-unstable	Neutral	Near neutral-stable	Stable	Very stable
4-5	0,78	0,65	0,69	1,00	1,21	0,78	0,51
5-6	1,12	1,26	0,86	1,00	0,96	0,83	0,63
6-7	0,94	0,94	0,83	1,00	0,94	0,90	0,87
7-8	1,04	1,04	0,97	1,00	0,98	1,27	-
8-9	0,92	0,85	0,89	1,00	0,90	-	-
9-10	-	1,11	0,85	1,00	-	-	-

Contrary to the results presented in Tables 2 and 3, the results in Table 4 show no clear trend with regard to stability dependence of the "small" scale turbulence energy level. Therefore, with the "small" scale turbulence energy level being roughly invariant with respect stability conditions, and the "large" scale turbulence energy level being highly dependent on ABL stability conditions, the present investigation shows that the DWM split in scales roughly "matches" the split in scales between the turbulence energy-containing range and the turbulence inertial subrange, thus in turn confirming the DWM stability conjecture.

3.3. Full-scale LiDAR measurements

The full-scale LiDAR measurements analyzed in this subsection relate to the same measuring campaign as described above. As a supplement to the sonic recordings, the Nordtank turbine was equipped with a pulsed LiDAR system mounted on a platform at the rear of the turbine nacelle, thus facilitating cross sectional scanning (i.e. cross sections perpendicular to the rotor axis) of the wake affected flow field behind the turbine. A detailed description of the experimental setup can be found in [6], where also the principle of extracting the mean wind speed characteristics from LiDAR measurements recorded outside the wake regime is described. In the present analysis, mean wind speeds associated with the LiDAR recordings are derived using this technique.

The wake characteristics in downstream cross sections are resolved as based on a Cartesian scan pattern consisting of 49 measurement points (i.e. 7×7). For various stability conditions, the basic idea is to resolve and compare the *wake deficit* characteristics in the meandering frame of reference (MFoR) as well as the *wake deficit dynamics*. This is performed for otherwise similar inflow conditions; i.e. mean wind speed and mean wind direction.

The wake deficit dynamics is obtained from "instantaneous" LiDAR cross sectional scans using the wake deficit tracking procedure introduced in [18]. With the wake deficit dynamics determined, it is straight forward to perform a transformation from the fixed frame of reference (FFoR), in which the measured wake affected flow is resolved, to the MFoR.

To obtain robust results, time series with a span ranging between 3 and 5 hours is used in this part of the investigation. To ensure a sufficient amount of data complying with the requirements, this means in turn that it is necessary to merge the former two unstable classes (i.e. "very unstable" and "unstable") into a new stability class denoted "unstable collapsed". Eventually three test cases, associated with low wind conditions and therefore pronounced deficits (i.e. high trust), are selected for this part of the analysis. The characteristics of the selected time series appear from Table 5 below. As expected the ambient turbulence level is increased with ABL stability conditions changing from stable over neutral to unstable.

Table 5. Overall characteristics of test cases.

Stability condition	Mean wind speed at hub height (U_0) [m/s]	Turbulence intensity at hub height [m/s]	Length of time series [hour]
Unstable collapsed	6.82	0.15	3.3
Neutral	7.03	0.14	3.2
Very stable	6.76	0.10	5.5

The results for the wake deficits, as expressed in the MFoR, are shown in Figure 2 for downstream distances ranging between 1D and 5D, where D denotes the rotor diameter. It is evident that the deficits expressed in the MFoR are almost invariant to the ABL stability conditions, thus in this respect confirming the conjecture on which the DWM modeling of non-neutral flow fields is based.

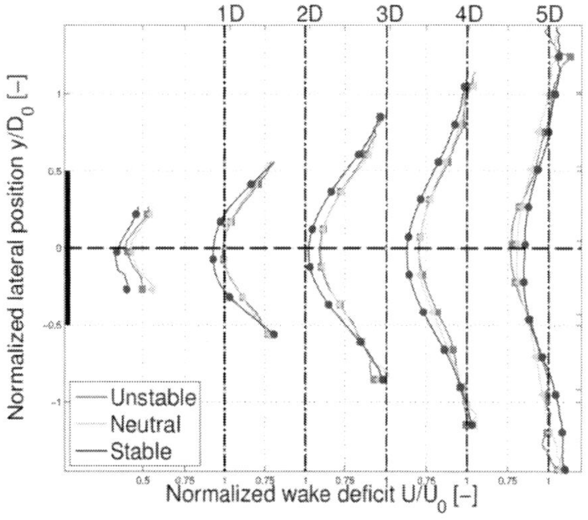

Figure 2. Normalized wake deficits in the MFoR depicted for three different stability conditions.

Turning to wake deficit dynamics, Table 6 shows the variance of the lateral wake center position for the investigated stability conditions. The lateral wake displacements are, in the context of DWM modelling , driven by the large scale lateral turbulent scales, and to facilitate direct comparisons with the inflow results given in Table 3, the results are normalized with respect to displacement variance associated with neutral conditions.

Table 6. Variance of the lateral wake center position normalized with respect to variance of the lateral wake center position associated with neutral conditions.

Downstream distance	Very unstable	Unstable	Near neutral-unstable	Neutral	Near neutral-stable	Stable	Very stable
3D	1,11	1,22	1,26	1,00	0,70	0,81	0,75
4D	1,09	1,36	1,10	1,00	0,59	0,57	0,49
5D	1,04	1,40	1,02	1,00	0,61	0,54	0,40

Qualitatively, also these results confirm the DWM conjecture stating that the wake meandering dynamics is driven by large scale turbulence structures, which in turn is highly dependent on ABL stability conditions. Although differences exists among values associated with *specific* stability classes, a quantitative comparison of the results in Table 2 (i.e. wind bin 6-7m/s) and Table 6 shows in addition that there is a reasonable agreement between the range of "large" scale variance stability dependence and the range of wake center lateral dynamics variance stability dependence, respectively, especially considering the fact that these results originate from not identical 10-minute recordings.

4. Performance of spectral tensor including buoyancy

Having rendered probable the validity of the DWM stability conjecture, we finally explore the performance of the generalized spectral tensor described in Section 2 for various ABL stability conditions, with the potential perspective of using this spectral tensor for the required DWM large scale turbulence modelling. Figure 3 shows a fit of the spectral tensor to the measured data presented in Figure 1. Full lines represent spectral fits, and dotted lines represent measured spectra.

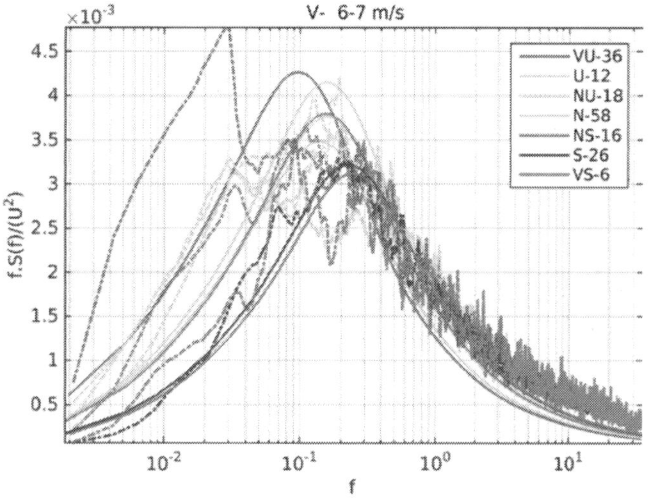

Figure 3. Spectral tensor fit to measured full-scale spectra of the lateral velocity component for various stability conditions and associated with mean wind speeds in the interval [6; 7]m/s.

Although the generalized spectral tensor is not meant to describe meso-scale effects (i.e. to include a *specific* modelling of such scales), the tensor is able to approximately capture the spectral "stability cascade" with increasing spectral length scales for increasing degree of ABL instability, and thereby paving the way for future detailed validation studies of the DWM model under non-neutral ABL stability conditions with focus on both wind turbine production and loading.

5. Conclusion

There is experimental evidence for a significant impact from atmospheric stability on wake affected flow fields in wind farms, which in turn affects both wind farm production and loading of the individual turbines. By analyzing an extensive amount of full-scale data based on both sonic measurements and LiDAR measurements, we have demonstrated the validity a conjecture stating that, in the context of DWM modelling, ABL stability impacts *only* the "large" turbulence scales constituting the meandering regime, whereas the "small" scale turbulence regime responsible for wake deficit expansion and attenuation in the MFoR can be considered invariant with respect to ABL stability conditions. Generalisation of the DWM model to non-neutral conditions is thus straight forward, and facilitated by the recent development of a spectral tensor including buoyancy effects.

Acknowledgments
The EUDP project "Impact of atmospheric stability conditions on wind farm loading and production" under contract 64010-0462 is acknowledged for financial support and thus for making this study possible.

References

[1] Hansen, K.S.; Barthelmie, R.; Ott, S. and Larsen, G.C. (2012). Park power deficit due to atmospheric stability. The Science of Making Torque from Wind, October 9-11, Oldenburg, Germany.
[2] Hansen, K.S.; Larsen, G.C. and Ott, S. (2014). Dependence of offshore wind turbine fatigue loads on atmospheric stratification. The Science of Making Torque from Wind 2012, June 17-20, Lyngby, Denmark.
[3] Larsen, G.C.; Larsen, T.J.; Mann, J.; Peña, A.; Hansen, K.S. and Madsen, H.Aa. (2009) The dependence of wake losses on atmospheric stability characteristics. EUROMECH Colloquium 508 on Wind Turbine Wakes, pp. 35–37.
[4] Keck, R.-E.; de Maré, M.; Churchfield, M.J.; Lee, S.; Larsen, G.C. and Madsen H.Aa. (2014). On atmospheric stability in the dynamic wake meandering model. Wind Energy, Volume 17, Issue 11, pages 1689–1710.
[5] Chougule, A. (2013). Influence of atmospheric stability on the spatial structure of turbulence. DTU Ph.D. thesis, 109p.
[6] Machefaux, E.; Larsen, G.C.; Koblitz, T.; Troldborg, N.; Kelly, M.; Chougule, A.; Hansen, K.S. and Rodrigo, J.S. (2014). Experimental and numerical study of the atmospheric stability impact on wind turbine wake. Accepted for publication in Wind Energy.
[7] Larsen, G.C.; Madsen, H.Aa.; Thomsen, K. and Larsen, T.J. (2008). Wake Meandering: A Pragmatic Approach. Wind Energy, Volume 11, Issue 4, pages 377–395.
[8] Madsen, H.Aa.; Larsen, G.C.; Larsen, T.J.; Troldborg, N. and Mikkelsen R. (2010). Calibration and Validation of the Dynamic Wake Meandering Model for Implementation in an Aeroelastic Code. Journal of Solar Energy Engineering, Vol. 132(4), 041014.
[9] Bingöl, F.; Mann, J. and Larsen G.C. (2010). LiDAR Measurements of Wake Dynamics, Part 1: One Dimensional Scanning. Wind Energy, Volume 13, Issue 1, pages 51–61.
[10] Larsen, T.J.; Madsen, H.Aa.; Larsen, G.C. and Hansen, K.S. (2013). Validation of the Dynamic Wake Meander Model for Loads and Power Production in the Egmond aan Zee Wind Farm. Wind Energy, Volume 16, Issue 4, pages 605–624.
[11] Larsen, G.C.; Larsen T.J.; Mann, J.; Peña, A.; Hansen, K.S. and Madsen, H.Aa. (2009). The dependence of wake losses on atmospheric stability characteristics. EUROMECH Colloquium 508 on Wind Turbine Wakes; 35–37.
[12] Mann, J. (1994). The Spatial Structure of Neutral Atmospheric Surface-Layer Turbulence. J. of Fluid Mech., 273, 141-168.
[13] von Kármán, T. (1948). Progress in the statistical theory of turbulence. Proc. Nat. Acad. Sci., 34:530–539.
[14] Kaimal, J.C. and Finnigan, J.J. (1994). Atmospheric Boundary layer Flows – Their structure and Measurement. Oxford University Press.
[15] Kaimal, J.C.; Wyngaard, J.C.; Izumi, Y. and Coté, O.R. (1972). Spectral characteristics of surface-layer turbulence. Q. J. R. Meteorol. Soc., 98(417):563–589. doi:10.1002/qj.49709841707.
[16] Mann, J. (1998). Wind field simulation, Prob. Eng. Mech. 13(4), pp. 269–282.
[17] Peña, A.; Gryning, S.E. and Mann J. (2010). On the length-scale of the wind profile. Quarterly Journal of the Royal Meteorological Society; 136(653); 2119–2131.
[18] Trujillo, J.; Bingöl, F.; Larsen, G.C. and Mann, J. (2011). Light detection and ranging measurements of wake dynamics. Part II: two-dimensional scanning. Wind Energy, Volume 14, Issue 1, pages 61–75.

Engineering models for merging wakes in wind farm optimization applications

E. Machefaux[1], G.C. Larsen[1] and J.P. Murcia Leon [1]
[1] DTU Wind Energy, Frederiksborgvej 399, 4000 Roskilde, Denmark
E-mail: ewma@dtu.dk

Abstract. The present paper deals with validation of 4 different engineering wake superposition approaches against detailed CFD simulations and covering different turbine interspacing, ambient turbulence intensities and mean wind speeds. The first engineering model is a simple linear superposition of wake deficits as applied in e.g. Fuga. The second approach is the square root of sums of squares approach, which is applied in the widely used PARK program. The third approach, which is presently used with the Dynamic Wake Meandering (DWM) model, assumes that the wake affected downstream flow field to be determined by a superposition of the ambient flow field and the dominating wake among contributions from all upstream turbines at any spatial position and at any time. The last approach developed by G.C. Larsen is a newly developed model based on a parabolic type of approach, which combines wake deficits successively. The study indicates that wake interaction depends strongly on the relative wake deficit magnitude, i.e. the deficit magnitude normalized with respect to the ambient mean wind speed, and that the dominant wake assumption within the DWM framework is the most accurate.

1. Introduction

Wind farm flow fields remain a challenging topic, which has attracted increasing attention during the last decade. Considerable progress in the fundamental understanding of such fields has been achieved through analysis of dedicated full-scale and wind tunnel experiments as well as through detailed CFD computations, which in turn has paved the way for development and validation of powerful engineering type of models preserving the essential physics of the problem while at the same time offering a substantial reduction in computational requirements. Computational demands is a critical parameter when it comes to design and optimization of wind farm layout in practice due to the huge number of wind farm flow conditions to be considered.

To remain of practical relevance, the commonly used approach in wind farm design codes is to combine single wake calculations with a recipe for wake superposition to determine the combined characteristics of the merged wake. The main challenge with predicting merged wakes is that, according to Crespo et al. [1], the ambient flow in which a downstream wake diffuses is fundamentally different than the one at upstream locations. Instead, today's commonly used wake superposition techniques rely on simple mathematical rules that are not justified by any physical means.

The present paper aims at evaluating the performance of 4 different wake superposition techniques. The benchmark is performed against results from detailed CFD simulations based on Large Eddy Simulation using the Actuator Disc rotor modeling technique under fully turbulent atmospheric flow. This model has been validated against merged wake lidar measurements as part of the study by Machefaux et al. in [14]. In the present research, merging wakes are studied with the most simple configuration, where two stall regulated rotors are aligned with the incoming wind, under different turbine spacings, ambient turbulence intensities and mean wind speeds.

One of the most simplistic wake overlapping approaches presently tested is a *linear superposition* of wake deficits, which was first introduced by Lissaman in [12]. As was argued by Crespo et al. [1], that this assumption fails for large perturbations where many upstream wakes are superimposed, as negative velocities may arise.

Thus, Katic et al. in [6] proposed to apply a linear superposition of the squares of the velocity deficits (*quadratic summation* rule). Various investigations showed that this method is beneficial in the case where many wakes are combined, giving better agreement with experimental results than the linear superposition for a small wind farm.

A power production benchmark proposed by Gaumond in [3] and based on the long term experimental observation at the Horns Rev offshore wind farm assessed the performance of these two simplistic approaches. It was found that the quadratic rule of Katic et al. in [6] associated to the standard industry N.O. Jensen wake model [5] may not be suitable for large wind farms, as it fails to capture properly the physic of merging wakes from neighboring rows of turbines.

This conclusion may initially appear contradictory with the observation by Katic [6], however it is not. The quadratic formulation of Katic at al. was tested for experimental data on a small wind farm, while the effect of neighboring wakes merging occurring at a certain downstream distance is only visible in large offshore wind farm as Horns Rev. In the latter observation, the quadratic rule failed to model deep array effects, as it reaches an equilibrium state (where the velocity deficit becomes constant) too quickly.

Recently, as part of the development of the wind farm topology optimization platform TOPFARM [7], two new wake superposition approaches were formulated. The first one is part of the in-house Aeroelastic code HAWC2 ([25]) coupled to the DWM model [8], where overlapping wakes are modeled using the so-called *dominant wake* approach. In this approach, the merged wake at a given location and at a given time in a wind farm is assumed to be characterized by the most dominant deficit of all upstream wakes contribution. In this way, the most attenuated and expanded upstream wakes contributes to the merged wake deficit more than what the conventional quadratic and linear summations would predict. This method proved robust to determine load and power production of the Egmond Aan Zee wind farm, as seen in the study by Larsen at al. in [11], and more recently, for the offshore Lillgrund wind farm [10]. The second technique, presently referred to as the *G.C. Larsen superposition model* [9], is based on a parabolic type of model, which combines wake deficits successively.

Among other observations, the present benchmark revealed that the performance of simple linear and quadratic summation approaches are highly dependent on the thrust force acting on the turbine.

2. Engineering models for predicting merging wakes

2.1. Introduction

In the present paper, we focus on the simplest merging wake configuration, where two turbines are aligned with the mean inflow direction. Specifically, this study aims at evaluating the performance of commonly used engineering models for single wake superposition, in order to predict both the mean deficit and the available aerodynamic power over the rotor area of a "fictitious" third turbine located downstream of the two wake generating rotors. The engineering models benchmark is based on CFD-LES simulations under sheared turbulent atmospheric inflow and covering a range of ambient configurations and turbine spacings, as described subsequently in Section 3.

A sketch of the investigated merging wake configuration is depicted in Fig. 1. The resulting wake deficits at the downstream location of a fictitious turbine $T3$ is obtained from superposition of single wake deficit generated by turbines $T1$ and $T2$, respectively. The inflow condition to the downwind turbine $T2$ varies with the chosen modeling approach, as will be detailed in Section 2.2.

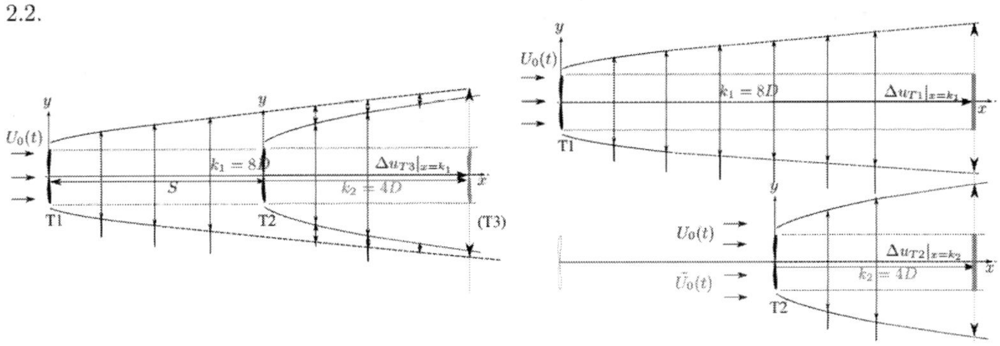

Figure 1. Sketch of the investigated wake overlapping situation. S denotes the turbine inter-spacing. The merged wake deficit (left figure) is decomposed into two single wake contributions: the wake of $T1$ at $k_1 = 8D$ downstream (top right figure); and the wake of $T2$ at $k_2 = 4D$ downstream (bottom right figure), where the origin of the x-axis is located at the most upstream rotor center. U_0 refers to the free stream velocity. Two different assumptions of the inflow conditions of turbine $T2$ are indicated: 1) using the mean wind speed U_0 or 2), using a reduced wind speed $\tilde{U}_0$ characterizing the disturbed incoming wind generated by the wake of $T1$.

We presently defined the wake velocity deficit in its normalized form, (Δu), as:

$$\Delta u = 1 - \frac{u_w}{U_0} \tag{1}$$

where u_w is the velocity in the wake and U_0 is the mean free stream velocity. The corresponding average deficit $\overline{\Delta u}$ over the rotor plane of area A is, in polar formulation, defined as:

$$\overline{\Delta u} = \frac{1}{A}\int_{-\pi}^{\pi}\int_{0}^{R} \Delta u r dr d\theta \tag{2}$$

The numerical evaluation of the integral in Eq. 2 is typically based on either: 1) a trapezoidal integration algorithm or 2); the Gauss integration technique as described in [9], where the averaged deficit is approximated using a forth order Gauss integration.
The corresponding available aerodynamic power P over the rotor plane is then obtained as:

$$P = \frac{1}{2}\rho A\left(U_0\left(1 - \frac{1}{A}\int_{-\pi}^{\pi}\int_{0}^{R} \Delta u r dr d\theta\right)\right)^3 = \frac{1}{2}\rho A\left((1-\overline{\Delta u})U_0\right)^3 \tag{3}$$

where U_0 is the mean free stream velocity. The determination of $\overline{\Delta u}$ varies with the tested engineering models as introduced subsequently.

2.2. The tested engineering models

2.2.1. Simple linear superposition

In this approach, the merged wake deficit at the downstream location corresponding to the $n^{th}+1$ fictitous rotor is obtained from linear superposition of the n upstream single wake deficits evaluated at the downstream location $x(n+1)$. Two assumptions on the characteristic of the inflow condition for the single wake calculations are investigated. The first modeling approach, as used by Lissaman et al. in [12], assumes that the ambient condition is the same for each wake generating turbine and equal to the free stream condition (U_0). The second approach, similar to the one resulting in the linearized CFD model FUGA [20], assumes that the ambient condition of each turbine is based on a reduced wind speed defined by the upstream wake affected inflow

($\tilde{U}_0(n)$). In the latter, the wake generated turbulence is neglected, and thus the turbulence intensity is constant and equal to the free stream one.

The average merged wake deficit resulting from the linear summation of n upstream velocity deficit contributions is then expressed as:

$$\overline{\Delta u}_{n+1} = \sum_{j=1}^{n} \left(\overline{\Delta u}_j|_{x(n+1)} \right) \tag{4}$$

Where in Eq. 4, each single wake contribution Δu_j is evaluated at the downstream location $x(n+1)$. The overbar denotes the averaging operator.

2.2.2. The "quadratic" approach
The quadratic superposition, as applied in the widely used PARK program [6], is obtained as the square root of the sums of the upstream single wakes deficits. Thus, the quadratic summation of n upstream velocity deficit contributions reads:

$$\overline{\Delta u}_{n+1} = \sqrt{\sum_{j=1}^{n} \left(\overline{\Delta u}_j|_{x(n+1)} \right)^2} \tag{5}$$

2.2.3. Multiple wakes in the Dynamic Wake Meandering approach
The superposition of wakes within the DWM model [8], [16] and [11], assumes that the wake affected downstream flow field is determined by a superposition of the ambient flow field and the *dominating* wake among contributions from all upstream turbines at any spatial position and at any time. This approach was motivated by an analysis of the wake deficit from 5 upstream turbines separated by 8 rotor diameters and calculated from the axis symmetric eddy viscosity model described in [16]. This model further assumes that the contributing upstream wakes are independent of each other, and thus, are determined under the same inflow characteristics (mean wind speed and turbulence field).
The superposed deficit in the DWM model is then calculated as:

$$\overline{\Delta u}_{n+1}(t) = MAX(\overline{\Delta u}_i(t))|_{x(n+1)} \tag{6}$$

where i is the upstream turbine index ranging from 1 to n in the park, and t denotes the time instant.

2.2.4. The G.C. Larsen wake superposition model
Finally, the G.C Larsen superposition model (GCL) described in [9] is evaluated in this study. We presently focus only on the wake-rotor interaction as fomulated in [9]. In this model, the instantaneous flow field at a given downstream turbine location x is obtained from the linear summation of all instantaneous single wake contributions at the same location x. Each of these single wake contributions are obtained from independent calculations conducted with an unsteady inflow field including the accumulated upstream wakes. As a result, this unsteady model, similarly to the DWM framework, takes into account the large scale wake meandering displacements.

The instantaneous wake velocity field $U(t)$ at the $n^{th}+1$ rotor in the turbine row is, in the *GCL* model, defined as:

$$U_{n+1}(t) = \overline{U_0} + \sum_{j=1}^{n} (U_j(t)|_{x(n+1)}^{\tilde{U}_j(t)} - \tilde{U}_j(t)) \tag{7}$$

In Eq. 7, each single wake contributions $U_j(t)$ to be superposed are determined at same downstream position $x(n+1)$ as where the resulting merged wake flow field is estimated, and simulated with the inflow field $\tilde{U}_j(t)$ described subsequently. The corresponding average deficit over the rotor area $\overline{\Delta u_{n+1}}(t)$ is calculated using Eq. 1 and Eq. 2.

The time dependent inflow condition $\tilde{U}_j(t)$ required for the simulations of the jth upstream single wake contribution is defined as:

$$\tilde{U}_j(t) = \overline{U_0} + \sum_{i=1}^{j-1} (U_i(t)|_{x(j)} - \overline{U_0}) \tag{8}$$

where $x(n)$ is the downstream distance from the most upstream rotor.

3. The merging wakes validation data

The present section aims at detailing the model and computational set up used to generate the merging wakes validation data.

3.1. Numerical model and computational set up

The wake velocity field was simulated using detailed unsteady CFD computations. The simulations were conducted using the incompressible Navier-Stokes solver EllipSys3D [18, 19, 23] and combines LES of the flow field with actuator disc modelling of the rotors. The atmospheric turbulence is modelled using the method presented by Troldborg et al. [27], where pre-generated Mann turbulence [17] is inserted in a cross-section upstream of the rotors. The used actuator disc model is based on the actuator shape model [21] combined with a blade element approach, where the aerodynamic loading is determined from the local flow conditions at the disc and 3D corrected aerofoil data as described by Réthoré et al. [22]. The LES model uses a sub grid scale model developed by Ta Phuoc [24], and the used solver parameters are identical to previous work [26, 15, 13].

The simulations are carried out in a regular Cartesian mesh of dimensions ($L_x = 24D_N$, $L_y = 14D_N$, $L_z = 14D_N$), where x, y and z denotes the streamwise, lateral and vertical coordinates, respectively, and D_N denotes the inserted turbine rotor diameter.

The simulated turbine is a stall regulated Nordtank 500 kW turbine, with a rotor diameter of 41 m and a hub height of 36 m. This turbine was chosen for practical reasons, as the rotational speed and the pitch is constant and therefore does not require any additional control. Furthermore, the present numerical model was validated for merging wake situations against full-scale lidar measurements conducted on the Nordtank turbine in [14].

The grid layout and the boundary conditions are in accordance with previous studies [26, 15]. The boundary conditions are as follows: Dirichlet conditions at the inlet and upper boundary ($x = 0$ and $z = 14D_N$), Neumann conditions at the outlet ($x = 24D_N$), symmetry condition at the bottom ($z = 0$), and cyclic boundary conditions on the sides ($y = 0$ and $y = 14D_N$).

In the region around and downstream of the rotors the grid spacing is equivalent to $dx = 60/D_N$ in the lateral and vertical direction, and $dz = 62/D_N$ in the stream-wise direction. The cells are stretched away from the wake region towards the domain boundaries, resulting in a total number of cells of 39.81 millions inside the domain. Each actuator disc is represented with a separate 2D polar grid with 94 radial and 180 angular elements.

The unsteady simulations are conducted using the time step $dt = 0.05s$. Fig. 2 shows an example of the coarse representation of the computational domain seen from the top, and the location of extraction planes.

3.2. Merging wakes parametric study

The present merged wake study involves a simple layout consisting of two Nordtank 500 kW stall regulated turbines (denoted $T1$ and $T2$) associated with a wide variety of ambient conditions. Specifically, two turbine inter-spacings are tested: a 'small' spacing of 4 rotor diameters,

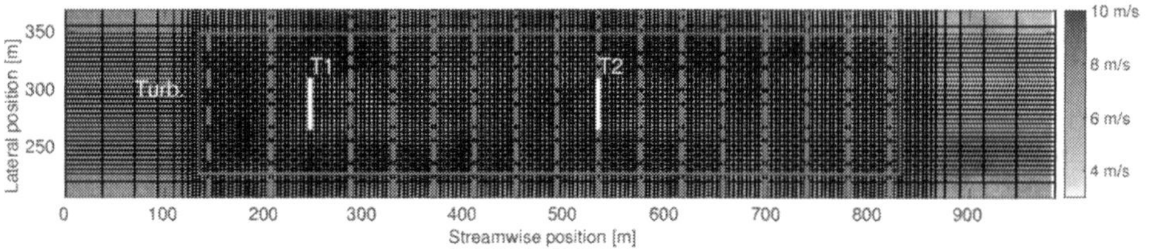

Figure 2. Coarse representation of the computational domain seen from the top. The white lines correspond to the wake of turbine $T1$ and $T2$ separated from 7 rotor diameters. Dash red lines are wake cross sections extracted at 5Hz and at a resolution of 1 m. The red rectangle shows a longitudinal extraction plane at hub height. The instantaneous streamwise velocity field is shown in the background where blue colors indicate regions of high wind speed and yellow colors show the wake region.

corresponding to a rather severe wake condition similar to the Middelgrunden wind farm outside Copenhagen harbor; and a 'larger' 7 diameters spacing, which is typical for offshore wind farm layouts in general. Additionally, two simulated ambient turbulence intensities are adopted, corresponding to typical offshore and onshore values of respectively 6% and 12%. Finally, four mean wind speeds are chosen covering different turbine loadings and thus various wake flow regimes. Table 1 summarizes the test matrix. The inflow wind profile is assumed neutral and modeled using a logarithmic law, where the roughness is equal to 5 cm. The friction velocity is adjusted to satisfy the prescribed hub height wind speed for all cases. Finally, synthetic Mann turbulence are generated using the same seed but scaled to match the required turbulence intensity and sheared inflow profile.

Table 1. Merging wake parametric study. Test cases matrix. The turbulence intensity corresponds to the free stream turbulence intensity I_{ref}, as resolved by the CFD-LES computations.

	5 m/s		**8 m/s**		**11 m/s**		**15 m/s**	
Merged wake: 4D spacing	6.3%	12.2%	6.3%	11.9%	6.3%	11.9%	6.4%	12.1%
Merged wake: 7D spacing	6.3%	12.2%	6.3%	11.9%	6.3%	11.9%	6.4%	12.1%

3.3. Simulation of the single wake contributions

The benchmark of merging wake models further requires, in addition to the simulations performed with both rotors, to determine the two turbine single wakes, independently. However, the tested wake accumulation models are based on various inflow field assumptions formulated in either instantaneous or average terms. This section therefore aims at describing the way each engineering model are related to the available LES computations.

3.3.1. Single wake definition in the DWM wake superposition framework

As described in [11], the DWM model assumes that the superposed single wake contributions are not influenced by other upstream wakes, and thus, can be simulated at the same unsteady inflow conditions. However, the DWM wake accumulation model is formulated in instantaneous terms (Eq. 6) so that the merged wake deficit at any downstream location is obtained from the instantaneous dominant wake among all upstream contributions. Therefore, it is crucial to ensure that the single wake contributions are simulated within the same large scale in the turbulence flow field, to ensure spatial and temporal correlation between the merged and the two independent single wake simulations.

The assumption of constant inflow for each of the turbines in the propagation row simplify the analogy with LES simulation. Specifically, only one single wake computations is required for the application of the DWM superposition, as the multiple turbine wakes can be analyzed by simply extracting the instantaneous velocity field at different downstream cross section, i.e.:

$$\Delta u_n(t)|_{x(n+1)}^{U_0(t)} \equiv \Delta u_{n+1}(t+dt)|_{x(n+2)}^{U_0(t)} \tag{9}$$

However, this analogy can be only applied after introducing a relevant time shit dt for each instantaneous deficit characterizing their respective distance from the wake emitting rotor, or more specifically from the upstream location where the Mann turbulence are inserted into the computational domain. The Taylor frozen turbulence hypothesis is adopted to quantify this time shift as $dt = S/\overline{U_0}$, where S is the turbine spacing and U_0 is the mean free stream velocity.

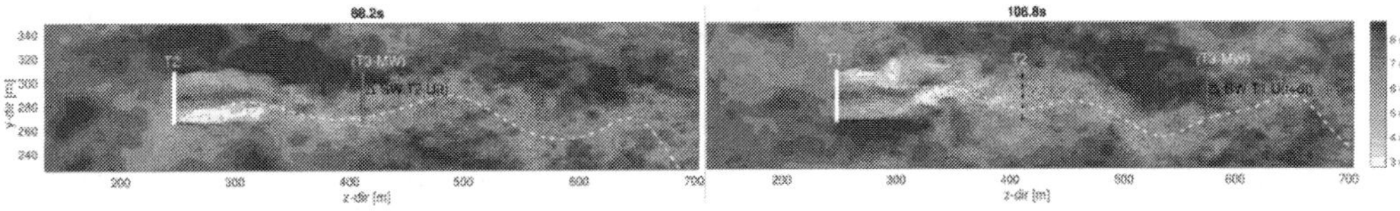

Figure 3. Dominant (instantaneous) wake analysis for the evaluation of the wake accumulation within the DWM framework. The two figures shows the instantaneous velocity contour at hub height at two different time instants. The dash lines represents the wake center obtained from applying the tracking procedure described in [28]. The time difference between the two snapshot is obtained from adopting the Taylor frozen turbulence hypothesis, illustrated by the large high velocity turbulence structure to the left of the two $T3$ locations.

As depicted in Fig. 3, the instantaneous "dominant" wake at the location of the fictitious turbine $T3$ (red section in Fig. 3) is conveniently obtained by analyzing the instantaneous wake of $T2$ at the time instant t and the instantaneous wake of $T1$ at the time instant $t + dt$, where dt is obtained from Taylor hypothesis. This has the advantage of reducing the number of computations in the present study.

3.3.2. First and second modeling approach of linear and quadratic summations

Meanwhile, other model such as FUGA [20] are based on linear summation of single wake deficits obtained at a different inflow conditions than the free stream to account for multiple wake overlapping. This implies first to perform single wake simulations of the most upstream turbine $T1$ at the same inflow conditions as the merged wake ones listed in Table 1.

Additionally, single wake computations of the downstream turbine $T2$ are conducted for a wide range of new inflow conditions to characterize the wake affected inflow. In the present approach, the average reduced wind speed over the rotor area of $T2$ is estimated from power curve reading based on the mean power produced by $T2$ in the merged wake simulation. The increased turbulence intensity generated by the inflow wake of $T1$ is neglected, and therefore the newly generated Mann turbulence field adapted to the reduced wind speed is scaled to reach the same turbulence level as for the merged wake simulation, however, inherently associated to a different turbulent realization. It is assumed that, considering a sufficiently long averaging time, this new realization of turbulence does not affect the average single wake deficits used in the present linear and quadratic summation. The resulting single wake test matrix is shown in Table 2.

3.3.3. The GCL wake superposition model

The analogy between the wake superposition as formulated in the GCL model and the available LES results is more challenging. As opposed to the DWM model, the unsteady inflow velocity of the downwind turbine is affected by the upstream wakes. Furthermore, $T2$ single wake simulations are performed under a different turbulence field, which is incompatible with the GCL formulation.

Table 2. Large Eddy Simulation of the single wake generated by turbine $T1$ and $T2$. Single wake simulations of $T2$ are performed at various mean wind speed characterizing the wake affected inflow generated by $T1$. The change of mean wind speed implies to generate a new set of Mann turbulence field, with the same turbulence level as for the single wake computation of $T1$ but with a new turbulent realization.

	6%	12%	6%	12%	6%	12%	6%	12%
SW $T1$, **4D-7D** U_0 (m/s)	5	5	8	8	11	11	15	15
SW $T2$, **4D,** $\tilde{U_0} \neq U_0$ (m/s)	3.95	4.5	6.45	7.24	9.22	10.05	12.84	14.02
SW $T2$, **7D,** $\tilde{U_0} \neq U_0$ (m/s)	4.32	4.85	7.04	7.62	9.88	10.36	13.04	14.19

From Eq. 7, the merged wake deficit for $n = 2$ rotors ($T1$ and $T2$) at the downstream distance corresponding to a third rotor $T3$ in the GCL superposition framework is given as:

$$\overline{\Delta u}_{T3}(t) = \overline{\Delta u}_{T1}(t)|_{x(T3)}^{U_0(t)} + \overline{\Delta u}_{T2}(t)|_{x(T3)}^{\tilde{U_0}(t)} \tag{10}$$

where the subscript $x(T3)$ corresponds to the downstream distance where the merged wake deficit is estimated, and the superscript U_0 and $\tilde{U_0}$ are the inflow velocity at the first and second rotors, respectively, as defined in Eq. 8. The first term on the right hand side is directly known from the single wake simulations of $T1$ as listed in Table 2. The last term is known from the reduced mean wind speed computations of $T2$, however, under a different turbulence field as previously discussed.

To circumvent this, we apply the equivalence in Eq. 9, so that the single wake computation of $T1$ simulated under the same turbulence field as for the merged wake, is used instead of the $T2$ simulation with the different turbulence realization. However, the difference of inflow conditions between $T1$ and $T2$ simulations requires a scaling of the instantaneous wake deficit.

This scaling procedure can be resumed into in three steps: 1) we first determine the two involved average deficit of $T1$ and $T2$ in their respective meandering frame of reference (MFoR) and at the downstream location of $T3$; 2), we correct each instantaneous velocity deficit of $T1$ by subtracting its average deficit in the MFoR and furthermore adding the average deficit of $T2$ in the MFoR. The final correction is similar to the one performed for the DWM model and is to apply a time shift calculated from Taylor frozen hypothesis on the previously obtained deficit to ensure spatial and temporal correlation of turbulence among the unsteady single wake deficits.

Once the average deficit in the MFoR is known from applying the tracking algorithm as in [28], step 2 gives:

$$\Delta u_{T2}(t)|_{x(T3)}^{\tilde{U_0}(\tilde{t})} \equiv \Delta u_{T1}(t)|_{x(T2)}^{U_0(t)} - \overline{\Delta u}_{T1}^{(M)}|_{x(T2)}^{U_0} + \overline{\Delta u}_{T2}^{(M)}|_{x(T3)}^{\tilde{U_0}} \tag{11}$$

In which the superscript (M) denotes the meandering frame of reference, and $x(T2)$ and $x(T3)$ are the downstream positions of $T2$ and $T3$, respectively.
Applying the time shift of step 3 gives:

$$\Delta u_{T2}(t)|_{x(T3)}^{\tilde{U_0}(t)} \equiv \Delta u_{T1}(t+dt)|_{x(T2)}^{U_0(t)} - \overline{\Delta u}_{T1}^{(M)}|_{x(T2)}^{U_0} + \overline{\Delta u}_{T2}^{(M)}|_{x(T3)}^{\tilde{U_0}} \tag{12}$$

4. Benchmark results and discussion

In this section, the error on the prediction of the average deficit and the available aerodynamic power over the rotor area and at the fictitious location of $T3$ is investigated. It is defined as:

$$\epsilon(CT) = 100 \cdot \left(\frac{Q_{model} - Q_{LES}}{Q_{model}} \right) \tag{13}$$

where Q refers to either integrated deficit or available aerodynamic power from Eq. 2 and 3, respectively. The results are shown as function of the average of the two turbine thrust coefficients for the four tested combinations of turbulence intensity and turbine spacing.

4.1. Linear and quadratic average models

As seen in Fig. 4, the way wakes accumulate is strongly dependent on the thrust force acting on the turbine. Low wind speed associated with a high thrust coefficient typically generates very pronounced wake deficits. Their resulting overlapped wake appears well captured by a quadratic summation with error of the order of 3-5 % for the higher CT configuration. Contrary, at high mean wind speed, the turbine is more 'transparent' to the flow, and linear summation of the very attenuated single wakes seems to be more representative of the merged wake deficit. This is consistent with a linear wake perturbation approach. At moderate wind speed, none of the summation technique is able to capture the merged wake deficit. The quadratic summation underestimates the deficit, whereas the linear summation overestimates it. Thus, averaging both contributions may enhance the agreement over the entire tested wind speed range, similarly to the Horns Rev wake losses study in [4].

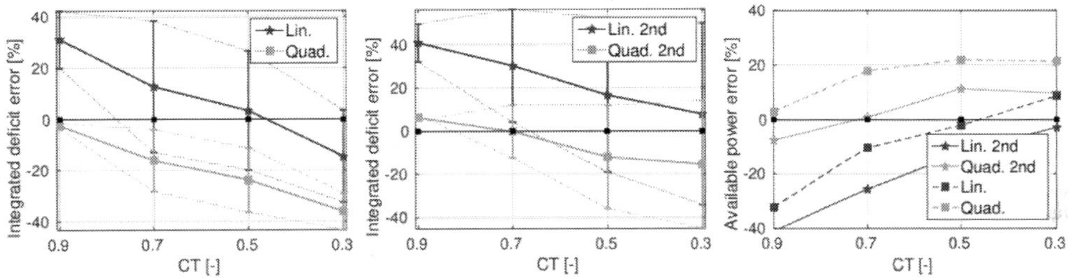

Figure 4. Error on the prediction of the integrated deficit from the steady linear and quadratic models and for the first and second modeling approach (respectively left and center plots). The available aerodynamic power at the downstream location of $T3$ is shown on the right plot. Error bars shows ± one standard deviation of the error for each thrust (mean wind speed) flow cases.

When considering the more realistic second modeling approach, the downstream single wake contributions is simulated at a reduced average wind speed characterizing the wake affected inflow conditions. Thus, the magnitude of the downstream wake generated by $T2$ is larger, as the CT coefficient of the downstream turbine increases with the reduced incoming wind. Therefore, the quadratic summation prediction error is reduced when using the second modeling approach.

The present results are consistent with earlier observations by Crespo et al. [1], which found that, when considering the superposition of two wakes abreast, that linear superposition of the small velocity deficits in the interference region is robust for predicting the resulting overlapped wake characteristics. Contrary, when considering the situation where two turbines are in a row, the linear superposition assumption overestimates substantially the velocity deficit. Thus, the quadratic summation of Katic et al. [6], proved to perform better for larger deficits. This conclusion is also consistent with the observations made by Gaumond in [3], the semi-empirical superposition law of Smith and Taylor [2] as well as with the wake-wake interaction formulated in the *G.C. Larsen superposition model* [9]

4.2. Dynamic Wake Meandering

Results of the DWM predictions are shown in Fig. 5. It is seen that the overall agreement is fair regardless of the thrust force acting on the turbine. Furthermore, the observed agreement is associated to a systematic overestimation of the available aerodynamic power. It is speculated if the assumption of constant inflow / advection velocity in the DWM framework can result in such power overestimation and therefore will require further investigation, in analogy with the study of the single wake advection velocity in [13].

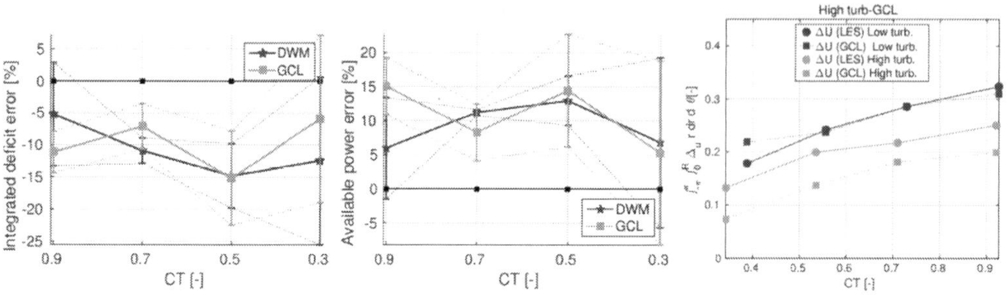

Figure 5. Error on the prediction of the integrated deficit (left) and the available aerodynamic power (middle) at the downstream location of $T3$, based on the Dynamic Wake Meandering and G.C Larsen wake superposition models. Error bars shows ± one standard deviation of the error for each flow case. (Right) Comparison between the G.C Larsen superposition model prediction of the integrated deficit over the rotor with LES results at the location of $T3$ for the low turbulent and high turbulent cases.

4.3. GCL

Finally, the predictions from the GCL model are shown in Fig. 5. This model shows similar overall performance to the DWM one. However, unlike the DWM predictions, the performance of the GCL model varies significantly with respect to the turbulence intensity. An excellent agreement is observed for the low turbulence case whereas the merged wake deficit is significantly underestimated when considering higher turbulence level, as seen in Fig. 5(right). A possible source of uncertainty may come from the implemented wake deficit scaling for the downwind turbine wake. The accuracy of this method can be deteriorated at high turbulence level, where the increased meandering magnitude can affect the performance of the wake tracking procedure and thus, the determination of the two scaling quantities in Eq. 12.

5. Conclusion

In the present work, merging wake engineering models have been evaluated against CFD-LES simulations using the in-house EllipSys3D flow solver for a range of mean wind speeds, turbine spacing and turbulence intensity. The study indicates that wake interaction depends strongly on the relative wake deficit magnitude, i.e. the deficit magnitude normalized with respect to the ambient mean wind speed. Furthermore, the dominant wake assumption within the DWM framework and the parabolic model of G.C. Larsen are both unsteady models capable of predicting the available aerodynamic power in the merged wake flow field with a good degree of accuracy. Simpler and widely used summation techniques such as the linear and quadratic techniques shows an acceptable performance as well, however, typically associated with a higher uncertainties as compared to unsteady models.

The present work will be extended to the study of large rows of turbine more representative of modern wind farm. A new model based on a weighted average combination of linear and quadratic techniques, where the turbine thrust coefficient is used as weighting factor, will be subsequently developed. Furthermore, the added wake turbulence will be quantified and modeled. Sideways wake merging, occurring when the accumulated wake of two turbine rows merged within the wind farm, will also be simulated and compare to the predictions from the wake to wake interaction formulation of the GCL model.

6. Acknowledgments

The major part of this research has been funded and carried out within the DSF FlowCenter project under the Danish Council for Strategic Research contract 2104-09-0026 and the EUDP project "Impact of atmospheric stability conditions on wind farm loading and production"" under contract 64010-0462.

References

[1] A. Crespo, J. Hernández, and S. Frandsen. Survey of modelling methods for wind turbine wakes and wind farms. *Wind Energy*, 1:1–24, 1999.
[2] Smith D. and Taylore G.J. Further analysis of turbine wake development and interaction data. In *Scientific proceedings*, pages 325–331. 13th BWEA Wind Energy Conf. Swansea, 1991.
[3] Mathieu Gaumond. Evaluation and Benchmarking of Wind Turbine Wake Models. 2012.
[4] G. Habenicht. Offshore wake modelling. Presentation at Renewable UK Offshore Wind, 2011.
[5] N. O. Jensen. A note on wind turbine interaction. Technical report, Risø-M-2411, Risø National Laboratory, Roskilde, Denmark, 1983.
[6] I. Katic, J. Høstrup, and N. O. Jensen. A simple model for cluster efficiency. *In proceedings of the European Wind Energy Association Conference and Exhibition (EWEC), Rome*, pages 407–410, 1986.
[7] G. C. Larsen, H. Aa. Madsen, T. J. Larsen, P.-E. Réthoré, and Peter Fuglsang. TOPFARM - A platform for wind farm topology optimization. *Book of Abstracts / Editor: Ivanell S. - Wake conference, Visvy, Sweden*, pages 40–42, 2011.
[8] G. C. Larsen, H. Aa. Madsen, K. Thomsen, and T. J. Larsen. Wake Meandering: A Pragmatic Approach. *Wind Energy*, 11:377–395, 2008.
[9] G.C. Larsen. From solitary wakes to wind farm wind fields a simple engineering approach. Risø-R-1727(EN). Technical report, DTU Wind Energy, Roskilde, Denmark,, 2009.
[10] T.J. Larsen, G.C. Larsen, H.Aa. Madsen, K. Thomsen, and S.M. Pedersen. Comparison of measured and simulated loads for the siemens swt 2.3 operating in wake conditions at the Lillgrund wind farm using HAWC2 and the DWM model. In *Scientific proceedings*, pages –. EWEC Offshore 2015 Copenhagen, 2015.
[11] Torben J. Larsen, Helge Aa. Madsen, Gunner C. Larsen, and Kurt S. Hansen. Validation of the dynamic wake meander model for loads and power production in the egmond aan zee wind farm. *Wind Energy*, 16(4):605–624, 2013.
[12] P.B.S. Lissaman. Energy effectiveness of arbitrary arrays of wind turbines. *Journal of Energy*, 3(6):323–328, 1979.
[13] E Machefaux, G C Larsen, N Troldborg, M Gaunaa, and A Rettenmeier. Empirical modeling of single-wake advection and expansion using full-scale pulsed lidar-based measurements. *Wind Energy*, 2014.
[14] E. Machefaux, G. C. Larsen, N. Troldborg, K.S. Hansen, N. Angelou, T. Mikkelsen, and J. Mann. Investigation of wake interaction using full-scale lidar measurements and large eddy simulation. 2015.
[15] E. Machefaux, N. Troldborg, G. C. Larsen, J. Mann, and H. Aa. Madsen. Experimental and Numerical Analysis of Wake to Wake Interaction in Wind Farms. In *Scientific proceedings*, pages 100–104. EWEC 2012 Copenhagen, 2012.
[16] H. Aa. Madsen, G. C. Larsen, T. J. Larsen, N. Troldborg, and Mikkelsen R. Calibration and Validation of the Dynamic Wake Meandering Model for Implementation in an Aeroelastic Code. *Journal of Solar Energy Engineering*, 2010.
[17] Jakob Mann. Wind field simulation. *Science*, 13(4):269–282, 1998.
[18] J. A. Michelsen. Basis 3d - a Platform for Development of multiblock PDE Solvers. Dept. of Fluid mechanics, DTU. *AFM 92-05*, 1994.
[19] J. A. Michelsen. *Block Structured Multigrid Solution of 2D and 3D elliptic PDE's.* AFM 94-05 - Department of Fluid Mechanics, Technical University of Denmark, 1994.
[20] Søren Ott. *Linearised CFD models for wakes*, volume 1772. 2011.
[21] P.-E. Réthoré and N. N. Sørensen. A discrete force allocation algorithm for modelling wind turbines in computational fluid dynamics. *Wind Energy*, 15:915–926, 2012.
[22] P.-E. Réthoré, P. van der Laan, N. Troldborg, F. Zahle, and N. N. Sørensen. Verification and validation of an actuator disc model. *Wind Energy*, 2013.
[23] N. N. Sørensen. *General Purpose Flow Solver Applied to Flow over Hills.* PhD thesis, Risø National Laboratory., 1995.
[24] L. Ta Phuoc, R. Lardat, M. Coutanceau, and G. Pineau. Recherche et analyse de modèles de turbulence de sous mailles adaptés aux écoulements instationnaires décollés. LIMSI, France. Technical report, LIMSI Report 93074, 2002.
[25] A.M. Hansen T.J.Larsen. *HAWC2 - User manual.* DTU-Risø-R-1597, 2007.
[26] N. Troldborg, G. C. Larsen, H. Aa. Madsen, K. S. Hansen, J. N. Sørensen, and R. Mikkelsen. Numerical Simulations of Wake Interaction between Two Wind Turbines at Various Inflow Conditions. *Wind Energy, 14, doi: 10.1002/we.433*, pages 859–876, 2010.
[27] N. Troldborg, J.N. Sørensen, R. Mikkelsen, and N.N Sørensen. A simple atmospheric boundary layer model applied to large eddy simulations of wind turbine wakes. *Wind Energy*, 2013.
[28] J. J. Trujillo, F. Bingöl, G. C. Larsen, J. Mann, and M. Kuehn. Lidar Measurements of Wake Dynamics, part 2: Two-dimensional scanning. *Wind Energy*, 14:61–75, 2011.

Wake Conference 2015 IOP Publishing
Journal of Physics: Conference Series **625** (2015) 012038 doi:10.1088/1742-6596/625/1/012038

Turbulent large-scale structure effects on wake meandering

Y-A Muller[1], C Masson[2] and S Aubrun[3]
[1,3]Univ. Orléans, INSA-CVL, PRISME EA-4229, Orléans, France [2] Département de génie mécanique, École de technologie supérieure, Montreal, Canada
E-mail: [1]yann-ael.muller@laposte.net, [2]christian.masson@etsmtl.ca, [3]sandrine.aubrun@univ-orleans.fr

Abstract. This work studies effects of large-scale turbulent structures on wake meandering using Large Eddy Simulations (LES) over an actuator disk. Other potential source of wake meandering such as the instablility mechanisms associated with tip vortices are not treated in this study. A crucial element of the efficient, pragmatic and successful simulations of large-scale turbulent structures in Atmospheric Boundary Layer (ABL) is the generation of the stochastic turbulent atmospheric flow. This is an essential capability since one source of wake meandering is these large - larger than the turbine diameter - turbulent structures. The unsteady wind turbine wake in ABL is simulated using a combination of LES and actuator disk approaches. In order to dedicate the large majority of the available computing power in the wake, the ABL ground region of the flow is not part of the computational domain. Instead, mixed Dirichlet/Neumann boundary conditions are applied at all the computational surfaces except at the outlet. Prescribed values for Dirichlet contribution of these boundary conditions are provided by a stochastic turbulent wind generator. This allows to simulate large-scale turbulent structures - larger than the computational domain - leading to an efficient simulation technique of wake meandering. Since the stochastic wind generator includes shear, the turbulence production is included in the analysis without the necessity of resolving the flow near the ground. The classical Smagorinsky sub-grid model is used.

The resulting numerical methodology has been implemented in OpenFOAM. Comparisons with experimental measurements in porous-disk wakes have been undertaken, and the agreements are good. While temporal resolution in experimental measurements is high, the spatial resolution is often too low. LES numerical results provide a more complete spatial description of the flow. They tend to demonstrate that inflow low frequency content - or large-scale turbulent structures - is an important parameter when simulating wake meandering and plays a significant role.

1. Introduction

Wind turbine wakes are critical for the design and operation of wind farms [1]. The wake velocity deficit contributes to decrease the wind farm energy production [2, 3]. Wakes should be considered during the design of wind farms in order to decrease energy losses associated with velocity deficit [4]. Wakes also increase turbulence intensities which influence the structural behavior and the life time of a wind turbine. Siting studies should consequently consider wakes during their analyses. Finally, the recovery rate of wake velocity deficit is highly dependent on the level of turbulence intensity [5]. Low turbulence intensities yield low recovery rate. This explains why wakes are critical for offshore developments.

Wakes can be considered as steady or unsteady phenomena. Analysing wake behavior as a steady phenomenon [6, 7, 8] typically leads to an overestimation of the velocity deficit. A more complete analysis is to consider wakes as non stationary flow structures [9, 10, 11, 12]. One of the key aspect of wind turbine wake dynamics is the phenomenon of meandering [13, 14, 15] which contibutes to fatigue loading. Measurements and observations of this phenomenon have been undertaken in situ [13, 15] and in wind tunnel [16, 17]. The causes of wake meandering were associated experimentally with the unsteady behavior of large-scale turbulent structures [18, 19] and the tip vortex instabilities in turbulent wake flows [20]. A pragmatic and efficient wake meandering model based on the effects of large scale (low frequency) structures has been proposed by Larsen et al. [10]. España et al. [18] have shown experimentally the importance of large scale turbulent structures on wake meandering behind porous disks.

When developing a simulation technique for wake meandering, an important aspect to consider is the flow model, including turbulence modeling, and the way wind turbine action on the flow is modeled with its associated level of details. Regarding flow and turbulence models, it ranges from basic momentum conservation to direct simulation of Navier-Stokes equations. Regarding the turbines, an actuator disk or actuator lines can be considered.Various combinations have been proposed and successfully applied [21, 22, 23]. Each combination was proposed in order to efficiently simulate a specific phenomenon (e.g. tip vortex shedding, blade tip losses, wakes). References [23, 24] are relevant to the work presented in this paper since they describe a simulation technique specifically developed to simulate unsteady wake behavior.

In this work, a simulation technique suitable for the analysis of large scale turbulent structures in Atmospheric Boundary Layer (ABL) is proposed and presented. A crucial element of this simulation technique is the generation of the stochastic turbulent atmospheric flow. This is an essential capability since one supposed source of wake meandering is these large - larger than the turbine diameter - turbulent structures.The unsteady wind turbine wake in ABL is simulated using Large Eddy Simulations (LES) in which the wind turbine is represented by an actuator disk. In order to dedicate the large majority of the available computing power in the wake, the ABL ground region of the flow is not part of the computational domain. Instead, mixed Dirichlet/Neumann boundary conditions are applied at all the computational boundary surfaces except at the outlet. Prescribed values for Dirichlet contribution of these boundary conditions are provided by the stochastic turbulent wind generator. This allows to simulate large-scale turbulent structures - larger than the computational domain - leading to an efficient simulation technique of wake meandering. Since the stochastic wind generator includes shear, the turbulence production is included in the analysis without the necessity of resolving the flow near the ground. The classical Smagorinsky sub-grid model is used. It has been determined that in such an implementation, this choice of sub-grid model is not critical since more than 80% of the kinetic energy of turbulence is resolved.

2. Mathematical model

2.1. Turbine representation

The turbine is modelled using an actuator disk of diameter D with an axial induction factor $a = 0.19$ corresponding to the measured value presented in [17]. This disk extracts momentum from the flow which produces a local deceleration. This constitutes an acceptable approximation of the wind turbine far wake, as described by Froude's theory [25]. This disk is introduced in the form of an internal boundary condition at the location where it is inserted, and where a pressure discontinuity is prescribed according to Eq.(1) but through which the velocity is continuous.

The imposed pressure discontinuity Δp is computed using the fluid density ρ, the instantaneous local wind speed at the disk location, U_d and the prescribed induction factor a:

$$\Delta p = 2a(1+a)\rho U_d^2 \tag{1}$$

2.2. Governing equations and turbulence modelling

The governing equations are the Navier-Stokes equations in which gravity and Coriolis terms are neglected. The LES approach with implicit filtering is adopted in this work. Consequently, only large-scale turbulent structures are resolved, and the smaller scale structures are modelled using the classical Smagorinsky model with a model constant C_s set to 0.167.

2.3. Synthetic turbulence generation

The stochastic generation of turbulent velocity series is a subject that has been approached by various authors [26, 27, 28], but the most commonly employed in the field of wind turbine simulation is probably that of Mann [29], which relies on the inversion of the spectral tensor of atmospheric turbulence.

In this work, the turbulent inflow is obtained by mean of a stochastic method similar to that of the aformentioned authors, which is described in [30]. The resulting series present realistic turbulence characteristics, especially regarding the largest scales of turbulence, which makes it particularly useful for the simulation of atmospheric flows. It makes use of an inverse wavelet transform to enable the simulation of sheared, non-homogeneous turbulent fields. It also introduces the concept of motion compensation in order to incorporate the temporal dynamic of the production and decay of the turbulent eddies.

3. Numerical method

All simulations are carried out using OpenFOAM [31]. The gradient terms are discretized using second order linear interpolations. The divergence terms are also discretized using second-order linear interpolations with the possibility of adding up to 20% of first-order upwinding to prevent any oscillations. This is the option *filteredLinear* in OpenFOAM. Time marching is done using backward Euler scheme. The pressure implicit split operator (PISO) algorithm [32] is used to handle pressure-velocity coupling. In OpenFOAM, it is available through *pisoFoam*. The systems of velocity equations are solved using linear algorithm (*smoothSolver*) while the geometric-algebraic multigrid algorithm (GAMG) is used for pressure equations.

3.1. Boundary conditions

Section 2.3 briefly described the methodology developed in order to produce realistic time series of turbulent velocity in ABL using a stochastic generator. In this section, the procedure to introduce these time series into the LES through appropriate boundary conditions is described.

There are two types of boundary conditions (BC) for differential equations: Dirichlet and Neumann types. For Dirichlet type of BC, the value of the variable is fixed at the boundary. For Neumann type of BC, it is the gradient of the variable that is fixed at the boundary.

Dirichlet and Neumann types of boundary conditions (BC) can be combined in a so-called hybrid BC. For a general dependent variable μ - which can be the pressure or velocity components - using f as the fraction of the BC associated with the Dirichlet's type of BC and the indices c and f to indicate the variable location at cell center and boundary face, respectively, the hydrid BC can be written as:

Dirichlet BC, prescribed value of μ: $\mu_{Dirichlet} = \mu$.

Neumann BC, prescribed $\partial\mu_f/\partial x$, x being the position: $\mu_{Neumann} = \mu_c + \underbrace{\frac{\partial \mu_f}{\partial x}}_{\text{specified}} (x_c - x_f)$.

Hybrid BC combining the previous two BCs:

$$\mu_{hybrid} = f\mu_{Dirichlet} + (1 - f)\mu_{Neumann} \tag{2}$$

The approach adopted in [23, 33, 24] consists in using steady-state BCs and turbulence is introduced through unsteady source terms in a vertical plane lying upstream of the rotor in the computational domain.

An alternative approach to introduce the turbulence is adopted in this work. It consists in imposing the fluctuating velocities at all boundaries of the computational domain except at the outlet. The velocity values are extracted from a spatially and temporally coherent synthetic velocity field, which is generated according to the stochastic method described in Section 2.3. This approach ensures a reasonable unsteady behavior of the flow in the entire computational domain: inspection of the numerical solutions reveals that the resolved turbulent kinetic energy near the boundary is similar to that which can be observed further inside the domain. Additionaly, it makes it possible to introduce, and study, turbulent structures larger than the computational domain. The largest turbulent scales in the synthetic velocity field may be larger than the size of the LES computational domain. Since the largest turbulent scales of the synthetic field are applied at the inlet, lateral, top and bottom boundaries, it ensures that the computational domain is swept by these large turbulent structures, and consequently, the effect of these large structures on the meandering of the wake is reproduced in the computational domain. However, coherence between the turbulent structures in the computational domain and the external atmospheric turbulence is needed. This is verified by comparing the time-spectra of the synthetic velocity and resolved velocity inside the LES as shown on Fig.4. It shows that the turbulence levels are similar for most frequencies, and especially low frequencies (large scales). Since the turbulence levels are similar, it is reasonable to assume that the simulated velocity in the domain is directly driven by the synthetic field, with a slight amount of discrepancy depending on the scale of the structures. This discrepancy is caused by the fact that the evolution of the simulated velocity field is obtained by solving the Navier-Stokes equations, whereas the evolution of the synthetic velocity field is the result of a simplified advection scheme relying mainly on Taylor hypothesis.

Consequently, the synthetic velocities are prescribed on all boundary faces of the computational domain, except at the outlet face which is orthogonal to the main wind speed direction. Zero gradient is applied on velocities at the outlet. For the lateral, top and bottom faces, the average mass flow rate is zero on each boundary face, but since the flow is turbulent, instantaneous flow rate may be non-zero.

During the initial phase of the development, a Dirichlet type boundary condition (*fixedValue* in OpenFOAM) was imposed both at inlet and lateral faces. The prescribed velocity values were given by the stochastic generator. This implementation allowed to obtain overall physically meaningful solutions, but with significant spurious spatial oscillations near cells adjacent to boundary faces with instantaneous outflow. This undesirable behavior of the solution is produced mainly by an inconsistency between the small-scale structures computed by the LES and the stochastic generator: the discrepancy between the synthetic velocity field and the simulated LES velocity field is mainly produced by the advection which is calculated differently in LES and synthetic turbulence generator. Only a minor part of this discrepancy can be attrubuted to the difference in resolution between synthetic and simulated turbulence.

While large-scale structures are typically coherent between the solutions of LES and stochastic generator, it is not the case for small-scale structures which experience a different evolution since their coherent time-scales are much smaller.

This inconsistency can be solved by using hybrid boundary conditions where the Dirichlet fraction f in Eq.(2) is taking values from 1 - for pure Dirichlet - to 0 - for pure Neumann - depending on the sign of the local wall normal velocity component (and thus the flow through the boundary face) of the instantaneous flow as given by Eqs.(3) and (4). It is to be noted that there is a value of f for each velocity component. This hybrid boundary condition is applied to the components of the velocity tangential to the wall, and is based on the value of the wall-normal component of the velocity. Its role is to loosen the specified value of the velocity at the boundary during outflows in order to prevent spurious spatial oscillations. Regardless of the direction of the flow, the wall-normal component of the velocity is always specified with a pure Dirichlet condition. When the normal component was not specified, a decrease of theaverage velocity of the flow in the main direction was observed due to an excess of mass exiting the computational domain.

Following this reasoning, Eq.(2) can be re-written in a vectorial form by replacing the scalar quantity f by the tensor F. Using tensor F allows the specification of a different value of the Dirichlet fraction for each velocity component expressed in term of the face normal n oriented toward the interior of the computational domain:

$$\text{for flux } \phi < 0 \quad F = \underbrace{1 * n \wedge n}_{\perp} + \underbrace{(I - n \wedge n) * 0}_{//} \tag{3}$$

$$\text{for } \phi \geq 0 \quad F = I \tag{4}$$

The operator $\wedge$ is the external product, resulting in a tensor. I is the identity matrix. Eq.(2) can be written as:

$$u = F * u_{Dirichlet} + (I - F) * u_{Neumann} \tag{5}$$

$$\text{with} \quad u_{Dirichlet} = u_{synthetic} \tag{6}$$

$$u_{Neumann} = u_c + \frac{\partial U_{ABL}}{\partial x} * (x_c - x_l) \tag{7}$$

The values of $u_{Dirichlet}$ at all boundary faces are given by the stochastic generator at proper time and location. Values of $u_{Neumann}$ at all boundary faces are determined from the velocity gradient computed from the averaged flow field as given by the ABL logarithmic expression. This hybrid boundary condition allows to maintain the flow rates at all boundaries as those given by the stochastic generator. Zero gradient is set at boundaries for pressure with a reference pressure prescribed at a given point in the computational domain.

3.2. Computational domain

The computational domain is illustrated in Fig. 1. The bottom face is not located at the ground, but at a height of $5m$ where turbulence is assumed to be completely developed. As such, the velocity is not zero at this bottom face. Coarser mesh resolutions (5.66m and 6.82m) were also tested. The results were very close. Furthermore, given the simplicity of the WT model used here, a finer mesh does not appear necessary. While we do focus on the large turbulent scales and do not make the wake recovery and diffusion the main priority of this study, it is important to point out that the values of the velocity on the centerline of the wake at 2D and 5D behind the rotor are in very good agreement with experimental data, as shown in Section 4.2.3. Computational domain parameters such as domain width Ly, domain height Lz, rotor diameter D, number of grid points N_{grid}, grid resolution Δ are reported in Table 1. The relatively small domain height adopted is the result of a compromise between computational cost and modelling accuracy. The choice of a relatively small domain height has been made to allow the use a uniform mesh to ensure that the implicit turbulence filtering scale does not change across the domain. Given such a constraint for the mesh, increasing the domain size while keeping the same resolution quickly increases the number of cells. Moreover it was decided to use a domain width greater than height so as to prioritize horizontal wake meandering over vertical, since the horizontal meandering shows a greater amplitude.

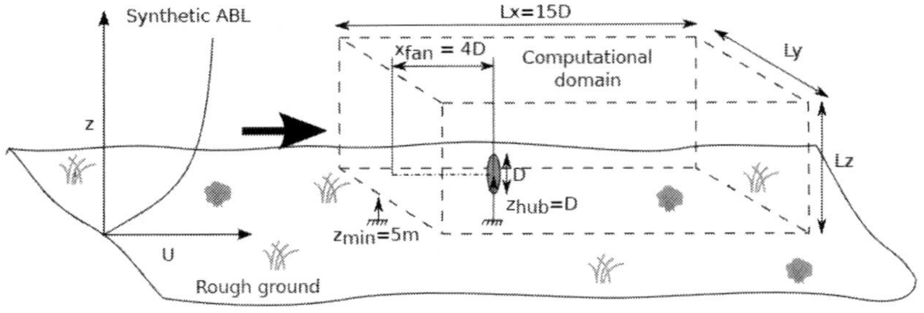

Figure 1. Computational domain

Case	$Lx\,(m)$	$Ly\,(m)$	$Lz\,(m)$	$\Delta\,(m)$	Nx	Ny	Nz	N_{grid}	$\Delta t\,(s)$
D80	1200	400	265	5.02	238	79	52	977704	0.45

Table 1. Computational domain parameters

3.3. Actuator disk implementation

The internal face at the disk location is determined by selecting a vertical plane in the grid in order to ensure that the disk is composed of existing grid cell faces. A grid face is considered to be part of the actuator disk when its center lies within a circle having the same diameter as the actuator disk. All grid faces having these caracteristics compose the discretized actuator disk on which the internal boundary condition is applied. Since the grid is cartesian, the discretized actuator disk is not a perfect circle. Its perimeter is of *stair* shape. With a grid resolution of 5m, there are about 16 cells radially over 1D. While this might seem coarse, an uniform Cartesian mesh is simple and has the advantage of ensuring that the implicit LES filter scale is homogeneous accross the whole domain. Through the actuator disk, only the pressure is discontinuous, the velocity is continuous. This does not create a singularity in the code, as it is implemented as a pair of coinciding cyclic boundaries internal to the domain.

In order to allow comparisons with experimental measurements in wind tunnel [17], a disk of $80\,m$ immersed in an ABL having a roughness height $z_0 = 0.02m$ is analysed. The actuator disk is located downstream of the inlet computational domain at a distance of $4D$. This distance is sufficient for the turbulence to reach an equilibrium state.

4. Results

4.1. ABL simulations

Figs.2 and 3 shows the vertical profiles of velocity and turbulent intensity at various longitudinal locations downstream of the inlet. The horizontal homogeneity is almost perfectly obtained, indicating that the proposed top and bottom BCs behavior is adequate. There is a noticable, but small, acceleration near the ground and a deceleration near the top region of the computational domain. Transition from acceleration toward deceleration occurs at about one third of the computational domain height.

The frequency weighted longitudinal velocity energy spectra in Fig.4 shows the presence of large-scale structures. The theoretical slope of $-2/3$ is also observed at high frequencies with high-frequency deviations most likely caused by contamination from the LES sub-scale filtering.

Fig.5 shows the contours of instantaneous velocity module in a vertical plane. The complete plane of the computational domain is presented showing that the proposed BC treatment prevents any spurious oscillation in the solution near the boundaries. When using pure

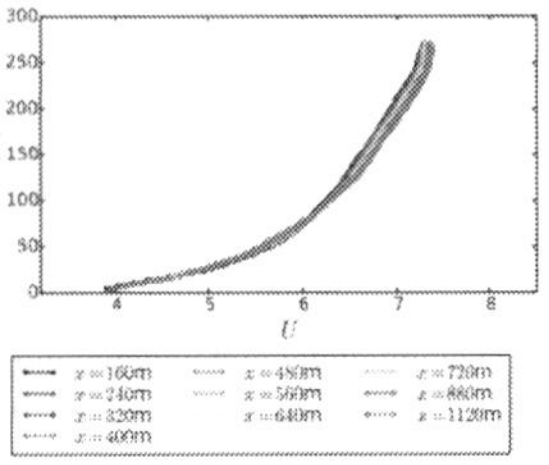

Figure 2. ABL velocity profiles for $z_0 = 0.02\,m$

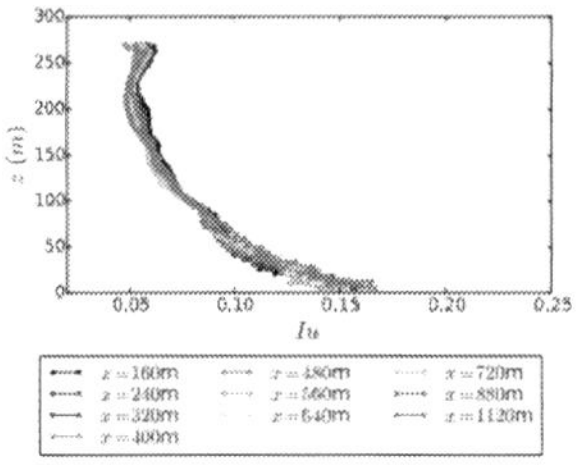

Figure 3. ABL turbulent intensity profiles for $z_0 = 0.02\,m$

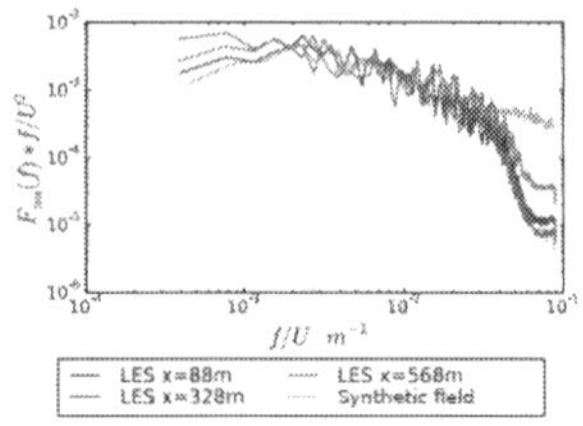

Figure 4. ABL longitudinal velocity energy spectra for $z_0 = 0.02\,m$

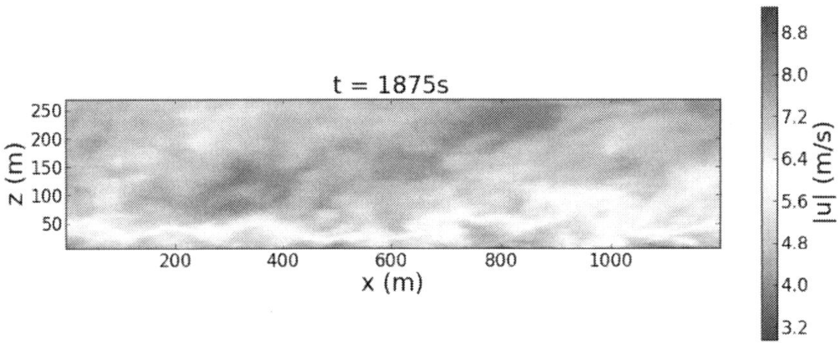

Figure 5. ABL instantaneous velocity field in median vertical plane for $z_0 = 0.02\,m$

fixedValue conditions the observed oscillations were stationnary, aligned with the boundaries on the whole outflow areas, and could extend up to about 5-10 cells in the computational domain. Their amplitude was of the order of 10% the average flow velocity, so not particularly large but definitely unphysical. Such spurious oscillations would clearly be seen on instantaneous as well as average velocity contours.

4.2. Wake simulations

4.2.1. Mean profiles The vertical profiles of the longitudinal velocity, Fig. 6 are consistent with the ABL velocity profiles previously presented. They exhibit an increase in velocity in the profile outside of the WT wake, which is caused mainly by the relatively small height of the computational size with respect to the disk diameter. This choice of computational domain height is however deemed acceptable, since the main goal of this work is to study the wake meandering. This decision is further supported by the results of a study of the sensitivity to the domain size and to the mesh resolution. The results of this study are reported in [30].

The values of velocity at the first cell near the ground at position z = 5 m are almost identical for all streamwise locations. This is due to the implemented boundary conditions, which let the tangential velocity be computed from the LES solution when the flow is leaving the computational domain, but set the velocity vector to values prescribed by the stochastic generator when the flow is entering. The value of the longitudinal velocity is effectively fixed 50 % of the time. In fact the probability density of the synthetic velocity normal component at the

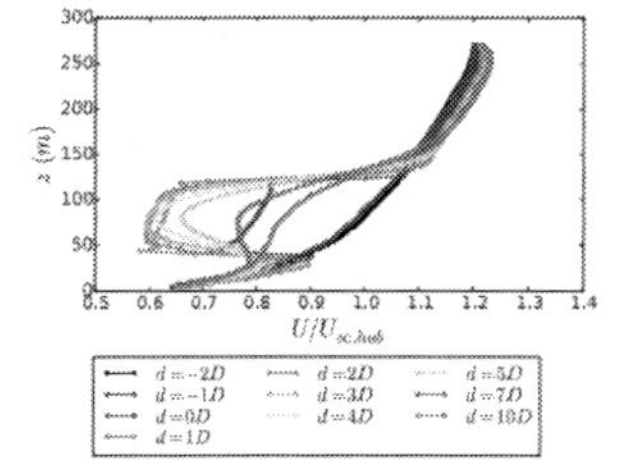

Figure 6. Longitudinal velocity profiles for $a = 0.19$ and $z_0 = 0.02\,m$

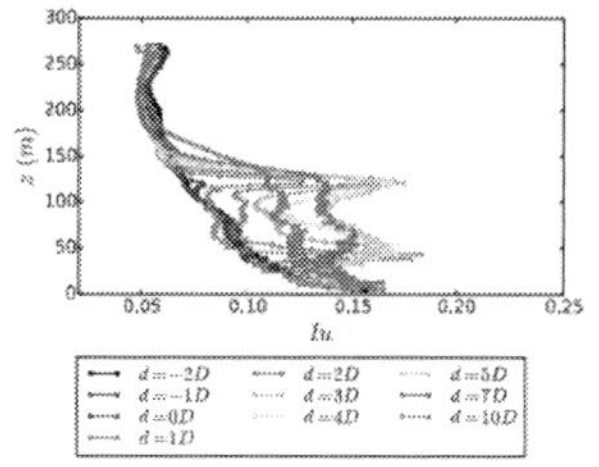

Figure 7. Turbulent intensity profiles for $a = 0.19$ and $z_0 = 0.02\,m$

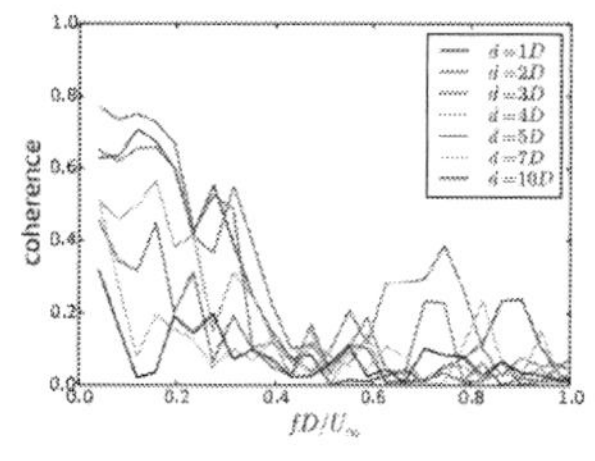

Figure 8. Wake coherence for $a = 0.19$ and $z_0 = 0.02\,m$

bottom boundary of the computational domain is centered on zero since the average velocity vector is tangential to the ground. This behavior will result in the prescription of the average velocity at the bottom boundary close to the average value of the synthetic field. The choice of the bottom face location of the computational domain ($z = 5\,m$ in this work) is critical. The value $z = 5\,m$ has been chosen since this elevation is of the order of magnitude of the mesh resolution, and the assumption of fully turbulent flow at the bottom boundary is valid only above a certain height. The results presented in Section 4.1 clearly show that this implemented boundary condition near the ground yield very good ABL velocity profiles close to the ground when using the traditional Smagorinsky model. When applying wall model at the ground, proper simulation of turbulence using LES near the ground needs a fine mesh, specific law of the wall and finally a turbulent model ables to handle the near wall region which is not the case with the traditional Smagorinsky model used in this work.

The implemented porous disk modeled is having a non-uniform pressure jump calculated from instantaneous local velocity. This is clearly not identical to an actuator disk having a uniform pressure. The current implementation is expected to be closer to an actual turbine, in that the average wind speed varies with height which produces a change in the average pressure jump.

Fig. 7 presents the vertical profile of the streamwise turbulence intensity at various downstream locations. The actuator disk is located at $x = 0D$. The value of the turbulence intensity in the middle part of the near wake is very close to the upstream undisturbed ABL value. This is to be expected since the main turbulence production mechanism is associated with velocity gradients which are significant only at the wake edges in the near wake region.

4.2.2. Meandering It is important to note that meandering is considered in this work as the time variation of the location of the wake velocity deficit centre caused by the large scale structures in the incident ABL. The unsteady simulation of the actuator disk immersed in the ABL allows the observation of wake time evolution. Fig. 9 presents the instantaneous flowfield in a vertical plane in which the turbine axis of rotation lies. It shows a variation of the wake trajectory which is typically associated with meandering as defined previously. The procedure adopted here to determine the instantaneous position of the wake is the same as the one described in Ref. [17]. It is different than the 2D Gaussian fitting procedure described in [19] but should yield similar results. It allows the calculation of wake position temporal series along transversal coordinates. The effets of averaged shear on velocities are removed when calculating z wake position in order to remove the effects of the undisturbed ABL velocity profile.

Fig. 8 shows coherence between the transverse velocity measured at the center of the actuator disk and the instantaneous location of the wake centerline on the y axis for different distances d on the x axis downstream of the rotor. The usual methods to improve the statistics have

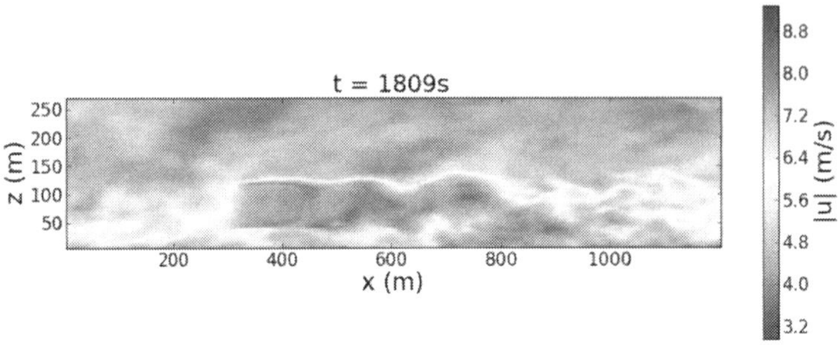

Figure 9. Wake instantaneous velocity field in median vertical plane for $a = 0.19$ and $z_0 = 0.02\,m$

been used where appropriate. The main reason why the spectra of Fig. 8 are noisy is that there are no symmetries in the flow that can be exploited to perform spatial averaging of the turbulence spectra measured in the wake of the turbine. Spectra from Fig. 8 looks noisier than those shown in Fig. 4. This is mainly associated the the scales which are of log type in Fig. 4 and linear in Fig. 8. The behavior of computed coherence shown in Fig.8 is very similar to the one computed from the observations [17] with high values of coherence at low frequencies and values close to zero at high frequencies. This change in behavior is occuring at a non-dimensional frequency around of $0.3 - 0.4$ (corresponding to a length scale of less than 3D), which is again in agreement with the wind-tunnel observations [17]. This frequency correponds to a wavelength of about three times the diameter of the rotor. The coherence is decreasing with the dowstream distance from the disk. The curves presented in Fig.8 are more noisy than the coherence obtained from the measurements. This is attributed to the reduced total number of computed samples/datapoints/timesteps in comparison with the experimental measurements. A better statistical convergence could certaintly be achieved using simulations with longer time periods, but considering the non-linear convergence rate with respect to the number of samples/datapoints/timesteps, the cost of these simulations becomes rapidly prohibitive. The physical time simulated to obtain the coherence is approximately 1 hr, which is already a long period in turbulent ABL simulations.

4.2.3. Validation The numerical results have been compared to experimental observations presented in Ref. [17]. More specifically, the following characteristics are compared:

- U_{2D}/U_∞ : ratio between the average wake centre velocity 2D dowstream and the upstream undisturbed velocity at hub height.
- U_{5D}/U_∞ : ratio between the average wake centre velocity 5D dowstream and theupstream undisturbed velocity at hub height.
- $\sigma_{wake,y}/D$: standard deviation of the wake centreline horizontal location at 5D downstream.
- $\sigma_{wake,z}/D$: standard deviation of the wake centreline vertical location at 5D downstream.

Numerical and experimental values are reported in Table 2.

Regarding non-dimensional velocities, the agreement with the measurements is almost perfect. However, the error on the standard deviation of the wake centre horizontal position is more than 20%. Nevertheless, these results shows that the meandering is qualitatively reproduced. The error on the vertical meandering is much larger. It is expected that some improvement

Case	U_{2D}/U_∞	U_{5D}/U_∞	$\sigma_{wake_y,5D}/D$	$\sigma_{wake_z,5D}/D$
exp. D80	0.604 (ref)	0.659 (ref)	0.17 (ref)	0.14 (ref)
D80	0.603 (0%err)	0.660 (0%err)	0.132 (22%err)	0.047 (66%err)

Table 2. Experimental/numerical comparisons

could be brought to these simulations by improving the synthetic turbulence generator: the spectral tensor model used in this work quite severely underestimates the fluctuations of the vertical components of the velocity, which we think is responsible for the large discrepancy found for the standard deviation of the vertical meandering. However, the integral scales from the numerical simulations and the measurements are similar (around $200m$ at $2.5D$). It is therefore suspected that there is a disagreement between the numerical and experimental turbulent structure lifespans.

One of the source of discrepancy is suspected to be the characteristics of the incident ABL turbulence. While the turbulence intensities are identical between experiments and simulations, the spectral tensor implemented in the stochastic generator is relatively simple and consequently does not reproduce some anisotropic characteristics of the wind tunnel turbulence. Using the spectral tensor of Mann [29] may improve the results of the simulations. Despite these limitations, the results shown are still useful given that there are relatively few actual results available in the literature on the subject of wake meandering. Moreover, it is believed that the proposed methodology is physically sound and that the current limitations can all be addressable in the near future.

5. Conclusion

This paper presents an application of stochastic temporal velocity series for the LES simulation of wind turbine wakes in atmospheric boundary layer. The actuator disk is modeled by means of a pressure jump dependent on the local unsteady velocity and the prescribed induction factor, which is the similar numerical version of the porous disk used experimentally.

The fundamental role of boundary conditions based on a directional hybrid formulation, which adopts the behavior of a Neumann or Dirichlet type boundary conditions depending on the direction of the flow at the boundaries, has been demonstrated. It is shown that this type of BC ensures at the same time a strict control of the mass flow rate in the computational domain, while preventing spurious spatial oscillations near the boundaries with locally exiting flow.

The study results show that the method is effective in producing longitudinally homogeneous velocity profiles in the entire computational domain. The turbulent intensity profiles are also exhibiting a relatively fully developed behavior.

The proposed method allows to model realistic atmospheric flows using LES, without requiring a precursor simulation which can be costly in term of computing resources and challenging to implement. Such a method could be useful in many engineering applications in the ABL.

The instantaneous horizontal or vertical position of the wake on a transverse section is determined from the instantaneous velocity deficit profile whose values are used to calculate a center of gravity of the position of the wake. The result is a time series of the transverse position of the wake center. Coherence between the wake velocity time series and the upstream transverse velocity reveals a very strong correlation of low frequency which decreases with increasing frequency and reaches a non-significant value for any wavelength of less than about two times the diameter of the rotor.

The relatively large values of the measured coherence suggest that the real-time forecasting of meandering from upstream transverse velocities is a realistic prospect which could be done, for example, from global stress measurements on the mast or wind turbine rotor.

The comparison with experimental results shows that the computed velocity deficits are quite satisfactory. The simulated horizontal meandering qualitatively reproduces the observed experimental behavior. The agreement of the numerical vertical meandering is less convincing. Improved synthetic turbulence from the stochastic generator should increase the agreement.

References

[1] Vermeer L J, Sørensen J N and Crespo A 2003 *Progress in Aerospace Sciences* **39** 467 – 510 ISSN 0376-0421
[2] Ainslie J F 1988 *J. Wind Engineering and Industrial Aerodynamics* **27** 213 – 224 ISSN 0167-6105
[3] Jensen N O 1983 A note on wind generator interaction
[4] Aubrun S 2012 *Int. J. Engineering Systems Modelling and Simulation* **4** 2–10
[5] Réthoré P E 2009 *Wind Turbine Wake in Atmospheric Turbulence* Ph.D. thesis Aalborg University
[6] El Kasmi A 2008 *Simulation de l'écoulement dans un parc d'éolienne situées sur un terrain accidenté* Ph.D. thesis ÉTS
[7] Leclerc C 1998 *Simulation numérique de l'écoulement tridimensionnel turbulent dans un parc éolien* Master's thesis École Polytechnique de Montréal
[8] Meyers J and Meneveau C 2012 *Wind Energy* **15** 305–317 ISSN 1099-1824
[9] Breton S P, Sibuet Watters C, Masson C, Gomez-Iradi S and Mundate X 2012 *Int. J. Engineering Systems Modelling and Simulation* **4** 11–26
[10] Larsen G C, Madsen H A, Thomsen K and Larsen T J 2008 *Wind Energy* **11** 377–395 ISSN 1099-1824
[11] Larsen T J, Madsen H A, Larsen G C and Hansen K S 2013 *Wind Energy* **16** 605–624
[12] Trujillo J J and Kühn M 2009 Adaptation of a lagrangian dispersion model for wind turbine wake meandering *Proc. of the EWEA Conf.* (Marseille, France: EWEA)
[13] Bingöl F, Mann J and Larsen G C 2010 *Wind Energy* **13** 51–61 ISSN 1099-1824
[14] España G, Aubrun S, Loyer S and Devinant P 2011 *Wind Energy* **14** 923–937 ISSN 1099-1824
[15] Taylor G J, Milborrow D J, McIntosh D N and Swift-Hook D T 1985 Wake measurements on the Nibe windmills *Proc. of 7th BWEA Wind Energy Conf.* (Oxford, UK: BWEA)
[16] España G 2009 *Étude expérimentale du sillage lointain des éoliennes à axe horizontal au moyen d'une modélisation simplifiée en couche limite atmosphérique* Ph.D. thesis Université d'Orléans
[17] Muller Y A, Aubrun S and Masson C 2015 *Experiments in Fluids* **Accepted**
[18] Espana G, Aubrun S, Loyer S and Devinant P 2012 *J. Wind Engineering and Industrial Aerodynamics* **101** 24 – 33 ISSN 0167-6105
[19] Trujillo J J, Bingöl F, Larsen G C, Mann J and Kühn M 2011 *Wind Energy* **14** 61–75 ISSN 1099-1824
[20] Medici D and Alfredsson P H 2006 *Wind Energy* **9** 219–236
[21] Leclerc C, Masson C, Ammara I and Paraschivoiu I 1999 *Wind Engineering* **23** 279–294
[22] Mikkelsen R 2003 *Actuator Disc Methods Applied to Wind Turbines* Ph.D. thesis Technical University of Denmark
[23] Troldborg N, Sørensen J N and Mikkelsen R 2009 *Actuator Line Modeling of Wind Turbine Wakes* ISBN 978-87-89502-80-9
[24] Keck R E, Mikkelsen R, Troldborg N, de Maré M and Hansen K S 2014 *Wind Energy* **17** 1247–1267 ISSN 0022-1120
[25] Froude R 1889 *Trans. Inst. Naval Architects* **30** 390
[26] Lee S, Lele S K and Moin P 1992 *Physics of Fluids A: Fluid Dynamics* **4** 1521 ISSN 08998213
[27] Klein M, Sadiki A and Janicka J 2003 *Journal of Computational Physics* **186** 652–665 ISSN 00219991
[28] Davidson L and Bilson M 2006 *Int. J. Heat and Fluid Flow* **27** 1028–1042 ISSN 0142-727X
[29] Mann J 1994 *J. Fluid Mechanics* **273** 141–168 ISSN 0022-1120
[30] Muller Y A 2014 *Étude du méandrement du sillage éolien lointain dans différentes conditions de rugosité* Ph.D. thesis Université d'Orléans
[31] OpenFOAM-Foundation URL `http://www.openfoam.org/`
[32] Issa R I, Gosman A D and Watkins A P 1986 *J. Computational Physics* **62** 66–82
[33] Troldborg N, Sørensen J N and Mikkelsen R 2007 *J. Physics: Conf. Series* **75** 012063 ISSN 1742-6596

A new analytical model for wind farm power prediction

Amin Niayifar and Fernando Porté-Agel

École Polytechnique Fédérale de Lausanne (EPFL), Wind Engineering and Renewable Energy Laboratory (WIRE), EPFL-ENAC-IIE-WIRE, 1015 Lausanne, Switzerland

E-Mails: Amin.niayifar@epfl.ch, Fernando.porte-agel@epfl.ch

Abstract. In this study, a new analytical approach is presented and validated to predict wind farm power production. The new model is an extension of the recently proposed by Bastankhah and Porté-Agel for a single wake. It assumes a self-similar Gaussian shape of the velocity deficit and satisfies conservation of mass and momentum. To estimate the velocity deficit in the wake, this model needs the local wake growth rate parameter which is calculated based on the local turbulence intensity in the wind farm. The interaction of the wakes is modeled by use of the velocity deficit superposition principle. Finally, the power curve is used to estimate the power production from the wind turbines. The wind farm model is compared to large-eddy simulation (LES) data and measurments of Horns Rev wind farm for a wide range of wind directions. Reasonable agreement between the proposed analytical model, LES data and measurments is obtained. This prediction is also found to be substantially better than the one obtained with a commonly used wind farm wake model.

1. Introduction

The energy field nowadays has reached a critical situation. Due to the limited fossil-based fuel resources and their negative impact on the environment and the climate, there is a growing urge to find alternative sources of energy which are clean, sustainable and cost-effective. Renewable energies play an important role in achieving these objectives. With the increasing global interest in renewable energies, wind energy is witnessing a continuous growth at an average annual rate of approximately 25%, as stated by the Global Wind Energy Council [1].

In wind farms, turbine wake effects are responsible for significant power losses, which can be as large as 25% of the total power output [2]. Accurate wind farm power prediction is of vital importance in order to minimize the impact of wakes on the downwind turbines. Various experimental, numerical and analytical studies have been carried out to investigate turbine wake effects. Accurate wake prediction and understanding could be achieved by using numerical methods such as large-eddy simulation (LES) and cutting edge experimental techniques. Although both experimental and numerical approaches provide accurate results, analytical models are preferred to use for wind farm optimization purposes due to their simplicity and low computational cost [3].

Several wind turbine wake models have been proposed to predict velocity deficit inside wind farms [4-8]. One of the most common wake models, which has been extensively used in the literature (e.g., [9]) as well as in commercial softwares (e.g., [10-14]), is the one proposed by Jensen [5]. It considers a top-hat shape for the normalized velocity deficit and states:

$$\frac{\Delta U}{U_\infty} = \frac{U_\infty - U_w}{U_\infty} = (1 - \sqrt{1 - C_T})/\left(1 + \frac{2k_{wake}x}{d_0}\right)^2, \tag{1}$$

where U_∞ is the undisturbed velocity, U_w the wake velocity, C_T the thrust coefficient of the turbine, k_{wake} the wake spreading parameter, d_0 the wind turbine diameter and x the distance behind the turbine. It should be mentioned that this model was derived using only mass conservation [8].

Frandsen et al. [7] later on proposed the following equation considering a top-hat shape for the velocity deficit and applying the conservation of mass and momentum to a control volume around the turbine:

$$\frac{\Delta U}{U_\infty} = \frac{1}{2}\left(1 - \sqrt{1 - 2\frac{A_0}{A_w}C_T}\right), \tag{2}$$

where A_0 is the area swept by the wind turbine blades and A_w is the cross-sectional area of the wake. Although these models are commonly used in the literature and commercial softwares, they tend to overestimate power prediction in the full-wake conditions and underestimate it in the partial-wake conditions due to the top-hat velocity deficit assumption.

Recently Bastankhah and Porté-Agel showed that a Gaussian wake model can provide substantially better results in the full-wake and partial-wake conditions [8]. They applied mass and momentum conservation to a control volume around one turbine, assuming a self-similar Gaussian profile for the velocity deficit, and derived the following expression:

$$\frac{\Delta U}{U_\infty} = \left(1 - \sqrt{1 - \frac{C_T}{8(k^*x/d_0 + \varepsilon)^2}}\right) \times exp\left(-\frac{1}{2(k^*x/d_0 + \varepsilon)^2}\left\{\left(\frac{z - z_h}{d_0}\right)^2 + \left(\frac{y}{d_0}\right)^2\right\}\right), \tag{3}$$

where x, y and z are streamwise, spanwise and vertical coordinates, respectively, and z_h is the hub height level. k^* denotes to the wake growth rate which is a function of thrust coefficient and local turbulence intensity [8]. Bastankhah and Porté-Agel also proposed the following expression for ε:

$$\varepsilon = 0.2\sqrt{\beta}, \tag{4}$$

where β is a function of C_T which is defined as:

$$\beta = \frac{1}{2}\frac{1 + \sqrt{1 - C_T}}{\sqrt{1 - C_T}}. \tag{5}$$

Inside a wind farm, depending on the direction of the wind, several wind turbines operate in the wakes of many up-wind turbines. To this effect, wind farm models predict wake effects using single wake models and apply superposition principles to deal with multiple wakes. Lissaman [4] used an analogy between the expansion of wind turbine wakes and the dispersion of point-source pollutants (e.g., from smoke stacks) in the atmospheric boundary layer to propose a model for the cumulative velocity deficit based on linear superposition of velocity deficits. The model states:

$$U_i = U_\infty - \sum_k (U_\infty - U_{ki}), \tag{6}$$

where U_i is the velocity at turbine i and U_{ki} is the wake velocity of turbine k at turbine i considering only those turbines whose wakes interact with turbine i. Later, Katic et al. [5] used the superposition of energy deficits instead of velocity deficits, to take into account the interaction of multiple wakes as follows:

$$U_i = U_\infty - \sqrt{\sum_k (U_\infty - U_{ki})^2}, \tag{7}$$

they assumed that the kinetic energy deficit of a multiple wake is equivalent to the sum of the energy deficits for individual wake at the down-wind position. Voutsinas et al. [6] followed the same approach as Katic et al. [5], but in order to estimate the energy deficit of each wake, they considered the difference between the inflow velocity at the turbine and the wake velocity as follows:

$$U_i = U_\infty - \sqrt{\sum_k (U_k - U_{ki})^2}. \tag{8}$$

Several studies (e.g., [15, 16]) have shown that an enhancement of turbulence intensity occurs inside wind farms with respect to the turbulence level of the incoming wind. Moreover, Bastankhah and Porté-Agel [8] showed that wake growth rate increases as turbulence intensity increases. To these effects, we expect that the assumption of constant wake growth rate, which has been extensively used in the common analytical wind farm models, is not realistic and instead in this paper, we propose an empirical equation for the local wake growth rate which is based on the local turbulence intensity. There have been a few research studies to model added turbulence intensity inside the wind farm [17-19]. These models usually rely on the thrust coefficient of wind turbines and ambient turbulence intensity to calculate added turbulence intensity at the wind turbine hub height. Added turbulence intensity is defined as:

$$I_+ = \sqrt{I_{wake}^2 - I_0^2}, \tag{9}$$

where I_{wake} is the turbulence intensity in the wake and I_0 the ambient turbulence intensity. Quarton and Ainsile [17] proposed the following empirical expression to predict the added turbulence intensity generated by a wind turbine:

$$I_+ = 4.8 C_T^{0.7} I_0^{0.68} (x/x_n)^{-0.57}, \tag{10}$$

where x_n is the length of the near-wake region which is defined as [20]:

$$x_n = \frac{\sqrt{0.214 + 0.144m}(1 - \sqrt{0.134 + 0.124m})}{(1 - \sqrt{0.214 + 0.144m})\sqrt{0.134 + 0.124m}} \frac{r_0}{(dr/dx)}, \tag{11}$$

where $m = \frac{1}{\sqrt{1-C_T}}$, $r_0 = d_0\sqrt{\frac{m+1}{2}}$ and dr/dx is defined by the following expression:

$$dr/dx = \sqrt{\left(\frac{dr}{dx}\right)_a^2 + \left(\frac{dr}{dx}\right)_m^2 + \left(\frac{dr}{dx}\right)_\lambda^2}, \tag{12}$$

where $\left(\frac{dr}{dx}\right)_a = 2.5I_0 + 0.005$, $\left(\frac{dr}{dx}\right)_m = \frac{(1-m)\sqrt{1.49+m}}{9.76(1+m)}$ and $\left(\frac{dr}{dx}\right)_\lambda = 0.012B\lambda$. B is the number of blades and λ is the tip speed ratio. Later, Hassan [18] suggested the following expression for the added turbulence intensity:

$$I_+ = 5.7C_T^{0.7}I_0^{0.68}(x/x_n)^{-0.96}\,. \tag{13}$$

Crespo and Hernandez [19], based on their numerical data, suggested the following empirical equation for the ranges of parameters, $5 < x/d_0 < 15$, $0.07 < I_u < 0.14$ and $0.1 < a < 0.4$ where a is the induction factor:

$$I_+ = 0.73a^{0.8325}I_0^{0.0325}(x/d)^{-0.32}\,. \tag{14}$$

In section 2, the analytical wind farm model is presented. Then, a description of case study (Horns Rev wind farm) is presented in section 3. The results are discussed and compared with LES, observed data and a commonly used analytical wind farm wake model. in section 4. The conclusion and future research are presented in section 5.

2. Description of the new analytical wind farm model

The new wind farm model applies the superposition of the velocity deficit to the recently proposed Gaussian wake model by Bastankhah and Porté-Agel [8]. Next, the details of the new wind farm model and its implementation are given.

2.1. *Analytical Model for the velocity deficit*

For each turbine in the farm, we apply the Gaussian wake model recently developed by Bastankhah and Porté-Agel [8] (equations (3), (4) and (5)). For single wakes, this model assumes a self-similar Gaussian distribution for the velocity deficit, which conserves mass and momentum. Since in this paper wind turbines operate in a range that thrust coefficient is approximately constant (Figure 3), we assume that k^* is only dependent to local turbulence intensity. Figure 1 shows the wake growth rate behind a V-80 turbine simulated with LES for a wide range of turbulence intensities of the incoming boundary layer flow at hub height [8]. Based on aforementioned numerical data we propose the following empirical equation for the range of conditions considered in this study ($0.065 < I_u < 0.15$) to calculate the wake growth rate:

$$k^* = 0.3837I + 0.003678, \tag{15}$$

where I is the local turbulence intensity at the rotor hub height.

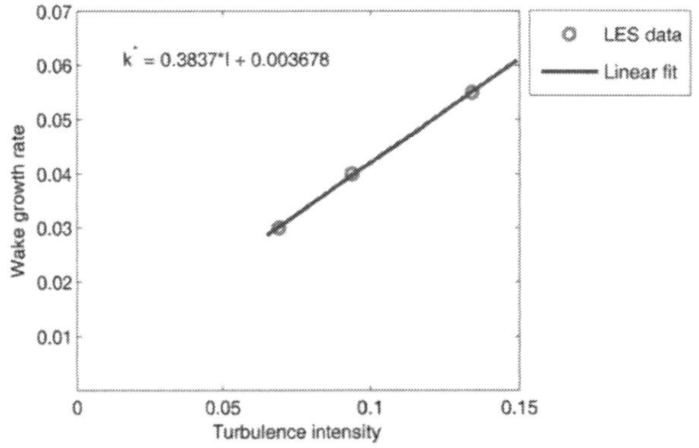

Figure 1. Wake growth rate for V-80 turbine in boundary layer flow with different turbulence intensities at hub height

The problem of multiple wake interactions is solved by applying a new approach of velocity deficit superposition. Previously Lissaman [4] used the velocity deficit superposition. He considered the difference between the undisturbed velocity and the wake velocity (equation (6)) which resulted in an overestimation of the velocity deficit [3]. Here, to calculate the velocity deficit, we instead propose to use the difference between the inflow velocity at the turbine and the wake velocity which is defined as:

$$U_i = U_\infty - \sum_k (U_k - U_{ki})\,. \tag{16}$$

As mentioned before, the single wake model conserves the mass and momentum. It is of vital importance that the velocity deficit superposition conserves mass and momentum. Lissaman [4] justified the linear superposition of the wakes using an analogy between turbine wakes and pollution plumes, whose Gaussian concentration distribution can be superimposed due to the linearity of the process. Since pollutant mass is conserved by applying superposition, in the same way superposition of velocity deficit conserves the linearized momentum deficit.

2.2. *Turbulence intensity model*

For the local turbulence intensity, we use a top-hat distribution with the wake diameter of 4σ, which has been derived empirically based on LES data [15]. σ is the standard deviation of the Gaussian-like velocity deficit and it is defined [8] as:

$$\sigma/d_0 = k^* x/d_0 + \varepsilon\,. \tag{17}$$

The enhancement of turbulence intensity for individual turbines is calculated from equation (14). Then, using equation (9) the local turbulence intensity can be found.

Extensive numerical and experimental studies have revealed that the turbulence intensity inside a wind farm increases and quickly reaches an equilibrium after 2-3 rows [15, 21]. A previous study by Frandsen and Thøgersen [22] has also shown that only the effect of neighboring turbines is important to predict the turbulence intensity in the wake of given turbine. To this effect, for every turbine, we consider only the added turbulence intensity resulting from the closeset upstream turbine whose wake has the most significant impact. It is defined as:

$$I_{+\,j} = \max\left(\frac{A_w 4}{\pi d_0{}^2} I_{+\,kj}\right), \tag{18}$$

where $I_{+\,j}$ is the added turbulence intensity at the turbine j, A_w the intersection between the wake (using equation (17)) and the rotor area, and $I_{+\,kj}$ the added turbulence intensity induced by the turbine k at the turbine j.

2.3. *Power prediction*

Power generated by each turbine is predicted using the power curve, which gives the power production as a function of wind speed. Here we use the data available for Vestas V-80 wind turbines and we fit a fifth degree polynomial.

3. Case description

The Horns Rev offshore wind farm is selected as a case study, since LES predictions [15, 23] and measurements [10, 24] are available to evaluate the performance of the new analytical wind farm model. The wind farm is located in the North Sea around 15 km off the westernmost point of Denmark. It has 160 MW total rated power capacity and consists of eighty Vestas V-80 wind turbines within an area of about 20 km^2. Each turbine has a rotor diameter of $d = 80\ m$ and a hub height of $H_{hub} = 70\ m$ (above sea level). Figure 2 shows a schematic of the Horns Rev wind farm layout where the wind turbines are positioned in a rhomboid shape with a minimum spacing of 7 rotor diameter between two consecutive turbines within 8 columns and ten rows (approximately 7° turned counterclockwise from North-South).

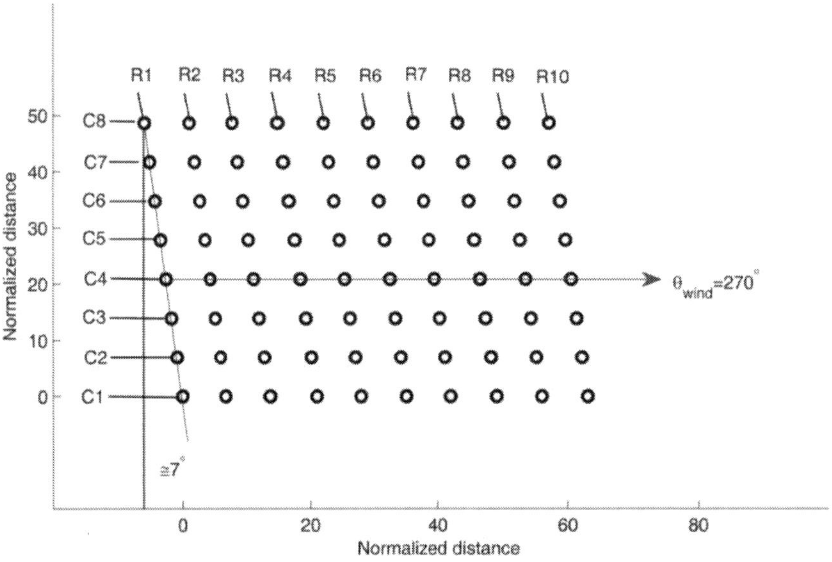

Figure 2. Layout of Horns Rev wind farm. Distances are normalized by the rotor diameter $d = 80$.

The analytical wind farm model needs wind turbine characteristics as an input (power curve and thrust coefficient) to predict the wind farm power. Here, we use the data that was used to validate LES by Wu and Porté-Agel [23]. Figure 3 shows the previously mentioned power curve and thrust coefficient for different wind speeds. The surface roughness for the farm is considered as $z_0 = 0.05\ m$. In this paper, the inflow wind condition is defined by a turbulence intensity of 7.7% at hub height and an average velocity of 8 ms^{-1} at the same height.

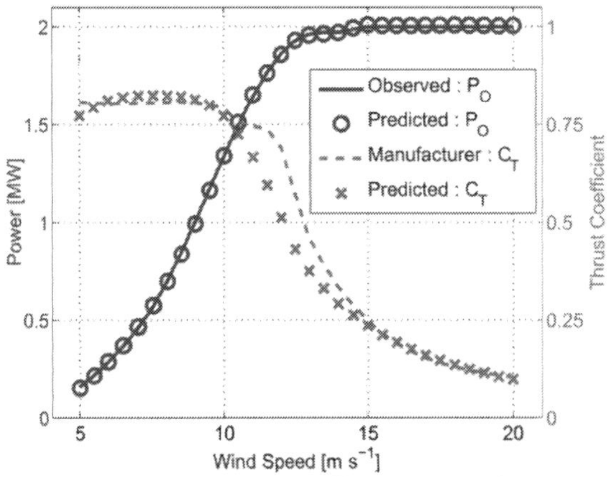

Figure 3. Measured and simulated power curve and thrust coefficient curves with respect to different range of wind speeds for the Vestas V-80 2 MW turbine, (Source: Wu and Porté-Agel, 2014)

4. Results and discussion

In this section, the new analytical wind farm model is used to simulate the turbine wakes and associated power losses in Horns Rev. The results are compared with available power measurements [10, 24] and LES data [15, 23], as well as predictions from an analytical top-hat wake model. The top-hat wake model is based on the model proposed by Katic et al [5]. In this top-hat wake model (e.g, WAsP), the wake growth rate is set to 0.04, which is the recommended value for offshore applications [25].

Figure 4 shows the simulated total normalized power output from the Horns Rev wind farm obtained from the new proposed model, LES [15] and the top-hat model [5] for a wide range of wind directions (from 173° to 353°). This allows the evaluation of the model over a variety of full-wake and partial-wake conditions. In this paper, the simulated power is normalized by the power of an equivalent number of stand-alone wind turbines operating in the same incoming wind condition. As it can be seen in Figure 4, while the top-hat model under predicts the normalized power, the proposed analytical model shows a good agreement with LES. A significant power loss of approximately 30% is seen when the wind farm is exposed to wind direction angles (173°, 270° and 353°) corresponding to full-wake conditions with small streamwise distances between consecutive wind turbines. Also, several local maxima (e.g., 185° and 340°) can be distinguished which correspond to large streamwise distances between turbines.

In order to evaluate the performance of different multiple wake superposition approaches, next we compare the normalized power output simulated with LES with the one obtained with the new analytical model using both velocity deficit superposition and energy deficit superposition. Figure 5 shows the normalized power output as a function of turbine row (averaged over columns 2, 3 and 4) in the wind farm. Here, for the calculation of velocity and energy deficits, we use the difference between the inflow velocity at the turbine and the wake velocity. In the case of using the difference between the undisturbed boundary-layer velocity and the wake velocity, due to the large number of turbine rows, it leads to negative velocities. As it can be seen in the Figure 5, velocity deficit superposition shows a good agreement with LES, while energy deficit superposition overestimates the normalized power by about 15%.

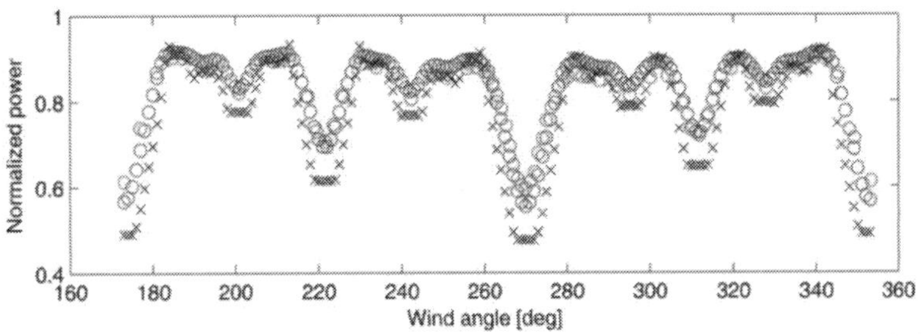

Figure 4. Distribution of the normalized Horns Rev wind farm power output obtained with the new analytical model and LES for different wind directions. Red circles correspond to LES, blue circles represent new analytical model and black crosses show top-hat model.

Figure 6 presents the normalized power output for three wind sectors (i.e., 270 ±5°, ±10° and ±15°) predicted by the new analytical model and WAsP, as well as the measurements [24]. From this figure it is clear that WAsP tends to overestimate the velocity deficit while the proposed model shows a very good agreement with the measurements. This result highlights the fact that assuming a top-hat shape for velocity deficit can lead to significant error in both full-wake and partial-wake conditions.

As mentioned before, the wake growth rate is a function of the local turbulence intensity. Therefore, in the analytical wind farm model we should predict the local turbulence intensity in the farm. Moreover, it is of vital importance to investigate the level of turbulence intensity in order to determine the impact of fatigue loading on the turbines. Figure 7 presents the level of turbulence intensity at hub height in the farm simulated using the models of Quarton [17], Hassan [18] and Crespo [19] as well as the LES predictions. From this figure it is clear that Quarton and Hassan models predict turbulence intensities that are much larger than both Crespo model and LES. As shown in the figure, Crespo model, which is used in the new analytical wind farm model, shows an acceptable agreement with LES.

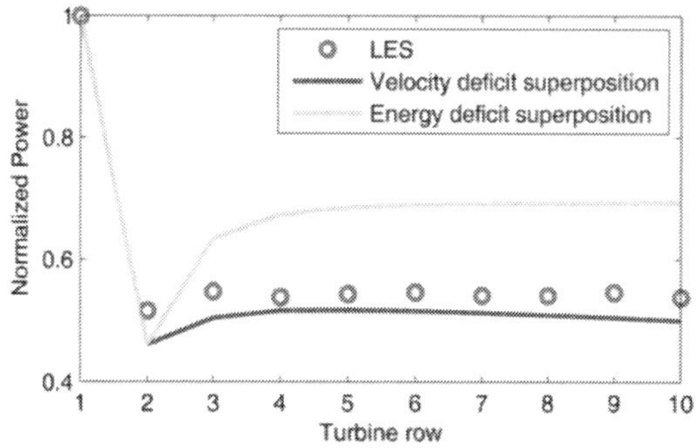

Figure 5. Comparison of the simulated power output for $\theta_{wind} = 270°$ by use of energy and velocity deficit superposition with LES data

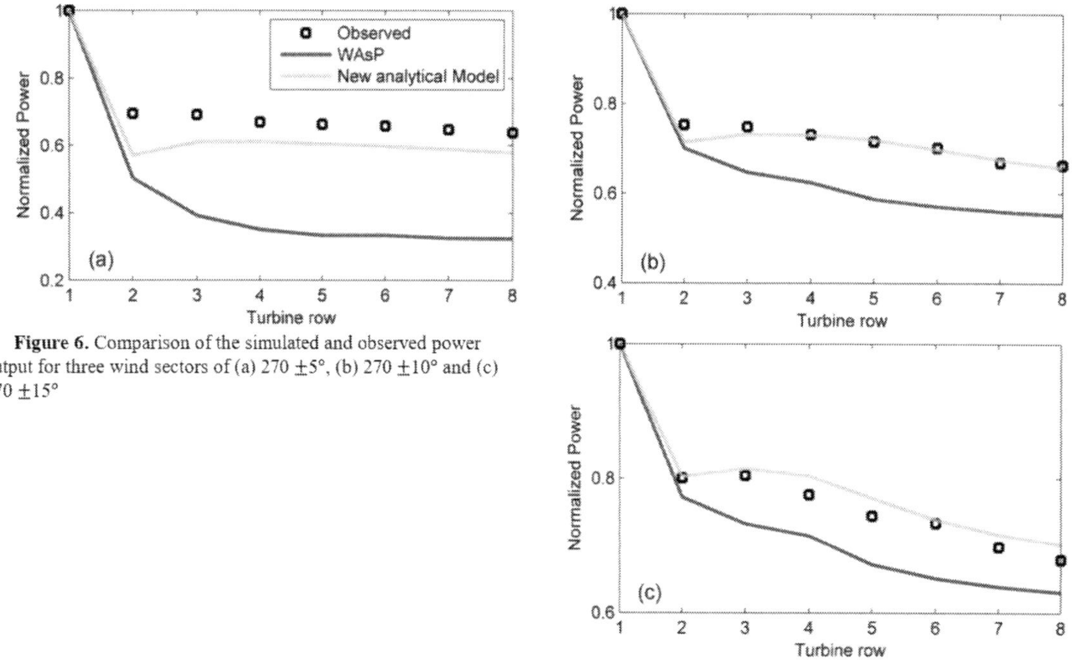

Figure 6. Comparison of the simulated and observed power output for three wind sectors of (a) 270 $\pm 5°$, (b) 270 $\pm 10°$ and (c) 270 $\pm 15°$

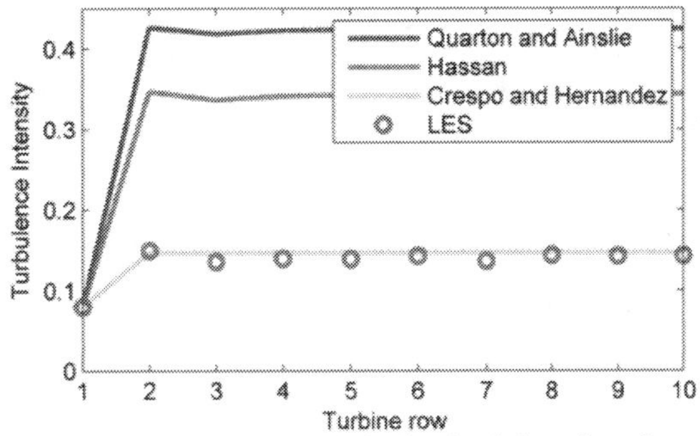

Figure 7. Comparison of the observed and simulated turbulence intensity at hub height

5. Conclusion

A new analytical model is proposed to estimate the power output of a wind farm. Individual wakes are modeled using a self-similar Gaussian distribution for the velocity deficit, which conserves mass and momentum. Wake velocity distribution is predicted by specifying local wake growth rate. In this study, we propose an empirical expression to calculate the wake growth rate based on the local turbulence intensity. The local turbulence intensity is predicted inside the farm by use of empirical models. A velocity deficit superposition is used to consider the effect of multiple wakes. The power curve is used to predict the power production from the wind turbines.

We select the Horns Rev wind farm as a case study since LES predictions and measurements are available for a wide range of inflow conditions. The comparison with the LES and observed data reveals that the obtained power prediction with the new analytical model is in reasonable agreement with the LES and experimental data. Also, it is observed that the power prediction of the new proposed model is substantially better than the one obtained by WAsP.

Future research will focus on developing an expression to calculate wake growth rate for a wider range of inflow conditions. Furthermore, the new analytical model can be used to design alternative layouts so as to maximize wind farm power output while minimizing the fatigue accumulation on the turbines.

Refersnces

[1] GlobalWind Energy Council. "GlobalWind Report: Annual Market Update 2012". Available online: http://www.gwec.net/wp-content/uploads/2012/06/Annual report 2012 LowRes.pdf (accessed on 30 June 2013).

[2] Barthelmie RJ, Pryor S, Frandsen ST, Hansen KS, Schepers J, Rados K, et al. Quantifying the impact of wind turbine wakes on power output at offshore wind farms. *J. Atmos. Oceanic Tech.* 2010;**27**(8):1302-17.

[3] Crespo A, Hernandez J and Frandsen S. Survey of modelling methods for wind turbine wakes and wind farms. *J. Wind energy.* 1999;**2**(1):1-24.

[4] Lissaman P. Energy effectiveness of arbitrary arrays of wind turbines. *J. Energy.* 1979;**3**(6):323-8.

[5] Katic I, Højstrup J and Jensen N, editors. A simple model for cluster efficiency. *EWEC*; 1986.

[6] Voutsinas S, Rados K and Zervos A. On the analysis of wake effects in wind parks. *J. Wind Eng.* 1990;**14**:204-19.

[7] Frandsen S, Barthelmie R, Pryor S, Rathmann O, Larsen S, Højstrup J and Thøgersen M. Analytical modelling of wind speed deficit in large offshore wind farms. *J. Wind energy.* 2006;**9**(1-2):39-53.

[8] Bastankhah M and Porté-Agel F. A new analytical model for wind-turbine wakes. *J. Renew.*

Energy. 2014;**70**:116-23.
[9] González JS, Rodriguez AGG, Mora JC, Santos JR and Payan MB. Optimization of wind farm turbines layout using an evolutive algorithm. *J. Renew. Energy*. 2010;**35**(8):1671-81.
[10] Barthelmie R, Larsen G, Frandsen S, Folkerts L, Rados K, Pryor S, Lange B and Schepers G. Comparison of wake model simulations with offshore wind turbine wake profiles measured by sodar. *J. Atmos. Oceanic Tech.* 2006;**23**(7):888-901.
[11] Thøgersen ML. , Wind PRO/PARK: introduction to wind turbine wake modelling and wake generated turbulence. Technical report. Niels Jernes Vej 10, DK- 9220 Aalborg, Denmark: EMD International A/S; 2005.
[12] Crasto G, Gravdahl A, Castellani F and Piccioni E. Wake modeling with the actuator disc concept. *J. Energy Proc.* 2012;**24**:385-92.
[13] GH WindFarmer theory manual. Garrad Hassan and Partners Ltd; 2009.
[14] version 1.3 ed. Openwind theoretical basis and validation. AWS Truepower, LCC; 2010.
[15] Porté-Agel F, Wu Y-T and Chen C-H. A numerical study of the effects of wind direction on turbine wakes and power losses in a large wind farm. *J. Energies*. 2013;**6**(10):5297-313.
[16] Abkar M and Porté-Agel F. Mean and turbulent kinetic energy budgets inside and above very large wind farms under conventionally-neutral condition. *J. Renew. Energy*. 2014;**70**:142-52.
[17] Quarton D and Ainslie J. Turbulence in wind turbine wakes. *J. Wind Eng.* 1990;**14**(1):15-23.
[18] Hassan U and Hassan G. A wind tunnel investigation of the wake structure within small wind turbine farms: Harwell Laboratory, Energy Technology Support Unit; 1993.
[19] Crespo A and Herna J. Turbulence characteristics in wind-turbine wakes. *J. Wind Eng Ind Aerodyn.* 1996;**61**(1):71-85.
[20] Vermeulen P, editor. An experimental analysis of wind turbine wakes. 3rd international symposium on wind energy systems; 1980.
[21] Abkar M and Porté-Agel F. A new wind-farm parameterization for large-scale atmospheric models. *J. Renew. Sustainable Energy Rev.* 2015;**7**(1):013121.
[22] Frandsen S and Thøgersen ML. Integrated fatigue loading for wind turbines in wind farms by combining ambient turbulence and wakes. *J. Wind Eng.* 1999;**23**(6):327-40.
[23] Wu Y-T and Porté-Agel F. Modeling turbine wakes and power losses within a wind farm using LES: An application to the Horns Rev offshore wind farm. *J. Renew. Energy*. 2015;**75**:945-55.
[24] Barthelmie RJ, Hansen K, Frandsen ST, Rathmann O, Schepers J, Schlez W, et al. Modelling and measuring flow and wind turbine wakes in large wind farms offshore. *J. Wind energy*. 2009;**12**(5):431-44.
[25] Barthelmie RJ and Jensen L. Evaluation of wind farm efficiency and wind turbine wakes at the Nysted offshore wind farm. *J. Wind energy*. 2010;**13**(6):573-86.

The impact of wake models on wind farm layout optimization

Jonas Schmidt and Bernhard Stoevesandt

Fraunhofer IWES, Ammerländer Heerstraße 136, 26129 Oldenburg, Germany

E-mail: jonas.schmidt@iwes.fraunhofer.de

Abstract. Results for nine gradient-based layout optimization runs of a wind farm with 25 turbines in flat terrain are presented, varying three different choices of the underlying wake model and three inflow scenarios. In all cases the AEP is maximised and the constraints are purely geometrical. A single inflow vector, a uniform wind rose and a realistic synthetic wind rose are studied, and the final layouts for the Jensen, the Ainslie and a CFD-based numerical wake model are compared. From this an estimate of the average variation of the turbine position due to the different wake models is obtained. All calculations were carried out with the in-house software *flapFOAM*.

1. Introduction

Wind farm layout optimization is crucial for advancing wind energy in Europe, since often either the available area at a site is limited or the meteorological conditions do not clearly prefer a single instance of inflow conditions. Wake situations may therefore be significant and should not be ignored during the process of layout optimization. For a recent review on the topic of wind farm optimization and more than 20 years of related research see [1], also [2, 3]. For a recent summary on the related topic of optimised wind farm control see [4].

The optimal wind farm layout depends on many site-specific parameters and constraints, especially for on-shore wind farms. Examples are limitations of the available area due to urban neighbourhood, land ownership, forest, inaccessible terrain or unsuitable soil, the requirement of an achievable and realisable cable network and grid connection, restrictions due to access roads, environmental constraints, etc. Such issues imprint on the objective function as well as the optimization constraints. The implementation of a multi-fidelity layout optimization software with complex objective functions beyond annual energy production (AEP) by DTU is described in [5].

In this work, however, we focus on the impact of the wake effect on the layout optimization of wind farms. It induces an interaction between the individual wind turbines and can therefore play a major role, especially at sites where the available area is limited and the wind rose is not dominated by a single wind direction. We therefore leave the above mentioned issues aside and idealise the problem by only considering the maximisation of the AEP of the wind farm.

Wind farm layout optimization relies on iterative methods, which are based on a large number of evaluations of the objective function. This involves the repeated calculation of the wake effect for each turbine, which may be expensive, depending on the method and the invoked models. While 0-equation wake models like the Jensen model [6] or the Frandsen model [7] are fast and therefore natural candidates for this purpose, more complex models like the Ainslie model [8] or the CFD-based numerical wake model [9] of the software *flapFOAM* require more numerical or algorithmic attention. However, these issues are purely technical and therefore either solved or solvable, in contrast to the principle complexity issue of full CFD simulations during run-time of the optimization routine.

It is the main purpose of this work to address the impact of the choice of wake model on the outcome of a layout optimization that purely focuses on the wake effect. Besides studying this question it also serves as a proof-of-principle for the coupling of the wind farm modelling code *flapFOAM*, that is being developed at the Fraunhofer Institute for Wind Energy and System Technology IWES, and the open source optimization tool box *Dakota* [10].

2. Methods

2.1. Wind farm modelling with flapFOAM

All wake modelling and wind farm calculations in this work are obtained with the software *flapFOAM*, which has been developed at Fraunhofer IWES since 2011. It was inspired by the software *FLaP*, which had been developed earlier at the University of Oldenburg (cf. [11]), without including code of the latter. *flapFOAM* is based on the idea of single-wake superposition, fully written in C++, and can read OpenFOAM simulation results. Its strictly modular structure allows the developer to extend and improve models independently of the core functionality of the code, and the user to select between a broad range of models and settings. The proof-of-principle of the numerical wake model based on CFD-RANS results was presented in [9], and progress on the inclusion of complex terrain effects was reported in [12]. A detailed description of the software will be given elsewhere, here we only provide a brief summary of the parts that are directly relevant for the current work.

The basic calculation algorithm of *flapFOAM* for the calculation of the wind vector v at a point p and the effective wind velocity at the i-th rotor $v^{(i)}_{\mathrm{rotor}}$ in the presence of wakes is sketched in Fig. 1 (left). The `CombinedWindWakeField` class contains the background wind field, which represents the wind field at the site without the wake effect, the deficit wind field, which provides all wake deficit data, the wake overlap rule, which is called `WakeAdditionModel`, and the `DownwindOrder`, which contains the information of the interdependence of the individual wind turbines.

Two different algorithms are implemented for the downwind order. The updating routine for the iterative model is sketched in the right panel of Fig. 1. In this case the respective wake deficits at any two turbines caused by one another are compared. From this the turbine dependencies and the global order of evaluation are determined. After the re-calculation of the effective wind speeds at the wind turbines, this process is repeated until convergence is reached. A simpler model determines a dominant wind direction and orders the turbines according to their corresponding coordinate with respect to this direction.

The crucial quantity for all wind farm calculations is the list of effective wind velocities at the rotors v_{rotor}. This quantity determines the thrust coefficient and therefore the wake strength for each wind turbine rotor. Depending on the `RotorModel` it is calculated as an average over the rotor disk, where the wake addition model specifies which power of the local wind speed is used as the integral kernel during the calculation of the mean.

The contribution of rotor i to the total local wind deficit, measured in a volume element of the disk that represents rotor j, depends on the current results of the dependencies of rotor j and $v^{(i)}_{\mathrm{rotor}}$, and on the selected `SingleWakeModel` for wind turbine i. The wake models are formulated with respect to the rotor reference frame, and their defining property is to return the velocity deficit at any point p, given the effective inflow velocity vector, the rotor model and the required constant model parameters. Several wake models are implemented in *flapFOAM*, and the three relevant for this work are briefly introduced in Section 2.3.

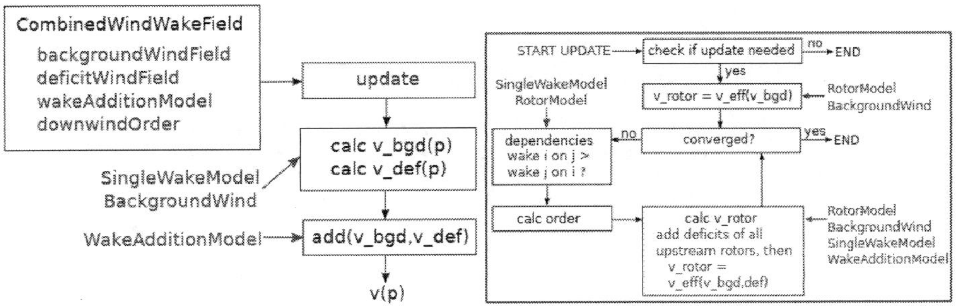

Figure 1. Sketch of local velocity calculation with *flapFOAM* (left) and the updating routine for the iterative downwind order model (right). The calculated velocities at the rotors, here denoted as `v_rotor`, is crucial for all wind farm calculations.

2.2. *Wind farm model*

The numerical wind turbine model that is considered in this work has $D = 120$ m rotor diameter, $H = 120$ m hub height and 2.53 MW nominal power. In what follows the effective wind speed at the rotor is obtained directly from the centre point of the disk, i.e., the simplest rotor model without disk integration is applied.

The model wind farm consists of 25 wind turbine of the above rotor type, and initially arranged in a regular 5×5 pattern of size 3×3 km^2. The available area for the layout optimization is a square of 8×8 km^2. In order to break the symmetry, the axes which define the initial wind farm layout are chosen as $(1, -0.01)$ and $(-0.01, 1)$ in the xy-plane. This is a simple way to obtain non-vanishing derivatives of the power yield in y-direction for wind vectors that are parallel to the x-axis, using a centred finite-difference gradient scheme[1].

2.3. *Wake models*

Three wake models are compared in this work, the Jensen, the Ainslie and a CFD-based wake model. The addition of wake deficits is preformed quadratically under the square root, and no partial wake or meandering models are applied.

2.3.1. Jensen model A classic wake model is the model by Jensen [6]. Its sole parameter is the constant wake expansion parameter, which determines the linear expansion of the wake radius with downstream distance, and is chosen as $k = 0.07$ throughout this work. The model furthermore ignores radial variations of the deficit inside the wake, and the application of mass conservation in a control volume then yields an inverse-square law with respect to distance for the axial deficit. The Jensen model is a zero-equation model, which makes it fast and therefore efficient for layout optimization applications.

2.3.2. Ainslie model The Ainslie model [8] is a 2-equation eddy-viscosity model, which is based on the reduction of the time-averaged Navier-Stokes equations by imposing axial symmetry. The eddy viscosity is the combination of an ambient turbulence contribution and a wake-induced term that depends on the centre line deficit. The Ainslie model is widely applied since it captures more of the flow dynamics than the Jensen model but is less complex than full RANS equations.

[1] Notice however that this is a matter of sensitivities and the chosen step size, rather than principle, and there are other options to achieve that goal. One example is the application of several optimisation runs with initial gradient step size of 5 D and subsequently smaller values.

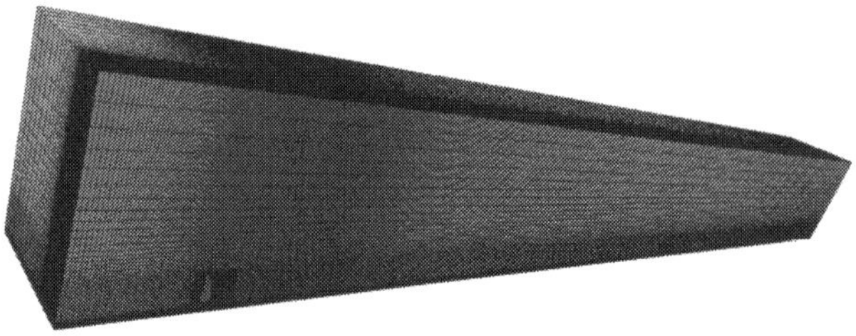

Figure 2. Cut through the mesh and the actuator disk (red).

In *flapFOAM* the Ainslie model is solved on-the-fly during run time on a specified 2D grid in the radial coordinate r and the disk orthogonal coordinate x, for discrete inflow wind speeds between which interpolation to user-specified numerical accuracy is applied. For this, a Crank-Nicolson discretization scheme is applied to yield a tri-diagonal matrix system, whose solutions are written to a data base in order to avoid redundant calculations.

Here the grid size of the regular finite difference grid in (x, r) was chosen as 9000×4500 m^2 with 900×450 grid points. The ambient turbulence intensity was chosen uniformly as 10%. All derivatives and interpolations are performed at second order numerical accuracy.

2.3.3. CFD wake model Basically the 3D-RANS equations applied to an isolated actuator disk define a (4+x)-equation wake model, where x represents the turbulence model equations. Hence it can be treated similar to the Ainslie model described above. However, due to the complexity of CFD simulations they are obtained before run time of *flapFOAM* and should span the range of inflow wind speeds of interest. Details of the implementation of a numerical wake model based on pre-calculated CFD-RANS results are given in [9].

For the purpose of this work we performed eight CFD-RANS simulations of a single uniform actuator disk in neutral stratification with OpenFOAM's *simpleFoam* solver (version 2.3.1), at inflow wind speeds 3, 5, 8, 10, 12, 15, 18 and 20 m/s at hub height 120 m.

The mesh has dimensions $8.8 \times 1.5 \times 1.0$ km^3 and was created using the IWES in-house tool *terrainMesher*, which is a follow-up of the open-source *terrainBlockMesher* [13]. It consists of 2.05 million cells, including the actuator disk with 1892 cells, cf. Fig. 2. The first cell height at the ground is 1 m and standard wall functions with roughness length $z_0 = 5$ cm were used. Both grading and refinement were applied to improve the resolution of the wake and the near-disk region.

We applied the k–ϵ–f_P turbulence model [14] with parameters as recommended there. Compared to the standard k–ϵ model this version includes a correction of turbulent viscosity that depends on the change of velocity gradients due to the presence of the actuator disk, enhancing the wake deficit. All boundary conditions at the inlet were obtained by a one-dimensional cyclic precursor run, given the mass flow according to a standard log-profile. The desired profile and the inflow wind speed at hub height were well matched by the results. All variables converged to residuals below 10^{-5} in all simulations. The converged results were mapped to a fully structured coarser grid with 277200 cells and minimal cell size of 10 m in order to speed up the optimization.

2.4. Inflow models

Inflow model specify for which set of inflow conditions the wind farm calculations are performed. In cases different from the single wind vector case, the obtained results are mean results over the corresponding background wind fields. Each individual contribution is therefore added with the associated statistical weight. Three different inflow cases are considered in this work.

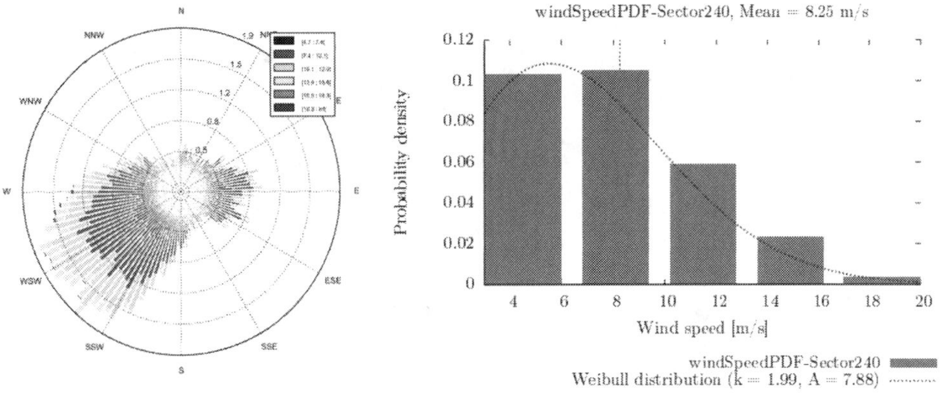

Figure 3. The wind rose (left) and the wind speed histogram for wind direction 240° (right).

2.4.1. Single wind vector The single vector inflow case (SWV) is defined by a uniform background wind with velocity vector $\mathbf{v} = (8, 0, 0)$ m/s, chosen in the linear regime of the considered turbine power curve. The SWV case has a large number of local and global minima, since there are many ways to arrange 25 wind turbines inside the 8×8 km^2 area such that they have almost free-stream conditions.

2.4.2. Uniform wind rose A uniform wind rose with 360 sectors (WR360U) and 8 m/s wind speed at hub height is used as a test scenario that has no trivial free-stream solutions. It can be expected that layout optimization results show the corresponding symmetry, possibly broken by the rectangular wind farm boundary.

2.4.3. Realistic wind rose This case is a synthetic non-uniform wind rose with 120 sectors (WR120R), each equipped with a wind speed histogram with maximal 5 bins in the range from 3 to 20 m/s, cf. Fig. 3. As for the WR360U scenario there are no free-stream solutions for the wind turbines. However, there is a strong bias from south-westerly wind directions. In total this inflow case comprises 510 non-vanishing wind inflow vectors, with mean wind speed 7.4 m/s and mean direction 230°.

2.5. Optimization

For efficient optimization, *flapFOAM* has been coupled to the open source optimization tool box *Dakota* (version 6.0.0), developed by Sandia National Laboratories [10]. The coupling has been achieved internally, i.e., in terms of linked libraries and the implementation of wrapper classes that communicate between the two software packages. This is efficient, because *flapFOAM* can call the optimization routine, and the slow process of reloading data into memory during optimization iterations is avoided.

Since we focus on the wake effect in this work, the objective function is the park efficiency without any further contributions. Here park efficiency is defined as the total power yield of the wind farm, normalized by the number of turbines and the power that an isolated turbine would produce in the specific inflow case. Notice that the latter varies depending on the scenario. For the chosen machine model it is 1.674 MW for the SWV and WR360U cases and 1.247 MW for the WR120R inflow case.

Parameter	Value
`method`	`conmin_mfd`
`max_iterations`	5000
`convergence_tolerance`	10^{-5}
`constraint_tolerance`	0.003
`max_function_evaluations`	10000
`scaling`	

Table 1. The parameters used for CONMIN optimization method, as given to Dakota. The `scaling` parameter activates the normalisation and requires no value in the `method` specification.

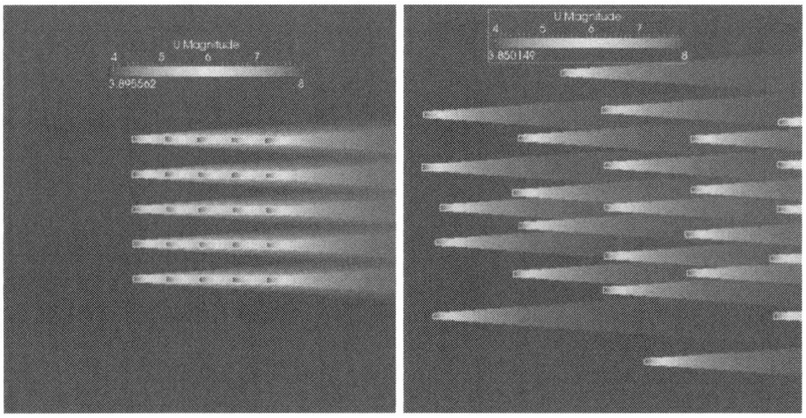

Figure 4. The SWV inflow case for the Jensen wake model. The left panel shows the initial condition, the right panel the final layout. The park efficiency improved from 68.5% to 97.5%.

Two types of optimization constraints apply. First, the positions have to respect the maximally available area for the wind farm, cf. Sec. 2.2, and second, a minimal inter-turbine distance of 500 m, i.e. 4.17 D, is required. For the considered 25 wind turbines within the square boundary there are 50 design variables and 400 constraints. During evaluation, the constraints are normalized with respect to characteristic scales.

All results presented in the following were obtained by the CONMIN optimization method, as provided by Dakota. This gradient-based constraint optimization algorithm was developed by Vanderplaats in 1973 [15] and has been applied to a vast number of problems ever since. It is limited to finding the nearest local minimum. Two variants are implemented in Dakota, a conjugate gradient optimization method (frcg) for unconstraint problems, and a method of feasible directions (mfd). The latter was chosen here, with parameters as listed in Table 1. It is based on the evaluation of gradients of the objective function and the constraints, from which a direction in the parameter space is determined. The variables are then altered according to that direction and a step size, until constraints are hit or the objective function does no longer improve. In that case a new feasible direction is determined, and the procedure is continued until convergence is reached. For all results a central finite-difference gradient scheme with step size 1 D was applied.

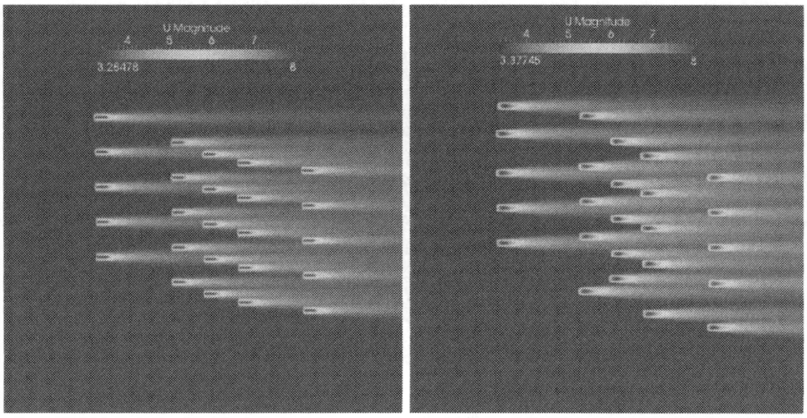

Figure 5. The final layout in the SWV inflow case for the Ainslie (left) and the CFD wake model (right). The park efficiencies improved from 65.3% and 61.9% in the initial configuration to 95.6% and 95.8%, respectively.

3. Results and discussion

3.1. Single wind vector

The initial and optimized layouts for the SWV inflow case and the Jensen wake model are shown in Fig. 4, for the Ainslie and CFD wake models the latter are shown in Fig. 5. The park efficiencies were increased by the optimization from 68.5%, 65.3% and 61.0% to 97.5%, 95.6% and 95.8% for the Jensen, Ainslie and CFD wake models, respectively.

For all wake models almost free-stream solutions were found by the CONMIN optimization method and the chosen initial conditions and parameters. Especially for the Jensen wake model, which has no radial dependence within the wake region, but also for the axisymmetric Ainslie wake model, this requires some tuning and is not automatic. For example, small step sizes may lead to changes of the objective function that are below the tolerance threshold, thereby terminating the optimization. Here slightly rotated initial grid axes were chosen to break the symmetry, cf. Sec. 2.2.

One obvious issue of the method is that it does not explore all of the available parameter space, but only the vicinity at each iteration. Hence the optimal park efficiency of 100% is not reached, and the turbines appear clustered and attached to the neighbour turbine's wake. For the Jensen model the clustering is less pronounced, since the relatively large choice of the opening angle yields a wide and constantly growing wake radius. In general the issue can be addressed by considering non-trivial rotor models or by increasing the sensitivity. However, for realistic applications the inflow will always involve a distribution of wind directions, which counteracts the effect.

3.2. Uniform wind rose

The left panel of Fig. 6 shows the initial mean wind speed at hub height for the Ainslie wake model and the WR360U inflow case. Clearly the layout is not the optimal configuration, since all turbines are located at positions where the mean wind speed is low compared to the outer regions. The right panel shows the optimized layout. Here no gradients of the mean wind speed are apparent beyond the turbine vicinities, as expected for the uniform wind rose case. One central turbine is surrounded by a ring of turbines, the remaining turbines are located near the wind farm boundary. This pattern is invariant under all common symmetries of the boundary

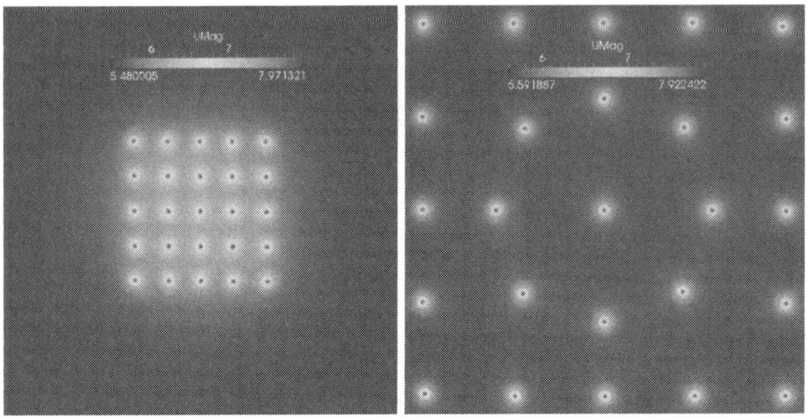

Figure 6. The WR360U inflow case for the Ainslie wake model. The left panel shows the mean wind speed at hub height for the initial condition, the right panel for the final layout.

Case	Wake	Initial FE	Final FE
WR360U	Jensen	88.0%	97.5%
	Ainslie	89.7%	92.2%
	CFD	87.3%	96.0%
WR120R	Jensen	93.1%	98.0%
	Ainslie	83.7%	98.3%
	CFD	92.1%	97.8%

WR360U	Ainslie	CFD
Jensen	1.0 D	4.1 D
Ainslie	–	3.2 D
WR120R	**Ainslie**	**CFD**
Jensen	4.7 D	4.0 D
Ainslie	–	1.7 D

Table 2. Left: Wind farm efficiencies (FE) for the WR360U and WR120R cases. Right: Average turbine distance between two final layouts, cf. Fig. 7.

square and the wind rose, as expected.

The wind farm efficiency is increased from 89.7% to 92.2% in case of the Ainslie model, cf. Table 2 (left). Notice that the park efficiencies are calculated with the same wake model that is used for the layout optimization. They should not be used directly for comparisons of the final layouts, since their relevance is a matter of validation and therefore model dependent. The issue of objective comparison of optimization results is beyond the scope of this paper and left for future work.

The final layouts for all three wake models and the WR360U inflow case are presented in the left panel of Fig. 7. They all show the expected symmetries, however, the turbine positions do not exactly coincide. As listed in Table 2 (right) the average distance between the positions of one particular turbine in two different final layouts is between 1.0 to 4.1 D. The Jensen and Ainslie models disagree little in the optimal layout, while the CFD based wake model yields a considerably different layout. This concerns the radius of the inner ring of turbines, which varies depending on the decrease of the wind deficit in the wake with distance from the rotor. Furthermore the outer turbines seem to follow a rounder shape for the CFD model case, with every second turbine shifted inwards from the boundary.

3.3. Realistic wind rose

The initial and final mean wind speed fields for the WR120R inflow case and the CFD wake model are shown in Fig. 8. The final layout is a distorted version of the result for the symmetric

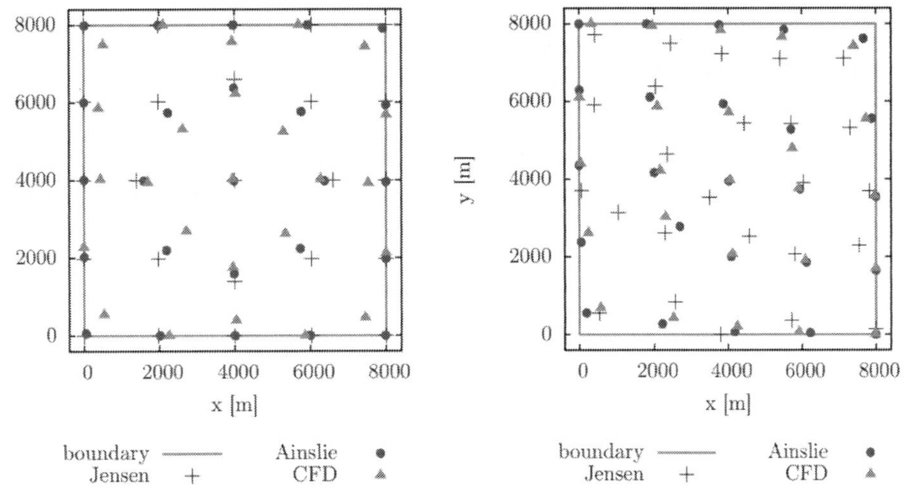

Figure 7. The final layouts for the WR360U (left) and WR120R (right) inflow cases and the three studied wake models.

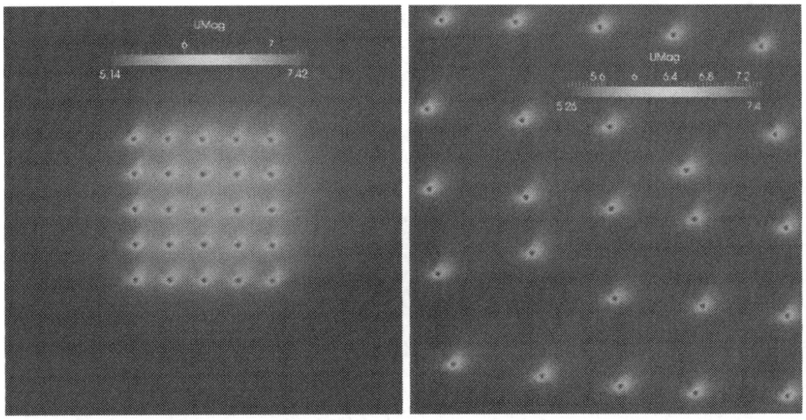

Figure 8. The WR120R inflow case for the CFD wake model. The left panel shows the mean wind speed at hub height for the initial condition, the right panel for the final layout.

WR360U case. A similar observation can be made for the Ainslie wake model, cf. Fig. 7 (right). Table 2 states that the final layouts of these two models coincides up to an average turbine position shift of 1.7 D.

The final layout result for the Jensen wake model is qualitatively different. It cannot be characterized as a distortion of the symmetric case, it rather shows features of a staggered layout orthogonal to the main wind direction south-west. The mean turbine position variation compared to the Ainslie and the CFD wake models is 4.7 and 4.0 D, respectively.

4. Conclusion

In this work we studied wind farm layout optimization of a wind farm with 25 wind turbines in flat terrain, using the software *flapFOAM*, which recently has been coupled to the optimization library *Dakota*. Three different wake models were compared, the Jensen wake model, the Ainslie wake model and a CFD-based numerical wake model that was obtained from pre-calculated actuator disk simulations with OpenFOAM. Layout optimization results were obtained with the gradient based CONMIN method for three different inflow cases, a single wind vector (SWV), a uniform wind rose with 360 sectors (WR360U) and a realistic, synthetic wind rose with 120 sectors and up to 5 wind speed bins per sector (WR120R).

The SWV inflow case has a large number of optimization minima, and for the three wake models three different solutions with nearly free-stream conditions for all turbines were found. The gradient based optimiser needs different degrees of tuning in order to avoid unfavourable blocking of a turbine in the bulk between neighbour wakes. Also the usage of non-purely local gradient based optimization methods can overcome this issue. Though the CFD wake model seems to be less sensitive to tuning, a quantitative comparison of the wake model impact is difficult for this inflow case due to the large number of solutions.

The uniform wind rose case has no solution with almost free-stream conditions. For all wake models final configurations were found that show the expected symmetries. The Jensen and Ainslie wake models agree on the turbine position up to 1.0 D on average, the CFD wake model yields variations of up to 4.1 D on average. This may be due to a significantly different decay behaviour of the wake deficit in this model.

For the WR120R inflow case the Ainslie and the CFD wake model are in good agreement, with an average variation of the turbine position of 1.7 D. Both models show a distortion of the symmetric solution that was found for the WR360U case. The Jensen model yields a qualitatively different solution that resembles a staggered layout with lines of turbines orthogonal to the main wind direction. The displacement compared to the other solutions is up to 4.7 D on average.

Our results demonstrate that wake models can have an impact on the outcome of wind farm layout optimization. Based on the our examples and a gradient based optimization method the magnitude of the effect can be estimated as an average variation of the turbine position in the range 1–5 D. An objective decision which wake model is the best for layout optimisation is difficult, since it involves the definition of a metric that ideally does not rely on engineering models and is validated for a large number of wind farm configurations.

References

[1] Herbert-Acero J F, Probst O, Rethore P E, Larsen G C and Castillo-Villar K K 2014 *Energies* **7** 6930–7016
[2] Samorani M 2010 The wind farm layout optimization problem Tech. rep. PowerLeeds School of Buisness
[3] Tesauro A, Rethore P E and Larsen G C 2012 *EWEA* (Copenhagen, Denmark)
[4] Knudsen T, Bak T and Svenstrup M 2014 *Wind Energy* n/a–n/a
[5] Rethore P E, Fuglsang P, Larsen G C, Buhl T, Larsen T J and Madsen H A 2014 *Wind Energy* **17** 1797–1816
[6] Jensen J O 1983 A note on wind generator interaction Tech. Rep. Risø-M-2411 Risø National Laboratory
[7] Frandsen S, Barthelmie R, Pryor S, Rathmann O, Larsen S, Højstrup J and Thøgersen M 2006 *Wind Energy* **9** 39–53
[8] Ainslie J F 1983 *BWEA Wind Energy Conference* (Oxford, UK)
[9] Schmidt J and Stoevesandt B 2014 *EWEA conference proceedings* (Barcelona, Spain)
[10] Adams e a 2009 Dakota, a multilevel parallel object-oriented framework for design optimization, parameter estimation, uncertainty quantification, and sensitivity analysis: Version 5.4 user's manual Tech. Rep. SAND2010-2183 Sandia Technical Report
[11] Lange B, Waldl H P, Guerrero A G, Heinemann D and Barthelmie R J 2003 *Wind Energy* **6** 87–104
[12] Schmidt J and Stoevesandt B 2014 *Journal of Physics: Conference Series* **524** 012136
[13] Schmidt J 2013 `https://github.com/jonasIWES/terrainBlockMesher.git` [Online; acc. 13-March-2013]
[14] van der Laan M P, Sorensen N N, Rethore P E, Mann J, Kelly M C, Troldborg N, Schepers J G and Machefaux E 2014 *Wind Energy* n/a–n/a ISSN 1099-1824
[15] Vanderplaats G N 1973 Tech. Rep. NASA TM X-62,282 NASA Technical Memorandum

Coupling of a free wake vortex ring near-wake model with the Jensen and Larsen far-wake deficit models

J W van Heemst[1], D Baldacchino[1], D Mehta[2,1] and G J W van Bussel[1]

[1]DUWIND - Delft University Wind Energy Research Institute, Kluyverweg 1, 2629 HS Delft, NL

[2]Energy Research Centre of the Netherlands, Westerduinweg 3, 1755 LE Petten, NL

E-mail: hans@vanheemst.com, d.baldacchino@tudelft.nl

Abstract. This paper presents a simple physical model to improve the currently used far-wake deficit models in the wind industry. The main improvement is deemed on the determination of the wake deficit in the near-wake. A Vortex Ring Model (VRM) is used to calculate the induced velocities in the near-wake, which are then coupled to the Jensen far-wake model and the Larsen far-wake model based on the concept of Eddy Viscosity (EV). The inviscid near-wake VRM is based on the shedding of discrete tip vortex rings released from a uniformly loaded actuator disc. The model is validated against wind tunnel measurements from experiments with a two-bladed turbine and a circular metal mesh with a uniform porosity to represent an actuator disc. The VRM shows a good agreement with the experimental data with respect to the wake deficit evolution. The VRM is coupled with two well-known engineering type far-wake models: the Jensen and Larsen wake deficit models. The results of the coupling of the VRM and the more elaborated Larsen far-wake model are compared against a 3D Large Eddy Simulation (LES) CFD model. This comparison shows the effect of different near-wake models on the development of centreline velocities in the far-wake. The centreline velocity deficit predicted by the VRM-Larsen model more closely matches LES calculations in comparison with the reference Larsen model.

1. Introduction

Most well-known engineering type wind farm wake models include a fairly simple approximation of the near-wake. This is often a combination of an assumed velocity deficit profile (e.g. square hat or Gaussian) and empirical data or a simple streamtube approach based on the conservation of momentum. Disadvantages of using empirical based near-wake models is the uncertainty from measurements that are already introduced at the start of the wake field and the requirement of calibrated coefficients corresponding to the applied case. An analytical, physics based model, independent of empirical constants and rotor blade parameters, could provide a better input to far-wake approximations. An approach by Segalini & Alfredsson [1] using a helical vortex approach is promising, but a simpler engineering type solution was sought.

Discrete vortex rings to describe the near-wake have been used by Baldacchino and van Bussel [2] to investigate the effect of yawed inflow and the existence of ground effects on the near-wake development. In the model, the turbine rotor is replaced by an actuator disc, which sheds discrete vortex rings. The rings are free to move, grow and interact. The resulting set of vortex rings can be used to describe a full 3D velocity field.

In the present paper, the vortex ring model (VRM) is combined with the Jensen [3] and Larsen [4] far-wake models. The Jensen model is selected because it is a widely used engineering model and has a relative simple implementation as compared to other wake models. As a second model, Larsen's Eddy Viscosity (EV) based far-wake model is selected because it is a more advanced model and offers compatibility with the VRM approach. The coupling could lead to a model with improved accuracy over the complete velocity field compared to the streamtube near-wake model, which is used in the original Larsen far-wake model. This paper discusses a stationary wake, but the VRM supports fluctuations in wind speed and wind direction [5], yawed flow [2] and dynamic loads on the actuator disc [6].

The approach to combine the VRM with the Larsen and Jensen far-wake models is discussed in this paper. The near wake vortex ring model is validated against wind tunnel measurements of (i) a two-bladed rotor model and (ii) a porous disc representing an actuator disc. The results of the VRM-Larsen model are compared to a Large Eddy Simulation (LES) CFD model, which is provided by the Energy Research Centre of the Netherlands (ECN). Lastly, a multiple wake simulation is briefly studied (velocity and power deficits) by applying the VRM-Larsen model to a case from the Horns Rev wind farm.

2. Derived numerical models

2.1. The vortex ring model

In the VRM, the wind turbine rotor is replaced by a uniformly loaded actuator disc, which sheds infinitely thin, discrete vortex rings (see figure 1). The rings are assumed to be inviscid and perfectly circular. The vortex rings are placed instantaneously in the free-stream after constant periods of time Δt. They have mutual influences on their displacement and expansion/contraction.

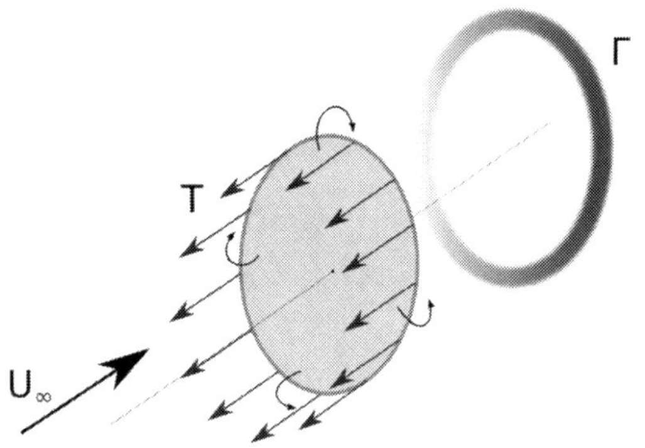

Figure 1. Schematic illustration of an actuator disc releasing a vortex ring

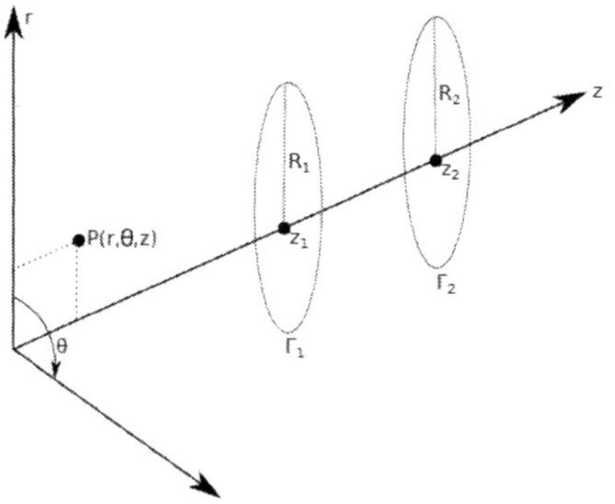

Figure 2. Coordinate system used on a pair of vortex rings in the free-stream

2.1.1. Derivation of the velocity field from a set of vortex rings The series of vortex rings can be used to calculate an axi-symmetric 3D velocity field. For an arbitrary point $P(r, \theta, z)$, the induced velocity from a vortex ring can be described by the set of equations displayed in equations 1 and 2, derived by Yoon and Heister [7] using the Biot-Savart law. These set of equations are valid on every point in the field, except on the vortex ring itself (where $R_i = r$ and $z = z_i$), which results in a singularity.

$$u_z = \frac{\Gamma_i}{2\pi\sqrt{[(z-z_i)^2 + (r+R_i)^2]}} \cdot \left[K(m) + \frac{R_i^2 - r^2 - (z-z_i)^2}{(z-z_i)^2 + (R_i - r)^2} E(m)\right] \tag{1}$$

$$u_r = \frac{\Gamma_i(z - z_i)}{2\pi r\sqrt{[(z - z_i)^2 + (r + R_i)^2]}} \cdot \left[K(m) - \frac{R_i^2 + r^2 + (z - z_i)^2}{(z - z_i)^2 + (R_i - r)^2}E(m)\right] \tag{2}$$

In these equations, $K(m)$ and $E(m)$ represent complete elliptic integrals of the first and second kind, where m is determined with $m = 4rR_i/((z - z_i)^2 + (r + R_i)^2)$.

2.1.2. An approach to determine vortex ring strength Van Kuik [8] derived a relation between the pressure jump Δp at the edge of an actuator disc and the shedding of vorticity in the flow. He used the continuity and momentum equations for inviscid and incompressible fluid to derive the following equation:

$$\frac{\partial \Gamma_{edge}}{\partial t} = \frac{\Delta p}{\rho} \tag{3}$$

Following the approach of Yu et al. [6], equation 3 is applied on a series of vortex rings. The pressure jump at the actuator disc is described as the thrust divided by the area. Using the definition of C_T, this can be written as:

$$\Delta p = \frac{T}{A} = \frac{\frac{1}{2}\rho C_T U_\infty^2 A}{A} = \frac{1}{2}\rho C_T U_\infty^2 \tag{4}$$

Substituting this into equation 3 gives the final expression of the vortex ring strength:

$$\Gamma_i = \frac{1}{2}U_\infty^2 C_T \Delta t \tag{5}$$

Here, Δt represents the period between the shedding of two vortex rings. Initially, Δt was linked to the revolution of a blade in an approach to simulate a realistic tip vortex pitch, constraining the ring radii by following an empirical wake expansion model as in [2]. The expansion model is not in good agreement with results from wind tunnel experiments, so the expansion model was disregarded and induced vortex ring expansion is used instead. This results in a much smaller value for Δt to guarantee the stability of the simulation when the vortex rings start interacting further down-stream. As a reference, a value of $\Delta t = 0.01s$ is used for the small rotor in the experiment discussed in section 3 and $\Delta t = 0.5s$ is used for the large rotor in the LES case discussed in section 5.

2.1.3. Numerical implementation of the free wake VRM The position of a vortex ring in the free wake, is determined from the axial velocity of the ring. This velocity is derived by superimposing the undisturbed wind speed, the self-induced velocity and the mutually induced velocity from other vortex rings. Following the approach by Yu et al. [6] and Baldacchino and van Bussel [2], the self-induced velocity is derived by determining the velocity at the centre of the ring by substituting $z = z_i$ and $r = 0$ in equations 1 and 2. This results in the simple equation $u_s = \Gamma/(2R)$. Baldacchino [9] compared this method with the more well-known Kelvin Theorem which takes into account the ring's core size a. His research showed that for small core sizes ($a/R < 0.01$), the differences between the two methods is negligible, thus the more simple approach is used since the rings are assumed infinitely thin. The mutually induced velocity of a vortex ring is determined by the total induced axial velocity on a arbitrary point on the vortex ring. An example of a pair of vortex rings is displayed in figure 2. In this case, the induced velocity on vortex ring 2 by vortex ring 1 is computed with:

$$u_{m,2} = \frac{\Gamma_1}{2\pi\sqrt{[(z_2 - z_1)^2 + (R_2 + R_1)^2]}} \cdot \left[K(m) + \frac{R_1^2 - R_2^2 - (z_2 - z_1)^2}{(z_2 - z_1)^2 + (R_1 - R_2)^2}E(m)\right] \tag{6}$$

This procedure is repeated for all n vortex rings, which concludes the derivation of the translational velocity of a single ring with:

$$u_i = U_\infty + u_s + u_m = U_\infty + \frac{\Gamma_i}{2R_i} + \sum_{j=1}^{n} u_{m,j} \tag{7}$$

The position of the vortex rings is then updated with a second order update scheme. The wake expansion is modelled by allowing the rings to expand and contract freely. The expansion of a ring is determined with the same method as the mutually induced axial velocity, except the radial velocity is now determined on the ring. The numerical timestep of the simulation is determined as $0.5\Delta t$.

The induced velocities result in vortex ring interaction. Figure 3 and 4 illustrate the distribution of vortex rings in the wake for two typical cases. The graphs display the effect of the thrust on vortex ring interaction. As expected, a higher thrust coefficient results in stronger vortex rings, which interact at an earlier stage compared to an actuator with a lower thrust. A converged solution is obtained by allowing the rings to travel far down-stream ($> 20D$) whereas only the velocities in the near-wake are analysed, far from the starting vortices.

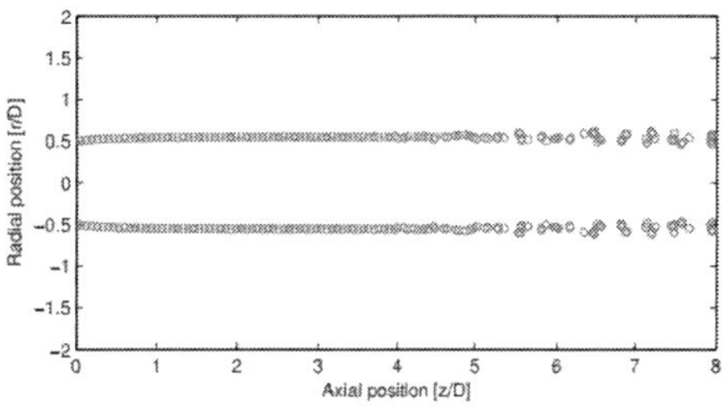

Figure 3. Simulated vortex ring distribution with $C_T = 5/9$

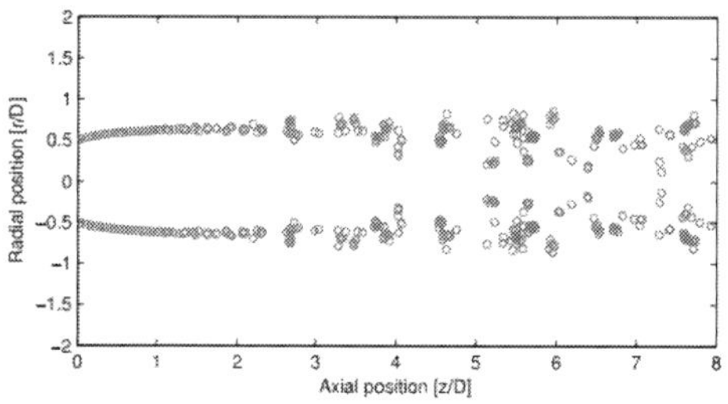

Figure 4. Simulated vortex ring distribution with $C_T = 8/9$

2.2. The Large Eddy Simulation model

The coupled VRM and far-wake models are compared with a LES based Energy-Conserving Navier-Stokes (ECNS) solver, which is a CFD code developed by ECN. It discretises the incompressible Navier-Stokes equations on a uniform Cartesian grid and solves them using a Finite-Volume approach. Central to the code is the use of energy-conserving schemes that guarantee the absence of numerical dissipation, not only through spatial-discretisation but also through temporal-integration [10]. With wind turbine or farm simulations, the presence of numerical dissipation is known to spuriously speed-up wake recovery, leading to the under prediction of wake deficit, which is averted in the ECNS code. To model the large, energy-containing scales of the flow, the ECNS uses the Smagorinsky model with the Smagorinsky constant C_S, set to 0.15 [11]. For simulating a single turbine, the inflow is assigned a simple Dirchlet condition, whereas the rest of the boundaries are outflow boundaries that prohibit back-flow. The turbines are modelled as uniformly loaded actuator disks that are introduced in the computational domain as body forces. Furthermore, the time-step is set to $0.01s$ to ensure that the solution is independent of the time-step. Further details regarding the LES code can be found in Mehta et al. [12].

3. Validation of the VRM with wind tunnel experiments

Experiments with a two-bladed rotor and an actuator disc represented by a porous mesh were carried out by Lignarolo et al. [13]. They provided the velocity field of the wake of the experiment for a comparison with the VRM. The experiments were conducted in the Open Jet Facility (OJF) at the Delft University of Technology. The experiments were conducted with a high thrust coefficient of $C_T = 0.93$, which is above normal operational conditions. This relatively high load stresses the model and approaches the limit. Higher values for C_T will result in instabilities with heavy ring interaction and rings travelling upstream. An overview of remaining parameters of the experiment is presented in table 1. Figure 5 shows velocity profiles at various distances

Table 1. Parameters of the wind tunnel experiment from Lignarolo et al. [13]

Parameter		Wind Turbine	Actuator
Diameter	D	0.6 m	0.6 m
Free-stream velocity	U_∞	4.7 m/s	4.7 m/s
Rotational frequency	ω	109.3 rad/s	-
Reynolds number (root)	Re_{cr}	32,000	-
Reynolds number (tip)	Re_{ct}	96,000	-
Reynolds number (diameter)	Re_D	188,000	188,000
Thrust coefficient	C_T	0.93	0.93
Tip speed ratio	λ	6.97	-
Turbulence Intensity	TI	0.5%	0.5%

down-wake from the rotor disk. Each graph shows the actuator and rotor measurements from the wind tunnel and the simulation results, in which a distinction is made between time-averaged (TA) and instantaneous (I) results. The time-averaged results show velocities averaged over two time periods $2\Delta t$, with a sample rate of 500 samples per period. This is done to verify if the instantaneous velocity field is representative for a time independent period. A condition is set to disregard a ring when it is within $0.02D$ of the axial position where induced velocities are determined to avoid the ring singularities. The instantaneous results were taken at the end of the simulation, where the vortex rings have reached a minimum distance of $20D$ and the exact axial position was taken between the closest two vortex rings. Wake deficits and expansion rates are in reasonable agreement between the VRM and the experimental results. The results of the instantaneous velocity field is in accordance with the time-averaged values. The instantaneous velocity field simulation has a much shorter computational period which is a significant advantage over the time-averaged method. Vortex ring expansion and interaction can be seen in figures 6 and 7, where the positions of the vortex rings at a fixed time instant are overlaid over the average velocity field of the experimental turbine and porous disc respectively. The display of strong vortex ring interaction seems to correspond with the destabilisation and growth of the wake shear layer, where the near-wake transitions to the turbulent far-wake. Further investigation would however be required to understand the relation between the mutual instabilities predicted by the dynamic (inviscid) vortex model and wake breakdown observations. Theoretically, the VRM should show a closer relation to the actuator disc when compared to the rotor, since the vortex rings are shed from an actuator disc and wake rotation is not modelled in the VRM. Figure 5 shows that the velocities from the porous disc experiment are closer to the VRM as compared to the 2-bladed turbine, but the deficit is still a bit underestimated.

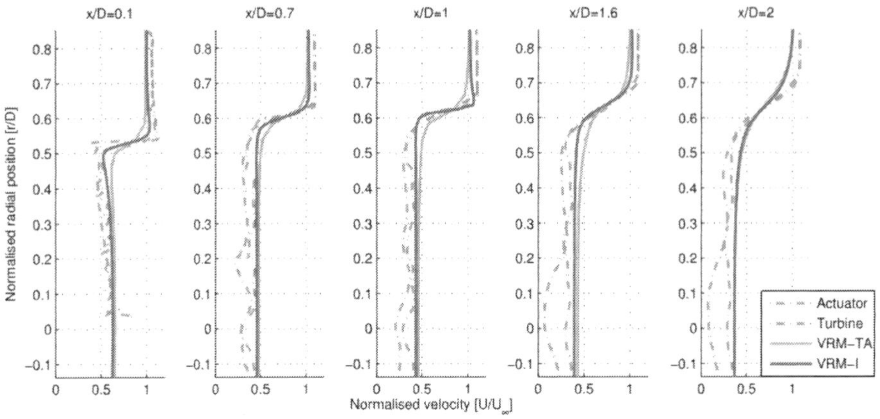

Figure 5. Comparison of the near-wake VRM with wind tunnel measurements showing velocity profiles of the porous disc, two-bladed turbine, time-averaged VRM and instantaneous VRM

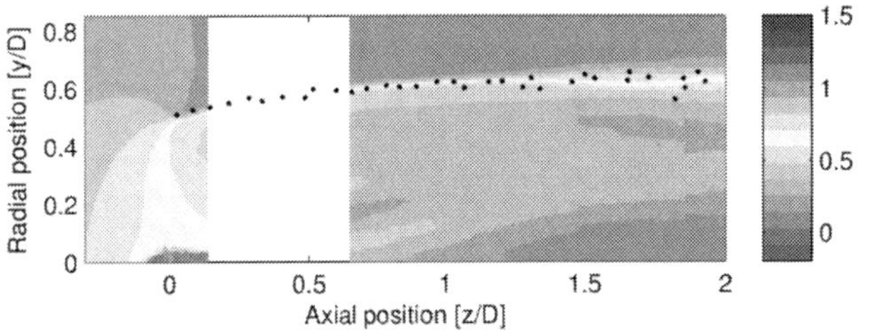

Figure 6. Time averaged normalised velocity field from the 2-bladed rotor. Overlayed are positions of the simulated vortex rings represented by black dots

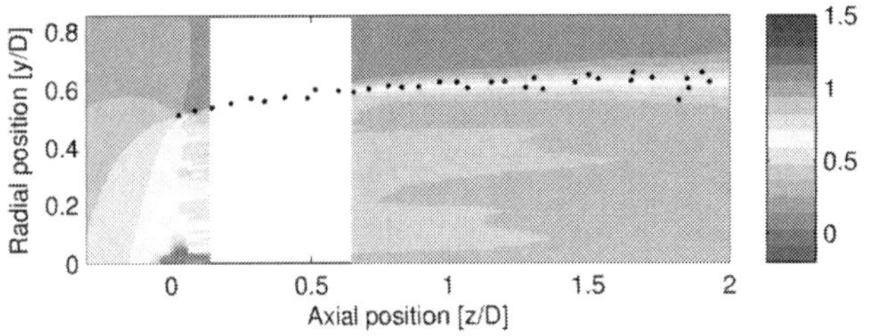

Figure 7. Time averaged normalised velocity field from the porous mesh. Overlayed are positions of the simulated vortex rings represented by black dots

4. Coupling the near and far-wake models

The VRM is coupled with the Jensen and Larsen wake deficit models and discussed in the following two sections. Results of the coupled simulation are discussed in section 5.

4.1. Coupling of the VRM with the Jensen model

The Jensen or PARK model is a engineering type far-wake model based on a linear expanding wake and the conservation of momentum [3], although Bastankhah and Porté-Agel have shown that it is actually only based on mass continuity [14]. The wake expansion is prescribed by the wake decay coefficient (WDC) k, which depends on the free-stream atmospheric turbulence and terrain roughness. The Jensen model does not distinguish the near- and far-wake. Although limited in physical representation, the model is popular due to its simplicity and low computational cost. For these reasons, the VRM is coupled to the Jensen model as a preliminary investigation. The VRM is used to determine the axial velocities at the end of the near-wake. It is found that at a distance of $2D$, the initial velocity deficit is converged when using $C_T = 8/9$, but this distance is of course sensitive to the operating C_T. This velocity deficit is then used to determine the disc averaged axial induction input to the Jensen model, resulting in an equal velocity at $2D$. The Jensen model is used to calculate the far-wake from $2D$ onwards, with the solved value for a and the given value for k. The initial expansion of the wake is fitted to match the expansion from the VRM. In summary, the VRM is used to determine the initial induction velocities and expansion for the Jensen model.

4.2. Coupling of the VRM with the Larsen EV model

The VRM was coupled to the wake deficit model of the Dynamic Wake Meandering (DWM) model introduced by Larsen et al [4]. The structure of the deficit model introduced in the original DWM, consists of three parts: a uniform inflow or start deficit from the upstream turbine, an initial expansion and wake deficit, determined by a near-wake model and finally, a far-wake model based on the EV turbulent mixing concept by Ainslie [15]. In the original model, near-wake velocity deficits and edge expansion is determined with a streamtube momentum model or a blade element momentum (BEM) model, if blade characteristics are known. This results in an initial velocity deficit at a position where the pressure has fully recovered, often taken to be $2D$ downstream. The disadvantage of the streamtube model is that the deficit is only determined at the end of the near-wake at an assumed position. Simultaneously, a hat profile is taken as an initial far-wake profile, whereas experimental results clearly show substantial thickening of the shear layer towards the end of the near-wake. The BEM model results in a more realistic wake profile but requires blade paramaters and pitch angles of the modelled wind turbine.

The Larsen model is well suited for coupling to the VRM, since the model explicitly distinguishes the near- and far-wake. In the present approach, the streamtube model is replaced by the VRM, from which the axial and radial velocities were determined at a position of $2D$. This velocity field is then extracted as an initial condition for the EV far-wake. In a multiple wake simulation, an area weighted average of the velocity on the down-wake rotor is determined from the results of the far-wake simulation. This weighted average is subsequently used to determine the strength of the vortex rings released from the second turbine. The induced velocities from the series of vortex rings are then added to the already calculated velocity field from the previous turbine in a region from $-2D$ towards $+2D$ with respect to the down-wake turbine. This results in a two-way interaction of the VRM and the Larsen EV model.

In Larsen's far-wake model, the turbulent mixing process is described with rotationally symmetric Navier-Stokes equations, where the pressure is assumed to be constant. The thin shear layer approximation is applied to the far-wake flow, i.e. radial flow derivatives are considered to be much larger then axial derivatives. This results in the momentum and continuity equations displayed in equation 8 and 9 in cylindrical coordinates.

$$U\frac{dU}{dz} + V\frac{dU}{dr} = -\frac{1}{r}\frac{d}{dr}(r\overline{uv}) \tag{8}$$

$$\frac{V}{r} + \frac{dV}{dr} + \frac{dU}{dz} = 0 \tag{9}$$

In these equations, U represents the velocity in the axial direction (z) and V the velocity in the radial direction (r). $\overline{uv}$ denote the temporal averaged fluctuating velocity components. Using the concept of EV, these Reynold stresses are expressed as:

$$-\overline{uv} = \nu_t\frac{dU}{dr} \tag{10}$$

The EV term ν_t consists of a laminar part, driven by the velocity difference across the shear layer and a turbulent part, based on the turbulence intensity, TI. It further includes filter functions for the build-up of turbulence and empirical constants. The latest derivation by Madsen et al. [16] and Keck et al. [17] was used in the present implementation of the Larsen model. The equations are discretised on a staggered uniform grid with a finite difference scheme, using an upwind scheme for the axial derivatives and a central scheme for the radial derivatives. This results in two non-homogeneous linear systems of equations for a single radial position.

5. Results of a VRM-Jensen and VRM-Larsen comparison with LES

The results for a single turbine case in laminar flow from the VRM-Larsen model are compared with the results from the LES model. For a single turbine, the domain is $12D, 8D, 8D$ in size and uses $120, 80, 80$ grid points with the actuator disk located at $4D$, to ensure a stable inflow. The simulation was run with a laminar inflow, i.e. turbulence in the domain was solely due to the actuator disks, in accordance with the tests carried out with the VRM-Larsen model. The LES simulations were carried out on the Delft University of Technology Beowulf cluster.

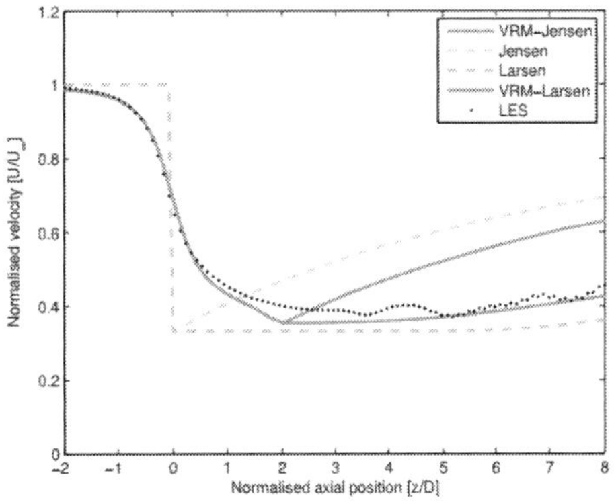

Figure 8. Normalised axial velocities along the centreline of the wake from the Jensen, Larsen, and LES simulations

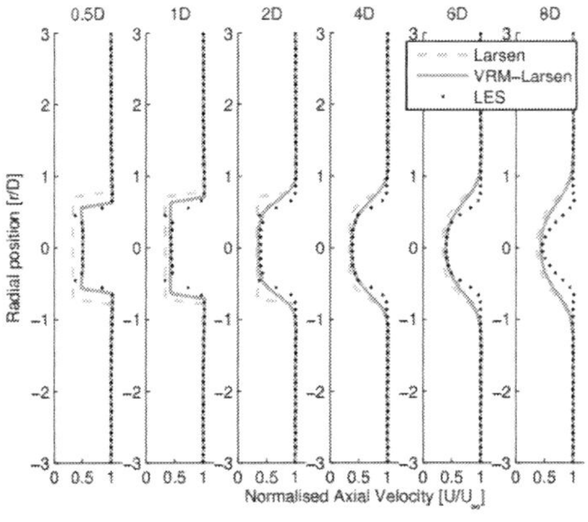

Figure 9. Wake profiles at various positions in the wake from the Larsen, VRM-Larsen and LES simulations

To approach the laminar inflow condition best in the Jensen model, a value for k of 0.03 was used, which is lower than the recommended low turbulence offshore value of 0.04. A value of 0 would simply result no wake recovery. The results can be seen in figure 8. The VRM-Jensen model generally results in a more gradual wake recovery as compared with the reference Jensen model. A lower value for k would perhaps result in closer fit with the LES, to which the model could be tuned to make it more suitable for laminar inflow. The coupling with the Jensen model was considered as a concept study and thus no further investigations were performed with it. A follow-up coupling with a more advanced model based on Jensen (e.g. the model derived by Bastankhah and Porté-Agel [14]) might lead to more favourable results.

The VRM-Larsen model shows a reasonable resemblance with the LES simulation in the near- and far-wake, as can be seen in figures 8 and 9. The streamtube near-wake model which was used in the Larsen model seems to overpredict the wake losses in the near-wake, which results in a lower centreline velocity in the far-wake.

The LES results show some unnatural behaviour along the edge of the rotor, where the velocity in the near-wake experiences a sudden drop which is visible in the first three wake

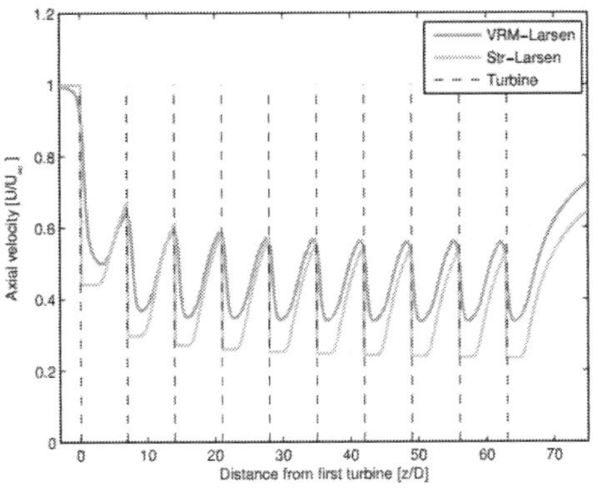

Figure 10. Velocities along the centreline for row E at Horns Rev in a full parallel wind direction

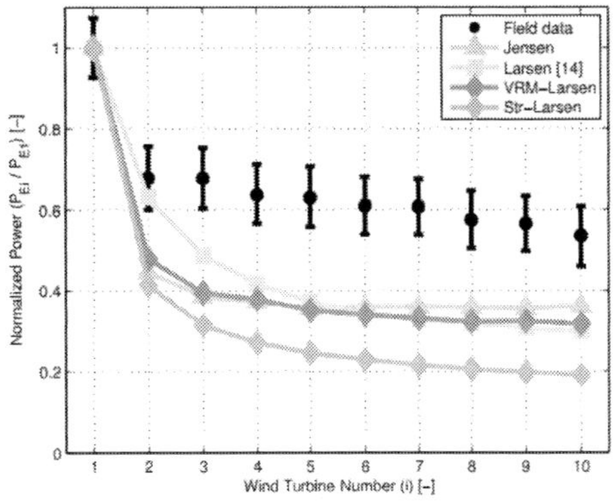

Figure 11. Normalised power in row E at Horns Rev. Field data taken from Gaumond et al. [18]

samples of figure 9. This is the result of the implementation of a circular actuator disc on a Cartesian grid of limited resolution. This could be circumvented by using a Gaussian distribution to apply the force more gradually on finite volumes, but this is not yet implemented in the ECNS code. A simulation with an increase in resolution on a smaller domain was carried out and showed a reduction of this effect. An increase of resolution on the same domain was not possible with the available computational power. The LES results also show a contraction of the wake at $8D$, which was not observed by the other models. This could be a side-effect of using laminar inflow or the used Smagorinsky model, which is dissipative in nature and possibly resulting in a quick dissipation of the turbulence created by the actuator disc.

The results from the LES calculations show that wake recovery due to turbine-generated turbulence in the far-wake starts later than the previously assumed distance of $2D$. Ideally, the VRM near-wake should be extended to a position between $2D - 4D$, however strong vortex ring interaction after $2D$ prohibited a converging velocity field in this region.

6. Discussion and recommendations from the LES results

The VRM-Larsen model has shown a reasonable accordance to a single wake LES simulation in laminar flow, with less than 1/100th of the required computational time. This encouraged further testing of the model and a multiple wake simulation was performed and compared with power deficit data from Horns Rev. A benchmarking case from Gaumond et al. [18] is used as a reference. The case involves a wind direction parallel to a wind farm row (270° West), with an uncertainty in the 10 minute average wind direction measurements of $\pm 2.5°$. The array consists of 10 Vestas V80 wind turbines, spaced at $7D$. The thrust coefficient of the wind turbines is taken as $C_T = 0.805$ in accordance with the wind turbine parameters from Hansen et al. [19] and is assumed constant between 5 and 9 m/s. The undisturbed wind speed was measured as $8m/s \pm 0.5m/s$.

The velocities along the centreline through the array of wind turbines is displayed in figure 10 for the Larsen and VRM-Larsen models. The figure shows influence of the VRM on the wake of the upwind turbine. The model captures the deceleration of the flow in front of the rotor, whereas the Larsen model only carries information down-stream.

Figure 11 shows the normalised power output of the turbines for the specified case. Included are the field measurements, results from the Jensen and the Larsen models as implemented by Gaumond and the VRM-Larsen and streamtube-Larsen models as implemented by the author. All models underpredict the power output of the wind turbines. Gaumond attributes this to uncertainty in the wind direction of the data, which is highly likely since it is 10 minute averaged and the western met mast is located relatively far from the wind farm, compared to the eastern met mast. The difference between Larsen by Gaumond and Streamtube-Larsen is initially surprising, since both models are based on the same EV model. However, the difference can be attributed to a number of uncertainties which are not addressed in detail in the paper by Gaumond. It is likely that the BEM near-wake model is used in Gaumond's Larsen simulation since the turbine parameters were known. In this near-wake model, the hub is disregarded and the airflow is allowed to pass without interruption, which results in higher velocities in the centre of the wake. The exact empirical constants used in Gaumond's EV-term are also unclear. These constants determine the relative influence from ambient turbulence in wake recovery and the results were found to be rather sensitive to variations in these empirical constants. The Larsen model by Gaumond might have been supplemented by correction models for turbine added turbulence or wake meandering from the DWM, which would have an effect on the power prediction. Lastly, the power prediction is perhaps averaged over different wind directions ($\pm 2.5°$), which is not implemented in the streamtube-Larsen and VRM-Larsen models.

Due to these uncertainties, the comparison was found to be inconclusive and further analyses of the model is required. Ongoing work will investigate a two-turbine case and compare the VRM-Larsen model with the LES calculations, in order to better understand discrepancies in the models and observations. This will also make for a fairer comparison, since all the parameters used in the LES are known. Such an approach also presents the opportunity to implement yawed inflow cases, ground effects and wake meandering.

7. Conclusion

The near-wake model based on ideal inviscid vortex rings has been validated against wind tunnel measurements. The model shows a reasonable to good agreement with the actuator disc represented by a porous mesh in terms of development of wake deficit and expansion.

The model is successfully coupled to the Jensen and Larsen far-wake models. The VRM-Jensen model shows a more gradual wake recovery and was developed as a proof of concept.

The VRM provides a good alternative for the streamtube momentum model that is used in the Larsen DWM model. The VRM provides the ability to describe an axi-symmetric near-wake velocity field. The resulting thicker shear layer and different velocities at the end of the near-wake influence the far-wake in a positive way with respect to LES results in a laminar flow, single turbine case. The VRM-Larsen model approaches the LES, whilst maintaining an engineering-type computational cost.

In future work, a two-turbine case will be designed for a LES comparison. This provides the opportunity to implement other wake effects in the near-wake, e.g. yawed inflow and ground effect or transfer functions from the LES or DWM, e.g. meandering in the far-wake and rotor added turbulence.

Acknowledgments

The authors would like to thank Lorenzo Lignarolo for sharing his results from the PIV measurements performed in the TU Delft OJF wind tunnel. Wei Yu is also acknowledged for her assistance in the implementation of the model.

References

[1] Segalini A and Alfredsson P H 2013 A simplified vortex model of propeller and wind-turbine wakes *Journal of Fluid Mechanics* **725** 91–116 ISSN 0022-1120

[2] Baldacchino D and van Bussel G 2014 Wind turbine wake stability investigations using a vortex ring modelling approach *Journal of Physics: Conference Series* **555** 012111 ISSN 17426596

[3] Jensen N 1983 A Note on Wind Generator Interaction Tech. Rep. Risø-M-2411 DTU Roskilde

[4] Larsen G C, Madsen H A, Torben J and Troldborg N 2008 Wake modeling and simulation Tech. Rep. Risø-R-1653(EN) DTU Roskilde

[5] de Vaal J B, Hansen M O L and Moan T 2014 Validation of a vortex ring wake model suited for aeroelastic simulations of floating wind turbines *Journal of Physics: Conference Series* **555** 012025 ISSN 1742-6588

[6] Yu W, Ferreira C J S, van Kuik G and Baldacchino D 2014 Vortex ring model analysis of an actuator disc with dynamic and non-uniform load *Paper in preparation*

[7] Yoon S S and Heister S D 2004 Analytical formulas for the velocity field induced by an infinitely thin vortex ring *International Journal of Numerical Methods in Fluids* **44** 665–672

[8] van Kuik G A M 2003 The edge singularity of an actuator disc with a constant normal load *22nd AIAA/ASME Wind Energy Symposium* (Reno) ISBN 1563475944 ISSN 0022-4502

[9] Baldacchino D 2012 *Horizontal Axis Wind Turbine (HAWT) wake stability investigations* Master's thesis Delft University of Technology

[10] Sanderse B 2013 *Energy-conserving discretization methods for the incompressible Navier-Stokes equations* Ph.D. thesis Eindhoven Univsersity of Technology

[11] Smagorinsky J 1963 General circulation experiments with the primitive equations I. The basic experiment *Monthly Weather Review* **91** 99–164 ISSN 0036-8075

[12] Mehta D, van Zuijlen A and Bijl H 2014 Energy-conserving schemes for wind farm aerodynamics *Journal of Physics: Conference Series* **524** 012157 ISSN 1742-6596

[13] Lignarolo L E M, Ragni D, Ferreira C J S and van Bussel G J W 2014 Kinetic energy entrainment in wind turbine and actuator disc wakes: an experimental analysis *Journal of Physics: Conference Series* **524** 012163 ISSN 1742-6596

[14] Bastankhah M and Porté-Agel F 2014 A new analytical model for wind-turbine wakes *Renewable Energy* **70** 116–123 ISSN 09601481

[15] Ainslie J 1988 Calculating the flowfield in the wake of wind turbines *Journal of Wind Engineering and Industrial Aerodynamics* **27** 213–224

[16] Madsen H A, Larsen G C, Larsen T J, Troldborg N and Mikkelsen R 2010 Calibration and Validation of the Dynamic Wake Meandering Model for Implementation in an Aeroelastic Code *Journal of Solar Energy Engineering* **132** 041014 ISSN 01996231

[17] Keck R E, Veldkamp D, Madsen H A and Larsen G C 2012 Implementation of a Mixing Length Turbulence Formulation Into the Dynamic Wake Meandering Model *Journal of Solar Energy Engineering* **134** 021012 ISSN 01996231

[18] Gaumond M, Réthoré P, Bechmann A, Ott S, Larsen G and Pena Diaz A 2012 Benchmarking of wind turbine wake models in large offshore wind farms *The Science of making Torque from Wind* (Oldenburg)

[19] Hansen K S, Barthelmie R J, Jensen L E and Sommer A 2012 The impact of turbulence intensity and atmospheric stability on power deficits due to wind turbine wakes at Horns Rev wind farm *Wind Energy* **15** 183–196

Numerical and Experimental Methods for Wake Flow Analysis in Complex Terrain

Francesco Castellani[1], Davide Astolfi[1], Emanuele Piccioni[1], Ludovico Terzi[2]

[1] Department of Engineering, University of Perugia, Perugia, Italy
[2] Renvico srl, Via San Gregorio 34, Milano 20124, Italy

E-mail: `francesco.castellani@unipg.it`

Abstract. Assessment and interpretation of the quality of wind farms power output is a non-trivial task, which poses at least three main challenges: reliable comprehension of free wind flow, which is stretched to the limit on very complex terrains, realistic model of how wake interactions resemble on the wind flow, awareness of the consequences on turbine control systems, including alignment patterns to the wind and, consequently, power output. The present work deals with an onshore wind farm in southern Italy, which has been a test case of IEA-Task 31 Wakebench project: 17 turbines, with 2.3 MW of rated power each, are sited on a very complex terrain. A cluster of machines is investigated through numerical and experimental methods: CFD is employed for simulating wind fields and power extraction, as well as wakes, are estimated through the Actuator Disc model. SCADA data mining techniques are employed for comparison between models and actual performances. The simulations are performed both on the real terrain and on flat terrain, in order to disentangle the effects of complex flow and wake effects. Attention is devoted to comparison between actual alignment patterns of the cluster of turbines and predicted flow deviation.

1. Introduction

The assessment of power output quality of wind farms operating in complex terrains is extremely challenging. The present-day abundance of SCADA control systems and the increasing computational power have provided a significant boost to the interpretation of wind farm performances. Basically, three are the main drivers: the wind flow itself, which might have rather extreme features on peculiarly complex sites, wake interactions and the response of the control system to the intertwining of these two phenomena. These challenges even force to revisit techniques which are well established for performance assessment of offshore wind farms. For example, in [1] it has been shown that, for onshore wind farms, farm efficiency analysis as a function of the wind direction (commonly known as polar efficiency) can not be tackled as in the offshore case [2, 3, 4, 5, 6] because it is not mathematically consistent. A novel definition is therefore proposed, which is by construction mathematically consistent; its physical interpretation is presented through three test cases, on terrains of increasing complexity (from very gentle to extremely challenging). The most complex wind farm analysed in [1] is also a test case of the IEA-Task 31 Wakebench project [7], aiming at developing a framework for the evaluation of wind farm flow models operating at microscale level, and it is also the object of this investigation. The terrain of this wind farm is indeed quite steep, with very high slopes

Journal of Physics: Conference Series **625** (2015) 012042 doi:10.1088/1742-6596/625/1/012042

(up to 60%) close to the turbines; also the layout is complex and large, resulting in non-trivial combination effects of complex wind flow and wake interactions. For some precious insight on wind flow on complex terrains, see for example [8] and [9].

The aim of the present paper is heavily combining the approaches above summarized with experimental SCADA data analysis. Wakes and topographical flow acceleration manifest in meandering wind, which the nacelle of the turbine under wake can not follow optimally. Therefore, clusters of turbines tend to behave as a whole, rather than a collection of individuality, and peculiar alignment patterns arise. In [10] it is shown, on the test case of two subclusters lying on very different terrains, that recurrent alignment patterns display huge performance deviations with respect to each other. In the present Paper, precisely one of the subclusters of turbines analysed in [10] is investigated, through numerical and experimental analysis. Further, the numerical simulations shall be performed both on the actual terrain and on flat terrain, in order to try to disentangle the effect of terrain-induced wind flow acceleration from pure wake effect.

The structure of the Paper is as follows: in Section 2 the test case wind farm is briefly described, in Section 3 the numerical methods and the SCADA data post-processing techniques are introduced. In Section 4 the results are collected, and finally in Section 5 some concluding remarks are sketched and further directions of the present work are proposed.

2. The Wind Farm

The layout of the test case wind farm is sketched in Figure 1: seventeen aerogenerators are installed on site, with 2.3 MW of rated power at around 12 m/s; the rotor diameter is 93 meters and the hub height is 80 meters. Six met-masts have been installed for site assessment: nowadays the only active, whose data shall be employed for the present work, is the one slightly north of the turbine SGM10, collecting data at hub height. From the topographic contour lines of Figure 1, the challenging complexity of the site is manifest: actually slopes up to 60% can be found in proximity of the turbines. The severity of the slopes has been quantified through the Ruggedness Index (RIX value), computed at a given point through the Wind Atlas Analysis and Application Program [11, 12] as the fractional extent of the surrounding terrain which is steeper than a certain critical slope. RIX values at turbine sites are displayed in Table 1. The site, which has been chosen as test case of the IEA-Task 31 Wakebench project [7], is also characterised by significant icing and lightning effects.

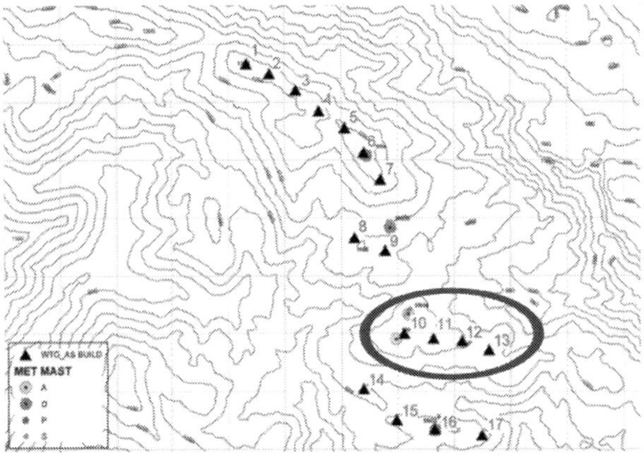

Figure 1. Wind farm layout.

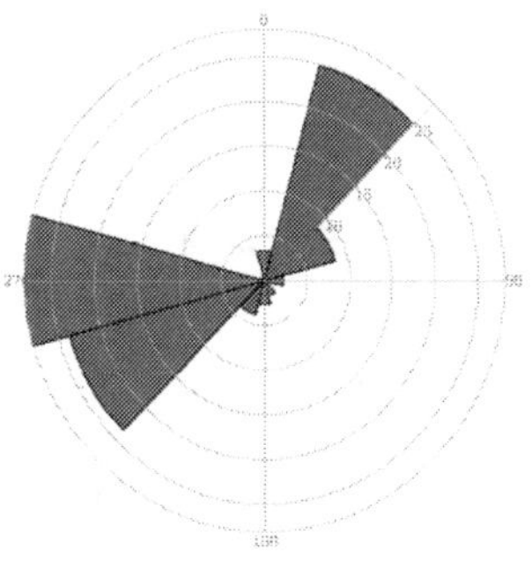

Figure 2. Wind rose: percentage of occurrence.

Journal of Physics: Conference Series **625** (2015) 012042 doi:10.1088/1742-6596/625/1/012042

Table 1. Ruggedness Index at turbine site.

Turbine	RIX value (%)
SGM10	24.9
SGM11	23.1
SGM12	21.7
SGM13	20.4

The analysis of the present Paper is focused on the subcluster SGM10-SGM13 of the wind farm, which is zoomed in Figure 3.

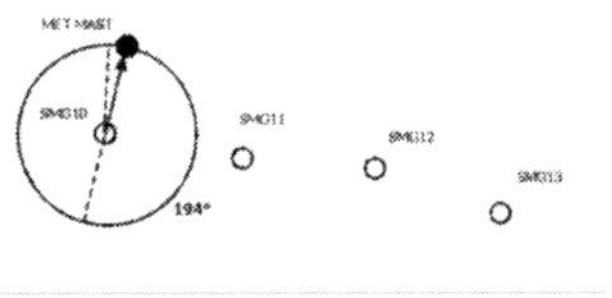

Figure 3. Layout of SGM10-SGM13 subcluster and met-mast

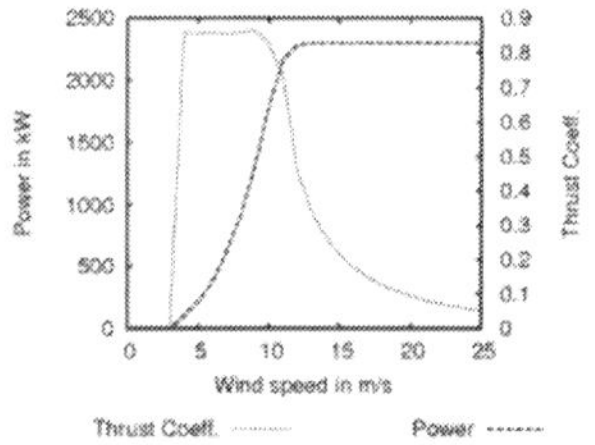

Figure 4. Wind turbine power and thrust curve.

The inter-turbine distance is such that considerable wake effects are expected: actually all the couples of nearby turbines of the subcluster are at around 2.8 rotor diameters one from the other. Further, the complexity of the terrain induces non-trivial wind flow, such that, from the very populated 270° sector (as shown in the wind rose of Figure 2), turbine SGM13, which is the most downstream, is by far the best performing [1]. Also, the met-mast lies very close to the cluster, and it can be considered a reliable landmark for undisturbed wind flow. On the grounds of the considerations above, the subcluster SGM10-SGM13, when the wind blows from the 270° sector, has been chosen as testing ground of the present work.

3. The numerical and experimental methods

The numerical simulations have been performed with the WindSim numerical tool [13]: this code is based on the numerical core PHOENICS, which solves the Reynolds Averaged Navier-Stokes (RANS) equations on a Cartesian grid; different solvers can be used and in the present work the General Collocated Velocity method (GCV) is employed for its capabality in reaching convergence on very steep terrains. The objective is simulating the behaviour of the SGM10-SGM13 cluster of turbines when there are certain conditions at met-mast. Two test cases have been chosen: 10 and 6 m/s at met mast. The inlet boundary is assumed to be a logarithmic profile blowing from 270°, whose intensity at the top of the boundary layer (assumed equal to 800 meters above ground level) is iteratively adjusted, in order to obtain 10 (6) m/s at hub height at met-mast position. The procedure is as follows: iterations are run, in order to provide an acceptable degree of convergence and stability. Subsequently, 100-iterations blocks are run. At the end of each of them, the target values are checked: in this case, wind intensity at met-mast. Deviation from the pre-run and from the expected results are evaluated, and boundary layer values are finely tuned for a new run, and so on. The two regimes have also been investigated by the point of view of free wind flow, producing a prediction for the undisturbed wind speed, wind direction and turbulence intensity at each turbine site. The presence of the rotors has been simulated through the Actuator Disc model [14, 15, 16, 17] both on the real terrain and on flat terrain. The turbulence model is RNG k-ϵ. It has been checked that it provides better convergence with respect to k-ω and modified k-ϵ, despite its inherent inability of optimally resembling the near-wake wind field. As shall arise throughout the discussion, the stream flow is driven mainly by the complexity of the terrain, and for this reason RNG k-ϵ provides better results. Similar considerations arise in [18], where the complex terrain of Askervein hill (also a test case of IEA-Task 31 Wakebench project [7]) is analyzed: it is shown that RNG k-ϵ is the computational scheme providing better performances. The computational domain for the AD model is divided in a coarse area and a refined area: the former is necessary to simulate the main wind field blowing in the mesoscale, the latter is needed for evaluating the interaction between the wind field and the turbines, modeled by the disk, on a local scale through the thrust coefficient. A grid independence study has been performed as a preliminary issue for the objectives of IEA-Task 31 Wakebench project [7]. For the present work, in particular, peculiar attention is deserved to grid independence of wind intensity at upstream turbine SGM10. The refined area employs a grid based on 22 subdivisions per turbine diameter. In [17] it is shown that results with the AD model do not depend on refined area grid resolution, as long as there are not less that 16 subdivisions per turbine diameter. The power production P is estimated through an integral post-processing method [16], using the local wind speed and the maximum pressure drop across the rotor area A (Equation 1).

$$P = \int_A v \cdot dp \cdot dA \simeq \sum_i \xi_i \Delta p_i v_i \tag{1}$$

The integral of Equation 1 is discretized using weights ξ_i associated to the grid resolution, Δp_i is the local pressure drop and v_i is the local wind speed. The weights ξ_i are normalized to 1 and are proportional to the ranking in relative distance of the i-th subdivision from the center of the diameter, because the higher the distance from the center of the disk, the higher the annular area spanned, and therefore the torque and the extracted power. The output of the AD model is therefore the active power, from which the undisturbed wind speed v_∞ is obtained through the power curve (see Figure 4), the turbulence intensity and the wind direction. The details of the computational grids are summarized in Table 2. The volume is described by terrain-following surfaces, close to the topographic surface, and flat surfaces at the top of the computational domain. Figures 5 and 6 show the computational grid at the level of terrain surface.

Table 2. The computational domains.

Model	Cells (X, Y, Z)
Free Flow	(279, 215, 30)
AD model + terrain	(256, 111, 61)
AD model + flat	(236, 103, 61)

The two simulation regimes have been chosen for the following reasons:

- 10 m/s is a critical value because, within one typical standard deviation, velocities associated to rated power are approached. The AD model clearly can not take into account control system dynamics approaching rated power, and therefore is expected not to reproduce brilliantly the interaction between rotor and wind flow.

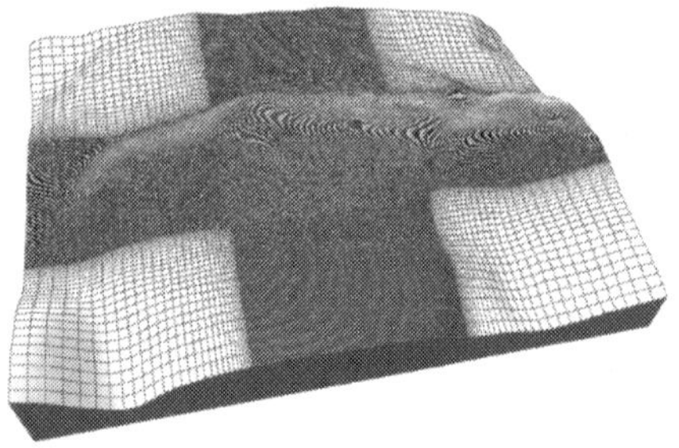

Figure 5. The computational domain: free flow model on actual terrain

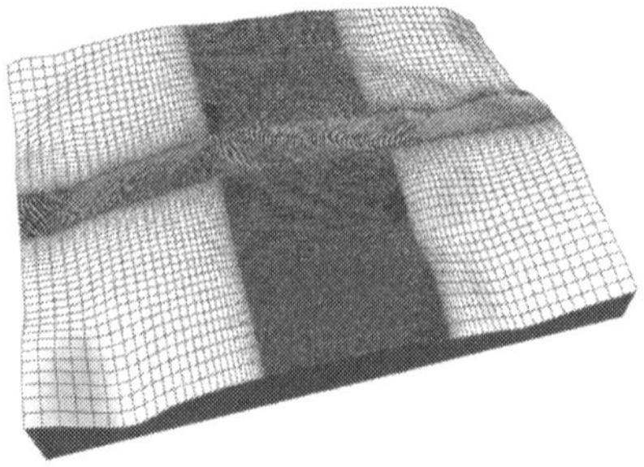

Figure 6. The computational domain: AD model on actual terrain

- 6 m/s has been instead chosen because, when the wind speed is of this order of magnitude, the thrust coefficient is high and the dynamics is strongly driven by wakes: it is therefore interesting to inquire how much of the actual performances is captured by a pure wake model, i.e. the AD on flat terrain.

SCADA data have been post-processed in order to reproduce the regimes simulated through the numerical models. Basically, three filters have been employed. First, the time steps are selected during which the subcluster of interest is producing output in unison. Subsequently, filters have been applied on the wind speed as measured by met-mast (correspondingly to the simulations regimes above), and on the orientation of the upstream SGM10 turbine ($270° \pm 5°$ direction). Thresholds for filtering have been estimated as follows: first, the relative error on each wind speed measurement has been computed. This data set has been averaged for a global estimate, amounting at 18%. Therefore, a conservative criterion has been chosen, which guarantees precise resembling to simulated regimes and fair data sets population: 10% of tolerance around the two wind speed values (10 and 6 m/s). Further, the AD model assumes disk orientation under orthogonal directions. The wind is expected to be somehow distorted at turbines SGM11, SGM12, SGM13 site, while within a certain extent it is not at turbine SGM10 site, since this machine is upstream. Similarly, while turbines SGM11, SGM12 and SGM13 are not expected to exactly orientate to the wind in the real world due to turbulence and complexity effects, turbine SGM10 within a certain extent is, because it is upstream when the wind blows from the 270° sector. Therefore it is reasonable to suppose that SGM10 satisfies the assumptions of the AD model and, correspondingly, to filter SCADA measurements in a way which resembles these assumptions. Since the order of magnitude of the standard deviation of nacelle position SCADA measurements is 5°, this value has been chosen as amplitude of tolerance on SGM10 orientation measurements, around the 270° centre. The filter on nacelle positions deserves a deeper discussion: actually it has been checked if the measurements are vexed by a systematic offset. The speed-up ratio between met-mast and the nearest turbine SGM10 has been computed and averaged against 1° amplitude of SGM10 nacelle position intervals: it is expected that it reaches a maximum along the direction joining SGM10 to the met-mast, because the met-mast would be under wake of turbine SGM10. From the layout it is computed that the inclination of the line joining SGM10 to met-mast is 194°. The measured speed-up displays its maximum at 198° and this means that there is a 4° systematic offset in SGM10 nacelle position measurements. The post-processing and the results displayed in Section 4 are to be intended as removed of this systematic bias. This procedure has been adopted, for couples of nearby turbines of the cluster, in order to check reliability of nacelle position measurements of each wind turbine. The results here on are to be intended as removed of this possible bias. On the data sets obtained filtering as described above, wind speed and direction and active power measurements have been averaged and compared against model predictions. The comparison of turbulence intensity model predictions against SCADA measurements deserves a deeper discussion: as pointed out in [19], the Turbulence Kinetic Energy (TKE) is calculated from a closure scheme in CFD models and it is non-trivial to reconstruct from it Turbulence Intensity (TI) and validate against measurements from cup anemometers. In [19] two reconstruction schemes are proposed, which act as lower and upper bounds for comparison against measurements: the approach is validated through the computation of the bias on a testing site in a costal area of Chile. In the present Paper, the following procedure has been adopted: turbulence intensity, both from models and experimental, has been computed at each turbine site referred to the amount of turbulence at met-mast site. The experimental turbulence intensity at met-mast has been computed filtering measurements according to the same criteria summarized above. This technique circumvents the issue of direct comparison of model and experimental TI and guarantess capability of appreciating the trend, moving from one turbine site to an other. The results are collected and discussed in the following Section 4.

4. Results

Figures 7, 8, 9 and 10 summarize the results for the test case of 10 m/s at met-mast and Figures 11, 12, 13 and 14 display the results for the 6 m/s at met-mast case. They are to be discussed together, because they share some basic features, albeit with non-trivial subtleties. First of all, from Figures 7 and 11 it arises that the free flow model fails in capturing the trend of speed intensity moving from SGM10 to SGM13: actually the model predicts a decreasing trend, while SCADA measurements show that a severe speed loss occurs moving from SGM10 to SGM11, but the trend inverts moving towards SGM12 and especially SGM13. Actually SGM13 displays a velocity even slightly higher that the upstream turbine SGM10, also because it is not quite in line with the other machines, and thus sees a much reduced wake. For this reason, actually, the most challenging testing ground of the numerical models is capturing the trend at turbines SGM11 and SGM12. The velocity (and consequently power) collapse moving from SGM10 to SGM11 is brilliantly captured by the AD model on real terrain. As concerns turbulence ratios, the AD model on flat terrain in general better resembles experimental data, especially in the 6 m/s case. When the wind intensity is 6 m/s, actually, the thrust coefficient is high and the turbulence dynamics is mainly driven by wakes. The intensity distribution instead primarily depends on the main stream flow and the amount of energy it moves: this is strongly guided by the terrain. Therefore, it is not surprising that the AD model on flat terrain better reproduces the local turbulence intensity distribution while the AD model on the real terrain best captures the features of wind speed, and therefore power, distribution. A stimulating subtlety is that the AD model on the real terrain at 10 m/s predicts almost a plateau in turbulence ratio, moving from SGM11 to SGM12, while the AD model on flat terrain predicts a severe increase from SGM11 to SGM12. Even though the orders of magnitude of the turbulence ratio predicted by the AD model on the real terrain are overestimated, it is interesting to notice that if one wants to understand the trend of the turbulence distribution, it becomes important to take into account terrain effects, as the wind intensity increases. It is somehow surprising that the AD model on flat terrain best captures the order of magnitude of turbulence intensity ratio: nevertheless remember it is a ratio between a standard deviation and an average, and therefore strong speed-ups to the wind flow might lead to an underestimate. The role of the terrain is fundamental also to understand the alignment patterns: even though nacelle anemometer is not experiencing the effect of the swirl, it is nevertheless useful to compare experimental data against the trend predicted by the models, also because the AD models are capable of encoding the global flow rather than the near-wake. It is expected that the free flow model can not simulate the real alignment patterns of turbines SGM11-SGM12-SGM13, because they do not experience the free stream. Particularly for this reason, the free flow model is included: it is interesting to compare how things are in the real world against how they would be if there was only the free stream and not the turbines. From Figure 10 and 14, it arises that the free flow model predicts a anticlockwise tendency moving from turbine SGM10 to SGM13. Actual nacelle orientations indeed display a similar trend, but with a much higher range of variability: 15° against 5°. These are brilliantly captured, especially at turbine SGM11, by the AD model on the real terrain. The general lesson is that in a complex site, as the chosen test case, the actual nacelle alignment patterns can be numerically estimated only if one takes into account also the terrain. The results demonstrate that the AD model is capable of reproducing a realistic flow deviation from the main stream, resembling the actual alignment patterns, even if it is based on the hypothesis of perfectly orthogonal rotors.

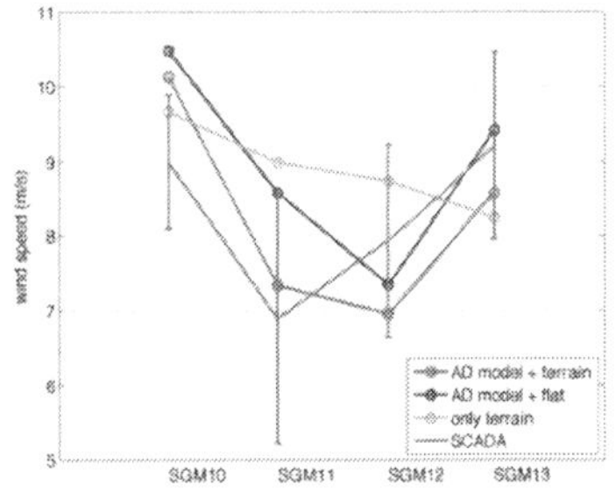

Figure 7. Models vs. experimental data: wind speed. 10 m/s

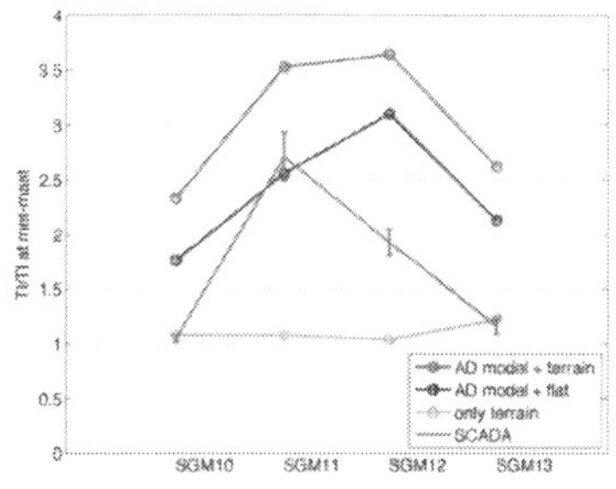

Figure 8. Models vs. experimental data: turbulence intensity relative to met-mast. 10 m/s

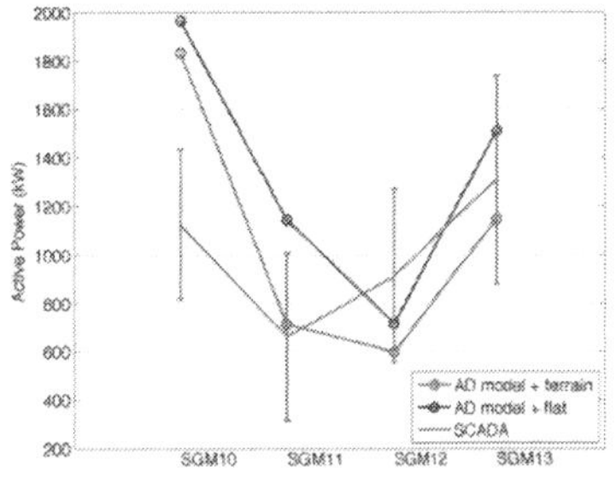

Figure 9. Models vs. experimental data: active power. 10 m/s

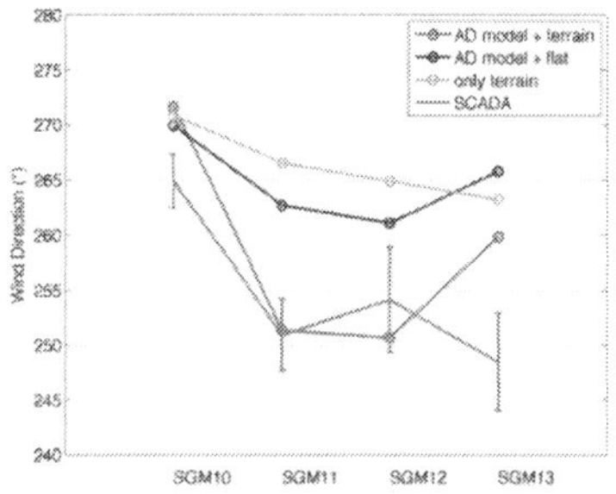

Figure 10. Models vs. experimental data: wind direction. 10 m/s

Figures 10 and 14 show that the wind blowing undisturbed from 270° is northward distorted by the terrain and the presence of the turbines. This basically explains why turbine SGM13, which is apparently the most downstream, actually is the best performing turbine. The flow of turbulence does not affect it and actually it displays almost the same turbulence ratio amount as the upstream turbine SGM10, while instead turbines SGM11 and SGM12 more or less double with respect to SGM10.

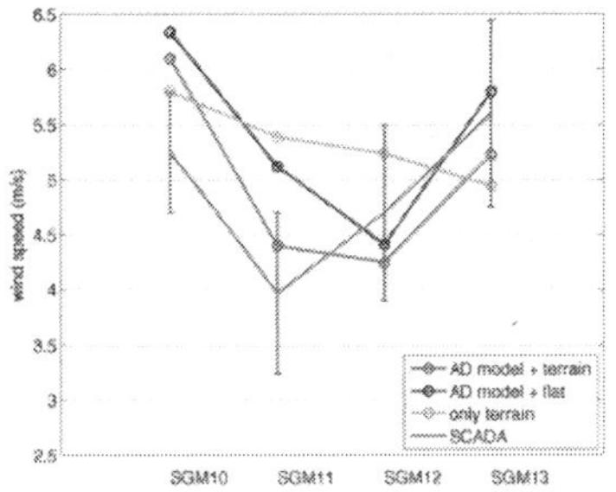

Figure 11. Models vs. experimental data: wind speed. 6 m/s

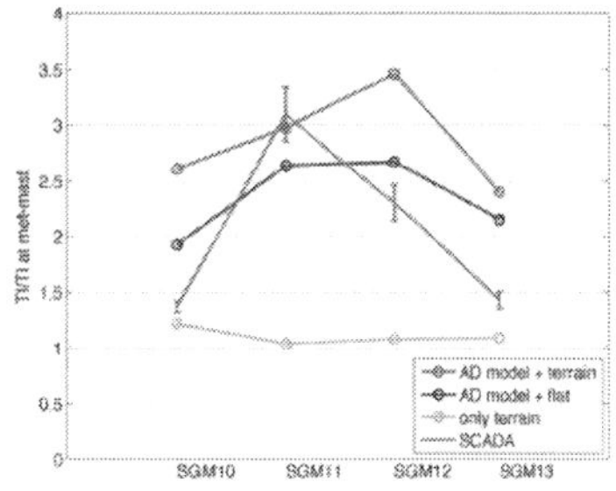

Figure 12. Models vs. experimental data: turbulence intensity relative to met-mast. 6 m/s

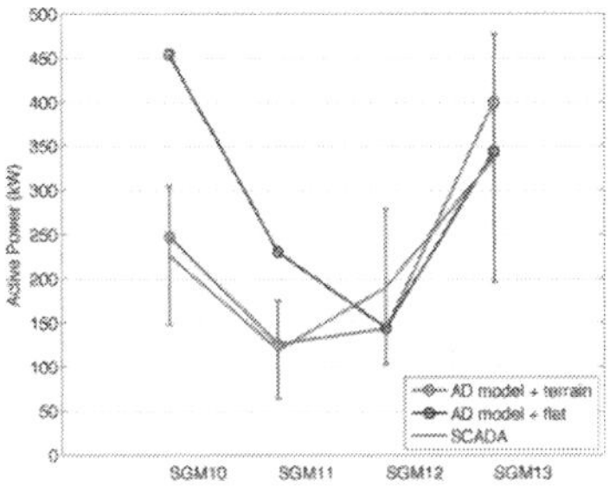

Figure 13. Models vs. experimental data: active power. 6 m/s

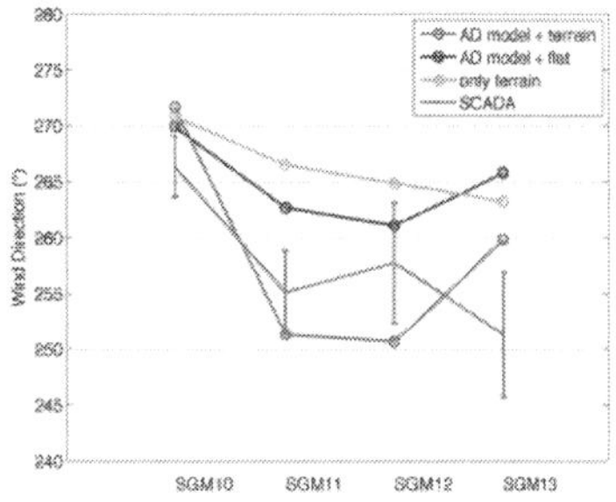

Figure 14. Models vs. experimental data: wind direction. 6 m/s

5. Discussion and Further Directions

In the present Paper, a cluster of turbines of an onshore wind farm sited in southern Italy has been analysed numerically and experimentally. The wind farm has been a test case of the IEA-Task 31 Wakebench project [7] because the very complex terrain conspires with wake effects, resulting in a very stimulating arena, which has also been addressed in [1] by the point of view of farm efficiency as a function of wind direction. The objective of the present work has been interpreting the performances of the selected cluster in terms of wake effects and topography, trying to disentangle them, or at least assessing their relevance. CFD Numerical modelling comes at hand: a free flow model and the AD model, for simulating the presence of the rotors, have been employed. The AD models are placed on the actual terrain and on flat terrain, in order to appreciate the different features of the predictions and comparing them against SCADA measurements. In order to have at hand comparable quantities on experimental and numerical sides, appropriate computational and data-mining techniques are employed, which are described in Section 3. The 270° wind direction sector has been chosen because it is the main wind direction (see Figure 2) and because it is a very relevant testing ground for the issues above. Two wind speed regimes have been chosen:

- 6 m/s at met mast, because the dynamics is expected to be dominated by wakes.
- 10 m/s at met-mast, because the topography effects are expected to become more relevant

and because the wind speed fluctuations might approach regimes associated to rated power. Therefore non-negligible influence of the control system in the machine dynamics is possible.

The results, displayed in Section 4, can be summarized as follows:

- The free flow model is unable to capture the trend of wind speed, moving from turbine SGM10 to SGM13. Actually it predicts decreasing wind speeds, while instead an abrupt fall occurs only from turbine SGM10 to SGM11, and the trend inverts at turbine SGM12 and especially SGM13.
- The AD model on the real terrain brilliantly captures the fall of wind speed from turbine SGM10 to SGM11. Further, it does not capture the rise at SGM12, but at least it predicts an almost flat trend. Instead, the AD model on the flat terrain overestimates the wind speed (and consequently the power) at turbine SGM11 and predicts a further decrease at turbine SGM12. This means that the terrain is fundamental in driving the main stream flow, transporting the energy, and that the terrain effects are crucial in order to interpret the performances of turbines SGM11 and SGM12. In other words, if there were only wakes, we should be expecting SGM12 to perform worst that SGM11, while it exactly does the opposite.
- The distribution of the turbulence intensity is best captured by the AD model on the flat terrain, especially in the 6 m/s case, when the dynamics is expected to be dominated by wake effects. The 10 m/s case is extremely instructive: even though the AD model on the real terrain overestimates the order of magnitude of turbulence intensity ratios, it best captures the trend at SGM12, with respect to the AD model on the flat terrain. This means that at high wind speeds, the effects of terrain are not negligible in understanding how tue turbulence intensity distributes.
- The AD model on the real terrain captures the main features of the actual turbine alignment patterns, while instead the AD model on the flat terrain and the free flow model do not capture the amplitude and the trend of northward wind flow distortion, which ultimately explains why performances of turbines SGM12 and SGM13 are far better than in an hypothetical flat terrain. The results demonstrate that the AD model reproduces realistic flow deviation from the main stream, resembling the actual alignment patterns, even if based on the hypothesis of perfectly orthogonal rotors.

The AD model by construction disregards rotational and near-wake effects, but it is a computationally accessible tool for investigating the global wind farm aerodynamic behaviour: it is therefore very appreciable that the AD model captures the main features of the wind speed trend, especially the abrupt fall at turbine SGM11. Yet, several are the possible developments of the present work in order to refine the predictions. The employed data set is an ensemble of 10-minute averages: yaw positions therefore by construction have uncertainty; further, even if data are filtered on a relatively narrow interval centered on 270°, the terrain is so complex that there possibly are several inflow and wake conditions. Since the nature of this source of information can not be circumvented, a more consistent comparison against numerical models can be obtained by encapsulating somehow uncertainty in the computational framework. In [20] a Gaussian weighting of simulation regimes is proposed and on the offshore Horns Rev test case it is shown that is sensibly leads to a more consistent framework for comparing models to reality. It would be valuable to push forward the research of the present work and include such approach, in particular because of the challenging complexity of our test site. Further, as shown in the preliminary analysis of [10], the alignment patterns to the wind in complex flow regimes have a huge impact on the amount of energy extracted: even though, as Figures 14 and 10 show, the actual alignment patterns averagely resemble the predictions from the AD model about the directional flow, an underworld of nacelle configurations exists, basically associated to the same wind conditions. Discriminating between them and inquiring which of them are energetically and-or mechanically favorable would be an impressive achievement to the issue of SCADA-based custom wind farm optimization and to the perspective of yaw active control systems [21].

References

[1] F. Castellani, D. Astolfi, L. Terzi, K.S. Hansen, and J.S. Rodrigo. Analysing wind farm efficiency on complex terrains. In *Journal of Physics: Conference Series*, volume 524, page 012142. IOP Publishing, 2014.

[2] R.J. Barthelmie, K.S. Hansen, and S.C. Pryor. Meteorological Controls on Wind Turbine Wakes. *Proceedings of the IEEE*, 101(4):1010–1019, April 2013.

[3] R.J. Barthelmie, S.C. Pryor, S.T. Frandsen, K.S. Hansen, J.G. Schepers, K. Rados, W. Schlez, A. Neubert, L.E. Jensen, and S. Neckelmann. Quantifying the impact of wind turbine wakes on power output at offshore wind farms. *Journal of Atmospheric and Oceanic Technology*, 27(8):1302–1317, 2010.

[4] K.S. Hansen, R.J Barthelmie, L.E Jensen, and A. Sommer. The impact of turbulence intensity and atmospheric stability on power deficits due to wind turbine wakes at horns rev wind farm. *Wind Energy*, 15(1):183–196, 2012.

[5] P. McKay, R. Carriveau, and D. S.-K. Ting. Wake impacts on downstream wind turbine performance and yaw alignment. *Wind Energy*, 16:221–234, March 2013.

[6] F. Port-Agel, Y.-T. Wu, and C.-H. Chen. A numerical study of the effects of wind direction on turbine wakes and power losses in a large wind farm. *Energies*, 6(10):5297–5313, 2013.

[7] J.S. Rodrigo, P. Gancarski, R.C. Arroyo, P. Moriarty, M. Chuchfield, J.W. Naughton, K.S. Hansen, E. Machefaux, T. Koblitz, E. Maguire, et al. Iea-task 31 wakebench: Towards a protocol for wind farm flow model evaluation. part 1: Flow-over-terrain models. In *Journal of Physics: Conference Series*, volume 524, page 012105. IOP Publishing, 2014.

[8] A. Makridis and J. Chick. Validation of a cfd model of wind turbine wakes with terrain effects. *Journal of Wind Engineering and Industrial Aerodynamics*, 123:12–29, 2013.

[9] M. Lee, S.H Lee, N. Hur, and C. Choi. A numerical simulation of flow field in a wind farm on complex terrain. *Wind and Structures*, 13(4):375, 2010.

[10] F. Castellani, D. Astolfi, and L. Terzi. Mathematical methods for performance evaluation of onshore wind farms through scada data mining. *Recent Advances in Energy, Environment and Financial Planning*, 2014.

[11] N.G. Mortensen, L. Landberg, I. Troen, and E. Lundtang Petersen. Wind atlas analysis and application program (wasp). 1993.

[12] A.J. Bowen and N.G. Mortensen. *WAsP prediction errors due to site orography*. 2004.

[13] P. Moreno, A.R Gravdahl, and M. Romero. Wind flow over complex terrain: application of linear and cfd models. In *European Wind Energy Conference and Exhibition*, pages 16–19, 2003.

[14] B. Sanderse. Aerodynamics of wind turbine wakes. *Energy Research Center of the Netherlands (ECN), ECN-E-09-016, Petten, The Netherlands, Tech. Rep*, 2009.

[15] P.M. Réthoré, N. Sørensen, and F. Zahle. Validation of an actuator disc model. In *2010 European Wind Energy Conference and Exhibition*, 2010.

[16] F. Castellani and A. Vignaroli. An application of the actuator disc model for wind turbine wakes calculations. *Applied Energy*, 101:432–440, 2013.

[17] G. Crasto, A.R. Gravdahl, F. Castellani, and E. Piccioni. Wake modeling with the actuator disc concept. *Energy Procedia*, 24:385–392, 2012.

[18] C. Peralta, H. Nugusse, S.P. Kokilavani, J. Schmidt, and B. Stoevesandt. Validation of the simplefoam (rans) solver for the atmospheric boundary layer in complex terrain. In *ITM Web of Conferences*, volume 2, page 01002. EDP Sciences, 2014.

[19] L. Casella, W. Langreder, A. Fischer, M. Ehlen, and D. Skoutelakos. Dynamic flow analysis using an openfoam based cfd tool: Validation of turbulence intensity in a testing site. In *ITM Web of Conferences*, volume 2, page 04002. EDP Sciences, 2014.

[20] M Gaumond, P-E Réthoré, Søren Ott, Alfredo Peña, Andreas Bechmann, and Kurt Schaldemose Hansen. Evaluation of the wind direction uncertainty and its impact on wake modeling at the horns rev offshore wind farm. *Wind Energy*, 17(8):1169–1178, 2014.

[21] P. Fleming, S. Lee, M. Churchfield, A. Scholbrock, J. Michalakes, K. Johnson, and P. Moriarty. The SOWFA super-controller: A high-fidelity tool for evaluating wind plant control approaches. Technical Report NREL/CP-5000-57175, National Renewable Energy Laboratory, Golden, CO, January 2013. To be presented at EWEA 2013 Vienna, Austria February 4-7, 2013.

WIND TURBINE WAKE INTERACTIONS; RESULTS FROM BLIND TESTS

Per-Åge Krogstad and Lars Sætran

Department of Energy and Process Engineering,
Norwegian University of Science and Technology NTNU, Trondheim, Norway

Abstract. Results from three "Blind test" Workshops on wind turbine wake modeling are presented. While the first "Blind test" (BT1, 2011) consisted of a single model turbine located in a large wind tunnel, the complexity was increased for each new test in order to see how various models performed. Thus the next "Blind test" (BT2, 2012) had two turbines mounted in-line. This is a crucial test for models intended to predict turbine performances in a wind farm. In the last "Blind test" (BT3, 2013) the two turbines were again mounted in-line, but offset sideways so that the rotor of the downstream turbine only intersected half the wake from the upstream turbine. This case produces high dynamic loads and strong asymmetry in the wake. For each "Blind test" the turbine geometry and wind tunnel environment was specified and the participants were asked to predict the turbine performances, as well as the wake development to five diameters downstream of the second turbine. For the first two tests axisymmetry could be assumed if the influence of the towers was neglected. This was not possible in BT3 and therefore only fully 3D methods could be applied. In all tests the prediction scatter was surprisingly high.

1. Background

The NREL full scale wind tunnel tests, completed in 2000, and the following blind tests by a large number of calculation methods (Simms et al. [1]) demonstrated how important it is to have detailed data to validate predictions against. The comparison showed that there was large uncertainty in the prediction methods, which called for more refined computer codes. Discussions of some of the issues that surfaced after the tests was presented in the special issue of *Wind Energy*, which was published in 2002 (Schreck [2]).

Since the NREL tests, significant improvements in turbine performance and wake development prediction methods have emerged. The progress in wake predictions over a decade of research may e.g. be found by comparing the review papers by Crespo et al. [3] and Sanderse et al. [4]. From simple analytical wake models based on classical theories, the standard is now to solve transport equations for turbulence properties or solving the turbulent field in time and space using large eddy simulation. But even at these levels of sophistication the results obtained rely heavily on assumptions incorporated by the modeler. This was clearly demonstrated in the comparisons of wake predictions performed by Cabezón et al. [5] using various turbulence transport equation models.

Many wake model studies have used full scale turbine wake data for validation. The disadvantages with this procedure is that the amount of data is sparse and initial and boundary conditions are not known in sufficient detail to allow one-on-one simulations to be made. In this respect wind tunnel studies are preferable, since the flow conditions may be specified at all levels required to set up a proper simulation. The major drawback is that the models operate at much lower Reynolds numbers than found at full scale. As part of the joint Norwegian research programs denoted Nowitech and Norcowe it was decided to set up a model experiment to be used as a data source when testing out turbulence models or developing new tools. Together Nowitech and Norcowe are involved in research on most aspects of offshore wind turbine technology and have about 35 PhD or PostDoc members, many involved with wind turbine aerodynamics. Before the data was released it was deemed a good idea to arrange open blind tests to find how the current models used would perform if only the turbine geometry was known.

The wind turbine model used for these tests was designed with the specific aim to form a test case for prediction methods. Therefore this is not a typical wind turbine layout, since measurement and wind tunnel restrictions, combined with test case challenges, had to be considered. For a test case it was important to use a simple geometry while also introducing effects that might be difficult to handle by a prediction method. This led to the design shown in figure 1. As may be seen from the photo, the turbine has 3 blades and the rotor sits on top of a stepped tower consisting of four cylinders of different diameters. The nacelle is also circular with a diameter of d = 90mm. It has an almost semi spherical hub cover at the front and back.

The model was tested in a wind tunnel that has a test section which is almost 12m long. The height is about 2m and the width almost 3m. For details, see one of the references, Adaramola & Krogstad [6] or Krogstad & Adaramola [7]. The rotor diameter is $D = 0.894$m and the centre of the rotor is located $z = 0.817$m above the floor level.

Figure 1. Model in the wind tunnel

The model was designed to operate at 10m/s with a tip speed ratio of TSR = 6, which is typical for a full scale turbine. In order to test the Reynolds number dependence of the blades, the performance of the turbine was measured over a wide tip speed range for reference velocities ranging from about U_{ref} = 7 to 15m/s. The power and thrust coefficients were found to be independent of Reynolds numbers for $U_{ref} > 9$m/s (see e.g. Krogstad & Adaramola [7]).

1.1. The blade geometry

To make the blade specification simple, the same airfoil was used all along the span. The airfoil selected is the 14% thick NREL S826 airfoil. It is originally intended to be used near the tip of a full scale turbine. The blades were made of aluminum and the maximum load on each blade was estimated to be about 15N. It was assumed that the blades had sufficient stiffness so that geometrical changes due to the loads could be neglected. Therefore this slender airfoil could be used also for the root sections. The airfoil is shown in figure 2.

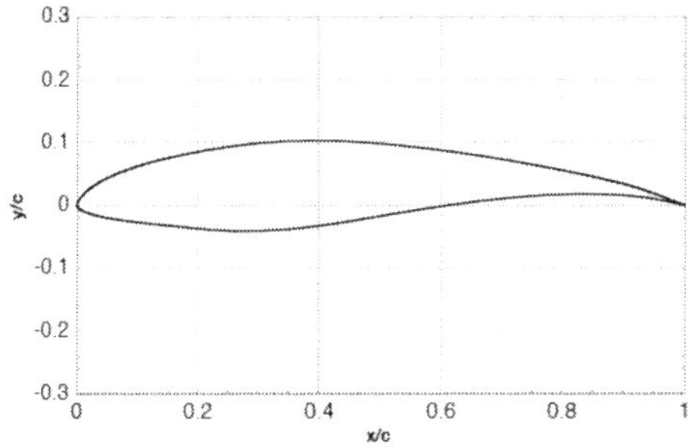

Figure 2. Shape of the NREL S628 airfoil

2. The BT1 test case

In this first "Blind test" the case consisted of the single model turbine shown in figure 1. Except for the geometry specifications and wind tunnel inlet conditions, no further information was given by the organizers. Thus, if the participants needed e.g. lookup tables to determine the load distributions on the blades, they would have to generate this information themselves. The way this was obtained would have impact on the results. Hence, the first compulsory output was predictions of the power and thrust coefficients for tip speed ratios (TSR) from TSR = 1 to 12. (The runaway TSR had been measured to be about TSR $\approx$ 11.5.)

The main focus of the blind test was however on the wake development behind the turbine. The participants were therefore asked to provide profiles of the velocity defect, $(U_{ref} - U)/U_{ref}$, and turbulent kinetic energy, $k^2/U_{ref}^2 = \frac{1}{2}(u_x^2 + u_y^2 + u_z^2)/U_{ref}^2$ at three downstream positions, $x/D = 1$, 3 and 5, measured from the plane of the rotor. The data should be provided for three operating conditions, TSR = 3, 6 and 10, i.e. in the fully stalled region, at design conditions and near the runaway point. The predictions were compared with measured thrust forces, power coefficients as well as velocity profiles. Full details of the measurements and predictions for BT1 may be found in Krogstad and Eriksen[8].

2.1. Participants and methods

Calculations were submitted by eight groups, some of whom supplied multiple predictions, giving a total of eleven submissions. In order to start the wake calculations, the power and the thrust coefficients of the turbine must first be estimated. Except for two groups, this data was obtained using a Blade Element Momentum method. Three simulations resolved the flow down to the level of the boundary layer on the blade and thereby avoiding the assumptions implied in the BEM methods. Below is a list of participants and the essence of their methods.

2.1.1. Agder Energy Production

J. A. Lund of Agder Energy Production used the software *openFoam* in combination with the $k - \omega$ SST turbulence model to calculate the wake flow. The rotor was modeled by means of an actuator disk where C_P and C_T were taken from a BEM model. It was documented that the numerical setup was grid-independent. The results will be labelled **Lund $k-\omega$ SST** in the plots.

2.1.2. Acona Flow Technology

E. Manger of Acona Flow Technology was the only participant who modeled the wind tunnel test case as it was performed. The entire model, including the rotor, tower and nacelle, located in the test section was meshed using 5.3×10^6 cells. A sliding mesh was used to allow the blades to rotate. The flow was solved using the *Ansys Fluent v.13-0* software and the $k-\omega$ SST turbulence model was incorporated in the solution. The results will be referred to as **Manger $k-\omega$ SST**.

J. Kvalvik of the same company performed a simulation using *openFoam*. Kvalvik used a cylidrical rotating grid containing the three rotor blades which was freely suspended in a short square box (about 2.5 diameters long) without the inclusion of the nacelle or tower. The turbulence model used was the one-equation model of Spalart and Allmaras and grid independence was demonstrated using up to 1.8×10^6 cells. Because the computational domain was relatively short, no wake data were submitted. The label used for these data is **Kvalvik**.

2.1.3. CMR Prototech

CMR Prototech submitted data, performed by T. Hansen, for a simulation that resolved the full flow details down to $y^+\approx 1$ on the actual blade geometry. The software package used was *STAR CCM+* and the turbulence was modeled using the $k-\omega$ SST turbulence model. Hansen assumed axisymetry and therefore did not include the tower. Only one blade sitting on the hub was modeled and periodic boundaries were applied to include the effects of the other blades. A grid refinement test with up to 23×10^6 cells was performed to demonstrate grid independence. His predictions will be denoted as **Hansen $k-\omega$ SST**.

2.1.4. DTU Mechanical Engineering

J. N. Sørensen and R.F. Mikkelsen at DTU, Denmark, did in principle provide 3 sets of simulation data, although two of these are closely related. The simplest was a BEM method for a turbine in an unbounded flow field (labelled **Sørensen & Mikkelsen BEM-nocorr**).

The model size is in the upper range of what may be considered a free field operation. Sørensen & Mikkelsen had therefore also estimated appropriate corrections for C_P and C_T due to the blockage caused by introducing a drag force in the tunnel. Thus the correction depends on the estimated C_T and at high thrust values corrections up to about 10% were made to the estimated free field C_T. The label used for these data is **Sørensen & Mikkelsen BEM-wtcorr**.

Finally the group delivered data for a combined actuator line / Large Eddy Simulation (LES) using a program called *EllipSys3D*. The forces on the flow as generated by the rotor were represented by rotating actuator lines and the rest of the flow was resolved using the Navier-Stokes equations on a very fine mesh. Only the sub-grid turbulence is modeled while the large scale turbulence is predicted directly from the Navier-Stokes equations. Hence, this solution would normally be considered very accurate, but the accuracy in the wake also depends on the modeling of the rotor loads from the BEM methods. Their predictions are labelled **Sørensen & Mikkelsen LES**.

2.1.5. GexCon

The GexCon group, consisting of J.A. Melheim, L. Sælen and M. Khalil, performed calculations using the software package $FLACS-Wind$ which is developed by GexCon. This is a transient CFD solver which in this case used the standard $k-\epsilon$ turbulence model. The computational domain was similar to the wind tunnel dimensions, but the increase in tunnel height to compensate for the growth of side wall boundary layers was not included. The rotor was represented as an actuator disk and the disk data was obtained using a BEM method. The effects of the tower were not included in the simulations. These predictions have been denoted **Melheim et al.** $k-\epsilon$.

2.1.6. National Institute of Advanced Industrial Science and Technology, Japan,

T. Kono from National Institute of Advanced Industrial Science and Technology, Japan, supplied the second set of data generated using LES with a software package called *Front Flow/red.* The rotor was simulated as an actuator disk where the data was taken from a conventional BEM simulation. The simulations were performed in a domain representing the wind tunnel test section and the effects of the tower and the nacelle were included by adding body forces. 2.4×10^6 cells were used. His data is labelled **Kono LES**.

2.1.7. NTNU

Data was submitted from the Department of Civil and Transport Engineering, NTNU, by L. Suja and P. Thomassen. The performance of the wind turbine was predicted using the aero-elastic code called *ASHES*, which makes use of the BEM theory. The method predicts C_P and C_T, as well as the radial distribution of the blade loadings, but is not able to provide information about the wake development. The results from this group will be labelled **Suja & Thomassen BEM**.

2.1.8. StormGeo / Univ. Stavanger

These simulations were submitted by S. Kalvig. She has used the software *openFoam* and represented the rotor as an actuator disk. C_P and C_T data for the disk was obtained from the measurements reported in Krogstad et al. [9]. The calculations were performed in a domain corresponding to the test section of the wind tunnel using a standard $k-\epsilon$ method, but the effects of the tower were not included. The predictions have been denoted **Kalvig** $k-\epsilon$.

Table 1. BT1 model overview

Name	Model type
Hansen $k-\omega$ **SST**	FR / SST
Manger $k-\omega$ **SST**	FR / SST
Kvalvik	FR / 1Eq
Kono LES	BEM-AD / LES
Sørensen & Mikkelsen LES	BEM-AL / LES
Kalvig $k-\epsilon$	BEM-AD / kEps
Melheim et al. $k-\epsilon$	BEM-AD / kEps
Lund $k-\omega$ **SST**	BEM-AD / SST
Sørensen & Mikkelsen BEM-nocorr	BEM-No tunnel walls / no wake calculations
Sørensen & Mikkelsen BEM-wtcorr	BEM-With tunnel walls / no wake calculations
Suja & Thomassen BEM	BEM / no wake calculations

2.2. BT1 model overview

The methods described above may be grouped according to the models used for the rotors and the equations included to solve turbulence field (see table 1). *BEM* implies that the blade loads are determined by a Blade Element Momentum method and tables of lift and drag coefficients,

AD means the rotor is approximated by an Actuator Disk, while *AL* is used for an Actuator Line model. *FR* means that a CFD code has been used to resolve the flow down to the rotor surface. The turbulence models have been denoted *SST*, *kEps* and *1Eq* for the $k-\omega$ SST, $k-\epsilon$ and Spalart-Allmares models, respectively. *LES* means that the wake is resolved using Large Eddy Simulations.

3. BT1 Results

3.1. Predicted performance

The power coefficient is shown in figure 3(a) and the thrust coefficient in 3(b). The symbols used are the same in all figures with measurements presented as the large black dots to distinguish them from the predictions. The predictions by the Hansen $k-\omega$ SST, Kvalvik Spalart-Allmaras and Manger $k-\omega$ SST methods are also represented by filled symbols to highlight the methods that solve the flow over the blades in detail, instead of relying in the BEM methods.

(a) 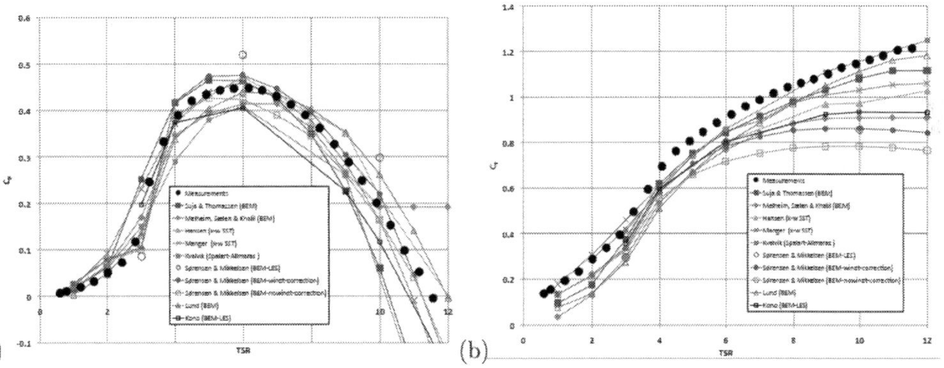 (b)

Figure 3. Turbine performance. (a) Power coefficient, (b) Thrust coefficient.

The spread in the power predictions (figure 3(a)) is surprisingly large, considering that most of the predictions use the same Blade Element Momentum theory as the basis. The turbine peak performance is generally predicted to within ±10% and most methods predict the peak performance where it should be, around the design tip speed ratio of TSR=6.

For the very low TSRs, where the blades are deeply stalled, the deviation from the measured power coefficient, C_P, is mostly small, while at very high TSRs there is significant spread in the predictions. Previous studies (Krogstad and Lund[10]) have shown that the first signs of stall occur around TSR=4 and it was therefore expected that the principal signs of departure from the measurements would be found here. But except for the predictions by Kvalvik, none of the methods appear to have special problems here. As the stall deepens, however, (at TSR=3 the blades are fully stalled) the spread becomes larger. But for the deep stall region for TSR < 3, all methods behave well. This suggests that the predictions of the airfoil data, that has to be performed before a BEM calculation can be made, have been successful for most participants, possibly with some inaccuracies in the transition region from the first signs of stall until the profile is completely stalled.

Looking at the thrust coefficient (figure 3(b)), it is seen that the general trend is that C_T is under-predicted by all models. There is considerable disagreement about its value for high TSR. This does not appear to have a systematic trend with respect to prediction methods in the sense that the BEM and CFD methods both appear to be distributed over the entire range of predicted values. It should also be noted that while the Sørensen & Mikkelsen BEM-nocorr method predicts the lowest C_T values for TSR > 5, this cannot be because it does not include the restrictions caused by the wind tunnel walls. The Suja & Thomassen BEM method, which also assumes infinite flow domain, is one of the best performers for TSR higher than the design condition.

The fully resolved simulations of the Kvalvik Spalart-Allmaras method perform excellent over the entire range of operations, which is remarkable since the computational domain is terminated only 2D downstream of the rotor and the predicted C_P was not particularly good. The predictions of Hansen $k - \omega$ SST, Kvalvik Spalart-Allmaras and Manger $k - \omega$ SST are in close agreement over the full range of TSR, with Kvalvik performing slightly better. At high TSR the Manger $k-\omega$ SST method performs slightly better than the one by Hansen $k-\omega$ SST in predicting C_T, suggesting that the latter method slightly under-predicts the airfoil performance at low angles of attack. Since the Hansen $k - \omega$ SST simulations use about five times as many grid points as Manger $k - \omega$ SST, and the turbulence model used is the same, this may suggest that Manger has been more successful in distributing the points wisely.

3.2. Wake data

The participants were asked to provide predictions of the spatial distribution across the wake at zero angle of yaw for three positions downstream of the rotor plane, at X/D = 1, 3 and 5, for three operating conditions, TSR = 3, 6 and 10. Space does not allow the results for all stations and all parameters to be shown here, so we restrict the presentation to the mean velocity and turbulent kinetic energy distributed along a horizontal line across the wake at the rotor axis elevation for the design condition, TSR=6, which should be the simplest case.

The first station for which data was requested was at X/D=1 (figure 4(a)). Here X is the distance measured from the plane of the rotor. The velocity defect measurements show an

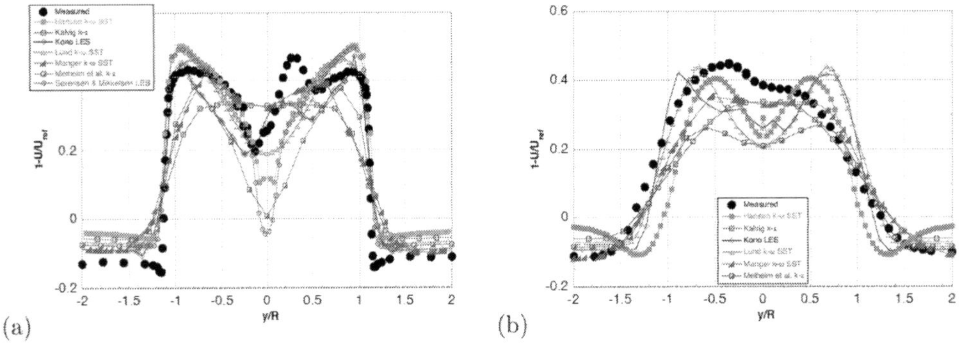

Figure 4. Mean velocity profiles along a horizontal line for TSR=6. (a) X/D = 1, (b) X/D = 5.

asymmetry in the form of a peak around $y/R \approx 0.2$. They are also present in the predictions of the two methods that include the full tower and nacelle geometry (Kono LES and Manger $k-\omega$ SST), although to a much smaller extent, and are therefore assumed to be due to the wake behind the tower. All predictions show the same shape with a dip in the centre. This is partly due to the flow acceleration caused by the displacement effect of the nacelle and partly because the nacelle was omitted in many methods, so there is no energy extraction here.

It is interesting to observe that the *STAR CCM+* predictions by Hansen, using the $k-\omega$ SST turbulence model, produce almost identical results to the LES predictions at this station. And so does the predictions by Lund $k-\omega$ SST, who uses a different solver but the same turbulence model. In fact these methods give a better collapse with the LES of Sørensen & Mikkelsen than the LES of Kono does.

The same comments that were made for $X/D = 1$ basically apply to the wake profiles at the last measurement station, $X/D = 5$ (figure 4(b)). (This is outside the calculation domain for the LES of Sørensen & Mikkelsen, so no results are available here.) The measurements have now started to develop a stronger velocity defect on the left hand side than on the right side. It is believed that this is a footprint of the tower wake which slowly rotates in the opposite direction to the rotor motion. This asymmetry is also to some extent captured by the predictions made by Manger $k-\omega$ SST and by Kono LES that incorporated the tower in the computational domain.

3.2.1. Turbulent kinetic energy profiles at TSR=6

The normalized turbulent kinetic energy at $X/D = 1$ along the horizontal line is shown in figur 5(a). (Note that the ordinate axis is logarithmic here since the range of the energy across the span is large and because the various prediction methods estimate quite different turbulence levels.) The turbulent kinetic energy distribution is characterized by strong peaks generated by the tip vortices from the blades. These peaks are captured by the predictions of Melheim et al. $k-\epsilon$, Kalvig $k-\epsilon$, Lund $k-\omega$ SST, and by Sørensen & Mikkelsen LES, although the levels are one order of magnitude lower than in the measurements. It is surprising that the LES of Kono did not pick up these peaks very well and gave so different turbulent energy from the LES predictions of Sørensen & Mikkelsen along the outer half of the wake. The profiles of Hansen $k-\omega$ SST have the correct shape, but the peak near the tip is more than two orders of magnitude too low. The measurements also indicate a high turbulence level behind the hub which could only be captured by the predictions where the hub of the model was included.

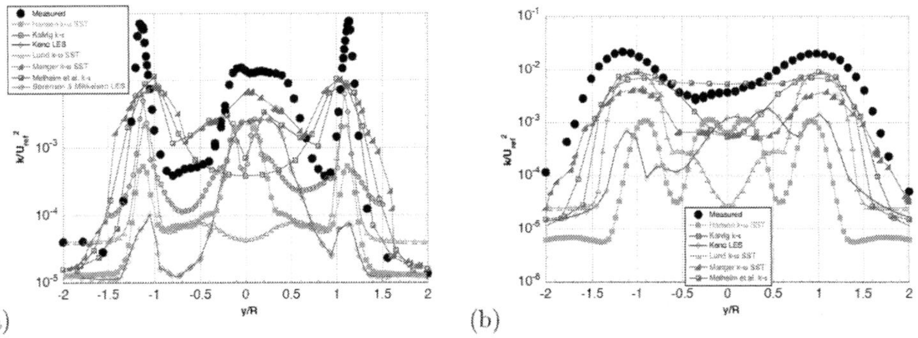

(a) (b)

Figure 5. Turbulent kinetic energy profiles for TSR=6. (a) $X/D = 1$, (b) $X/D = 5$.

Figure 5(b) show the distributions at $X/D = 5$. Surprisingly, the simple actuator disk / $k-\epsilon$ model of Kalvig is now performing very well as it gives the right shape as well as roughly the correct level. The predictions by Manger $k-\omega$ SST also show the correct shape, although the turbulent kinetic energy level is low. But the method of Melheim et al. $k-\epsilon$ is the one that appears to represent the measurements the best. The other methods, which all predict roughly the same k levels, show very ragged distributions, indicating that the spanwise turbulent diffusion is underestimated.

It is hard to select a best method or turbulence model from the presented wake and turbulence distributions due to the large variations in the results. But most predictions are at least able

Journal of Physics: Conference Series **625** (2015) 012043 doi:10.1088/1742-6596/625/1/012043

to predict the general shape of the velocity defect profiles as well as the location of the tip vortices and how they grow downstream. Overall it appears that the LES method of Sørensen & Mikkelsen (which also produced good predictions at $X/D = 3$, not shown here) is producing the most consistent results, but this comes at considerable computational costs.

4. The BT2 test case

The overall impressions from the first "Blind test" was that many of the methods were predicting the flow around a single wind turbine quite well. There was a demand for a more challenging case and so in 2012 a second "Blind test" was organised. This time two similar turbines were mounted in line so that the downstream turbine operated entirely within the wake of the first. The two turbines used identical blades, so except for a small difference in nacelle diameters, the two rotors were identical. But the first test had shown the importance of the tower wake, as it affected the wake symmetry, so a bulkier tower was used on the upstream turbine. Again it should be pointed out that the purpose of these blind tests was to generate test cases for prediction methods and not to simulate state of the art turbine designs.

Figure 6. Blind test 2 setup

The setup of BT2 is shown in figure 6. The upstream turbine T_1 was positioned $2D$ from the test section inlet. It was verified that this location is sufficiently far downstream for the operation of the turbines not to affect the uniformity of the inlet velocity profile. The rotor of downstream turbine T_2 was positioned $S = 3D$ downstream from the upstream turbine rotor. Both turbines rotate the same way and had the same hub height. The turbines were alternately positioned on a high accuracy six component aerodynamic balance in order to allow the thrust on the rotor to be measured separately. The thrust of the tower and nacelle was measured and subtracted from the thrust of the turbine, to yield the thrust of the rotor alone. Each turbine was fitted with a torque transducer and a speed of rotation sensor on the shaft mounted directly behind the rotor. The power generated by both turbines could be measured simultaneously. The measurement and predictions for BT2 are fully documented in Pierella et al.[11].

Again the invitation document for the "Blind test 2" workshop included all the geometrical details of the turbines, and the inlet conditions, but nothing more. The task given was to predict the performances of both turbines and the wake development downstream of T_2. Obviously this gave a handicap for modelers who had not been entirely successful in predicting the wake behind the first turbine in BT1, but the philosophy was that now that the wake data behind the upstream turbine had been made available they could improve their predictions.

Three test cases were defined. For all three the upstream turbine was operated at its design condition (TSR = 6). In Case A the downstream turbine was operated at TSR = 4 (close to its peak performance), Case B had TSR = 7 for the downstream turbine (close to runaway) and for Case C TSR was 2.5 (rotor operating in the stall mode). For both turbines the upstream reference velocity was used for scaling.

4.1. Participants and methods at BT2
There were eight participants at BT1 and the number was the same for BT2, except that there was a few of the BT1 groups dropping out and a few new coming in.

Manger from Acona participated with the same CFD model as last time (see 2.1.2 above for details).

Khalil and Sælen of GexCon again participated with their $FLACS-Wind$ method (described in section 2.1.5 above) and produced two sets of predictions. One was for the turbines in the wind tunnel test section where the walls were treated as no-slip boundaries. This will be labeled **GexCon-sim1**. In the other set of data the walls were removed and the simulation emulates an unbounded domain. This will be denoted **GexCon-sim2**.

Mikkelsen and Sarmast from DTU and KTH delivered LES predictions (section 2.1.4 above).

The wind turbine group of Agder Energy Production (section 2.1.1) had in the meantime been reorganized into a company called Meventus, but the method used was the same. Therefore the label used in the next set of plots has been changed to **Meventus**.

The new participants and their methods were as follows:

4.1.1. CMR Instrumentation
Hallanger and Sand of CMR Instrumentation used an in-house CFD code called *Music.* The code was developed at CMR, and uses a standard steady state $k-\epsilon$ model for turbulence closure with the coefficients proposed by Launder and Spalding[12].

The discretized equations are solved on a co-located grid using the SIMPLE algorithm from Ferziger and Perić[13]. The rotors were modeled by an actuator disc with rotation, where the 2D airfoil data were estimated using *XFOIL* (see Drela[14]). A sub-grid turbulence model is added, in order to represent the additional turbulence generation by the wind turbine, adapted from models for flow in large rod bundles by Sha and Launder[15]. The total number of grid points for the case was $3.9 \cdot 10^5$. The walls were included in the simulations, while the towers of the turbines were not. The hubs were taken into account as a flow resistance, but the axial extent of the nacelles was not represented.

The results from Hallanger and Sand will be labeled **CMR**.

4.1.2. de Vaal - NTNU Marintek
De Vaal performed his calculations using *Fluent v. 14.0.* Axisymmetry was assumed, and the turbine actuator disc was included via a BEM model. The 2D airfoil data were calculated using the Q^3uick software implemented by García[16], and used the experimental airfoil performance data set that had been generated the **DTU** contributors since the previous "Blind test". The grid was composed of 508×236 cells in the streamwise and radial directions, respectively. 80 of the radial cells were used along the rotor span. The tower and the nacelle were not simulated, but the wind tunnel walls were taken into account as no-slip surfaces. The turbulence closure model was the Reynolds Stress model (RSM), as implemented in *Fluent.*

These simulations are denoted **DeVaal**.

4.1.3. University of Puerto Rico
Leonardi and Martinez Tossas, from the University of Puerto Rico, used OpenFOAM to perform a LES simulation with a standard Smagorinsky sub-grid scale model ($C_S = 0.1$). The

turbine rotors were simulated as actuator lines. The walls, roof and floors were included in the simulation as no-slip surfaces, while the towers and the nacelles of the turbines were omitted. High Reynolds number airfoil data from Somers[17] were used as input for the simulations, and the *AirfoilPrep* tool from NREL [18] was used to correct the airfoil data for rotational effects.

The results from University of Puerto Rico will be labeled **Leonardi**.

4.1.4. Middle East Technical University, METU, Turky

The results from METU were produced by Uzol and Sezer Uzol. The authors used the free-wake code *AeroSIM+*, a 3D unsteady free-wake vortex panel method. For the turbines, the blades are discretized using quadrilateral panel elements and vortex ring elements are placed within each element. During an unsteady run, at each time instant, the influence coefficients of surface and wake panels are calculated using the induced velocity values from each vortex segment on the collocation points by means of the Biot-Savart law. Then the blades are moved and a new solution is calculated for the next time step. No tower or hub were modeled, while the boundaries of the wind tunnel were treated as no-slip surfaces.

The results from METU will be labeled **Uzol**.

4.2. BT2 model overview

The methods, grouped according to the models used for the rotors, and the equations applied to solve turbulence field are listed in table 2.

Table 2. BT2 model overview

Name	Model type
Acona	FR / SST
Meventus	FR / SST
CMR	BEM-AD / kEps
De Vaal	BEM-AD / RSM
GexCon-sim1	BEM-AD-With tunnel walls / kEps
GexCon-sim2	BEM-AD-No tunnel walls / kEps
DTU-KTH	BEM-AL / LES
Leonardi	BEM-AL / LES
Uzol	Vortex Panel Method

5. BT2 Results

5.1. Predicted performance

In the following plots, the marker symbols used are indicative of the rotor modelling techniques: a square means that the rotor was fully resolved, a triangle is used when the rotor was modeled as an actuator disc and a circle when an actuator line model was used. The vortex panel method of Uzol is different from all the others and is indicated by a five pointed star.

The mandatory predictions for the upstream turbine at its design condition (TSR = 6) show a surprisingly large scatter (±20%) in the power coefficient (see Figure 7(a)). It was not expected because the results from BT1 were closer to the measurements and the data had been available for a year so an improvement rather than a worsening was expected. Also experiments at the correct Reynolds numbers had been performed on 2D versions of the airfoil both at DTU, Denmark, and METU, Turkey. However, some participants preferred to provide their own data sets using software such as *XFOIL*.

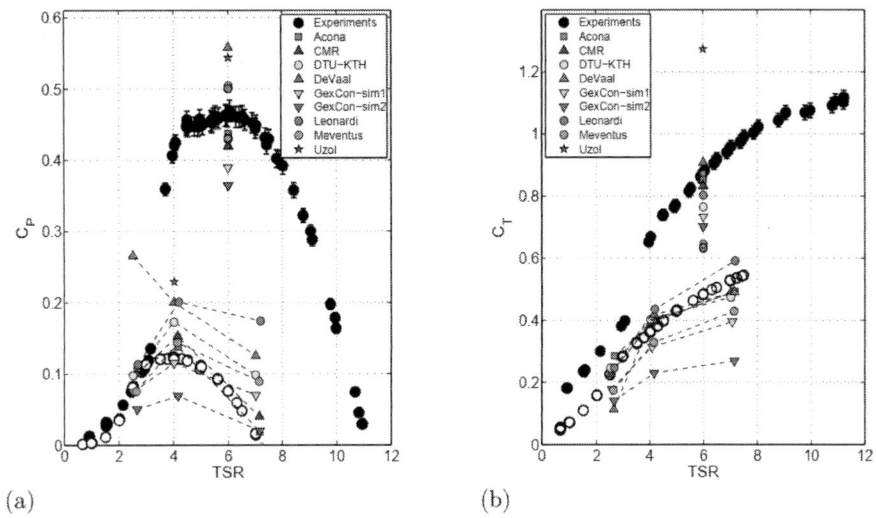

Figure 7. Experimental C_P of T_1 (•) and T_2 (∘) vs. simulations (grey symbols). (a) Power coefficient, (b) Thrust coefficient.

When the same airfoil data sets were used in different BEM models, as in the case of de Vaal and DTU-KTH, the predicted values varied significantly, suggesting that the predictions depend just as much on how the data was implemented as on the quality of the data. In this case de Vaal used an actuator disk model while DTU-KTH used the actuator line model. Judging from the predicted power coefficients, the more complicated actuator line method appear to give better estimates than the actuator disk. Acona's predictions, the only one this time to resolve the flow down through the boundary layers to the rotor surface and including the turbine structure, performed quite well, as also found in the BT1 blind test.

Figure 7(b) shows a similar scatter for the thrust coefficients. The free vortex method of Uzol gave far too high thrust, while all the other models either matched or underpredicted the rotor thrust compared to the measurements. Acona's results again gave the best agreement. The actuator disc predictions from de Vaal and CMR matched closely the thrust of the rotor. It was less successful in the power prediction, suggesting that the airfoil performance data they used are very different. All the actuator line methods had a tendency to underpredict the thrust, similarly to what happened in the Blind Test 1. The GexCon-2 simulation, which did not include the walls, show lower power and thrust values than the one with walls (GexCon-sim1), indicating a significant effect of the solid blockage on the rotor performance.

The power coefficients predicted for T_2 (Figure 7(a)) show more scatter than for T_1. The downstream turbine had its peak efficiency at $\lambda_2 = 4$ (Case A), while in Case B, at $\lambda_2 = 7$, the turbine was close to the runaway point. In Case C ($\lambda_2 = 2.5$) the power produced by the turbine was significantly lower than at the optimum point, and the blades were most likely heavily stalled. Despite the low incoming velocity and the consequent low local Reynolds number experienced by T_2, the experimental performance curve of T_2 does not indicate any of the irregularities observed at low Reynolds numbers for the single turbine. The downstream turbine was probably less sensitive to Reynolds number effects now due to the high turbulence level in the wake of the upstream turbine.

Except for the predictions of de Vaal, all the simulations correctly predicted the peak performance at $\lambda_2 \approx 4$. At $\lambda_2 = 4$, GexCon-sim1 performed best, closely matching the experimental results, while the simulation without the walls (GexCon-sim2) predicted a significantly lower performance. The fully resolved simulation from Acona performed well, overestimating only marginally the power production. Despite using the same 2D airfoil data, DTU-KTH and de Vaal had again different power results, with the actuator line model being lower and more accurate. This result is remarkable, since a higher thrust prediction for the first turbine means less kinetic energy availabe to the second turbine, and in this sense the prediction from DTU-KTH should have been higher than the ones from de Vaal. This suggests that de Vaal's model overestimated the turbulent diffusion in the wake and hence the velocity recovery rate. The free wake method from Uzol dramatically overestimated the downstream turbine power production, as it did for the upstream turbine.

The thrust coefficients for T_2, Figure7(b), are mostly under-predicted, but closer to the experimental data for low TSRs. The CMR method performs very well throughout the whole tip speed range, being in close agreement with the experiments. Again the Acona model produced high quality predictions under all conditions, while the other models showed high variability and sensitivity to the operating conditions of the turbines.

For discussions of the $\lambda = 2.5$ and 7 cases, please see Pierella et al. [11] for full details.

5.2. BT2 wake data

The fact that the upstream turbine has its peak performance at TSR ≈ 6 and the downstream turbine at TSR ≈ 4, suggests that since the rotors are identical the effective wake velocity across the rotor area seen by T_2 is about 67% of the incoming velocity, U_{ref}. However, in order to avoid any uncertainties due to an interpreted reference velocity for T_2, it was decided to scale all data with the same upstream velocity, U_{ref}.

5.2.1. Mean velocity profiles at TSR=6 for T_1 and 4 for T_2.

Here we show the mean velocity profiles at $X = 1D$ and $X = 4D$ for Case A, when both turbines were operating at their peak performance. For the other cases the reader is referred to Pierella et al. [11] where all cases are documented. The experimental mean velocity profile at $X = 1D$ (black dots in figure 8(a)) was slightly asymmetric, primarily due to the wake generated by the tower of the second turbine. The velocity outside of the wake was 20% higher than the reference velocity due to the blockage effect of the turbines, which produces an overall drag coefficient of about 1.2. Therefore, the modellers who underestimated the thrust for both turbines, like Meventus, predicted a lower velocity increase outside of the wake. Furthermore, not including the walls in the simulation also led to a deviation in the prediction of the undisturbed velocity, as demonstrated by the 10% difference between GexCon-1 and GexCon-2. At the centre line the models that did not include the nacelle failed spectacularly by producing only minor velocity defects or even a speedup in the centre region. Thus only the predictions by Acona and CMR reproduced the velocity profile shape here.

At $X = 4D$ turbulent diffusion has smoothed out the details of the initial conditions, which has improved the agreement between the predictions and the measurements (figure 8(b)), even in the cases where the turbine geometry have not been faithfully represented. The asymmetry from the first station has been shifted due to the wake rotation so that the traces of the tower has moved over from the negative Z to the positive Z side. The wake covers almost the whole wind tunnel width and the profile has a nearly Gaussian shape. The method of CMR performed very well across the whole span, while Acona's predictions indicated still a very strong velocity defect near the centre of the wake, suggesting an underestimation of the turbulent diffusion. This was confirmed by low turbulence levels predicted at all downstream distances (see next section).

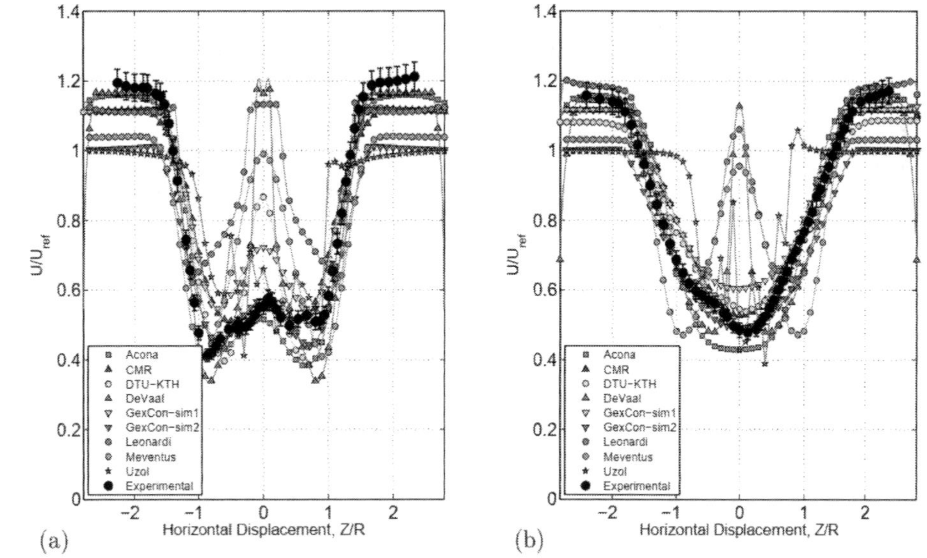

Figure 8. Case A, horizontal velocity profiles. (a) $X = 1D$, (b) $X = 4D$.

While the LES method of DTU-KTH predicted the wake very well across most of the span, the LES from Leonardi and the RANS-RSM from de Vaal performed quite well in the outer part of the wake and in the shear layer, but the velocity peak at the centre line was still over 100% too high, again suggesting that the spanwise diffusion was strongly underestimated. The free wake code from Uzol appears to have had convergence problems going from $X = 4D$, predicting a rather noisy profile at $X = 4D$ which was narrower than at $X = 1D$.

5.2.2. Turbulent kinetic energy profiles at TSR=6 for T_1 and 4 for T_2.

The profiles of the streamwise normal Reynolds stress, u'^2, at the same stations are shown in figures 9(a) and (b). Only u'^2 was measured using a single wire hot wire probe. Many of the simulations resolved the Reynolds stress tensor, but for those who used k transport equations the isotropic approximation $u'^2 = 2k/3$ has been used in the presentations.

Close to the rotor ($X = 1D$), the experimental streamwise Reynolds stress profile, figure 9(a), features three distinct peaks, two in the outer wake shear layer and one close to the wake centre line, slightly skewed to the positive side of the Z axis. The two outer peaks are characteristic of the tip vortices while the centre peak comes from the nacelle and the tower wakes and is therefore asymmetric. While the scatter in the predictions for BT1 was so large that logarithmic ordinate axes had to be used, all the models predicted a turbulence intensity of the same order of magnitude as in the measurements for this case. The LES from DTU-KTH is seen to have produced oscillations at this position. This is particularly evident near the centre, where strong turbulence has been generated in the shear layer produced by the centre line velocity overshoot. The tip vortices in the Acona simulations also produced two stress peaks but the level is much too low. It was suggested by the author that this may be related to problems in the interpolation of the turbulent kinetic energy at the interface between the revolving rotor mesh and the background grid. However, there are obvious problems also away from the interface. The method did not predict the peak at the wake centre line observed

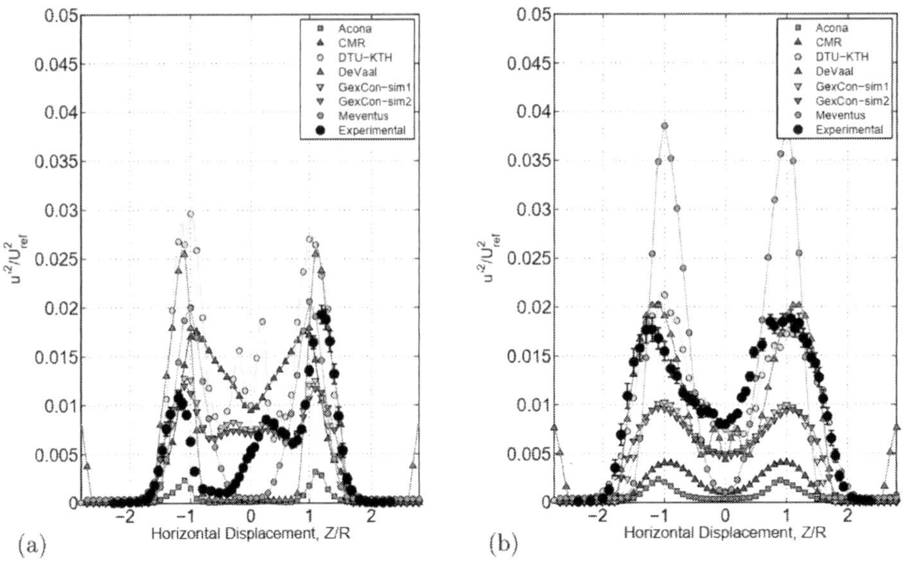

Figure 9. Case A, horizontal profiles of u'^2/U_{ref}^2. (a) $X = 1D$, (b) $X = 4D$.

in the measurements, despite the fact that the computations included the entire model in the simulations.

Most of the methods predicted too high turbulence intensity in the tip vortex region at this first position. However, this changed rapidly as the flow developed downstream, so that at the last station ($X = 4D$) about half of the methods strongly underestimated the stress levels (figure 9(b)). Here turbulent diffusion has spread a considerably amount of turbulent energy across the wake both in the predictions and measurements and so the profile shape is virtually identical for all methods, but the peak levels are significantly different. All the 2-equations models are now predicting too low turbulence intensities, which seems to confirm that such turbulence models are too dissipative to correctly represent a complicated anisotropic turbulent flow. The actuator line LES from DTU-KTH and RSM from de Vaal agree quite well with the experimental data all across the wake. They employed the same airfoil data for their rotor predictions, but incorporated totally different descriptions of the turbulent transport process.

6. The BT3 test case

In the last "Blind test", organised in 2013, the same two turbines from BT2 were used, but they were offset from the centre line by about $D/4$ either way so that the downstream turbine rotor only intersected half the wake of the first turbine. This is probably a more realistic operating condition for turbines in a wind farm and constitutes additional challenges compared to the BT2 case. The second rotor now experiences a highly non-uniform inflow with velocities that change rapidly around the periphery. This adds important hysteresis effects on the flow over the blades which may make the steady state data bases for the blade characteristics questionable. Also it introduces serious dynamic forces on the blades, but these could not be measured in the present setup.

Again the focus of the test was on the performance of the downstream turbine and the wake generated behind it. While the first two "Blind tests" had been for turbines exposed to

Figure 10. Upstream turbine and grid used in BT3.

a uniform, low turbulence inflow conditions, an additional case was generated for this "Blind test", where a background turbulence level typical for atmospheric conditions were generated using a large scale grid. Figure 10 shows the grid upstream of turbine T_1. The turbulence field generated by this grid has been extensively documented, see e.g. Davidson and Krogstad[19]. At the position of the upstream turbine rotor the flow is uniform and the turbulence level was $T_u = 10\%$, decaying to about 5% at the location of the second turbine rotor plane.

6.1. Participants and methods at BT3

Due to the increased complexity of this "Blind test" and because axisymmetry could no longer be assumed even in the simplest modelling, the number of participants for this test dropped to 5. Acona, DTU-KTH, GexCon and CMR contributed predictions with the same methods applied in BT2 . New contributions were submitted by CD-adapco. Their details are as follows:

6.1.1. CD-adapco.

S. Evans from CD-adapco, London, UK, provided predictions using their own software Star-CCM+. Around the rotors, cylindrical domains were created so that the arbitrary sliding interface functionality of STAR-CCM+ could be used to model the rotor motion. A polyhedral mesh was created containing 14 million mesh elements with the boundary layer on the blades being resolved down to a Y^+ value of less than 2, somewhat dependent on the operating conditions. Predictions were provided using the $k - \omega$ SST Detached Eddy Simulation (DES) model with curvature corrections. The mesh was created in such a way that the DES model operated in LES mode in the wake regions of the flow. The simulation ran in a two-step approach. First the case was run using a steady approach with multiple rotating frames. After the simulation was considered to be sufficiently initialised, it was switched to transient simulation. For a description of how the software is used for rotating flows, see Medonca et al.

[20].

6.2. *BT3 model overview*

The solution methods for the rotors and the turbulence equations used for BT3 are listed in table 3.

Table 3. BT3 model overview

Name	Model type
Acona	FR / k-ω-SST
CD-adapco	FR / k-ω-DES
CMR	BEM-AD / kEps
GexCon	BEM-AD / kEps
DTU-KTH	BEM-AL / LES

7. BT3 Results

Due to space limitations and to allow comparisons with BT2, only the non-grid case will be presented here. For full documentation the reader should refer to Krogstad et al.[21].

7.1. *Predicted performance*

As in the BT2 cases, the operating conditions for T_1 was kept constant at a tip speed ratio of TSR = 6. The speed of rotation for T_2 was then varied in order to cover the required TSR range.

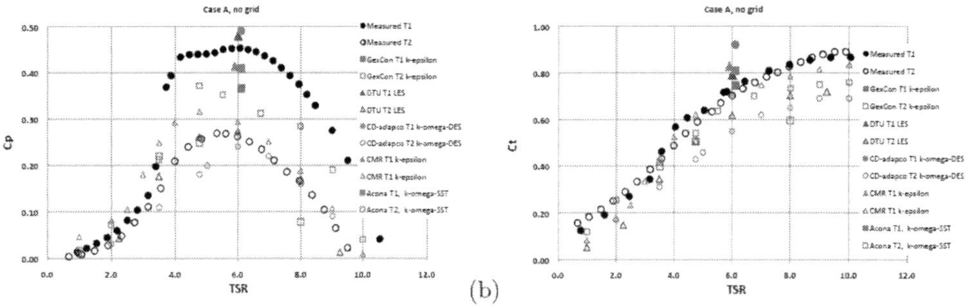

(a) (b)

Figure 11. Turbine performances, Case A, no turbulence grid. (a) Power coefficient, (b) Thrust coefficient.

The scatter in the power predictions for the upstream turbine is about the same as in the two previous "Blind tests". This is surprisingly large, considering that most of the participants have already preformed predictions for these turbines a couple of times before and know the performance curves. Compared to the BT2 test case, the energy output of T_2 has now almost doubled and is stretched out over a much broader TSR operating range. It may be noted that for TSR < 3, the T_2 power production follows the same trend independent of the amount of wake it operates in. This is because the blades are completely stalled and the time needed to recover from the stall appears not to be sufficient as it passes in and out of the wake. Since the lift and drag coefficients are virtually independent of angle of attack in the fully stalled region, the operating condition is not essential for the amount of energy produced.

The predictions for the second turbine vary by more than 50 % in some regions, but this is understandable, since we here have the most complicated test case so far. However, some of the predictions for T_2 are in fact very good at high tip speeds, e.g. the curves obtained by CMR Instrumentation and CD-adapco for TSR > 6. The T_2 power predictions by DTU are excellent for the entire TSR range. An oddity in the C_P predictions is that the only two methods that fully resolve the flow over the blades all the way down through the boundary layers (i.e. Acona and CD-adapco) produce so dramatically different results for TSR < 9.

Thrust coefficient measurements (figure 11(b)) show that except for a short region around TSR = 4, the C_T data for both turbines are very similar. Intuitively the velocity fields seen by the two rotors are expected to be very different with the downstream turbine being exposed to a lower momentum. Despite this, the coefficients are the same when scaled with the same parameters, showing that the physical forces are in fact almost identical. It is conjectured that the velocity increase seen by T_2 in the external flow outside the wake from T_1 may have been sufficient to compensate for the momentum loss in the wake, to produce very similar turbine drag. The general trend is that C_T is mainly over-predicted for the T_1 turbine, but under-predicted for T_2. GexCon has matched the force on turbine T_1 very well, but underpredicts the forces on T_2 significantly at TSR=8. The differences between the Acona and CD-adapco predictions are again quite evident, especially for TSR < 6 where the Acona model is doing very well.

7.2. Wake data

As an example of the wake complexity behind the second turbine and the predictions obtained, the data for $TSR_1 = 6$ and $TSR_2 = 4.75$ will be presented. These are the conditions when both turbines operate at their peak performances. Results for the other cases ($TSR_1 = 6$ and $TSR_2 = 3.5$ and 8) may be found in Krogstad et al.[21].

7.2.1. Mean velocity profiles at TSR=6 for T_1 and 4 for T_2.

The first station for which data was requested was at X/D=1 downstream of the rotor plane of T_2. The mean velocity profiles measured along a horizontal line parallel to the floor at the elevation of the rotor axis are shown in figure 12(a), for the non-turbulent case. Also, data was requested for X/D=3 and these are shown in figure 12(b). The centre of turbine T_1 is located at y/R = -0.45 and the rotor extends from y/R = -1.50 to y/R = +0.61. T_2 has its centre at y/R = +0.45 with a rotor that spans from y/R = -0.55 to y/R = +1.45. These are indicated by the arrows above each plot to help the interpretation of the data.The outer edges of the combined wake are quite evident, but the central part is a mix of influences from the two wakes.

Compared to the wake behind a single turbine operating at its best performance, this is a much more complicated wake and very far from the Gaussian profiles used by simple models. However, the sharp gradients are quickly smeared out by diffusion so that the profile at X/D=3 is already much smoother.

Except for the CMR $k-\epsilon$ model, all methods have reproduced the general shape of the profile at $X/D = 1$ (figure 12(a)). The GexCon method performs very well, avoiding the oscillations found in the predictions of Acona, CD-adapco and to some extent in the DTU results. As the flow develops downstream the strong gradients disappear in all models except in the predictions by Acona. This model predicted a profile shape at $X/D = 3$ (figure 12(b)) which is very similar to the one at $X/D = 1$, suggesting too low turbulent diffusion. Again, the GexCon model was very close to the measurements, while the diffusion in the $k - \epsilon$ model of CMR is definitely too strong, making the profile almost Gaussian here. In general the other methods predict too strong velocity defects at this station.

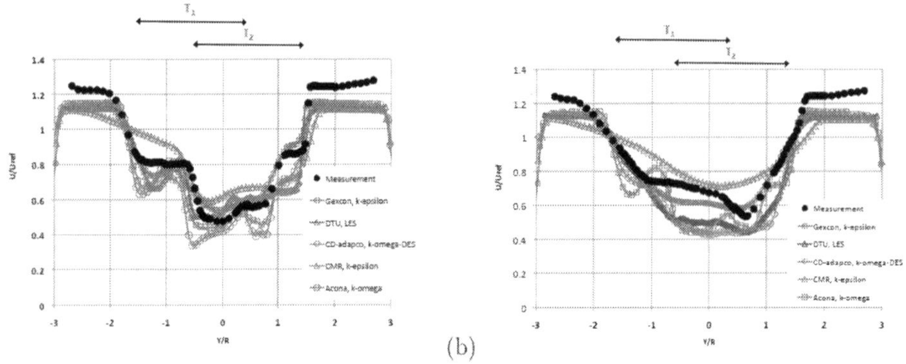

(a) (b)

Figure 12. Mean velocity profiles for TSR$_1$=6 and TSR$_2$=4.75 at (a) $X/D = 1$ and (b) $X/D = 3$. Positions of the rotors are indicated above the figures.

7.2.2. $< u_x^2 >$ stress profiles at TSR$_1$ = 6 and TSR$_2$ = 4.75.

The streamwise normal Reynolds stresses are shown in figure 13(a) for $X/D = 1$ and in figure 13(b) for $X/D = 3$. Again the simplification $< u_x^2 > \approx 2k/3$ was used for the k models. For this "Blind test" all Reynolds stresses were measured at a few stations and the excellent performance of the simplification was documented (see Krogstad et al.[21] for details).

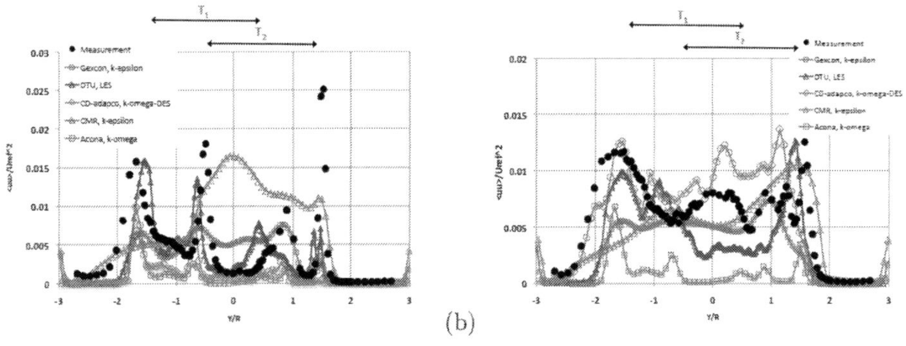

(a) (b)

Figure 13. Turbulent stress $< u_x^2 >$ for TSR$_1$=6 and TSR$_2$=4.75 at (a) $X/D = 1$ and (b) $X/D = 3$.

The complexity of the wake is very evident in the stress profiles. Very distinct energy peaks in the many tip vortices may be observed. All but the CMR method have predicted these peaks and perhaps not surprisingly the LES methods have done so most successfully. The turbulence level generated by the Acona method is very low at both stations suggesting that the kinetic energy generation from the shear layers has been underestimated. This supports the conjecture that the very small change in mean velocity profiles between the two stations was due to incorrect turbulence prediction.

The turbulence introduced by the grid significantly affected the wake data (not shown here). Partly by making the incoming wake smoother and partly by increasing the diffusion in the second wake and thereby removing many of the steep gradients. Together these had a significant effect on both the mean velocity and kinetic energy profiles.

8. Concluding remarks

The performance of model turbines and the wake formed by the rotors have been predicted for three test cases of increasing complexity. The methods comprised all levels of sophistication that are being used today, ranging from simple standard Blade Element Momentum methods to Large Eddy Simulations.

In most studies reported in the literature, where the aim is to sort out the best turbulence model to be used, the same operator uses the same program and grid distribution while incorporating different models in an attempt to best reproduce a reference distribution. This could be experimental results or other reliable references. Blind tests like those reported here take away the target and therefore opens up for the uncertainty connected with the model operator. This is more like the environment a design engineer works in. Using the code available he/she tries to set up the calculation to the best of his/her skills. This, of course, makes it difficult to objectively sort out a best prediction method but gives valuable information about the state of the art of the computations that are routinely made for wind turbines.

The spread in the results was probably larger than was initially expected. The uncertainty in the predicted turbine performance is of the order of $\pm 10\%$ near the design conditions for the simplest case of a single turbine in a uniform flow, but increases considerably when it was operating in the wake of another. Especially when only part of the wake hits the rotor.

To compute the wake flow, standard $k-\epsilon$ models, $k-\omega$ SST models and even large eddy simulations were used. There is no obvious winner when it comes to the turbulence model used. There is as much variation within the groups using the $k-\omega$ SST model as there is between these and the other models applied, underlining the dependence of the skill of the model developer or user on the results. Unfortunately the picture was not much clearer with the LES simulations either. However, it appears safe to say that using actuator line theory to predict the initial rotor condition gives better wake predictions than the actuator disk. In general this gave results that were more in line with the experimental data.

Wind tunnel size limited the downstream distance that the wakes could be measured to about 5D for the single turbine and 3 to 4D for the two turbine cases. It would have been interesting to see how the methods would perform down to, say $X/D = 10$, since it appears that some of the methods have problems accounting for the spanwise turbulent diffusion and therefore retain spanwise inhomogeneity over much longer distances than are indicated in the measurements and the LES calculations.

It is apparent from this exercise that when it comes to wind farm planning, the estimates of wake development has not reached the reliability that is necessary for accurate estimates of turbine interactions. However, the situation may not be as bad as some of the tests reported here may suggest. Most of the tests were performed in homogeneous and virtually turbulence free conditions. It was found that when adding free stream turbulence the agreement between measurements and predictions improved. In shear flow conditions the differences may be even smaller since both shear and background turbulence tend to increase turbulence diffusion, making the wakes more Gaussian.

References

[1] Simms D, Schreck S, Hand M and Fingersh L 2001 *National Renewable Energy Laboratory, Colorado, USA* **NREL/TP-500-29494** 1–44
[2] Schreck S 2002 *Wind Energy* **5** 77–257
[3] Crespo A, Hernández J and Frandsen S 1999 *Wind Energy* **2** 1–24
[4] Sanderse B, van der Pijl S and Koren B 2011 *Wind Eneergy* **14** 799–819
[5] Cabezón D, Migoya E and Crespo A 2011 *Wind Energy* **14** 909–921

[6] Adaramola M and Krogstad P Å 2011 *Renewable Energy* **36** 2078–2086
[7] Krogstad P Å and Adaramola M 2012 *Wind Energy* **15** 743–756
[8] Krogstad P Å and Eriksen P 2013 *Renewable Energy,* **50** 325–333
[9] Krogstad P Å, Karlsen J and Adaramola M 2010 *17 Australasian Fluid Mechanics Conference, Auckland, New Zealand* **Paper 024** 1–4
[10] Krogstad P Å and Lund J 2012 *Wind Energy* **15** 443–457
[11] Pierella F, Krogstad P Å and Sætran L 2014 *Renewable Energy* **70** 62–77
[12] Launder B and Spalding D 1974 *Computer Methods in Applied Mechanics and Engineering* **3** 269 – 289
[13] Ferziger J H and Perić M 1999 *Computational methods for fluid dynamics* vol 2 (Springer Berlin etc)
[14] Drela M 1989 Xfoil: An analysis and design system for low reynolds number airfoils Tech. rep. NREL
[15] Sha W and Launder B 1979 *Argonne Lab. Rep. AN-77* **73**
[16] García N R 2011 *Unsteady Viscous-Inviscid Interaction Technique for Wind Turbine Airfoils* Ph.D. thesis
[17] Somers D 1999 Design and experimental results for the s825 airfoil Tech. rep.
[18] Hansen C 2009 Nwtc design codes (airfoilprep)
[19] Davidson P and Krogstad P Å 2008 *J. Fluid Mechanics,* **602** 287–302
[20] Mendonca F, Baris O and Capon G 2012 *Proc. ASME Turbo Expo 2012: Power for Land, Sea and Air,* **GT2012** 1–10
[21] Krogstad P Å, Sætran L and Adaramola M 2015 *J. Fluids and Structures* **52** 65–80

Wake Conference 2015 IOP Publishing
Journal of Physics: Conference Series **625** (2015) 012044 doi:10.1088/1742-6596/625/1/012044

Atmospheric stability assessment for the characterization of offshore wind conditions

J. Sanz Rodrigo[1], E. Cantero[1], B. García[1], F. Borbón[1], U. Irigoyen[1], S. Lozano[1], P.M. Fernandes[1], R.A. Chávez[1]

National Renewable Energy Centre (CENER)
Calle Ciudad de la Innovación 7, 31621-Sarriguren, Spain

E-mail: jsrodrigo@cener.com

Abstract. Based on the Fino-1 offshore met mast database, different instrument set-ups and methodologies for stability characterization have been tested using non-dimensional numbers like the gradient and bulk Richardson number, and their equivalences with the Obukhov parameter $\zeta = z/L$, which can be measured locally with the use of a sonic anemometer. These equivalences depend to a large extent on the suitability of empirical stability functions obtained in horizontally-homogeneous conditions. The bulk Richardson number method, based on Grachev and Fairall (1997) empirical function, is the least demanding measurement method for stability characterization offering a more practical approach to wind farm designers than using the sonic method. Alternatively, the AMOK method, used by FUGA wake model and also based on the bulk Richardson number, assumes surface-layer theory and avoids using stability functions, which results in a more robust formulation. A 9-class stability classification based on Sorbjan and Grachev (2010) is used to generalize the categorization of wind conditions. Based on flux-profile analysis it was concluded that unfortunately the local ζ is not sufficient to describe the scaling behaviour of the stable boundary layer. Indeed, larger wind shear than predicted by classical onshore stability functions is found, probably as a result of lower boundary layer depths.

1. Introduction

Offshore wind power (onshore as well) is challenging many traditional wind resource assessment methodologies that have traditionally relied on a rather simplistic description of the wind climate distribution based on velocity and wind direction. Large wind farms are immersed in the marine atmospheric boundary layer (ABL) where wind conditions are modulated by atmospheric stability throughout the entire operating range of wind turbines. Low turbulence intensities due to the very low aerodynamic roughness of the sea surface are responsible for slow decaying wakes that last much longer than in onshore conditions [1].

Traditional wind farm design tools, developed for onshore wind deployment, typically assume neutral atmospheric stratification resulting in higher turbulence intensities and larger wake decay coefficients than in offshore conditions [2]. As a result these models will systematically underestimate wake effects in offshore conditions.

High fidelity wind assessment tools need to account for a thermally stratified turbulent structure of the marine ABL to produce accurate wind farm models. The addition of a third dimension to the description of the wind climate based on atmospheric stability will produce a more detailed spectrum of wind conditions which will allow a more complete wind farm design eventually leading to improved performance. On the other hand, it is necessary to make sure that the added complexity is not detrimental to the robustness and uncertainty of the design methodology. Hence, best-practice guidelines will be produced to guide wind analysts on the characterization of atmospheric stability and its influence on the wind profile.

2. Stability assessment from measurements

2.1 The Obukhov length method from sonic anemometer measurements

Local-scaling theory [3] generalizes Monin-Obukhov (M-O) surface-layer theory [4] to the ABL. It states that any dimensionless turbulence characteristic will only depend on a reduced set of scales which, by dimensional analysis, lead to the Obukhov length scale L:

$$L = -\frac{u_*^3}{\kappa \frac{g}{\Theta_0} \overline{w\theta}} \tag{1}$$

where g is the gravity, u_* is the friction velocity, Θ_0 is the surface temperature and $\overline{w\theta}$ is the kinematic heat flux. The dimensionless height $\zeta = z/L$ is used as stability parameter ($\zeta < 0$ indicates unstable, $\zeta > 0$ stable and $\zeta = 0$ neutral conditions). The Obukhov length is obtained from sonic anemometers using eddy-correlation techniques. The result is a local evaluation of stability based on turbulent fluxes averaged over periods from minutes to one hour to account for the energy in the microscale turbulence range. Since the Obukhov length method based on sonic measurements does not require any hypothesis, produces a local value of stability and is the most widely used stability parameter in ABL theories, we will use the ζ parameter as our target stability metric and the sonic method as our benchmark to compare other profile-based methods.

According to local-scaling theory, any dimensionless turbulent characteristic will depend solely on this parameter evaluated at any given height within the ABL. For example, the nondimensional wind shear:

$$\phi_m(\zeta) = \frac{\kappa z}{u_*} \frac{\partial U}{\partial z} \tag{2}$$

is equal to 1 in neutral conditions, where $\kappa = 0.41$ is the von Karman constant and U is the horizontal velocity. In non-neutral conditions several stability functions $\phi_x(\zeta)$ can be found in the literature, being the ones from Dyer [5], based on flux-profile measurements in Kansas, the most widely used:

$$\phi_x(\zeta) = \begin{cases} (1 - a_x\zeta)^{p_x} & \zeta < 0 \\ 1 + b_x\zeta & \zeta \geq 0 \end{cases} \tag{3}$$

with $a_m = a_h = 16$, $b_m = b_h = 5$, $p_m = -1/4$ and $p_h = -1/2$, where the subscript m is related to the wind speed shear and h to the temperature shear. Integrating (2) between the surface roughness z_0 and a given height yields the well known logarithmic wind profile:

$$U = \frac{u_*}{\kappa}\left[\ln\left(\frac{z}{z_0}\right) - \psi_m(\zeta)\right] \tag{4}$$

where $\psi_m(\zeta)$ is the integrated stability function for momentum [6]. In offshore conditions the Charnock relation links the roughness length to the shear stress [7]:

$$z_0 = C_{ch} \frac{u_*^2}{g} \tag{5}$$

where the Charnock constant C_{ch} ranges from 0.008 to 0.06 in the literature [8]. This constant can obtained by fitting equation (4) in neutral conditions. Peña and Gryning [8] found a value of 0.012 from measurements at Horns Rev and Sanz Rodrigo [9] found a value of 0.006 at Fino-1, both met-masts located in open-sea conditions in the North Sea. Lange et al [10] found a value of 0.03 at Rodsand, 11 km from the coast in the Baltic Sea. This higher value is related to a dependency of the surface roughness with the fetch (distance to the coast). A fetch-dependent model is proposed by [11].

2.2 The gradient Richardson method from profile measurements

The gradient Richardson number is the ratio of turbulent kinetic energy generation by buoyancy to that generated by shear. Using eddy-viscosity assumptions, the turbulent fluxes are defined in terms of velocity and temperature gradients to yield:

$$Ri(\zeta) = \frac{\frac{g}{\Theta_0}\frac{\partial \Theta}{\partial z}}{\left(\frac{\partial U}{\partial z}\right)^2} = \zeta \frac{\phi_h(\zeta)}{\phi_m^2(\zeta)} \tag{6}$$

By iteration in (6) it is possible to derive ζ from Ri if suitable stability functions are available. Alternatively, there are empirical functions that relate directly ζ and Ri like Sorbjan and Grachev [12].

$$\zeta = \begin{cases} Ri & \zeta \le 0 \\ \dfrac{Ri}{0.9} \dfrac{(1+300Ri^2)^{9/4}}{(1+250Ri^2)^{3/2}} & \zeta > 0 \end{cases} \tag{7}$$

Computing the Richardson number gradients using finite differences between sensors (finite-differences *FD* method) will typically result in biased assessments due to a sensor resolution of about the same order of the temperature differences. This method shall only be used in connection with differential temperature measurements. Alternatively, gradients can be obtained by fitting a logarithmic function of the type X = A log(z) +B to the observational data (profile-fit *PFIT* method). This method is recommended if more than 3 sensors are available in order to produce a more robust local estimate of *Ri*. Nevertheless, the accuracy of this method depends, to a large extent, on the adequacy of the underlying empirical relationship to the target site conditions.

2.3 The bulk Richardson method from profile measurements

A more robust estimate of stability is the bulk Richardson number Ri_b, defined in terms of a temperature difference and a single velocity level:

$$Ri_b(\zeta) = \frac{gz\Delta\Theta}{\Theta_0 \overline{U}^2} \tag{8}$$

where the height z is taken here as the mean height between the two levels of temperature. Grachev and Fairall [13] derived a relationship between the Ri_b and ζ for offshore conditions:

$$\zeta = \begin{cases} \dfrac{C_1 Ri_b}{1-C_2 Ri_b} & \zeta > 0 \\ C_1 Ri_b & \zeta \le 0 \end{cases} \tag{9}$$

with $C_1 = 10$, $C_2 = 5$, implying a critical bulk Richardson $Ri_b = 1/C_2 = 0.2$. $\Delta\Theta$ in (8) is derived from the air to water temperature difference, where the water temperature should be as close to the surface as possible and z in (9) is therefore half the air temperature height. Similar to the Ri method, $\Delta\Theta$ should be based on differential temperature measurements or well calibrated absolute temperature sensors to avoid biased Ri_b.

The definition of the Ri_b number in (8) is generally applicable to any velocity or temperature heights, resulting in a variation of Ri_b with height. In connection to wind resource assessment methodologies, the resulting value from (9) in absolute terms is only meaningful if the same definition is used in the flow model. For practical reasons, we could take two alternatives: 1) to refer to surface-layer stability by measuring velocity and temperature at low heights (<20 m), equivalent to the experimental conditions of Grachev and Fairall, or 2) to refer to a "hub-height stability" by measuring velocity and temperature at hub-height. The first option will be more meaningful to implement in surface-layer models, while the latter is more appropriate in the context of wind energy applications. We typically speak of hub-height wind speed and turbulence to classify wind conditions so we could do the same for stability and refer it to hub-height.

Ott and Nielsen [14] propose a method, also based on the bulk Richardson number, that eliminates the dependency on empirical functions by assuming surface-layer M-O theory and near-neutral conditions ($\zeta < 0.1$). Then the relation between the Ri_b and ζ is:

$$\zeta \approx Ri_b \frac{(\log z_u / z_0)^2}{\log z_t / z_0} \tag{10}$$

where z_u and z_t are the heights above sea level of the velocity and temperature sensors and z_0 is obtained by iteration using the M-O logarithmic velocity profile (4) combined with the Charnock relation (5).

3. The Fino-1 test case

The Fino-1 offshore platform is located 45km off the Borknun Island in Germany (54.014°N, 6.5905°E). A 100-m mast fully equipped with meteorological and oceanographic instruments is measuring since 2003 [15]. Profile measurements of velocity at 8 levels (33, 40, 50, 60, 70, 80, 90 and 100 m), temperature and relative humidity at 5 levels (33, 40, 50, 70 and 100m) are averaged at 10min intervals. Three sonic anemometers located at 40, 60 and 80m measure at 10 Hz the 3D velocity components. Sonic anemometer measurements are available for the year 2006 although the 60 m sensor only works during the first half of the year.

Sonic anemometer measurements are rotated to the mean streamline coordinate system using the planar fit method [9][16][17]. In order to reduce the scatter from flux measurements, both the profile and the flux data are averaged to an hourly database. Only open water conditions are considered, without mast distortion effects, reducing the analysis to the wind direction sector 225°±45°. The analysis filters out velocities outside of the typical operating conditions of wind turbines (4 to 25 m/s). The quality control on the data and a detailed description of the flux-profile method can be found in [9].

4. On the validity of local-scaling offshore

Based on flux-profile analysis [9] it is possible to test whether general ABL local-scaling theory work offshore similarly to onshore conditions. This would allow extending the applicability of onshore stability functions over homogenous flat terrain like (3) to the offshore environment.

Figure 1 shows the non-dimensional wind shear derived from the three sonic levels based on (2), compared to the stability function from Dyer (3) obtained from onshore conditions. The local wind shear is obtained from a profile fit to the cup velocity measurements as it is done for the *Ri* method. If local scaling theory would apply all the data points from the three levels would align along the same trend. While the unstable and near-neutral regimes are well described by this function, the stable regime shows a larger wind shear, with larger deviation from Dyer's function as we look to levels closer to the sea surface. This indicates that, local scaling theory is not sufficient to account for the vertical structure of the stable marine boundary layer at Fino-1.

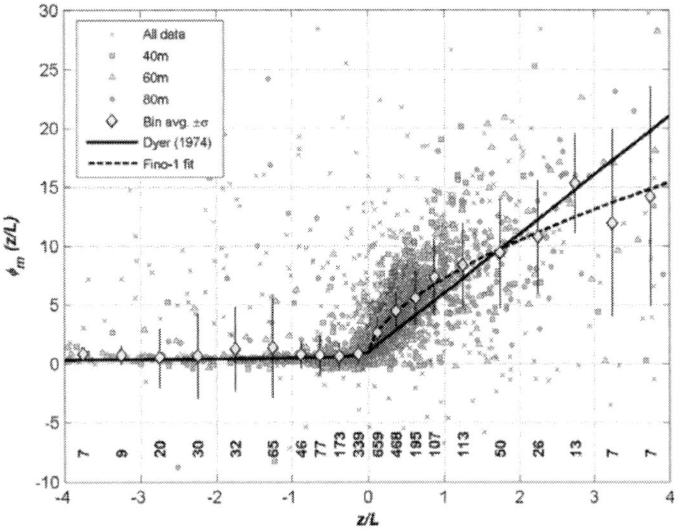

Figure 1: Non-dimensional wind shear at 40, 60 and 80 m compared to Dyer's stability function (3) and Fino-1 best fit curve (11). "All data" shows the initial dataset before applying quality control filters to the flux measurements. Yellow dots are bin-averages considering the three measurement levels together.

Other multi-scale similarity theories [18] might be more appropriate to account with the Coriolis and free-atmosphere stability effects of the ABL, which influence the height of the boundary layer that can be particularly shallow in offshore conditions. The surface-layer can be also affected by the state of the sea. Smedman et al. [19] show that M-O theory is only valid in growing sea conditions, i.e. when wind is increasing and the waves are created by the local wind and move slower than the wind (wave age lower than 0.8). In these conditions the undulated surface is equivalent to a roughness length as predicted by the Charnock dependency on the friction velocity. In swell conditions, when long waves travel faster than the wind due to a distant storm, the wind profile is not logarithmic and develops a kink at some height above the wave surface.

Nevertheless, for practical reasons, boundary layer models in wind energy applications rely on M-O theory to predict the wind profile in offshore conditions. This is not only for the convenience of using models developed previously for onshore conditions but also due to the adequacy to the available measurements from resource assessment campaigns, which do not include measurements of the wave regime or the free-atmosphere stability. For the same reasons we shall also simplify the

Journal of Physics: Conference Series **625** (2015) 012044 doi:10.1088/1742-6596/625/1/012044

analysis by considering only atmospheric stability, based on the local ζ parameter, as the most important scale for the thermally stratified boundary layer over relevant wind turbine operating conditions. Then, we can fit a specific offshore stability function based on stability bin averages from the three sonic levels (Figure 1). Instead of the linear behavior predicted by Dyer's function, the stable range follows a somewhat parabolic shape, similar to the function found by Duynkerke [20] in the 200-m Cabauw mast (Netherlands):

$$\phi_m(\zeta) = 1 + b_m\zeta\left(1 + \frac{b_m}{a_m}\zeta\right)^{a_m - 1} \tag{11}$$

with constants $a_m = 0.8$ $b_m = 5$. The Fino-1 case fits to this function with coefficients $a_m = 0.6$ $b_m = 30$.

5. Equivalence among stability assessment methods

Figure 2 shows the regression of both Ri_b-based methods for estimating ζ versus the sonic method as reference at two levels of the Fino-1 mast, 40 and 80 m. They both show reasonably good performance although larger spread is observed when using air temperature measurements far from the surface layer. This is to be expected since the Fuga method is based on surface-layer theory and the Grachev and Fairall empirical expression was also obtained from measurements close to the sea surface.

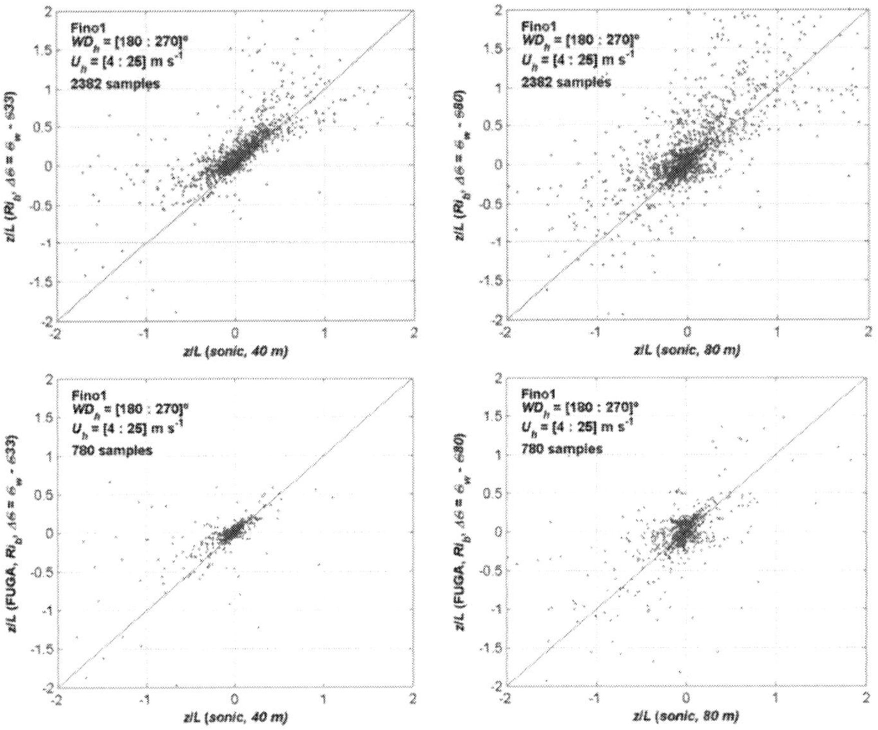

Figure 2: Regression of the ζ parameter obtained from the Grachev and Fairall (top) and FUGA (bottom) methods versus the sonic method using the 33 m (left) and 80 m (right) air temperature levels at Fino-1.

6. Atmospheric stability classification

It is relevant to define stability classes that can allow us to categorize wind conditions. Gryning et al. [21] define a seven-category classification based on the Obukhov length. This definition is somehow ambiguous since the Obukhov length depends on the height above ground level when measuring beyond the surface layer. A more generalized categorization is based on the dimensionless stability parameter ζ associated to a given height. Based on this parameter, Sorbjan and Grachev [12] identify four regimes in the stable boundary layer: "nearly-neutral" ($0 < \zeta < 0.02$); "weakly-stable" ($0.02 < \zeta < 0.6$); "very-stable" ($0.6 < \zeta < 50$); and "extremely stable" ($\zeta > 50$). In the "near-neutral" regime, Ri is a linear function of ζ, while in the "very stable" regime the Ri is an exponential function of ζ.

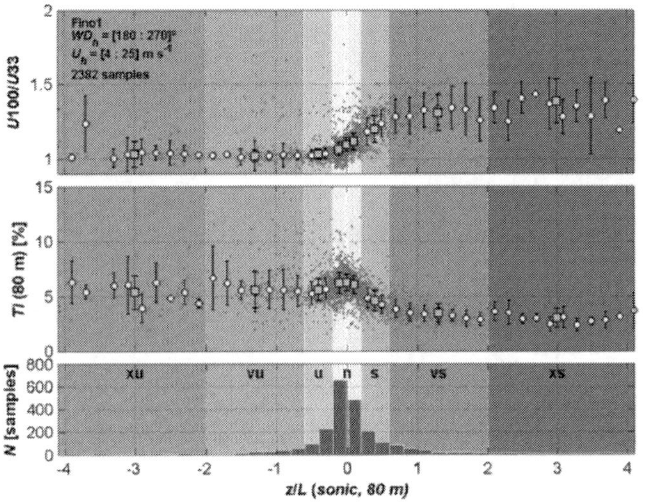

Figure 3: Wind shear (top), turbulence intensity (middle), and number of samples (bottom) vs sonic stability at Fino-1 (80 m). The blue and red shading indicate the different stability classes as given in Table 1. Yellow dots represent bin-averages and blue squares stability class-averages.

Table 1: Classification of atmospheric stability (symmetric for the unstable range)

Category	Range
near-neutral (n)	$0 < \zeta < 0.02$
weakly stable (ws)	$0.02 < \zeta < 0.2$
stable (s)	$0.2 < \zeta < 0.6$
very stable (vs)	$0.6 < \zeta < 2$
extremely stable (xs)	$\zeta > 2$

Figure 3 shows the wind conditions (wind shear and turbulence intensity) as a function of the sonic stability parameter at 80 m at Fino-1. The yellow dots show bin-averaged quantities based on a uniform bin-width of 0.2. The histogram of the stability parameter indicates that the most relevant

Journal of Physics: Conference Series **625** (2015) 012044 doi:10.1088/1742-6596/625/1/012044

range is $|\zeta| < 2$. Beyond $|\zeta| > 2$ very few profiles are found, showing very large scatter. The blue square dots indicate the ensemble mean values of each stability class following Sorbjan and Grachev classification in the stable range and assuming a symmetric classification in the unstable range for simplicity. The "extremely un/stable" regime limit is shifted to $|\zeta| = 2$ in order to avoid contamination of the large scatter found at the high ends of the scale to the "very un/stable" class. An additional limit is added at $|\zeta| = 0.2$ to add higher resolution in the most frequent stability range. Hence, a nine-class symmetric stability classification is defined (Table 1).

The ensemble averages of the stability classes follow fairly well the variability of the wind conditions for the relevant range of stabilities in Fino-1. The wind shear is roughly constant in unstable conditions and progressively increases in neutral to stable conditions reaching a plateau in very stable conditions.

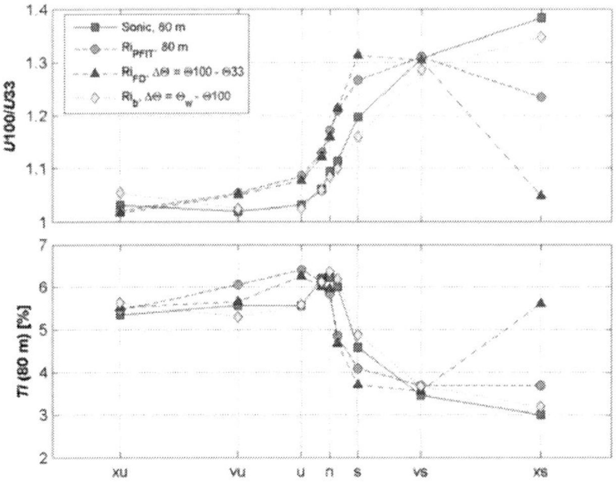

Figure 4: Class-averaged wind conditions at the open-sea sector of Fino-1 (225° ± 90°) as a function of the stability parameter ζ based on different stability assessment methods.

Figure 4 shows the class-averaged wind conditions (wind speed ratio between 100 and 33 m, *U100/U33*, and turbulence intensity at 80 m, *TI*, based on the ratio of the standard deviation to the mean wind speed) at the open-sea sector of Fino-1 as a function of the stability class for different stability assessment methods of determining the ζ parameter. The *xs* class shows large discrepancies due to poor statistics from low number of samples with large measurement uncertainties. It is noticed the good equivalence of the Ri_b and the sonic methods, compared to the gradient *Ri* methods that tend to overestimate the wind shear. In fact, the *Ri* methods produce a very distinct distribution of stabilities compared to that of Figure 3 from the sonic. The histogram from the *Ri* method has a larger number of unstable cases and a wider range of stabilities. As a result Figure 4 shows an offset of stability classes compared to the sonic method.

A local maximum of turbulence intensity is noticed for neutral conditions, a distinct behaviour in the offshore environment, caused by the increased roughness in connection with growing sea at high wind speeds. In fact, for *wu* and *ws* conditions, that take a very substantial share of the data, we find the same level of turbulence in this case. Hence, there is no monotonic decrease of turbulence with stability and therefore we cannot take turbulence intensity as a proxy of stability. In this particular case, the wind shear would be a better proxy since there is a more distinct change when we move from unstable to stable conditions.

7. Stability distributions

Figure 5 shows the distribution of atmospheric stability against wind speed based on the four different methods of stability classification described below. It is observed how stability dominates the wind climate over operational wind turbine speeds. The wind climate is roughly divided into 60% of unstable range and 40% of stable range.

Taking the sonic method as a benchmark (Figure 5a), it is observed that the bulk methods reproduce reasonably well the stability distribution in the stable range. The unstable range overpredicts the number of extremely unstable conditions. This is due to situations with Ri_b much larger than the critical value of 0.2 that are shifted to the unstable regime following Grachev and Fairall function (8). These situations are found when the water-air temperature differences are large (more than 1 °C) and velocities are low. On the other hand, the method based on the gradient Richardson (Figure 5b) produces a distribution with too frequent extremely stable and unstable cases.

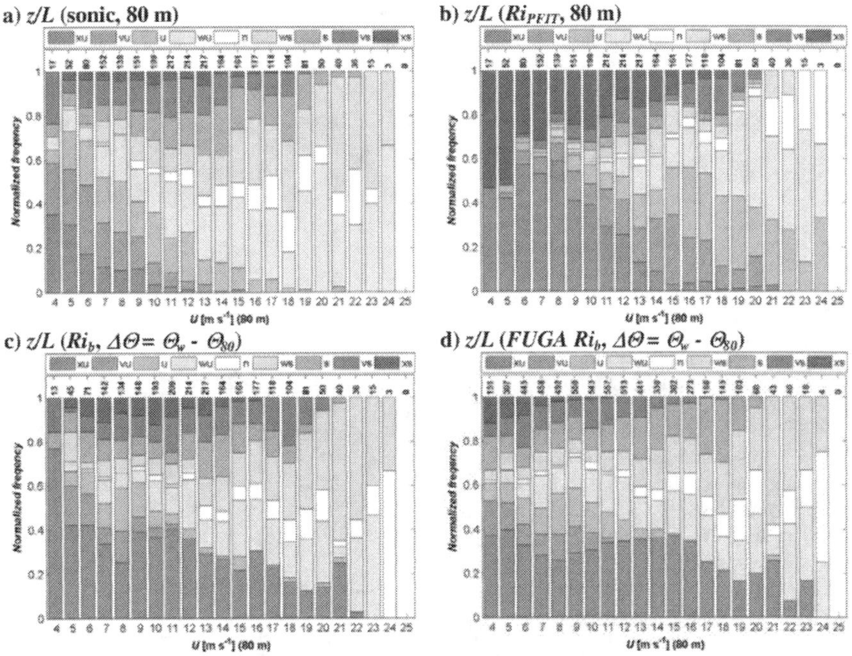

Figure 5: Distribution of stability with wind speed based on different stability parameters: a) sonic method, b) *Ri PFIT* method, c) Ri_b method, d) Ri_b *FUGA* method

8. Conclusions and discussion

It is generally understood by the wind energy community that the next generation of wind farm tools will include atmospheric stability as a design driver. While model developers are already exploring this possibility, it remains difficult to deal with this extra dimension if suitable stability measurements are not routinely acquired in resource assessment campaigns. This work attempts to provide some guidance as to how to perform these measurements that are particularly relevant in the offshore environment. Due to the inherent difficulty of the stability assessment methods, it is convenient to adopt certain conventions when it comes to measuring and defining stability. This will narrow down the stability assessment methods to a few practical ones that can be more consistently used in connection to state-of-the-art models.

The bulk Richardson method seems to be a robust approach to characterize stability offshore, requiring the minimum level of instrumentation while still providing good performance in the characterization of mean wind conditions compared to the sonic method, more elaborate and costly. The method requires differential temperature measurements or well calibrated absolute temperature sensors in order to reduce as much as the bias in the evaluation of potential temperature gradients. The sonic method however remains the only way of measuring local stability without the use of empirical functions or theoretical assumptions. In the long term, this instrument shall be the preferred add-on to mast or remote sensing campaigns to characterize not only stability but other turbulence quantities that are relevant for wind farm and turbine design models.

The universality of onshore stability functions has been demonstrated to be invalid in the stable marine boundary layer. It is important to use existing offshore masts, where flux-profile analysis can be done based on several sonic anemometers, to calibrate stability functions for the offshore environment and verify similarity theories including wave conditions. Based on this experimental approach, more accurate parameterizations will be included in mesoscale and microscale wind farm models to account for stable conditions. It is under these conditions that array efficiencies are the lowest and where wake effects take longer to dissipate due to the lower ambient turbulence conditions.

Acknowledgements
This work has been carried out with the support from three projects: the Spanish national project INNPACTO-EMERGE, a project from Carbon Trust for the U.K. Offshore Wind Accelerator and the "MesoWake" Marie Curie International Outgoing Fellowship (PIOF-GA-2013-624562). The availability of Fino-1 measurements from the German Hydrographic Institute (DSH) and the German Wind Energy Institute (DEWI) is greatly appreciated too.

References
[1] Frandsen A, Barthelmie R, Rathmann O, Jørgensen HE, Badger J, Hansen K, Ott S, Réthoré P-E, Larsen SE, Jensen LE 2007 Summary report: The shadow effect of large wind farms: measurements, data analysis and modeling. Risø National Laboratory report, Risø-R-1615(EN)
[2] Barthelmie RJ, Frandsen ST, Nielsen NM, Pryor SC, Rethore PE and Jørgensen HE 2007 Modelling and measurements of power losses and turbulence intensity in wind turbine wakes at Middelgrunden offshore wind farm *Wind Energy* **10** 217-228
[3] Nieuwstadt F.T.M. 1984 The Turbulent Structure of the Stable, Nocturnal Boundary Layer *J. Atmos Sci.* 41: 2202-2216
[4] Monin AS and Obukhov AM 1954 Basic laws of turbulent mixing in the atmosphere near the ground *Tr. Akad. Nauk. SSR* **151** 163-187
[5] Dyer AJ 1974 A Review of Flux Profile Relationships *Boundary Layer Meteorol.* **7** 363-372
[6] Paulson CA 1970 The mathematical representation of wind speed an temperature profiles in the unstable atmospheric surface layer *J. Appl. Meteorol.* **9** 857-861
[7] Charnock H 1955 Wind stress on a water surface *Q. J. Roy. Meteor. Soc.* **81** 639-640
[8] Peña A and Gryning S-E 2008 Charnock's Roughness and Non-dimensional Wind Profiles Over the Sea *Boundary-Layer Meteorol.* **129** 479-495
[9] Sanz Rodrigo J 2011 Flux-profile characterization of the offshore ABL for the parameterization of CFD models *EWEA Offshore 2012*, Amsterdam, The Netherlands, November 2011
[10] Lange B, Larsen S, Højstrup J and Barthelmie R 2004 The influence of thermal effects on the

wind speed profile of the coastal marine boundary layer *Boundary Layer Meteorol.* **112** 587-617
[11] Lange B, Larsen S, Højstrup J and Barthelmie R 2004 Importance of thermal effects and sea surface roughness for offshore wind resource assessment *J. Wind Eng. Ind. Aerod.* **92** 959-988
[12] Sorbjan Z and Grachev AA 2010 An Evaluation of the Flux-Gradient Relationship in the Stable Boundary Layer *Boundary-Layer Meteorol* **135** 385-405
[13] Grachev AA and Fairall CW 1997 Dependence of the Monin-Obukhov Stability Parameter on the Bulk Richardson Number over the Ocean *Jouranl of Applied Meteorology* **36** 406-414
[14] Ott S and Nielsen M 2014 Developments of the offshore wind turbine wake model Fuga. *DTU Wind Energy report* E-0046, January 2014
[15] Neumann T, Nolopp K, Riedel M, Strack M, Herklotz K, Stein J 2004 One Year Operation of the First Offshore Wind Research Platform in the German Bight – FINO I. German Wind Energy Conference DEWEK 2004, Wilhelmshaven, Germany, 2004
[16] Tautz S, Lange B, Heinemann D 2004 Correction of the heat and momentum flux measurements with the ultrasonic anemometers at the FINO I offshore meteorological mast for flow distortion and mounting effects. German Wind Energy Conference DEWEK 2004, Wilhelmshaven, Germany, 2004
[17] Wilczak JM, Oncley SP and Stage SA 2001 Sonic anemometer tilt correction algorithms *Boundary-Layer Meteorol.* **99** 127-150
[18] Zilitinkevich S and Calanca P 2000 An extended similarity-theory for the stably stratified atmospheric surface layer. *Quart. J. Roy. Meteorol. Soc.*, **126** 1913-1923
[19] Smedman A-S, Högstrom U, Bergström H, Johansson C, Sjöblom A and Larsén XG 2003 New findings concerning the structure of the marine atmospheric boundary layer over the Baltic Sea – possible implications for wind energy installations. *Wind Engineering* **27** 431-447
[20] Duynkerke PG 1999 Turbulence, Radiation and Fog in Dutch Stable Boundary Layers *Boundary Layer Meteorol.* **90** 447-477
[21] Gryning S-E, Batchvarova E, Brümmer B, Jørgensen H and Larsen S 2007 On the extension of the wind profile over homogeneous terrain beyond the surface layer *Boundary-Layer Meteorol.* **124** 251–268